# Lecture Notes in Electrical Engineering 1137

The book series *Lecture Notes in Electrical Engineering* (LNEE) publishes the latest developments in Electrical Engineering—quickly, informally and in high quality. While original research reported in proceedings and monographs has traditionally formed the core of LNEE, we also encourage authors to submit books devoted to supporting student education and professional training in the various fields and applications areas of electrical engineering. The series cover classical and emerging topics concerning:

- Communication Engineering, Information Theory and Networks
- Electronics Engineering and Microelectronics
- Signal, Image and Speech Processing
- Wireless and Mobile Communication
- Circuits and Systems
- Energy Systems, Power Electronics and Electrical Machines
- Electro-optical Engineering
- Instrumentation Engineering
- Avionics Engineering
- Control Systems
- Internet-of-Things and Cybersecurity
- Biomedical Devices, MEMS and NEMS

For general information about this book series, comments or suggestions, please contact leontina.dicecco@springer.com.

To submit a proposal or request further information, please contact the Publishing Editor in your country:

**China**

Jasmine Dou, Editor (jasmine.dou@springer.com)

**India, Japan, Rest of Asia**

Swati Meherishi, Editorial Director (Swati.Meherishi@springer.com)

**Southeast Asia, Australia, New Zealand**

Ramesh Nath Premnath, Editor (ramesh.premnath@springernature.com)

**USA, Canada**

Michael Luby, Senior Editor (michael.luby@springer.com)

**All other Countries**

Leontina Di Cecco, Senior Editor (leontina.dicecco@springer.com)

**** This series is indexed by EI Compendex and Scopus databases. ****

Yong Qin · Limin Jia · Jianwei Yang ·
Lijun Diao · Dechen Yao · Min An
Editors

# Proceedings of the 6th International Conference on Electrical Engineering and Information Technologies for Rail Transportation (EITRT) 2023

## Rail Transportation Operation Management Technologies

Springer

*Editors*
Yong Qin
Beijing Jiaotong University
Beijing, China

Limin Jia
Beijing Jiaotong University
Beijing, China

Jianwei Yang
Beijing University of Civil Engineering and Architecture
Beijing, China

Lijun Diao
Beijing Jiaotong University
Beijing, China

Dechen Yao
Beijing University of Civil Engineering and Architecture
Beijing, China

Min An
University of Salford
Salford, UK

ISSN 1876-1100 ISSN 1876-1119 (electronic)
Lecture Notes in Electrical Engineering
ISBN 978-981-99-9310-9 ISBN 978-981-99-9311-6 (eBook)
https://doi.org/10.1007/978-981-99-9311-6



This Springer imprint is published by the registered company Springer Nature Singapore Pte Ltd.
The registered company address is: 152 Beach Road, #21-01/04 Gateway East, Singapore 189721, Singapore

# Contents

# Study on the Compilation of Train Diagram and Transportation Capacity of Urban Rail Transit

Guangjian Zhang(✉) and Lizhu Zhang

School of Automobile and Transportation, Tianjin University of Technology and Education, Tianjin 300222, China
Zgjcgx@163.com, zlzjcb@163.com

**Abstract.** The train diagram of urban rail transit is the concentrated embodiment of the transportation organization of the whole urban rail transit system. By analyzing the passenger flow data of a metro line in a city, the section passenger flow in peak hours and the maximum section passenger flow in time period in the whole day are calculated from the OD matrix. The number of running trains and the running interval are also calculated, and fine-tuning is carried out, so as to get the full-time driving plan. On the basis of determining the train routing, stop plan, turn back plan and vehicle operation plan, the train diagram shall be prepared. In order to ensure the subway operation enterprises to arrange transportation production reasonably, through the analysis of subway line capacity and train capacity, the transportation capacity is analyzed and studied.

**Keywords:** Train diagram · OD matrix · Full-time driving plan · Transportation capacity

## 1 Introduction

The dispatch and command work of urban rail transit is an important part of the operation and management of urban rail transit systems. The important foundation of its work is that the train operates well according to the train diagram. The train diagram is a graphical explanation of the spatial and temporal relationship of train operation. It interprets the order in which each train occupies the section, train operation time, the arrival and departure or passing time of each station, the dwell time at the station and the turnaround time at the turnaround station and train routing and the inbound and outbound time at the vehicle depot. It is an important foundation for subway operation enterprises to organize train operations [1]. Zhou Zhengduo [2] et al. established an optimization model for redundant time layout of train operation diagrams of high-speed railway. It has positive significance in ensuring the order of transportation organization and improving the quality of passenger transportation services. Wang [3] et al. established a state space model that considers the evolution of train departure time and changes in passenger capacity inside the train. Jovanović [4] et al. fixed the train interval operation time and dwell time, and abstracted the optimal allocation of train interval redundancy time as a


Y. Qin et al. (Eds.): EITRT 2023, LNEE 1137, pp. 1–11, 2024.
https://doi.org/10.1007/978-981-99-9311-6_1

knapsack problem. Shi Xiaojun [5] analyzed the elements and steps of preparation of train operation diagram based on his practical experience in preparing Tianjin Metro No.1, and summarized the preparation techniques. Xu Ruihua [6] et al. analyzed the design concept and overall structure of computer compilation of train operation diagrams for urban rail transit, and studied a series of issues related to the operation and management of urban rail transit systems. Du Yijia [7] et al. analyzed the functional characteristics of train operation diagram compilation systems for domestic and foreign railway networks, providing a platform and tool for the development of train operation diagram compilation systems for networked railway lines. Rong Jian [8] conducted research on management optimization strategies for train operation diagram preparation to improve the level of train operation diagram preparation. Guo, H. [9] built a mathematic model to characterize the time-varying demand on passengers arriving at the platform of a metro station linking to an intercity railway station. Zhang Kun [10] briefly analyzed the factors and strategies of subway transportation capacity to improve the transportation efficiency of urban rail transit. Kang Liujiang [11] achieved collaborative compilation of network operation diagrams.

The above literature has not studied the most basic method of compiling train diagrams. So, the peak hour OD matrix (7:00~8:00) of a certain line in a city is given. The process of compiling train operation diagrams is calculated and studied. In general, the given 14 stations are named A, B, C, D, E, F, G, H, I, J, K, L, M and N. The average station spacing is 1543 m, with one end being the depot and the other end being the parking lot. A $\rightarrow$ N is the downlink, and N $\rightarrow$ A is the uplink.

## 2 Technical Route

Compiling a train diagram is a complex task that requires maintaining certain principles. On the premise of ensuring safety and reliability, the train speed is increased and the running time of the train is reduced for the convenience of passengers. Appropriate turn back methods are selected to fully utilize train and line capabilities. For economic reasons, the number of vehicles in operation should be minimized as much as possible [12]. The technical route for preparing train operation diagram is shown in Fig. 1.

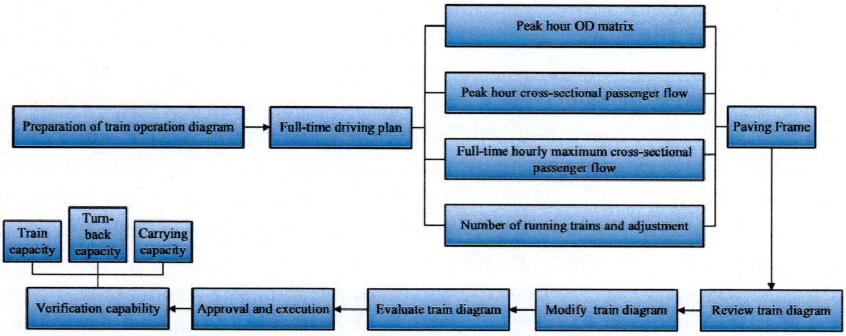


**Fig. 1.** Compilation process of train operation diagram

## 3 Process Exploration

### 3.1 Calculation of Cross-Sectional Passenger Flow

According to the peak hour OD matrix, the cross-sectional passenger flow during peak hours can be calculated. As shown in Table 1 below. The number of downlink people and the number of uplink people in the table refer to the cross-sectional passenger flow during peak hours.

$$P_j = P_i - P_x + P_s \tag{1}$$

where, $P_j$ is passenger flow of the jth cross section. $P_i$ is passenger flow of the ith cross section. $P_x$ is number of passengers getting off at the station. $P_s$ is number of passengers boarding at the station.

**Table 1.** Peak hour (7:00~8:00) cross-sectional passenger flow (unit: person)

| The number of downlink people | The number of downlink people boarding | The number of downlink people getting off | Station name | The number of uplink people boarding | The number of uplink people getting off | The number of uplink people |
|---|---|---|---|---|---|---|
| 13461 | 13461 | 0 | A | 0 | 11486 | 11486 |
| 19474 | 6145 | 132 | B | 105 | 5232 | 16613 |
| 24294 | 5174 | 354 | C | 382 | 5027 | 21258 |
| 26764 | 3273 | 803 | D | 859 | 4347 | 24746 |
| 30387 | 4703 | 1080 | E | 1180 | 4290 | 27856 |
| 31145 | 3617 | 2859 | F | 2189 | 3443 | 29110 |
| 27282 | 7064 | 10927 | G | 11784 | 5637 | 22963 |
| 26376 | 2907 | 3813 | H | 6519 | 2464 | 18908 |
| 23012 | 1928 | 5292 | I | 4057 | 1622 | 16473 |
| 19875 | 1485 | 4622 | J | 4830 | 998 | 12641 |
| 14263 | 1041 | 6653 | K | 3236 | 534 | 9939 |
| 9018 | 716 | 5961 | L | 3829 | 288 | 6398 |
| 4475 | 103 | 4646 | M | 3895 | 330 | 2833 |
| | 0 | 4475 | N | 2833 | 0 | |

There is a peak period, which occurs at the F-G section. The other sections gradually decrease from the peak section to both sides. Therefore, it can be inferred that two ends of the line are in the suburbs, with less passenger flow, and the middle is in the city center. The commercial community and other areas have a larger passenger flow, and most people are in the city center.

### 3.2 Division of Working Day Periods

The operating time of urban rail transit systems in China is generally 18 h. From 5:00 to 23:00. Due to the fact that the OD matrix is obtained from research on weekdays, only the weekday train diagram and capability are discussed. The division of time periods and the distribution proportion of full-time hourly maximum cross-sectional passenger flow are shown in Table 2 below. From Table 1, it can be seen that the maximum section passenger flow during peak hours is 31145 passengers, which can be calculated as full-time hourly maximum cross-sectional passenger flow.

**Table 2.** Full-time hourly maximum cross-sectional passenger flow

| Time | Distribution ratio (%) | Passenger flow | Time | Distribution ratio (%) | Passenger flow (%) |
|---|---|---|---|---|---|
| 5:00–6:00 | 17 | 5295 | 14:00–15:00 | 58 | 18064 |
| 6:00–7:00 | 45 | 14015 | 15:00–16:00 | 69 | 21490 |
| 7:00–8:00 | 1 | 31145 | 16:00–17:00 | 85 | 26473 |
| 8:00–9:00 | 75 | 23359 | 17:00–18:00 | 62 | 19310 |
| 9:00–10:00 | 5 | 15573 | 18:00–19:00 | 42 | 13081 |
| 10:00–11:00 | 53 | 16507 | 19:00–20:00 | 34 | 10589 |
| 11:00–12:00 | 63 | 19621 | 20:00–21:00 | 29 | 9032 |
| 12:00–13:00 | 6 | 18687 | 21:00–22:00 | 26 | 8098 |
| 13:00–14:00 | 6 | 17441 | 22:00–23:00 | 16 | 4983 |

### 3.3 Determination of Train Formation Plan

In urban rail transit, three types of vehicle A, B, and C are specified. According to the situation, Type A vehicles are selected. The commonly used 6-vehicle formation scheme in urban rail transit has been adopted. That means six A-type vehicles are coupled to form a train. Length of each carriage of type A vehicle is 24.4 or 22.8 m. Its width is 3 m. Vehicle capacity is 310 persons.

### 3.4 Preparation of A Full Day Driving Plan

The full day driving plan is the plan for the number of trains that will operate in each hour during the operating hours of the urban rail transit system. The daily operation tasks of urban rail transit lines are specified.

$$n_i = \frac{P_{\max}}{P_{train}\beta} \tag{2}$$

where, $n_i$ is number of hourly trains operating throughout the day. $P_{train}$ is number of train personnel quota. $P_{\max}$ is one-way maximum cross-sectional passenger flow. $\beta$ is full load rate of line section.

$$\beta = \frac{P_{\max}}{C_{\max}} \times 100\% \tag{3}$$

where, $C_{\max}$ is the transportation capacity of the line during peak hours.

The value of $\beta$ can be determined based on different time periods. During peak hours of passenger flow, it can be greater than 1, and at other times it can be lower than 1. Its function can determine the basic balance of the number of trains running, without causing frequent train entry and exit. Therefore, 110% is used during peak hours, and 90% is used during other operating hours. The number of train personnel quota during peak and off-peak hours is respectively 2046 and 1674.

The number of trains running is calculated, and the departure interval also needs to be calculated. The calculated driving interval during a certain period of time may be relatively long. If it is too long, the waiting time for passengers will be increased, and the quality of service will be reduced. Therefore, considering all factors, the driving interval should be between 9:00 and 21:00. Generally, it should not exceed 6 min. During off-peak operating hours, it is generally not recommended to exceed 10 min. The calculation results and the final number of trains are shown in Table 4 below. As shown in Table 3 below

$$t_{interval} = 60/n_i \tag{4}$$

where, $t_{interval}$ is interval time of trains running.

### 3.5 Determination of Train Stop Time

The main purpose of subway trains stopping at stations is to carry out passenger transportation operations, such as passengers boarding and disembarking. When calculating the train dwell time, it is necessary to minimize the train dwell time and improve transportation efficiency while meeting the requirements of train dwell operations to the greatest extent possible.

$$t_{station} = \frac{h\left(P_{boarding} + P_{disembarking}\right)t_{boarding}(disembarking)}{nmd} + t_{onandoff} - t_{\Delta} \tag{5}$$

Where, $P_{boarding}$, $P_{disembarking}$ is number of people boarding or disembarking at stations during peak hours. $T_{boarding(disembarking)}$ is the average time required for each passenger to board or disembark (Acceptable 1.2 s). $T_{on\ and\ off}$ is time for opening and closing the door (Acceptable 18 s) [13]. $t_{\Delta}$is overlapping time between opening and closing doors and passenger boarding (Acceptable 3 s). $h$ is peak hour coefficient (Acceptable 1.15). $n$ is number of trains operating during peak hours (Acceptable 16). $m$ is number of train formations (Acceptable 6). $D$ is number of doors on each side of each vehicle (Acceptable 5).

According to the above formula, the stopping time of the train can be calculated. To consider a certain amount of surplus, an upward floating treatment should be carried out, usually an integer multiple of 5. As shown in Table 4 below.

**Table 3.** The number of hourly trains operating throughout the day

| Time | Interval time of trains running after adjustment | Number of running trains | Number of running trains after adjustment | Time | Interval time of trains running after adjustment | Number of running trains | Number of running trains after adjustment |
|---|---|---|---|---|---|---|---|
| 5:00–6:00 | 10 min | 4 | 6 | 14:00–15:00 | 5 min 30 s | 11 | 11 |
| 6:00–7:00 | 6 min 42 s | 9 | 9 | 15:00–16:00 | 4 min 36 s | 13 | 13 |
| 7:00–8:00 | 3 min 45 s | 16 | 16 | 16:00–17:00 | 4 min 36 s | 13 | 13 |
| 8:00–9:00 | 4 min 18 s | 14 | 14 | 17:00–18:00 | 5 min | 12 | 12 |
| 9:00–10:00 | 6 min | 10 | 10 | 18:00–19:00 | 6 min | 8 | 10 |
| 10:00–11:00 | 6 min | 10 | 10 | 19:00–20:00 | 6 min | 7 | 10 |
| 11:00–12:00 | 5 min | 12 | 12 | 20:00–21:00 | 6 min | 6 | 10 |
| 12:00–13:00 | 5 min | 12 | 12 | 21:00–22:00 | 10 min | 5 | 6 |
| 13:00–14:00 | 5 min 30 s | 11 | 11 | 22:00–23:00 | 10 min | 3 | 6 |

**Table 4.** Stopping time at each station

| Station | Number of passengers boarding and disembarking | Stopping time (s) | Correction of stop time (s) | Station | Number of passengers boarding and disembarking | Stopping time (s) | Correction of stop time (s) |
|---|---|---|---|---|---|---|---|
| A | 13461 | 53.70 | 55 | H | 6720 | 34.32 | 35 |
| B | 6277 | 33.05 | 35 | I | 7220 | 35.76 | 40 |
| C | 5528 | 30.89 | 35 | J | 6107 | 32.56 | 35 |
| D | 4076 | 26.72 | 30 | K | 7694 | 37.12 | 40 |
| E | 5783 | 31.63 | 35 | L | 6677 | 34.20 | 35 |
| F | 6476 | 33.62 | 35 | M | 4749 | 28.65 | 30 |
| G | 17991 | 66.72 | 70 | N | 4475 | 27.87 | 30 |

### 3.6 Determination of Interval Operation Time

Nowadays, trains generally run under the moving block signal system. The average operating speed of the section is 50 km/h. According to the distance between stations, the running time between trains is calculated as shown in Table 5.

**Table 5.** Train interval operation time

| Interval | Interval distance (m) | Running time (s) | Interval | Interval distance (m) | Running time (s) |
|---|---|---|---|---|---|
| A—B | 1569 | 113 | H—I | 1258 | 91 |
| B—C | 1235 | 89 | I—J | 1496 | 108 |
| C—D | 1458 | 105 | J—K | 1789 | 129 |
| D—E | 1469 | 106 | K—L | 1691 | 122 |
| E—F | 2103 | 152 | L—M | 1854 | 134 |
| F—G | 1256 | 91 | M—N | 1395 | 101 |
| G—H | 1479 | 107 | | | |

### 3.7 Determination of Train Turnback Time

The turn back operation of trains on this line at destination stations A and H belongs to the post station turn back operation. Train stopping at platform 1 handles passenger disembarking operations within a certain stopping time. After completion, the train enters turn back line 2 from the main line, and the shunting route can be processed. The train is within the turn back line, and the drivers of the front and rear trains change their heads. After staying for a certain period of time, the previous train leaves the station. At the same time, the switch is opened on the main line and the shunting signal is open, entering the downlink platform 3, and the train completes the turn back operation.

The calculation formula is:

$$h_{departure} = t_{leave} + t_{operation}^{exit} + t_{confirm} + t_{departure} + t_{stop} \tag{6}$$

where, $t_{\text{leave}}$ is the time when the departing train leaves the station block zone (Acceptable 1.5 min). $t_{operation}^{exit}$ is time for handling exiting turn back line (Acceptable 0.5 min). $T_{confirm}$ is time for confirming the signal (Acceptable 0.5 min). $T_{departure}$ is time the train runs from the station's turn back line to the departure line (Acceptable 1.5 min). $T_{stop}$ is train stop time including passenger boarding (Acceptable 1 min).

Therefore, the final turnback time of the train is 5 min, or 300 s.

### 3.8 Preparation of Vehicle Allocation Plan

To complete the passenger transportation task, the urban rail transit system must maintain a certain number of vehicles. According to usage, it can be divided into three categories: operating vehicles, maintenance vehicles and backup vehicles.

For operational vehicles:

$$N = \frac{n_{peak\ hours} \theta_{train} m}{3600} \tag{7}$$

where, $N$ is Number of operating vehicles. $N_{peak\ hours}$ is number of operating trains during peak hours. $\theta_{train}$ is train turnover time. $m$ is number of train formations.

Among them:

$$\theta_{train} = \sum t_{running} + \sum t_{station} + \sum t_{turnback\ station\ stop} \tag{8}$$

where, $t_{running}$ is the sum of the running time of trains traveling back and forth between different sections on the line. $T_{station}$ is the total stopping time of each intermediate station for a train traveling back and forth on the line. $T_{turnback\ station\ stop}$ is the total dwell time of the train at the turnaround station.

Based on the above dwell time, interval operation time and turnaround time, adding an additional 20 s for train start and 10 s for parking, train turnover time $\theta_{train}$ is 4306 s namely 71.77 min. So, the number of operating vehicles is 115. The number of maintenance vehicles and the number of backup vehicles are generally determined based on 10% of the number of operating vehicles. So the total number of maintenance vehicles and backup vehicles is 23. The final total number of vehicles was determined to be 138.

## 4 Preparation of Train Diagram

For the drawing of train diagrams, manual drawing can be used. But with the continuous development of computer technology, major simulation software has been launched one after another. For example, there is Matlab, AutoCAD, etc. Nowadays, China's urban rail system adopts the internationally advanced Automatic Train Control (ATC). This system can lay out train diagrams. Therefore, ATC Train Dispatch Simulation System of Urban Rail Transit is used for development.

Based on the data determined above, they are input into the software. Train diagram is drawn. The results are shown in Fig. 2 (partial view).

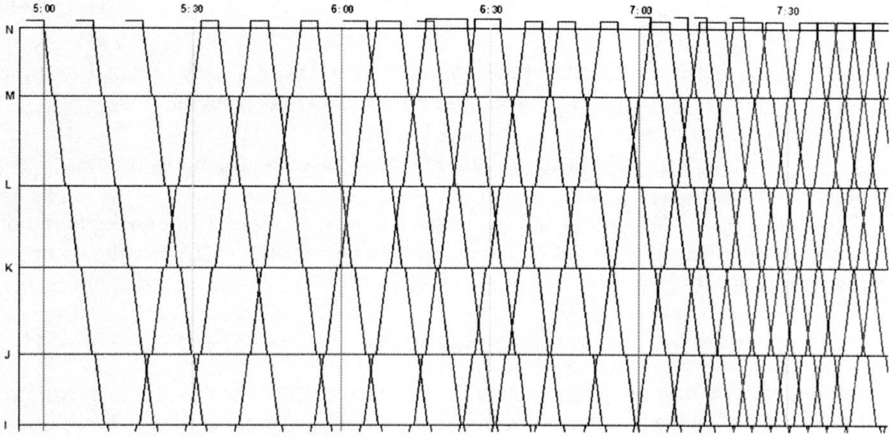


**Fig. 2.** Partial enlarged view of train diagram

### 4.1 Exploration of Train Transportation Capacity

For train passing capacity, that is the maximum number of trains that can pass in a certain direction within 1 h on a certain route.

$$n_{\max} = \frac{3600}{t_{interval}} \tag{9}$$

where, $n_{\max}$ is the maximum number of trains that the line can pass through within 1 h. $t_{\text{interval}}$ is minimum train interval.

For this line, the moving block control principle is adopted. Under the moving block system, the tracking point of the following train is the tail of the preceding train. Tracking interval of moving block train:

$$I = \frac{V_{\max}}{3.6a} + 3.6 \times \frac{L_F + L_A}{V_{\max}} \tag{10}$$

where, $V_{\max}$ is maximum running speed of the train (Acceptable 80 km/h). $a$ is train braking deceleration (Acceptable 1.3 m/s$^2$). $L_F$ is distance of train travel within the confirmation signal and braking response time $L_F = (30\,s + 1.5\,s) \times \frac{50000\,m}{3600\,s} \approx 438\,m$. $L_A$ is the safe distance from the target stopping point of the following train to the starting point of the track circuit occupied by the preceding train (Acceptable 200 m).

After analysis, it can be concluded that the minimum tracking interval of the train is 108 s.

For turn back capability, it refers to the maximum number of trains that can be turned back at a turn back station within a unit hour.

$$n_{turnback} = \frac{3600}{h_{depart}} \tag{11}$$

where, $n_{turnback}$ is the maximum number of trains that can turn back within 1 h at the turn back station. $H_{depart}$ is Time for train turn back departure interval.

So, the turnback capacity of this line is 12 trains.

For train capacity, it refers to the amount of space that a train can transport passengers when fully loaded.

$Traincapacity = Numberofpassengerspervehicle \times Numberoftrainformations$ For the transportation capacity of the line, it is equal to the product of the passing capacity and the train capacity.

$$p = n_{\max} m p_{train} \tag{12}$$

where, $p$ is maximum conveying capacity per unit time of the line. $N_{max}$ is the maximum number of trains that the line can pass through within 1 h. $m$ is number of train formations. $P_{train}$ is vehicle personnel quota.

So, the transportation capacity of this line is 63240 people.

## 5 Conclusion

By peak hour OD matrix for a specific route through the city, peak hour cross-sectional passenger flow and full-time hourly maximum cross-sectional passenger flow are calculated. So, the full-day driving plan can be obtained. The departure interval has been fine-tuned, resulting in a feasible full-day driving plan. The stopping time and interval running time are calculated ulteriorly. After determining the turn back method, turn back time, train routing, stop plan, and vehicle allocation plan, ATC simulation software is used to draw train diagrams to obtain the planned train diagram, and transportation capacity is explored. When preparing the train diagram, it is necessary to conduct a comprehensive analysis of the system. Comprehensive planning should be achieved, constantly summarizing experience to improve drawing quality. In the later stage, further research should be conducted on the crew plan, which refers to the use of the plan by crew members.

**Acknowledgment.** This work was supported by China university industry- university research innovation fund-Beichuang Teaching Assistant Project (Phase II) (2021BCB02003). The authors are grateful to the anonymous reviewers for their constructive comments on the paper.

## References


1. Cao, J., Ying, X.: Urban Rail Transit Dispatch and Command Work, 1st edn. China Communications Press, Beijing (2017). (in Chinese)
2. Zhou, Z., Zhou, L.: Optimization of buffer time layout of high speed railways' train working diagram for minimizing passengers' arrival delay. Railw. Transp. Econ. **45**(2), 25–32 (2023). (in Chinese)
3. Wang, X., Xin, T.P., Wang, H.W., et al.: A generative adversarial network based learning approach to the autonomous decision making of high-speed trains. IEEE Trans. Veh. Technol.Veh. Technol. **71**(3), 2399–2412 (2022)
4. Jovanović, P., Kecman, P., Bojović, N., et al.: Optimal allocation of buffer times to increase train schedule robustness. Eur. J. Oper. Res.Oper. Res. **256**(1), 44–54 (2017)
5. Shi, X.: Probing into the compiling of train operation schedule for mass rail transit. Urban Rapid Rail Transit **21**(2), 24–27 (2008). (in Chinese)
6. Xu, R., Jiang, Z., Zhu, X., et al.: Key problems in computer compilation of train diagram in UMT. Res. Urban Rail Transit (5), 31–35 (2005). (in Chinese)
7. Du, Y., Chen, D., Lv, H., et al: Function of collaborative compilation system of train working diagram for urban rail transit network. Railw. Transp. Econ. **44**(10), 99–104,119 (2022). (in Chinese)
8. Rong, J.: A study on the optimization of train working diagram compilation and management. Railw. Transp. Econ. **41**(2), 27–31 (2019). (in Chinese)
9. Guo, H., Bai, Y., Hu, Q., Zhuang, H., Feng, X.: Optimization on metro timetable considering train capacity and passenger demand from intercity railways. Smart Resilient Transp. **3**(1), 66–77 (2021)
10. Zhang, K.: Analysis of the influencing factors and strategies for the transportation capacity of subway rail transit. Jushe (30), 129 (2017). (in Chinese)
11. Kang, L., Zhu, X., Sun, H., et al.: Last train timetabling optimization and bus bridging service management in urban railway transit networks. Omega-Int. J. Manag. Sci. **84**(4), 31–44 (2019)
12. Mao, B.: Operation and Management of Urban Rail Transit System, 2nd edn. China Communications Press, Beijing (2017). (in Chinese)
13. Miao, Q., Pan, Z.: Research of urban rail transit train dwelling time at station. Res. Urban Rail Transit **20**(06), 37–40 (2017). (in Chinese)

Check for updates

# Research on High-Speed Railway Timetable Rescheduling Based on Genetic Algorithm

Xinyi Du[1,2], Li Wang[1,2(✉)], and Xianghao Wang[1,2]

[1] School of Traffic and Transportation, Beijing Jiaotong University, Beijing 100044, China
{duxinyi,wangli,xianghaow}@bjtu.edu.cn

[2] Beijing Engineering Research Center of Urban Traffic Information Intelligent Sensing and Service Technologies, Beijing 100044, China

**Abstract.** This paper presents an effective approach to address the challenge of rapidly restoring train operation order and automatically generating scheduling optimization schemes during abnormal events. The study focuses on high-speed railway, and establishes a high-speed railway train operation adjustment model, with the primary objective of minimizing the weighted total arrival delay time while satisfying various time and capacity constraints. To overcome the NP-hard nature of the problem, we propose a genetic algorithm approach with an appropriate coding mode, fitness function, crossover, and mutation rules. To validate the proposed model and algorithm, we use the operational data from the Beijing-Shanghai high-speed railway. The results of comparing the advantages and disadvantages of the genetic algorithm with the interval-only accelerated operation method demonstrate the feasibility and effectiveness of genetic algorithm, which can provide decision support for dispatching high-speed railway train operation scheduling.

**Keywords:** Timetable Rescheduling · High-speed Railway · Genetic Algorithm

## 1 Introduction

According to the Statistical Bulletin of China State Railway Group Co., Ltd. in 2021 [1], by the end of 2021, the mileage of China's high-speed rail operation had exceed 40,000 km, and the number of trains had reached 4,153, ranking first in the world. Adjustment of train operation is also called real-time scheduling, which means that under emergency conditions, according to the length of delay time and the number of trains affected by delay propagation, the normal operation of delayed trains can be resumed and the influence of delay propagation can be reduced as much as possible, so as to ensure the high punctuality rate of trains.

Emergency events can be divided into slight disturbance and severe disturbance [2]. Train real-time scheduling optimization under slight disturbance is mainly aimed at the scene where the train has slight initial delay, and the model complexity increases exponentially with the complexity of the study scene, which cannot solve the problem of large-scale serious disturbance [3]. Shuguang Zhan [4] proposed a train operation


Y. Qin et al. (Eds.): EITRT 2023, LNEE 1137, pp. 12–19, 2024.
https://doi.org/10.1007/978-981-99-9311-6_2

adjustment model based on event activity network for the complete or partial interruption of the interval. Ruhu Gao et al. [5] proposed an optimization algorithm of running diagram based on ADMM (Alternating Direction Method of Multipliers). Zhiwen Liao et al. [6] proposed a train operation adjustment model based on priority. Considering transfer, Pengcheng Wen et al. [7] established a linear integer programming model of train operation adjustment suitable for the last train period. Considering passenger transfer, Pu Zhang et al. [8] established a multi-objective mixed integer linear programming model to minimize the total train delay time and the number of passengers who failed to transfer. Considering severe initial delay and interval speed limit conditions, Mingming Wang et al. [9] constructed a multi-objective nonlinear train scheduling optimization model based on event activity network. Based on global conflict analysis, Jun Zhang et al. [10] constructed a bi-objective programming model for schedule adjustment, which was solved by the algorithm of Pareto optimality and Nash equilibrium. Mingming Wang et al. [11] proposed a hybrid heuristic algorithm of genetic and particle swarm optimization to solve the train operation adjustment model under slight initial delay, but the gap between the feasible solution and the optimal solution in an acceptable time was 19.6%. Ming Lei et al. [12] applied the idea of co-evolutionary genetic algorithm to solve the train operation adjustment model.

## 2 Definition of Problem

Train real-time scheduling problem is a NP-hard problem, in order to simplify the research, the following assumptions are put forward:

(1) Taking a single railway line as the research object, trains in different directions of line run independently, and one of directions is selected for research.
(2) Assuming that the number of arrival and departure lines and staff at each station are sufficient.

## 3 Model Construction

### 3.1 Model Parameters & Decision Variables

Before building the model, it is necessary to define the known parameters and unknown variables. The symbolic meanings of sets, indexes, parameters and variables used in mathematical models are shown in Table 1.

### 3.2 Model Construction

In this paper, the goal is to minimize the weighted total arrival delay time, which mainly measures the total delay time of trains arriving at each station.

$$\min z = \sum_{k \in K} \sum_{u \in U} a_k \times \left(A_u^k - A\prime_u^k\right) \tag{1}$$

In order to ensure the feasibility of the train operation adjustment scheme, the adjustment should be carried out under certain constraints.

**Table 1.** The meanings of the parameters and decision variables

| Symbols | Meanings |
|---|---|
| $K$ | Set of train services, indexed by $k$, $k \in \{1, 2, \ldots, n\}$ |
| $U$ | Set of stations, indexed by $u$, $u \in \{1, 2, \ldots, m\}$ |
| $\tau_{\text{start}}$ | Additional time caused by starting at station |
| $\tau_{\text{stop}}$ | Additional time caused by stopping at station |
| $t_{\text{arr}}$ | The headway of arrival at station |
| $t_{\text{dep}}$ | The headway of departure from station |
| $T_u^k$ | Minimum dwell time of train $k$ at station $u$ |
| $A_u^{'k}$ | Scheduled arrival time of train $k$ at station $u$ |
| $D_u^{'k}$ | Scheduled departure time of train $k$ from station $u$ |
| $A_u^k$ | Actual arrival time of train $k$ at station $u$ |
| $D_u^k$ | Actual departure time of train $k$ from station $u$ |
| $r_{u,u+1}$ | Minimum interval time of the same track occupied by adjacent train |
| $I$ | Time interval between trains spaced by automatic block signals |

- The Constraint of Dwell Time

The dwell time of trains at each station must be longer than the minimum dwell time.

$$D_u^k - A_u^k \geq T_u^k, k \in K, u \in U \tag{2}$$

- The Constraint of Interval Time of Track

Adjacent trains need to meet the constraint of minimum interval time of the same track.

$$\begin{cases} \left|A_u^{k+1} - A_u^k\right| \geq I \\ \left|D_u^{k+1} - D_u^k\right| \geq I \\ k \in K, u \in U \end{cases} \tag{3}$$

- The Constraint of Departure Time

The passenger train can only depart from the railway station after ensuring that all passengers who purchase tickets have completed boarding and landing, and it is usually not allowed to depart earlier than the scheduled departure time.

$$\begin{cases} D_u^k \geq D{\prime}_u^k \\ D_u^{k+1} - D_u^k \geq t_{\text{dep}} \\ A_u^{k+1} - A_u^k \geq t_{\text{arr}} \\ k \in K, u \in U \end{cases} \tag{4}$$

- The Constraint of Minimum Interval Time

In order to ensure the safety of train operation, the running speed of trains cannot exceed the maximum running speed specified according to the line design conditions, that is, it meets the requirement of minimum interval time of the same track occupied by adjacent train.

$$A_{u+1}^{k} - D_{u}^{k} - \tau_{\text{start}} - \tau_{\text{stop}} \geq r_{u,u+1}, k \in K, u \in U \tag{5}$$

- The Constraint of Crossing

Trains in the same direction cannot appear in the same space at the same time. Even at the station, one arrival and departure line can only be occupied by one train at the same time.

$$\begin{cases} D_{u}^{k} > D_{u}^{g} \Rightarrow A_{u}^{k} > A_{u}^{g} \\ k \in K, u \in U, k \neq g \end{cases} \tag{6}$$

## 4 Model Solving

Genetic algorithm is a random search algorithm and can solve combinatorial optimization problems well. The basic idea of solving the problem of timetable rescheduling by genetic algorithm is as follows:

- Chromosome Coding

In order to make chromosomes correspond to the adjustment results better, this paper adopts integer coding, the gene is the number of delayed trains, and the chromosome length is the number of all delayed trains.

- Population Initialization

The departure sequence of 50 delayed trains is randomly generated, and the randomly generated departure sequence is likely to be an infeasible solution. The method adopted is to directly eliminate the infeasible solution by using constraints.

- Constructing Fitness Function

The fitness function designed in this paper is the reciprocal of the total weighted arrival delay time of the next station of the adjustment station, that is, the higher the chromosome fitness with the smallest total weighted arrival delay time when the train arrives at the next station according to the adjusted departure sequence.

$$\begin{cases} f_{u}^{i} = \frac{1}{\sum_{k \in K'} a_{k} \times \left(A_{u+1}^{k} - A_{u+1}^{\prime k}\right)} \\ i = 1, 2 \ldots n \end{cases} \tag{7}$$

- Choice

The better parental chromosomes were screened out by selection operation, and the chromosomes with low fitness were eliminated. According to the principle that the higher the fitness of the parental chromosome, the greater the probability of inheritance to the next generation, this paper uses roulette to select the better parental chromosome.

$$P_i = \frac{f_u^i}{\sum_{i=1}^{n} f_u^i} \times 100\% \tag{8}$$

- Crossing

Because the chromosomes encoded in this paper are the departure sequence of delayed trains, no matter what way the two parental chromosomes cross, it will inevitably produce infeasible solutions, so the diversity of population depends on mutation operation.

- Mutation

In order to make the new individual produced feasible solution, the genes of the two positions are randomly exchanged, so as to increase the diversity of population and prevent falling into local optimal solution.

- Stop

When the iteration number reaches the maximum iteration number, the adjustment is stopped, which is manually input, and the default maximum iteration number is 100 generations.

# 5 Example Analysis

## 5.1 Example Description

There are 24 stations along the Beijing-Shanghai high-speed railway (see Fig. 1). Because TianjinXi Station is not on the main line of Beijing-Shanghai high-speed railway, this paper selects the remaining 23 stations except TianjinXi Station as the research object.

## 5.2 Delay Condition Setting

This paper selects 19 trains from BeijingNan to Shanghai Hongqiao from 7:00 to 12:00 one day. In order to facilitate the subsequent adjustment of train departure sequence, the trains are numbered as 1~19 from small to large according to the departure sequence of the departure station. When scheduling the train timetable, it can be found that G1, G3, G5, G7, G9 and G11 trains have few stops, high slope of train running line, high travel speed and overtake other trains at some stations. Therefore, it is determined that these six trains are all high-grade trains and the remaining 13 trains are low-grade trains. The adjustment weight of high-grade trains is 0.3, and that of low-grade trains is 0.2.

Assuming that all trains start on time, the research scope of this paper is the delay of arrival of a single train, that is, the delay of a single train in 22 sections of Beijing-Shanghai high-speed railway. Set the delay situation as G113 arrives at CangzhouXi Station 16 min later, and use only interval accelerated operation and genetic algorithm to output adjustment results for this situation.

**Fig. 1.** Beijing-Shanghai high-speed railway route map

### 5.3 Analysis of Results

Train G113 was originally scheduled to arrive at CangzhouXi Station at 9:30, but actually arrived at 9:46. The subsequent high-grade train G5 was originally scheduled to cross G113 at CangzhouXi Station at 9:40. However, due to the delay of the train, the train G5 was forced to slow down in TianjinNan-CangzhouXi Section. The delayed time of arriving at CangzhouXi Station is the actual arrival time of train G113 plus the minimum arrival interval. As scheduled, the train G5 does not need to stop at CangzhouXi Station. Train G113 runs on time between CangzhouXi Station and DezhouDong Station, and resumes on time at DezhouDong Station at 10:12. Train G5 runs in the next interval and resumes punctual operation at Zaozhuang Station at 11:07. There are 2 late trains involved in this adjustment, with a total arrival delay of 60.15 min and 6 late stations. All late trains resume punctual operation before arriving at the terminal. The scheduled train timetable and the actual train timetable are adjusted by interval acceleration method and genetic algorithm (see Fig. 2).

The adjustment result is that the total arrival delay time is 504.05 min only by interval acceleration operation, while the total arrival delay time is 67.15 min by genetic algorithm adjustment, which is 86.7% less than the delay time of only interval acceleration operation; The number of delay stations only for interval acceleration operation is 20, and the total number of stations with delay is adjusted to 6 by genetic algorithm. Therefore, the calculation result of genetic algorithm is better, which is mainly reflected in the better results of total arrival delay time, total number of late trains and terminal delay time, and the output result is also more stable. Under the condition of small delay time, all delayed trains can resume punctual operation, ensuring that the total delay time and adjustment time are within an acceptable range, and the objective function value is better.

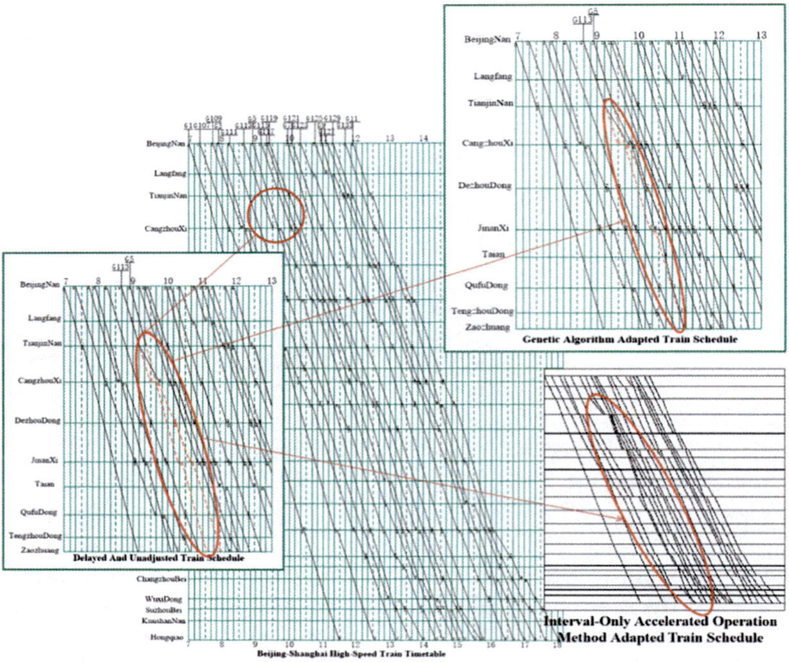


**Fig. 2.** The best rescheduled time-distance diagram based on the proposed GA and interval-only accelerated operation method respectively.

## 6 Conclusion

In this paper, aiming at minimizing the total weighted arrival delay of trains and satisfying various time interval constraints, a train operation adjustment model is established, and genetic algorithm is used to solve the model. By calculating the arrival and departure time of delayed trains at each delayed station, their arrival and departure times are determined by using constraints, so that all the generated solutions are feasible solutions, which improves the operation efficiency and shortens the time for finding satisfactory solutions. Compared with the adjustment scheme of only accelerating operation in interval, the calculation result of genetic algorithm is better, which is mainly reflected in the better results of total arrival delay time, total number of late trains and terminal delay time, and the output result is also more stable.

**Acknowledgements.** This study is funded by the National Key Research and Development Program of China (Grant No. 2022YFB4300603) and the National Natural Science Foundation of China (72001021).

## References


1. China State Railway Group Co., Ltd.: Statistical Bulletin 2021 [EB/OL]. People's Railway News (01 May 2022)
2. Cacchiani, V., et al.: An overview of recovery models and algorithms for real-time railway rescheduling. Transp. Res. Part B-Methodol. **63**, 15–37 (2014)
3. Quaglietta, E., et al.: The ON-TIME real-time railway traffic management framework: a proof-of-concept using a scalable standardised data communication architecture. Transp. Res. Part C-Emerg. Technol. **63**, 23–50 (2016)
4. Zhan, S.-G.: Real-time high speed train rescheduling considering the multiple units connecting. J. Transp. Eng. Inf. **18**(02), 1–9, 38 (2020). (in Chinese)
5. Gao, R., Niu, H.: Study on additional train timetable algorithm in flexible manner based on ADMM approach. J. China Railw. Soc. **43**(02), 21–29 (2021). (in Chinese)
6. Zhiwen, L., Qin, Y., Wang, M.: Research on high-speed railway operation adjustment model based on priority. Smart Resilient Transp. **4**(1), 12–21 (2022)
7. Wen, P., Zhao, P., Yao, X., Zhang, P.: Timetable rescheduling during last train period in high-speed rail network considering passenger transfer connections. J. China Railw. Soc. **44**(12), 1–9 (2022). (in Chinese)
8. Zhang, P., Zhao, P., Qiao, K., Wen, P.: Research on high-speed railway timetable rescheduling considering passenger transfer. J. China Railw. Soc. **44**(2), 8–15 (2022). (in Chinese)
9. Wang, M., Wang, L., Xu, X., Qin, Y.: Real-time train dispatching on high-speed railway under severe primary delays and speed limit. J. China Railw. Soc. **44**(1), 8–16 (2022). (in Chinese)
10. Zhang, J., Ye, Y.: Timetable rescheduling model of high-speed railway based on global conflict resolution. J. China Railw. Soc. **45**(2), 1–12 (2023). (in Chinese)
11. Wang, M., Wang, L., Xu, X., Qin, Y., Qin, L.: Genetic algorithm-based particle swarm optimization approach to reschedule high-speed railway timetables: a case study in China. J. Adv. Transp., 1–12 (2019)
12. Ming, L.E.I., Xuelei, M.E.N.G.: High-speed railway operation adjustment based on co-evolutionary genetic algorithm. J. Railw. Sci. Eng. **14**(6), 1137–1145 (2017)

# An Optimization Method of Suburban Railway Stopping Scheme Under the Need of Running Multiple Train Types

Rudong Yang[1,3,4](✉), Ming Xu[2,3], Xiang Wu[1,3,4], Zhenhao Fei[1,3], and Ruihua Xu[3,4]

[1] CASCO Signal Co., Ltd., Shanghai 200072, China
{yangrudong,wuxiang,feizhenhao}@casco.com.cn
[2] Shanghai Suburban Railway Operation Co., Ltd., Shanghai 201102, China
13162622612@126.com
[3] Shanghai Engineering Research Center of Railway Intelligent Dispatching Command System, Shanghai 200072, China
rhx@tongji.edu.cn
[4] Tongji University, Shanghai 201804, China

**Abstract.** With the development of regional integration, suburban railway need to interconnect with national railways and operate networked within itself. Passenger flow in the region show the following trends: increasing cross-line passenger volume, uneven spatial-temporal distribution, and the inability of local trains in meeting fast access needs. To adapt to those trends, the suburban railway need to operate multiple train types, especially express and local trains. In this paper, we take the stopping scheme of express trains as the research object, take the total travel time of passengers and the operation cost of railway enterprises as objectives, establish the optimization model, design an improved ant colony algorithm based on computer simulation, so as to calculate the number of stops and specific stopping station of express trains. Finally, taking Shanghai suburban railway Airport Connecting Line as an example, the optimized operation scheme is running 3 express trains and 6 local trains in peak hour, the stopping station of the express train are SLN, DJQ and PDJC. The total travel time of the optimized scheme is reduced by 1.8% compared with the direct trains, despite the increase in operation cost. The case proves that optimizing the stopping scheme of express trains can save the overall travel time of passengers while meeting the transportation demand, proves the effectiveness of the model and algorithm, improves the service quality of suburban railway.

**Keywords:** Suburban Railway · Stopping Scheme · Express and Local Train · Improved Ant Colony Algorithm

## 1 Introduction

In recent years, China's suburban railways have flourished, especially in metropolitan areas such as the Yangtze River Delta and the Guangdong-Hong Kong-Macao Bay Area, where thousands of kilometers of suburban railway lines are planned to meet the needs


Y. Qin et al. (Eds.): EITRT 2023, LNEE 1137, pp. 20–30, 2024.
https://doi.org/10.1007/978-981-99-9311-6_3

of passengers for fast access to travel and commute in the face of the expanding urban spatial scales. At the same time, National Comprehensive Transportation Planning and various policy development plans have put forward the policy of integrating the four networks of rail transportation (the four networks of rail transportation include national trunk railway network, intercity railway network, suburban railway network and the urban rail transit network) [1], suggesting that suburban railway lines need to meet the requirement of interconnection with national railways and the networked operation mode within suburban railway network, which means that certain lines should run multiple types of trains, including the trains from the national railway lines, interline or crossline trains from other suburban railway lines and the express/local trains of this certain line, so as to meet the diverse travel needs as well as to guarantee the service level.

The stopping scheme is directly related to the passenger travel time and service level. The train stopping scheme is formulated according to the technical characteristics of the line and stations, travel demand, organization mode and other factors, which determine the train's stopping stations to ensure that the transportation capacity provided on the line could well matches the passenger travel demand. The appropriate formulation of train stopping scheme is the basis for improving the level of passenger transportation services and achieving efficient utilization of line capacity. Therefore, a series of factors including passenger demand, station conditions, station service capacity, train travel speed, and section passing capacity need to be considered in the process of stopping scheme making.

In this paper, the train stopping mode of the suburban railway is divided into two types, local train and express train, in the local train stopping mode, the train stops at each station of the line, each station can get the same service density, which helps to collect and disperse passenger flow around the station and improve the efficiency of passenger travel and transfer along the line, but will cause a corresponding decrease in transport efficiency between important stations, which will also lead to a waste of travel time and capacity utilization. In the express train stopping mode, train stops at important transportation hubs, core areas stations, etc. The service level around the stations is different, and the train travel speed is improved by reducing stops, which helps to realize the fast transportation demand of passengers, but to a certain extent, it affects the travel efficiency of passengers along the line.

Many scholars at home and abroad have studied the optimization of train stopping scheme, Suh [2] constructed a stopping scheme optimization model based on OD data, inter-station distance and other factors for urban rail transit in Seoul, South Korea, and the solution results showed that the express/local train mode during peak hours can save passengers' travel time in general although it makes some passengers' waiting time increase. Jamili [3] studied the inter-stop train operation mode of urban rail transit under uncertainty conditions and proposed a robustness-based model of urban rail transit express/local train stopping scheme. Zeng [4] studied the operation of express/local trains on the basis of short and long routes, and constructed a single-objective optimization model with passenger travel time and a dual-objective nonlinear integer optimization model with passenger travel cost and business operation cost, respectively, but the maximum train operating speed in the model did not exceed 80 km/h. Su [5] constructed an alternative set of train stopping schemes based on the analysis of passenger travel demand and used simulated annealing algorithm to solve the HSR train running scheme. Yusin Lee [6] analyzed the problem of express/local trains in Japanese suburban railway and analyzed the importance of running express/local train running schemes for the influence of passenger travel time and passenger flow induced increase. Also, there are some papers that study the effect of operating express/local trains on line capacity [7, 8]. Ding [9] pointed out that the current urban lines have spatial and temporal uneven passenger flow characteristics, and studied the interrelationship between the factors influencing fast and slow trains to study the line capacity by setting different combinations of express and local trains. Zhang [10] focuses on the effect of the number of stops along the suburban line, which reduces the efficiency and attractiveness of the line, and investigates the operation of express/local trains to improve the line's capacity. In summary, the current research on the compilation of suburban railway operation scheme is more in-depth on the train operation route and operation scheme, but there is a lack of research on the train type, operation quantity and stopping scheme under the condition of multi-mode rail transit cooperative operation. Therefore, in order to match the actual situation of suburban railway under the integration of urban agglomerations, it is necessary to study and develop the optimization model of train stopping scheme for suburban railway, so as to better optimize the train operation scheme of urban railroads.

In view of the above situation, this paper proposes an optimization method of stopping scheme of suburban railway for running multiple types of trains. Based on the actual organization requirements, the corresponding optimization model for the stopping scheme of express trains is established with the total travel time of passengers and the operating profit of enterprises as the objectives, so as to determine the number of stops and specific stops stations of express trains, and design an ant colony algorithm based on computer simulation.

The remainder of this paper is organized as follows. In the next section the train stopping scheme's optimization model is constructed which include the parameters, Objective function and Constraints. In the third section, the designed ant colony algorithm is introduced. In the fourth part the proposed methodology is applied to the Shanghai suburban railway Airport Connecting Line as a case study. The final section summarizes the paper.

## 2 Model

In this section, the optimization model of the train stopping scheme is proposed.

### 2.1 Model Assumption

The passengers will firstly take the trains whose travel time are shorter according to the express train's operation proportion if the trains will pass the target station. If the express train doesn't pass the target station, then choose the local train.

Only two train types, express and local train, are considered in this paper. The running speed is the same, and the difference is the station stop time and extra time for train's starting and stopping because of the different train stopping plans.

## 2.2 Indices and Parameters

See Table 1.

**Table 1.** Indices and parameters

| Notation | Description |
|---|---|
| $n$ | Station index, 1,2,3,…, $n$ |
| $Q_{ij}$ | Local train's travel time between station i and station j, include the station stop time and extra time for train's starting and stopping |
| $L_{ij}$ | Distance between station i and station j |
| $t_{extra}$ | The extra time for train's starting and stopping when the train stop at one station |
| $I$ | The minimum interval time between two trains |
| $\mu$ | Train type, if $\mu = 1$, it is express train; if $\mu = 2$, it is local train |
| $s$ | The length of each periodic period |
| $T_S$ | The total travel time of all passengers during the period S |
| $P_{ijs}$ | The passenger volumes travel from station i to station j during period $s$ |
| $m_x$ | Various operating cost of train; $m_1$ stands for fixed train cost, $m_2$ stands for the train's operation cost per kilometer, $m_3$ stands for train stopping cost, $m_4$ stands for passenger fare per kilometer |
| $\lambda$ | The proportion of passengers who take the express trains |
| $t_i$ | The train's stop time at station i |
| $Z_i$ | If $Z_i = 0$, it means the express train won't stop at station i; if $Z_i = 1$, it means the express train will stop at station i |
| $C$ | The lower limit of train attendance |
| $N_s^{\mu}$ | The number of type $\mu$ trains operating during period $s$ |

## 2.3 Modeling

The formulation of the suburban railway stopping scheme should be based on meeting the travel needs of passengers, focusing on providing passengers with faster and more convenient transportation services, improving transportation efficiency, reducing travel time, and reducing the operating costs of suburban railway enterprises. Therefore, the objective function of train stopping plan should be to maximize the total travel time of passengers and the operating income of the enterprise.

**Objective function**

$$minT_s = \sum_{i=1}^{n}\sum_{j=1}^{n}\left[\begin{array}{l} Z_iZ_jP_{ijs}\left[\lambda\left(Q_{ij} + \sum_{x=i+1}^{j-1}(Z_x - 1) * t_{extra} + \frac{s}{N_s^1}\right) + (1-\lambda)\left(Q_{ij} + \frac{s}{N_s^2}\right)\right] \\ +(1 - Z_iZ_j)P_{ijs}\left(Q_{ij} + \frac{s}{N_s^2}\right) \end{array}\right] \quad (1)$$

$$minZ_s = \sum_{i=1}^{n}\sum_{j=1}^{n} m_4 P_{ijs} L_{ij} - \left[ (m_1 + m_2 L_{1n}) \cdot \left(N_s^1 + N_s^2\right) + \sum_{u=1}^{2}\sum_{j=2}^{n-1} m_3 Z_j N_s^u \right] \quad (2)$$

**Constraints**

$$Z_i Z_j\left[\lambda * CN_s^1 + (1-\lambda) * CN_s^2\right] + \left(1 - Z_i Z_j\right) * CN_s^2 \geq P_{ijs} \quad (3)$$

$$\frac{P_{ijs}}{Z_i Z_j\left[\lambda * CN_s^1 + (1-\lambda) * CN_s^2\right] + (1 - Z_i Z_j) * CN_s^2} \geq c \quad (4)$$

$$\frac{s}{N_s^1 + N_s^2} \geq I \quad (5)$$

$$\frac{N_s^1}{N_s^1 + N_s^2} \leq \lambda \quad (6)$$

$$N_s^2 \geq 1, s = 1, 2, \ldots, S \quad (7)$$

$$x_1 = 1, x_n = 1 \quad (8)$$

$$i, j = 1, 2, \ldots, n \quad (9)$$

The objective function (1) is to minimize the total travel time of passengers in each time period s, including train's running time, waiting time in the station and extra stopping time. When there are express trains running between station i and station j, passengers shall choose the express train in a certain proportion, which shall be no less than the proportion of the express train in all the operating train. The objective function (2) is to maximize the operating income of the railway enterprise in period s, which is composed of fixed cost of train operation, variable cost of train operation(include) and ticket income (which is determined by factors such as passenger volume, the tariff per kilometer, etc.).

Constraint (3) indicates that the transport capacity should meet the demand in the time period s; Constraint (4) indicates that the full load rate of the train should not be lower than the set value during period s; Constraint (5) indicates that the departure interval of the local train and express train should be bigger than the minimum interval time; Constraint (6) indicates that if passengers choose the express trains according to a certain proportion during the period s, the proportion shall not be less than the proportion of the express train in all trains; Constraint (7) indicates that at least one local train should be operated during the period s, while constraint (8) indicates that the express train need to stop at both the starting and ending stations.

## 3 Algorithm

The combination of station nodes in the suburban railway stopping scheme is complex, so the corresponding ant colony algorithm is studied and combined with computer simulation to search and traverse the feasible solutions for different stopping schemes of express trains, so as to find the optimal stopping scheme.

### 3.1 Algorithm Description

The problem of designing train stopping scheme for suburban railway can be considered as a directed empowerment map G, $G = (N, A)$, set N = {1,2,3..., n} is the stopping station of the train, set $A = (i, j|i, j \in N, i < j)$ is the express train's stopping station connection.

In this paper, we design an ant colony algorithm to study the optimization of train stopping scheme of suburban railway and simulate the stopping scheme of express trains. In the ant colony algorithm, each ant starts from the first station and selects the next node according to pseudo-random, and the next node serial number j satisfies greater than or equal to the previous node serial number i. The ant moves until the end station n and puts the selected node into the solution set S after the movement is completed. Each element in the solution set S at this point is a feasible stopping solution for the train.

### 3.2 Heuristic Factor Construction

The heuristic factor represents the strength of the role of the randomness factor in the search process of the ant colony, which can inspire the ants to choose a better node and indicates the relative importance of the trajectory. As the key to the solution of the ant colony algorithm, the heuristic factor is particularly important to determine, and the total passenger saving time between selected stopping stations will be used as the heuristic factor in this paper.

$$\theta(i,j) = \theta_{ijs} \tag{10}$$

In Eq. (10):

$$\theta_{ijs} = \sum_{k=i+1}^{j} (t_k + t_{extra}) \cdot \sum_{r \in s} \lambda \cdot P_{rjs} - \sum_{m=1}^{r-1} \lambda \cdot P_{rjs}\left(t_m + t_{extra} + \frac{s}{N_s^2} - \frac{s}{N_s^1}\right) \tag{11}$$

In the Eq. (11), $\sum_{r \in s} \lambda \cdot P_{rjs}$ is the number of passengers when the train stop at station r, j; $\sum_{k=i+1}^{j} (t_k + t_{extra}) \cdot \sum_{r \in s} \lambda \cdot P_{rjs}$ indicates the total passenger time saved when the express train stops at station r; $\sum_{m=1}^{r-1} \lambda \cdot P_{rjs}\left(t_m + t_{extra} + \frac{s}{N_s^2} - \frac{s}{N_s^1}\right)$ represents the increase in passenger waiting time corresponding to an express train stop at station r.

### 3.3 Pheromone Update Rules

The pheromone concentration indicates the number of times a path is selected; the denser the pheromone of a path, the higher its probability of being selected, and the pheromone concentration of a path with higher selection increases over time. In the ant colony algorithm studied in this paper, the pheromone is updated globally, and the best path is updated after all feasible solutions are traversed.

The dissipation degree of pheromone is denoted by ρ, which responds to the disappearance level of pheromone. Taking too large a value will affect the randomness and global optimality; taking too low a value will lead to a lower convergence rate, which is generally taken between [0.2,0.5]. The pheromone concentration dissipation rules on each path are:

$$\tau_{\omega+1}(i,j) = \rho\tau_{\omega+1}(i,j) + \Delta\tau_{\omega}(i,j) \tag{12}$$

The pheromone concentration update rule of the ant colony is as follows, $\Delta\tau(i,j)$ denotes the pheromone increment:

$$\Delta\tau(i,j) = \sum_{k=1}^{n} \Delta\tau_{\omega}^{k}(i,j) \tag{13}$$

The pheromone concentration update rule on the walk path of a single ant is:

$$\Delta\tau_{\omega}^{k}(i,j) = \begin{cases} \frac{Q}{T_k}, \textit{if the k} - \textit{th ant passes through}(i,j) \\ 0, \textit{the other situations} \end{cases} \tag{14}$$

where $Q$ is the pheromone constant, which is the total number of pheromones after traversing once; $T_k$ is the value of the optimization objective function for the k-th scenario.

### 3.4 Target Conversion Rules

The model has two objective functions: the minimum value of total passenger travel time (objective function 1) and the maximum value of enterprise operation revenue (objective function 2). In each iteration of the ant colony algorithm, when the traversal is completed once, the value of objective function 1 of each ant, $T_s$, and the minimum value of objective function 1 of all ants, $T_{s,min}$, are converted by dimensionless conversion, $T_s^{'} = \frac{T_{s,min}}{T_s}$; the same process is performed for objective function 2 to obtain $Z_s^{'} = \frac{Z_s}{Z_{s,max}}$. The final solution $\varphi = \omega_1 T_s^{'} + \omega_2 Z_s^{'}$, where $\varphi$ is the bi-objective function evaluation value, and $\omega_1$, $\omega_2$ are the weight coefficients, which are determined by the social value and economic benefits brought by each objective, thus indicating the importance of the objective function. The weight coefficients in this model are set according to the relevant historical experience of suburban railway, and the total passenger travel time weight coefficient $\omega_1$ is higher.

### 3.5 Algorithm Steps

Step 1: Importing passenger flow OD data, section running time, station stopping time data, etc. for the study time period, and setting the corresponding parameters, initial stopping scenarios $X_0$ and the solution $\omega = 1$;

Step 2: Based on the input OD data, the corresponding section passenger flow is calculated, and the number of trains, departure interval and other information is determined based on the train full capacity index and the section passenger flow;

Step 3: Update the heuristic factor (formula (10)) based on the calculated number of trains and departure interval information;

Step 4: Construct the corresponding feasible solution for the stopping scheme and add the feasible solution to the result set by traversal and complete one traversal;

Step 5: All feasible solutions are evaluated comprehensively according to the value of the objective function, and the solution with the best value is obtained and recorded as the new large station train stop solution, $X_{\omega}^{'}$;

Step 6: If the algorithm reaches the maximum number of iterations, then output the optimal solution $X_{\omega}^{'}$, otherwise make $\omega = \omega + 1$ and skip to Step2.

# 4 Case Study

## 4.1 Related Data

This paper takes Shanghai suburban railway Airport Connecting Line (in the first stage, the operation section is from Shanghai Hongqiao Station (SHHQ) to Shanghai East Railway Station (SHD)) as an instance for verification, Fig. 1 is the line's schematic map. The Airport Connecting Line is under construction and will open in 2024. According to the feasibility report, the section distance and running time information are shown in Table 2. If a train skip one station, it will save 2 min because of the station stop time and extra time for train's starting and stopping. There are two types of EMU (Electric Multiple Unit), 4 carriages and 8 carriages, and the rated capacity of each carriage is 768. The operating cost information is shown in Table 3. The passenger OD (origin and destination) data in the rush hour is listed in Table 4 and section passenger volumes are shown in Table 5.

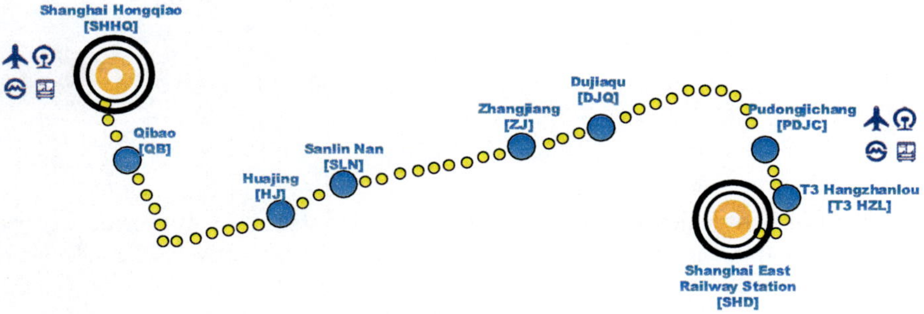


**Fig. 1.** The schematic map of Airport Connecting Line.

## 4.2 The Calculation of Stopping Scheme

According to the feasibility report of Shanghai suburban railway Airport Connecting Line, set the proportion of express train ridership $\lambda = 0.8$, set the pheromone factor

**Table 2.** Train route information of the Airport Connecting Line

| Operation section | Stopping type | Distance (km) | Travel time (min) | Station stop time+extra time (min) |
|---|---|---|---|---|
| Shanghai Hongqiao-Shanghai East Railway Station | Local train | 68.66 | 48 | 2 |

**Table 3.** Operating cost of the Airport Connecting Line

| Fixed train cost (yuan/hour) | Operation cost of train(yuan/km) | Train stopping cost(yuan/station) | Passenger fare (yuan/km) |
|---|---|---|---|
| 5000 | 120 | 200 | 0.45 |

**Table 4.** Passenger OD volumes of Airport Connecting Line in the rush hour

| OD Volume | SHHQ | QB | HJ | SLN | ZJ | DJQ | PDJC | T3 HZL | SHD |
|---|---|---|---|---|---|---|---|---|---|
| SHHQ | 0 | 928 | 1063 | 2109 | 1001 | 1092 | 188 | 104 | 484 |
| QB | 0 | 0 | 1133 | 474 | 93 | 151 | 22 | 12 | 28 |
| HJ | 0 | 0 | 0 | 96 | 287 | 892 | 44 | 24 | 62 |
| SLN | 0 | 0 | 0 | 0 | 1280 | 712 | 131 | 73 | 176 |
| ZJ | 0 | 0 | 0 | 0 | 0 | 756 | 141 | 78 | 57 |
| DJQ | 0 | 0 | 0 | 0 | 0 | 0 | 72 | 40 | 53 |
| PDJC | 0 | 0 | 0 | 0 | 0 | 0 | 0 | 220 | 789 |
| T3 HZL | 0 | 0 | 0 | 0 | 0 | 0 | 0 | 0 | 438 |
| SHD | 0 | 0 | 0 | 0 | 0 | 0 | 0 | 0 | 0 |

$\alpha = 1$, heuristic function factor $\beta = 1$, pheromone constant $Q = 1$, and the number of ants set to 100, and calculate 100 times after iteration respectively, the calculated operation plan is running 3 express trains/h and 6 local trains/h in the peak hour, and the results of the stopping scheme are shown in Table 6.

From the results of 100 iterations, Solution 4 with 3 stopping stations appears 64 times, while Solution 3, Solution 2 and Solution 1 appear 19, 11 and 6 times respectively, which indicates that Solution 4 is the better train stopping option, the total travel time of Solution 4 is reduced by 1.8% compared with the direct trains in Solution, despite the increase in operation cost. The result also verifies the feasibility of the ant colony algorithm.

**Table 5.** Section Passenger volumes of Airport Connecting Line in the rush hour

| Section name | Morning peak | | Evening peak | |
|---|---|---|---|---|
| | SHHQ-SHD | SHD-SHHQ | SHHQ-SHD | SHD-SHHQ |
| SHHQ-QB | 5206 | 5850 | 6518 | 2887 |
| QB-HJ | 5933 | 3478 | 2784 | 3944 |
| HJ-SLN | 5116 | 3074 | 1546 | 3921 |
| SLN-ZJ | 7462 | 2763 | 1147 | 6166 |
| ZJ-DJQ | 5554 | 3087 | 1769 | 3733 |
| DJQ-PDJC | 1890 | 1987 | 993 | 367 |
| PDJC-T3 HZL | 2273 | 3056 | 1785 | 808 |
| T3 HZL-SHD | 2143 | 3290 | 1883 | 691 |

**Table 6.** The calculation result of stopping scheme of the 100 iterations (all the trains depart from SHHQ)

| Solution No | Number of express train stops | Stopping scheme | Operating revenue of enterprise(yuan) | Total passenger travel time(min) | Solution appearances times |
|---|---|---|---|---|---|
| 1 | 1 | PDJC | −10781.3 | 336043.5 | 6 |
| 2 | 2 | DJQ, PDJC | −11329.2 | 335283.8 | 11 |
| 3 | 3 | HJ, DJQ, PDJC | −12063.9 | 334385.1 | 19 |
| 4 | 3 | SLN, DJQ, PDJC | −12103.3 | 330273.2 | 64 |

From the calculation results, it can be seen that the overall travel time of passengers is less when the express train makes 3 stops compared to 2 stops or 1 stop, but the increase in the number of stops will reduce the operating revenue of the enterprise, because the operating cost of the enterprise is greater than the ticket revenue of passengers in all combinations of stopping station in the model, so the operating revenue of enterprise is negative in the model results. From the OD and section passenger flow data, it can be seen that the passenger flow between the stops of the express train is large, and by optimizing the stopping scheme of the express train, it can play a role in reducing the travel time of passengers. At the same time, taking into account the demand of passengers of suburban railway to board and alight at intermediate stations and interchange into urban areas, it is more reasonable to set 3 stations for the intermediate station stopping scheme, which not only ensures the travel time but also meets the travel demand of passengers.

By operating express trains, the entire travel time will also be reduced, saving about 6 to 8 min of travel time, the passenger time between SHHQ and PDJC can be further compressed to 30 min, the SHHQ to SHD can be compressed to 40 min. The operation of the express train can achieve a fast and close connection and is also conducive to the development requirements of the two airports to fully integrate the resources.

## 5 Conclusion

This paper proposed an optimization method of suburban railway stopping scheme to meet the demand of running different kinds of trains, and could also meet the travel time and passenger flow volume requirements. And designed an improved ant colony algorithm based on computer simulation. Finally, the case study of Shanghai Suburban Railway Airport Connecting Line is analyzed, and obtain the optimized stopping scheme for the express trains, including the number of stops and specific stations, which verifies the effectiveness of the method.

**Acknowledgements.** This work is supported by Science and Technology Research and Development Program of China National Railway Group Corporation: Research on System Architecture and Key Technologies of Railway Dispatching Command Management and Control System (P2022X013).

## References

1. Outline of the Fourteenth Five-Year Plan for National Economic and Social Development of the People's Republic of China and the Vision 2035, http://www.gov.cn/xinwen/2021-03/13/content_5592681.htm. Accessed 13 Mar 2021. (in Chinese)
2. Sogin, S.L., Caughron, B.M., Chadwick, S.G.: Optimizing skip stop service in passenger rail transportation. In: Joint Rail Conference 2012:501. Philadelphia, Pennsylvania, USA (2012)
3. Jamili, A., Aghaee, M.P.: Robust stop-skipping patterns in urban railway operations under traffic alteration situation. Transp. Res. Part C: Emerg. Technol. **61**, 63–74 (2015). Elsevier, New York
4. Zeng, C., Hou, Y., Luo, Q., et al.: Optimal scheme of train stopping at express-local train station in condition of full-length and short-run routes. Urban Mass Transit **22**(7), 63–66 (2019). (in Chinese)
5. Su, H., Tao, W., Hu, X.: A line planning approach for high-speed rail networks with time-dependent demand and capacity constraints. Math. Probl. Eng. (2019). Hindawi, London, United Kingdom
6. Lee, Y., Chen, C.Y.: A heuristic for the train pathing and timetabling problem. Transp. Res. Part B: Methodol. **43**(8–9), 837–851 (2009). Elsevier, New York
7. Xin-Miao, Z., Ai-Bing, X., Lu, L.I.U.: Carrying capacity of stations under express/local operation of urban rail transit. J. Transp. Syst. Eng. Inf. Technol. **17**(06), 200–206 (2017). (in Chinese)
8. Zhang, C., Han, B.M., Zhang, Q.: Optimization methods of express and local train skip stop scheme on airport rail line. Urban Rapid Rail Transit **28**(05), 67–70 (2015). (in Chinese)
9. Ding, X., Zhang, S., Liu, Z., et al.: The analysis and calculation method of urban rail transit carrying capacity based on express-slow mode. Math. Probl. Eng., 1–9 (2016). Hindawi, London, United Kingdom
10. Li, S., Zhang, X., Chen, J., et al.: Study on express-slow routes mode of foreign subway. In: Proceedings of the 2013 International Conference on Electrical and Information Technologies for Rail Transportation (EITRT2013), vol. II. Springer, Berlin Heidelberg (2014)

# Urban Rail Transit Track Area Extraction Using LiDAR

Shen Tuo[1], Zhou Jinhuang[1], Xie Yuanxiang[1], Deng Chenxin[2(✉)], and Qian Yanzuo[1]

[1] Shanghai Key Laboratory of Rail Infrastructure Durability and System Safety, Shanghai 201804, China
st8250@163.com, {2133387,2233390,1933341}@tongji.edu.cn
[2] Shanghai Zegao Electrical Engineering Technology Co, Ltd., Shanghai, China
dengchenxin@shzegao.com

**Abstract.** The intrusion of obstacles into the track area is a significant issue that impacts the safety of urban rail transit. The obstacle detection technology based on LiDAR (LIght Detection and Ranging) meets the detection requirements of urban rail transit. However, under the limitations of current LiDAR performance, existing methods can only extract the track area up to 100 m, which is insufficient for the speed and braking distance of urban rail transit trains. In light of the characteristics of urban rail transit lines, a new method is proposed to indirectly extract the track area by detecting references parallel to the rail, i.e., tunnel walls, protective walls, and sound barriers. By means of border points selection, clustering, and curve fitting, two curves of the references on both sides are obtained. A track centerline is obtained from the two curves and is expanded to extract the track area according to size specifications. The proposed method in this paper is evaluated in a variety of scenes, including tunnel, elevated, and ground sections of urban rail transit lines. The result shows that the proposed method successfully extracts the track area within 150 m.

**Keywords:** LiDAR · Track area · Extraction method · Urban rail transit

## 1 Introduction

The intrusion of obstacles into the track area is an important issue affecting the safety of urban rail transit. The collision between trains and obstacles has caused major operational safety accidents. On November 15, 2017, 27 passengers and 2 staff were injured in a subway rear-end accident at Yuqun Station of Singapore Metro during the morning rush hour. On December 25, 2018, a train struck a pedestrian who entered the track area without permission in Shanghai Metro Line 3, resulting in the death of the person. On March 18, 2019, two MTR trains collided, injuring two drivers and causing serious damage to the trains. On the other hand, according to *Railway applications - Automated urban guided transport (AUGT) - Safety requirements* (IEC62267), fully automated trains with GoA4 (Grades of Automation 4) should be operated without drivers [1]. It is urgent for developing an on-board active obstacle detection technology.


Y. Qin et al. (Eds.): EITRT 2023, LNEE 1137, pp. 31–46, 2024.
https://doi.org/10.1007/978-981-99-9311-6_4

A train operates along fixed rails, so only obstacles that intrude into the track area may cause a collision accident. For the same reason, the train can only take braking measures when there is an obstacle in its path. To avoid accidents or minimize the speed in case of a collision, the on-board active obstacle detection technology should have a long detection range and be able to estimate the distance of obstacles. Trains, pedestrians, falling stones, equipment, and other potential non-specific objects in the track area will pose a threat to safety and should be detected as obstacles. The train operates early and late and passes through tunnels and outdoor sections, so the detection technology should be able to work stably under various lighting conditions. The obstacle detection technology based on LiDAR (LIght Detection And Ranging) meets the above requirements. However, due to the limitation of scanning accuracy, the point cloud collected by LiDAR is sparse at a long distance, making it difficult to extract the track area to determine whether a potential obstacle is intruding.

In light of the urgent need for the on-board active obstacle detection technology and the difficulties of current research, based on the characteristics of urban rail transit lines, this paper proposes a LiDAR-based track area extraction method, which can be applied to the LiDAR-based obstacle detection to meet the requirements and help improve safety.

## 2 Related Works

### 2.1 The On-Board Active Obstacle Detection Technology

The main sensor types used in the on-board active obstacle detection technology include: vehicle-vehicle communication [2, 3], vision [4, 5], millimeter wave radar [6] and LiDAR [7]. Among these, the first one senses other trains through vehicle-vehicle communication to avoid collisions between trains. This kind of technology has a long detection range and accurate estimation of the distance between trains and is less affected by external environmental conditions. However, it is not able to detect pedestrians, falling stones, and other obstacles that do not establish communication with trains. Videos and images contain rich and intuitive information, so vision-based technology is widely researched. Especially with the development of deep learning, vision-based technology performs excellently in detecting known obstacles. However, cameras are highly sensitive to lighting conditions. In particular, when a train enters and leaves tunnels, the drastic light change will make cameras difficult to work normally. In addition, vision-based technology is difficult to estimate the distance of obstacles. Stereo vision is limited by the width of train vehicles [8] and the distance estimation method based on deep learning is only able to cope with known obstacles [9]. Millimeter wave radar works stably in various external environmental conditions such as darkness, rain, and fog, and completes speed and distance measurements at the same time. However, due to its large operating wavelength, the millimeter wave radar performs poorly in detecting small targets such as falling stones and toolboxes. LiDAR works in any light conditions, performs well in distance measurement, and does not ignore small objects in scan lines. Recently, the scanning accuracy of LiDAR has been improving while the costs of equipment have been decreasing. After all, LiDAR is suitable for the on-board active obstacle detection. Table 1 shows the comparison of the above technologies.

**Table 1.** Comparison of obstacle detection technologies

| Performance | | Vehicle-Vehicle Communication | Vision | Millimeter Wave Radar | LiDAR |
|---|---|---|---|---|---|
| Detection range | | Long | Medium | Long | Medium |
| Target distance estimation | | Strong | Weak | Medium | Strong |
| Detecting capability | Trains | Strong | Strong | Strong | Strong |
| | Medium obstacles | – | Strong | Medium | Strong |
| | Small obstacles | – | Medium | Weak | Strong |
| Performance in bad conditions | Darkness | Strong | Weak | Strong | Strong |
| | Rainy weather | Strong | Weak | Strong | Medium |
| | Snow or fog | Strong | – | Strong | Weak |

## 2.2 LiDAR-Based Track Area Extraction Method

The track area is slightly wider than the zone between the two rails where intruding objects will collide with the train. Extracting the track area is the first step of obstacle detection.

Some studies use other sensors to extract the track area. Zhang [10] projects point clouds to images by calibrating the camera with LiDAR and extracts the track area using vision methods. This method relies on vision methods to complete an accurate segmentation of the track area, which is weak to lighting and weather conditions. Guo [11] builds a track map in advance with historical GPS data and uses real-time GPS information to match the location of trains in the track map to extract the track area. This method requires real-time accurate GPS information, which is difficult to work effectively in urban rail transit with multiple tunnels and buildings.

The track area can be obtained by extending the zone between two rails to both sides in a certain proportion. Stein [12] uses template matching to detect the notch of rails in the light rail and combines it with gauge information to prevent misjudgment. Geng [13] compares the features of LiDAR point cloud in a railway environment, and finds that features such as roughness, anisotropy, and curvature can be used to separate rail points from surrounding objects. Considering the vertical height difference between the rail surface and the ground, Chen [14] proposes methods to find potential rail points by detecting the height difference among points in small neighborhoods. Guo [15] detects potential rail points based on the feature that outer grounds near the rails miss points due to laser beams being blocked by the rail. Ren [16] proposes a rail detection method for low-density point cloud, which uses relative height, orientation and rail gauge to roughly select rail points and detect rails through a series of filtering steps. However, these methods all require a certain density of rail points in the point cloud and only work for a range of up to 100 m under the limitations of current LiDAR performance. The detection range of 100 m is far from sufficient for the speed and braking distance of trains in urban rail transit. Therefore, this paper aims to increase the extraction distance of the track area.

In addition to direct detection of rails, in urban rail transit lines, references parallel to the rails e.g., tunnel walls in tunnel sections, protective walls and sound barriers in elevated and ground sections, provide feasibility for increasing the extraction distance of the track areas. Thus, this paper proposes a new method of indirectly extracting the track area by detecting the references using LiDAR. So that the detection range of obstacles can be increased to meet the demand for safe and efficient operations in urban rail transit.

# 3 Feasibility Analysis and Process

## 3.1 Feasibility Analysis of Track Area Extraction Based on References

LiDAR emits laser beams at a certain frequency and receives reflections from objects in the surrounding environment. The emission angle and the time difference between emitting and receiving of the laser beams are used to determine the direction and distance of the points formed by objects. To fully scan the surroundings, LiDAR emits laser beams in all directions using mechanical rotation, resonant mirrors, and other techniques [17]. However, due to the limitation of minimum angle difference between laser beams, the points formed by the same object become fewer as the distance increases. Furthermore, the rail has a small surface and a poor reflection angle, which causes the points of the rail to be lost at closer distances.

Most sections in urban rail transit have references parallel to the rails. To illustrate, tunnel walls exist in tunnel sections and protective walls and sound barriers exist in elevated and ground sections. These references have a larger surface and a better reflection angle, so their points exist at longer distances.

Figure 1 shows a raw 3D point cloud collected by Livox Tele-15 LiDAR in an elevated section and the 2D point cloud projected to the horizontal plane. As seen in Fig. 1, the rails are recognizable up to about 80m and completely lost at more than 100 m. In contrast, the references still have several points at about 150 m and are recognized as a clear curve in the 2D point cloud. The references are detected well when the rails are lost, so the method proposed in this paper to indirectly extract the track area by detecting the reference is feasible.

## 3.2 Track Area Extraction Method Based on References

### 3.2.1 Overview of the Main Process

According to the analysis in Sect. 3.1, Fig. 2 shows the position relationship of detectable references and rails, track centerline, and track area on the horizontal plane. They are parallel to each other, so after fitting the curve of the references, the track area is obtained by translation and expansion.

Figure 3 shows the main process of the track area extraction method proposed in this paper. Firstly, the Lidar data is pre-processed and projected to the horizontal plane to get a 2D point cloud. Then, the border points formed by the reference on both sides

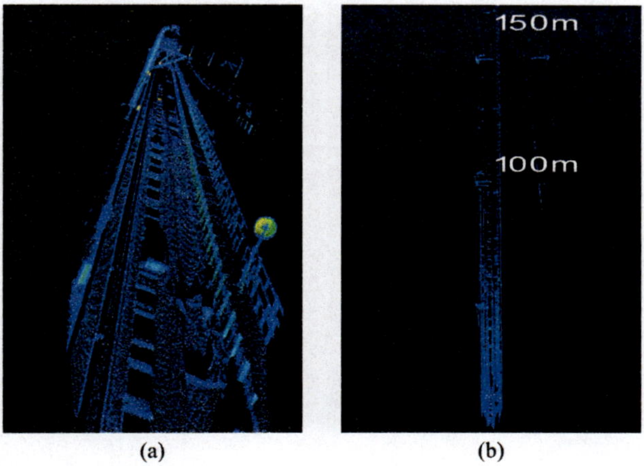


**Fig. 1.** The point cloud collected by Livox Tele-15 LiDAR in an elevated section. (a) The raw 3D point cloud. (b) The 2D point cloud projected to the horizontal plane.

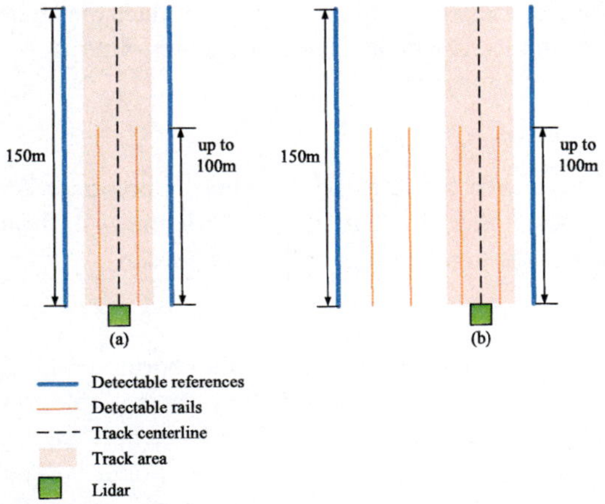


**Fig. 2.** The position relationship of detectable references and rails, track centerline, and track area on the horizontal plane. (a) Single line section. (b) Double line section.

are selected, the wrong points are removed by clustering, and the curve of each border is fitted. Finally, a better border curve is selected, and the track centerline is obtained by translating the better border curve and is expanded to obtain the track area.

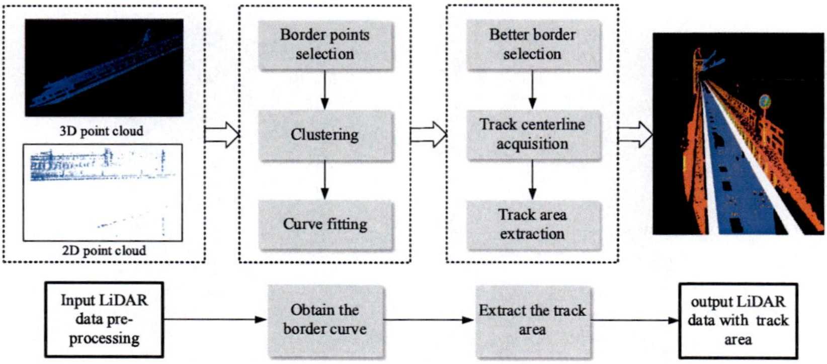


**Fig. 3.** The main process of the track area extraction method.

### 3.2.2 Details of the Process

(1) *data pre-processing.*

The pre-processing of the point cloud is mainly to reduce the algorithm consumption and to obtain a high-quality 2D point cloud, including three steps of voxel filtering, horizontal projection, and cropping.

(2) *Border points selection.*

In order to fit the curve of the reference, the border points on both sides should be selected. For ease of description, the left and right references in front of the train in the point cloud are noted as the upper border and lower border.

(3) *Border points clustering.*

The border points clustering is mainly to get real border points. Due to the limited height of protective walls and sound barriers, they are not able to block all the laser beams. As a result, some wrong points formed by large buildings or trees outside the references are also selected in step (2). The raw border points will be fitted as incorrect curves. It is necessary to cluster the raw border points and remove the wrong points.

(4) *Border curve fitting.*

Using the real border points, curve fitting is performed to obtain the border curves. These two border curves represent the reference on both sides.

(5) *Track Area Extraction.*

After obtaining the two curves for the upper and lower borders, it is necessary to select a better curve from them. Figure 4 shows a cloud point at a bend. As the train turns to the right, the left references have a better reflection angle than the right, making the points of the upper border denser. In straights, there is also a difference in the quality of the upper and lower borders due to different types of references on both sides, such as taller sound barriers on one side and protective walls on the other. Therefore, a certain rule is needed to select the better curve, which will be described in detail in Sect. 4.

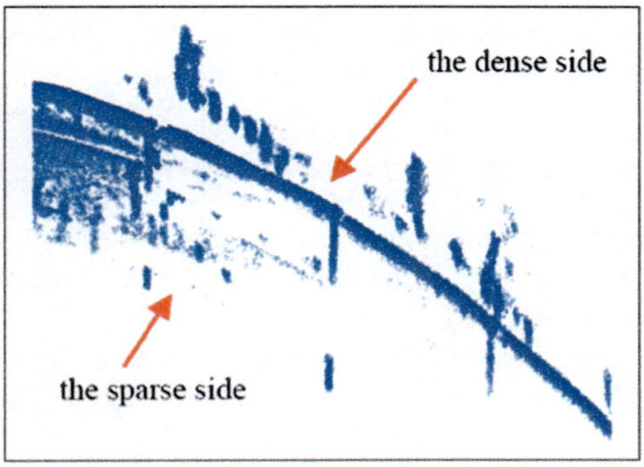


**Fig. 4.** The cloud point at a bend.

After the better border curve is selected, the track centerline is obtained by translating it a certain length according to the prior knowledge about urban rail transit. Finally, the track area is obtained by extending a certain length from the track centerline to both sides.

## 4 Track Area Extraction Method Based on References

### 4.1 Data Pre-processing

The density of the point cloud decreases with distance, and each frame contains large numbers of points. In the voxel filtering method, the point near the center of a voxel grid is used to represent all points in this grid, which makes the density uniform and reduces the total number of points.

The mentioned references are vertical to the ground. By projecting the point cloud to the horizontal plane, the task of fitting the reference surface is transferred to fitting a curve, which makes better use of the points and reduces the algorithm consumption. For an accurate description, a coordinate system is established in the 2D point cloud by defining the center of the train front as the origin, the forward of the train as the X-axis, and the left direction as the Y-axis, as shown in Fig. 5.

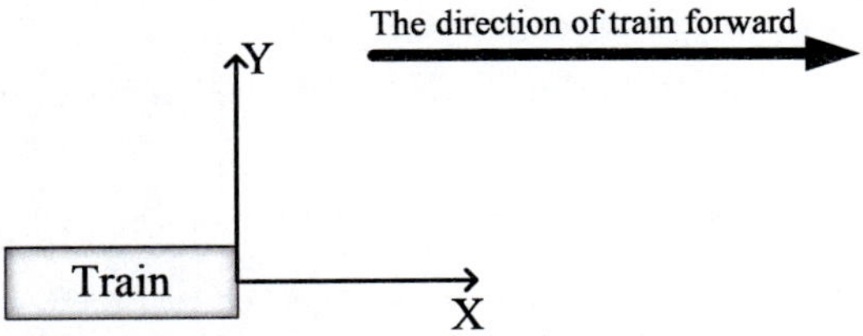


**Fig. 5.** Coordinate system in the 2D point cloud.

Due to the limitation of the field of view, references near the train are not sensed. On the other hand, as the maximum detection range of LiDAR, the point cloud exceeding a certain distance is too sparse to be used. Therefore, the point cloud is cropped and the points with x-coordinate in a certain range are retained.

### 4.2 Border Points Selection Based on Fixed-Length Segmentation

After the pre-processing of the point cloud, to select the borders points fully, this paper proposes a method to select the border points based on fixed-length segmentation, drawing on the idea of differentiation.

As shown in Fig. 6, the point cloud is segmented with a fixed length along X-axis, and the border points in each segment are selected. The detailed steps are described as follows:

At first, in a segment with a fixed length of $a$, find $k$ points with the largest y-coordinate, and calculate their average value as the average height of the upper border. Similarly, find $k$ points with the smallest y-coordinate to calculate the average height of the lower border.

Secondly, extend from the average height of the upper (lower) border by length $b$ downward (up) to obtain a selected area. All points in that area (orange area) are selected.

Thirdly, perform the above operations on each segment to select all border points.

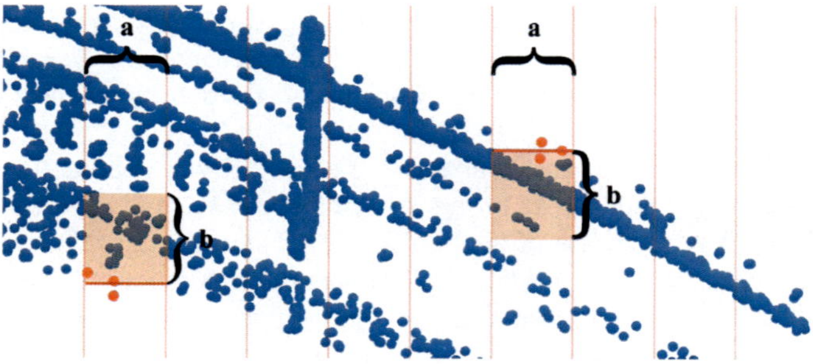


**Fig. 6.** Border points selection based on fixed length segmentation.

With the above method, the selection effect of the border points is shown in Fig. 7.

### 4.3 Border Points Clustering Based on DBSCAN

The purpose of the border points clustering is to get the real border points. DBSCAN (Density-Based Spatial Clustering of Applications with Noise) is a classical clustering algorithm [18]. The neighborhood radius *EPS* and the number threshold *MinPts* are the two main parameters in DBSCAN. A point is defined as a core-point if the number of points in its circular neighborhood with *EPS* reaches *MinPts*, otherwise as a non-core-point. The relationship from a core-point to other points in its neighborhood is directly

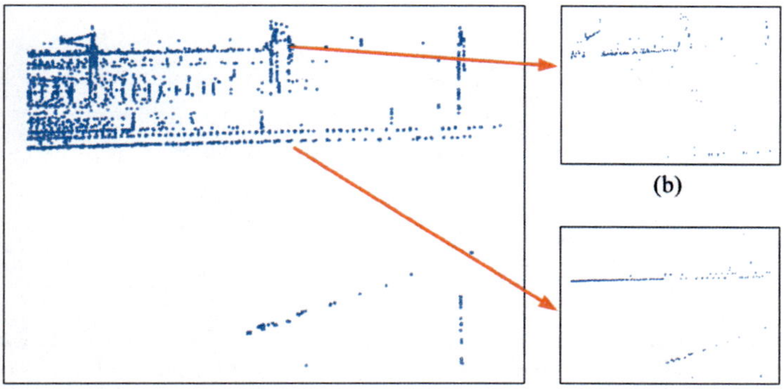

(a) (b) (c)

**Fig. 7.** The effect of border points selection. The fixed length $a = 1m$, the extended length $b = 2m$, and the number of sampling points $k = 3$. (a) Original 2D point cloud. (b) Points of the upper border. (c) Points of the lower border.

density-reachable. If there exists a chain of points satisfying directly density-reachable from a point $P$ to a point $Q$, then the relationship from $P$ to $Q$ is density-reachable. If there exists a core-point $S$ with a point $P$ and a point $Q$, so when $P$ and $Q$ are both density-reachable from $S$, that the relationship between $P$ and $Q$ is density-connected. A cluster contains points that are density-connected with each other. Points that do not belong to any cluster are noise.

As shown in Fig. 8, the radius of red circles is *EPS* and *MinPts* is 4. $P_2$ and $P_3$ are core-points because the number of points in each neighborhood is 5, while the number of $P_1$, $P_4$ and $P_5$ is less than *MinPts*, thus $P_1$, $P_4$ and $P_5$ are non-core-points. For ease of description, the relationships from $P$ to $Q$ is noted as $P \rightarrow Q$. With the above rules of DBSCAN, $P_2 \rightarrow P_1$, $P_2 \rightarrow P_3$, $P_3 \rightarrow P_2$ and $P_3 \rightarrow P_4$ are directly density-reachable. Because there exists a chain $P_2 \rightarrow P_3 \rightarrow P_4$ satisfying directly density-reachable, $P_2 \rightarrow P_4$ is density-reachable. Because there exists a core-point $P_2$ and $P_2 \rightarrow P_1$ as well as $P_2 \rightarrow P_4$ is density-reachable, the relationship between $P_1$ and $P_4$ is density-connected. Based on the same rule, the relationship among blue points as well as green points is density-connected, so they are clustered as two clusters. The black points are not density-reachable from any core-point and therefore become noise points.

In this study, the cluster of the real border points is a thin strip, so DBSCAN are suitable for the clustering. The real border points are stretched along the X-axis. To pay more attention to the connection of points in the X-axis when clustering to better separate the real border points from the wrong points, the y-coordinate of each point is amplified before clustering.

Figure 9 shows the clustering effect of the raw border points of the lower border in Fig. 7 using DBSCAN. It can be seen that the real border points (blue) and the wrong points (orange, etc.) are separated properly.

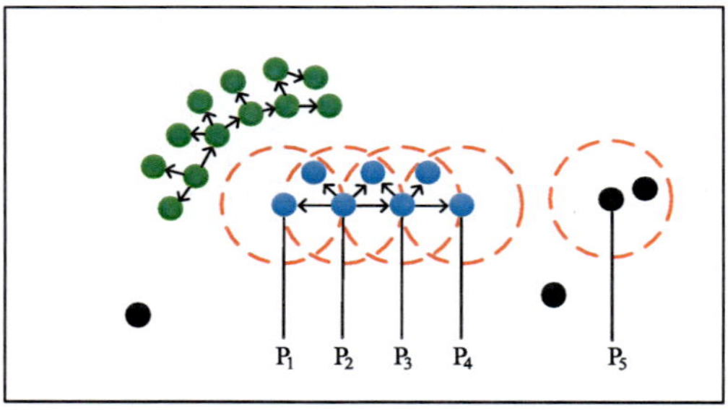


**Fig. 8.** Schematic diagram of DBSCAN clustering, where the radius of the red circle is *EPS* and *MinPts* is 4. The blue points and green points are two clusters and the black points are noise points.

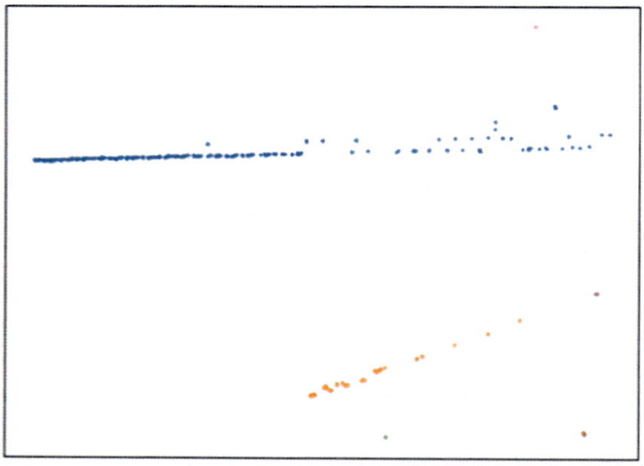

**Fig. 9.** The clustering effect of the raw border points using DBSCAN, where the neighborhood radius *EPS* is as 20 m, the number threshold *MinPts* is as 20% of the total, and the magnification of y-coordinate is 20. The blue cluster contains the real border points.

### 4.4 Border Curve Fitting Based on Least Squares

Least squares is the most common method to solve curve fitting problem. When fitting a polynomial curve, let a polynomial of degree t

$$y = f(x) = A_1x^t + A_2x^{t-1} + \cdots + A_tx + A_{t+1} \tag{1}$$

For a point $(x_i, y_i)$, there is the predicted value of y

$$y_i^{'} = f(x_i) \tag{2}$$

For all points, there is the sum of squared errors

$$Q = \sum_{i=1}^{n} \left(y_i^{'} - y_i\right)^2 = \sum_{i=1}^{n} (f(x_i) - y_i)^2 \tag{3}$$

The sum of squared errors $Q$ is a function with the polynomial coefficients ($A_1, A_2, \ldots, A_{t+1}$) as independent variables. The curve fitting problem is transformed into an optimization problem that minimizes $Q$. Least squares is used to solve this optimization problem to obtain the best coefficients.

The degree of the polynomial needs to be determined in advance to get a reasonable result. The urban rail lines have restrictions such as minimum turning radius and have no continuous sharp reverse curves, so a polynomial of degree 2 is able to fit the border curve well. The effect of using the least squares to fit the curve of a border at the bend is shown in Fig. 10.

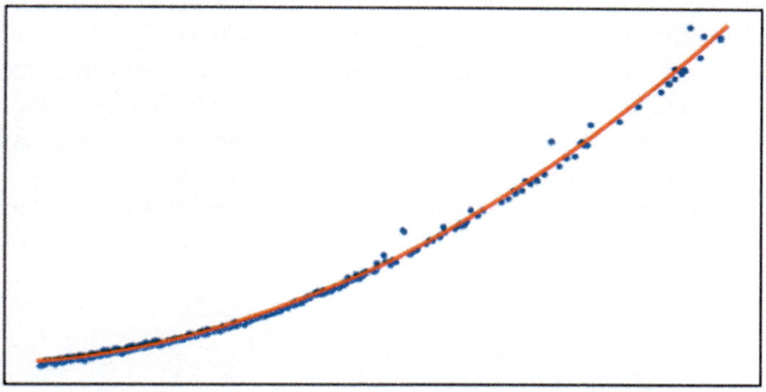

**Fig. 10.** The effect of using the least squares to fit the curve

### 4.5 Track Area Extraction Based on the Prior Knowledge of Urban Rail Transit

#### 4.5.1 Better Border Selection

RMSE (Root Mean Square Error) measures the effect of curve fitting by calculating the difference between the points and the fitted curve, as Eq. 4

$$RMSE = \sqrt{\frac{1}{n}\sum_{i=1}^{n}\left(y_i' - y_i\right)^2} \tag{4}$$

where the meaning of each variable on the right is consistent with Eq. 3.

In this paper, the better border is selected by combining the number of border points and the RMSE of fitted curves. Equation 5 shows the selection rule.

$$S = \begin{cases} A, n_A > n_B \times \beta \textbf{ or } (n_A \le n_B \times \beta \textbf{ and } RMSE(A) \le RMSE(B)) \\ B, n_B > n_A \times \beta \textbf{ or } (n_B \le n_A \times \beta \textbf{ and } RMSE(A) > RMSE(B)) \end{cases} \tag{5}$$

where $S$ is the better border, $A$ is the upper border, $B$ is the lower border, $n_A$ is the number of points on the upper border, $n_B$ is the number of points on the lower border, $\beta$ is a preset constant.

#### 4.5.2 Track Centerline Obtaining and Track Area Extraction

In the data pre-processing, the points whose x-coordinate is smaller than a preset value (for ease of description, the preset value is noted as $x_l$) are removed. To obtain the track centerline, there are three ways as follows: (a) Use a rail extraction method to detect the rails at a close distance, obtain the track center at $x = x_l$, , and translate the better border to overlap with it. (b) Preset a length from the track centerline to the references and translate the better border by this length. (c) Use the average value of y-coordinate at $x = x_l$ for the upper and lower border as the track center and translate the better border to overlap with it. The first way is the most accurate but requires extra algorithm consumption. The second way relies on a suitable preset length, but different references have different real distances to the centerline, for instance, sound barriers are more outward compared to protective walls. The third way is not suitable for double line sections but can reduce the impact of different reference types.

Therefore, this paper combines way (b) and way (c) to adaptively obtain the track centerline. The average heights of the upper and lower border in segment $(x_l, x_l + 1)$ is reused from chapter 4.2. If the difference between the two average heights is less than the preset threshold $l$ then use way (c), otherwise use way (b).

The effect of using the above method to select the better border and acquire the track centerline is shown in Fig. 11.

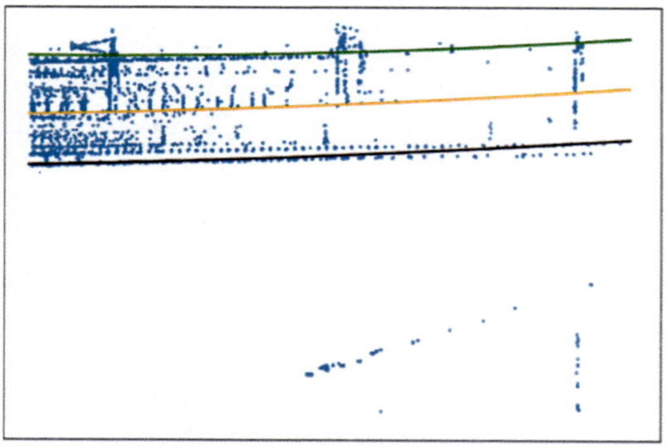

**Fig. 11.** The effect of better border selection and track centerline acquirement, where the green line is a fitted upper border curve, the red line is a fitted lower border curve, the black line is the selected better border curve, and the orange line is the track centerline.

After obtaining the track centerline, according to size guides in relevant specifications, e.g., gauge and train width, the track area is extracted by extending a certain length from the track centerline to both sides.

Taking Shanghai as an example, according to "Technical Specifications of Urban Rail Transit Projects", this paper sets the preset threshold $l$ as 6 m, and the extension length on both sides as 1.3 m.

# 5 Case Study

## 5.1 Experimental Environment

In the experiment, LiDAR was mounted on a cart at a height of 1.5 m from the ground. Static and dynamic experiments are conducted in the stations and intervals of Shanghai Metro Line 11. In the dynamic experiment, LiDAR moves forward with the cart at a speed of about 5 km/h. Experimental photos are shown in Fig. 12, and the detailed sections of each experiment are shown in Table 2.

(a) (b)

**Fig. 12.** The experimental photos. (a) The experimental equipment installation. (b) The experimental site.

**Table 2.** Detailed locations of experiments

| Experiment type | Experimental spot | Section Type |
|---|---|---|
| Static | Station A | Elevated section |
| | Station B | Tunnel section |
| Dynamic | Station A to Station C interval | Tunnel, elevated, and ground section |

In the experiments, we use cubes with a side length of 0.5 m and staffs with a height of 1.7 m as obstacles. Two cases are conducted where the obstacles were placed inside and outside the track area. Since these obstacles have high reflectivity and easily recognizable shapes, we can use them as position markers.

The LiDAR chosen in the experiment is Livox Tele-15, and its main parameters are shown in Table 3.

## 5.2 Result and Discussions

The performance of the proposed method cannot be measured by the overlap rate between the extracted track area and real track area because it is difficult to manually label the track area due to the sparse or even lost points in the point cloud at long distances.

**Table 3.** The main parameters of Livox Tele-15 LiDAR

| Parameter | value |
|---|---|
| Laser wavelength | 905 nm |
| Field of view | 14.5° × 16.2° |
| Range precision | 2 cm |
| Angular precision | <0.03° |
| Beam divergence | 0.12° × 0.02° |
| Data volume | 240,000 points per second |

Obstacles, as well as trackside markers, can be considered as position markers with known relationship to the actual position of the track area. Therefore, the performance of the proposed method is evaluated by checking whether the position relationship between the extracted track area and the mentioned position markers is consistent with the true situation.

Table 4 shows the results of extracting the track area within 150 m, which shows that the proposed method has a high accuracy in various experimental scenes.

**Table 4.** Experimental results of each experiment

| Experimental scenes | | Frame | Accuracy |
|---|---|---|---|
| static experiment | Tunnel | 973 | 99.6% |
| | Elevated | 909 | 94.5% |
| dynamic experiment | | 2000 | 91.6% |

In Table 4, the accuracy in tunnel sections is higher than that in elevated sections, because tunnel walls form more points than the sound barriers and the protective walls, and there is no interference from the points outside the tunnel walls in tunnel sections. The accuracy in the dynamic experiment is lower than that in the static experiment, because the same object is scanned at different moments and its points in the point cloud deviates. By analyzing incorrect results, we find that there exist some trees close to the reference in outdoor sections, which makes the extraction results of the track area deviate. Figure 13 shows the effect of the track area extraction method. As seen in Fig. 13, the track area extracted by the method in this paper is consistent with the rail alignment, and the extraction distance reaches 150 m.

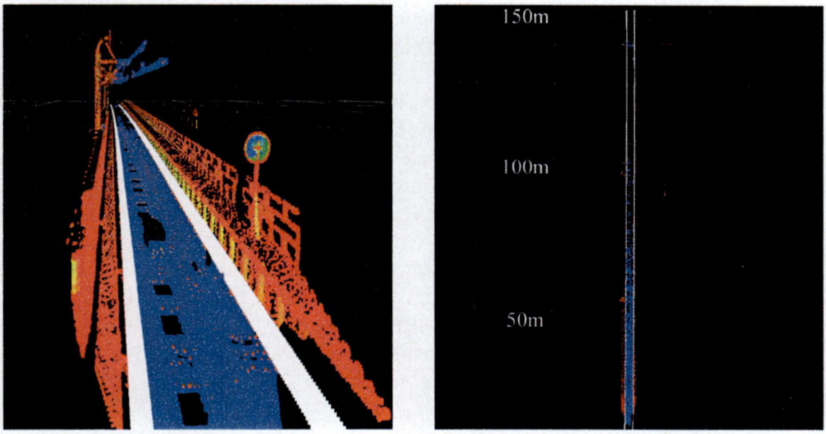


(a) (b)

**Fig. 13.** The effect of the proposed track area extraction method. The track area is the blue part inside the two white lines. (a) A front stereo view. (b) A horizontal projection view.

## 6 Conclusion

LiDAR has a limited detection range for rails, and the points of rails are almost lost at distances exceeding 80 m. The effective range of the methods to extract the track area by detecting rails is short, which makes it difficult to meet the requirements of the braking distance. The tunnel walls, protective walls, and sound barriers have a consistent alignment with the rails and are more accessible to LiDAR. In this paper, a method is proposed to extract the track area by detecting the mentioned references. The experimental results show that the proposed method has an effective extraction range of 150 m. However, the proposed method uses a 2D point cloud to extract the track area on the horizontal plane, lacking height information. Further work will investigate the extraction of a 3D track area to better cope with uphill and downhill sections.

**Acknowledgments.** This work was supported by National Key Research and Development Project of China (2022YFB4300501), Science and Technology Commission of Shanghai Municipality (23DZ2204900), and The Natural Science Foundation of Shanghai (21ZR1423800).

## References

1. Commission, I.E.: Railway applications—automated urban guided transport (AUGT)—safety requirements. In: IEC62267: International Standard (2009)
2. Shen, T., Song, H.: A new movement authority based on vehicle-centric communication. Wirel. Commun. Mob. Comput.. Commun. Mob. Comput. **2018**(2018), 1–10 (2018)
3. Song, H., Shen, T., Wang, W.: Train-centric communication-based close proximity driving train movement authority system. IEEE Intell. Transp. Syst. Mag. **10**(3), 22–34 (2018)
4. Ristić-Durrant, D., Franke, M., Michels, K.: A review of vision-based on-board obstacle detection and distance estimation in railways. Sensors **21**(10), 3452. (2021)

5. Huang, S., Yang, L., Tao, T., et al.: A Method of Tram Obstacle Intrusion Detection and Track Recognition Based on Instance Segmentation. Shanghai Highway, no.161(02), pp. 89–94+103+168 (2021). (in Chinese)
6. Zeng, X., Xiong, Q., Wang, Y., et al.: Modern tram auxiliary safety protection technology based on obstacles detection. J. Tongji Univ. (Nat. Sci.) **47**(01), 64–70 (2019). (in Chinese)
7. Shen, T., Qian, Y., Xie, L., et al.: Obstacle detection algorithm of fully automatic train considering reflection intensity. J. Tongji Univ. (Nat. Sci.) **50**(01), 6–12 (2022). (in Chinese)
8. Assaf, E.H., von Einem, C., Cadena, C., et al.: High-precision low-cost gimballing platform for long-range railway obstacle detection. Sensors **22**(2), 474 (2022)
9. Ristić-Durrant, D., Haseeb, M.A., Franke, M., et al.: Artificial intelligence for obstacle detection in railways: project smart and beyond. In: Dependable Computing-EDCC 2020 Workshops: AI4RAILS, DREAMS, DSOGRI, SERENE 2020, Munich, Germany, September 7, 2020, Proceedings 16, pp. 44–55. Springer International Publishing (2020)
10. Zhangyu, W., Guizhen, Y., Xinkai, W., et al.: A camera and LiDAR data fusion method for railway object detection. IEEE Sens. J. **21**(12), 13442–13454 (2021)
11. Guo, S.: Research on Detection Method of Obstacle in Front of Train Operation Based on Radar Measurement Data. Lanzhou Jiaotong University (2020). (in Chinese)
12. Stein, D., Spindler, M., Kuper, J., et al.: Rail detection using lidar sensors. Int. J. Sustain. Dev. Plan. **11**(1), 65–78 (2016)
13. Geng, Y., Wang, Z., Jia, L., et al.: Comparison of lidar point cloud features in railway environment. In: 2020 IEEE International Conference on Computational Intelligence and Virtual Environments for Measurement Systems and Applications (CIVEMSA), pp. 1–6. IEEE (2020)
14. Chen, C.: Accurate Modeling of Railway Tracks Based on MLS Point Cloud. Southwest Jiaotong University (2019). (in Chinese)
15. Guo, Z., Cai, B., Jiang, W.: A railway track detection method using LiDAR. J. Cent. South Univ. (Sci. Technol.) **51**(02), 560–566 (2020). (in Chinese)
16. Ren, Y., Ai, C., Lu, P., et al.: An automated rail extraction framework for low-density LiDAR data without sensor configuration information. IEEE Sens. J. **22**(13), 13234–13243 (2022)
17. Kimoto, K., Asada, N., Mori, T., et al.: Development of small size 3D LIDAR. In: 2014 IEEE International Conference on Robotics and Automation (ICRA), pp. 4620–4626. IEEE (2014)
18. Ester, M., Kriegel, H.P., Sander, J., et al.: A density-based algorithm for discovering clusters in large spatial databases with noise. In: kdd, vol. 96(34), pp. 226–231 (1996)

# Evaluating First-Train-Timetable-Network Performance in Urban Rail Transit with Percolation Theory

Tianlei Zhu, Zhiao Ma, and Xin Yang(✉)

State Key Lab of Rail Traffic Control and Safety, Beijing Jiaotong University, Beijing 100044, China
xiny@bjtu.edu.cn

**Abstract.** As modern large cities heavily rely on urban rail transit (URT) systems as their most crucial transportation infrastructure to meet daily commuting demands, the irrational design of the first train timetable and the lack of effective connections between different lines could result in excessively long travel times for passengers. Therefore, it is necessary to explore the flow organization mechanism of the URT system. In this paper, based on the temporal network theory, the concept of the First-Train-Timetable-Network (FTTN) is proposed, and the FTTN organization mechanism of Beijing and Shanghai's URT systems is studied based on the percolation theory. We discovered an important parameter that can characterize the rationality of the first train timetable arrangement, namely, the critical time $t_c$. Additionally, we observed the FTTN's percolation transition phenomenon and evaluated the network performance. The results show that the FTTN model proposed in this study, together with the URT network performance evaluation method, can effectively reflect the actual situation of the URT network, providing valuable insights for URT operators to formulate the first train timetable.

**Keywords:** Urban rail transit · Percolation theory · Network performance · First train timetable

## 1 Introduction

Urban rail transit (URT) is the backbone of public transportation systems in megacities, responsible for meeting people's daily commuting demands [1]. The URT industry in China has seen rapid development in the past decade, with 21 cities adding URT systems from the end of 2017 to the end of 2022, a growth rate of 63.64%, and an operating mileage of 10,291.95 km, a growth rate of 104.95%. However, as the size of the URT network continues to expand, the network structure becomes increasingly complex [2], and the development of train timetables becomes more challenging [3]. Particularly for the first train of the day, passengers with this need are often more concerned about travel time costs. If the first train timetable is unreasonable, passengers may have to wait for an extended period, which will cause significant inconvenience to them. To solve this problem, in-depth research on the operational mechanism of URT is needed to explore the


Y. Qin et al. (Eds.): EITRT 2023, LNEE 1137, pp. 47–54, 2024.
https://doi.org/10.1007/978-981-99-9311-6_5

formation laws of the traffic network. This will help us better understand the network's performance and provide suggestions for the operation of URT to develop better train timetables that meet actual demands, reduce passenger travel costs, and improve service levels.

This paper summarizes the current status and limitations of the application of temporal network theory and percolation theory in the field of transportation. Based on temporal network theory, the concept of First-Train-Timetable-Network (FTTN) and its construction method are proposed, and the concepts of time-varying connectivity and time-varying network efficiency are used to measure the performance of FTTN. We studied the URT systems of Beijing and Shanghai, and observed the percolation transition phenomenon of the FTTN and a parameter called "critical time", which can reflect the rationality of the first train timetable. The performance of FTTN in the two cities was measured. The results showed that FTTN can reflect the actual operation of the first train quite well, and the critical time $t_c$ and other metrics can accurately characterize the performance of FTTN. This provides suggestions for the development of train timetables and has practical implications for urban rail transit operation.

## 2 State of the Art

### 2.1 Temporal Network and Its Metrics

Temporal networks have emerged to address the deficiency of static networks in expressing dynamic factors in the network, and have been widely applied in various fields. Eckmann et al. [4] studied nonlinear and self-organizing phenomena in email networks based on the theory of temporal networks. Adar et al. [5] analyzed the spread of computer viruses on blog networks. Han et al. [6] established a dynamic protein network model and studied the interactions between proteins in organisms. In addition, many scholars have studied the measurement indicators of temporal networks, such as the fastest path, network diameter, characteristic temporal path length, closeness centrality, and temporal correlation coefficient.

There have been numerous studies on temporal networks, but the application of temporal network theory to the transportation field is still not comprehensive enough, especially in the establishment of URT network models using temporal network theory. There is a lack of research on how to reasonably apply the core ideas of temporal networks to various modes of transportation networks, or the establishment of temporal networks for multimodal transportation.

### 2.2 Percolation Theory in Transportation Network

Percolation theory can well reflect the evolution process and some self-organizing characteristics of networks, and it has been applied in the field of transportation. Wang et al. [7] established an agent-based traffic model and studied the network flow and percolation characteristics. Li et al. [8] studied the percolation transition phenomenon in the Beijing road traffic network and demonstrated the effectiveness of percolation-based bottleneck identification. In addition, Zeng et al. [9, 10] conducted further research on urban road traffic networks based on percolation theory. However, these studies mainly focused on urban road traffic networks, while research on URT networks and aviation networks is relatively scarce.

The application of percolation theory in the field of transportation is still in its infancy and has mainly focused on the domain of road traffic. However, road traffic is a free-flow network, while URT networks and aviation networks are non-free-flow networks. How to use percolation theory to characterize the properties of non-free-flow networks is a significant issue that deserves attention. There is also limited research on using percolation theory to evaluate the merits and demerits of URT train timetables.

## 3 Methodology

### 3.1 Construction of FTTN Model

In URT system, passengers are unable to travel between stations even if the physical line structure is intact when there is no train in operation. If we regard the stations in the URT system as nodes, there are no edges in the network, and there is no path between any two distinct nodes. Only when there is a train running on a certain line can passengers move on that line. Based on this idea, we propose the construction method for FTTN, which follows the following rules.

- For a certain moment, if there is no first train passing through between two adjacent stations, it is considered that there is no connecting edge between the two stations.
- For a certain moment, if there has been a first train passing between two adjacent stations, it is considered that there is a directed and unweighted edge between the two stations.

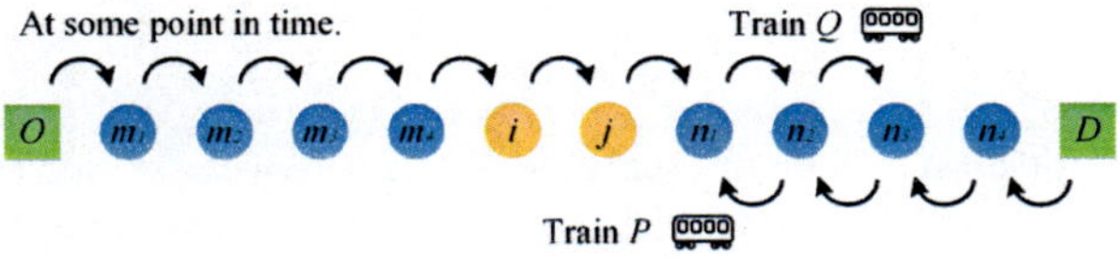


**Fig. 1.** FTTN network model diagram

When the first train does not pass through a station, passengers may have difficulty reaching their destination within a reasonable travel time, and therefore there is no path between the two points. Once the first train passes through, passengers can take the first or the next train within a reasonable waiting time. As the first train often runs early and has a low passenger demand, the impact of passenger flow on the network is minimal, and there is no situation where passengers cannot board due to congestion. Therefore, once the first train passes through, the edges of the network are assumed to be unweighted and will not be failed which caused by heavy passenger volume. URT trains have up and down directions, so the edges are directed. As shown in Fig. 1, if the first train from station i to station j has already passed through station j, then there is an edge from i to j, and if the first train from station j to station i has not yet passed through station i, there is no edge from j to i.

### 3.2 Metrics of URT Network Performance

Evaluation metrics for static networks often do not consider network changes and are therefore usually not suitable for directed unweighted networks. To characterize the properties of the FTTN network, two indicators that can evaluate dynamic directed unweighted networks were selected in this paper, namely time-varying network connectivity ($TVNC^t$) and time-varying network efficiency ($TVNE^t$).

$TVNC^t$ represents the URT network's reachability during a specific time period, and is an important indicator of network performance in the face of station or line failures. A higher $TVNC^t$ value indicates better passenger travel success rates and improved network performance. To calculate this metric, we have enhanced the connectivity index proposed by Schneider et al. [11], and the formula is shown below:

$$TVNC^t = \frac{1}{Nt}\sum_{t_0=0}^{t} G(t_0), 0 \le t_0 \le t \tag{1}$$

where t is the end point-in-time of the research, $t_0$ represents the current time, $G(t_0)$ is a function which represents the number of nodes in the giant component at time $t_0$, and N is the number of nodes in the network.

Network efficiency is a widely used performance metric for evaluating networks. It is calculated as the average reciprocal of the shortest path length between any two points in the network. A smaller shortest path length results in a larger reciprocal value, indicating a higher network efficiency. Scholars have studied the temporal variation of network efficiency, referred to as $TVNE^t$. To calculate this metric, we use the evaluation method proposed by Tang et al. [12], which is as follows:

$$TVNE^t = \frac{1}{N(N-1)}\sum_{i,j} \frac{1}{d_{ij}(t_0)}, 0 \le t_0 \le t \tag{2}$$

where $d_{ij}(t_0)$ is a function which represents the distance between station i and j. It should be noted that since the FTTN network is an unweighted network, we assume that the distance between two adjacent stations is 1. If there is no reachable path between two points, then $d_{ij}(t_0)$ is considered infinite and $1/d_{ij}(t_0) = 0$. Additionally, $\forall i, j \in N$ and $i \neq j$ and the meaning of $d_{ij}$ is different from the meaning of $d_{ij}$ in directed networks.

## 4 Cases Study

### 4.1 Data Description

This paper analyzes FTTN networks in Beijing and Shanghai using data that includes topological information of the lines and stations, as well as timetables of the first train. The data is obtained from official websites of Beijing and Shanghai subways and processed by removing incomplete data, deducing train routings, and extracting interval travel times. The data set includes 45 lines, 779 stations, 133 first train timetables, and 915 interval travel time data points, with a minimum time unit of 1 min.

## 4.2 Percolation Transition in FTTN

Due to the fact that URT trains are divided into up and down directions, and the concept of connected components only applies to undirected graphs, we adopt the concept of weak connectivity in directed graphs proposed by Nicosia et al. [13] here, which means that if there is a directed edge between two adjacent nodes *i* and *j*, regardless the direction of the edge, the two points are considered weakly connected. This concept is also consistent with the actual situation, because for passengers, if they are waiting for the first train in the downward direction at a certain time point, even if only the first train in the upward direction is passing by, the passengers can still wait for the first train in the downward direction within a reasonable time range.

The percolation transition phenomena in the URT systems of Beijing and Shanghai are shown in Fig. 2. Since the earliest departure time of the first train in each city is different, we take the earliest departure time of the first train in each city as the starting time, that is, $t = 0$. As the first trains of each line depart, the FTTN network gradually becomes complete. Figure 2(a) to (c) show the network states of the Beijing URT's FTTN at $t = 30, 45$, and 60, respectively. Nodes of the same color represent a cluster, and here we color the largest four clusters, while the smaller clusters are shown in gray. The FTTN network of the Shanghai URT is shown in Fig. 2(d) to (f).

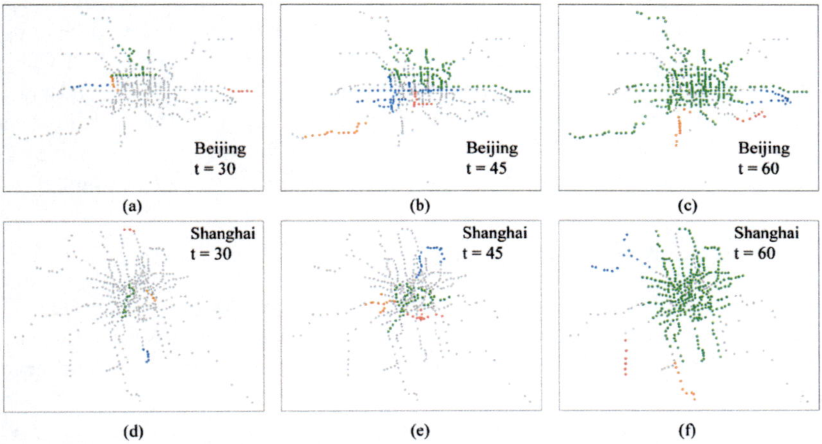


**Fig. 2.** Percolation transition in Beijing & Shanghai FTTN network

Here we can observe that, in the FTTN networks of Beijing and Shanghai URT, within the first 30 min of operation, most of the nodes in the network are unreachable and many clusters are composed of first departures from stations around the network. After 45 min of operation, approximately half of the stations in the Beijing URT network are reachable, whereas the number is a little lower in the Shanghai URT network, and several large clusters form in both networks. After 60 min of operation, most of the stations in both cities' FTTN networks merge into a single cluster, with only some stations located far from the network center remaining unreachable. This is due to the consideration of actual passenger flow demand in the development of the first departure timetables, with some lines having low passenger demand and thus later first departures.

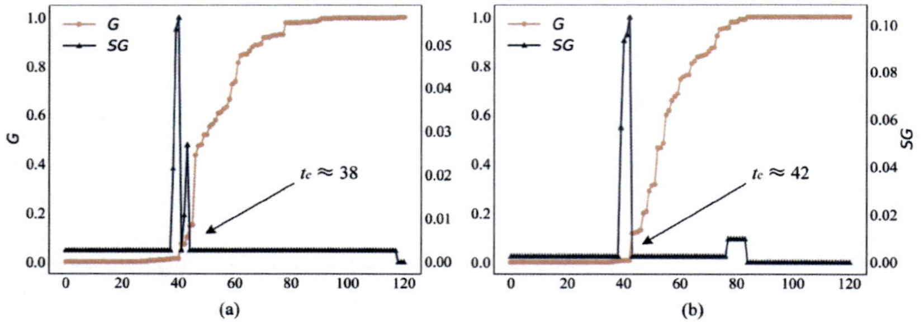


**Fig. 3.** Identification of $t_c$ in (a) Beijing URT system. (b) Shanghai URT system

Additionally, we identified trends in G and SG, which represent the proportion of nodes in the largest and second largest connected components in the network, respectively. We incorporate the concept of percolation thresholds into practical significance, meaning that FTTN network connectivity suddenly increases at a critical time $t_c$, resulting in higher network reachability and a sudden increase in the number of accessible stations for passengers. We refer to this percolation threshold as the key time $t_c$ in this paper. Figure 3 clearly illustrates the relationship between G and SG in the FTTN networks of the two cities, with Fig. 3(a) representing the $t_c$ for Beijing URT and Fig. 3(b) representing the $t_c$ for Shanghai URT. Beijing URT network's $t_c$ is 38 min, whereas Shanghai URT network's $t_c$ is 42 min, indicating that the first departure timetable design for Beijing URT system is slightly better than that for Shanghai URT system.

### 4.3 Performance Assessment of FTTN

Here, we calculated the two indicators mentioned earlier. As our constructed FTTN considers dynamic traffic flow, these two indicators can comprehensively evaluate the performance of the FTTN from the perspective of both connectivity and network topology.

The results of the $TVNC^t$ value and average $TVNE^t$ are shown in Table 1, which indicates that the $TVNC^t$ of the Beijing URT network is slightly higher than that of the Shanghai URT network. According to the calculation results of the $TVNE^t$ in Fig. 4, it can be seen that there is almost no difference in the $TVNE^t$ between the two cities in the early stage of operation. However, with the continuous increase in operation time, the $TVNE^t$ of the Beijing URT network begins to exceed that of the Shanghai URT network around 50 min after operation. The results of these two indicators indicate that the first train timetable design of the Beijing URT is more reasonable. Compared to the Shanghai URT, the integration time required for the FTTN network is shorter, and passengers can complete their travel demands with a higher probability in a shorter time after their travel. Therefore, it can be considered that the first train timetable design of the Beijing URT is slightly better than that of the Shanghai URT.

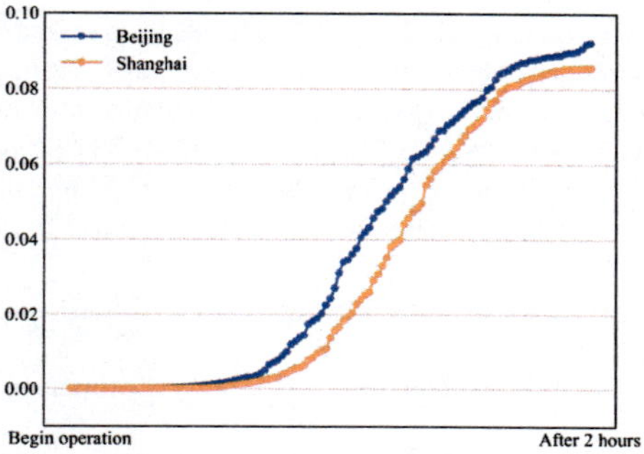


**Fig. 4.** The trend of $TVNE^t$

**Table 1.** Comparation of different metrics

| Performance metrics | Beijing | Shanghai | Percentage difference |
|---|---|---|---|
| $TVNC^t$ | 0.55634 | 0.53642 | 3.71% |
| Average $TVNE^t$ | 0.03660 | 0.03098 | 18.14% |

## 5 Conclusions

This article is the first to apply percolation theory and temporal network theory to the evaluation of train timetables in URT. The proposed evaluation model for first-train timetables can effectively assess the performance of specific cities' first-train timetables, providing timetable developers with a new perspective. In addition, we also constructed an FTTN model and observed percolation transitions on the FTTN. During the percolation transition, we gave the percolation threshold practical significance and named it the critical time, which can accurately reflect the sudden change in network properties during the FTTN integration process within this framework.

In future research, the relationship between train timetable and passenger demand should be considered. Reasonable first train timetables should be developed based on actual passenger needs and operational costs, and more indicators that can accurately characterize the performance of the first train timetable should be proposed.

**Acknowledgement.** This research was supported by the National Key R&D Program of China (No. 2020YFB1600702), the National Natural Science Foundation of China (Nos. 72071015, 72288101).

## References

1. Ma, Z., Yang, X., Wu, J., Chen, A., Wei, Y., Gao, Z.: Measuring the resilience of an urban rail transit network: A multi-dimensional evaluation model. Transp. Policy **129**, 38–50 (2022)
2. Du, Y., Wang, H., Gao, Q., Pan, N., Zhao, C., Liu, C.: Resilience concepts in integrated urban transport: a comprehensive review on multi-mode framework. SRT **4**, 105–133 (2022)

3. Guo, H., Bai, Y., Hu, Q., Zhuang, H., Feng, X.: Optimization on metro timetable considering train capacity and passenger demand from intercity railways. SRT **3**, 66–77 (2021)
4. Eckmann, J.-P., Moses, E., Sergi, D.: Entropy of dialogues creates coherent structures in e-mail traffic. Proc. Natl. Acad. Sci. U.S.A. **101**, 14333–14337 (2004)
5. Adar, E., Adamic, L.A.: Tracking information epidemics in Blogspace. In: The 2005 IEEE/WIC/ACM International Conference on Web Intelligence (WI'05), pp. 207–214. IEEE, Compiegne, France (2005)
6. Han, J.-D.J., et al.: Evidence for dynamically organized modularity in the yeast protein–protein interaction network. Nature **430**, 88–93 (2004)
7. Wang, F., Li, D., Xu, X., Wu, R., Havlin, S.: Percolation properties in a traffic model. EPL **112**, 38001 (2015)
8. Li, D., et al.: Percolation transition in dynamical traffic network with evolving critical bottlenecks. Proc. Natl. Acad. Sci. U.S.A. **112**, 669–672 (2015)
9. Zeng, G., et al.: Switch between critical percolation modes in city traffic dynamics. Proc. Natl. Acad. Sci. U.S.A. **116**, 23–28 (2019)
10. Zeng, G., et al.: Multiple metastable network states in urban traffic. Proc. Natl. Acad. Sci. U.S.A. **117**, 17528–17534 (2020)
11. Schneider, C.M., Moreira, A.A., Andrade, J.S., Havlin, S., Herrmann, H.J.: Mitigation of malicious attacks on networks. Proc. Natl. Acad. Sci. U.S.A. **108**, 3838–3841 (2011)
12. Tang, J., Musolesi, M., Mascolo, C., Latora, V.: Temporal distance metrics for social network analysis. In: Proceedings of the 2nd ACM Workshop on Online Social Networks, pp. 31–36. ACM, Barcelona Spain (2009)
13. Nicosia, V., Tang, J., Musolesi, M., Russo, G., Mascolo, C., Latora, V.: Components in time-varying graphs. Chaos **22**, 023101 (2012)

# Probabilistic Power Flow Analysis of All-Parallel AT Traction Power Supply System

Lan Yuan and Lyu Xiaoqin(✉)

School of Electrical Engineering, Southwest Jiaotong University, Chengdu 610031, Sichuan Province, China
xiaoqin93@163.com

**Abstract.** In order to evaluate the impact of train randomness and volatility on the traction power supply system, this paper firstly establishes the probability model of the train based on the probability distribution function of the train position from the time-position curve obtained by the train traction calculation. Then, considering the multiconductor characteristic and real structure of all-parallel autotransformer (AT) traction power supply system with a traction substation and two power supply arms, the probabilistic power flow calculation of the train-network system under the actual line parameters is carried out using the Monte Carlo simulation method. The results show that the probability model of the train and the probabilistic power flow calculation method can well analyze the probability distribution of the output power of the substation, the voltage amplitudes of the train, contact wires, rails and feeders under different train operation conditions.

**Keywords:** All-parallel AT traction power supply system · Traction calculation · Probabilistic power flow · Monte Carlo simulation

## 1 Introduction

With the development of high-speed railways in China, the proportion of traction load in power system is increasing. High speed trains, the traction load, have various operation conditions, such as traction and braking, and have the characteristics of randomness and volatility. When trains are running, in addition to the uncertain influence on node voltages of power grid [1], due to the complex network of traction power supply system [2], the node voltages of the traction network will also be uncertainly influenced.

Power flow analysis of traction power supply system can be briefly divided into deterministic power flow [3] and probabilistic power flow [1, 4–6]. For probabilistic power flow, the traction load in Refs. [1] and [4] was modeled from the low-voltage side of the traction substation, so the power flow distribution inside the traction power supply system cannot be analyzed. Reference [5] constructed the position probability distribution of the train, and used Monte Carlo simulation to calculate the train voltage and the substation supply power. Reference [6] established the train probabilistic model and carried out the probabilistic power flow calculation. However, the traction power supply system has been greatly simplified and the multi-conductor characteristics of the


Y. Qin et al. (Eds.): EITRT 2023, LNEE 1137, pp. 55–62, 2024.
https://doi.org/10.1007/978-981-99-9311-6_6

traction network has not been considered. Reference [7] calculated the probability power flow of urban rail traction power supply system which cannot be directly applied to the railway traction power supply system.

In this paper, the probabilistic model of the train and the node admittance matrix of the train-network system with multi-conductor characteristic including a traction substation and its two power supply arms and the train are established. Then, based on an actual line parameters, the probability power flow of the all-parallel AT traction power supply system is calculated by Monte Carlo simulation method.

## 2 Power Flow Calculation Model of the Real Train-Network System

### 2.1 Train Traction Load Model Based on Traction Calculation

The active power of the train at time $t$ can be calculated according to different operation conditions, as shown in Eq. (1), where the unit of active power $P_t$ and auxiliary power $P_f$ is $MW$, $\eta$ is the comprehensive efficiency (including gear machine efficiency, motor efficiency, inverter efficiency), $F_t$ and $B_t$ are the traction and regenerative braking force when the train speed is $v_t$, and the calculation formula is given in reference [8]. In addition, when the train power factor is $\cos\varphi$, the train reactive power is $Q_t = \frac{P_t}{\tan\varphi}$.

$$P_t = \begin{cases} \frac{F_t v_t}{3.6\times 1000\times\eta} + P_f, \textit{Traction} \\ P_f, \textit{Coasting drifting} \\ \frac{-B_t v_t \eta}{3.6\times 1000} + P_f, \textit{Brake} \end{cases} \tag{1}$$

Under the constant speed control strategy [9], considering the three operation conditions of traction, regenerative braking and coasting drifting and the traction/regenerative braking characteristics of the train, taking into account the line parameters such as ramp, bend, tunnel and speed limit, the speed-time curve, position-time curve, power-time curve and condition-time curve of the train can be obtained by train traction calculation.

### 2.2 Modeling of All-Parallel aT Traction Power Supply System

The all-parallel AT traction power supply system is shown in Fig. 1. The traction transformer adopts Vx connection (220/55 kV). The up and down traction network consists of contact wire (T1/T2), rail (R1/R2), positive feeder (F1/F2) and protect wire 1 & 2.

The admittance matrix of each component of the traction network is derived from references [10] and [11]. The contact wire and the feeder in the neutral section are disconnected, and the rail is connected. The train is regarded as a power source model, and the power of the train at each simulation moment can be obtained from the train traction calculation model in Sect. 2.1. Only the train section and the substation section contain the injected current vector. The node voltage equation of the whole system can be obtained as follows:

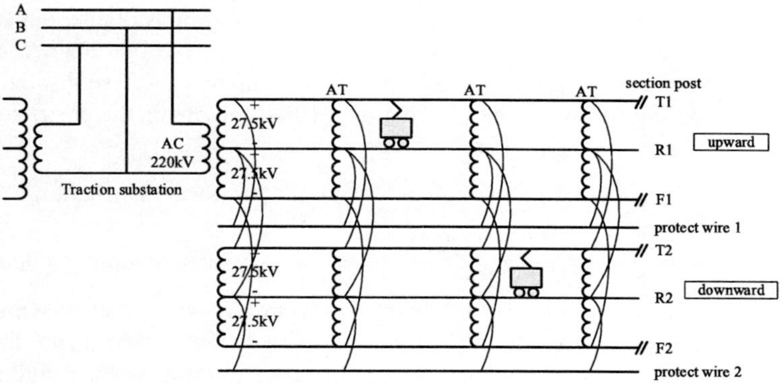


**Fig. 1.** Structure of all-parallel AT-fed power system

$$\begin{bmatrix} \mathbf{Y}_1+\mathbf{Z}_1^{-1} & -\mathbf{Z}_1^{-1} & & & & \\ -\mathbf{Z}_1^{-1} & \mathbf{Z}_1^{-1}+\mathbf{Y}_2+\mathbf{Z}_2^{-1} & -\mathbf{Z}_2^{-1} & & & \\ & -\mathbf{Z}_2^{-1} & \mathbf{Z}_2^{-1}+\mathbf{Y}_3+\mathbf{Z}_3^{-1} & -\mathbf{Z}_3^{-1} & & \\ & & \ddots & \ddots & \ddots & \\ & & & & \mathbf{Z}_{N-2}^{-1}+\mathbf{Y}_{N-1}+\mathbf{Z}_{N-1}^{-1} & -\mathbf{Z}_{N-1}^{-1} \\ & & & & -\mathbf{Z}_{N-1}^{-1} & \mathbf{Z}_{N-1}^{-1}+\mathbf{Y}_N \end{bmatrix} \begin{bmatrix} \mathbf{U}_1 \\ \mathbf{U}_2 \\ \mathbf{U}_3 \\ \vdots \\ \mathbf{U}_{N-1} \\ \mathbf{U}_N \end{bmatrix} = \begin{bmatrix} \mathbf{I}_1 \\ \mathbf{I}_2 \\ \mathbf{I}_3 \\ \vdots \\ \mathbf{I}_{N-1} \\ \mathbf{I}_N \end{bmatrix} \quad (2)$$

# 3 Probabilistic Power Flow Algorithm for All-Parallel aT Traction Power Supply System

## 3.1 Probability Model of Train Traction Load

The results of probabilistic power flow of traction power supply system are mainly affected by train power. In order to calculate the probabilistic power flow of traction power supply system, the probabilistic model of train traction load should be established first. When the train is running in a section, the train power is related to the position. As long as the probability distribution of the train position is constructed, the train power can be obtained through the train power-position curve.

The longer the train passes through a certain position, the greater the probability that the train is at this position. Therefore, the time-position curve $t = g(X)$ of the train can be used to construct the position probability distribution of the train. Since the position-time function $X = g^{-1}(t)$ of the train is monotonous and non-decreasing, the probability distribution function of the train position [7] is as follows:

$$F(X_1) = P(X \leq X_1) = P(g^{-1}(t) \leq g^{-1}(T_{X_1})) = P(t \leq T_{X_1}) = \frac{T_{X_1}}{T} = \frac{g(X_1)}{T} \quad (3)$$

where $T$ is the total running time of the train between the two stations, and $T_{X_1}$ is the time required for the train to run to the location $X_1$.

In fact, the analytical formula of $t = g(X)$ is generally not available, but the train time-position curve composed of a series of discrete points can be obtained by train traction calculation or measured data. Monte Carlo simulation of these discrete points and random sampling by inverse transformation method can obtain the position sample of the train. The specific steps [7] are as follows:

**Step 1.** Generate a pseudo-random number $r$ that obeys uniform distribution $U$ (0,1).

**Step 2.** If $\frac{g(X_{j-1})}{T} = \sum_{i=1}^{j-1} P(X_i) < r \leq \sum_{i=1}^{j} P(X_j) = \frac{g(X_j)}{T}$, $X_j$ is the sampling value.

The up and down lines of the all-parallel AT traction power supply system have a lateral connection line at the AT, so the train departure in the up and down directions cannot be regarded as an independent event. If there are multiple trains running in the two stations, the first train position in one direction (up or down) is sampled first. The positions of other trains in this direction and all trains in the other direction are determined by the train diagram.

### 3.2 Probabilistic Power Flow Algorithm Based on Monte Carlo Simulation

Firstly, based on the train traction load calculated by traction calculation, the probability distribution function of train position is constructed, and the first train position sample is obtained by Monte Carlo sampling. The other train positions are determined by the train diagram, and the train power is determined by the train power-position relationship. Then, the system node admittance matrix is established based on the multi-conductor chain circuit model of the all-parallel AT traction power supply system. According to the system node voltage equation, the train injection current and node voltage are iteratively calculated, and the results of multiple sampling experiments are counted. Finally, the kernel density estimation method is used to obtain the probability distribution of output power of substation, voltage amplitudes of the train, contact wires, rails and feeders.The specific steps of probabilistic power flow algorithm based on Monte Carlo simulation are shown in Fig. 2.

## 4 Case Study

The power supply diagram of the traction power supply system is shown in Fig. 3, including one traction substation, two power supply arms, two section posts, two ATs and two stations. All line data are referred to [9]. The train adopts CRH2A EMU, with comprehensive efficiency of 0.85, lagging power factor equals 0.97, auxiliary power of 490 kVA, short-circuit capacity of power system of 4 GVA, rated capacity of traction transformer of (40 + 40) *MVA*, Vx connection, short-circuit impedance ratio of 10.5%, leakage reactance of autotransformer of j0.45 Ω, transverse connection resistance of 1e−6 Ω, rail grounding resistance of 100 Ω/km, and iteration accuracy of 1e−4 kV.

### 4.1 Train Traction Calculation

From the train traction calculation, the time-position curve, power-position curve and condition-position curve of the up train running between station A and B are shown in Fig. 4.

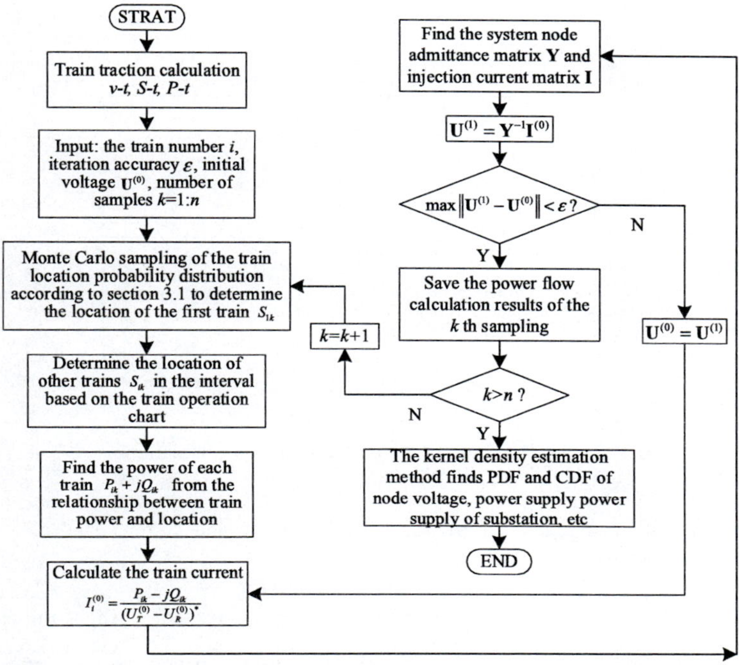


**Fig. 2.** Probabilistic power flow algorithm flow based on Monte Carlo simulation

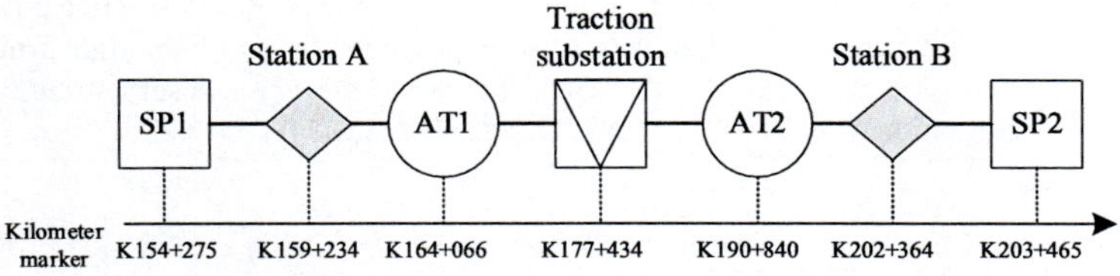


**Fig. 3.** Power supply schematic

## 4.2 Probabilistic Power Flow Calculating Results

A train runs on the up line between station A and B. Monte Carlo sampling 10000 times takes 20.4 s, and the average number of power flow calculation iterations per sampling is 3 times. The mean and standard deviation of train voltage amplitude, train injection current, left arm feeder current, right arm feeder current, active power of the left arm of the substation, reactive power of the left arm of the substation, active power of the right arm of the substation, reactive power of the right arm of the substation are shown in Table 1.

Combined with Fig. 4 and Table 1, it can be seen that under this line parameters, because the end point of the train interval operation is on the right arm of the traction substation, train regenerative braking conditions are mostly concentrated on the right

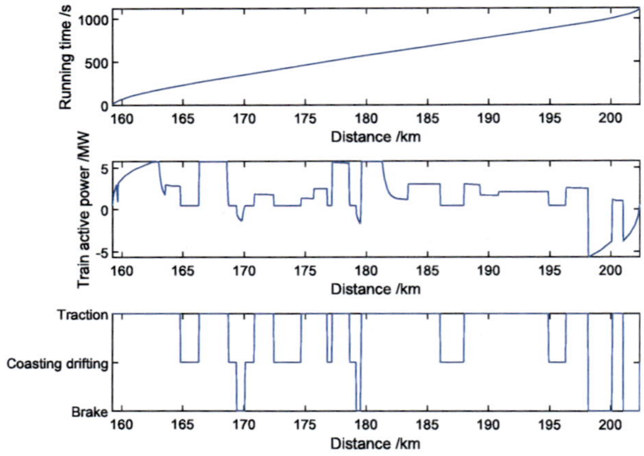


**Fig. 4.** The result of the up train traction calculation

**Table 1.** The mean and standard deviation of train voltage and current, feeder current and substation power supply

| Parameter | Train voltage amplitude/kV | Train injection current/kA | Left arm feeder current/kA | Right arm feeder current/kA |
|---|---|---|---|---|
| $\mu$ | 27.44 | 0.07 | 0.05 | 0.02 |
| $\sigma$ | 0.15 | 0.10 | 0.07 | 0.08 |
| Parameter | Active power of the left arm of the substation/MW | Train injection current/kA | Active power of the right arm of the substation/MW | Reactive power of the right arm of the substation/MVar |
| $\mu$ | 2.50 | 0.65 | 1.25 | 0.33 |
| $\sigma$ | 3.87 | 1.02 | 4.19 | 1.04 |

arm of the traction substation. When the train is regeneratively braked, the active power of the train supplied by the traction substation is negative, so the mean value of the active power of the right arm is also lower than that of the left arm. Moreover, due to the high power factor of the train, the average reactive power supplied by the traction substation is low.

The average voltage of each section of different conductors is shown in Fig. 5. SP1, AT1, TS left, TS right, AT2, SP2 represent section post 1, AT1 section, substation left arm power supply section, substation right arm power supply section, AT2 section, section post 2 respectively. And T1, T2, R1, R2, F1, F2 represent the up and down contact wire, the up and down rail, the up and down feeder. According to Fig. 5, the train runs on the up line, so the voltage of the up contact wire is lower than that of the down contact wire. The voltage of the up rail is higher than that of the down rail, and the voltage of the up rail is the highest. The voltage of the up and down feeder is not much different. For other sections except the train section, due to the parallel connection of the up and down lines, the voltages of up and down contact wires, rails, and feeders are equal. Since the train is basically in the right arm of the traction substation when regenerative braking occurs, and the regenerative braking will cause the network voltage to rise, the contact wire and feeder voltage of each section of the right arm of the traction substation ( TS right, AT2, SP2) are higher than the left arm of the traction substation.

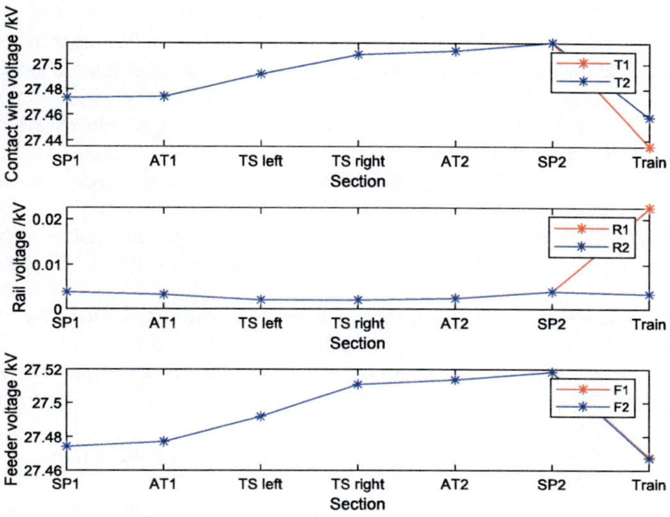


**Fig. 5.** The voltage mean value of each section of different conductors when the train runs under different operation conditions

## 5 Conclusion

In this paper, the probability characteristics of traction network node voltages when trains are running on the basis of establishing the probabilistic model of trains under all operation conditions are analyzed to evaluate the influence of train randomness and fluctuation on the traction network voltages. Because when the train is in regenerative braking condition, the train mainly runs in the right supply arm of the traction substation, the average active power supplied by the right supply arm is smaller than that of the left supply arm. And the mean value of the contact wire and feeder voltages of each section in the right supply arm are higher than those of the left supply arm. Moreover, the mean value of the up contact wire voltage of the train section is lower than that of the down contact wire voltage, and for the mean value of the rail voltages, the voltage where the train is located is the highest.

## References

1. Che, Y., Lyu, X., Wang, X.: Probabilistic analysis of impact of traction load power on power grid considering correlation. Proc. CSEE **42**(09),3250–3261 (2022). (in Chinese)
2. Xiaoqin, L., Xiaoru, W., Yulong, C., Ruiqing, F.: Eigenvalue-based harmonic instability analysis of electrical railway vehicle-network system. IEEE Trans. Transp. Electrif. **5**(3), 727–744 (2019)
3. Hu, H., He, Z., Wang, J., et al.: Power flow calculation of high-speed railway traction network based on train-network coupling system. Proc. CSEE **32**(19), 101–108 (2012). (in Chinese)
4. Che, Y., Zhao, Q., Chen, B., Zhu, J.: Time sequence probabilistic influence of traction load with small sample of newly high-speed railway on power grid. In: 18th International Conference on AC and DC Power Transmission, China, pp. 504–510 (2022)
5. Ho, T.K., Chi, Y.L., Wang, J., et al.: Probabilistic load flow in AC electrified railways. IEE Proc.-Electr. Power Appl. **152**(4), 1003–1013 (2005)
6. Chen, Y., Chen, M., Cui, M., et al.: Voltage unbalance probability pre-assessment of electrified railways with uncertain traction load. IEEE Trans. Transp. Electrif. (2022)
7. Liu, W., Li, Q., Tang, B., Chen, M.: Probabilistic load flow for urban rail traction power supply based on Monte Carlo simulation. J. Southwest Jiaotong Univ. **45**(4), 561–567 (2010). (in Chinese)
8. He, X.,Peng, X., Zhou, Y., Xiao, J.: Dynamic traction load of a multi-mass EMU braking model. J. Southwest Jiaotong Univ. **48**(05), 878–884 (2013). (in Chinese)
9. Wang, K.: Research on Energy Consumption Assessment and Optimization for Traction Power Supply System of High-Speed Railway. Southwest Jiaotong University, Chengdu (2020). (in Chinese)
10. Wu, M.: Uniform chain circuit model for traction networks of electric railways. Proc. CSEE **30**(28), 52–58 (2010). (in Chinese)
11. Lyu, X., Wang, X.: State-space model of high-speed railway traction power-supply system. Proc. CSEE **37**(03), 857–869 (2017). (in Chinese)

# Fully Automatic Dispatch of Coupling and Uncoupling Based on Interoperable Flexible Formation

Sheng Wang(✉) and Bo Zheng

CASCO Signal Ltd., Shang Hai 200070, China
wangsheng@casco.com.cn

**Abstract.** Urban rail transit has the characteristics of tides and different spatial passenger flow distributions. Many technologies are raised and researched recently. Based on its operational demand characteristics, fully automatic online coupling and uncoupling can solve some of the contradictions between transportation capacity and energy consumption, headway and service quality. Based on the inter-operable protocol and flexible formation, the dynamic identification of coupling trains is designed based on the precise location messages from ZC. Meanwhile, with the definition of coupling and uncoupling service trips in the timetable, the detail about the fully automatic dispatch process of coupling and uncoupling is carried out in this paper. At the same time, relevant operational suggestions are provided for some abnormal scenarios.

**Keywords:** Flexible formation · Coupling · Uncoupling · Full automation operation

## 1 Introduction

Urban rail transit exhibits significant differences in passenger flow distribution in different spaces and times. In traditional operation models, based on passenger flow conditions, organization methods such as running with different routes and varying headways during morning and evening rush hours are often used [1]. With the emergence of coupling technology, it can better solve the contradiction between transportation capacity and energy consumption, headway and service quality [1]. Online coupling and uncoupling refers to a transportation organization technology that achieves coordination between passenger flow demand and actual capacity by flexibly changing train formation in the mainline area, based on the characteristics of passenger flow in different intervals and time periods, while ensuring high train service frequency [2, 3]. At present, there is increasing research on coupling, and in addition to traditional coupling technologies, virtual coupling and dynamic coupling are also under in-depth research [4]. They are proposed to further increase railway line capacity by reducing the headway from absolute brake distances to relative brake distances, however, there are still many challenges to solve before its practical implementations [5].


Y. Qin et al. (Eds.): EITRT 2023, LNEE 1137, pp. 63–70, 2024.
https://doi.org/10.1007/978-981-99-9311-6_7

To meet the requirements of fully automatic coupling and uncoupling for online flexible formation, it is necessary to identify the coupling trains and automatically manage the coupling and uncoupling with timetable. The main purpose of this article is to study the automatic train supervision system (ATS) based on interoperable protocols, how to automatically manages the entire process of coupling and uncoupling through the service defined in timetable.

## 2 Identification of Flexible Formation Coupling Trains

The identification of coupled trains is a prerequisite for managing coupling and uncoupling. For existing fixed formation technologies, such as Shanghai Line 16 [6], ATS only needs to generate train objects based on the pre-defined fixed formation relationship after obtaining the coupling status of the on-board report through the relationship of the formations configured in static data [7]. The combination relationship between the two units is pre-defined; For flexible formation, if there are a large number of online vehicles, this "random" combination cannot be predefined through enumeration, and it is necessary to dynamically generate new train objects according to certain principles based on the coupling relationship between different units.

The display and tracking of precise train location on the ATS layout use the AP from ZC. Based on this, to ensure the consistency of information sources, the identification of coupling trains in this article also relies on the ZC location message. The dynamic calculation of coupling formation is carried out through the train formation identification field in the AP message, which needs to include the following information:

(1) The coupling status of train units
Head and tail are not coupling; Head is coupling while tail is not; Tail is coupling while head is not; The coupling status is unknown.
(2) Number of coupling train formation
The number of train units within the coupling formation (this article only studies the scenario of 2-units coupling).
(3) Train unit sequence within the coupling formation
When in a coupling state, the train units are arranged sequentially from the beginning.

Based on the above content, the basic principle for calculating coupling formation when ATS receives AP information from a unit is:

(1) If the coupling status of this train is "Head and tail are not coupling", the train formation is recognized as a single one;
(2) If the coupling status of this train is "Head is coupling while tail is not", obtain the head unit ID from the train unit sequence within the coupling formation. If the head unit status is "Tail is coupling while head is not", combine the two successfully matched train units into a coupling train, and generate a new train unit number and train object at the same time;
(3) If the coupling status of the front and rear ends of this train is "Tail is coupling while head is not", obtain the tail car ID from the train unit sequence within the coupling formation. If the status of the tail car is "Head is coupling while tail is not", combine the two successfully matched train units into a coupling train, and generate a new train unit number and train object at the same time;

To ensure the uniqueness of the train formation identification number within ATS, the train formation number is a fixed number of digits. Assuming a single train formation number is three digits and displayed in six digits within ATS, for example, if a single train unit number is "001", the train formation number maintained in ATS is "000001". Then, when train "001" is coupled with train "002", the new train object generated is "001002" (using a combination of small formation in front and large formation in back).

## 3 Definition of Coupling and Uncoupling in Timetable

In order to achieve fully automatic coupling and uncoupling task control, when preparing the timetable, it is necessary to pre-define the coupling and uncoupling tasks based on operational requirements. The defined content includes: the type of formation to perform the service (whether it is a long formation coupling train); The task type of this trip at the terminal (uncoupling, be coupled, to couple).

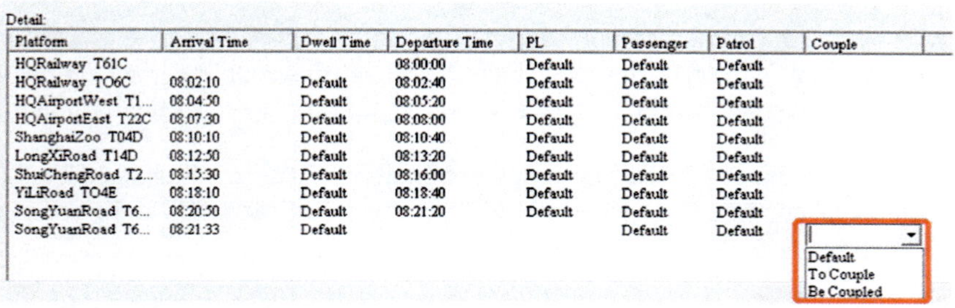
Detail:

| Platform | Arrival Time | Dwell Time | Departure Time | PL | Passenger | Patrol | Couple |
|---|---|---|---|---|---|---|---|
| HQRailway T61C | | | 08:00:00 | Default | Default | Default | |
| HQRailway TO6C | 08:02:10 | Default | 08:02:40 | Default | Default | Default | |
| HQAirportWest T1... | 08:04:50 | Default | 08:05:20 | Default | Default | Default | |
| HQAirportEast T22C | 08:07:30 | Default | 08:08:00 | Default | Default | Default | |
| ShanghaiZoo T04D | 08:10:10 | Default | 08:10:40 | Default | Default | Default | |
| LongXiRoad T14D | 08:12:50 | Default | 08:13:20 | Default | Default | Default | |
| ShuiChengRoad T2... | 08:15:30 | Default | 08:16:00 | Default | Default | Default | |
| YiLiRoad TO4E | 08:18:10 | Default | 08:18:40 | Default | Default | Default | |
| SongYuanRoad T6... | 08:20:50 | Default | 08:21:20 | Default | Default | Default | |
| SongYuanRoad T6... | 08:21:33 | Default | | | Default | Default | |



**Fig. 1.** Definition of coupling service type in timetable

As shown in Fig. 1: If the single formation train is required to couple at the terminal, the last record in this trip should be set "be coupled" or "to couple" based on the characteristics of the coupling task. The "be coupled" train will first arrive at the coupling area and stop steadily before waiting for the 'to couple' train to collide.

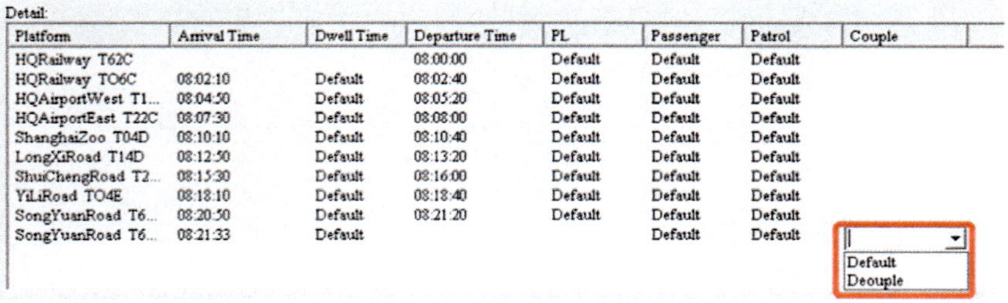
Detail:

| Platform | Arrival Time | Dwell Time | Departure Time | PL | Passenger | Patrol | Couple |
|---|---|---|---|---|---|---|---|
| HQRailway T62C | | | 08:00:00 | Default | Default | Default | |
| HQRailway TO6C | 08:02:10 | Default | 08:02:40 | Default | Default | Default | |
| HQAirportWest T1... | 08:04:50 | Default | 08:05:20 | Default | Default | Default | |
| HQAirportEast T22C | 08:07:30 | Default | 08:08:00 | Default | Default | Default | |
| ShanghaiZoo T04D | 08:10:10 | Default | 08:10:40 | Default | Default | Default | |
| LongXiRoad T14D | 08:12:50 | Default | 08:13:20 | Default | Default | Default | |
| ShuiChengRoad T2... | 08:15:30 | Default | 08:16:00 | Default | Default | Default | |
| YiLiRoad TO4E | 08:18:10 | Default | 08:18:40 | Default | Default | Default | |
| SongYuanRoad T6... | 08:20:50 | Default | 08:21:20 | Default | Default | Default | |
| SongYuanRoad T6... | 08:21:33 | Default | | | Default | Default | |



**Fig. 2.** Definition of uncoupling service type in timetable

As shown in Fig. 2: If the coupling formation train is required to uncouple at the terminal, the last record in this trip should be set "uncouple" and this trip should be also marked as long formation.

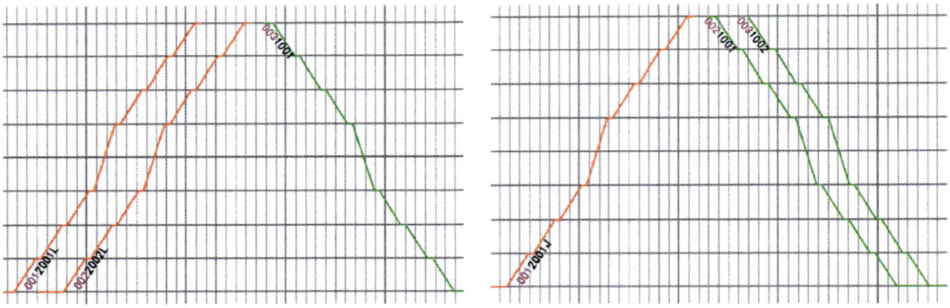

(a) coupling trips. (b) uncoupling trips.

**Fig. 3.** Coupling and uncoupling trips

As shown in Fig. 3(a): After two single formation trips coupling, they are merged into a new service number. Set a special character "L" at the end of the train number with a coupling task.

As shown in Fig. 3(b): After the coupling formation trip uncoupling, it shall be split into two independent new service numbers. Set a special character "J" at the end of the train number with an uncoupling task.

## 4 The Dispatch Process of Fully Automatic Coupling and Uncoupling

In the topology diagram of the layout, it is necessary to predefine the coupling and uncoupling area. The areas in mainline can be divided into two categories: Type 1, which can be stabled for each short formation train; Type 2, only used for coupling and uncoupling operations and single formation stabling in this area.

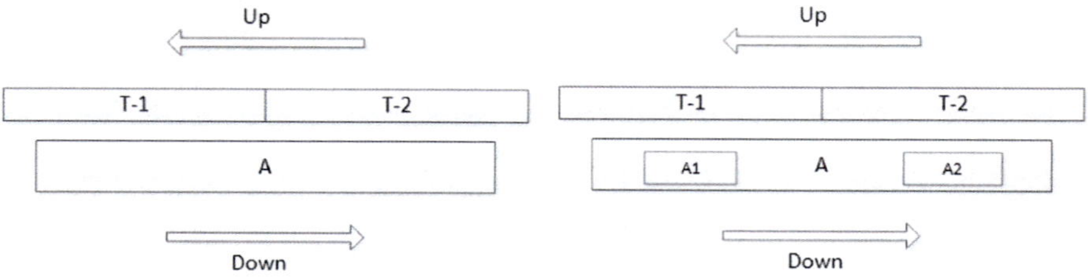


(a) area type 1. (b) area type 2.

**Fig. 4.** Coupling and uncoupling area type

As shown in Fig. 4(a), there is a coupling and uncoupling area that can meet the stabling requirement for each short formation. Track T is the coupling and uncoupling area designed for the system. We split it into T-1 and T-2, which are associated with two single formation stopping areas A1 and A2, respectively. At the same time, a large stopping area A is defined within the system as the coupling formation train stabling area which is the logical area of the system and is transparent to users.

A destination of A1 or A2 can be set for single formation trains for stabling, and the trains will stop on the corresponding T-1 or T-2 tracks; For coupling trains which is upward running, destination A1 needs to be associated (at this time, ATS sends stabling area A according to the train coupling status). If A2 is assigned incorrectly, it will be rejected. Similarly, for downward running trains, destination A2 needs to be assigned, otherwise it will be rejected.

As shown in Fig. 4(b), the area is only used for coupling and uncoupling operations and single formation trains' stabling. Similarly, in this scenario, T needs to be split into T-1 and T-2, but at this point only one stopping area A is associated. When the ATS sends the coupling or uncoupling operation command to the train, the stopping area are all A, and the train selects the stopping point based according to the coupling or uncoupling task in the command.

This type of coupling and uncoupling area is applicable for stopping only one train (which can be single or coupling formation), such as turn back areas and passenger platforms designed as coupling and uncoupling area.

## 4.1 The Dispatch Process of Coupling

In brief, the dispatch process of coupling is divided into four stages: be coupled command, to couple command, holding command, and new service association [8, 9].

As shown in Fig. 5(a), Stage 1: When the train (group number 000001, trip number 2001, destination A2) stops steadily at or departs from the platform, check whether the last node of the current trip is "be coupled". If it is TRUE and the current platform is the last one before the coupling area, and the train is also at the "CBTC control level", the "be coupled" command shall be automatically sent. When the next coupling area is Type 1, the coupling stopping area is A2. When it is Type 2, the coupling stopping area is A. The train automatically runs to the "be coupled" area A according to the control command and electronic map, stopping steadily and waiting for the "to couple" train to collide.

As shown in Fig. 5(b), Stage 2: When the train (train set 000002, trip number 2002, destination A2) stops steadily at or departs from the platform, check whether the last node of the current trip is "to couple". If it is TRUE and the current platform is the last one before the coupling area, and the train is at the "CBTC control level", and the train with "be coupled" operation mode described in stage 1 has stopped in the coupling area, the "to couple" command shall be automatically sent. When the next coupling area is Type 1, the stopping area shall be set to A2, and when it is Type 2, the stopping area shall be set to A. If the train is in "to couple" service and the coupling route meets the setting conditions, the route shall be automatically triggered.

As shown in Fig. 5(c), Stage 3: ATS shall hold the coupling command for the "be coupled" train and the "to couple" train until the process is successful or failed. Before

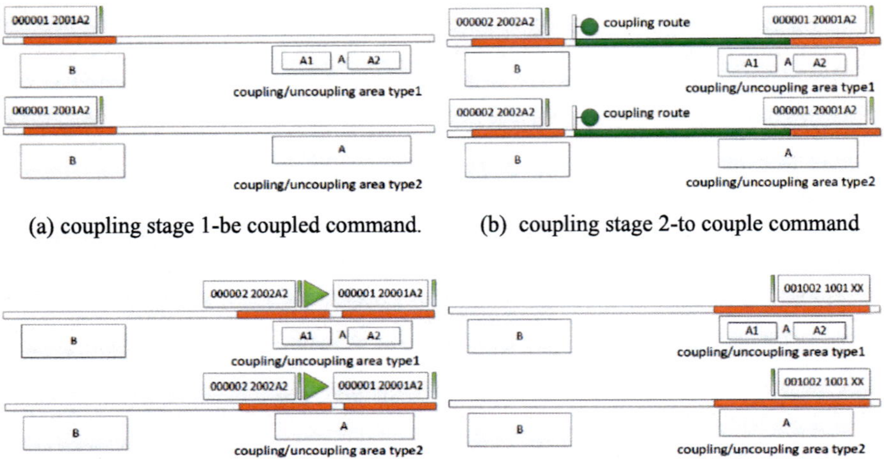


(a) coupling stage 1-be coupled command. (b) coupling stage 2-to couple command

(c) coupling stage 3-holding command. (d) coupling stage 4- new service association.

**Fig. 5.** The dispatch process of coupling

the coupling is completed, the tracking of train unit formation shall still be kept with single formation which is reported from ZC's AP.

As shown in Fig. 5(d), Stage 4: When the coupling process is completed, two train units with new coupling status in the AP information shall be reported by ZC. One unit's status is "Head is coupling while tail is not" and another one's status is "Tail is coupling while head is not". ATS automatically combines the two train units into a new train object, and the generated new train group number is 001002 with the principles in Chap. 2. According to the definition of the timetable (as shown in Fig. 2), check the trip started from coupling area and associate with the service number for the new coupling formation train. As for the departure platform, it depends on the coupling area type. If the coupling area is Type 1, the departure platform area is determined to be A1 or A2 based on the location of the main control end after coupling; If the coupling area is Type 2, then the departure area is A.

After the coupling command is sent, when there is an abnormal situation that requires adjusting the operation to cancel the coupling task, the method of canceling the coupling command can be used to terminate the execution of the coupling task. If the "be coupled" or "to couple" command has been sent and the coupling is cancelled, ATS will no longer hold the coupling command for the train, and the train will terminate the coupling process on its own; If the train operation with coupling tasks in the timetable is cancelled before the "be coupled" or "to couple" command is set, ATS will no longer prepare the commands, and the coupling route shall not be automatically triggered.

### 4.2 The Dispatch Process of Uncoupling

As for the dispatch process of uncoupling, it is divided into three stages: uncoupling command, holding command, and new service association [8, 9].

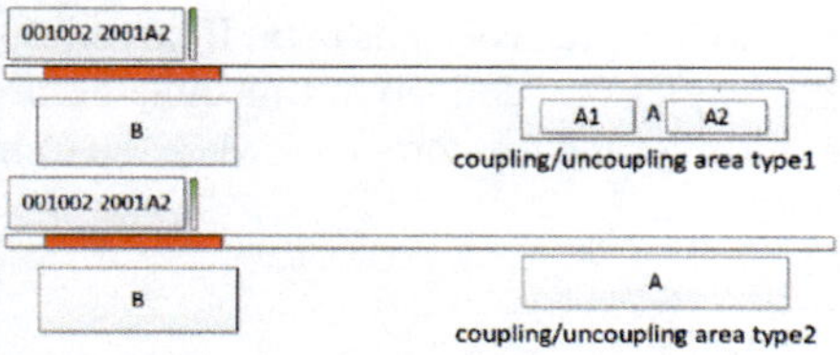


(a) uncoupling stage 1-uncoupling command.

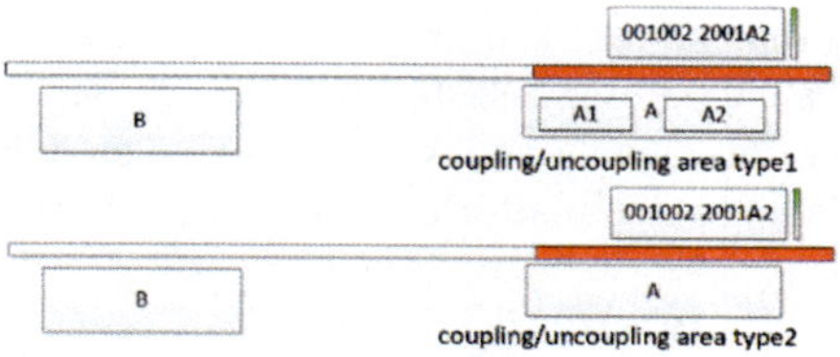


(b) uncoupling stage 2-holding command

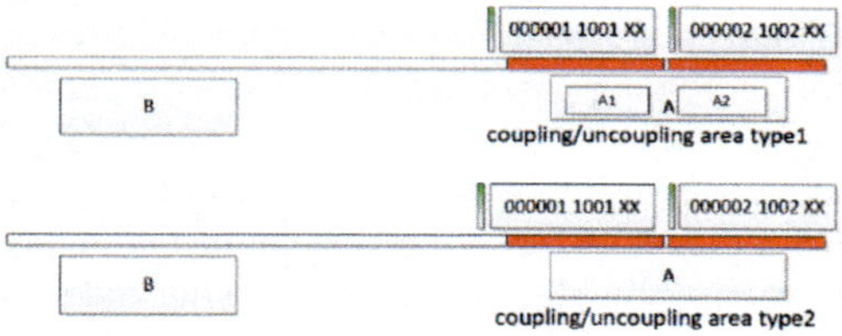


(c) uncoupling stage 3- new service association.

**Fig. 6.** The dispatch process of uncoupling

As shown in Fig. 6(a), Stage 1: When the train (group number 001002, trip number 2001, destination A2) stops steadily and accurately at or departs from the platform, check whether the last node of the trip is "uncoupling". If it is TRUE and the current platform is the last one before the uncoupling area, and the train is at the "CBTC control level", the uncoupling command shall be automatically sent. The stopping area shall be A regardless of whether the coupling and uncoupling area type is Type 1 or Type 2.

As shown in Fig. 6(b), Stage 2: The train arrives at the destination to begin the uncoupling process. ATS shall hold the uncoupling command for the train until the process is completed. During this time, the AP information reported from ZC is still a coupling unit, and ATS keeps the display status of coupling formation.

As shown in Fig. 6(c), Stage 3: When the uncoupling process is completed, the AP information of two units reported from ZC shall be transformed into independent two single train units. ATS automatically deletes the internal coupling train object and generates two independent single formation trains 000001 and 000002. According to the definition of the timetable (as shown in Fig. 2), the trip starting from the stopping areas of T-1 and T-2 shall be checked. For type 1 of coupling and uncoupling area, the train stabling at location A1 shall be associated with the trip whose start point is T-1 in the timetable, and the train stabling at location A2 shall be associated with the trip whose start point is T-2 in the timetable; For coupling and uncoupling area type 2, two trips shall share the same starting in timetable. Based on the relative locations of the trains' up and down running direction after uncoupling and the departure sequence, the early departing train shall be associated first, and then the late departing one.

If the uncoupling command is in progress, when there is an abnormal situation that requires adjusting the operation to cancel the uncoupling command, the execution of the uncoupling task can be terminated by using the method of canceling uncoupling command. Once the uncoupling process is cancelled, ATS will no longer hold the uncoupling command for the train, and the train shall terminate the process on its own; If cancelling the uncoupling task for the train whose uncoupling task is defined in timetable before the command is set, ATS will no longer automatically send the command when the train arrives at the platform in front of the uncoupling area.

## 5 Conclusion

The dispatch of fully automatic coupling and uncoupling not only involves the control of train tasks, but also needs to consider the collaboration of various external interface specialties, such as passenger information system for predicting train formation, platform announcement system for guiding passengers to suitable position, and inhibition about PSD interfaces, and so on [10].

The fully automatic coupling and uncoupling operation that meets the requirements of flexible formation has gradually transitioned from the laboratory debugging stage to the engineering application stage. The dispatch of coupling and uncoupling described in this article is only based on the mainline timetable, and has not yet involved the coupling and uncoupling process of the fully automatic depot. With the gradual application of this technology, there will definitely be more demands in subsequent practice, in order to continuously improve our design to adapt to different operational scenarios.

Meanwhile, with the in-depth research of virtual coupling technology, how to apply dispatch management technology based on flexible formation of physical coupling and uncoupling to virtual coupling scenarios is also a new topic that needs to be studied in the future [11].

## References

1. Zhaoling, L., Yezhi, Y.: Key technologies of flexible marshalling in signaling system for urban rail transit. Urban Rail Transit **18**(12), 72–77 (2021). (in Chinese)
2. Tang, Y., Ma, B.: Research on the flexible car-set operation organization of urban rail transit. J. Railw. Eng. Soc. **31**(8), 96–100 (2014). (in Chinese)
3. Fan, H., He, Y.: Flexible marshalling scheme and function realization of urban rail transit train. Urban Mass Transit **24**(9), 200–203 (2021). (in Chinese)
4. Wu, Q., Ge, X., Zhu, S., Cole, C., Spiryagin, M.: Physical coupling and decoupling of railway trains at cruising speeds: train control and dynamics. Int. J. Rail Transp. (2023). https://doi.org/10.1080/23248378.2023.2169963
5. Yang, Y., Yan, F.: Research on train dynamic coupling strategy based on distributed model predictive control. In: Journal of Physics: Conference Series, vol. 2183, p. 012029 (2022)
6. Wang, D., Huang, Q.: Application of flexible train formation on rail transit FAO lines. Urban Mass Transit **22**(S2), 102–105 (2019). (in Chinese)
7. Liu, L.,Yu, X., Kang, L.: Research on automatic dispatching management for online coupling and uncoupling of subway trains. Rail Transit **2**, 59–61 (2022). (in Chinese)
8. Wei, L., Yan, Y., Wang, H.: Research on online coupling and uncoupling technology of metro trains in FAO. Technol. Innov. Appl. **12**(28), 1–4 (2022). (in Chinese)
9. Ronghua, H.: Application of flexible urban rail transit train formation technology. Urban Mass Transit **25**(11), 143–147 (2022). (in Chinese)
10. Zhiyuan, Q.: Optimization of Shanghai metro line 16 online coupling and uncoupling functions. Urban Mass Transit **11**, 103–106 (2022). (in Chinese)
11. Felez, J., Vaquero-Serrano, M.A.: Virtual coupling in railways: a comprehensive review. Machines **11**(5), 521 (2023)

# Research on Soil Potential Distribution Caused by Metro Stray Current

Tang Yuhang[1(✉)], Yu Kun[1], Ni Yanru[1], Cheng Xinxiang[1], Zeng Xiangjun[1], and Han Wei[2]

[1] Changsha University of Science and Technology, Changsha 410000, China
843368522@qq.com, chxinee028@stu.csust.edu.cn
[2] State Grid Taiyuan Power Supply Company, Taiyuan 030000, China

**Abstract.** The harm of stray current in urban rail transit is becoming increasingly serious, and the corrosion of buried metal pipelines and DC bias of grounding transformers caused by it have attracted attention from various aspects. Soil potential distribution is a common entry point for metal corrosion and DC bias problems. Therefore, from the perspective of urban soil potential distribution, this article constructed a four-layer ground grid return current model and deduces the voltage distribution of steel rails, and constructed a urban soil potential distribution model. Analyzed the interference range and degree of stray current during subway operation. The results indicated that an increase in the DC resistance of the rail will significantly increase the soil potential, increase the range of stray current influence, and increase the risk of DC bias and rail corrosion in the substation. Increasing the rail-ground transition resistance can effectively reduce the risk of stray current hazards in nearby soil areas. The larger soil resistivity also has a certain limiting effect on the propagation of stray currents.

**Keywords:** Stray current · Urban rail transit · Soil potential distribution · Metal corrosion · DC bias

## 1 Introduction

At present, urban rail transit in China is booming, bringing great convenience to people's travel but inevitably generating the harm of stray current [1]. DC traction is commonly used in urban rail transit in China. Due to the inability of the rail to achieve complete insulation to the ground, some current will leak into the ground, which is called stray current [2]. Stray current can cause potential fluctuations in nearby soil [3], invading nearby buried metal pipelines and steel structures, causing electrochemical corrosion [4], cause DC bias in the neutral grounding transformer of nearby substations [5], seriously endangering the safe and stable operation of urban power grids [6].

To study the impact of stray current on urban power grids and buried metals, starting from the distribution of urban soil potential, and analyzing the interference range and degree of stray currents during subway operation, it is first necessary to model the subway traction system. At present, the modeling of the DC traction system for dual


Y. Qin et al. (Eds.): EITRT 2023, LNEE 1137, pp. 71–79, 2024.
https://doi.org/10.1007/978-981-99-9311-6_8

terminal power supply in urban rail transit in China is relatively mature [7], and the four-layer grounding network circuit model is widely accepted by scholars [8]. Based on this, scholars have conducted numerical calculations of stray currents [9], rail potential distribution [10], and research on the characteristics of soil potential distribution [11]; Simulating the distribution of urban soil potential is a common entry point for addressing issues such as buried metal corrosion, DC bias in substations, and evaluating the range and degree of stray current interference, which has attracted scholars' attention. Li Wenfeng proposed a vertical layered soil model and derived the potential distribution generated by point current sources using the complex image method [12]. S. Lin established a soil model based on the finite element method and analyzed the impact of stray current on soil potential [13].

This article used the finite element simulation software COMSOL Multiphysics (COMSOL Corporation, USA) and built an urban soil potential simulation model based on real operating parameters of urban rail transit, including Changsha Metro Line 1 and Line 2. Calculate the rail voltage of the subway line and couple it as a line voltage source with the urban soil model to achieve simulation of urban soil potential distribution; We have studied the influencing factors of urban soil potential distribution, including rail DC resistance, rail-drainage network transition resistance, and soil resistivity, which have certain guiding significance for evaluating the interference degree and impact range of stray currents.

## 2 Ground Grid Reflux Model

China's urban rail transit mainly adopts dual ended DC power supply, which means that the traction power substations(TPS) at both ends of each section of the line supply power to the subway. Changsha Metro Line 1 includes nine TPS, and Line 2 includes ten TPS. Therefore, the two lines are divided into 17 sections by the TPS. The rail voltage distribution of 17 sections of the line was calculated using a four-layer ground grid reflux model, which was equivalent to a line voltage source imported into the urban soil model built by COMSOL to simulate the diffusion of stray current in the urban soil environment and achieve the simulation of urban soil potential.

The schematic diagram of the DC traction system of urban rail transit is shown in Fig. 1, where the parts that affect the distribution of rail voltage and stray current include rail DC resistance $R_G$, rail-drainage network transition resistance $R_{g1}$, drainage network DC resistance $R_P$, drainage network-structural steel bar transition resistance $R_{g2}$, structural steel bar DC resistance $R_J$, structural steel bar-soil transition resistance $R_{g3}$, and soil DC resistance $R_D$. The above parts constitute a four-layer grounding network reflux model, as shown in Fig. 2. Divide the entire system vertically into many finite elements and derive a $\pi$ type equivalent circuit to replace the distribution parameters of each finite element. From this, a network composed of concentrated parameters is constructed to establish node voltage groups and solve rail potential, current, and stray current.

In Fig. 2, I1 and I2 are the traction currents provided by the TPS at both ends to the locomotive.

To simplify the calculation, the drainage network, structural steel bars, and transition resistance are equivalent to the track earth transition resistance $R_g$, resulting in simplified

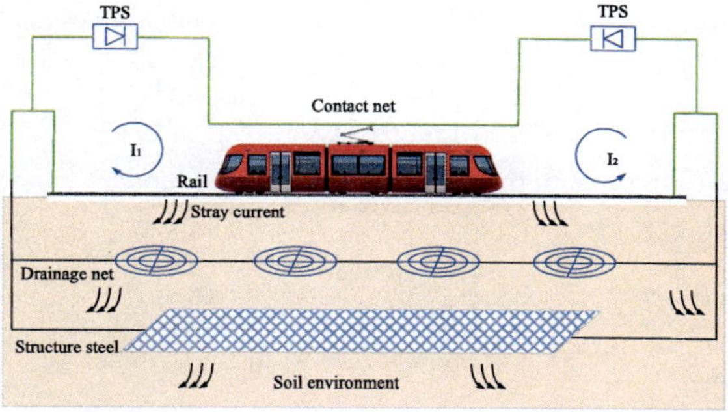


**Fig. 1.** The schematic diagram of the DC traction system.

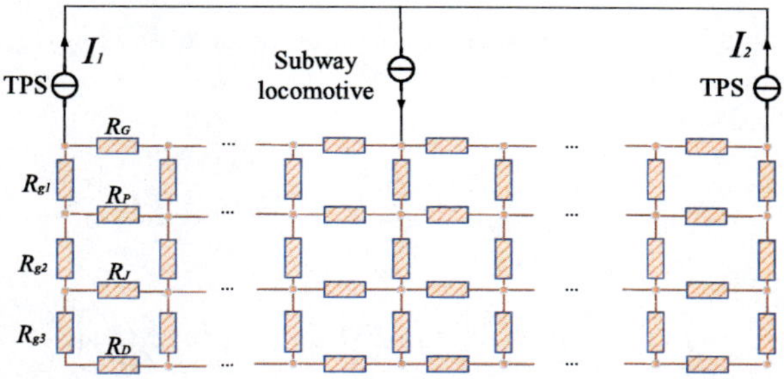


**Fig. 2.** The schematic diagram of the four-layer grounding network reflux model.

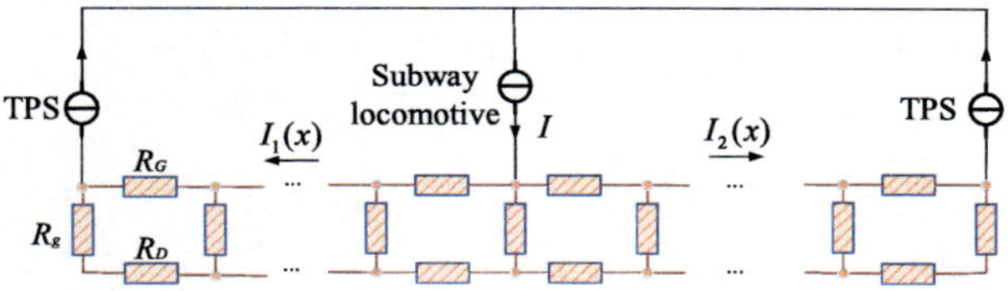


**Fig. 3.** The schematic diagram of simplified ground grid model.

circuit as Fig. 3. A subway locomotive runs on a $L$ long subway line with TPS at both ends. The subway locomotive divides the line into two sections of length $L_1$ and $L_2$, respectively. In Fig. 3, $I_1(x)$ represents the rail current in the first section of the line with length $L_1$, and $I_2(x)$ represents the rail current in the second section of the line with length $L_2$. According to Kirchhoff's law:

$$\frac{d}{dx}\begin{bmatrix} I_G(x) \\ U_G(x) \end{bmatrix} = \begin{bmatrix} 0 & 1/R_{GD} \\ R_G + R_D & 0 \end{bmatrix}\begin{bmatrix} I_G(x) \\ U_G(x) \end{bmatrix} + \begin{bmatrix} 0 \\ I \cdot R_D \end{bmatrix} \tag{1}$$

$$I_D(x) = I - I_G(x) \tag{2}$$

For the above non-homogeneous differential equation system, its general solution is:

$$\begin{bmatrix} I_G(x) \\ U_G(x) \end{bmatrix} = \begin{bmatrix} a_1 \ a_2 \\ a_3 \ a_4 \end{bmatrix} \begin{bmatrix} C_1 e^{\lambda_1 x} \\ C_2 e^{\lambda_2 x} \end{bmatrix} + \begin{bmatrix} f_1 \\ f_2 \end{bmatrix} \tag{3}$$

For the above equation system, there are the following boundary conditions:

The first line:

$$\begin{cases} I_{G1}(0) = \frac{L_2}{L} I \\ I_{G1}(L_1) = \frac{L_2}{L} I \end{cases} \tag{4}$$

The second line:

$$\begin{cases} I_{G2}(L_1) = \frac{L_1}{L} I \\ I_{G2}(L) = \frac{L_1}{L} I \end{cases} \tag{5}$$

If n metro locomotives are running on a metro line with length $L$, and their positions are $d_1$~$d_n$, as shown in Fig. 4, the traction current of each train is $I_1$~$I_n$, then according to the superposition theorem:

At position of 0~$d_1$:

$$I_G(x) = -\sum_{i=1}^{n} \sum_{j=1}^{2} \left[ a_j C_j(I_i) e^{\lambda_j (d_i - x)} + f_j(I_i) \right] \tag{6}$$

$$U_G(x) = \sum_{i=1}^{n} \sum_{j=1}^{2} \left[ a_{j+2} C_j(I_i) e^{\lambda_j (d_i - x)} + f_j(I_i) \right] \tag{7}$$

At position of $d_1$~$d_2$:

$$I_G(x) = \sum_{j=1}^{2} \left[ a_j C_j(I_1) e^{\lambda_j (d_i - x)} + f_j(I_1) \right] - \sum_{i=2}^{n} \sum_{j=1}^{2} \left[ a_j C_j(I_i) e^{\lambda_j (d_i - x)} + f_j(I_i) \right] \tag{8}$$

$$U_G(x) = \sum_{i=1}^{n} \sum_{j=1}^{2} \left[ a_{j+2} C_j(I_i) e^{\lambda_j (d_i - x)} + f_j(I_i) \right] \tag{9}$$

and so on, at position of $d_{n-1}$~$d_n$:

$$I_G(x) = \sum_{i=1}^{n-1} \sum_{j=1}^{2} \left[ a_j C_j(I_i) e^{\lambda_j (x - d_i)} + f_j(I_i) \right] - \sum_{j=1}^{2} \left[ a_j C_j(I_n) e^{\lambda_j (d_n - x)} + f_j(I_n) \right] \tag{10}$$

$$U_G(x) = \sum_{i=1}^{n-1} \sum_{j=1}^{2} \left( a_{j+2} C_j(I_i) e^{\lambda_j (x - d_i)} \right) + \sum_{j=1}^{2} \left( a_{j+2} C_j(I_n) e^{\lambda_j (d_n - x)} \right) \tag{11}$$

## 3 Line Modeling Methods

COMSOL is a finite element simulation software that supports three-dimensional modeling and coupling of multiple physical fields. The current (ec) module in COMSOL can be used to calculate the electric field, current, and potential distribution in conductive media. The calculated rail voltage of each section of the line is equivalent to the line voltage source applied to the soil model, which can intuitively simulate the soil potential distribution caused by stray current propagation in the soil model. As shown in Fig. 4, an example of a soil model with a size of 4 * 4 *1 km was constructed. The subway line in this model consists of three TPS, one single car section, and one double car section, with a total of three subway locomotives. Calculated the rail voltage distribution of the line using the parameters shown in Table 1, as shown in Fig. 4.

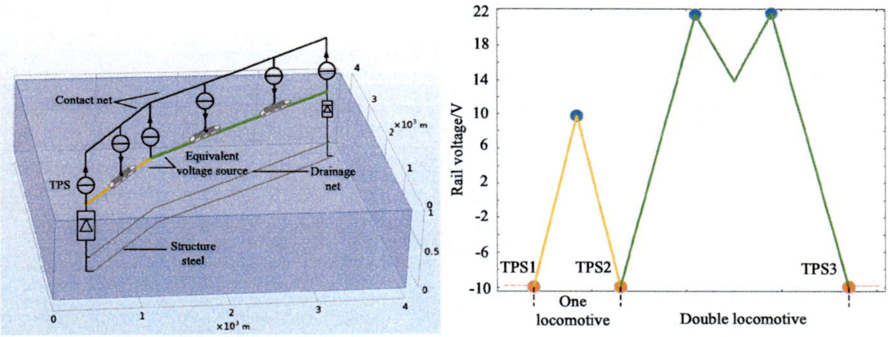


(a) subway line model. (b) Rail voltage.

**Fig. 4.** The schematic diagram of subway line model and rail voltage

**Table 1.** Rail voltage calculation parameters

| Parameter (unit) | Value | Parameter (unit) | Value |
|---|---|---|---|
| $R_G$ (Ω/km) | 0.026 | $I_Q$ (A) | 2000 |
| $R_P$ (Ω/km) | 0.66 | $R_{g2}$ (Ω · km) | 15 |
| $R_J$ (Ω/km) | 0.24 | $R_{g3}$ (Ω · km) | 15 |
| $R_D$ (Ω/km) | 0.001 | $\rho$ (Ω · km) | 0.01 |
| $R_{g1}$ (Ω · km) | 15 | | |

Figure 5 shown the simulated surface potential distribution and soil potential isosurface distribution of the soil model. When the traction current is equal to 2000 A, the extreme values of the rail potential in the two sections are 9.8 V, −9.8 V, and 21.6 V, respectively. The maximum value of the rail potential appeared at the locomotive position, while the minimum value appeared at the TPS. The distribution of soil potential is manifested by the radiation of the steel rail position towards the surrounding area,

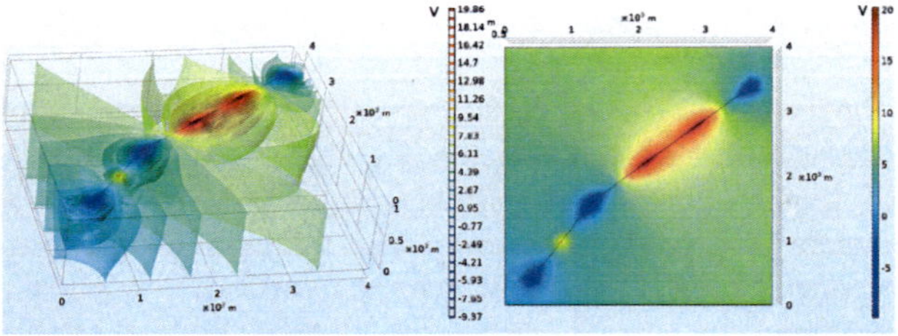

**Fig. 5.** Simulation result

and the range of each equivalent value near the subway line is relatively small. The potential fluctuation near the subway line is more severe, and is greatly affected by the stray current of the subway.

## 4 Factors Affecting the Distribution of Soil Potential

Based on the modeling method mentioned in the previous section, a soil potential simulation model for Changsha Metro Lines 1 and 2, as shown in Fig. 6, was constructed. The model has a length, width of 30 km, depth of 5 km. Other parameters are consistent with the example in the previous section. The rail voltage distribution of subway lines is calculated from actual operating parameters.

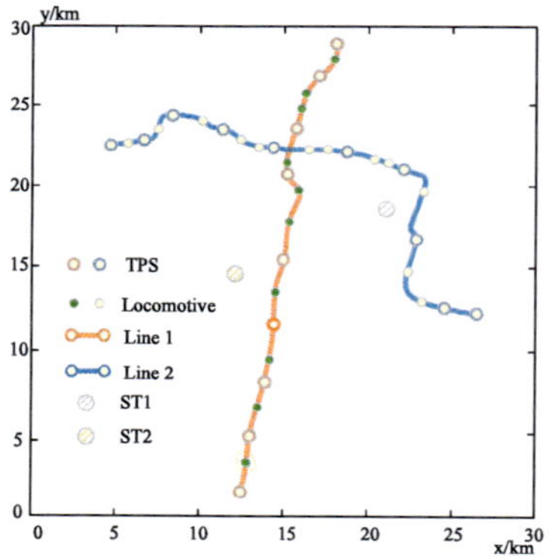


**Fig. 6.** The schematic diagram of Changsha Metro Lines 1 and 2.

During the dynamic operation of multiple trains, the main factors affecting the soil potential distribution are the rail DC resistance, transition resistance, and soil resistivity. Select two 220 kV substations in Changsha, ST1(21.1, 17.9) and ST2(12.4, 14.6), which are severely affected by stray current interference, and simulate the changes in soil potential under different parameters. In order to explore the impact of various parameters on the soil potential distribution and provide some reference for DC bias suppression, this article discussed the use of a single variable principle, with each subway operation time of 100 s and the initial positions of each locomotive still shown in Fig. 6.

### 4.1 The Influence of Rail DC Resistance on Siol Potential Distribution

Maintain $R_{g1} = 15\ \Omega \cdot \text{km}$, $\rho = 0.1\ \Omega \cdot \text{km}$, and change the $R_G$ to 0.013 Ω/km, 0.026 Ω/km, and 0.039 Ω/km, respectively. Obtained the changes in soil potential distribution as shown in Fig. 7. During the entire operation of the subway, the time periods for significant potential changes in the two substations are 0 s to 30 s (acceleration process) and 65 s to 90 s (braking process), during which the distribution of stray current is affected by changes in traction current. Within 30 s to 65 s, the train is in a coasting state and the traction current remains basically unchanged. During this period, the potential changes in the two substations are due to changes in the locomotive position, which affects the distribution of stray current.

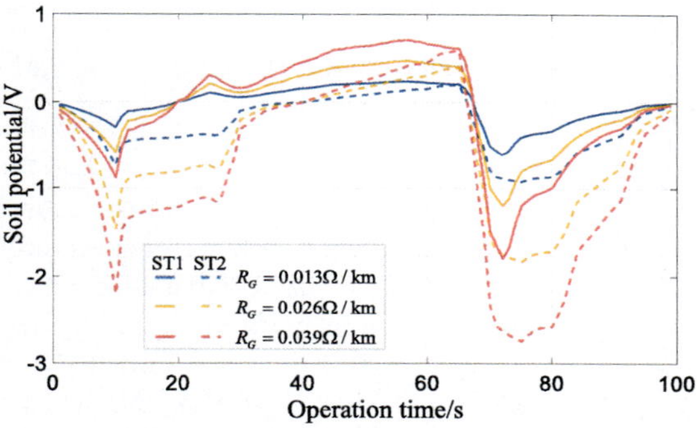


**Fig. 7.** Influence of $R_G$ on soil potential of 2 substations

Changing $R_G$ will not change the trend of soil potential changes. When it increases from 0.013 Ω/km to 0.026 Ω/km and 0.039 Ω/km, the soil potential and potential fluctuations in the two substations also increase by about 2 or 3 times the original. The increase in the DC resistance of the steel rail will significantly increase the soil potential, increase the range of stray current influence, and exacerbate the risk of DC bias and rail corrosion in the substation.

### 4.2 The Influence of Rail-Ground Transition Resistance on Soil Potential Distribution

Maintain $= R_G$ 0.026 Ω/km, $\rho = 0.1\ \Omega \cdot \text{km}$, change the $R_{g1}$ to $R_{g1} = 5\ \Omega \cdot \text{km}$, $R_{g1} = 15\ \Omega \cdot \text{km}$, $R_{g1} = 25\ \Omega \cdot \text{km}$. The simulation results are shown in Fig. 8. The change is negatively correlated with the change in nearby soil potential. Increasing from 15 to about 25 reduces the maximum, minimum, and potential fluctuation amplitude of the two substations to 90% of their original values; Reducing $R_{g1}$ from 15 $\Omega \cdot \text{km}$ to 5 $\Omega \cdot \text{km}$ increases the maximum, minimum, and potential fluctuation amplitude of the soil potential of the two substations to the original 108%. Increasing $R_{g1}$ (such as increasing the resistivity of the rail coating) can effectively reduce the risk of stray current hazards in nearby areas.

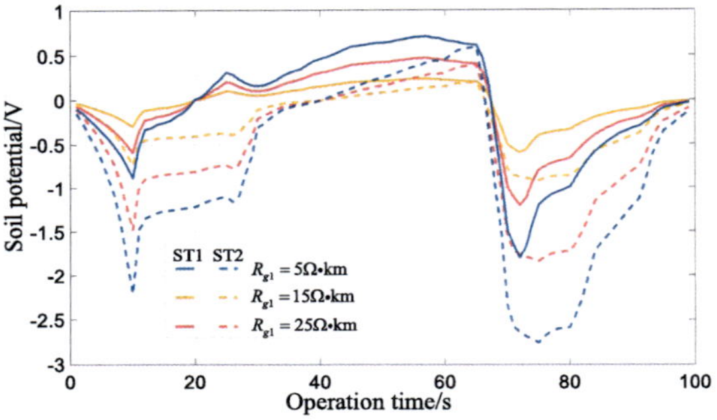


**Fig. 8.** Influence of $R_{g1}$ on soil potential of 2 substations

### 4.3 The Influence of Soil Resistivity on the Distribution of Soil Potential Distribution

Maintain $= R_G$ 0.026 $\Omega$/km, $R_{g1} = 15\ \Omega \cdot$ km, change the $\rho$ to $\rho$=0.05 $\Omega \cdot$ km, $\rho = 0.10$ $\Omega \cdot$ km, respectively. The simulation results are shown in Fig. 9. As the soil resistivity increases, the soil potential fluctuations of the two substations slightly increase, but the impact is extremely weak. At the 35th second of the simulation run. The change in soil resistivity will not affect the total amount of stray current leaked by the steel rail, but will affect the propagation range of stray current in the soil. Compared to = 0.1 and = 0.05, the range of each potential equivalent is wider, indicating that when the soil resistivity decreases, the stray current will propagate further and the soil potential amplitude will also slightly decrease.

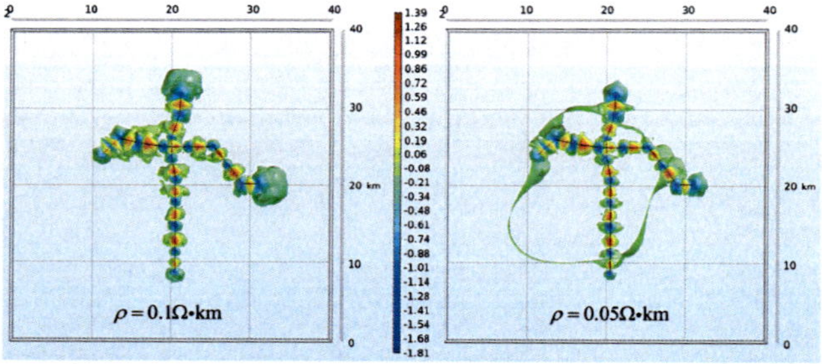


**Fig. 9.** Influence of $\rho$ on soil potential of two substations

## 5 Conclusion

(1) Changing the DC resistance of the steel rail will directly affect the stray current value that the steel rail leaks into the ground. Excessive DC resistance of the steel rail can exacerbate the DC bias and corrosion of the substation.

(2) The change in transition resistance between the rail and the drainage network is negatively correlated with the change in nearby soil potential. The increase in transition resistance between the steel rail and the drainage network reduces the stray current leaked by the steel rail, thereby reducing the risk of nearby areas being harmed by stray current.

**Acknowledgments.** This research was supported by the Science and Technology Project of the State Grid Corporation of China: Research on the coupling influence and suppression method of grid DC bias caused by metro stray current (5100-202199276A-0-0-00).

## References

1. Fichera, F., Mariscotti, A., Ogunsola, A.: Evaluating stray current from DC electrified transit systems with lumped parameter and multi-layer soil models. In: Eurocon, pp. 1187–1192 (2013)
2. Wang, A., Lin, S., et al.: Evaluation model of DC current distribution in AC power systems caused by stray current of DC metro systems. IEEE Trans. Power Deliv. **36**(1), 114–123 (2021)
3. Hen, R., Yang, Y., Jin, T.: A hierarchical coordinated control strategy based on multi-port energy router of urban rail transit. Prot. Control Mod. Power Syst. **7**(2), 201–212 (2022)
4. Zhou, H., Zhang, J., Li, J., et al.: Simulation analysis of interference of subway stray current. J. East China Jiaotong Univ. **39**(06) (2022). (in Chinese)
5. Ni, Y., Yu, K., Zeng, X., et al.: A correlation analysis on transformer DC bias current caused by metro stray current. J. Electr. Power Sci. Technol. **36**(06), 136–143 (2021). (in Chinese)
6. Wu, G., Xiao, L., Zhang, F., et al.: Simulation analysis of distribution characteristics of urban rail transit stray current in power system. South. Power Grid Technol. **13**(10), 39–43 (2019). (in Chinese)
7. Wang, A., Lin, S., Li, J., et al.: Stray current simulation model of the long line of DC metro systems. High Volt. Eng. **46**(04), 1379–1386 (2020). (in Chinese)
8. Yu, K.: Simulation Research on Stray Current in Subway Based on CDEGS. Southwest Jiaotong University (2015). (in Chinese)
9. Pang, Y.: Research on Stray Current in Urban Rail Transit. Southwest Jiaotong University (2008). (in Chinese)
10. Yan, M., Li, X., Wang, K., et al.: Research and simulation of potential and stray current distribution on subway rail. J. Beijing Inst. Pet. Chem. Technol. **21**(01), 37–41 (2013). (in Chinese)
11. Wang, Y., Huang, Y., Peng, C., et al.: Evaluation model for dynamic interference range of subway stray current based on surface potential gradient. J. Beijing Jiaotong Univ. **44**(03), 30–36 (2020). (in Chinese)
12. Li, W., Yang, H., Xu, F.: The effect of vertical layered soil on surface potential under a three-dimensional composite layered soil model. J. Sichuan Univ. (Eng. Sci. Ed.) **44**(05), 161–167 (2012). (in Chinese)
13. Lin, S., Zhou, Q., Lin, X., et al.: Infinitesimal method based calculation of metro stray current in multiple power supply sections. IEEE Access **8**, 96581–96591 (2020)

# Research on Hybrid Hub-and-Spoke Transportation Network of High-Speed Rail Express

Yansen Pei[1] and Xiaoning Zhu[1,2](✉)

[1] School of Traffic and Transportation, Beijing Jiaotong University, Beijing 100044, China
{peiyansen,xnzhu}@bjtu.edu.cn

[2] Key Laboratory of Transport Industry of Comprehensive Transportation Theory (Beijing Jiaotong University), Ministry of Transport, Beijing 100044, China

**Abstract.** In recent years, the rapid development of China's economy has driven the booming logistics industry, and then puts forward higher requirements for the ability and efficiency of the express industry. High-speed rail express, as a new mode of transportation, will play an important role in the improvement of China's logistics capacity. At present, most studies on high-speed rail express network are based on the hub-and-spoke network. This paper firstly analyzes the advantages and disadvantages of fully connected network and pure hub-and-spoke network, and then proposes the necessity of constructing hybrid hub-and-spoke network in high-speed rail express. Then, with the goal of minimizing the total cost of network transportation, an integer programming model is established to construct a hybrid hub-and-spoke network of high-speed rail express, considering the constraints of node operation ability and number of hubs, and a numerical experiment is designed to solve and verify the model. The influence of number of hubs on the optimal target value is analyzed, and the proportion of OD flow taking different transportation modes is calculated. Finally, the optimal target values of fully connected network, pure hub-and-spoke network and hybrid hub-and-spoke network are compared, and it is found that the use of hybrid hub-and-spoke network in the fast freight transportation of high-speed rail can not only exert the scale effect, but also effectively reduce the detour transportation to save the transportation cost.

**Keywords:** High-Speed Rail Express · Operation Network · Hybrid Hub-and-Spoke Network · Integer Programming

## 1 Introduction

In recent years, Chinese economy has been booming. In the Internet era, e-commerce has obtained an unprecedented development environment, logistics industry is a strong cornerstone of its development, and the express industry as the pillar of the logistics industry, more need to keep pace with the times, in order to adapt to the economic development brought about by the higher demand for express goods transportation. High-speed rail express is a full-course transportation service for small items with advanced


Y. Qin et al. (Eds.): EITRT 2023, LNEE 1137, pp. 80–87, 2024.
https://doi.org/10.1007/978-981-99-9311-6_9

logistics requirements by utilizing railway transportation resources such as high-speed passenger units and high-speed confirmation trains. Compared with traditional express delivery, high-speed rail express is more timely, punctual, safe, stable and environmental friendly, especially in the medium and long distance transportation, high-speed rail express can play its unique advantages. In the 14th Five-Year Plan for the Development of Modern Logistics, the state clearly proposes to promote the steady development of railway (high-speed rail) express, improve the railway (high-speed rail) express network, and innovate the railway (high-speed rail) express service, so as to improve the service capacity and efficiency of the logistics industry, adapt to the diversified logistics needs, and promote the development of green logistics.

However, high-speed rail express entered the market late, with low brand awareness and recognition. Although the high-speed rail network has covered most of the large and medium-sized cities in the country, the number of business outlets of high-speed rail express in the covered cities is not large [1]. Meanwhile, due to the restrictions of railway transport capacity, organization mode and other factors, it has not yet become the main form of rapid freight transport in China. Its development is still in the stage of exploration and innovation. At present, the research on the layout of high-speed railway express operation network in the academic circle focuses on the layout of large-scale high-speed railway express base [2]. Ma Chunshan et al. [3], aiming at maximizing the income of railway enterprises, constructed an integer programming model for the location and scale optimization of high-speed railway express base based on service network, and adopted a heuristic iterative algorithm framework to solve the problem. Wang Jinzhi [4] constructed the distribution center location index system, and on this basis, combined with the basic principle and application process of fuzzy clustering method, established the location model of high-speed rail express distribution center with the existing railway freight station as the alternative location. Sun Huijun et al. [5] established a two-layer programming model for the location of logistics distribution center. Shen Hao [6] established a P-median hub location model with coverage distance restriction, and solved the model through a hybrid intelligent algorithm combining genetic algorithm and particle swarm optimization algorithm. In terms of hub location selection, many scholars have also conducted researches on enhancing the robustness of location selection results. Xu Shuangsong et al. [7] abstracted the optimization model of high-speed rail fast cargo transportation network into a mixed integer quadratic cone programming problem by using distributed robust optimization method under the framework of hub-spoke network, and proposed corresponding outer approximation (OA) algorithms. Fu Chenyi et al. [8] proposed a two-stage robust optimization model with demand-related uncertainty set for bike sharing system, which maximizes daily returns and enhances the robustness of the system.

After determining the physical nodes of high-speed rail express, the nodes need to be classified and connected into a network to form a high-speed rail express operation network based on the consideration of node operation capacity, line transportation capacity and delivery time of goods in an appropriate way. Deng Xiaozhen [9] pointed out that China's express transportation demand network is a large-scale hub-and-spoke network, analyzed the spatial distribution characteristics of China's express demand, selected 35 stations from 223 existing high-speed rail express operation stations as candidate nodes of high-speed rail express special train network, and built a three-node transportation network of high-speed rail express through K-means clustering model. He Kaixian [10] established mixed integer programming models for zero-load operation network optimization of hybrid hub-and-spoke highway with no section capacity restriction and with section capacity restriction respectively. In addition, some researches focus on freight flow distribution and operation organization model of high-speed rail express. Bin Song et al. [11], aiming at the lack of effective integration of high-speed rail express trunk transportation and urban end transportation, studied the optimization of high-speed rail express service path based on urban end transportation and two-dimensional loading. Zhu Endong et al. [12] used the modeling method of space-time network to optimize vehicle flow paths, marshalling plans and train operation plans. Fuzzy evaluation and analytic hierarchy process (AHP) are used to evaluate and analyze the operation mode of high-speed express in railway development, and through node stratification and regional division, a hub-and-spoke transport service network based on node splitting is constructed. Li Gaobo [13] constructed a hybrid hub-and-spoke multimodal transport hub network model considering service time requirements and hub capacity constraints.

According to the results of literature review, it can be seen that most studies on high-speed rail express in China are based on hub-and-spoke network, but there are few studies on hybrid hub-and-spoke network. Aiming at minimizing the costs of network transportation, this paper constructs a hybrid hub-and-spoke network for high-speed rail express freight transportation, and compares it with fully connected network and pure hub-and-spoke network.

## 2 Problem Statement

The hub-and-spoke network collects OD flows to one or two hubs and then transports them to the destinations. Compared with the fully connected logistics network, the hub-and-spoke logistics network can generate scale economy of transportation between hubs. As shown in Fig. 1(a), in a fully connected logistics network, the transport demand of a single line is generally small, and the cargo load rate is often not high. Moreover, if the logistics demand between two points has obvious directivity, the situation of empty cars returning may occur, resulting in huge losses of logistics benefits. In the hub-and-spoke logistics network, as shown in Fig. 1(b), OD flows are generally gathered in one or two hub stations for transportation, forming the transportation scale benefits between hub stations. In addition, the hub-and-spoke network can reduce the transportation resources needed and promote the integrated utilization of logistics resources.

However, hub-and-spoke logistics network also has disadvantages. For example, the hub-and-spoke network will inevitably bring detour costs for cargo transportation, and the transport route is often not the shortest path. Moreover, due to the diminishing economies of scale, when the transport demand between two non-hub nodes reaches a certain level, the scale benefits brought by the centralized transport through trunk lines may hardly offset the costs brought by the detour transport. At this time, for the two nodes. The hub-and-spoke transport network will no longer be practical. In addition, the centralized transportation of goods at the hub station will bring node logistics costs, including not only the narrow economic cost, but also the processing time cost and waiting time cost of goods.

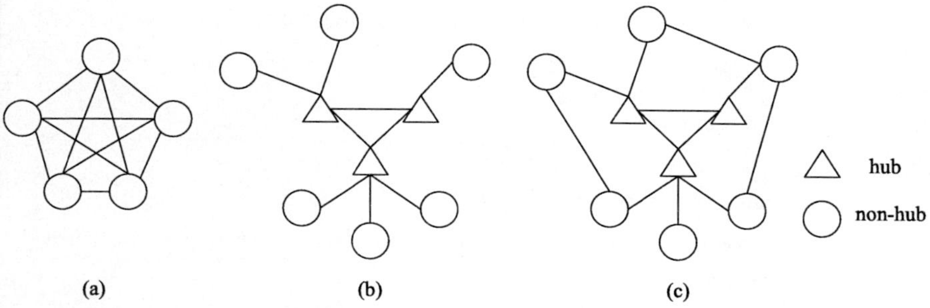


**Fig. 1.** Schematic diagram of logistics network

For hub-and-spoke logistics network, scale effect and detour transportation cost are the most fundamental pair of contradictions, so we need a suitable way to deal with this problem, which can not only make full use of the scale effect, but also reduce the detour transportation cost to the greatest extent. In this context, hybrid hub-and-spoke network emerges. As shown in Fig. 1(c), it allows some non-hub nodes to carry out through-transport of goods on the basis of pure axle-spoke network, so as to avoid large detoured transport costs. It is the combination of pure hub-and-spoke network and fully connected network, and is a good method to solve the above contradictions. Therefore, aiming at minimizing the costs of network transportation, this paper constructs a hybrid hub-and-spoke network for fast freight transportation of high-speed rail express, and compares it with fully connected network and pure hub-and-spoke network.

## 3 Model Formulation

### 3.1 Model Assumption

This paper makes the following assumptions about the research questions:

- The demands for goods transportation between all OD pairs are known.
- Goods between a pair of OD points can't be transported through different routes.
- The transport network allows any OD flow to carry out single-point transit, two-point transit and through-transport.
- The scale effect is only generated on the transport routes between hubs.
- Any non-hub node can belong to more than one hub, that is, the goods of any non-hub node can transit through more than one hub.
- The calculation of node cost includes fixed cost and non-fixed cost, assuming that the non-fixed cost is proportional to the flow through the node.

### 3.2 Optimization Model

In this model, specified number of hubs in given nodes are firstly chosen, and then transportation modes (direct connection transportation, single-hub transshipment or double-hub transshipment) and transportation paths for all OD flows are selected, so as to minimize the total transportation cost of the network.

**Table 1.** Notation of the optimization model.

| set, parameters and scalars | |
|---|---|
| $i, j, k, m \in N$ | index of nodes |
| $H_i$ | fixed construction cost of Hub $i$ |
| $f_{ij}$ | cargo flow from node c to node $j$ |
| $c_{ij}$ | transportation cost per unit goods per unit distance between node $i$ and node $j$ |
| $d_{ij}$ | the distance between node $i$ to node $j$ |
| $\alpha$ | the discount factor generated by scale effect |
| $p$ | the specified number of hubs |
| $F_k$ | the maximum cargo flow capacity of hub $k$ |
| variables | |
| $y_k$ | is 1 if node $k$ is selected as the hub |
| $x_{ij}$ | is 1 if the goods between node $i$ and node $j$ choose direct connection transportation |
| $z_{ijkm}$ | is 1 if the goods between node $i$ and node $j$ are transported through hubs $k$ and $m$, $k = m$ refers to single-hub transshipment, $k \neq m$ refers to double-hub transshipment |

The complete notation are described in Table 1.
The IP model is formulated in the following:

$$min \sum_{i \in N} H_i y_i + \sum_{(i,j) \in N} f_{ij} c_{ij} x_{ij} d_{ij} + \sum_{(i,j,k,m) \in N} \alpha f_{ij} c_{km} z_{ijkm} d_{km} \quad (1)$$

$$\sum_{k} y_k = p, \quad \forall k \in N \quad (2)$$

$$z_{ijkm} \leq y_k \cdot y_m, \quad \forall i, j, k, m \in N \quad (3)$$

$$\sum_{(k,m) \in N} z_{ijkm} + x_{ij} = 1, \quad \forall i, j \in N \quad (4)$$

$$\sum_{(i,j,k) \in N} f_{ij} \cdot z_{ijkm} \leq F_m \cdot y_m, \quad \forall m \in N \quad (5)$$

$$x_{ij}, y_k, z_{ijkm} \in \{0, 1\}, \quad \forall i, j, k, m \in N \quad (6)$$

Function (1) means to minimize the total cost of transportation network, including fixed cost of hub location, direct connection transportation cost and transshipment cost. Constraints (2) specifies the number of hubs in the network; Constraints (3) means that only after node $k$ and node $m$ are determined as hubs can the goods transit through the two nodes. When $k = m$, it is a single-hub transshipment; when $k \neq m$, it is a double-hub transshipment; Constraints (4) means that the cargo flow between node $i$ and node $j$

can only choose one between transit transportation and direct connection transportation. Constraints (5) represents the capacity constraint of hub $m$. Constraints (6) is the value description of decision variables.

## 4 Numerical Experiment

In this paper, a numerical experiment containing 15 nodes is designed, from which 5 hub nodes need to be selected, and the transportation mode of each group of OD flows (single-hub-transshipment, double-hub-transshipment or direct connection transportation) is determined. The model is solved by Gurobi 10.0.1 on a PC with 2.20 GHz CPU and 16 GB RAM, and the optimal solution was obtained within 30 s in all experiments.

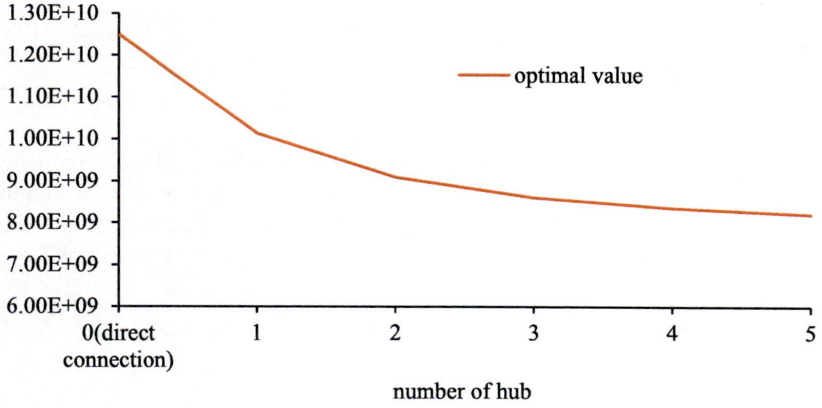


**Fig. 2.** The optimal target value under different number of hubs

Experiments with different number of hubs are conducted to study the influence of the number of hubs on the optimal target value, and the results are shown in Fig. 2.

As can be seen from Fig. 2, compared with 0 hubs (fully connected network), the establishment of hybrid hub-and-spoke network can effectively reduce logistics costs, and the optimal target value decreases with the increase in the number of hubs, because with the increase in the number of hubs, each OD flow has a greater possibility to choose a more favorable path, thus reducing the total logistics cost. At the same time, we can notice that the decline trend of the target value slows down gradually with the increase of the number of hubs. This is because on the premise that the total transportation demand remains unchanged, the marginal benefit brought by the number of hubs to a certain extent will no longer increase. Almost all OD flows have adopted the optimal path theoretically. It can be seen that the number of hubs is not the more the better. When choosing the number of hubs, we should consider the actual transportation demand to make a correct decision.

The statistics of OD flow choosing different transport modes under different number of hubs are shown in Fig. 3.

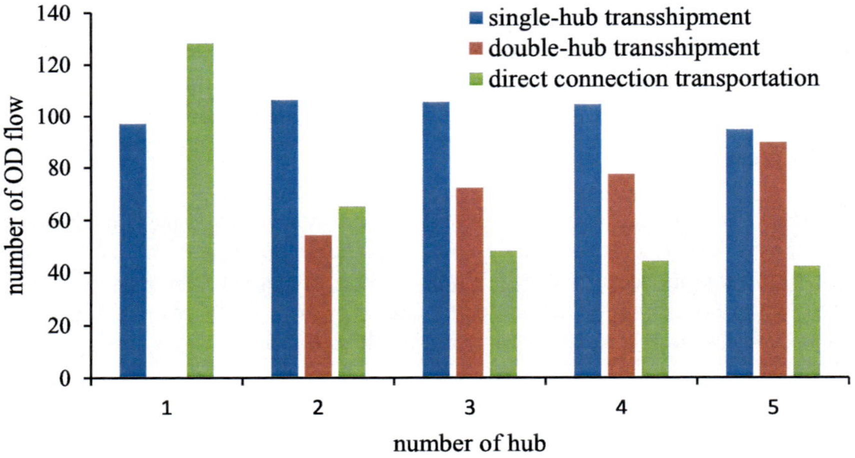


**Fig. 3.** OD flow transport mode

As can be seen from Fig. 3, with the increase in the number of hubs, the number of OD flows choosing direct connection transportation gradually decreases, and more OD flows choose single-hub transshipment or double-hub transshipment mode, which reduces the transportation cost with scale effect of the hub-and-spoke network.

**Table 2.** Table captions should be placed above the tables.

| Network | Total cost |
|---|---|
| fully connected network | 1.25E+10 |
| pure hub-and-spoke network | 8.43E+09 |
| hybrid hub-and-spoke network | 8.29E+09 |

In the pure hub-and-spoke network, the transportation path of each OD flow is determined. Under the condition of the same number of hubs (5 hubs), the total transportation cost is compared with the hybrid hub-and-spoke network and fully connected network, as shown in Table 2. It can be seen that the hybrid hub-and-spoke network can indeed further save the transportation cost.

## 5 Conclusions

In this paper, the advantages and disadvantages of fully connected network, pure hub-and-spoke network and hybrid hub-and-spoke network are analyzed. The hybrid hub-and-spoke network can not only exert the scale effect of hub-and-spoke network, but also effectively reduce the cost of detour transportation. Then, with the goal of minimizing the total transportation cost, an integer programming model was established to construct the hybrid hub-and-spoke high-speed rail express transportation network. Finally, a numerical experiment is designed to solve and verify the model, and the influence of the number of hubs is analyzed. The advantages of the hybrid hub-and-spoke transportation network are verified by comparing the three transportation networks.

**Acknowledgments.** This work is supported by the Joint Funds of the National Natural Science Foundation of China (No. U2034208).

## References

1. Zhou, Y.: market positioning and operation strategy research of CRHE express. Railw. Econ. Res. **168**(04), 41–45 (2022). (in Chinese)
2. Wang, H.Q., Ding, X.D., Huo, J.Y.: Operation network layout method and case study of high-speed railway express cabinet. Railw. Freight Transp. **41**(03), 48–52+58 (2023). (in Chinese)
3. Ma, C.S., Liao, Z.W.: An optimization method of high-speed rail express base location selection and scale based on service network. J. Railw. Sci. Eng. **20**(02), 463–472 (2023). (in Chinese)
4. Wang, J.Z.: Location of Distribution Center for High-Speed Railway Express Based on Railway Freight Stations, p. 70. Beijing Jiaotong University (2022). (in Chinese)
5. Sun, H.J., Gao, Z.Y., Wu, J.J.: A bi-level programming model and solution algorithm for the location of logistics distribution centers. Appl. Math. Model. **32**(4), 610–616 (2008)
6. Shen, H.: Study on Location Selection of High Speed Rail Express Hub Based on the Value of Freight Travel Time, p. 69. Southwest Jiaotong University (2021). (in Chinese)
7. Xu, S.S., Liu, K.L.: Hub-and-spoke network optimization method of China railway highspeed express based on distributed robust optimization. Chin. J. Manag. Sci., 1–11 (2022). (in Chinese)
8. Fu, C.Y., Zhu, N., Ma, S., et al.: A two-stage robust approach to integrated station location and rebalancing vehicle service design in bike-sharing systems. Eur. J. Oper. Res. **298**(3), 915–938 (2022)
9. Deng, X.Z.: Research on the Optimization of Express Transportation Network of High-Speed Railway Under the Evolution of Express Transportation Demand Space, p. 153. Southwest Jiaotong University (2021). (in Chinese)
10. He, K.X.: Optimization of Hybrid Hub-and-Spoke Network for Less-than-Truckload Transportation, p. 66. Central South University (2014). (in Chinese)
11. Bin, S., Zhou, S.Q., Zhang, D.Z., Zhang, T.: Research on optimization problem of the service path of high speed railway express with the consideration of the last-mile distribution in cities and two-dimensional loading. Railw. Transp. Econ. **45**(03), 53–61 (2023). (in Chinese)
12. Zhu, E.D., Crainic, T.G., Gendreau, M.: Scheduled service network design for freight rail transportation. Oper. Res. **62**(2), 383–400 (2014)
13. Li, G.B.: Design of Hybrid Hub-and-Spoke Intermodal Hub Network with Capacity Constraints, p. 75. Chang'an University (2014). (in Chinese)

# Optimization of Train Operation Control Based on Soft Actor-Critic Deep Reinforcement Learning Algorithm

Zhuyuan Lan and Huiqin Pei(✉)

School of Electrical and Automation Engineering, East China Jiaotong University, Nanchang 330013, China
{zhuyuanlan,peihuiqin}@ecjtu.edu.cn

**Abstract.** With the rising interest in the application of artificial intelligence methods in society, the focus on the use of deep reinforcement learning in railway transportation networks has grown significantly. This study aims to explore the optimization problem in train operation control and proposes a deep reinforcement learning approach based on the Soft Actor-Critic algorithm as a training framework. The objective is to achieve energy efficiency and on-time arrivals. By considering various factors such as train speed limits, energy consumption, and operational efficiency, a reward function is designed to guide the train in learning appropriate operating strategies. Through the integration of a realistic train dynamics model and mechanical characteristics, simulation experiments validate the feasibility of this approach in achieving energy savings and punctual arrivals.

**Keywords:** Train Control Optimization · Energy Efficiency · Soft Actor-Critic Algorithm

## 1 Introduction

The optimization of train operation control is a critical task within the railway transportation system, aiming to enhance efficiency, safety, and energy utilization. Traditional methods face challenges when confronted with the complexity of dynamic systems, uncertain operational conditions, and the need for real-time handling of unexpected situations. To address these challenges and strengthen train operation control, this paper presents innovative approach based on the Soft Actor-Critic (SAC) deep reinforcement learning algorithm.

Previous studies have made significant contributions to the field of train operation control. Lin et al. [1] integrated traffic path optimization and train information planning using a simulated annealing algorithm, improving the overall system performance. Zhao and Dai [2] developed an event-triggered adaptive control method for multiple high-speed trains, considering deceptive attacks on bottleneck sections. Wang et al. [3] proposed a comprehensive energy-efficient train operation method based on spatiotemporal velocity network, effectively optimizing energy utilization. Huang et al. [4]


Y. Qin et al. (Eds.): EITRT 2023, LNEE 1137, pp. 88–96, 2024.
https://doi.org/10.1007/978-981-99-9311-6_10

presented a data-driven model and ensemble algorithm based on machine learning for optimizing discrete train speed distribution in urban rail transit. Zhao et al. [5] applied the GA-PSO algorithm to optimize energy-efficient driving of trains in urban rail transit systems. Liu et al. [6] introduced an intelligent traction control method for permanent magnet maglev trains based on model predictive fuzzy PID control and online optimization. Su et al. [7] developed a model predictive control based on nonlinear safety balancing distance for virtual coupled trains on gradient terrain. Zhang et al. [8] proposed an event-driven deep reinforcement learning approach for intelligent train operation, effectively controlling train operations in dynamic environments. Li et al. [9] achieved approximate cost-optimal energy management for hydrogen-powered train units using the double Q-learning algorithm. Meng et al. [10] applied deep reinforcement learning to energy-efficient train operation for autonomous driving, demonstrating its effectiveness in optimizing train operations. Liao et al. [11] proposed a deep reinforcement learning method for rescheduling energy-oriented train timetables under disturbances. Liu et al. [12] developed a DQN-based intelligent control method for heavy-loaded trains on long downhill sections. Guo et al. [13] focused on optimizing subway timetables considering intercity railway train capacity and passenger demand. Zhiwen et al. [14] investigated a priority-based high-speed railway operation adjustment model to improve train operation efficiency. Su et al. [15] presented an energy-saving method for cooperative train control based on multi-agent reinforcement learning. Shang et al. [16] introduced a deep reinforcement learning method with a reference system to handle constraints in energy-efficient train control. The proposed method in this paper is inspired by their work. Wang et al. [17] applied machine learning techniques to intelligent operation of heavy-loaded trains, effectively addressing the data imbalance issue. Zhou et al. [18] focused on developing intelligent train operation algorithms that combine expert knowledge with reinforcement learning techniques. Zhu et al. [19] proposed a method using deep reinforcement learning to optimize the performance of communication-based train control systems. Additionally, Zhu et al. [20] introduced an eco-driving algorithm for trains by allocating energy and adopting the Q-learning method. These studies have provided valuable insights and techniques for the development of the optimization framework proposed in this paper.

In this study, we employ the SAC deep reinforcement learning algorithm to optimize the control of train operations. This approach combines the strengths of deep learning and actor-critic methods to enhance the decision-making process and refine the control over train operations. The SAC algorithm represents a cutting-edge deep reinforcement learning technique that has demonstrated impressive efficacy across various control problems. Specifically tailored for optimizing train operation control, the SAC algorithm adeptly handles the continuous action space critical for governing the output force, acceleration, and velocity of trains. Leveraging a neural network-based actor-critic architecture, the SAC algorithm learns the optimal strategy for controlling train operations by iteratively updating the network parameters.

In summary, this paper introduces a novel method for optimizing train operation control based on the Soft Actor Critic deep reinforcement learning algorithm. By integrating advanced deep reinforcement learning techniques, this method holds significant promise in revolutionizing railway transportation systems, elevating efficiency, safety, and mitigating energy consumption.

Empirical evaluation using real-world train operation data will provide valuable insights into the effectiveness and practicality of this method in enhancing train operation control. The results obtained from this evaluation will showcase the potential of the proposed approach in reducing energy consumption and comprehensively improving the performance of train operations.

## 2 Train Operation Trajectory Control Problem

The dynamics of train operation involve a complex interplay of various factors. By incorporating insights from previous studies on train operation control and applying Newton's second law, this paper presents a formulated train dynamics model represented by Eqs. (1) and (2):

$$\frac{dv}{dx} = \frac{\mu_1 F(v)_t - \mu_2 F(v)_b - F(v)_{basic} - F(v)_{add}}{M(1+\tau)} \tag{1}$$

$$\frac{dt}{dx} = \frac{1}{v} \tag{2}$$

Here, $F(v)_t$ represents the train's traction force, $F(v)_b$ denotes the train's braking force, $F(v)_{basic}$ signifies the basic resistance encountered during train operation, and $F(v)_{add}$ refers to the additional resistance caused by factors like curves and gradients. The coefficients $\mu_1$ and $\mu_2$ represent the train's traction and braking coefficients, respectively.

Furthermore, the basic resistance can be calculated using the formula:

$$F(v)_{basic} = av + bv^2 + cv^3 \tag{3}$$

The traction force and braking force of the train are influenced by the mechanical characteristics of the train, while the additional resistance is determined by the design of the train's operating route. The basic resistance, on the other hand, is affected by factors such as weather conditions. Empirical values are typically assigned to these forces in this study. The traction/braking characteristic curve employed in this paper is depicted in the accompanying Fig. 1.

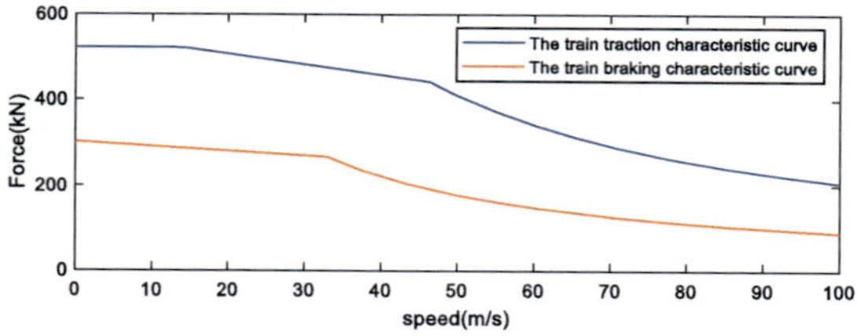


**Fig. 1.** The train traction/braking characteristic curve

During train operation control, the optimization of traction and braking forces assumes a pivotal role in ensuring both safe arrival and minimal energy consumption.

Through precise calibration of these forces, the train can achieve optimal acceleration and deceleration, thereby promoting efficient energy utilization and upholding safety throughout the entire journey.

In this research, a time step of 1 s is adopted. The displacement of the train within each time step is denoted as dx, and the force exerted by the train at the present time is represented as $F$. To quantify the overall energy consumption of the train throughout the entire operational phase, the following equation is employed:

$$\mathrm{E} = \int_0^T Fvdt \tag{4}$$

Furthermore, this study establishes predefined criteria for on-time arrival, parking accuracy, and maximum speed limit of the train. These criteria are quantified by the following equations:

$$\left|\mathrm{t}_{ar} - \mathrm{t}_{plan}\right| \leq \mathrm{t}_{ae} \tag{5}$$

$$\left|\mathrm{d}_{ar} - \mathrm{d}_{plan}\right| \leq \mathrm{d}_{ae} \tag{6}$$

$$\mathrm{v}_t \leq \mathrm{v}_{limit} \tag{7}$$

Equation (5) represents the permissible discrepancy between the actual arrival time of the train and the scheduled arrival time. Equation (6) specifies the acceptable margin of error between the actual stopping position of the train and the designated stopping position. Finally, Eq. (7) establishes the requirement that the train's speed must not exceed the prescribed speed limit.

## 3 Deep Reinforcement Learning Methodology

Deep Reinforcement Learning (DRL) is an amalgamation of deep learning and reinforcement learning techniques that empowers agents to learn and make decisions in complex environments. By harnessing the capabilities of deep neural networks, DRL methods are able to approximate the value function or policy function of agents.

In recent years, DRL methods have experienced remarkable growth and have garnered considerable attention due to their ability to handle high-dimensional input spaces and learn directly from raw sensory inputs, such as images or sensor data. The applications of DRL span diverse domains, encompassing fields such as robotic control, AI training for gaming competitions, and various realms of automated control.

In this study, the chosen DRL method for train operation control is the Soft Actor-Critic (SAC) algorithm. SAC combines the advantages of actor-critic methods with entropy regularization, facilitating the learning of both policy and value functions. Its performance in handling tasks within continuous action spaces has been extensively validated across multiple domains.

By employing the SAC algorithm for train operation control, the aim is to optimize the traction and braking forces of the train, thereby achieving efficient and secure operation.

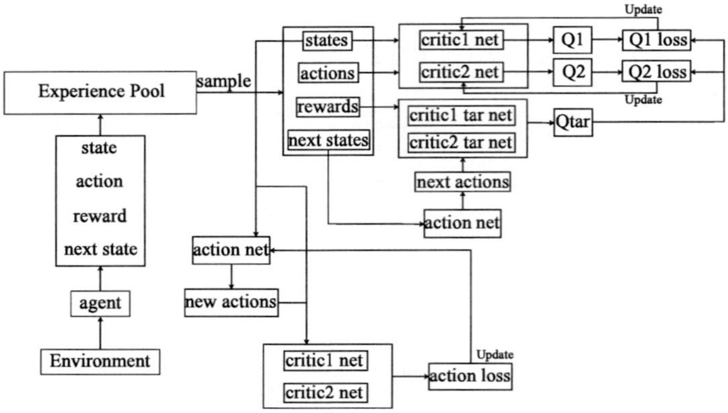


**Fig. 2.** The SAC Algorithm Learning Framework

The SAC algorithm empowers the agent to learn and adjust its control policy based on environmental feedback, enabling the train to make intelligent real-time decisions.

The learning framework of the SAC algorithm is depicted in Fig. 2.

The SAC algorithm's network architecture comprises several essential components:

Policy Network (Actor Network): The policy network takes the current state as input and produces a probability distribution over continuous actions. Its purpose is to learn an optimal policy that generates the most suitable actions given a specific state.

Q Networks (Critic Networks): The SAC algorithm employs two Q networks to estimate the action-value function (Q-function). These networks take both the state and action as input and output corresponding Q-values. Their role is to evaluate the quality of actions and guide the learning process.

Target Q Networks (Target Critic Networks): Alongside the Q networks, SAC incorporates two target Q networks. These networks possess the same structure as the Q networks but use distinct sets of parameters. They provide stable Q-value estimates used for computing the target values during training.

Temperature Parameter: SAC introduces a temperature parameter to strike a balance between exploration and exploitation. It modulates the entropy of the action distribution, promoting exploration. The temperature parameter undergoes updates during training to discover an appropriate balance between exploration and exploitation.

The interaction among these components establishes the learning framework of the SAC algorithm. The policy network generates actions, the Q networks estimate Q-values, the target Q networks supply target values for updating the Q networks, and the temperature parameter influences the exploration-exploitation trade-off. The entire network architecture undergoes iterative training to refine the policy and Q-function parameters.

The formulation of an appropriate reward function during the train operation control process holds significant significance as it serves as a guiding principle for the train to acquire suitable operational strategies. The reward function must be customized to address specific objectives, encompassing considerations of safety, energy efficiency, and timeliness.

Therefore, in this study, the immediate reward function is carefully crafted according to the following formulation:

$$r_{im} = r_v + r_d + r_{si} \tag{8}$$

Among them, $r_v$ represents a velocity-related reward function, and its assignment is as follows when it falls below the target velocity (the average speed multiplied by the coefficient ω):

$$r_v = v_t/\omega v_{ave} \tag{9}$$

When the velocity exceeds the target speed, the assignment of $r_v$ is as follows, in order to incentivize the train to maintain a speed close to the target:

$$r_v = 2 - v_t/\omega v_{ave} \tag{10}$$

In order to encourage the train to progress towards the target point, the reward function related to position is defined as follows:

$$r_d = 1 - |d_t - d_{plan}| \tag{11}$$

When the remaining distance of the train is less than one-ninth of the planned distance, it is desirable to encourage the train to slow down and stop. In this case, the value of $r_v$ is set to 0, and the reward function for stopping is defined as follows:

$$r_{si} = 1 - (v_t/v_{ave}) \tag{12}$$

To mitigate the array of issues caused by an overly dense reward function, this paper proposes a method to appropriately reduce the density of the reward function by introducing reward step.

When the train reaches the terminal state, that is, when it arrives at the destination, the reward function is defined as follows:

$$\begin{aligned} r_{done} = {} & \rho_1\big(1 - \big(t_{ar} - t_{plan}\big)/t_{plan}\big) + \rho_2(1 - (v_{ar} - v_{ave})/v_{ave}) + \\ & \rho_3\big(1 - \big(d_{ar} - d_{plan}\big)/d_{plan}\big) + \rho_4(1 - (E_{ar} - E_{PSO}/E_{PSO})) \end{aligned} \tag{13}$$

Where $\rho_1$, $\rho_2$, $\rho_3$, and $\rho_4$ are adjustable coefficients, Ear represents the cumulative energy consumption of the train upon arrival, and $E_{PSO}$ denotes the cumulative energy consumption obtained using the particle swarm optimization algorithm.

To ensure the fidelity of train control in practical settings, our approach takes into account the impact of the train's actual mechanical system. The actions in our method are represented within a range of [−1, 1], which corresponds to the braking/traction coefficient. These coefficients are then multiplied by the maximum braking force and maximum traction force available at the current velocity. This enables the algorithm to control the train in a manner that closely emulates real-world conditions. By incorporating these considerations, our approach strives to enhance the authenticity of train control in practical scenarios.

## 4 Experimentation

To validate the feasibility of the proposed method in this article, simulation verification was carried out using the actual basic resistance coefficient of the Wuhan-Changsha South section of the Wuhan-Guangzhou Line. Specifically, the values chosen for the parameters were a = 5.2, b = 0.038, and c = 0.00112.

The following table shows the settings for the experimental environment (see Table 1):

**Table 1.** Experimental environment

| Software Platform | Programming Language | Learning framework | Operating System |
|---|---|---|---|
| Pycharm | Python3.9 | Pytorch | Windows10 |

The SAC algorithm parameter settings are presented in the following table (see Table 2):

**Table 2.** The SAC algorithm parameters

| Experience Pool Size | Sample Size | Discount Factor | Temperature Parameter |
|---|---|---|---|
| 18000 | 1800 | 0.99 | $log_2 0.01$ |

The following table presents the train-related parameters used in the experiments (see Table 3):

**Table 3.** The train-related parameters

| Type | Mass | Operating Mileage | Operating Time |
|---|---|---|---|
| CRH380AL | 890 t | 2000 m | 188 s |

After conducting 2000 episodes of iteration, we obtained the reward function and the speed curve of the train, which are depicted in Fig. 3.

It is evident from the observation that the reward function shows a steady increase and gradually stabilizes around a certain value.

In the end, we have obtained a train trajectory with a cumulative energy consumption of 179480.5 J and a travel distance of 1999.7 m. The Particle Swarm Optimization algorithm yielded an energy consumption value of 181933.4 J, while the SAC algorithm's result slightly outperformed it. This unequivocally demonstrates the effectiveness of this approach.

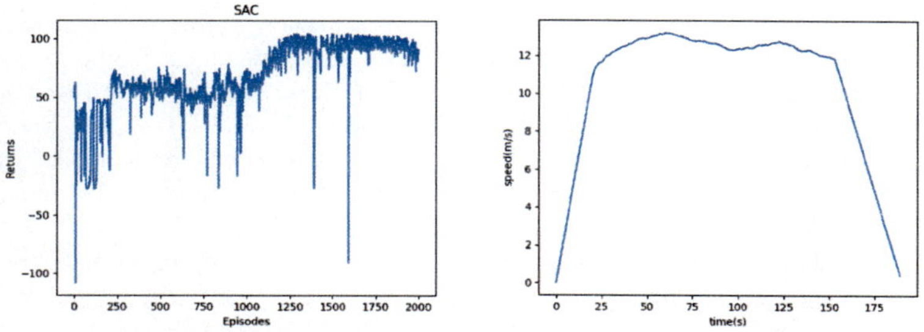


(a) Reward function curve of SAC (b) The speed curve of train

**Fig. 3.** The result of iteration

## 5 Conclusion

In summary, this study presented a reinforcement learning-based approach for optimizing train control. By employing the SAC algorithm, an optimal policy was learned to adjust traction and braking forces, leading to safe and energy-efficient train operations. Experimental results demonstrated the effectiveness of the proposed approach in achieving punctual arrivals, precise stopping, and adherence to speed limits.

Future research can explore several directions. Firstly, extending the approach to incorporate additional factors such as passenger comfort and train stability would contribute to more comprehensive train control strategies. By integrating these aspects into the reinforcement learning framework, a more holistic approach can be developed. Secondly, investigating the scalability and generalizability of the approach across different train systems and operating conditions would be valuable. This analysis would provide insights into the adaptability of the proposed approach in various real-world scenarios.

**Acknowledgments.** This work was supported by the Natural Science Foundation of P. R. China for Young Scientists under Grant 62103145; the Key Research and Development Project of Jiangxi Provincial Technology Department under Grant 20202BBEL53018; and the Graduate Student Innovation Special Fund Project of Jiangxi Province under Grant YC2022-s485.

## References

1. Lin, B., Zhao, Y., Lin, R., et al.: Integrating traffic routing optimization and train formation plan using simulated annealing algorithm. Appl. Math. Model. **93**, 811–830 (2021)
2. Zhao, H., Dai, X.W.: Event-triggered adaptive control for multiple high-speed trains with deception attacks in bottleneck sections. Inf. Sci. **547**, 470–481 (2021)
3. Wang, X., Tang, T., Su, S., et al.: An integrated energy-efficient train operation approach based on the space-time-speed network methodology. Transp. Res. Part E: Logist. Transp. Rev. **150**, 102323 (2021)
4. Huang, K., Wu, J., Yang, X., et al.: Discrete train speed profile optimization for urban rail transit: a data-driven model and integrated algorithms based on machine learning. J. Adv. Transp. **2019**, 7258986 (2019)

5. Zhao, L., Peng, J., Wang, J., et al.: Optimization control of energy-efficient driving for trains in urban rail transit based on GA-PSO algorithm. In: 4th International Conference on Electrical and Information Technologies for Rail Transportation, Novel Traction Drive Technologies of Rail Transportation, pp. 777–785. Springer, Singapore (2020)
6. Liu, Y., Fan, K., Ouyang, Q.: Intelligent traction control method based on model predictive fuzzy PID control and online optimization for permanent magnetic maglev trains. IEEE Access **9**, 29032–29046 (2021)
7. Su, S., She, J., Li, K., et al.: A nonlinear safety equilibrium spacing-based model predictive control for virtually coupled train set over gradient terrains. IEEE Trans. Transp. Electrif. **8**(2), 2810–2824 (2021)
8. Zhang, L., Zhou, M., Li, Z.: An intelligent train operation method based on event-driven deep reinforcement learning. IEEE Trans. Ind. Inform. **18**(10), 6973–6980 (2021)
9. Li, Q., Meng, X., Gao, F., et al.: Approximate cost-optimal energy management of hydrogen electric multiple unit trains using double Q-learning algorithm. IEEE Trans. Ind. Electron. **69**(9), 9099–9110 (2021)
10. Meng, X., Wang, H., Lin, M., et al.: Deep reinforcement learning for energy-efficient train operation of automatic driving. In: 8th International Conference on Computer Science and Network Technology, pp. 123–126. IEEE, Hangzhou (2020)
11. Liao, J., Yang, G., Zhang, S., et al.: A deep reinforcement learning approach for the energy-aimed train timetable rescheduling problem under disturbances. IEEE Trans. Transp. Electrif. **7**(4), 3096–3109 (2021)
12. Liu, W., Su, S., Tang, T., et al.: A DQN-based intelligent control method for heavy haul trains on long steep downhill section. Transp. Res. Part C: Emerg. Technol. **129**, 103249 (2021)
13. Guo, H., Bai, Y., Hu, Q., et al.: Optimization on metro timetable considering train capacity and passenger demand from intercity railways. Smart Resilient Transp. **3**(1), 66–77 (2021)
14. Zhiwen, L., Qin, Y., Wang, M.: Research on high-speed railway operation adjustment model based on priority. Smart Resilient Transp. **4**(1), 12–21 (2022)
15. Su, S., Wang, X., Tang, T., et al.: Energy-efficient operation by cooperative control among trains: A multi-agent reinforcement learning approach. Control Eng. Pract. **116**, 104901 (2021)
16. Shang, M., Zhou, Y., Fujita, H.: Deep reinforcement learning with reference system to handle constraints for energy-efficient train control. Inf. Sci. **570**, 708–721 (2021)
17. Wang, X., Li, S., Tang, T., et al.: Intelligent operation of heavy haul train with data imbalance: a machine learning method. Knowl.-Based Syst. **163**, 36–50 (2019)
18. Zhou, K., Song, S., Xue, A., et al.: Smart train operation algorithms based on expert knowledge and reinforcement learning. IEEE Trans. Syst. Man Cybern. Syst. **52**(2), 716–727 (2020)
19. Zhu, L., He, Y., Yu, F.R., et al.: Communication-based train control system performance optimization using deep reinforcement learning. In: IEEE Trans. Veh. Technol. **66**(12), 10705–10717 (2017)
20. Zhu, Q., Su, S., Tang, T., et al.: An eco-driving algorithm for trains through distributing energy: a Q-Learning approach. ISA Trans. **122**, 24–37 (2022)

# Rolling Element Localized Fault Diagnosis Method for Train Bearings Based on Orthogonal Bi-axial Signals

Yongliang Bai[1,2](✉), Hai Xue[1], Jiadong Meng[1], and Jiangtao Chen[1]

[1] School of Mechanical Engineering, Lanzhou Jiaotong University, Lanzhou 730070, China
yongliangbai@126.com

[2] Beijing Key Laboratory of Performance Guarantee on Urban Rail Transit Vehicles, Beijing University of Civil Engineering and Architecture, Beijing 100044, China

**Abstract.** This paper proposes a rolling element fault diagnosis method for train bearings based on orthogonal bi-axial vibration signals, which addresses the complex and difficult-to-diagnose rolling element fault signals due to the lack of reference signals and the high diagnostic difficulty. Firstly, the coupling form of orthogonal bi-axial vibration signals in rolling element fault signals is analyzed. Secondly, a reference setting method based on orthogonal bi-axial signals is proposed. Finally, the effectiveness and advancement of the proposed algorithm are verified by processing localized rolling element faults on a laboratory-scale test rig and collecting signals at different speeds. This method provides a certain reference and theoretical guidance for the diagnosis of rolling element faults in train bearings under complex service environments.

**Keywords:** Train Vibration · Orthogonal Biaxles · Rolling Element fault

## 1 Introduction

As the mileage and vehicle inventory of China's rail transit continue to increase, ensuring the safe operation and accurate maintenance of trains has become a key research focus in the industry. Bearings are one of the key rotating components in urban rail transit trains, and accurate maintenance and fault diagnosis can have an important impact on the intelligence and greenization of rail transit. However, the harsh working conditions of axle box bearings in urban rail transit trains make them prone to local faults in various components, such as the cage, inner and outer rings, and rolling elements [1, 2]. According to statistics, rolling element localized faults account for a high proportion of bearing failures in urban rail transit vehicles, especially under high-speed and heavy-load operating conditions, where slight faults in rollers can easily lead to fault expansion and large-area localized faults on the working surface of rolling elements [3, 4]. Therefore, it is necessary to analyze the characteristics of this type of fault signal, study its fault vibration signal characteristics, and use innovative diagnostic algorithms to diagnose rolling element localized faults in axle box bearings of urban rail transit vehicles.


Y. Qin et al. (Eds.): EITRT 2023, LNEE 1137, pp. 97–105, 2024.
https://doi.org/10.1007/978-981-99-9311-6_11

Scholars in the industry have conducted relevant research on the diagnosis of rolling element localized faults. Li Xianglong [5] showed through statistical data from actual operating routes that rolling element localized faults account for 85.1% of failures in rail vehicles, emphasizing the importance of diagnosing faults in the working surface peeling and cracking of rollers. Peng Fan [4] conducted dynamic analysis and running-in experiments on rolling element faults in rail vehicles, verifying the diagnosis of tapered roller bearing roller faults based on wavelet denoising and HHT. Cheng Yao [6] studied a local fault diagnosis method for train rolling elements based on deconvolution and verified it using experimental signals. Zhang Azhong [7] characterized the dynamic characteristics of rolling element localized faults under different speed conditions. Jiang Fucai [8] conducted bearing roller fault diagnosis based on resonance demodulation using vibration signals from actual line-operated vehicle bearings. Although these studies have achieved good diagnostic effects, they are all based on single-directional axle box vibration signals, which have limitations such as incomplete fault information acquisition and lack of reference signals. Therefore, it is necessary to develop more effective localized rolling element fault diagnosis methods for axle box bearings in rail transit vehicles by utilizing bi-axial vibration signals based on their anisotropy.

# 2 Methodology

## 2.1 Traditional Single-Axis SGMD Algorithm

The Symplectic Geometry Dode Decomposition (SGMD) algorithm was proposed by Haiyang Pan in 2019 [9]. The key idea of this algorithm is to solve the eigenvalues of the Hamilton matrix based on the symplectic matrix similarity transformation, and further reconstruct the Symplectic Geometry Components (SGCs) using the corresponding eigenvectors. For a complex signal, SGMD decomposition can achieve adaptive parameter setting and obtain a certain number of SGCs. This decomposition algorithm can eliminate the influence of noise components and adaptively process signals to obtain single-component SGCs, showing excellent decomposition ability and good anti-interference performance in vibration signal decomposition while keeping the input time signal unchanged.

The core of the SGMD algorithm is the symplectic matrix similarity transformation, which solves the eigenvalues of the target matrix Hamilton matrix, and uses its corresponding eigenvectors to reconstruct SGCs, thereby adaptively decomposing complex signals while denoising them and obtaining several SGCs [10]. For discrete vibration signals $x(i) = (x_1, x_2, \cdots, x_n)$, the decomposition process based on SGMD is as follows:

(1) *Phase space reconstruction*

The multi-dimensional signal is reconstructed based on Takens embedding, and the trajectory matrix $\mathbf{X}$ is constructed by using one-dimensional input time series delay topology algorithm.

$$\boldsymbol{X} = \begin{bmatrix} x_1 & x_{1+\tau} & \cdots & x_{1+(d-1)\tau} \\ \vdots & \vdots & & \vdots \\ x_m & x_{m+\tau} & \cdots & x_{m+(d-1)\tau} \end{bmatrix} \tag{1}$$

Where, $d$ is the embedding dimension, $\tau$ is the delay time, $m = n - (d-1)\tau$. Here, the embedding dimension determines the final number of SGCs, and its value criterion is

$$d = \begin{cases} round(n/3) & f_{\max}/F_s < 10^{-3} \\ 1.2 \times (F_s/f_{\max}) & f_{\max}/F_s \geq 10^{-3} \end{cases} \tag{2}$$

(2) *Symplectic geometric matrix transformation*

The goal is to get the Hamilton matrix, which can extract the time-displacement similarity of its own sequence according to the trajectory matrix, so as to obtain the covariance symmetric matrix

$$A = X^T X \tag{3}$$

Then, Hamilton matrix $M$ is represented based on matrix

$$M = \begin{bmatrix} A & 0 \\ 0 & -A^{\mathrm{T}} \end{bmatrix} \tag{4}$$

After obtaining the Hamilton matrix, it is desirable to square the matrix ($N = M^2$) to satisfy the Hamilton matrix property. It can be obtained from the above content and satisfy the characteristics of Hamilton matrix, so it can realize the symplectic orthogonal matrix $Q$ representation

$$\boldsymbol{Q}^{\mathrm{T}}\boldsymbol{N}\boldsymbol{Q} = \begin{bmatrix} B & R \\ 0 & B^{\mathrm{T}} \end{bmatrix} \tag{5}$$

(3) *Diagonal averaging*

The dimension of the original single component obtained from the above steps is $\mathbf{Z}_{\mathrm{i}}$, and the component combination can be further arranged. By definition $1 \leq i \leq d$, $1 \leq j \leq m$, can be obtained. Therefore, the corresponding diagonal averaging transformation matrix is expressed as

$$y_k = \begin{cases} \frac{1}{k}\sum\limits_{p=1}^{k} z^*_{p,k-p+1} & 1 \leq k \leq d^* \\ \frac{1}{d^*}\sum\limits_{p=1}^{d^*} z^*_{p,k-p+1} & d^* \leq k \leq m^* \\ \frac{1}{n-k+1}\sum\limits_{p=k-m^*+1}^{n-m^*+1} z^*_{p,k-p+1} & m^* \leq k \leq n \end{cases} \tag{6}$$

Further, based on the diagonal averaging operation, a single component combination $Y = Y_1 + Y_2 + \cdots + Y_d$ can be obtained.

(4) *Single component reconstruction*

In the signal component group obtained based on the decomposition algorithm, the main component is located at the front end of the matrix element. Therefore, the first original single-component signal is taken as the basis for feature analysis and comparison with the remaining components. The components with high similarity are selected and reconstructed to obtain the first signal component. Then, according to the reconstructed components, the residual matrix is obtained, and the residual signal is obtained by summing the residual matrix. Normalized mean square error (NMSE) is calculated by comparing the residual term with the initial signal.

$$NMSE = \frac{\sum_{e=1}^{n} g^h(e)}{\sum_{e=1}^{n} x(e)} \tag{7}$$

Where, $h$ represents the number of iterations in the loop. When $NMSE < th$, $th = 0.01$, the corresponding iterative algorithm is terminated. Otherwise, the initial matrix repetition operation based on the residual matrix is performed until the cutoff condition is met

$$x(n) = \sum_{h=1}^{N} SGC^h(n) + g^{(N+1)}(n) \tag{8}$$

## 2.2 Improved Orthogonal Biaxial SGMD Algorithm

It can be seen from the above that after the signal is decomposed into one-dimensional components, the signal needs to be reconstructed. In the process of bearing vibration signal decomposition, the anisotropy of vertical and axial vibration signal of axle box bearing needs to be taken into account. Therefore, based on the regional fault characteristics of bearing rollers, this paper chooses vertical vibration signal as the main component. Taking the axial vibration signal and the reference signal as $S_v(n)$ and $S_a(n)$. The orthogonal biaxial SGMD decomposition and reconstruction are carried out. If the vertical vibration signal and the axial vibration signal are respectively, then the corresponding SGCs component $SGC_a(n)$ and $SGC_v(n)$ can be obtained after the SGMD decomposition

$$\begin{cases} SGC_a(n) = \sum_{h=1}^{N} SGC_a^h(n) + g_a^{(N+1)}(n) \\ SGC_v(n) = \sum_{h=1}^{N} SGC_v^h(n) + g_v^{(N+1)}(n) \end{cases} \tag{9}$$

As can be seen from Eq. 9, the corresponding components obtained are all sets of multiple components. Therefore, the most important components in and are selected respectively for signal reconstruction, that is, the components with the highest kurtosis value in the component set are extracted respectively for reconstruction, and finally the orthogonal biaxial fusion signal $S_c(n)$ is obtained

$$S_c(n) = SGC_a^{h1}(n) + SGC_v^{h2}(n)\Big|_{SGC^h(n)=\max(Kurtosis(SGC^h(n)))} \tag{10}$$

With the orthogonal biaxis SGMD signal decomposition and reconstruction method, the feature extraction and diagnosis of regional roller faults can be realized more effectively.

# 3 Verification of Equal Scale Bogie Model in Laboratory

## 3.1 Introduction of Experimental Equipment

In order to verify the effect of the orthogonal two-axis SGMD signal processing algorithm proposed in this paper, fault processing and signal acquisition were carried out in the laboratory of equal scale rail vehicle bogies, and the diagnostic effect of regional roller faults of bearings was compared and verified by the obtained fault vibration signals. The proportional test bench used is shown in Fig. 1.

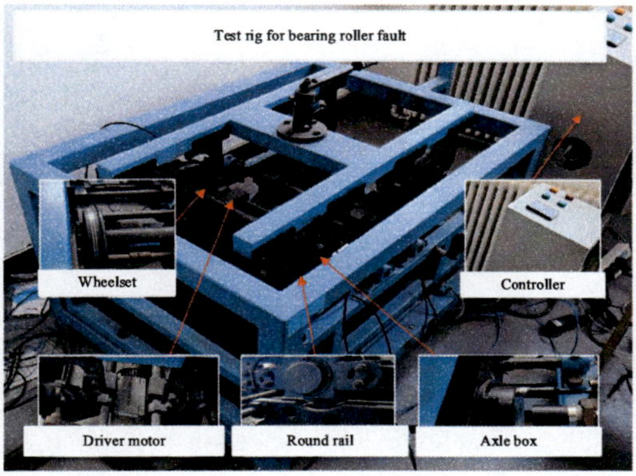


**Fig. 1.** Laboratory proportional bearing failure simulation test bench

Based on the concept of similarity simulation, the test bench scales key rotating component parameters such as wheelset diameter and axle box bearing size by 1:5 to realize dynamic simulation of wheelset-axle box fault vibration. The proportional test bench is composed of front and rear wheelsets, proportional axle boxes, circular track, LabView programmable speed controller and other key components. The upper cantilever beam is used to restrict the proportional bogie. The corresponding acquisition system structure is shown in Fig. 2.

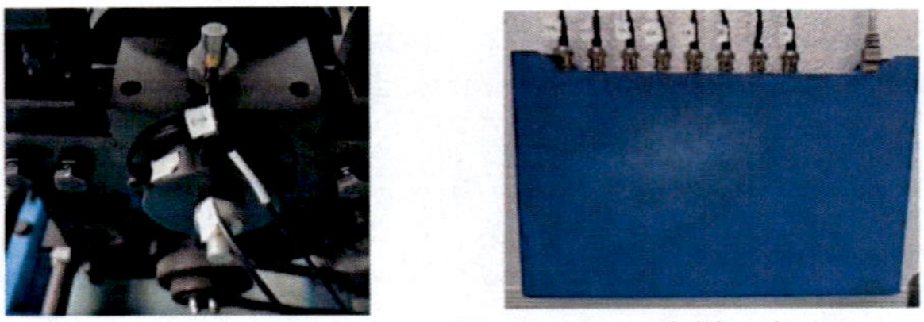

(a) ICP acceleration sensor (b) signal acquisition instrument

**Fig. 2.** Signal acquisition system

### 3.2 Setting of Experimental Conditions

In order to fully verify the effect of the orthogonal two-axis SGMD algorithm proposed in this paper, two bearing states were set respectively in healthy condition and regional fault condition. Among them, the bearing type was 351306 double-row tapered roller bearing. The fault simulation was realized by machining, and the fault size was 3.5 mm wide (see Fig. 3).

Healthy Regional fault

**Fig. 3.** Bearing condition setting

According to the Nyquist sampling law and fault characteristic frequency calculation, the corresponding sampling frequency is set to 20 K, the signal sampling time length is 1 s, and the corresponding analog bogie linear speed is 40 km/h. ICP acceleration sensor is used to collect and pick up the vertical and lateral acceleration signals of the axle box respectively.

### 3.3 Verifying Regional Roller Faults

Healthy bearing vibration signals and regional fault vibration signals were picked up respectively, and their envelope spectra were analyzed. The corresponding time domain and spectrum envelope results were shown in Fig. 4.

As can be seen from Fig. 4, in the time domain characterization, periodic impact signals exist in the corresponding time domain waveforms regardless of health condition or fault condition, and the periodicity of the impact signals is basically the same, but the amplitude of each periodic impact signal is quite different, indicating the necessity of studying orthogonal biaxial signals. At the same time, it can be found from the

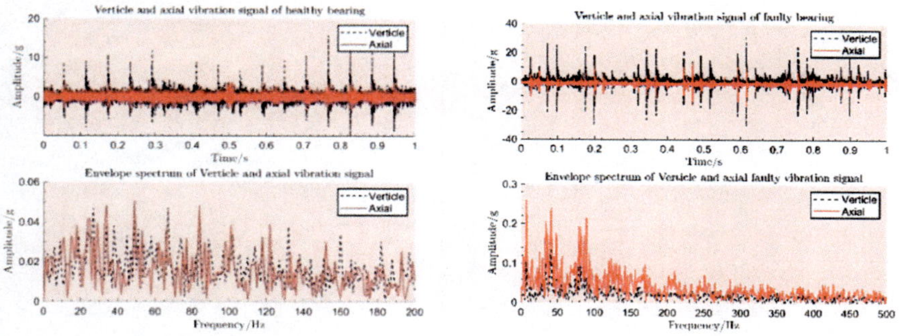


**Fig. 4.** Time domain and spectral waveform of healthy bearing

envelope spectrum that the axial signal has less interference than the vertical signal, but the corresponding fault characteristic frequency can not be found from the original signal envelope spectrum. Therefore, it is necessary to carry out signal processing based on orthogonal biaxial SGMD decomposition and reconstruction.

For regional bearing roller faults, SGMD decomposition of vertical signal and axial signal is carried out respectively, and kurtosis values of each component are calculated. The results are shown in Fig. 5.

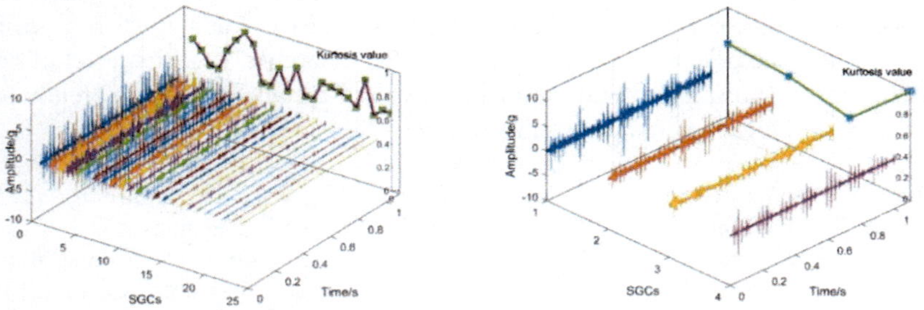


(a) Vertical signal SGMD decomposition and kurtosis calculation (b) Axial signal SGMD decomposition and kurtosis calculation

**Fig. 5.** Orthogonal biaxial signal SGMD decomposition

The most kurtosis component in vertical signal SGGs and the most kurtosis component in axial signal SGGs are reconstructed, and envelope spectrum analysis is performed on the reconstructed signal, and the results are shown in Fig. 6.

As can be seen from Fig. 6, compared with the original vertical and axial vibration signals, the envelope spectrum of the signal after decomposition and reconstruction by orthogonal biaxial SGMD shows obvious even order frequency doubling of regional roller fault characteristic frequency *Froller* (*2*Froller, 4*Froller, 8*Froller*). The diagnosis target of regional roller fault is realized.

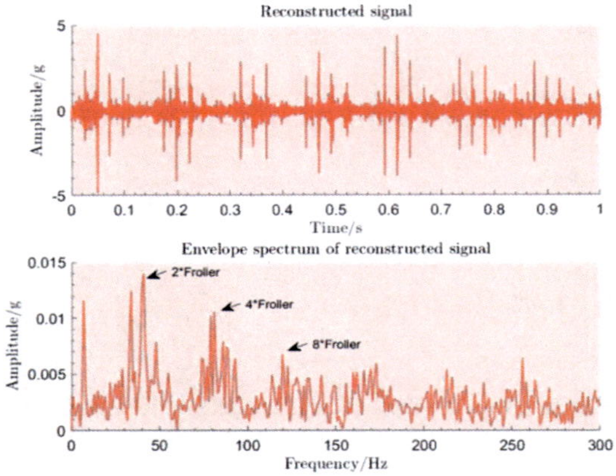


**Fig. 6.** Reconstruct the signal time domain signal and its envelope spectrum

# 4 Conclusion

Aiming at the anisotropic characteristics of vertical and axial vibration signals of rail vehicles, this paper proposes the decomposition and reconstruction of regional fault signals of orthogonal biaxial bearing rollers based on SGMD, and uses the proportional fault simulation test bench to collect data and verify the effect. The results show that the proposed method can effectively diagnose the regional roller fault of axle box bearing of rail vehicles. The proposed method can enrich the theoretical research level of bearing roller fault diagnosis algorithm, and can provide a certain method reference for the service performance guarantee of rail vehicles.

**Acknowledgments.** This work is supported by the Young Scholars Science Foundation of Lanzhou Jiaotong University (2022048), the Natural Science Foundation of Gansu Province (23JRRA909),the Gansu Provincial Education Department University Teachers' Innovation Fund Project (2023A-043), and the Science and Technology Projects of Gansu Province under Project (21JR7RA305).

# References

1. Cao, H.R., Fan, F., Zhou, K., He, Z.J.: Wheel-bearing fault diagnosis of trains using empirical wavelet transform. Measurement **82**, 439–449 (2016). (in English)
2. Bai, Y.L., Yang, J.W., Wang, J.H., Li, Q.: Intelligent diagnosis for railway wheel flat using frequency-domain Gramian angular field and transfer learning network. Ieee Access **8**, 105118–105126 (2020). (in English)
3. Fu, W.L., Shao, K.X., Tan, J.W., Wang, K.: Fault diagnosis for rolling bearings based on composite multiscale fine-sorted dispersion entropy and SVM with hybrid mutation SCA-HHO algorithm optimization. Ieee Access **8**, 13086–13104 (2020). (in English)
4. Fan, P.: Research on Characteristic Signal Extraction and Diagnosis Method of Bearing Fault of High-speed Train. Master's Thesis, Southwest Jiaotong University (2018)

5. Xianglong, L.: Failure analysis and preventive measures of rollers of newly manufactured 197726 bearings. Roll. Stock **04**, 41–42 (2002)
6. Yao, C.: Research on Rolling Bearing Fault Diagnosis and Condition Monitoring Technology for High-Speed Trains. Ph.D. Thesis, Southwest Jiaotong University (2019)
7. Azhong, Z.: Research on Dynamic Characteristics of Axlebox Bearing Failure Conditions of High-Speed Train. Master's thesis, Southwest Jiaotong University (2021)
8. Fucai, J.: Discussion on common problems of on-board monitoring devices in locomotive data analysis applications. Gansu Sci. Technol. **23**, 37–41 (2021)
9. Pan, H.Y., Yang, Y., Li, X., Zheng, J.D., Cheng, J.S.: Symplectic geometry mode decomposition and its application to rotating machinery compound fault diagnosis. Mech. Syst. Signal Process. **114**, 189–211 (2019). (in English)
10. Yan, X., Liu, Y., Jia, M.: A fault diagnosis approach for rolling bearing integrated SGMD, IMSDE and multiclass relevance vector machine. Sensors **15**, 1–6 (2020)

# Modeling of Hybrid EMUs and Rule-Based Energy Management Strategy

Dongfan Nie[1(✉)], Chao Li[1], Ke Su[2], Yuxin Yang[2], and Jie Chen[1]

[1] School of Electrical Engineering, Beijing Engineering, Research Center of Electric Rail Transportation, Beijing Jiaotong University, Beijing 100044, China
{21121465,19117012,jiechen}@bjtu.edu.cn

[2] The Fourth Company of Beijing Metro Operation Co., Beijing, China
suke6132991@163.com

**Abstract.** In recent years, railway transportation has developed rapidly, and in order to cope with sustainable development, hybrid power systems (HPS) are applied in rail transportation to achieve energy saving and environmental protection. In the de-sign of hybrid electric multiple units (EMUs), most of them use a combination of diesel generator (DG) sets and energy storage devices to provide energy for the load from multiple power sources. In this paper, firstly, the HPS is modeled, and the DG, power battery and traction system are established. Secondly, a rule-based energy management strategy is designed, which formulates the operation rules of the HPS according to the operating characteristics and status of the power battery and DG set, which can be used to ensure the reliable operation of the system with low fuel consumption under various operating conditions. Finally, simulation experiments are conducted to verify the effectiveness and reliability of the proposed method.

**Keywords:** Hybrid power system (HPS) · Electric multiple units (EMUs) · Energy management strategy · Rule-based control strategy

## 1 Fi Introduction

With the development of HPS in automobile [1], hybrid power has also been widely used in rail transiport [2]. In the design of hybrid EMUs, most of them adopt the structure of combining DG set and energy storage device [3], and carry out power distribution of two or more power sources through energy management system [4]. Energy management strategy has an important impact on the running performance of trains. Energy management strategy can make the HPS run stably, reduce the fuel consumption of train operation, improve the system efficiency, and achieve the purpose of energy saving. Energy management strategies fall into two categories: rule-based and optimization-based. The rule-based control strategy is based on engineering experience, which makes the rules in advance, simple control and easy to implement. Control methods based on optimization can be divided into global optimization and instantaneous optimization. Global optimization includes dynamic programming algorithm [6], Pontriagin minimum


Y. Qin et al. (Eds.): EITRT 2023, LNEE 1137, pp. 106–113, 2024.
https://doi.org/10.1007/978-981-99-9311-6_12

principle [7] and other methods, while instantaneous optimization includes model predictive control algorithm [8] and equivalent fuel consumption method [9]. This kind of control strategy can adjust the internal energy distribution mode of the system adaptively according to the driving condition of the vehicle, but it also has some problems, such as heavy calculation and difficult real-time application.

## 2 Hybrid EMUs Model

Most hybrid EMUs adopt the structure of DG and energy storage device which combine the two power sources. The power distribution of the two power sources is carried out through energy management strategy. The construction of HPS model [10] includes DG, power battery pack, traction system, etc., which can be used to verify the implementation of energy management control strategy, as shown in Fig. 1.

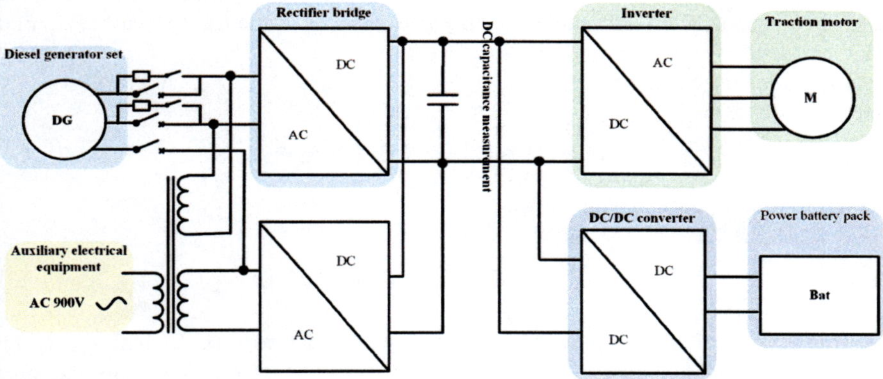


**Fig. 1.** Hybrid power systems topology.

### 2.1 Diesel Generator Set

DG is powered by diesel fuel, which is composed of diesel engine and generator. At present, the more commonly used engine and its speed control system is composed of speed controller, accelerator actuator and diesel engine. The basic structure is shown in Fig. 2.

The stator side of permanent magnet synchronous generator contains three-phase symmetrical winding, and there is relative movement between stator and rotor during operation. After simplified analysis, the three-phase Park transformation can be used to get the generator model.

The stator voltage equation is:

$$\begin{cases} u_d = R \times i_d + \frac{d\varphi_d}{dt} - \omega_e \times \varphi_q \\ u_q = R \times i_q + \frac{d\varphi_q}{dt} + \omega_e \times \varphi_d \end{cases} \quad (1)$$

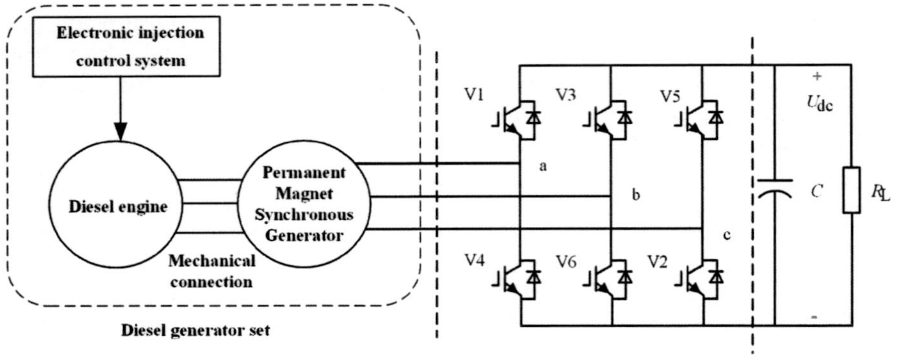


**Fig. 2.** Topology of DG.

Among them, $i_d$ and $i_q$ is the current of stator d and q axis., $u_d$ and $u_q$ is the voltage of stator d and q axis, $R$ for the stator armature resistance, $\omega_e$ is the rotor electrical angular velocity.

The torque equation is:

$$T_e = \frac{3}{2}p \times i_q\left[\left(L_d - L_q\right) \times i_d + \varphi_f\right] \tag{2}$$

Mechanical motion equation of motor:

$$J\frac{d\omega_m}{dt} = T_e - T_L - B\omega_m - T_f \tag{3}$$

Among them, $J$ is the moment of inertia, $B$ is the damping coefficient, $\omega_m$ is the mechanical angular velocity, $T_L$ is the load torque, $T_e$ is the electromagnetic torque, start is greater than the friction torque motor, is less than the static friction torque motor.

### 2.2 DC/DC and Battery Pack

The HPS uses DC/DC and power batteries. Control the charging and discharging of batteries through DC/DC. The circuit topology is shown in Fig. 3.

DC/DC converter is used for energy control and plays a role in stabilizing DC voltage. Conduction and shutdown of IGBT of the bidirectional Buck/Boost circuit are independently controlled and operated separately, respectively in Buck and Boost modes in different energy flow directions.

In Buck operation mode, the inductor current increases linearly, as shown in the formula:

$$\mathrm{L}\frac{di_L}{dt} = U_{dc} - U_{bat} \tag{4}$$

where, $i_L$ is the inductance current, $U_{dc}$ is DC side voltage, $U_{bat}$ is the terminal voltage of the battery.

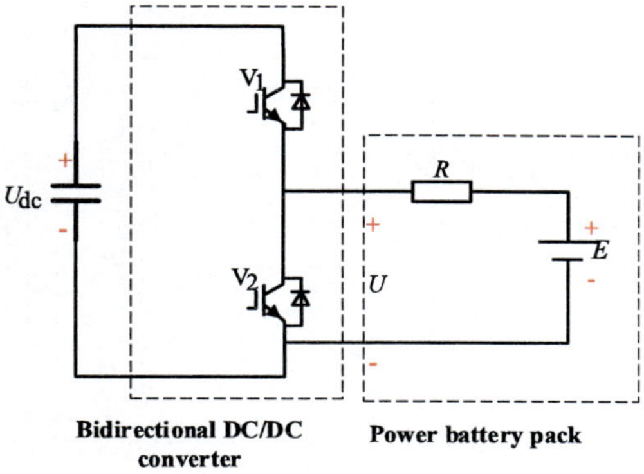


**Fig. 3.** DC/DC and battery pack topology.

In Boost operating mode, the inductor current linearly decreases, as shown in the formula:

$$L\frac{di_L}{dt} = -U_{bat} \tag{5}$$

The state of charge of a battery is related to charging and discharging, as shown in the formula:

$$SOC(t) = SOC_0 - \frac{\int I(t)dt}{C_n} \tag{6}$$

where, $SOC_0$ is the initial value of battery SOC, $C_n$ is the battery capacity.

The open circuit voltage (OCV) of the battery has a positive correlation with the state of charge of the battery, as shown in. Because the battery SOC is too high or too low, the open circuit voltage of the battery will change rapidly, so in the application, the battery SOC should vary between 10% and 90%.

### 2.3 Traction System

The traction system includes a traction inverter and a load motor, as shown in Fig. 4.

When constructing the HPS model, the load is equivalent to a load motor. The traction system is composed of load and DC/AC inverter.

## 3 Optimal Control of Energy Management

The HPS is allocated reasonably through energy management control strategies. This can ensure stable system operation and reduce fuel consumption. The control strategy is to develop corresponding rules in advance based on experience. The rule-based instantaneous power control is mainly controlled by controlling the speed of DG and battery status to realize power distribution and control of battery SOC.

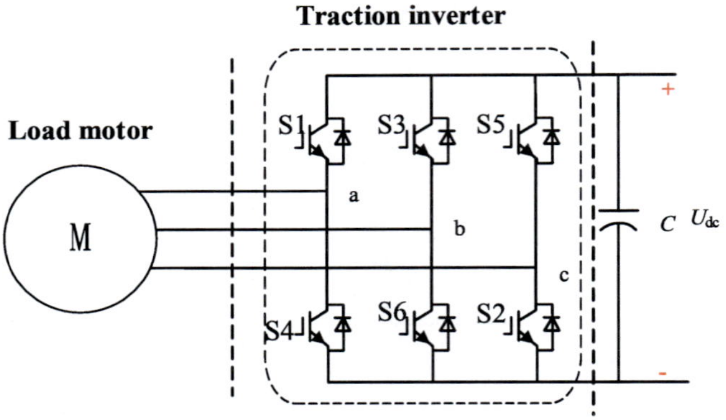


**Fig. 4.** Topological structure of traction system.

### 3.1 Diesel Generator Speed Control

When operating under traction conditions, DG sets adopt hysteresis control to adjust the speed. When operating under other conditions except traction conditions, due to the predictability of working conditions, DG sets adopt fixed speed control.

Hysteresis control ensures that the DG set can respond quickly and stabilize quickly within the hysteresis range, as shown in Fig. 5.

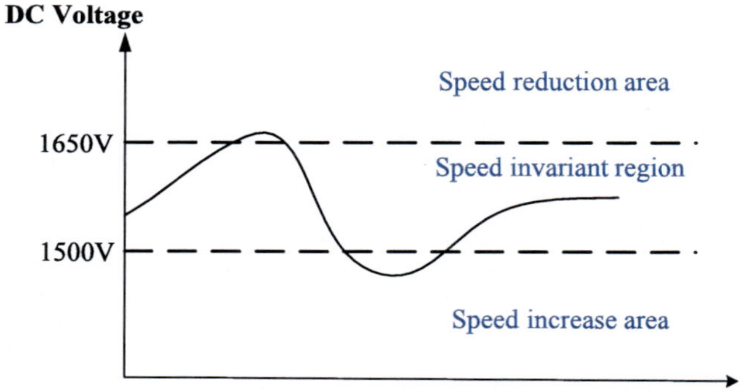


**Fig. 5.** Hysteresis control of generator.

Under other conditions, DG speed is set according to the operating condition and the change of DC side voltage. The setting values are shown in Table 1.

### 3.2 Droop Control of Power Battery Pack

The charge and discharge control of battery pack in traction condition adopts the double loop control completed by DC/DC converter. The battery control is shown in Fig. 6.

**Table 1.** Rule Control of DG

| Condition | | Speed of DG |
|---|---|---|
| Coasting condition | Charge | 1700 rpm |
| | Not charging | 1500 rpm |
| Braking conditions | Electric braking is normal | 1400 rpm |
| | Electric brake failure | 1500 rpm |
| Parking conditions | Charge | 1700 rpm |
| | Not charging | 600 rpm |

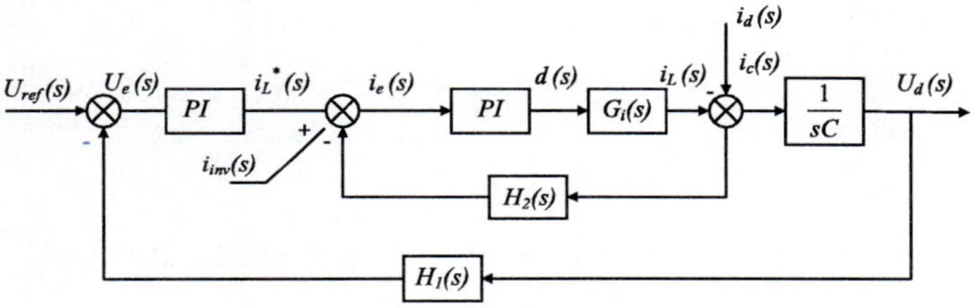


**Fig. 6.** Dual closed-loop control.

The target voltage of the outer voltage loop is set as the sagging voltage of the DG. Target voltage expression of the DC side can be obtained by fitting the load power and DC side voltage:

$$U_{dc} = 1832 \times \frac{N_{dg}}{2000} - 0.4P_{dg} \tag{7}$$

where, $N_{dg}$ is the speed of the diesel engine, $P_{dg}$ is the output power of the DG.

The dual closed-loop control strategy of voltage outer loop and current inner loop can achieve control of the battery, so as to achieve the goal of stabilizing intermediate DC voltage.

## 4 Simulation Result

The hybrid system model is simulated in Matlab, and the rule-based energy management control strategy is added to the system to simulate the traction, idle running, braking, parking and other conditions of the hybrid EMUs.

Power distribution results are shown in. The curve in the figure shows that the DG and battery can reasonably provide energy to the load motor (see Fig. 7).

The target speed and actual speed of the DG are shown in. It can be seen that the speed of the generator can operate according to the rules (see Fig. 8).

The target voltage and actual voltage of the DC side are shown in. The DC side voltage can remain stable during the whole operation (see Fig. 9).

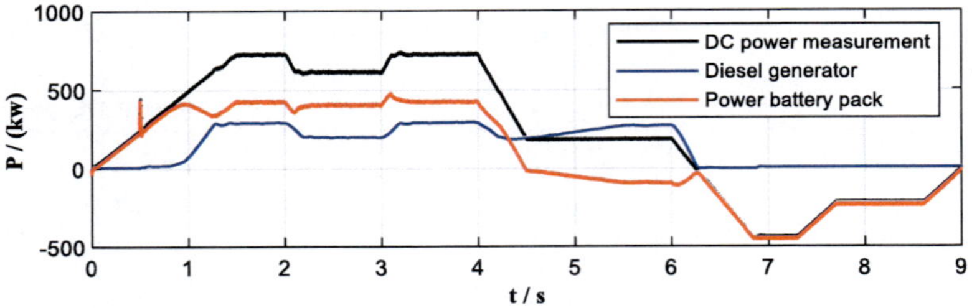


**Fig. 7.** Power configuration.

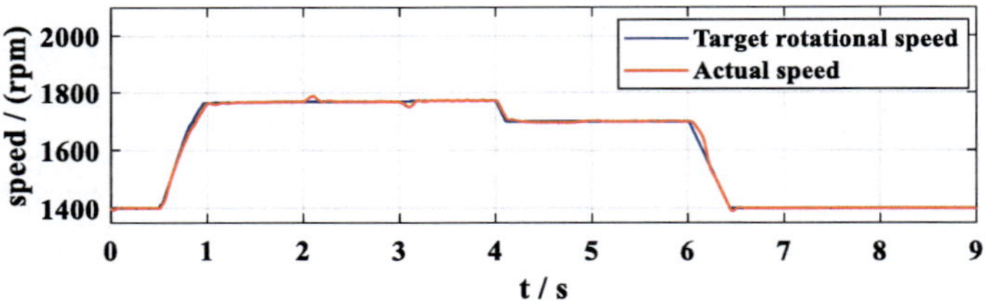


**Fig. 8.** Target speed and actual speed of DG.

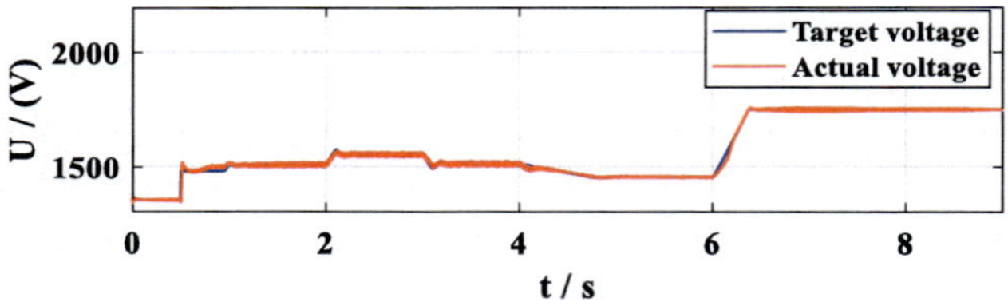


**Fig. 9.** DC side reference voltage and actual voltage.

## 5 Summarize

The HPS composed of DG and power battery can rationally allocate power source under the control of energy management. The HPS is modeled and divided into DG, power battery pack, DC/DC, load motor, etc. The two power sources are connected to the DC side to provide energy for the load. By means of rule-based energy management control strategy, the speed control of DG and droop control of power battery pack are added to the HPS, and the two power sources are reasonably distributed. The control strategy proposed in this paper is verified by Matlab simulation, which can reasonably allocate two power supplies. Meanwhile, it can achieve stable control of HPS and reduce fuel consumption to save energy.

**Acknowledgment.** Project supported by the Fundamental Research Funds for the Central Universities.

## References


1. Yongqi, Y., Yingyu, Z.: Hybrid electric vehicle energy management strategy based on rules strategy overview. Car (5) times, 10–12 (2023). (in Chinese)
2. Li: a study of China's present situation and development trend of high speed EMU. Information science and technology of China (10), 105–106 (2015). (in Chinese)
3. Shibuya, H., Kondo, K.: Designing methods of capacitance and control system for a diesel engine and EDLC hybrid powered railway traction system. IEEE Trans. Ind. Electron. **58**(9), 4232–4240 (2011). https://doi.org/10.1109/TIE.2010.2100332
4. Xiuchun, Z., Ge, G.: Hybrid electric vehicle energy management strategy research were reviewed. J. Autom. **42**(3), 321–334 (2016). (in Chinese)
5. Car Aerial: Hybrid power train configuration and energy management strategy research. Master degree, Beijing Jiaotong University (2022). (in Chinese)
6. Tang, W., Wang, Y., Jiao, X., Ren, L.: Hierarchical energy management strategy based on adaptive dynamic programming for hybrid electric vehicles in car-following scenarios. Energy **265**, 126264 (2023). https://doi.org/10.1016/j.energy.2022.126264
7. Kim, N., Cha, S., Peng, H.: Optimal control of hybrid electric vehicles based on Pontryagin's minimum principle. IEEE Trans. Control Syst. Technol. **19**(5), 1279–1287 (2011). https://doi.org/10.1109/TCST.2010.2061232
8. Lü, X., et al.: Hybrid electric vehicles: a review of energy management strategies based on model predictive control. J. Energy Storage **56**, 106112 (2022). https://doi.org/10.1016/j.est.2022.106112
9. Dengfeng, S., Chen, W., Yu, H.-S., Tong, Z.: Easily show families: composite power split hybrid electric vehicle energy management strategy research. Automot. Eng. **39**(1), 15–22, 27 (2017). (in Chinese)
10. Gengyi, L.: Hybrid electric locomotive traction system modeling and energy management research. Master degree, Beijing Jiaotong University (2021). (in Chinese)

# Research on Parameter Sensitivity About Singular Characteristics of Contact Force of High-Speed Railway

Jian Zhang[1](✉), RuiKai Niu[2], Wenzheng Liu[2], and Qing han[1]

[1] School of Robot Engineering, Yangtze Normal University, Chongqing 408100, China
20190020@yznu.edu.cn

[2] School of Electrical Engineering, Beijing Jiaotong University, Beijing 100044, China

**Abstract.** The alteration of contact force of pantograph- catenary (CFPC) at the dropper points (DP) is amplified remarkably when the speed of train is enhanced, and the collecting current character of pantograph- catenary took a turn for the worse. The chain-catenary/pantograph simulation model was established by adopting the finite element method (FEM) in the research to probe optimizing method of CFPC, which is analyzed that is obtained by calculation. First, the CFPC ridge-line on both sides of the DP is located by using the wavelet-ridge method to analyze singularity of the CFPC used by the Lipschitz exponent (LE). The conclusion is drawn that the CFPC reduction- rate on the right hand of the DP is larger than the increasing-rate on the left hand. Then, Sobol method is used to quantitatively analyze the relation between the CFPC singularity index and catenary-parameter. The results show that the singularity of CFPC can be reduced by increasing the contact wire tension and decreasing messenger wire tension, and the effect of catenary tension on the singularity at the DP is greater than the linear density of contact wire and messenger wire.

**Keywords:** Contact force of pantograph- catenary (CFPC) · Singular Characteristics · Sensitivity · Wavelet

## 1 Introduction

Electric energy can be gotten by means of pantograph sliding forward along the contact wire while the train is moving. The collecting current character of pantograph- catenary is appraised by the CFPC that is impacted by the structure of catenary, which varies along with working of the train. The alteration of contact force at the DP increases sharply as increasing train speed so far as to show up the pantograph-catenary offline that causes the worse collecting current quality.

Many scholars have made a lot of research and achieved a lot of results to reform the collecting current quality between pantograph and catenary. For example, Kim J, Lee J et al. analyzed the relation between contact wire deflection, catenary stiffness and contact force to analyzing the problem of rising pantograph-catenary off-line when the train moves on at great high speed [1, 2]. M. Bocciolone, F. Resta, M. Zhu et al. analyzed


Y. Qin et al. (Eds.): EITRT 2023, LNEE 1137, pp. 114–122, 2024.
https://doi.org/10.1007/978-981-99-9311-6_13

effect of aerodynamic lifting force on CFPC under the action of double-pantograph as the establishing the simulation system of pantograph-catenary a foundation [3–6]. In this research, the most academics analyzed the CFPC based on the average value, standard deviation, offline rate in a certain section as a whole. However, the CFPC at the DP becomes worse as the operating speed increases. As a result, it is very necessary to discuss the variation law of CFPC at the DP.

The singular characteristics of CFPC of the elastic chain and simple chain suspension catenary was unpacked in the previous study [7, 8]. However, there is no research on the quantitative study of the singular characteristics of CFPC. Therefore, the CFPC ridges on either side of the DP is located by using the wavelet-ridge method, and CFPC change was analyzed by the LE in this study. Then the Sobol that is the method of global sensitivity analysis is adopted to study quantitatively the relation between LE and catenary-parameter.

## 2 The Coupled Pantograph -Catenary Model

In this study, we adopted the Euler-Bernoulli beam element to establish the catenary model that is used to emulate the particle motion of the catenary [9, 10]. The equation of motion of the contact wire particle is shown below:

$$m_c \frac{\partial^2 u_c}{\partial t^2} + \frac{\partial^2}{\partial x^2}\left(EI_c \frac{\partial^2 u_c}{\partial x^2}\right) - \frac{\partial}{\partial x}\left(T_c \frac{\partial u_m}{\partial x}\right) + k_d(u_m - u_c)\delta(x - x_n) = F\delta(x - vt) \tag{1}$$

And the equation of motion of the messenger wire particle is shown below:

$$m_m \frac{\partial^2 u_m}{\partial t^2} + \frac{\partial^2}{\partial x^2}\left(EI_m \frac{\partial^2 u_m}{\partial x^2}\right) - \frac{\partial}{\partial x}\left(T_m \frac{\partial u_m}{\partial x}\right) + k_d(u_m - u_c)\delta(x - x_n) + k_s u_m(x - x_s) = 0 \tag{2}$$

where $m_c$ and $m_m$ shows the element mass of the contact wire and messenger wire respectively, $EI_c$ and $EI_m$ shows the bending stiffness of the contact wire and messenger wire respectively, $T_c$ and $T_m$ shows the contact wire tension and messenger wire respectively, $u_c$ and $u_m$ shows the displacement of contact wire and messenger wire respectively, $x_n$ shows the distance in a straight line between the DP and the motor point, $x$ shows the position of the motor point, $x_s$ shows the distance in a straight line between the support point and the motor point, $t$ shows the elapsed time, and $v$ shows the operating speed, $k_d$ shows the dropper stiffness, $k_s$ shows the equivalent stiffness of the supporting device, $\delta$ is the impulse function, $F$ shows the CFPC.

The simulation model of pantograph was established by adopting three-mass model [11], which equation is shown in Eq. (3).

$$\begin{cases} m_1 \ddot{y}_1 + c_1 \dot{y}_1 + k_1 y_1 - c_1 \dot{y}_2 - k_1 y_2 = -F(t) \\ m_2 \ddot{y}_2 + (c_1 + c_2)\dot{y}_2 + (k_1 + k_2) y_2 - c_2 \dot{y}_3 - k_2 y_3 - c_1 \dot{y}_1 - k_1 y_1 = 0 \\ m_3 \ddot{y}_3 + (c_2 + c_3)\dot{y}_3 + (k_2 + k_3) y_3 - c_2 \dot{y}_2 - k_2 y_2 = 0 \end{cases} \tag{3}$$

where, $m_1$, $m_2$ and $m_3$ shows the mass of pantograph head, upper frame and lower frame respectively, $k_i$ shows equivalent stiffness ($i = 1,2,3$), $c_i$ shows equivalent damping, $F(t)$ shows the CFPC, $F_0$ shows the static lifting force of pantograph.

We establish the catenary model by using the catenary structure of the high-speed railway of Beijing-Shanghai and Beijing-Tianjin, and single-slide pantograph is established, which is shown in Figs. 1 and 2.

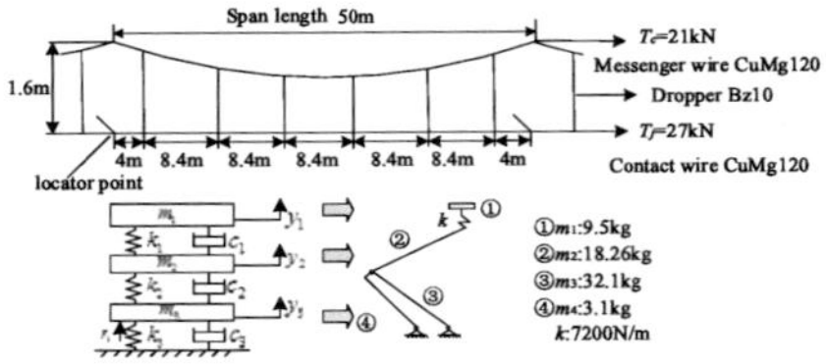


**Fig. 1.** Simple chain catenary

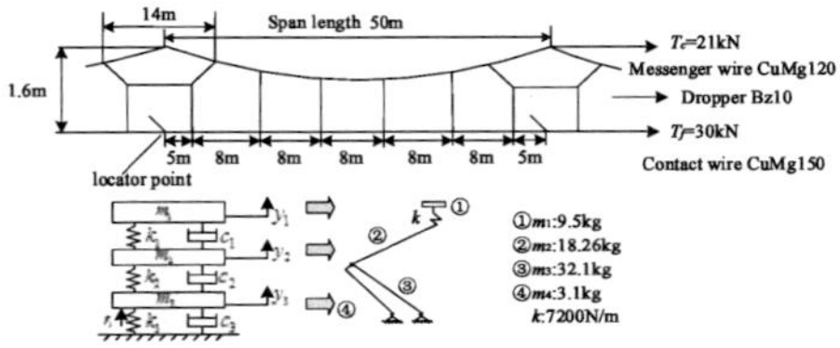


**Fig. 2.** Elastic chain catenary

## 3 The Method for Extracting Singular Signal

The mutation rate of the CFPC at the DP is intensified along with the improvement of the operating speed, and the collecting current character of pantograph -catenary takes a turn for the worse. A combination of wavelet ridge based on wavelet transform and LE is adopted to extract the change rate of CFPC on either side of the DP in this research to assay the singularity of CFPC. The CFPC curve of the simple chain catenary is shown in Fig. 3. Based on Gaussian wavelet transform, the wavelet coefficient diagram is displayed in Fig. 4 is obtained.

The calculation process of LE is displayed in Fig. 5, and the contact force LE is calculated, which is displayed in Fig. 6.

The LE of CFPC in one span both of simple chain catenary and elastic chain catenary is different in Fig. 6. And the LE on the left hand of the DP is outweigh than that on the right hand. It implies that the increasing-rate of CFPC on the DP left hand is under than decline rate on the DP right hand.

## 4 Parameter Sensitivity Analysis for the Singular Characteristics of CFPC

The method of Sobol is used to analyze quantitatively the singularity sensitivity of CFPC.

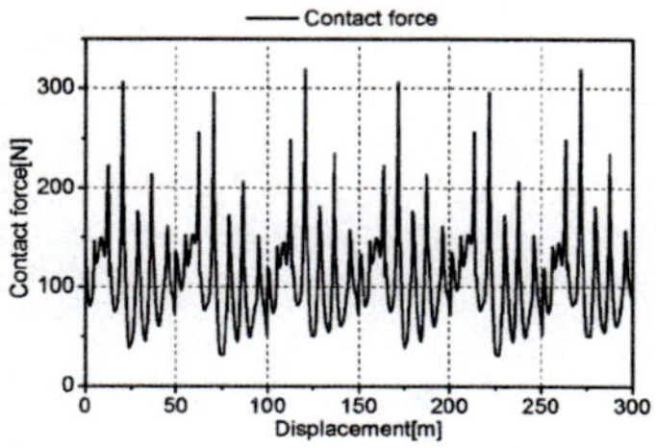


**Fig.3.** The curve of CFPC

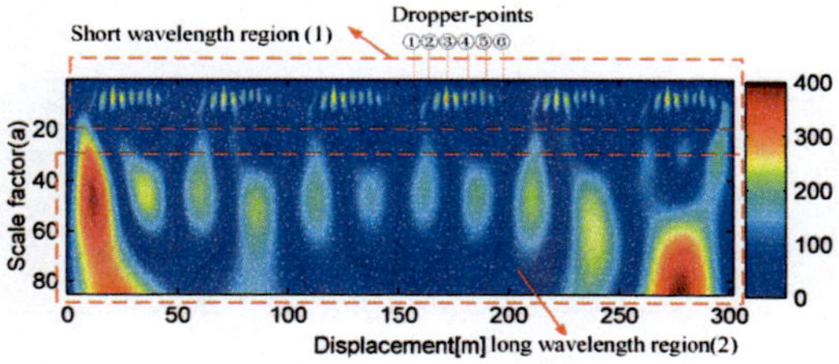


**Fig. 4.** Wavelet coefficient diagram of CFPC

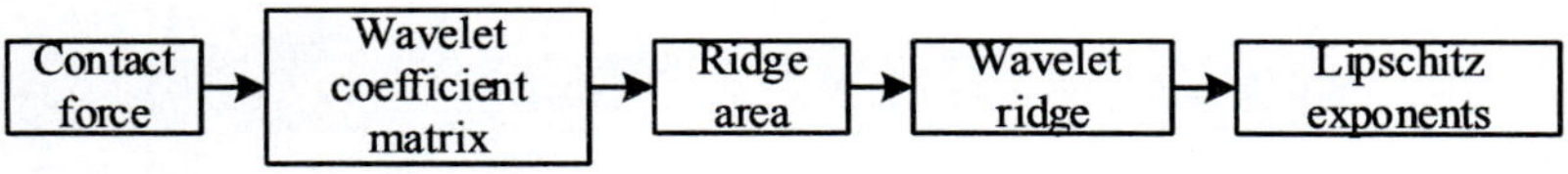


**Fig.5.** The extraction process of singular signal

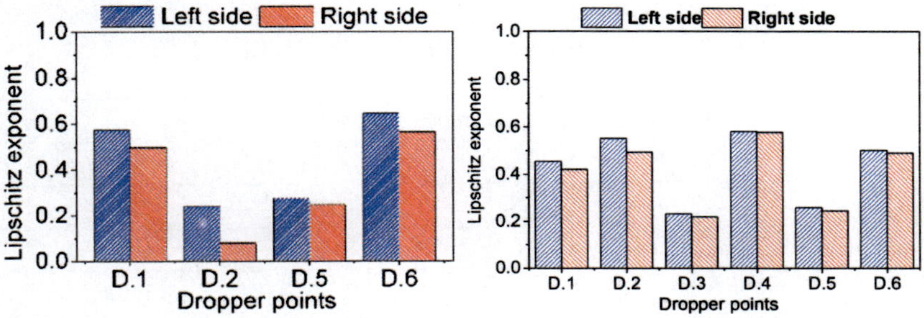


(a) Simple chain catenary (b) Elastic chain catenary

**Fig. 6.** LE of the CFPC

### 4.1 Sobol Method

Firstly, the method of central composite design is used [12], as displayed in Fig. 7. And the parameter combination of catenary model is devised for simulation calculations. The structural arguments of the catenary in Sect. 2 were regarded as the underlying value and set as a central point to analyze parameter sensitivity as displayed in Tables 1 and

2. Then the results obtained using Sobol method is analyzed. And that the influence of all parameters of catenary on the CFPC at the DP is obtained.

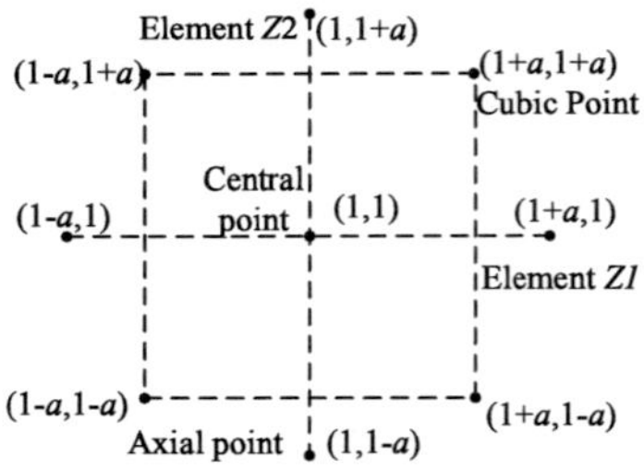


**Fig.7.** The method of ventral composite design

**Table 1.** Arguments of simple chain catenary

| Point | The design arguments of catenary | | | |
|---|---|---|---|---|
| | $T_c$/kN | $T_j$/kN | $m_c$/kg/m | $m_j$/kg/m |
| Point (1 + a) | 23.1 | 29.7 | 1.188 | 1.188 |
| Central point | 21 | 27 | 1.08 | 1.08 |
| Point (1 − a) | 18.9 | 24.3 | 0.972 | 0.972 |
| Point (a) | 25.2 | 32.4 | 1.296 | 1.296 |
| Point (−a) | 16.8 | 21.6 | 0.864 | 0.864 |

**Table 2.** Arguments of elastic chain catenary

| Point | The design arguments of catenary | | | |
|---|---|---|---|---|
| | $T_c$/kN | $T_j$/kN | $m_c$/kg/m | $m_j$/kg/m |
| Point (1 + a) | 23.1 | 33 | 1.188 | 1.474 |
| Central point | 21 | 30 | 1.08 | 1.34 |
| Point (1 − a) | 18.9 | 27 | 0.972 | 1.206 |
| Point (a) | 25.2 | 36 | 1.296 | 1.608 |
| Point (−a) | 16.8 | 24 | 0.864 | 1.072 |

Wherein, $T_c$ and $T_j$ show the tension of messenger wire and contact wire respectively, $m_c$ and $m_j$ show the linear density of messenger wire and the linear density of contact wire respectively. The relation between the structural arguments and CFPC using the Sobol is displayed in Eq. (4) [13].

$$f(X) = f(x_1, x_2, \ldots, x_n), x_i \sim U(0, 1), i = 1, 2, \ldots n \quad (4)$$

wherein $x_i$ shows the catenary arguments ($n = 4$) and $f(X)$ shows the evaluating indicator of the collecting current character, which is shown in Eq. (5) that is decomposed into the sum of sub-items:

$$f(x_1, x_2, \cdots x_n) = f_0 + \sum_{i=1}^{n} f_i(x_i) + \sum_{1\leq i<j\leq n}\sum f_{ij}(x_i, x_j) + f_{12\cdots n}(x_1, x_2, \cdots x_n) \tag{5}$$

where $f_i$ $(x_i)$ and $f_{ij}$ $(x_i, x_j)$ and other parameters are the subkeys of $f$ $(x_1, x_2, \ldots, x_n)$, which satisfies $\int_0^1 f_{1,2\cdots s}(x_1, x_2, \cdots x_s)dx_k = 0$, $(1 \leq k \leq s)$, and $f_0$ means the constant. Each subkey in the Eq. (5) is calculated as displayed below:

$$f_0 = \int f(x_1, x_2 \cdots x_n)dx_1 dx_2 \cdots dx_n \tag{6}$$

$$f_i(x_i) = \int f(x_1, x_2 \cdots x_n)dx_1 \cdots dx_{i-1}dx_{i+1} \cdots dx_n - f_0 \tag{7}$$

$$f_{ij}(x_i, x_j) = \int f(x_1, x_2 \cdots x_n)dx_1 \cdots dx_{i-1}dx_{i+1} \cdots dx_n - f_0 - f_i(x_i) - f_j(x_j) \tag{8}$$

The calculation equation for sensitivity coefficient is displayed below:

$$S_{1,2\cdots n} = D_{1,2\cdots s}/D \tag{9}$$

The sensitivity coefficient can satisfy the relation that is displayed below:

$$\sum_{i=1}^{n} S_i + \sum_{1=i}\sum_{1<j=n} S_{ij} + S_{12\cdots n} = 1 \tag{10}$$

wherein $D_i = \int f_I^2(x_i)dI$ and $D_{1,2,\cdots s} = \int f_{1,2,\cdots s}^2(x_1, x_2, \cdots x_s)dx_1\, dx_2 \cdots dx_s$ denote the partial variance. Furthermore, $D = \int f^2(x_1, x_2, \cdots x_n)dx_1 dx_2 \cdots dx_n - f_0^2$ denotes the total variance. At the same time, $I_i$ shows the 1st-order sensitivity coefficient, $S_{ij}$ shows the 2nd-order sensitivity coefficient. And the magnitude of the sensitivity coefficient determines the degree to the parameter affects the CFPC.

From the above analysis, it can be seen that the functional relation between the design arguments and the CFPC is the basis for calculating the sensitivity coefficient. An outstanding functional relation between the design arguments and the CFPC in Sect. 3 is found, which shows obvious nonlinear characteristics. As a consequence, the 2nd-order nonlinear regression equation is used to express the relationship between the parameters and CFPC. $f(x)$ is predicted as the nonlinear model, which is displayed below.

$$f(x) = F_0 + a_1x_1 + a_2x_2 + a_3x_3 + a_4x_4 + b_1x_1x_2 + b_2x_1x_3 + b_3x_1x_4 + c_1x_2x_3 + c_2x_2x_4 + d_1x_3x_4 \tag{11}$$

wherein $f(x)$ shows the singular characteristics target. $x_1, x_2, x_3, x_4$ denote respectively the tension of catenary, the linear density of the catenary. $a_1, a_2, a_3, a_4, b_1, b_2, b_3, c_1$, $c_2$ and $d_1$ are the regression coefficients of the equation.

## 4.2 Sensitivity Analysis

The sensitivity of simple chain catenary parameters to the contact- force singularity at each DP is displayed whilst the train is working at a speed of 300 km/h, and the tension and linear density of catenary are changed respectively. The sensitivity coefficient is shown below.

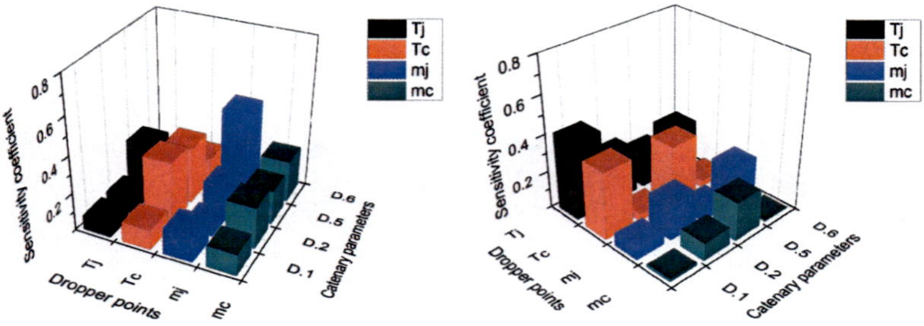


(a) Left hand (b) Right hand

**Fig.8.** LE sensitivity of CFPC of simple chain catenary

According to Fig. 8, we can see that the sensitivity of the CFPC LE on the left and right hands of the DP of a simple chain catenary to the tension and density of the catenary is different. Among them, the LE of the CFPC on the left hand of the DP ② and ⑤ in Fig. 8(a) is highly sensitive to the tension and density of the contact wire. That is to say, adjusting the tension and density of the contact wire can significantly change the mutation rate of the CFPC on the right hand at the DP.

The sensitivity of the CFPC singularity about the elastic chain catenary parameters is displayed in Fig. 9 when the train works at a of 350 km/h with changing the tension and linear density of catenary.

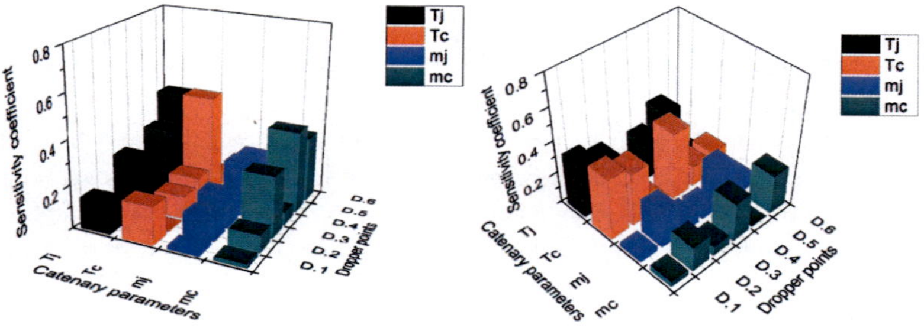


(a) Left hand (b) Right hand

**Fig.9.** LE sensitivity of CFPC of elastic chain catenary

As displayed in Fig. 9 (a), the sensitivity of LE in descending order is contact wire tension, messenger wire tension, contact wire linear density and messenger wire linear density. It can be seen that the sensitivity coefficients of contact wire tension, messenger wire tension are higher on the right side of each DP in Fig. 9 (b). In summary, the singularity of the tension of catenary at each DP is more obvious.

In summary, the singularity of the CFPC at the DP is greatly affected by the tension of catenary.

## 5 Conclusions

In this study, the method of combining wavelet-ridge and LE are employed to locate the position of the DP and extract mutant rate of contact to analyze the singularity. And the results are obtained as follows:

(1) The wavelet ridge can achieve the localization and extraction of LE, and quantitative analysis of the LE of CFPC can be achieved by using the Sobol sensitivity analysis method.
(2) The sink rate of the CFPC on the right hand at the DP of the simple and elastic catenary is greater than the increasing- rate of the CFPC on the left hand.
(3) Effect of the tension of catenary on the singularity for CFPC on both sides at the DP is more significant than the linear density of catenary.

**Acknowledgment.** This research is supported by the Science and Technology Project Affiliated to the Education Department of Chongqing (KJQN202201437 and KJZD-K202201404) and the Natural Science Foundation of Chongqing (cstc2020jcyj-msxmX0801 and CSTB2022NSCQ-MSX1199).

## References

1. Kim, J., Chae, H., Park, B., et al.: State sensitivity analysis of the pantograph system for a high-speed rail vehicle considering span length and static uplift force. J. Sound Vib. **303**(3–5), 405–427 (2007)
2. Lee, J., Kim, Y., Paik, J., et al.: Performance evaluation and design optimization using differential evolutionary algorithm of the pantograph for the high-speed train. J. Mech. Sci. Technol. **26**(10), 3253–3260 (2012)
3. Bucca, G., Carnevale, M., Collina, A., et al.: Adoption of different pantographs' preloads to improve multiple collection and speed up existing lines. Veh. Syst. Dyn. **50**(sup1), 403–418 (2012)
4. Facchinetti, A., Gasparetto, L., Bruni, S.: Real-time catenary models for the hardware-in-the-loop simulation of the pantograph–catenary interaction. Veh. Syst. Dyn. **51**(4), 499–516 (2013)
5. Zhu, M., Zhang, S., Jiang, J., et al.: Enhancing pantograph-catenary dynamic performance using an inertance-integrated damping system. Veh. Syst. Dyn. **60**(6), 1902–1932 (2022)
6. Vesali, F., Rezvani, M.A., Molatefi, H.: Analysis of conceptual similarities and differences of wave speed and critical speed in the overhead catenary system. Measurement **176**(3), 109164 (2021)
7. Zhang, J., Yi, J., Liu, W.: Study of time-frequency characteristics of contact force signal and parameter sensitivity of catenary of high-speed railway. IEEE Trans. Instrum. Meas. **71**(1), 1–11 (2022)

8. He, F., Zhang, J.: Research on time–frequency characteristics of contact force based on wavelet transform. IET Electr. Syst. Transp. **12**, 287–297 (2022)
9. Zou, D., Zhang, W.H., Li, R.P., Zhou, N., et al.: Determining damping characteristics of railway-overhead-wire system for finite-element analysis. Veh. Syst. Dyn. **54**(7), 902–917 (2016)
10. Yong, H.C., Lee, K., Park, Y., Kang, B., et al.: Influence of contact wire pre-sag on the dynamics of pantograph–railway catenary. Int. J. Mech. Sci. **52**(11), 1471–1490 (2010)
11. Park, T.J., Han, C.S., Jang, J.H.: Dynamic sensitivity analysis for the pantograph of a high-speed rail vehicle. J. Sound Vib. **266**(2), 235–260 (2003)
12. He, W., Xue, W., et al.: Optimization Experiment Design Method and Data Analysis. Chemical Industry Publishing House, Beijing (2012)
13. Sobol, I.M.: Global sensitivity indices for nonlinear mathematical models and their Monte Carlo estimates. Math. Comput. Simul **55**(1), 271–280 (2014)

# Risk Factors of Streetcar Intersections Based on the FISM Method

## A Case Study of Jiaxing Streetcar Line T1


Congcong Chen[1,2], Yufei Jin[3], Weiwei Song[2](✉), and Yuling Ye[2]

[1] School of Rail Transportation, Shandong Jiaotong University, Ji'nan 250357, China
[2] The Key Laboratory of Road and Traffic Engineering of the Ministry of Education, Tongji University, Shanghai 201804, China
woshisww@163.com, yyling71@163.com
[3] Air Traffic Management College, Civil Aviation Flight University of China, Guanghan 618300, China
ww0806@foxmail.com



**Abstract.** This paper highlights the utilization of the Fuzzy Interpretive Structural Modeling (FISM) method to analyze the interrelationships among influencing factors. By identifying the top influencing factors, including other traffic participants, the number of intersecting roads, performance of other traffic vehicles, streetcar lane priority, and streetcar drivers, the research provides a comprehensive understanding of the factors directly impacting streetcar operations at intersections. Based on the findings, several practical management strategies are proposed. These include strengthening safety education and standardizing driver behavior through increased driver rotation frequency and regency response, and public participation. By implementing these strategies, the paper aims to ensure the safe and efficient operation of streetcars at urban intersections, contributing to a sustainable and reliable urban transportation system.



**Keywords:** Streetcar intersections · Influencing factors · FISM · Management strategies · Traffic safety


## 1 Introduction

The streetcar systems, serving as rapid transit systems, has garnered attention in numerous cities. The characteristics such as expedious construction periods and cost-effectiveness when compared to subway systems. It is worth noting that streetcars function as low-capacity systems, often operating in conjunction with urban roadways and engaging in shared use of road rights. Consequently, the intricate nature of traffic organization at intersections within urban road networks is intensified, thereby elevating safety risks. As a result, a comprehensive understanding of the safety factors pertinent to streetcar intersections becomes imperative to facilitate informed decision-making and effectively address the challenges posed by their operation.



Y. Qin et al. (Eds.): EITRT 2023, LNEE 1137, pp. 123–131, 2024.
https://doi.org/10.1007/978-981-99-9311-6_14

Streetcars, as a form of ground-based public transport, occupy space on the road that is also used by other vehicles, leading to conflicts with other traffic. Intersection areas, in particular, witness a high frequency of streetcar accidents [1].

While there are numerous articles addressing the study of urban intersections, the causes of frequent road traffic accidents are often attributed to factors such as people, vehicles, and traffic management [2–4]. However, research specifically focusing on the safety operation of streetcar intersections is insufficient. Only a few articles analyze the safety of streetcar intersections by examining existing data in both temporal and spatial dimensions or conduct corresponding research on conflict areas at streetcar intersections [5, 6].

Furthermore, the Fuzzy Interpretation Structure Model (FISM) is commonly employed in project evaluation, such as highway construction or management modes. Previous studies have concluded that factors such as land expropriation issues and ambiguous construction and management regulations significantly impact project construction [7, 8]. Although FISM has been used sparingly in transportation, it has found application in areas such as passenger traffic optimization and assessing risks in multimodal transport networks [9, 10].

This study is based on the current situation of streetcar intersections in Jiaxing City. Through methods such as literature review, expert surveys, and case studies, the key factors influencing streetcar intersection risks are determined. The fuzzy explanatory structure model (FISM) is then employed to illustrate the relationships among these factors. Furthermore, taking the intersection scenarios of Jiaxing streetcar Line T1 as an example, recommendations for risk prevention and control in streetcar intersections are proposed.

## 2 Current Status of Jiaxing Streetcar

### 2.1 Current Status of Jiaxing Streetcar Planning and Construction

The Jiaxing Streetcar (Phase I) project encompasses the T1 and T2 lines, spanning a total operating length of 15.6km and comprising 20 stations, with an estimated total investment of 2.85 billion yuan. The first phase of the T1 Line stretches from Anle Road Station to Jiaxing South Station, covering a distance of 13.8km, with an average inter-station distance of 0.986km. Similarly, the first phase of the T2 Line operates from Yuehebei Station to Huancheng South Road Station, covering a distance of 1.8km, with an average inter-station distance of 0.45km. The T1 Line have commenced operations on June 25, 2021, with a total of 16 stations and an operating mileage of 13.8 km. The future planned network of the Jiaxing streetcar construction consists of three lines: T1, T2, and T6, with a combined length of 35.7 km. The comprehensive network encompasses seven lines, spanning a total length of 98 km.

Notably, the majority of intersections and urban road intersections along the streetcar lines adopt a level intersection mode, characterized by hybrid right of way, which introduces more risk factors and complex compositions compared to dedicated right of way. Therefore, analyzing and studying the risk factors at streetcar intersections holds significant practical importance.

### 2.2 Typical Accidents of Jiaxing Streetcar

During the trial operation, two accidents occurred where a tram collided with a car. Moreover, from January 15 to 20, 2022, within a span of five days, six vehicles illegally obstructed streetcars in Jiaxing, violating traffic laws.

### 2.3 Example of Intersection

Through field investigation, it was found that there are 30 intersections along the existing streetcar T1lines in Jiaxing City. For the purpose of analysis, 8 intersections were selected. The selected intersections are presented in the following Table 1.

**Table 1.** Example of intersection.

| Serial number | Intersection name |
|---|---|
| 1 | Zhongshan East Road and Jiahe Road intersection |
| 2 | Intersection of Jiahe Road and Qinjian Road |
| 3 | Intersection of Fonggong Road and Luli Street |
| 4 | Spinners Road and Ling Kung Tong Road |
| 5 | Spinners Road and Central Road South |
| 6 | Spinner Road and Yuquan Road |
| 7 | You boxing Road and Commercial Avenue |
| 8 | Youquan Road and Qingfeng Road |

Observations were made on the traffic volume at eight intersections during peak hours in the morning and evening. Furthermore, the streetcar passing patterns at these 8 intersections can be broadly categorized into two types: turning and going straight. To provide a visual representation, the streetcar crossing diagram (Figs. 1, and 2) is presented, where the yellow arrow indicates the presence of streetcars. Corresponding data for each intersection is provided in Table 2 and Fig. 3.

According to analysis, the intersection of Fanggong Road and Zhonghuan South Road has the highest traffic volume. During the evening peak hours, the conflicting traffic volume at the intersection of Fanggong Road and Luli Street, Zhongshan East Road and Jiahe Road, Fanggong Road and Central South Road, and Fanggong Road and Youquan Road significantly increased, reaching about 1000 vehicles.

## 3 Study of the Risk Factors of Streetcar Intersections

### 3.1 Identification of Risk Factors of Streetcar Intersections

The preliminary factor list was extracted through field investigation and literature review. Seven experts were invited to participate in a questionnaire survey regarding the relationship between the influencing factors. Following the second round of interviews, 15 key influencing factors were identified and summarized (Table 3).

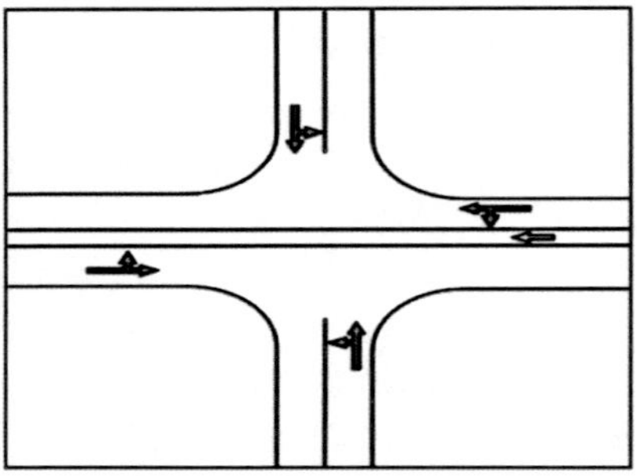

**Fig. 1.** Streetcar intersection diagram (going straight).

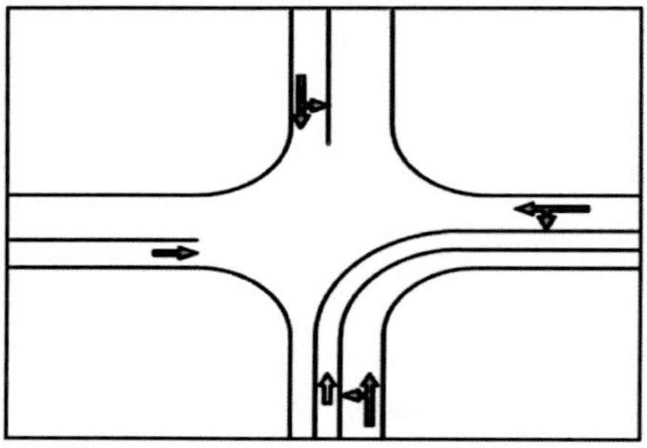

**Fig. 2.** Streetcar intersection diagram (turning).

**Table 2.** Number of vehicles at intersection Conflicts (including non-motor vehicles).

| Serial number | 1 | 2 | 3 | 4 | 5 | 6 | 7 | 8 |
|---|---|---|---|---|---|---|---|---|
| Morning peak periods | 827 | 822 | 56 | 559 | 1512 | 402 | 909 | 414 |
| Evening peak periods | 1008 | 1008 | 31 | 625 | 1042 | 657 | 1057 | 372 |

## 3.2 Development of the FISM Model

### Establish the correlation strength matrix

This study categorizes the intensity of interaction between risk factors into six levels. Let A be a binary fuzzy relation matrix of risk factors, denoted as $A = \left(a_{ij}\right)_{n\times n}$, where $a_{ij}$ represents the judgment result of the interaction between risk factor $a_i$ and $a_j$. It is

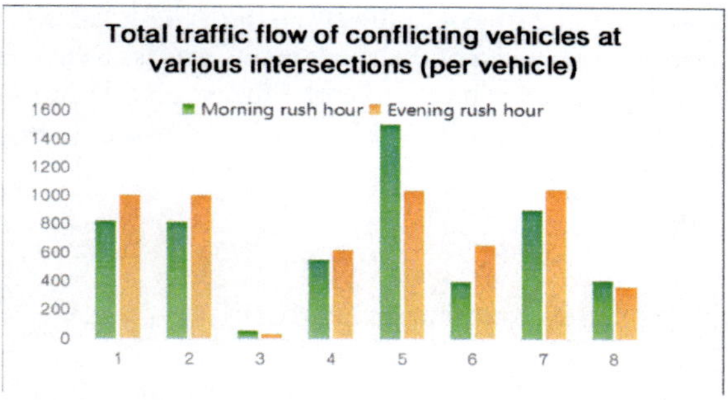


**Fig. 3.** Total morning and evening peak flow at each intersection.

**Table 3.** Risk factors of streetcar intersections.

| Classifying | Factor | Risk Factor |
|---|---|---|
| External risk factor | $S_1$ | Type of intersection |
| | $S_2$ | Other traffic participants |
| | $S_3$ | Number of lanes at an intersection |
| | $S_4$ | Performance of other traffic vehicles |
| | $S_5$ | Allocation of right of way |
| | $S_6$ | Traffic volume conflicting with streetcar operation direction |
| Internal risk factor | $S_7$ | Streetcar running speed |
| | $S_8$ | Streetcar service frequency |
| | $S_9$ | Streetcar signal priority |
| | $S_{10}$ | Trolley lane priority |
| | $S_{11}$ | Streetcar driver |
| | $S_{12}$ | Streetcar vehicle factor |
| Other risk | $S_{13}$ | Natural disaster |
| | $S_{14}$ | Social risk factor |
| | $S_{15}$ | Social environmental factors |

defined as follows:

$$a_{ij} = \begin{cases} 0.0,\ R_i\ on\ R_j\ No\ impact \\ 0.1,\ R_i\ on\ R_j\ Very\ low\ impact \\ 0.3,\ R_i\ on\ R_j\ low\ impact \\ 0.5,\ R_i\ on\ R_j\ Medium\ impact \\ 0.7,\ R_i\ on\ R_j\ High\ impact \\ 1.0,\ R_i\ on\ R_j\ Very\ high\ impact \end{cases} \tag{1}$$

Seven experts were invited to assess the strength of the relationship between pairs of risk factors. The arithmetic mean of their judgments was taken to quantify the degree of influence between the factors and construct the fuzzy relationship matrix. The fuzzy membership function (refer to Eq. (2)) was selected, and based on this, the correlation strength matrix $B$ was constructed.

$$b_{ij} = a_{ij} / \left(a_{i\cdot} + a_{\cdot j} - a_{ij}\right) \tag{2}$$

Calculate the fuzzy correlation strength matrix B:

$$B = \begin{bmatrix}
0 & 0.05 & 0.16 & 0.09 & 0 & 0.20 & 0 & 0 & 0 & 0.14 & 0 & 0 & 0 & 0 & 0.15 \\
0 & 0 & 0 & 0 & 0 & 0 & 0 & 0 & 0 & 0 & 0 & 0 & 0 & 0 & 0 \\
0 & 0 & 0 & 0 & 0 & 0 & 0.06 & 0 & 0 & 0 & 0 & 0 & 0 & 0 & 0 \\
0 & 0 & 0 & 0 & 0 & 0 & 0.14 & 0 & 0 & 0 & 0 & 0 & 0 & 0 & 0 \\
0 & 0.16 & 0 & 0 & 0 & 0 & 0 & 0.30 & 0 & 0 & 0 & 0 & 0 & 0 & 0.16 \\
0 & 0 & 0.12 & 0 & 0 & 0 & 0.38 & 0 & 0.14 & 0.16 & 0 & 0.07 & 0 & 0 & 0 \\
0 & 0.05 & 0 & 0.18 & 0 & 0 & 0 & 0 & 0.23 & 0 & 0 & 0.57 & 0 & 0 & 0 \\
0 & 0.07 & 0.11 & 0 & 0 & 0 & 0.16 & 0 & 0 & 0 & 0 & 0 & 0 & 0 & 0 \\
0 & 0 & 0 & 0 & 0 & 0 & 0 & 0 & 0 & 0.30 & 0.12 & 0 & 0 & 0 & 0 \\
0 & 0.08 & 0 & 0 & 0 & 0 & 0.13 & 0 & 0 & 0 & 0 & 0 & 0 & 0 & 0 \\
0 & 0 & 0 & 0 & 0 & 0 & 0.11 & 0 & 0 & 0 & 0 & 0 & 0 & 0 & 0 \\
0 & 0 & 0 & 0 & 0 & 0 & 0.11 & 0 & 0.16 & 0 & 0 & 0 & 0 & 0 & 0 \\
0 & 0.15 & 0 & 0 & 0 & 0 & 0 & 0 & 0 & 0 & 0.45 & 0.14 & 0 & 0.19 & 0 \\
0 & 0.17 & 0 & 0 & 0 & 0 & 0 & 0 & 0 & 0 & 0.28 & 0 & 0 & 0 & 0 \\
0 & 0.04 & 0 & 0 & 0 & 0 & 0.05 & 0.2 & 0 & 0 & 0.08 & 0 & 0 & 0 & 0
\end{bmatrix}$$

**Correlation strength matrix Booleanisation to obtain the adjacency matrix**

Since each element in the fuzzy correlation strength matrix B is a real number in the range of [0,1], a threshold $\lambda$ is required to filter the matrix, such that:

$$c_{ij} = \begin{cases} 1, b_{ij} \geq \lambda \\ 0, b_{ij} < \lambda \end{cases} \tag{3}$$

The value of lambda determines the precision of system division. Based on this, the analysis and decision-making objectives are combined with the current situation. Through multiple experiments, it has been observed that if the value of lambda is greater than or less than 0.15, it leads to undesirable consequences in the lower-level factor, rendering it impractical. Therefore, a threshold value of $\lambda = 0.15$ is established. Utilizing

Eq. (3), the adjacency matrix $C$ is computed:

$$C = \begin{bmatrix} 0&0&1&0&0&1&0&0&0&0&0&0&0&0&1 \\ 0&0&0&0&0&0&0&0&0&0&0&0&0&0&0 \\ 0&0&0&0&0&0&0&0&0&0&0&0&0&0&0 \\ 0&0&0&0&0&0&0&0&0&0&0&0&0&0&0 \\ 0&1&0&0&0&0&0&1&0&0&0&0&0&0&1 \\ 0&0&0&0&0&0&1&0&0&1&0&0&0&0&0 \\ 0&0&0&1&0&0&0&0&1&0&0&1&0&0&0 \\ 0&0&0&0&0&0&1&0&0&0&0&0&0&0&0 \\ 0&0&0&0&0&0&0&0&0&1&0&0&0&0&0 \\ 0&0&0&0&0&0&0&0&0&0&0&0&0&0&0 \\ 0&0&0&0&0&0&0&0&0&0&0&0&0&0&0 \\ 0&0&0&0&0&0&0&0&1&0&0&0&0&0&0 \\ 0&1&0&0&0&0&0&0&0&0&1&0&0&1&0 \\ 0&1&0&0&0&0&0&0&0&0&1&0&0&0&0 \\ 0&0&0&0&0&0&0&1&0&0&0&0&0&0&0 \end{bmatrix}$$

The reachability matrix $M$ can be obtained by performing Boolean operations on the adjacency matrix C. When all the products are equal, the availability matrix $M$ is obtained:

$$M = (C+I)^{k+1} = (C+I)^{k} \neq (C+I)^{k-1} \neq \cdots \neq (C+I) \tag{4}$$

According to Eq. (4), the accessibility matrix $M$ is calculated:

$$M = \begin{bmatrix} 1&0&1&1&0&1&1&1&1&1&0&1&0&0&1 \\ 0&1&0&0&0&0&0&0&0&0&0&0&0&0&0 \\ 0&0&1&0&0&0&0&0&0&0&0&0&0&0&0 \\ 0&0&0&1&0&0&0&0&0&0&0&0&0&0&0 \\ 0&1&0&1&1&0&1&1&1&1&0&1&0&0&1 \\ 0&0&0&1&0&1&1&0&1&1&0&1&0&0&0 \\ 0&0&0&1&0&0&1&0&1&1&0&1&0&0&0 \\ 0&0&0&1&0&0&1&1&1&1&0&1&0&0&0 \\ 0&0&0&0&0&0&0&0&1&1&0&0&0&0&0 \\ 0&0&0&0&0&0&0&0&0&1&0&0&0&0&0 \\ 0&0&0&0&0&0&0&0&0&0&1&0&0&0&0 \\ 0&0&0&0&0&0&0&0&1&1&0&1&0&0&0 \\ 0&1&0&0&0&0&0&0&0&0&1&0&1&1&0 \\ 0&1&0&0&0&0&0&0&0&0&1&0&0&1&0 \\ 0&0&0&1&0&0&1&1&1&1&0&1&0&0&1 \end{bmatrix}$$

**Using ISM to Hierarchize Influencing Factors**

Based on the interaction between risk factors, the ISM diagram of risk factors at streetcar intersections is obtained, as depicted in the Fig. 4.

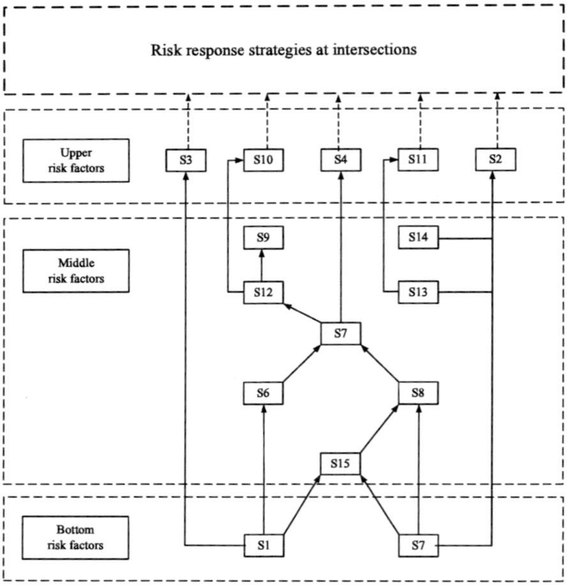


**Fig. 4.** ISM map of risk factors at streetcar intersections

The primary influencing factors, namely $S_2$, $S_3$, $S_4$, $S_{10}$, and $S_{11}$, have a direct impact on the operation of trams at intersections. The intermediate influencing factors consist of the integrated factors $S_{14}$, $S_{12}$, $S_{13}$, $S_7$, $S_6$, $S_8$, and $S_{15}$. The bottom influencing factors consist of $S_1$ and $S_5$. These factors are considered to have the most critical and fundamental influence on the operation of trams at intersections.

## 4 Conclusion and Suggestion

In this paper, the FISM method is employed to analyze the influencing factors and their hierarchical relationships at streetcar intersections. The interdependencies between these factors are established. From the obtained diagram, it is determined that $S_1$ (Type of intersection) and $S_5$ (Allocation of right of way) constitute the bottom factors in the ISM hierarchy and significantly impact the safe operation of streetcars. $S_2$ (Other traffic participants), $S_3$ (Number of lanes at an intersection), $S_4$ (Performance of other traffic vehicles), $S_{10}$ (Trolley lane priority), and $S_{11}$ (Streetcar driver) directly influence the safe passage of streetcars at intersections.

Based on the research results, the following suggestions are proposed for streetcar intersections in Jiaxing City:

(1) Enhance safety education and standardize driver behavior. This can be achieved by increasing driver rotation frequency, reducing the number of operations, and properly managing their work schedules to ensure drivers' alertness. Implement measures such as early alcohol testing before departure and regular psychological

counseling to enhance driver conduct and operating standards, thereby reducing unsafe behaviors.

(2) Strengthen legal oversight. Shift the government's approach to punishing motor vehicle and non-motor vehicle drivers from administrative means to legal means. Illegal lane changes, random lane snatching, and encroachment by other vehicles pose significant safety risks to streetcars and are major contributors to collision accidents. Strengthen cooperation with traffic management departments from the start of operations, intensify awareness campaigns, and enforce stricter penalties for illegal behaviors by other vehicles to reduce traffic accidents.

(3) Establish dedicated streetcar lanes and prioritize streetcar signals and right-of-way. Install signage, traffic lights, section isolation guardrails, and other infrastructure to ensure the exclusive right of way for streetcars, separating them from other modes of transportation. It is noted that some intersections are located in densely populated areas such as schools and shopping centers, and the streetcar lines intersect with multiple channels and four-way intersections, making it challenging to establish dedicated streetcar lanes initially. Therefore, it is recommended to prioritize streetcar signals at intersections and provide disguised priority for the streetcar line.

In summary, the management of streetcars at urban intersections requires a comprehensive approach involving traffic control, traffic safety, driver training, traffic supervision, information management, emergency management, and public participation. Through comprehensive, scientific, and effective management strategies, the safe, smooth, and efficient operation of streetcars at urban intersections can be ensured.

## References


1. Zhen, S., Qi, P., Jiashen, C.: Statistics and analysis of modern tram operation accidents in France, Australia and the United States. Urban Rail Transit Res. (03), 67–70+152 (2019). (in Chinese)
2. Jiajia, Q., Zhenkui, L., Jiabao, L.: Research on road traffic safety risk based on accident tree. Value Eng. **34**(16), 192–195 (2015). (in Chinese)
3. Chenggong, Y.: Safety evaluation and spatial-temporal characteristics analysis of signalized intersection. Qinghai Transp. Sci. Technol. **135**(06), 50–53 (2018). (in Chinese)
4. Haixing, W., Guiping, X., Lei, N.: Research on traffic conflict simulation at signal-controlled intersection based on traffic volume. China Saf. Sci. J. (03), 23–25+2 (2004). (in Chinese)
5. Qi, S., Yong, F., Hua, H., et al.: Collision risk assessment model of streetcar and motor vehicle traffic at plane signal-controlled intersection. China Saf. Prod. Sci. Technol. **16**(S1), 147–151 (2020). (in Chinese)
6. Żochowska, R., Klos, M.J., Soczówka, P.: The analysis of traffic safety on the intersections of roadways and tram tracks. Roads Bridg.-Drogi i Mosty **20**(1), 41–56 (2021)
7. Bhatt, N., Sarkar, D.: Evaluation of success and risk factors for highway project performance through integrated analytical hierarchy process and fuzzy interpretive structural modelling. Int. J. Constr. Manag. **20**(6), 653–665 (2020)
8. Pramudya, A.A., Wibowo, A.: Ranking risks of BOT toll road investment projects in Indonesia using fuzzy interpretive structural modelling. Constr. Econ. Build. **22**(4), 59–80 (2022)
9. Jian, W., Yang, C.: Optimization method of passenger traffic structure based on fuzzy interpretation structure model. Traffic Inf. Saf. **35**(04), 112–118 (2017). (in Chinese)
10. Yan-Hu, H., Ya-Fang, G.: Research on risk structure model of multimodal transportation network based on FISM. J. Highw. Transp. Sci. Technol. **35**(09), 152–158 (2018). (in Chinese)

# A Method for Optimizing the Configuration of Fuel Cell Hybrid Trains

Chang Rui[1], Cao Yufeng[2], Liu Yang[2], and Liu Jianqiang[1](✉)

[1] Institute of Power Electronics and Electric Traction, Beijing Jiaotong University, Beijing 100044, China
{21121411,liujq}@bjtu.edu.cn
[2] China Energy Railway Equipment Co., Ltd, Beijing 100091, China
11590069@ceic.com

**Abstract.** Capacity configuration is an important step in the design of hybrid trains, which directly affects the operational stability and economy of the train. In this study, a capacity configuration method for fuel cell hybrid trains considering train performance and economy is proposed. The characteristics of train demand energy and power under acceleration, cruising, and hill-climbing conditions, as well as the characteristics of the corresponding power sources, are analyzed to establish the corresponding allocation principles. The constraints of the train under different acceleration, cruising, and range conditions are studied, and the differences in power system mass and other aspects due to different capacity configuration combinations are considered, and each configuration combination is verified in a differentiated manner. The optimization objective of this study is to minimize the total cost, and the energy management strategy of Pontryagin's minimization principle(PMP) is used in comparing the fuel cell consumption, and a particle swarm algorithm is used to obtain its optimal solution. The cost differences and configuration differences arising from the different operation methods are analyzed in terms of whether the initial and final values of SOC change in conjunction with the operating road conditions.

**Keywords:** Hydrogen fuel hybrid system · Capacity configuration · Joint optimization · Power performance · Particle swarm optimization algorithm

## 1 Introduction

Power supply sizing is important to the design of hybrid trains. It affects the train's performance, including its driveability, fuel economy, and total cost. Fuel economy has been a key factor in the use of hybrid trains [1]. For fuel cell hybrid systems, power supply size has a greater impact on cost than other types of hybrid vehicles today due to the expensive fuel cell system.

Some authors tend to rely on performance metrics when designing hybrid systems and consider only the power performance metrics that must be met by the hybrid system during the design process. References [2] configure the power of the fuel cell system according to the power demand of the vehicle when it is running at its maximum


Y. Qin et al. (Eds.): EITRT 2023, LNEE 1137, pp. 132–139, 2024.
https://doi.org/10.1007/978-981-99-9311-6_15

speed, with the auxiliary power system supplementing the remaining power. Reference [3] determines the power mixing and system power supply parameters under the constraints of meeting the maximum speed, acceleration time per 100 km, and hill-climbing capability requirements. Reference [4] studied the parameter matching of fuel cell/battery/supercapacitor hybrid power systems by combining the characteristics of three power sources.

Some researchers tend to ignore the vehicle's energy management system (EMS) and concentrate on optimizing the size of vehicle components to reduce the computational burden [5]. Wu et al. [6] proposed a method to optimize the size of vehicle components using a parallel chaotic optimization algorithm (PCOA). Cai et al. [7] proposed a flowchart-based dimensioning method. Eren et al. [8] studied the multi-objective optimal sizing of a hybrid electric vehicle.

Several researchers optimize dynamical systems by using EMS based on optimized strategies. Reference[9] Co-optimization of two-source configuration and fuzzy control parameters using an improved Elitist Non-dominated Sorting Genetic Algorithm (NSGA-II) with elitist strategy. Reference [10] proposes an improved convex optimization (CVX) algorithm that synergistically optimizes capacity allocation and energy management strategies to obtain a bi-objective Pareto front for the quality and energy consumption of hybrid energy storage systems. Jain et al. [11] used a multi-objective generalized algorithm to investigate the optimal component dimensions of trains, including fuel economy and vehicle performance and found several Pareto optimal solutions for the vehicle.

## 2 A Joint Optimization-Based Configuration Method

The performance of a fuel cell hybrid is determined by three aspects: drive cycle, powertrain parameters, and energy management strategy. In this study, we build an optimization problem and solve it to find the optimal powertrain parameters for a specific line operation.

### 2.1 Establishment of Constraints

The combination of capacity configurations needs to meet the vehicle's power performance metrics, including cruising speed, acceleration time, and range.

#### 2.1.1 Power Performance Constraints

The vehicle power performance constraint depends on the demand for vehicle acceleration capability, top-speed running capability, and range performance. The dynamics equation of the train can be expressed as Eq. (1).

$$mgfv\cos\alpha + C_{\mathrm{D}}Av^3/21.15 + \delta mvdv/\mathrm{d}t + mgv\sin\alpha = P_{\mathrm{m}}\eta_{\mathrm{T}}\eta_{\mathrm{md}} \tag{1}$$

where $m$ is the weight of the car, $g$ is the acceleration of gravity, $f$ is the friction coefficient, $v$ is the running speed, $\alpha$ is the running slope of the train, $C_D$ is the wind resistance coefficient, and $A$ is the windward area.

Considering the maximum running speed, the acceleration is 0 at this time, and usually, the maximum speed indicator is considered in the case of a flat and straight road, the gradient is 0. The solution gives as Eq. (2).

$$\begin{cases} v_{\max} = \left[\sqrt{1/4p^2 + 1/27q^3} + p/2\right]^{1/3} - \left[\sqrt{1/4p^2 + 1/27q^3} - p/2\right]^{1/3} \\ p = 21.15\left(P_{m\_\max}\eta_T\eta_{md}\right)/(C_D A) \\ q = 21.15(mgf)/(C_D A) \\ P_{m\_\max} = \min\left(P_{\max\_allowed}, P_{fcmax}\eta_{dc} + V_{\min}(V_{ocv} - V_{\min})/R_{bat} - P_{aux}\right) \end{cases} \tag{2}$$

where $P_{m\_\max}$ denotes the smaller of the maximum power that the power system can provide or the maximum power of the motor.

Figure 1(a) represents the power constraint scenario for the power system at different maximum speeds and the power increases as the maximum operating speed increases.

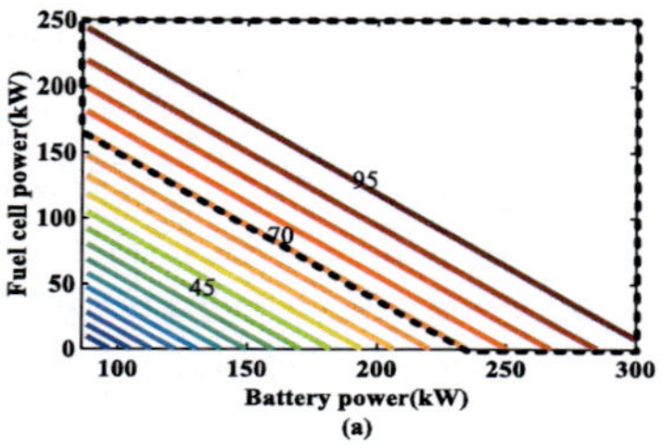


**Fig. 1.** Binding relationship diagram. (a) Maximum allowable speed constraint Domain. (b) Accelerated Time-Constrained Domain

The minimum acceleration time is calculated as Eq. (3), and the acceleration time to the target speed is calculated by discrete means:

$$\begin{cases} v(k) = F(k)\Delta t/(\delta m) + v(k-1) \\ u(0) = 0 \\ F(k) = T_{m\_\max}(k)i_g\eta_T/r - mgf - C_D A v(k-1)^2/21.15 \\ T_{m\_\max}(k) = \min\left(T_{m\_\max,allowed}, P_{m\_\max}(k)\eta_{md}/\omega(k)\right) \end{cases} \tag{3}$$

where F represents the combined force on the vehicle, Tm_max represents the lesser of the allowable motor torque or the torque in the constant work region, ig represents the transmission ratio, and r represents the axle radius.

Figure 1(b) represents the relationship between the acceleration time from 0 to 70 mph and the fuel cell power, and battery power, with the power demand increasing as the acceleration time decreases.

### 2.1.2 Fuel Cell Minimum Power Constraint

As the current cost of hydrogen fuel cells and hydrogen is relatively expensive, the constraint of power performance only will lead to the smaller fuel cell in the configuration result, so the lower limit of the power constraint of the fuel cell is chosen in this paper as the power that can bear the train running at a uniform speed at the maximum speed.

### 2.1.3 Checking of Feasible Domains

**Figure** 2 shows the relationship between the points of the feasibility check and the constraint domain. a) is the combination of points outside the feasible domain, b) and c) is the combination of points inside the feasible domain, and d) is the different combination of points leading to the change of the feasible domain.

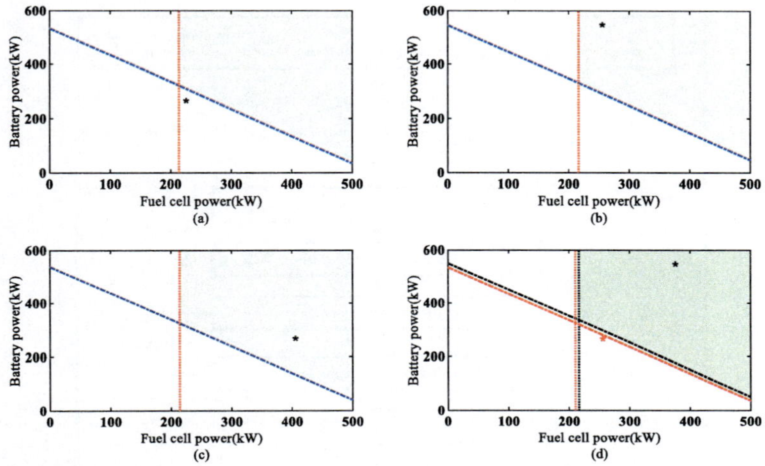


**Fig. 2.** Capacity allocation point and feasible domain relationship diagram. (a) Combination(1,220) check. (b) Combination(2,250) check. (c) Combination(1,400) check. (d) Comparison of (1,250) and (2,370) check fields.

## 2.2 Total Cost Optimization Function

The goal of optimization is to minimize the total cost of the whole system. The total cost of the power system includes the initial acquisition cost and the operating cost incurred during operation, and the operating cost mainly considers the cost of hydrogen in operation. The following optimization function can be established:

$$\begin{aligned} &\min_{\vec{x}} J = J_{\text{ope}} + J_{\text{fix}} \\ &\text{s.t.}\vec{x} = \left(N_{\text{bat\_p}}, P_{\text{fcnetmax}}\right)^{\text{T}} \in U \end{aligned} \tag{4}$$

where $J_{fix}$ is the acquisition cost, $J_{ope}$ is the operation cost and $x$ is the optimization vector, which contains the maximum net power of a fuel cell $P_{FCNetmax}$ and the number of batteries in parallel $N_{batp}$.

## 2.3 Joint Optimization Method

Combining the effects of operating lines and energy management methods on capacity allocation, this paper adopts a joint optimization method to solve the capacity allocation problem with total cost as the optimization objective. The block diagram is shown in Fig. 3. For different capacity configuration combinations, firstly, according to the aforementioned constraint establishment, check whether the combination points are located in the feasible domain.

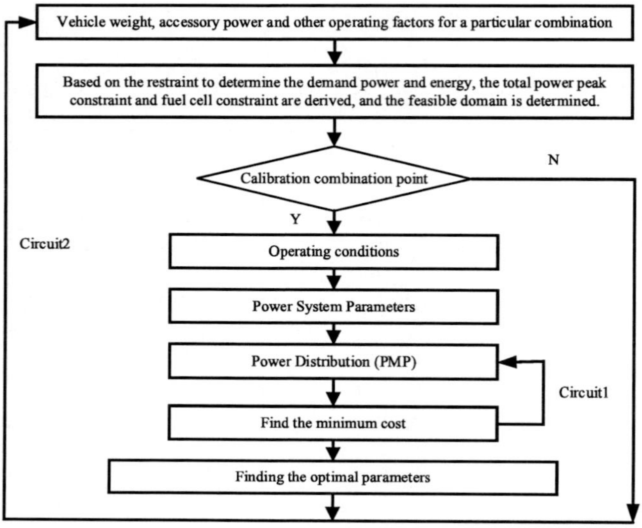


**Fig. 3.** Co-Optimization block diagram

To measure the change in demand power caused by the change in powertrain weight due to the capacity configuration approach, the vehicle weight can be considered as fuel cell weight, DCDC weight, battery weight, and other body weight that does not vary with the capacity configuration, and for different combinations of capacity configurations, the whole vehicle weight is calibrated based on the baseline weight.

Due to the change of powertrain, the vehicle weight changes from $m$ to $m'$, and the power demand changes from $P_{dmd}$ to $P'_{dmd}$.

$$P'_{\mathrm{dmd}} = P_{\mathrm{aux}} + \frac{P_{\mathrm{dmd}} - P_{\mathrm{aux}}}{m} m' \quad (5)$$

The calculation is located in the feasible domain of the configuration point for the weight of the car, the impact of accessory power situation, in this paper to the weight of the car AW2, according to the minimum time operation mode in a specific line operation derived from the power as the power demand benchmark $P_{dmdbase}$, when the power system configuration is shown in Table 1.

The role of optimization-based energy management is to allocate power between the two sources to achieve minimum operational hydrogen consumption. Optimization

**Table 1.** Optimization of baseline parameters

| Parameter Name | Value |
|---|---|
| Fuel cell power rating(kW) | 243 |
| Fuel cell weight (t) | 0.675 |
| DC/DC weight(t) | 0.108 |
| Power Battery Weight(t) | 2.3 |
| Baseline of battery parallel number | 2 |

function expressed as Eq. (6).

$$J = \min \sum_{k=0}^{N-1} C_{fc,k} + C_{bat,k}$$
$$s.t. \begin{cases} |SoC_0 - SoC_{end}| \leq \Delta SoC \\ SOC_L < SOC < SOC_H \\ P_{fc,\min} < P_{fc} < P_{fc,\max} \\ P_{bat,\min} < P_{bat} < P_{bat,\max} \end{cases} \quad (6)$$

Using the PMP algorithm, the optimal hydrogen consumption for a single run can be derived.

When analyzing the total cost, the two main aspects of "initial cost + operating cost" are considered. The price of battery is RMB 1500/kWh, and the price of fuel cell is RMB 2000/kW. As a result of the analysis of fuel cell and hydrogen prices, the fuel cell is replaced once every 6 years, and the battery is replaced once every 8 years. 2000 h is the estimated annual operating time, and 12000 h is the fuel cell life of the train.

Using optimization search of capacity configuration combinations by particle swarm algorithm. Particle swarm optimization (PSO) algorithm belongs to a type of evolutionary algorithm, which starts from the perspective of randomized solutions and finds the optimal solution by iteration.

## 2.4 Optimization Results

Figure 4(a) shows the variation of the total cost with fuel cell power and the number of batteries in parallel in the condition that the end value of SOC does not change, and (b) shows the variation of the total cost in the condition that the end value of SOC declines. (c) and (d) show the variation of one-way hydrogen consumption mass with fuel cell power and the number of batteries in parallel. When the operation mode is fully powered by fuel cells and the initial and final values of SOC need to be maintained at the same value, the particle swarm algorithm can lead to the optimal configuration of the combination of (1,363), which is located in the feasible domain and achieves the smallest commissioning cost of 478.473 Million for six years.

Since the price of one unit of electricity generated by hydrogen is more than the current feed-in tariff, the operation method using battery consumption maintenance, i.e., allowing the battery SOC end value to fall to the lower boundary, will result in a significant cost reduction. **Table** 2 lists its cost and changes.

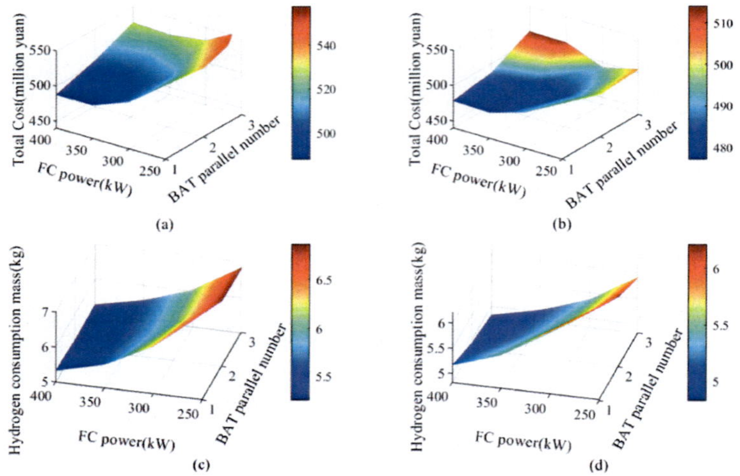


**Fig. 4.** Trend of total cost and hydrogen consumption with capacity configuration. (a) Total cost when SOC does not drop.(b) Total cost when SOC drops. (c) Hydrogen consumption when SOC does not drop. (d) Hydrogen consumption when SOC drops.

Table 2 illustrates the trend of the capacity configuration due to the operation mode, i.e., when the SOC of the battery in the end-state of operation is allowed to decrease more, the optimal combination of capacity configuration tends to increase the number of batteries in parallel and decrease the rated power of the fuel cell, and the total cost will decrease with the increase of the energy borne by the battery.

**Table 2.** Optimal costs for different operating methods

| SOC reduction per 20km | Optimal cost combination | Total cost (million yuan) |
|---|---|---|
| 0% | 1,363 | 478.473 |
| 2% | 1.13,348.5 | 468.267 |
| 4% | 2.28,293 | 452.137 |
| 6% | 3,236 | 421.993 |

## 3 Conclusion

1. In this paper, the relationship between the index requirements and power for accelerated operation, cruising operation, and range is studied to derive a range of power system constraints on bus power and energy.

2. In this paper, the joint operating conditions, energy management strategy, and capacity configuration are optimized in three parts and the optimal solution is found by PSO.
3. In this paper, we study the cost changes due to two different operation methods: maintaining the SOC end value and reducing the SOC end value, and conclude that reducing the SOC end value can effectively reduce the total cost, and the optimal configuration combination will tend to increase the battery capacity and reduce the fuel cell power.

**Acknowledgments.** This work was supported by research fund for the scheme research program of advanced rail transit comprehensive test and research base of China Energy Railway Equipment Co., Lt(TZKY-21-16) and Beijing Natural Science Foundation(L201017).

## References

1. Tiaozhong, W.: Application of hydrogen energy and supercapacitor in modern trams. Electromechanical Eng. Technol. **49**(09), 185–187 (2020). (in Chinese)
2. Yu Ming. Preliminary study on fuel cell vehicle power system parameter matching and control strategy. Shanghai Jiaotong University, 2011. (in Chinese)
3. Huang, M., Wen, P., Zhang, Z., et al.: Research on hybrid ratio of fuel cell hybrid vehicle based on ADVISOR. Int. J. Hydrogen Energy **41**(36), 16282–16286 (2016)
4. Yang, K., Danxiu, D., Wang, J., et al. Parameter matching and control strategies for electric buses with multiple energy sources [J]. J. Automot. Saf. Energy Conserv. **11**(3), 362–370 (2020). (in Chinese)
5. Xu, L., Ouyang, M., Li, J., Yang, F., Lu, L., Hua, J.: Optimal sizing of plug-in fuel cell electric vehicles using models of vehicle performance and system cost. Appl. Energy **103**, 477–487 (2013)
6. Wu, X., Cao, B., Li, X., Xu, J., Ren, X.: Component sizing optimization of plug-in hybrid electric vehicles. Appl. Energy **88**, 799–804 (2011)
7. Cai, Q., Brett, D.J.L., Browning, D., Brandon, N.P.: A sizing-design methodology for hybrid fuel cell power systems and its application to an unmanned underwater vehicle. J. Power. Sources **195**, 6559–6569 (2010)
8. Eren, Y., Gorgun, H.: An applied methodology for multi-objective optimum sizing of hybrid electric vehicle components. Int. J. Hydrogen Energy (2014). https://doi.org/10.1016/j.ijhydene.2014.12.024
9. Hao-Nan, M.: Research on Collaborative Optimization of Capacity Configuration and Intelligent Energy Management Strategy for Energy Storage Trams[D]. Beijing Jiaotong University (2021). https://doi.org/10.26944/d.cnki.gbfju.2021.003075 (inChinese)
10. Kun, A.X., Zhong Ping, Y., Wang et al.: Pareto solution set for capacity configuration of hybrid energy storage system for trams based on improved con-vex optimization algorithm[J]. J. Electr. Eng. Technol. **35**(14), 3116–3125 (2020). https://doi.org/10.19595/j.cnki.1000-6753.tces.190712 (in Chinese)
11. Jain, M., Desai, C., Kharma, N., Williamson, S.S.: Optimal powertrain component sizing of a fuel cell plug-in hybrid electric vehicle using multi-objective genetic algorithm. In: 35th Annual Conference of IEEE, IECON 2009, Porto, Portugal, November 03–05, pp. 3741–6

# Magnetic Levitation System Control Based on Improved LADRC

Ziwei Wu[1,2], Kuangang Fan[1,2,4(✉)], Xuetao Zhang[2,3], and Weichao Li[1,2]

[1] School of Electrical Engineering and Automation, Jiangxi University of Science and Technology, Ganzhou 341000, China
kuangangfriend@163.com
[2] Key Laboratory of Magnetic Suspension Technology of Jiangxi Province, Ganzhou 341000, China
[3] School of Mechanical and Electrical Engineering, Jiangxi University of Science and Technology 341000, Ganzhou, Jiangxi, China
[4] Ganjiang Institute of Innovation, Chinese Academy of Sciences, Ganzhou 341000, China

**Abstract.** In response to the problem of poor anti-interference and stability caused by the complex structure of the magnetic levitation ball system, an improved sparrow search algorithm (ISSA) combined with linear active disturbance rejection control (LADRC) is adopted to solve the above problems. Firstly, a mathematical model of magnetic levitation ball was established based on the research object of magnetic levitation ball; Secondly, an improved sparrow search algorithm is used to adjust the parameters of the linear active disturbance rejection controller (LADRC) to achieve stable control of the magnetic levitation ball; Finally, PID, LADRC, and ISSA-LADRC were introduced for simulation analysis and experimental verification. The final results indicate that ISSA-LADRC has strong stability and anti-interference ability.

**Keywords:** Magnetic levitation · Improved sparrow search algorithm · LADRC · PID · Optimization · Anti-interference

## 1 Introduction

In recent years, magnetic levitation technology [1–3] has been applied in many fields due to its frictionless, lossless, and low-power characteristics. Due to the open loop instability, uncertainty, and complex nonlinear characteristics of magnetic levitation systems, its practical application has become more difficult. Therefore, how to achieve its stable control [4–6]. Has been a research topic for researchers in this field.

Professor Han Jingqing proposed the auto disturbance rejection control (ADRC) control mode [7]. ADRC successfully retained the advantages of PID and made up for the shortcomings of PID. By attributing various unpredictable disturbances in the controlled object to the total disturbance, and then expanding the state observer for real-time evaluation and correction, the controlled object has good performance [8]. Due to the initial nonlinear form of self-disturbance rejection control, which had many


Y. Qin et al. (Eds.): EITRT 2023, LNEE 1137, pp. 140–148, 2024.
https://doi.org/10.1007/978-981-99-9311-6_16

parameters, the problem of difficult parameter tuning arose. Gao Z adopted a frequency scale approach to connect the adjustable parameters of the controller with the bandwidth, solving the problem of difficult parameter tuning in the early stage of self-disturbance rejection [9]. Wei [10] proposed using cuckoo birds and selecting appropriate fitness functions to tune LADRC, which is a simple and convenient method to use. Ouyang et al. [11] proposed an adaptive LADRC method that exhibits good dynamic performance.

From the above content, it can be seen that the parameter tuning of the controller has played an extremely important role in theoretical exploration and practical applications.

# 2 Introduction of Magnetic Levitation Ball System

## 2.1 Establishment of Mathematical Model for Magnetic Levitation Ball System

The magnetic levitation system consists of a laser displacement ranging sensor, a steel ball, an electromagnet, a power amplifier, and a control terminal. The working principle of this device is to drive electromagnetic force through output current to achieve suspension. Figure 1. The structural diagram of the magnetic levitation system is as follows:

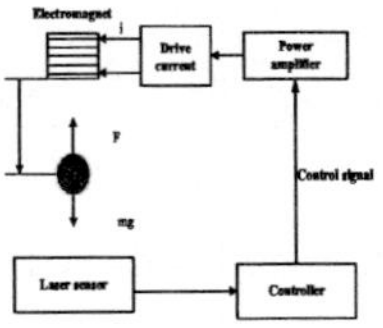


**Fig. 1.** Magnetic levitation ball system

The model expression of the maglev ball system can be obtained from Fig. 1 as follows:

$$\begin{cases} m\dfrac{d^2x(t)}{dt^2} = F(x,i) + mg, \\ F(x,i) = \dfrac{\mu_0 N^2 S}{4}\left(\dfrac{i}{x}\right)^2 \\ F(i_0,x_0) + mg = 0 \\ u(t) = Ri(t) + L_1\dfrac{di(t)}{dt} \end{cases} \tag{1}$$

In formula (1), $F(i,x)$ is the suction force of the magnet, and $F(i_0,x_0)$ is the suction force generated by the steel ball exiting the magnet at the equilibrium point.

According to Eq. (1), the suction force $F(i,x)$ of the magnet on the steel ball is expanded using Taylor's formula at equilibrium point $(i_0,x_0)$.

$$F(x,i) = F(x_0,i_0) + k_i(i - i_0) + k_x(x - x_0) + oR(x) \tag{2}$$

$k_i$ is the stiffness coefficient for the current, and $k_x$ is the stiffness coefficient for the suspension height. Therefore, by calculating the partial derivative of the electromagnetic attraction at the equilibrium point, we can obtain:

$$k_i = \left.\frac{\partial F}{\partial i}\right|_{(x_0,i_0)} = \frac{\mu_0 N^2 S i_0}{2x_0^2} \tag{3}$$

$$k_x = \left.\frac{\partial F}{\partial x}\right|_{(x_0,i_0)} = -\frac{\mu_0 N^2 S i_0^2}{2x_0^3} \tag{4}$$

In Eq. (2), $oR(x)$ is the higher-order term of electromagnetic force $F(i, x)$. Since the magnetic attraction at the equilibrium point is equal to the gravity of the steel ball, therefore:

$$F(i_0, x_0) + mg = 0 \tag{5}$$

The system (1) can be represented as:

$$m\frac{d^2x}{dt^2} = k_i(i - i_0) + k_x(x - x_0) \tag{6}$$

Bringing formulas (3) and (4) into (6) yields:

$$m\frac{d^2x(t)}{dt^2} = \frac{\mu_0 N^2 i_0 S}{2x_0^2} i - \frac{\mu_0 N^2 i_0^2 S}{2x_0^3} x \tag{7}$$

Applying the Laplace transform to both sides of the Eq. (7) yields:

$$s^2X(s) = \frac{1}{m}\frac{\mu_0 N^2 i_0 S}{2x_0^2} I(s) - \frac{1}{m}\frac{\mu_0 N^2 i_0^2 S}{2x_0^3} X(s) \tag{8}$$

Due to the presence of a power amplifier in a single point magnetic levitation ball device:

$$\mathrm{G_1(s)} = \frac{U(s)}{I(s)} = \mathrm{K_a} \tag{9}$$

Substituting formula (9) into formula (8) yields:

$$s^2X(s) = \frac{1}{m}\frac{\mu_0 N^2 \mathrm{i_0} S}{2\mathrm{x_0^2 K_a}} U(s) - \frac{1}{m}\frac{\mu_0 N^2 \mathrm{i_0^2} S}{2\mathrm{x_0^3}} X(s) \tag{10}$$

The transfer function expression of the single point magnetic levitation ball device can be obtained from Eq. (10) as follows:

$$G(s) = \frac{\frac{1}{m}\frac{\mu_0 N^2 i_0 S}{2x_0^2 K_a}}{s^2 + \frac{1}{m}\frac{\mu_0 N^2 i_0^3 S}{2X_0^3}} \tag{11}$$

The parameters of the system are shown in Table 1.

**Table 1.** Maglev system parameters

| Physical meaning | Parameter value | Physical meaning | Parameter value |
|---|---|---|---|
| Ball quality $m(kg)$ | 0.094 | Static inductance $L_1(Hm)$ | 135 |
| Coil turns $N$ | 2450 | Ball radius $r(m)$ | 0.0125 |
| Equilibrium current $i_0(A)$ | 0.5752 | Coil resistance $R(\Omega)$ | 13.8 |
| Balance height $x_0(m)$ | 0.01 | Permeability $\mu_0(H/m)$ | $4\pi \times 10^{-7}$ |
| Magnetic pole area $S(m^2)$ | $\pi \times 10^{-4}$ | | |

# 3 Linear Active Disturbance Rejection Control

## 3.1 LADRC Controller Design

Linear active disturbance rejection control (LADRC) consists of the following three parts: tracking Differentiator (TD), line state error feedback control rate (LESF), and linear state extended observer (LESO). Tracking Differentiator (TD) is used to reduce overshoot. The linear state error feedback control law (LESF) is responsible for receiving the error signal $e_1$ calculated by the LESO and TD modules, and differentiating this signal to obtain the differential error signal $e_2$, and calculating the initial control signal $u_0$ based on these two indicators. The LESO module is the most important module in LADRC, which can accurately evaluate the system state and total interference without knowing the precise model of the controlled object. The control structure of LADRC is shown in Fig. 2.

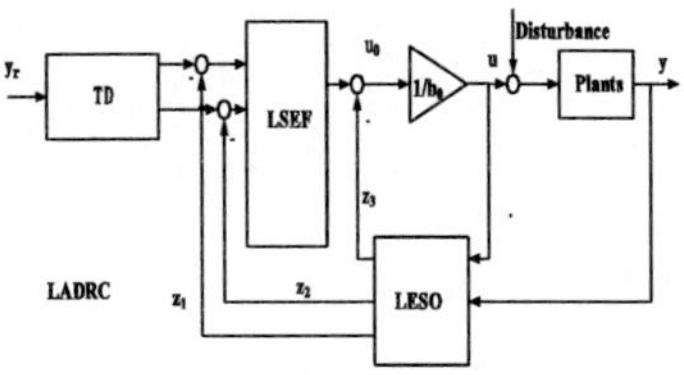


**Fig. 2.** Linear active disturbance rejection control structure

(1) Tracking Differentiator (TD)

The function of a differentiator is to resolve the conflict between overshoot and response speed, with the specific expression as follows:

$$\begin{cases} e_1(t) = y_r(t) - z_1(t) \\ e_2(t) = \dot{y}_r(t) - z_2(t) \end{cases} \tag{12}$$

Substitute $e_2(t) = -z_2(t)$ for $e_2(t) = \dot{y}_r(t) - z_2(t)$ in formula (12) to avoid signal jitter caused by changes in the set target value. Therefore, the following formula can be obtained:

$$\begin{cases} e_1(t) = y_r(t) - z_1(t) \\ e_2(t) = -z_2(t) \end{cases} \tag{13}$$

(2) Linear state error feedback control law (LSEF)

LESF consists of a combination of proportion and differentiation (PD combination). Therefore, the expression for the PD module is as follows:

$$u_0(t) = k_p e_1(t) + k_d e_2(t) \tag{14}$$

$k_p$ and $k_d$ are proportional coefficient and differential coefficient, respectively, which are related to the control bandwidth $\omega_c$, and their expressions are $k_p = \omega_c^2$, $k_d = 2\omega_c$.

(3) Linear Extended state observer Module (LESO)

LESO can effectively estimate the state variables of controlled objects. The expression is as follows:

$$\begin{cases} e(t) = z_1(t) - y(t) \\ z_1(t+h) = z_1(t) + h(z_2(t) - \beta_1 e(t)) \\ z_2(t+h) = z_2(t) + h(z_3(t) - \beta_2 e(t) + b_0 u) \\ z_3(t+h) = z_3(t) + h(-\beta_3 e(t)) \end{cases} \tag{15}$$

where, $\beta_1$,$\beta_2$ and $\beta_3$ are the adjustable gains of the linear extended state observer, which are related to the observation bandwidth $\omega_o$, and can be expressed as: $\beta_1 = 3\omega_0$, $\beta_2 = 3\omega_0^2$, $\beta_3 = \omega_0^3$, $y$ are the output of the system, and $u$ is the output control quantity of the LADRC controller. $z_1$, $z_2$ and $z_3$ evaluated $x_1$, $x_2$ and the total interference to the system, respectively.

# 4 Improved Sparrow Search Algorithm

## 4.1 Improved Sparrow Search Algorithm

Due to the fact that the sparrow search algorithm may fall into local optima, this article improves the discoverer's update formula. The specific improvements are as follows:

$$w{=}w_{\max} - \frac{t(w_{\max} - w_{\min})}{iter_{\max}} \tag{16}$$

where $w_{\max}$ is the maximum value of the weight factor; $w_{\min}$ is the minimum value of the weight factor; $t$ is the current number of iterations.

The improved discoverer update formula is as follows:

$$X_{\mathrm{i,j}}^{\mathrm{t+1}} = \begin{cases} X_{i,j}^t + w \cdot \zeta \cdot \left| X_{i,j}^t - f_g^t \right|, & R_2 < ST \\ X_{i,j}^t + Q \cdot L, & R_2 \geq ST \end{cases} \tag{17}$$

The random numbers $\zeta \in [0, 1]$ and $f_g^t$ are the optimal fitness values of the $t$ generation population.

# 5 LADRC is Optimized by Improving Sparrow Search Algorithm

## 5.1 LADRC is Optimized by Improving Sparrow Search Algorithm

Use the improved sparrow search algorithm to adjust the three parameters ($b_0$,$\omega_c$, $\omega_o$) of linear autoimmunity. Here, the ITAE error index is used as the fitness function, which combines the improved sparrow search algorithm with linear auto disturbance rejection, and its expression is:

$$\text{ITAE} = \int_0^t t|e_y|d\tau = \int_0^t t|y_r - y_{out}|d\tau \tag{18}$$

The specific optimization principle is shown in Fig. 3.

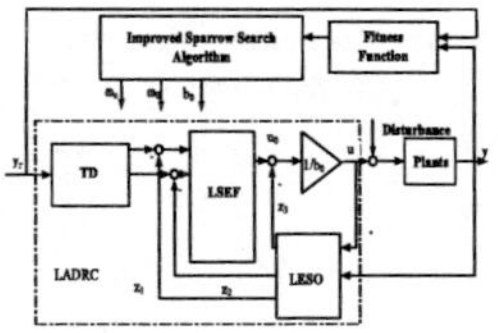


**Fig. 3.** Improved sparrow search algorithm optimization LADRC structure diagram

# 6 Simulation Analysis

In order to better reflect the control effect, error standards are used to evaluate the accuracy of the results. The two selected error indicators are error integral indicator *IAE* and root mean square error *RMSE*. Their mathematical expressions are as follows:

$$\text{IAE} = \int_0^t |e_y|d\tau = \int_0^t |y_r - y_{out}|d\tau \tag{19}$$

$$\text{RMSE} = \sqrt{\frac{1}{t}\int_0^t e_y^2 d\tau} = \sqrt{\frac{1}{t}\int_0^t (y_r - y_{out})^2 d\tau} \tag{20}$$

Here $y_r$ is the set suspension height and $y_{out}$ is the actual suspension height.

After optimizing the sparrow search algorithm, the three parameters of LADRC are as follows: $b_0$= 5.2658,$\omega_c$= 912.2154,$\omega_o$= 859.5641.

## 6.1 Simulation Analysis

From the results in Fig. 4(a), it can be seen that the PID generates excessive overshoot. The required adjustment time is 3.072 s, with an overshoot of up to 18.75%. The overshoot of LADRC is small, only 2.21%, but its adjustment time reaches 1.345 s. Only ISSA-LADRC meets the requirements of stable control, with overshoot of only 1.85%

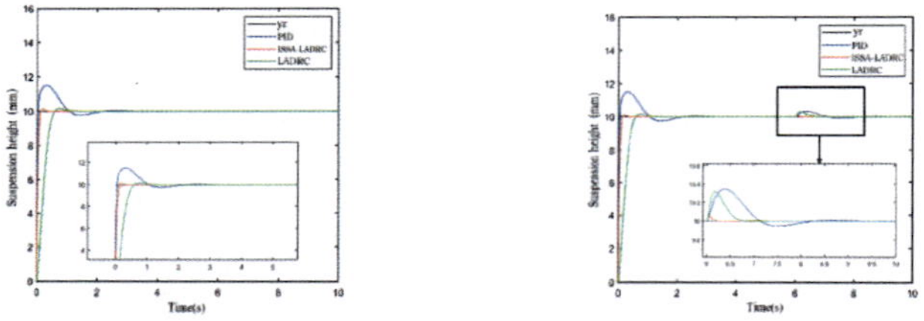

(a) Step response. (b) Step response with interference.

**Fig. 4.** Step response

and adjustment time of only 0.589 s. Compared with PID and LADRC, the adjustment time of ISSA-LADRAC has decreased by 80.83% and 56.21%, respectively.

The *IAE* error indicator of PID is $1.286 \times 10^{-3}$, while the *IAE* error indicator of LADRC is $2.240 \times 10^{-3}$. However, the *IAE* error indicator of ISSA-LADRC is $3.770 \times 10^{-4}$. Compared with PID and LADRC, the *IAE* error indicator of ISSA-LADRC has decreased by 70.68% and 83.17%. The *RMSE* error index of PID is $7.305 \times 10^{-4}$, and the *RMSE* error value of LADRC is $3.751 \times 10^{-4}$. At this time, the *RMSE* error index of ISSA-LADRC is only $1.479 \times 10^{-5}$. Compared with PID and LADRC, the *RMSE* error index of ISSA-LADRC has decreased by 97.98% and 96.06%, respectively. On the contrary, only ISSA-LADRC has much smaller errors than PID and LADRC, indicating that ISSA-LADRC has good dynamic performance.

In order to compare the anti-interference performance of the three control methods, a signal with an amplitude of 5 was added as interference at the 6th second, and the anti-interference effect is shown in Fig. 4(b). From Fig. 4(b), it can be seen that after adding step interference with a value of 5, The fluctuation distances of PID, LADRC, and ISSA-LADRC are 0.375mm, 0.312mm, and 0.071mm, respectively. Compared with PID and LADRC, the fluctuation amplitude of ISSA-LADRAC under interference conditions decreased by 81.07% and 77.24%, respectively.

## 7 Experimental Verification

The components of this experiment platform are: STM32F10 embedded system, steel ball, A/D module, electromagnet, drive module, laser displacement sensor, computer control terminal. The specific platform is shown in Fig. 5:

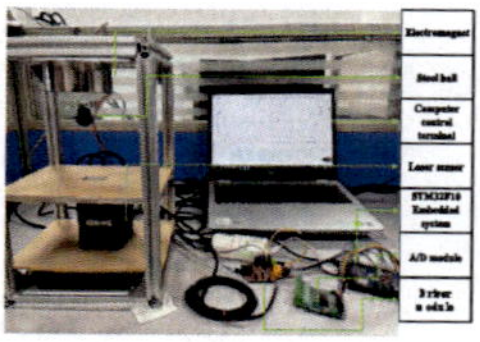

**Fig.5.** Experimental platform building

The waveform display is using vofa+, and the programming is using Keil uvision5 software. The sampling frequency of this experimental platform is 1000.

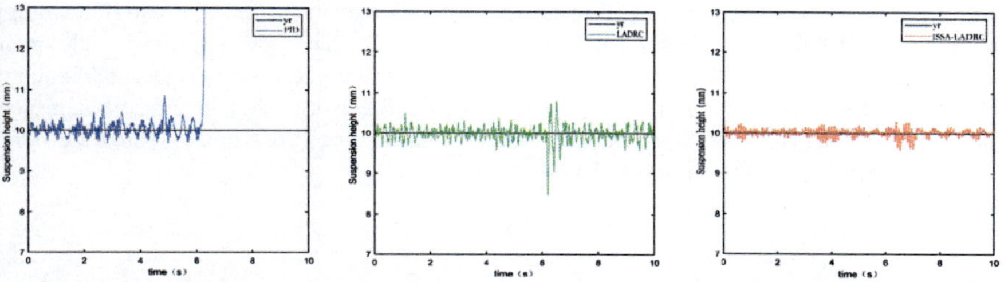


(a) PID control effect (b) LADRC control effect (c)ISSA- LADRC control effect

**Fig. 6.** Experimental effect of three control methods

As shown in Fig. 6., the errors between the actual suspension height of PID, LADRC, and ISSA-LADRC without interference and the set suspension height are $\pm 0.640$, $\pm 0.632$, and $\pm 0.301$, respectively. Compared with PID and LADRC, the errors of ISSA-LADRC are reduced by 52.97% and 52.37%, respectively. After adding interference, the PID lost control and fell off, unable to recover to the set suspension height. The fluctuation distances of LADRC and ISSA-LADRC after interference were 1.475mm and 0.423mm, respectively. Compared with LADRC, the fluctuation distance of ISSA-LADRC after interference decreased by 71.32%.

## 8 Conclusion

(1) When the input signal is a step signal, although the PID control has a fast response speed, the overshoot is too large. Although the overshoot of LADRC control is smaller than PID, its response speed is slow and it takes longer to stabilize the suspension. ISSA-LADRC has almost no overshoot, and the adjustment time is the shortest. Compared to PID and LADRC, the adjustment time of ISSA-LADRC has decreased by 80.83% and 56.21%.

(2) When adding step interference with an amplitude of 5, the fluctuation distances of PID, LADRC, and ISSA-LADRC under interference are 0.375 mm, 0.312 mm, and 0.071 mm, respectively. Compared with PID and LADRC, the fluctuation distance of ISSA-LADRC under interference decreased by 81.07% and 77.24%, respectively.

(3) In experimental verification, when no interference was added, compared with PID and LADRC, the actual suspension height and balance position errors of ISSA-LADRC were reduced by 52.97% and 52.32%, respectively. After adding interference, the magnetic levitation ball cannot recover to its equilibrium position under PID control. Compared to LADRC, the deviation distance of ISSA-LADRC after interference decreased by 71.32%. Therefore, ISSA-LADRC has better stability and anti-interference performance.

**Acknowledgments.** This research was funded by the Central Guided Local Science and Technology Funding Project of the Science and Technology Department of Jiangxi Province(Cross-regional Cooperation, 20221ZDH04052), the 03 Special Project and 5G Program of the Science and Technology Department of Jiangxi Province (No. 20193ABC03A058), the Program of Qingjiang Excellent Young Talents in Jiangxi University of Science and Technology (JXUSTQJBJ2019004), and This work was supported by the Key Research and Development Plan of Ganzhou (industrial field) (No. [2019]60), a grant from the Research Projects of Ganjiang Innovation Academy, Chinese Academy of Sciences. (No. E255J001), and the cultivation project of the State Key Laboratory of Green Development and High-Value Utilization of Ionic Rare-Earth Resources in Jiangxi Province (20194AFD44003).

## References

1. Han, H., Kim, D.: Magnetic levitation. Springer, Amsterdam, The Netherlands (2016)
2. Lee, H., Kim, K., Lee, J.: Review of maglev train technologies. IEEE Trans. Magn. **42**(7), 1917–1925 (2006)
3. Deng, Z., Liu, Z., Li, H., et al.: Development status and prospect of maglv train. J. Southwest Jiao Tong Univ. **57**(3), 455–474, 530 (2022) (in Chinese)
4. Yang, J., Shi, H., Hu, H., et al.: ADRC algorithm for single-point suspension system. J. Ordnance Equip. Eng. **41**(4), 167–171 (2020). (in Chinese)
5. Gao, Q., Yan, H., Zou, H., et al.: Magnetic levitation using diamagnetism: Mechanism, applications and prospects. Sci. China Technol. Sci. **64**(1), 44–58 (2021)
6. Shi, Y., Peng, X., Qu, L., et al.: High overload maneuver control of ramjet missile based on active disturbance rejection control. J. Ordnance Equip. Eng. **42**(7), 13–18 (2021)
7. Han, J.: Nonlinear design methods for control systems. IFAC Proc. Vol. **32**(2), 1531–1536 (1999)
8. Han, J.: From PID to active disturbance rejection control. IEEE Trans. Industr. Electron. **56**(3), 900–906 (2009)
9. Sun, Y., Xu, Q., Qian, H., et al.: Adaptive sliding mode control of maglev system based on RBF neural network minimum parameter learning method. Measurement **141**, 217–226 (2019)
10. Wei, L., Fan, K., Zhou, X., Hu, L., et al.: Linear active disturbance rejection control of magnetic levitation system based on cuckoo search. In: 11th IEEE Data Driven Control and Learning Systems Conference, DDCLS 2022, pp. 151–157 (2022)
11. Ouyang, Q., Fan, K., Liu, Y., et al.: Adaptive LADRC parameter optimization in magnetic levitation. IEEE Access **2021**(9), 36791–36801 (2021)

# Synergia Evaluation of Multi-type Rail Transit Operation Oriented to the Four-Railway-Network Integration


Gehui Liu[1], Ling Liu[1,2](✉), Jun Liu[1], Qing Li[1], and Huiru Zhang[3]

[1] Beijing National Railway Research & Design Institute of Signal & Communication Group Co. Ltd., Beijing 100070, China
{627LL,junliu,liqing}@crscd.com.cn

[2] Beijing Engineering Technology Research Center of Operation Control System for High Speed Railways, Beijing 100070, China

[3] Beijing Mass Transit Railway Operation Co. Ltd., Beijing 100044, China
zhanghuiru@bjtu.edu.cn



**Abstract.** The four-railway-network integration is the policy guidance of multi-type rail transit for intensive resource usage, transportation efficiency enhancement, and regional society and economy development. The synergia improvement concerning multi-type rail transit is an important rising of the service level of regional rail transit system and guidance for urbanization progress. Based on the summary of the current status and policy directions of the synergetic development, the concept and key indicators evaluating the degree of synergia are introduced. Four main synergetic patterns are distinguished referring to the typical cases, including independent operation, passenger service integration, cross-line operation, and regional coordination. Detailed patterns and evaluation indicators with respect to the passenger service integration and cross-line operation are further put forward respectively according to the improvement of passenger service and the difficulty of connection implementation. The proposed theoretical methods and indicators provide the firm support and active reference for the synergetic development of multi-type rail transit systems.



**Keywords:** Multi-type rail transit · Synergy pattern · Four-railway-network integration · Cross-operation operation · Transportation organizations


## 1 Introduction

With the rapid development of social economy and technology, rail transit has a promoting trend of multi-type and network-level operation, becoming an important support for urban construction and guiding spatial configuration in current urbanization progress. By the end of 2022, 155,000 km of the national railways were in operation, including 42,000 km of high-speed railway. A total of 55 cities within mainland China operated urban rail transit, including 8,012.85 km of metro, 1,223.35 km of suburban express, and 564.75 km of tram lines [1]. The productive expansion of multi-type rail transit



Y. Qin et al. (Eds.): EITRT 2023, LNEE 1137, pp. 149–163, 2024.
https://doi.org/10.1007/978-981-99-9311-6_17

networks well serves the high-density passenger demand, but the lack of linkage among different rail types and the independence of operation constraints further improvement of transportation efficiency and service quality. It is necessary for different rail transit networks to innovate from standalone operation to collaboration for purpose of efficient utilization of multi-type networks. Synergy is a comprehensive indicator to describe the level of collaboration among multi-type rail transits by combining a series of aspects such as information interactions, ticket mode, cross-line operation, and emergency management. The analyzing and controlling of the synergy status of multi-type rail transit networks is helpful to promote the connection and coordination within the large-scale, complex, multi-type regional rail networks.

The existing research concerning the multi-type rail transit synergy contains three main streams. The first stream focuses on the theory of synergy. [2] summarized the four levels and five development modes of multi-type rail transit synergy. Information exchange and cross-line connectivity are two key routes for collaboration promotion. [3] proposed the main integration patterns of multi-type rail transit networks and put forward three levels of standards consistence, service composition, and technology integration. [4] introduced a framework on characteristics of multi-type transportation networks, focusing on connectivity and resilience of comprehensive urban rail networks.

The second stream follows the implementation of synergy. [5] analyzed the technical requirements of cross-line operation involving the infrastructure, signals, trains, auxiliary lines, and power supply. The features of travel planning, ticketing, and service integration were also summarized. [6] raised the development strategy in terms of space, demand, and supply of multi-type rail transit, including overall planning, hub junction improving, integrated service, and station-city interaction. [7] compared the technical standard between express and ordinary lines of urban rail transit. Three modes of transportation organization are concluded, including cross-line within the same modes, cross-line between diverse modes, and transfer (i.e., not cross-line) between different modes. [8] advised the development routes related to management structure, network planning, and mode selection on synergy improvement.

In the third stream, which mainly pays attention to operation synergy, existing studies widely optimize the cross-line timetable for two connected metro lines [9, 10]. [11] arranged a mathematical optimization model to coordinate the fairness of transfer and direct passengers. A comprehensive passenger and freight scheduling model was proposed in [12], which simultaneously obtained the train timetable and cargo allocation. [13] came up with a novel train service scheduling model in the scenario of multi-type rail transit systems integration and interconnection.

According to existing research, the complex transportation organization within single type of rail transit, such as skip-stop operation, multi-vehicle trains, and cross-line operation, has attracted a lot of attention. Concerning the synergy of multi-type rail transits, most studies discuss the scope, development target, and implementation technology of the synergy, focusing on the construction and renovation of the network. Several indicator structure and classification approaches have been designed to evaluate the level of synergy, but the more systematic and comprehensive methods required further investigation. In this study, the classification and indicators describing multi-type rail network synergy are proposed based on the well-rounded analysis of the synergy requirement.

The synergy analysis method at the theoretical level is observably improved for possible application in cross-line operation, connection renovation, and network planning.

In the rest of the study, Sect. 2 explains the composition of multi-type rail transit and the overview of synergy development. Detailed synergy patterns and indicator structure are proposed and represented with actual cases in Sect. 3. Section 4 concludes the whole study and denotes the further research direction.

## 2 Multi-type Rail Transit Networks

Rail transit is the crucial support for city daily operation and urban group development. Constructing a multi-type rail transit system to support the high-quality urban cluster development is a major socio-economic problem. In various stages of planning, construction, and operation, however, independence of multi-type rail transit is insufficient to match the requirements of social and economic development in the aspects of functional division, network layout, transfer hub, and passenger services. To promote the rail transit coordination, the National Development and Reform Commission proposed the "Four-railway-network integration" of main-line railway, intercity railway, suburban railway, and urban rail transit. Then, the China Urban Rail Transport Association and China National Rail Co. Ltd. issued related executive programs in 2020, highlighting the construction of a modern integrated hub with smooth connections. The "Four-railway-network integration" has been listed in the National 14-th Five Year Plan issued in 2021, which claims the connectivity improvement of infrastructure in the metropolitan area. The integration and synergy of multi-type rail transit network is beneficial to improving the passenger service level, reducing the travel cost, and optimizing the public transportation network from the supply side.

According to the function and service areas, the rail transit system in China is classified into four categories, namely main-line railway (including high-speed and ordinary railway), intercity railway, suburban railway, and urban rail transit. The technical indicators, functional orientation, and service range of each system are shown in Table 1 and Fig. 1.

(1) Main-line railway, belonging to the national transportation support system, constitutes the uniform network among provincial administrative districts. The main-line railway serves the passenger and cargo transport mainly by the long and medium distance, requiring the high level of station serve and ticket management.
(2) Intercity railway, belonging to the city group rail transit framework, connects several cities within the group and main-line railway networks. The main service object is the short and medium distance passenger flow within the city group and key towns. Some peripheral clusters are also taken into account.
(3) Suburban railway, pertaining to the metropolitan rail transit system, always connects the main-line railway and urban rail transit and applies a closer technical standard to the main-line railway. The suburban railway mainly serves passengers between the core area between suburbs, satellite cities, and towns. At present, the concept and function of the suburban railway in China are not distinct. Figure 1 shows that the suburban trains can operate on independent lines or on main-line railways.

(4) Urban rail transit, usually denoting metro, including light rail and tram, is an important public transportation mode in the core of the metropolitan area, supporting the passenger corridor of the city. The range of service is long and medium distance within the city, providing punctual, reliable, and convenient passenger transportation through high-frequency operation.

**Table 1.** Technical indicators of multi-mode rail transit

| Type | Service range | Design speed (km/h) | Travel speed (km/h) | Mean station spacing (km) |
|---|---|---|---|---|
| Main-line railway | National range (≥150 km) | 200–350 | >160 | 30–60 |
| Intercity railway | City group (50–150 km) | 120–200 | >120 | 10–20 |
| Suburban railway | Metropolitan area (30–50 km) | 100–160 | >50 | 2–5 |
| Urban rail transit | Urban area (5–30 km) | 80–100 | 20–40 | 0.8–1.5 |

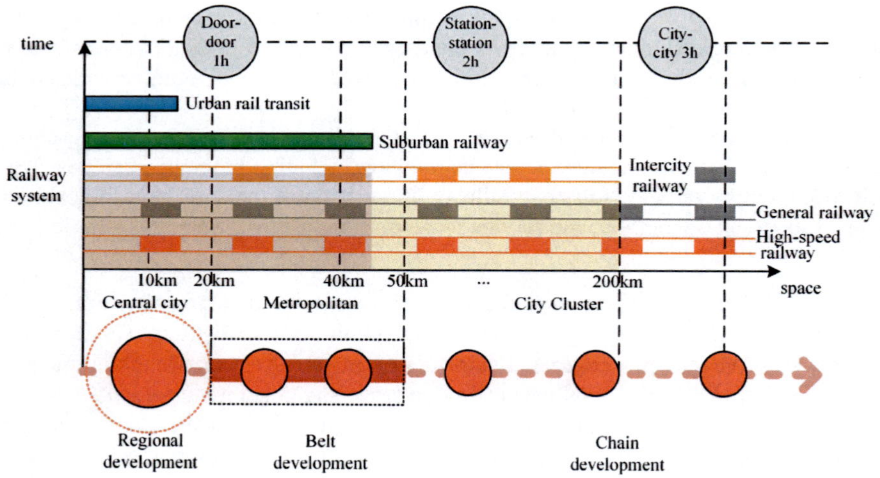


**Fig. 1.** Composition and function of multi-type rail transit system

Multi-type rail transit synergy is in line with the policy orientation of "Four-railway-network integration", which refers to a multi-level rail transportation organization with explicit service functions and compatible operations, based on various rail transit networks, supported by seamless connection and efficient rail transit hubs, and supplemented with flexible and diversified transportation implementation. The goal of synergy is the efficient and intensive utilization of different rail transit resources, the remarkable improvement of passenger travel efficiency, and prominent support for the public transportation development in metropolitan.

According to the service radius listed in Table 1, the main-line and intercity railways with a wide service range have to connect to inner city hubs. The spatial convergence of multi-type rail transit networks is located in the core of the city and becomes the focus of the rail transit synergy. In addition, the scale and structure of the city group have an important influence on railway layout, which should be regarded as an important factor in the synergy. The synergy of multi-type rail transit has to consider the passenger scale of each type of rail transit and transfer requirements among different types. Only when the rail transit network has a large scale, it is necessary to coordinate passenger service among different types. Furthermore, it is only in the case of a certain proportion of the transfer flows that the transfer arrangement needs to be optimized in order to improve the transfer efficiency.

Cross-line train operation is the key to realizing the coordination of multi-type rail transit. The cross-line allows trains to run across different lines, avoiding the time-consuming process of changing trains at stations. Through the cross-line operation, passengers are not restricted by the original physical separation and transfer mode between lines, thus realizing the effective integration of multi-type rail network resources. It is important to note that the cross-line operations are technically demanding and need not only infrastructure, trains, signals, communication, etc., but also high-performance operational management coordination.

## 3 Synergy Evaluation of Multi-type Rail Transit

Safety, efficiency, cost reduction, and high-perforce services are the common goals of different types of rail transit systems. With the large-scale network operation, more attention should be paid to the synergy of the network planning and transportation organizations. According to the integration of information interactions, ticket mode, connectivity, dispatching, and timetable organization, four synergy patterns are proposed to describe multi-type rail transit coordination, namely standalone operation, service integration, cross-line operation, and regional collaboration. The four synergy levels range from low to high, also from independent operation among different lines to coordination at the planning stage. Higher levels of synergy mean more accessible and convenient travel for passengers.

### 3.1 Synergy Pattern of Standalone Operation

Different urban rail transit lines operate separately with no connectivity and less information interaction under the standalone operation level. Passengers have to go through a complicated transfer process, especially transferring between different types of rail transit lines, where a time-consuming egress and access station walking path is required. Improvements in passenger service with the low level of synergy depend in large part on flexible transportation organizations. As shown in Fig. 2, Philadelphia Subway operates multiple modes of trains that save travel time for long-distance travelers. Similar flexible operation modes, including multiple routes, skip-stop trains, regional trains, and straight trains, provide a reference for cross-line operation. The common Y-shape line and co-operating on same lines are generally included in the single line category, as the main and branch lines have the same operation company.

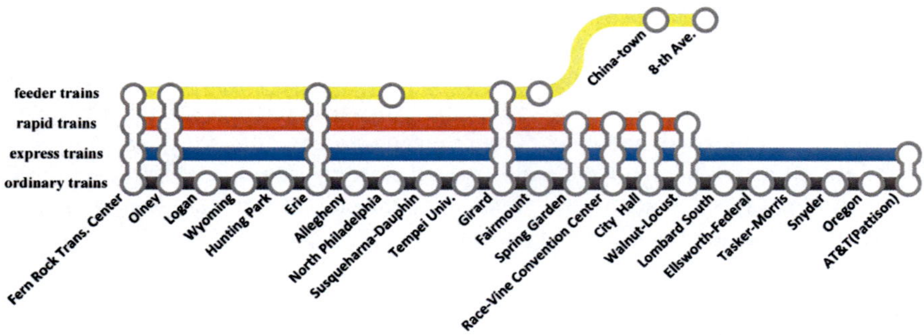


**Fig. 2.** Four operation mode of trains in Broad Street (Orange) Lines in Philadelphia Subway

### 3.2 Synergy Pattern of Service Integration

The service integration starts improving the passenger service between different lines that remain standalone operation. Rapid passenger transfer and basic information sharing are two typical characteristics. The service integration is further subdivided into three levels, including convenient transfer, ticket consistence, and passenger-oriented train operation.

**Convenient transfer**. Shortening the transfer time is of great significance for passengers requiring cross-line travel. The most effective way is to replace the outside-station or station-hall transfer with the platform transfer, which has been extensively applied in urban rail transit systems. Figure 3 illustrates the transfer design between Line 2 and Line 4 of Wuhan Metro. Four-direction platform transfer is realized by consecutive two transfer stations.

In the case of non-recognition of ticket and security check between different rail types, passengers have to walk through a complete egress and access station process. Therefore, reducing the transfer time between different types of rail transit is more obvious for coordinating rail transit networks. A typical example is the Xipu Station in Chengdu, China, as shown in Fig. 4, which is a joint station with platform transfer for urban rail transit and suburban railways.

**Ticket consistence**. Ticket consistence between different rail transit types helps the seamless transfer and travel with one ticket. The ticket consistence is usually actuated between different operation companies, usually exists in adjacent urban rail transit networks, such as the connection between of Line 1 of Shaoxing Metro and Line 5 of Hangzhou Metro, and Line 2 of Foshan Metro to the high-speed Guangzhou South Station.

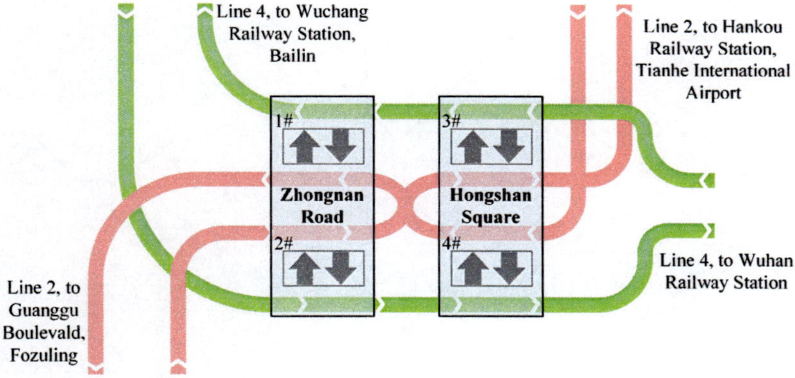


**Fig. 3.** Platform transfer between Line 2 and 4 of Wuhan Metro

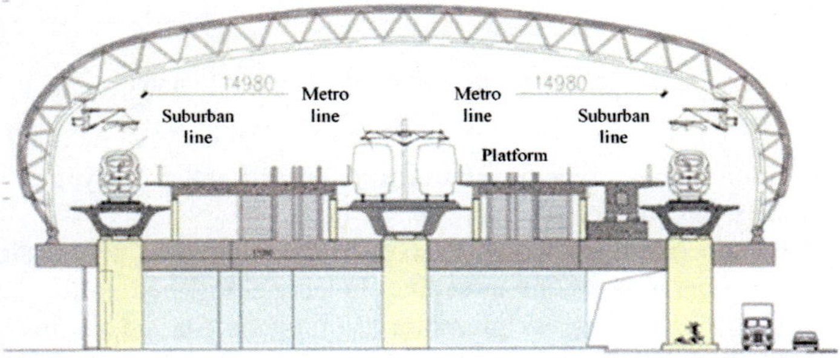


**Fig. 4.** Structure graph of Xipu station in Chengdu, China

The consistence of tickets among various rail transit types is far more complex than single type. The zone charging as shown in Fig. 5 is a solution by dividing the whole rail transit types by geographical areas. The underground, overground, docklands light rail, and national rail stations in London are all allocated into the areas for a unified charging.

**Passenger-oriented train operation.** Train operation coordination relies on the interaction of information, including line status, passenger status, and risk events. The typical approach is cooperating the timetable of different lines at stations or hubs. The mainline and intercity railway easily produce the massive passenger flow towards to the urban rail transit, appealing the transportation capacity increasing by adding trains or turning-around operations to relieve the crowded passengers.

### 3.3 Synergy Pattern of Cross-line Operation

The cross-line operation is an important feature of multi-type rail transit synergy, which needs infrastructure connectivity, transportation capacity sharing, coordinated scheduling, and integrated emergency control. Compared with the cross-type operation, the

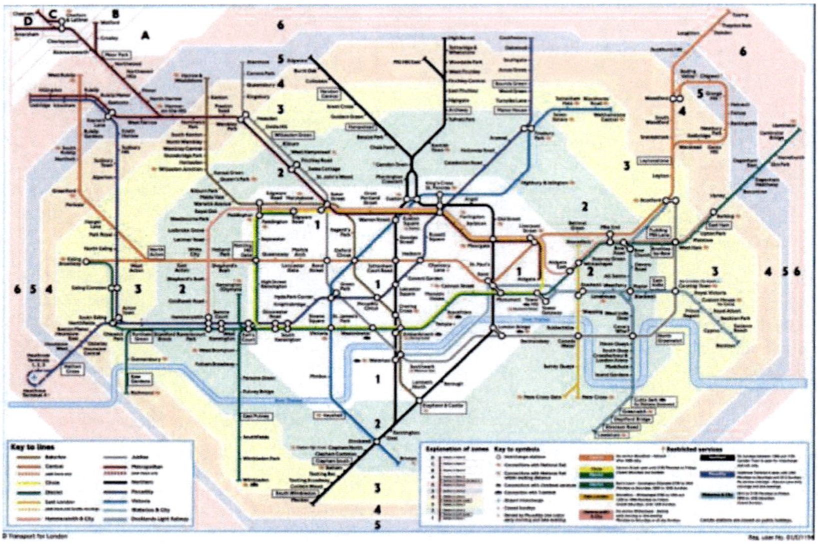

**Fig. 5.** Zone charging map of rail transit in London

cross-line operation within a type of rail transit requires smaller renovation works and is easy to achieve a high degree of synergy.

The cross-line operation has made observed progress in urban tail transit. Figure 6 shows a cross-line operation among three lines in Chongqing metro, including Line 4, Line 5, and Circle Line. The express train starts at Tangjiatuo Station on Line 4, and then successively passes through the Circle line and Line 5, finally stops at Tiaodeng Station on Line 5. Express trains can save passengers up to 33 min of travel time compared to ordinary trains. The implementation process of the three-line connectivity involves many specialties such as power supply, communication, and signal. A uniform signal and train control system is newly set up to achieve the cross-line operation.

In practice, it is easier to execute the cross-line operation with less difference in trains, infrastructure, facilities, signal facilities. For example, the Jiangtiao Line, a suburban railway in Chongqing applies the As-type trains, which can be powered by AC 25 kV or DC 1.5 kV contact network. The Line 5 of Chongqing Metro, physically connected with Jiangtiao Line, runs As-type trains powered by DC 1.5 kV contact network. The compatibility of the two lines on hardware and software determines the practicability of cross-line operation.

### 3.4 Synergy Pattern of Regional Collaboration

The regional collaboration means the cross-line operation and coordinated transportation management among various rail transit types on a large spatial scale, achieving the results on standard recognition, infrastructure connectivity, and integrated transportation service.

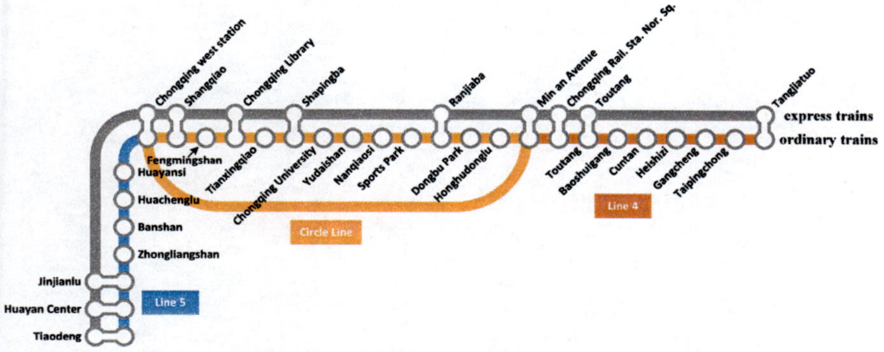


**Fig. 6.** Cross-line operation among Line 4, Line 5, and Circle Line of Chongqing Rail Transit

**Network-level operation collaboration.** The cross-line operation is developed from local lines to a large-scale network, further radiating throughout the metropolitan area and city group. A typical example is the Tokyo metropolitan area network, a multi-layered rail network based on Shinkansen, JR national rail, private rail, metro, and medium volume rail.

(1) Shinkansen high-speed railway and JR national railway undertake the long and medium distance passenger transportation between major cities. The Shinkansen railway is independent from other rail transits. The JR national railway serving intercity travel, overlaps with the JR commuter lines mainly serving the travel within the metropolitan area.
(2) Suburban railway of Tokyo metropolitan consists of JR commuter lines and private railways, having a total length of 2,000 km and a spatial radial shape. The suburban railway strengthens links between downtown and 26 satellite cities. These lines mainly connect to the JR Yamanote Line for transfer.
(3) The urban rail transit system contains subway and medium volume rail. As shown in Fig. 7, the 13 metro lines are all connected to JR Yamanote Line and converge with JR and private lines into major hubs, such as Ikebukuro, Shinjuku, and Shibuya Stations.

One of the representative features of rail synergy in Tokyo is "line integration". The proportion of suburban trains running through the metro lines is more than 60%. Apart from Ginza Line, Marunouchi Line (and its feeder line), and Hanzomon Line, 10 metro lines operate cross-line trains connecting 18 suburban lines, as listed in Table 2.

Another typical feature is the sophisticated transportation hub and transfer system. There are 112 transfer nodes in various transportation systems, such as metro, private railway, and JR railway, of which 27 of the 29 stations on the Yamashiro Line are transfer stations. The large number of transfer nodes and the balanced space layout effectively disperse the transfer passenger flow and reduce the pressure resulting from the massive passengers.

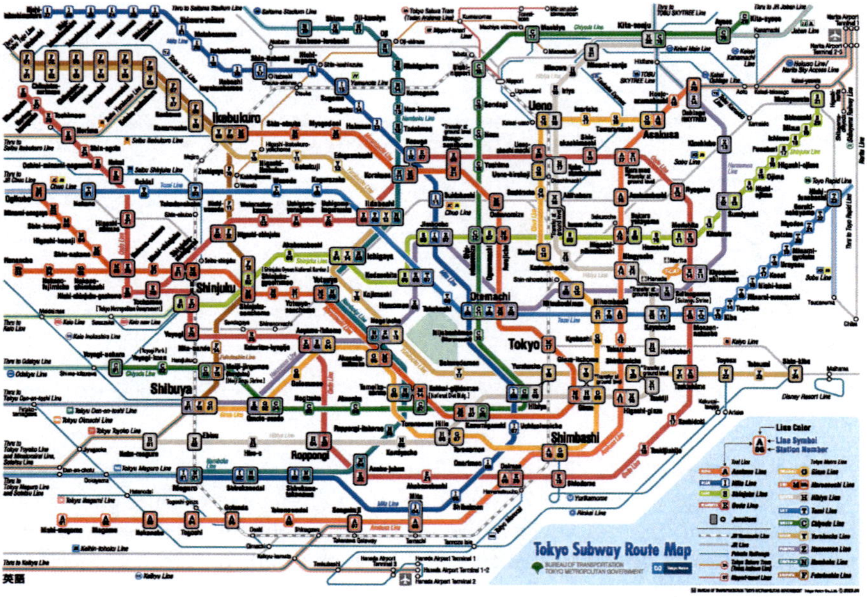


**Fig. 7.** Route map of Tokyo Subway

**Synergy at the planning state**. The coordinated planning scheme for the rail network is a favorable support for the operation synergy. For example, Japan considered the cross-line operation when designing the metro and suburban lines since 1951. With the exception of Shinkansen high-speed railway, other types of rail transits adopt a consistent or compatible system format, which lays the foundation for the current network-level flexible and coordinated operation. At present, China high-speed railway is built according to the same standard, ensuring the integration of new-built lines and existing lines. The urban rail transit lines with different planning period, construction series, technical indicators are relatively difficult to develop the operation synergy. A great deal of technical improvement and verification is required for the synergy improvement. The comprehensive synergy relies on the boundary reduction among different rail transit networks by promoting the consistence of system format, transportation organization, information interaction, jurisdiction division, and security control, finally providing key conditions for operation synergy at the planning stage.

### 3.5 Indicator Structure for Synergy Evaluation

The judgment indicator structure for the proposed four-level synergy is provided in Table 3. The sublevel of the pattern of service integration (i.e., level 2) and cross-line operation (i.e., level 3) is explained in detail.

**Table 2.** Cross-line operation among metro and suburban lines in Tokyo (partial)

| Line | Number of stations | Length (km) | Suburban lines with cross-line operation | Cross-line operation segments and length (km) |
|---|---|---|---|---|
| Asakusa Line | 20 | 18.3 | Keisei Oshiage Line | Oshiage - Narita Airport, Higashi-narita 72.8 |
| | | | Hosuku Line | Oshiage - Inba-Nihon-Idai 32.3 |
| | | | Keihin Kyukou Dentetsu | Sengakuji - Higashi-narita 73.4 |
| | | | Shibayama Railway Line | Oshiage - Higashi-narita 72.8 |
| Hibiya Line | 21 | 20.3 | Tobu Isesaki (Skyfree) Line | Kita-senju - Tobu-dobutsu-koen 33.9 |
| | | | Tokyu Toyoko Line | Naka-meguro - Hiyoshi 11.9 |
| Tozai Line | 23 | 30.8 | JR Chuo Line | Nakano - Mitaka 9.4 |
| | | | Sobu Line | Nishi-funabashi - Tsudanuma 6.1 |
| | | | Toyo Rapid Line | Nishi-funabashi - Toyo-katsutadai 16.2 |
| Mita Line | 27 | 26.5 | Tokyu Meguro Line | Meguro - Musashi Kosugi Eki 9.1 |
| Chiyuda Line | 20 | 24 | Odakyu Line | Yoyougi-uehara - Hon-Atsugi, Karakida 52.5 |
| | | | JR Joban Line | Ayase - Toride 29.7 |

**Table 3.** Indicator structure for synergy of multi-type rail transits

| Synergy level (pattern) | Cross-line operation | Information interaction | Standard recognition | Ticket management | Transfer | Dispatchment and emergency coordination | Facility sharing | Timetable collaboration | Planning collaboration |
|---|---|---|---|---|---|---|---|---|---|
| 1 (standalone operation) | × | × | × | Not consistent | Station-hall transfer | × | × | × | × |
| 2 (service integration) | × | Basic passenger and emergency information | × | Almost consistent | Platform transfer | Station level | × | Transfer-oriented | × |
| | Sub-level | Massive passenger information interaction | Ticket consistence with the same operational company | | | Ticket consistence with different operational companies | | | Timetable collaboration |
| | 2.1 | × | √ | | | × | | | × |
| | 2.2 | × | √ | | | √ | | | × |
| | 2.3 | √ | √ | | | √ | | | √ |
| 3 (cross-line operation) | √ | Frequent interaction | √ | Consistent | Not required | Line level | √ | Integrated scheduling | Partial collaboration |
| | Sub-level | Number of indicators reached | | | Indicator | | | | |

*(continued)*

**Table 3.** (*continued*)

| Synergy level (pattern) | Cross-line operation | Information interaction | Standard recognition | Ticket management | Transfer | Dispatchment and emergency coordination | Facility sharing | Timetable collaboration | Planning collaboration |
|---|---|---|---|---|---|---|---|---|---|
| | 3.1 | 0 | | (1) Cross-line between different rail transit types<br>(2) Cross-line between different operational companies<br>(3) Consistent on-train facilities<br>(4) Cross-line without drivers shift change<br>(5) Two-direction cross-line operation<br>(6) Multiple routes or stop schemes for cross-line trains | | | | | |
| | 3.2 | 2 | | | | | | | |
| | 3.3 | 4 | | | | | | | |
| | 3.4 | 6 | | | | | | | |
| | 3.5 | Unattended cross operation (achieving GoA3 operation level) | | | | | | | |
| 4 (regional collaboration) | √ | Full interaction | √ | Consistent | Not required | Regional level | √ | Integrated scheduling | Full collaboration |

## 4 Conclusion

Promoting the synergy of multi-type rail transit is the requirement for public transportation system and city cluster development. This study summaries the current status of multi-type rail transit in China and proposed the level of synergy on account of the information interaction, ticket mode, cross-line operation, dispatching system, and transportation management. Four patterns of synergy, including standalone operation, service integration, cross-line operation, and regional collaboration, are concluded based on the actual case studies. The core of the synergy is the matching the passenger demand and transportation supply. The crucial approach is the efficiency improvement at transfer stations and cross-line operation.

The synergy evaluation is an explicit direction for modern rail transit development, guiding the regional network capacity expansion, transportation resources allocation optimization, and four-railway-network integration. The detailed classes and indicators of synergy should be refined in the further study for a specifical direction on rail transit projects.

**Acknowledgement.** This work is supported by National Key R&D Program of China (2022YFB4300500).

## References

1. China Association of Metros: Overview of urban rail transit in China mainland in 2022 (2023) (in Chinese)
2. Peng, Q., Luo, J., Wen, W., Wen, C.: Study on regional multi-standard rail transportation organization coordination modes. J. Transp. Eng. Inf. **18**(4), 1–10+22 (2020) (in Chinese)
3. Zhao, S., Ji, Y., Yang, L., Li, D.: Integration mode and selection decision-making method of suburban railway and other rail transit. Urban Rapid Rail Transit. **35**(4), 106–113+134 (2022) (in Chinese)
4. Du, Y., Wang, H., Gao, Q., Pan, N., Zhao, C., Liu, C.: Resilience concepts in integrated urban transport: a comprehensive review on multi-mode framework. Smart Resilient Transp. **4**(2), 105–133 (2022)
5. Jiang, Z., Su, M., Li, H., Xu, S.: Connotation and technical features of rail transit interconnection. Urban Rapid Rail Transit. **34**(3), 46–51 (2021). (in Chinese)
6. Zhang, Y., Lin, Y.: Research on the development strategy of "Four networks "Integration" of different urban rail transits. Railw. Econ. Res. **168**(4), 1–5 (2022). (in Chinese)
7. Liu, J., Wang, M., Zhou, C., Zhang, J., Li, Qing.: Study on coordinated mode and multi-discipline coordinated scheme for urban rapid and urban traditional rail transits. Railw. Signal. Commun. Eng. **19**(3), 59–64 (2022) (in Chinese)
8. Wang, M., Wang, Z., Liu, J.: Analysis of synergetic and integrative development of regional rail transit. Railw. Signal. Commun. Eng. **15**(5), 80–84 (2018). (in Chinese)
9. Guo, H., Bai, Y., Hu, Q., Zhuang, H., Feng, X.: Optimization on metro timetable considering train capacity and passenger demand from intercity railway. Smart Resilient Transp. **3**(1), 66–77 (2021)
10. Li, M., Li, H.: Optimal design of subway train cross-line operation scheme based on passenger smart card data. Sustainability **14**, 6420 (2022)

11. Sun, L., Lu, H., Xu, Y., Kong, D., Shao, J.: Fairness-oriented train service design for urban rail transit cross-line operation. Physica A **606**, 128124 (2022)
12. Hörsting, L., Cleophas, C.: Scheduling shared passenger and freight transport on a fixed infrastructure. Eur. J. Oper. Res. **306**, 1158–1169 (2023)
13. Lin, L., Meng, X., Cheng, X., Ma, Y., Xu, S.: Train service plan design under the condition of multimodal rail transit systems integration and interconnection. J. Adv. Transp. **2023**, 8275191 (2023)

# Development Status, Main Issues, and Policy Suggestions for Transportation Science Popularization Bases

Ben Niu, Jing Dong(✉), Wenhao Shang, and Xiaofei Li

China Academy of Transportation Sciences, Beijing 100029, China
b.n@qq.com, dongjing797@qq.com

**Abstract.** Science popularization bases are crucial venues for conducting science popularization activities, serving as public platforms for providing scientific education services, and constituting an essential component of science and technology popularization work, with distinct characteristics of public welfare. This article, based on extensive and in-depth research on transportation science popularization bases, comprehensively analyzes the current development status and main existing issues in aspects such as regional distribution, base types, supporting units, accreditation institutions, resource conditions, and science popularization activities of industry science popularization bases. It puts forward policy recommendations for the development of transportation science popularization bases from six perspectives: strengthening the mission of science popularization development, optimizing the macro-layout of science popularization bases, promoting innovative development of science popularization bases, enhancing the development of science popularization resources, strengthening the construction of science popularization talent teams, and optimizing the accreditation of national transportation science popularization bases.

**Keywords:** Scientific and technological popularization · Science popularization bases · Transportation · Policy suggestions

## 1 Introduction

Technological innovation and science popularization are the two wings of innovative development. Science popularization is an essential component of a nation's innovation system, and strengthening it is an inevitable requirement for promoting an innovation-driven development strategy. The Party Central Committee and the State Council have always attached great importance to science popularization. In 2016, General Secretary Xi Jinping made an important statement at the National Science and Technology Innovation Conference: "Technological innovation and science popularization are the two wings of innovative development, and science popularization must be placed on an equal footing with technological innovation." [1] In the same year, the State Council issued the "13th Five-Year National Science and Technology Innovation Plan" [2], elevating science popularization to the level of national strategic construction. In September


Y. Qin et al. (Eds.): EITRT 2023, LNEE 1137, pp. 164–173, 2024.
https://doi.org/10.1007/978-981-99-9311-6_18

2022, the General Office of the CPC Central Committee and the General Office of the State Council issued the "Opinions on Further Strengthening Science and Technology Popularization in the New Era" [3], proposing requirements such as "improving the layout of science popularization infrastructure, promoting the balanced development of national science popularization infrastructure, and encouraging the construction of science popularization bases with regional, industrial, and disciplinary characteristics."

In recent years, the Ministry of Transport has increasingly emphasized industry science popularization and taken the construction of science popularization bases as an important focus. In 2020, the Ministry of Transport and the Ministry of Science and Technology jointly issued the "National Transportation Science Popularization Base Management Measures" [4] (hereinafter referred to as "Measures"), initiating the identification of national transportation science popularization bases and actively guiding the high-quality development of industry science popularization bases. At present, the transportation industry has initially formed a group of distinctive science popularization bases, playing a vital role in industry science popularization work. However, issues still exist, such as insufficient awareness of the importance of science popularization bases, imbalanced regional development, single science popularization display methods, inadequate personnel in science popularization, lack of distinct features in science popularization activities, and the need to enhance science popularization creation capabilities. To address these issues, the research team conducted a specialized investigation and, based on the analysis of the current situation and existing problems, proposed policy recommendations to accelerate the development of transportation science popularization bases.

## 2 Development Status of Transportation Science Popularization Bases

To comprehensively understand the current development status of transportation science popularization bases, the research team conducted extensive and in-depth investigations on 291 bases across 31 provinces in China through methods such as questionnaires, telephone surveys, focused group discussions, and on-site visits. The analysis of the relevant information is as follows:

### 2.1 Overall Situation of Science Popularization Bases

In terms of regional distribution, all 31 provinces in the country have established transportation science popularization bases, but there are significant disparities between provinces. Shandong, Beijing, and Guangdong, which rank in the top three in terms of the number of bases, each have more than 20; eight provinces, including Zhejiang, Jiangsu, and Shanghai, have more than 10; and the remaining 20 provinces have fewer than 10, with Jilin and Tibet having the least at only one each. Divided into three regions, the eastern region has 169 bases, accounting for 58.1% of the total; the central region has 55, accounting for 18.9%; and the western region has 67, accounting for 23%. It can be seen that the distribution of transportation science popularization bases is closely related

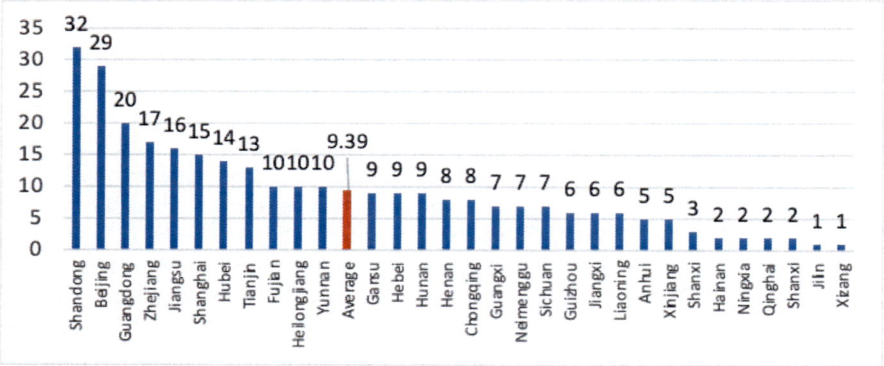


**Fig. 1.** Number of transportation science popularization bases in 31 provinces across the country

to the level of regional economic development, with the number of bases in economically better-off areas significantly higher than in other areas (Fig. 1).

Among the 333 prefecture-level cities in the country, only 93 have transportation science popularization bases, accounting for 27.9% of the total number of prefecture-level cities. This means that more than two-thirds of prefecture-level cities in China do not have transportation science popularization bases. Meanwhile, these bases are mainly concentrated in economically developed provincial capital cities. Among the 27 provincial capital cities nationwide, 25 have established bases, with the exceptions of Changchun and Haikou. Among the 306 non-provincial capital cities, 77.8% do not have a science popularization base. It can be seen that the distribution of transportation science popularization bases varies greatly between cities, far from meeting the requirements for balanced development (Table 1).

**Table 1.** Distribution of transportation science popularization bases in prefecture-level cities in China

| Number and proportion of prefecture-level cities | National | Of which: | |
|---|---|---|---|
| | | Provincial capital cities | Non-provincial capital cities |
| Total number of prefecture-level cities | 333 (100%) | 27 (100%) | 306 (100%) |
| Number of cities with bases | 93 (27.9%) | 25 (92.6%) | 68 (22.2%) |
| Number of cities without bases | 240 (72.1%) | 2 (7.4%) | 238 (77.8%) |

Based on the recognition of science popularization bases, a total of 181 science popularization bases have been recognized by government departments at all levels, scientific and technological associations, industry associations (associations), and other organizations. Among them, there are 35 national-level science popularization bases, including 30 National Transportation Science Popularization Bases jointly recognized by the Ministry of Transport and the Ministry of Science and Technology, and 5 National Education Science Popularization Bases recognized by the China Association for Science and Technology. There are also 64 science popularization bases recognized by transportation industry associations such as the China Highway Society, the Navigation Society, and the Railway Society. Furthermore, 15 science popularization bases have been recognized by provincial or municipal local science and technology authorities. In addition, 8 and 2 science popularization bases have been recognized by local science and technology associations and local industry associations (associations), respectively. Moreover, there are 57 museums recognized or registered by the State Administration of Cultural Heritage. Some of the bases have also obtained recognition from two or more channels. It can be seen that a favorable situation has been formed in the transportation science popularization bases, with joint efforts from government departments at all levels and the participation of industry-leading scientific and technological resources.

### 2.2 Resources and Conditions of Science Popularization Bases

Regarding the venues for science popularization, the average area of each base is 45,601 square meters, including indoor venues and outdoor spaces. The average area of indoor venues in science popularization bases is 5,902 square meters, with nearly half of the bases equipped with audiovisual auditoriums, which have an average area of 430 square meters. It is evident that transportation science popularization bases have relatively good indoor venues and outdoor space conditions, basically meeting the requirements of the "Methods" for venues with an area larger than 1,000 square meters.

In terms of science popularization team building, as of 2020, transportation science popularization bases have a total of 7,275 full-time and part-time science popularization personnel, with an average of 25 personnel per base, representing an increase of 8.7% compared to 2019. In terms of the composition of full-time and part-time science popularization personnel, there are a total of 2,910 full-time personnel, averaging 10 per base, and 4,365 part-time personnel, averaging 15 per base, with a full-time to part-time personnel ratio of 1:1.44. According to statistics, in 2020, the ratio of full-time to part-time personnel in the national science popularization workforce was 1:6.28, which is much higher than the ratio of full-time to part-time personnel in transportation science popularization bases.

In terms of investment in popular science funding, the total investment in transportation and popular science funding in 2020 increased by 139.0% compared to 2019. Examining the total investment in popular science funding across various regions, the eastern region contributed the most, with investment proportions of 56.9% and 64.6% in 2019 and 2020, respectively. The western region had relatively lower investments, with proportions of only 4.6% and 6.9% in the two years, but the growth rate was faster, reaching 258.3%. In terms of the average investment in each science popularization base in 2020, the national average investment was 4.45 million RMB. Among them, the central region had the highest investment, reaching 5.35 million RMB; the eastern region had a larger number of bases, and the average investment remained relatively high at 4.95 million RMB; while the western region invested 1.68 million RMB.

### 2.3 Popular Science Activities Implementation

Regarding the opening status of science popularization bases, the proportion of bases with an opening time of "more than 30 days" in 2019 and 2020 reached 92.2% and 94.5% of the total bases, respectively. Moreover, about half of the science popularization bases had an annual opening time of "more than 300 days," meaning they were essentially open year-round under normal circumstances. Due to the impact of the pandemic, the number of science popularization bases with an opening time of "more than 300 days" in 2020 decreased by 10% compared to 2019. From the perspective of the reception situation at science popularization bases, the annual average number of visitors to transportation science popularization bases in 2020 was 23,811. Among them, the proportion of bases receiving more than 5,000 visitors annually reached 54%. In terms of the implementation of popular science activities, the average number of activities held at each science popularization base was 46, 29, and 26 times in 2019, 2020, and 2021, respectively. Due to the impact of the pandemic, the number of popular science activities in 2020 and 2021 decreased significantly. However, many bases innovatively carried out various online activities through "Internet + popular science," maintaining a frequency of at least one activity every two weeks.

### 2.4 Popular Science Works and Publicity Channels

In terms of popular science works creation, transportation science popularization bases have produced 5,043 popular science articles, with a total readership of 10.87 million, averaging 2,155 reads per article. They have published 308 popular science books with a circulation of 12.95 million copies, averaging 42,000 copies per book. Additionally, they have produced 81,166 popular science videos, with a total viewership of 171.71 million, averaging 2,116 views per video. As short video applications continue to develop, popular science videos have become the primary direction for popular science creation. Although the average viewership of popular science videos is not significantly different from that of popular science articles, the number of videos is 16 times that of articles. Although the creation of popular science books is more difficult than other types due to the requirement of a relatively complete content system and the need for text popularization and illustrations, the individual dissemination effect of published books far exceeds the other two categories.

In terms of popular science publicity channels, the transportation science popularization bases' publicity channels include websites, WeChat official accounts, short video platform accounts, Weibo, and others, with a total of 304 different types of publicity channels. Among them, the number of WeChat official accounts and websites is the highest, while the number of short video platform accounts is relatively lower.

## 3 Main Issues in the Development of Transportation Science Popularization Bases

From the current situation analysis, transportation science popularization bases in various regions have become an important social force in carrying out industry popular science work. However, as the public's scientific literacy continues to improve, the demand for the quantity, quality, and service level of science popularization bases is increasing. The problems in the development of science popularization bases have become more prominent, mainly reflected in the following six aspects:

### 3.1 Insufficient Understanding of the Importance of Science Popularization Bases, and the Need for Strengthened Policy and Financial Support

The new era and new situation require the development of transportation popular science undertakings to significantly enhance social mobilization capabilities and mobilize and gather industry-leading popular science resources on a larger scale and to a greater extent. However, the entire transportation industry's understanding of the connotation, positioning, and function of science popularization bases is not clear and accurate enough, and the recognition of the platform and carrier role of science popularization bases in popular science undertakings is insufficient. At the same time, science popularization bases also lack the necessary policy and financial support.

Currently, besides national transportation science popularization bases, provincial transportation authorities and industry associations (associations) also carry out some regional or professional science popularization base identification work, but lack more robust policy support. In terms of financial support, the popular science funds of the Ministry of Science and Technology and the China Association for Science and Technology have been included in the fiscal budget, but the Ministry of Transportation has not yet established a special popular science fund, making it difficult to play a "catalyst" guiding role within the industry.

### 3.2 Insufficient Total Number of Science Popularization Bases, and Uneven Development Among Regions

In terms of the total number, the number of transportation science popularization bases is far from meeting the actual needs of popular science work, and a multi-level, multi-tiered, and specialized science popularization base system that meets the needs of different regions and target groups has not yet been formed. At the same time, the development of science popularization bases is closely related to the level of regional economic development. Science popularization bases are mainly concentrated in economically developed eastern provinces and provincial capitals of central and western provinces, with a smaller number of bases in western regions, and their driving and radiating capabilities are significantly weaker.

### 3.3 The Means of Popular Science Exhibitions Are Single, and the Ability to Provide Popular Science Services Needs to Be Improved

Currently, transportation science popularization bases focus on the construction of infrastructure venues, with the exhibition content not closely integrated with technological innovation and development. The creative level of scientific display design is not high, with the primary means of display being panels and physical models as supplementary. Better examples include video demonstrations, but truly interactive and immersive exhibits are rare. After the completion of some science popularization bases, there is a lack of attention to subsequent investment and management, with facilities not being updated in a timely manner and mechanisms not being effectively executed, affecting the sustainable development of science popularization bases.

### 3.4 The Total Number of Popular Science Personnel is Insufficient, and the Overall Quality Needs Improvement

The popular science personnel team is the life and vitality of the science popularization base, the planners and implementers of popular science activities, and the providers and practitioners of popular science services. The full-time and part-time popular science personnel of transportation science popularization bases are insufficient in total, with an unreasonable structure, relatively outdated knowledge structure, and room for improvement in the quality of scientific display program research, visit route design, popular science activity planning, and on-site explanations, affecting the improvement of service capabilities of science popularization bases.

### 3.5 Popular Science Activities Lack Distinctive Features, and There Are Fewer Branded Events

In addition to the visiting experience of museum exhibits, organizing regular or irregular popular science activities is an important means for science popularization bases to promote the dissemination of scientific and technological knowledge. At present, popular science activities organized by transportation science popularization bases have strong homogeneity, insufficient exploration of local, field-specific, and base-specific resources, lack distinctive features, and have a limited number of well-known branded events, resulting in limited social influence and reach.

### 3.6 Popular Science Creation Capabilities Need to Be Improved, and an Atmosphere for Active Creation Has not yet Formed

The creation and publication of original popular science works are core aspects of popular science work and an important embodiment of the intrinsic competitiveness of science popularization bases. It is well-known that creating popular science works requires a high level of professional knowledge and popular science creation skills. However, the current industry management departments and relevant units lack incentive policies and measures for popular science creation, and popular science works are relatively weak in terms of title evaluation compared to scientific research achievements. This makes it difficult to attract a large number of researchers and popular science workers to actively engage in popular science creation, and an atmosphere for actively carrying out popular science creation has not yet formed.

# 4 Policy Recommendations for Accelerating the Development of Transportation Science Popularization Bases

Under the strategic deployment of national science popularization work, such as the "Opinions on Further Strengthening Science and Technology Popularization Work in the New Era," and the important arrangements for the development of scientific and technological innovation in the transportation field, such as the "Outline of Long-term Development Plan for Science and Technology Innovation in the Transportation Field (2021–2035)," the industry's science and technology popularization work faces new opportunities and challenges. At the same time, it also puts forward higher requirements and standards for the development of industry science popularization bases. In view of the main problems in the development of transportation science popularization bases, the following policy recommendations are proposed:

## 4.1 Strengthen the Mission of Science Popularization Development and Enhance the Recognition of the Importance of Science Popularization Bases

Implement the important instructions of General Secretary Xi Jinping on science popularization work, organize relevant government departments and enterprises in the industry to deeply study and implement the "Opinions on Further Strengthening Science and Technology Popularization Work in the New Era," focus on the important tasks of building a strong transportation nation and promoting scientific and technological innovation, and enhance the strategic tasks and mission responsibility of transportation science popularization development. Promote the construction of a large-scale science popularization pattern in which the whole industry participates. Deeply understand the platform and position functions of science popularization bases, enhance the recognition of the importance of science popularization bases, and effectively play their important role in cultivating the scientific spirit, training scientific and technological innovation talents, and creating a social innovation atmosphere.

## 4.2 Optimize the Macro-Layout of Science Popularization Bases and Achieve Balanced Development of Science Popularization Public Services

Rely on transportation science and technology resources, such as teaching, research, production, and services, support and encourage various entities, such as universities, research institutes, vocational and technical schools, production and operation units, science and technology venues, and research and innovation platforms, to actively participate in the construction of science popularization bases. Continuously optimize the macro-layout of transportation science popularization bases, gradually form a science popularization base development system that covers all levels and categories, with a reasonable layout, led by national transportation science popularization bases, jointly participated by local transportation authorities and industry associations, and promotes the balanced development of transportation science popularization public services.

### 4.3 Promote the Innovative Development of Science Popularization Bases and Comprehensively Improve the Science Popularization Service Capabilities

First, strengthen the deep integration of science popularization and scientific and technological innovation, vigorously promote the popularization of industry science and technology resources, and closely combine the sharing of large scientific research instruments and equipment, the dissemination of frontier science and technology policies, and the promotion of advanced scientific and technological innovation achievements. Second, give full play to the science popularization function of scientific and technological innovation platforms, reasonably set science popularization tasks and assessment indicators in platform construction and open projects, and fully tap their social benefits. Third, strengthen the hardware and software design of science popularization venues, carry out targeted exhibition space layout, facility and equipment development, visit route design according to the science popularization audience, and add science popularization facilities that are available for public participation, experience, and interaction, mainly for the dissemination and learning of knowledge, to improve the interest and participation of science popularization. Fourth, make full use of information technology, promote the development of science popularization informatization, and vigorously develop online science popularization.

### 4.4 Strengthen the Construction of a Popular Science Talent Team and Improve the Volunteer Service System

Optimize the policy environment for the development of popular science talents, streamline the career development channels for popular science workers, and rationally formulate the professional title evaluation standards for full-time popular science workers. Strengthen the sense of responsibility and honor of high-end scientific and technological talents in participating in popular science, and actively play the role of scientific researchers and engineering technicians as the main force in popular science communication. Carry out extensive popular science capacity training, relying on universities, research institutions, and popular science venues to strengthen the cultivation and utilization of professional popular science talents. Strengthen the construction of popular science volunteer service organizations and teams, and cultivate a combined team of professional and amateur talents with excellent quality and wide coverage for popular science work.

### 4.5 Enhance the Development of Popular Science Resources and Create Distinctive Popular Science Brands

On the basis of the normalization of the opening of science popularization bases, actively carry out thematic popular science activities with distinctive features and outstanding results, combined with the characteristics, advantages, and actual needs of the public in the base. Encourage science popularization bases to create original popular science works such as articles, books, and videos, increase the promotion of excellent popular science works, and continuously improve the originality of popular science works. Strengthen the development of creative cultural products with popular science themes, realizing the organic integration of functionality, knowledge, and artistry.

### 4.6 Optimize the Recognition of National Transportation Science Popularization Bases and Scientifically Guide the Comprehensive Development of Industry Science Popularization Bases

Continuously optimize the application, review, recognition, and evaluation of national transportation science popularization bases, expand the application scope, scientifically determine the review and evaluation indicators and standards, strictly standardize the recognition workflow, and promote construction and management through evaluation. Enable the industry to deeply understand the connotation, positioning, and functions of science popularization bases, attract more social entities to actively participate in the construction and development of science popularization bases and the popular science cause, give full play to the demonstration, guidance, and driving role of national transportation science popularization bases, and thus scientifically guide the comprehensive development of industry science popularization bases.

## References


1. The N.P.C. of P.R.C. website, http://www.npc.gov.cn/npc/kgfb/202209/67860e9c845546008cd6c99e417b5c30.shtml, last accessed 2022/12/20 (in Chinese)
2. The State Council The P.R.C. website, http://www.gov.cn/zhengce/content/2016-08/08/content_5098072.htm, last accessed 2022/12/21. (in Chinese)
3. The State Council The P.R.C. website, http://www.gov.cn/zhengce/2022-09/04/content_5708260.htm, last accessed 2022/12/23. (in Chinese)
4. Ren, F., Zhang, Y., Liu, G.: Introduction to the science popularization industry. Springer Nature (2021)

# Simulation-Based Resource Allocation for Railway Container Terminals

Jing Tang[1], Jialu Yan[1], Zhaoyuan Sun[1], and Li Wang[1,2](✉)

[1] School of Traffic and Transportation, Beijing Jiaotong University, Beijing 100044, China
{20252051,20252052,20252054,liwang}@bjtu.edu.cn

[2] Key Laboratory of Transport Industry of Comprehensive Transportation Theory (Beijing Jiaotong University), Ministry of Transport, Beijing 100044, China

**Abstract.** The railway container terminal is an essential pillar of the intermodal transport network. Although many researches using mathematical models have conducted in this field, most of which only focus on the local operation instead of the global system. Based on this, this project proposes the study of simulation and optimization. Taking Wuhan Railway Container Terminal as an example, FlexSim software is used to design a complete container terminal operation scheduling simulation process and to explore the impact of different resource allocations. It provides scientific support for the selection of operation mode and resource allocation of railway container terminals.

**Keywords:** Train container terminal · Resource allocation · Simulation

## 1 Introduction

Railway container terminal for the railway container transport system to provide transit succession service, which is mainly applicable to the situation when the horizontal distance between the customer and the terminal operation area is far away, and the container is realized with the help of the external container truck for the cargo flow transfer work between the two [1]. Due to the existence of a certain between the two operation areas, this kind of transshipment for the container public-rail transport unified coordination organization also put forward higher requirement.

At present, the research on railway container terminal at home and abroad mainly takes mathematical models as the entry point, among which mixed integer programming and heuristic learning are frequently applied.

Based on the "gantry crane-truck" cooperative loading and unloading scheme, Wang Xiaohan et al. [2] constructed a multi-objective nonlinear mixed integer programming model with heuristic rules, genetic algorithm and simulated annealing algorithm with the objectives of maximizing the work balance rate and minimizing the work cost. Similarly, Martin Tschöke et al. [3] combined the "gantry crane-truck" mixed integer programming with efficient dynamic programming heuristic. Analogizing traveling salesman problem, Konrad Stephan et al. [4] derived the complexity results by differentiating the transfer direction of the container transfer, etc. Peng Guo et al. [5] utilized the constraint


Y. Qin et al. (Eds.): EITRT 2023, LNEE 1137, pp. 174–181, 2024.
https://doi.org/10.1007/978-981-99-9311-6_19

programming logic-based benders decomposition (LBBD). Kai Feng et al. [6] applied deep reinforcement learning to gantry cranes, introducing an element of uncertainty to make them more relevant to real-world situations.

However, these mathematical modelling methods have limitations in carrying out overall system optimization for such an intricate discrete-event dynamic system. The experimental results from Johannes Karder et al. [7] have showed a slight delay in the adaption of gantry crane speeds when constructing an open-ended genetic algorithm to achieve dynamic warehouse operations.

Moreover, the computer simulation methods are the best choice to solve such problems. Mei Sha et al. [8] simulated an integrated container handling system to determine ratios between quay yard and intra-terminal transfer equipment, which has shown to be practical and effective.

In general, there is less research related to the application of computer simulation to resource allocation at railway container terminals. Based on the validity of the use of computer simulations in similar scenarios, the project proposes the study of simulation optimization.

# 2 Design Principle

## 2.1 Design Idea

The main research content of this project is the simulation process design of railway container terminal operation scheduling. At the stage, we take Wuhan Wujiashan Container Terminal as an example to build a simulation model, focusing on the railway container terminal operation mode and resource coordination problems, exploring to what extent it would be affected by the changes in the number of trucks, loading and unloading lines and gantry cranes.[9] The design idea of the work is shown in Fig. 1.

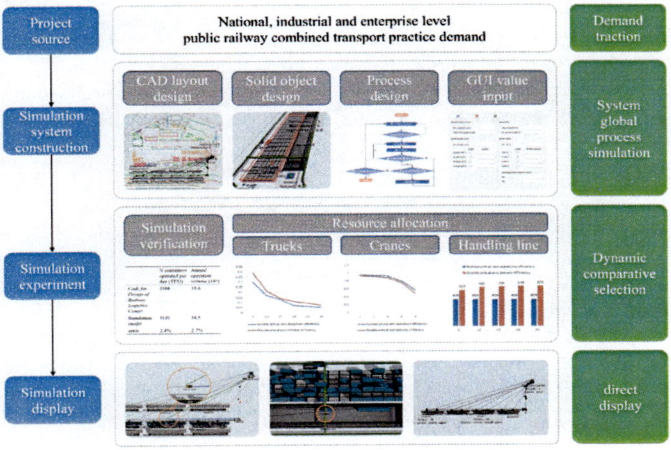


**Fig. 1.** Diagram of design idea

## 2.2 Research Methodology

The current discrete event simulation modelling software includes FlexSim, Witness, Arena and eM-Plant, etc., where FlexSim software has superb 3D dis-play effect, powerful data statistical analysis function that can be applied to sim-ulation modelling in multiple fields.

The basic modelling objects in FlexSim are around nine categories: Fixed Re-sources, Task Executers, Travel Networks, Conveyors, Visual, A*Navigation, AGV, Fluid, Dashboards, among which the following four categories are mainly used in the simulation of major objects in a railway container terminal.

(1) Fixed Resources: including Source to simulate the arrival of the container trains, Sink to indicate the departure of container trains, queue referring to yard to store containers.
(2) Task Executers: including container trucks (Truck), container truck task al-locators (Task executors), gantry cranes (Cranes) and other transport equipment.
(3) Travel Networks: including Network Node, Traffic Control, etc. to control the container truck path to guide the object.
(4) Dashboards: Use the Dashboards module to analyze the data of the model operation, including WIP, Throughput, State Pie, Zone Statistics and other graphical units.

The steps are as follows:

1. Layout design:

Referring to the functional partition of Wuhan Railway Container Terminal, the simulation model is laid out and placed physically. Our simulation area is shown in Fig. 2, which mainly contains the main yard and loading/unloading line, truck route, etc.

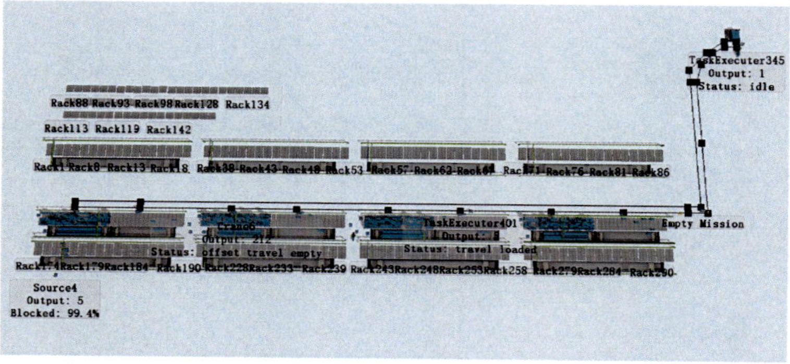


**Fig. 2.** Railway container terminal simulation model diagram

2. Process analysis:

The container operation of the railway container terminal includes: loading, unloading, storing in the yard, transporting away from the yard; the shuttle train operation of the railway container terminal includes: shuttle train arrival, shuttle train waiting for unloading, shuttle train waiting for loading, shuttle train departure. A diagram of the relevant event triggering relationships in the simulation model of a specific railway container terminal is shown in Fig. 3.

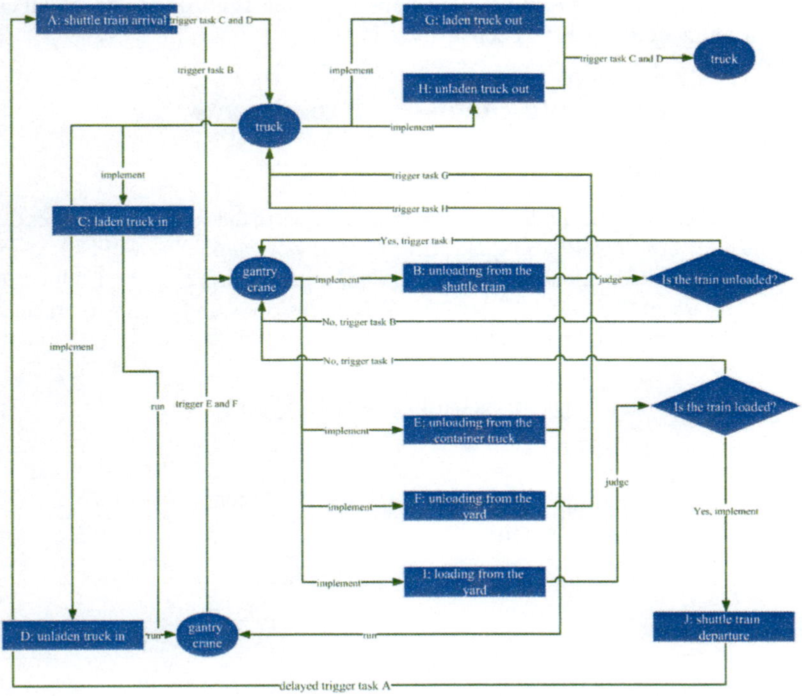


**Fig. 3.** Railway container terminal operation flow trigger mechanism diagram

3. Physical connections:

Using entity connections and intermediate port connections of FlexSim, a logical flow design is implemented to ensure orderly operations between the devices and to plan the transport routes for both unladen and laden tasks of the trucks.

4. Parameter calibration:

As a complex discrete system simulation, the dynamic operation process of it needs to consider more practical factors in the process of system operation. Collected and collated the research data and empirical values of the parameters such as station interval time, the number of loading and unloading lines and the capacity of the class train in advance, they have been used as inputs.

5. Model running:

Set the simulation time and run the model.

6. Analysis of results:

By repeatedly adjusting the parameters in steps, data such as utilization of orbital gantry crane in different contexts have been analyzed.

# 3 Model Validation and Simulation Experiments

## 3.1 Model Validation

To verify the validity of the model, the parameter settings in the Railway Logistics Centre Design Specification (Q/CR9133-2016) (in Chinese) were applied to validate the model for a train loading and unloading line of 1050m, a gantry crane span of 40m without cantilever, a yard of 7 rows, 154 bays, 1078 box positions and 5 gantry cranes. The simulation results are shown in Table 1.

**Table 1.** Simulation Validates Results

| Heading level | Number of daily work containers (TEU) | Number of annual work containers ($10^4$TEU) |
|---|---|---|
| Railway Logistics Center Design Specifications | 1104 | 33.6 |
| Simulation Model | 1131 | 34.5 |
| Errors | 2.4% | 2.7% |

## 3.2 Simulation Experiments

Based on the simulation model, the project conducted experiments related to equipment resource allocation.

**Analysis of the effect of the number and speed of gantry cranes on operational capacity**

Using the control variable method, the effect of different numbers of gantry cranes on operational efficiency was investigated and the results are shown in Fig. 4(a). The trend in efficiency changes is clearly not linear. The lower the number of gantry cranes, the higher the overall operational rate under the premise of meeting the capacity. It is worth noting that with the same number of gantry cranes, the efficiency is higher when delivering double arrival and departure of containers than the normal of that.

### Analysis of the influence of the number of container trucks on the operation capacity

With the other simulation model parameters set unchanged, the number of container trucks is adjusted to compare and analyze the operating rate of trucks under the double containers arrival and departure efficiencies, and the results are shown in Fig. 4(b). The higher the container arrival and dispatch efficiency, the higher the utilization rate of the container truck; the more the number of container trucks, the lower the operation efficiency of the container truck when the container arrival and dispatch efficiency is certain. In addition, the container trucks are more efficient in the case of double containers than when they are normal.

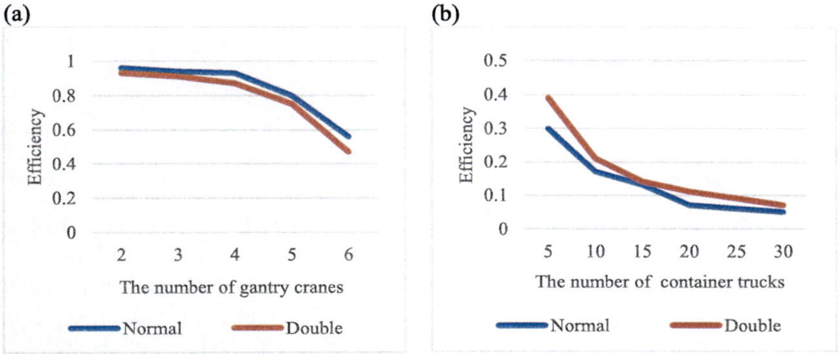


**Fig. 4.** Operation efficiency with normal and double arrival and departure of containers of (a) gantry cranes, (b) container trucks

### Optimal equipment configuration for different arrival and departure volumes

In the actual process, the number and frequency of container arrivals and departures vary. And depending on the actual situation, the terminal requires different equipment arrangement, adjust the container arrival and dispatch situation to explore the following optimal configuration, [10] the results are shown in Fig. 5.

To conclude, in a single line, when containers are arriving and departing at normal efficiency, a minimum number of trucks is sufficient to meet the container transport requirements. As the efficiency of container arrival and departure increases, the demand for the number of trucks also increases. The results of the adjustment and optimization show that when the number of trucks is within ten in a single loading line, the increase in the number of trucks has a significant effect on the capacity and throughput of the railway working area.

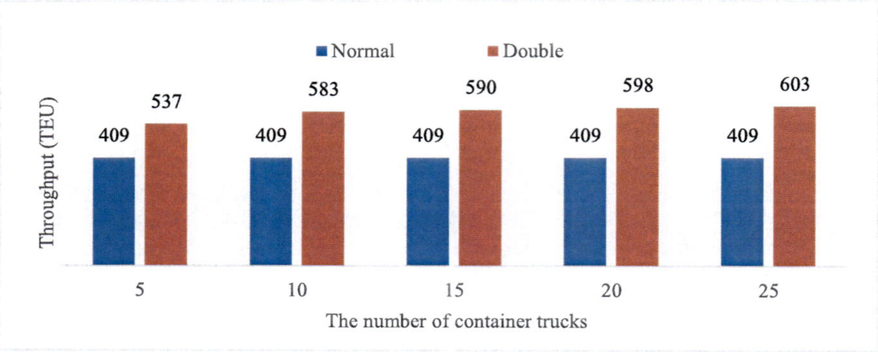


**Fig. 5.** Throughput with normal and double arrival and departure of containers

# 4 Innovative Features

The innovative features of this work are mainly reflected in the following three aspects:

1. Based on the national transportation development strategy, the work identifies the simulation process, model description and modeling steps for the actual needs of railroad container transportation, applies the computer simulation to the railroad container hub, and conducts a dynamic study from the overall situation of Wuhan Wujiashan railroad container center station, breaking through the limitations of mathematical modeling and other methods.
2. The validity of the simulation model is verified according to the relevant design specifications of the railway container hub, and a number of experiments related to the configuration of equipment resources such as gantry crane and collector truck are conducted using the model to find the optimal configuration, and conclusions with reference significance are drawn.

# 5 Conclusion

This work uses simulation and optimisation techniques, combined with the railway container terminal operation process, taking Wuhan Wujiashan railway container centre station as an example, and develops a railway container terminal operation simulation system, simulating the operation efficiency of terminal under different resource allocation conditions, and realising the dynamic simulation visualisation effect and data analysis. The research results of this topic can provide decision support for the planning, design and construction of railway container yards, as well as provide a basis for the configuration of facilities and equipment in railway container operation. In addition, the simulation system developed in this topic can be used as an experience and demonstration platform for railway container transport courses, which has good application prospects and promotion space.

## References


1. MI Tengfei, F.: Analysis of influencing factors of resource allocation of railway container terminal. China Water Transp. **19**(9), 65–66 (2019) (in Chinese)
2. Wang Xiaohan, F., Jia Yulin, S, Cai Jiaxin, T.: Joint optimization of resource allocation and operation scheduling at railway container terminals. Control. Decis. **36**(12), 3063–3073 (2021) (in Chinese)
3. Tschöke Martin, F., Boysen Nils, S.: Container supply with multi-trailer trucks: parking strategies to speed up the gantry crane-based loading of freight trains in rail yards. OR Spectr. **40**, 319–339 (2018)
4. Stephan, K., Boysen, N.: Crane scheduling in railway yards: an analysis of computational complexity. J. Sched. **20**(5), 507–526 (2017)
5. Guo, P.F., He, X.S., Luan, Y., Wang, Y.T.: Logic-based Benders decomposition for gantry crane scheduling with transferring position constraints in a rail–road container terminal. Eng. Optim. 53(1), 86–106 (2021)
6. Feng, K.F., Yang, L.S., He, D.T., et al.: A study on deep reinforcement learning-based crane scheduling model for uncertainty tasks. High Temp. Mater. Process. **41**(1), 469–481 (2022)
7. Karder, J.F., Beham, A.S., Werth, B.T., et al.: Integrated machine learning in open-ended crane scheduling: learning movement speeds and service times. Procedia Comput. Sci. **200**, 1031–1040 (2022)
8. Sha, M.F., Notteboom, T.S., Zhang, T.T., et al.: Simulation model to determine ratios between quay, yard and intra-terminal transfer equipment in an integrated container handling system. J. Int. Logist. Trade **19**(1), 1–18 (2021)
9. Wang Li, F.: Key Resources Scheduling Optimization Theory and Method of Railway Container Terminal, 1st edn. Beijing Jiaotong University Press, Beijing (2015)
10. He Xun, F.: Optimization study on synchronous transfer scheduling in railway container terminals. In: Southwest Jiaotong University (Master's Thesis) (2019)

# Research on Virtual Coupling Technology for Vehicle-to-Vehicle Communication Delay

Yiheng Chen[1](✉), Jinbai Zou[1], Xiaoyong Wang[2], Zhepu Xu[1], and Ziran He[1]

[1] School of Railway Transportation, Shanghai Institute of Technology, Shanghai 201418, China
azure_yiheng@163.com
[2] CASCO Signal Co., Ltd., Shanghai 200072, China

**Abstract.** The virtual coupling technology based on vehicle-to-vehicle communication can significantly shorten the distance between trains and improve the efficiency of the railway network, which is at the forefront of rail transit research. As a critical link in the virtual coupling system, vehicle-to-vehicle communication profoundly affects the coordinated operation of the trains under the virtual coupling state. Under the background of the shorter distance between trains and more complex train control technology, the virtual coupling system has more strict requirements for communication delay. This paper discusses the effect of communication delay on the virtual coupling of trains, which can provide theoretical support for vehicle-to-vehicle communication system performance design. In the first stage, a train dynamics model is proposed, and the model predictive control (MPC) method is used to realize the virtual coupling among multiple trains. In the second stage, using this dynamics model, the effect of communication delay on the virtual coupling train control system is analyzed through a simulation experiment. Finally, according to the experimental conclusion, the communication delay will not only cause the response lag of the following train but also aggravate the vibration phenomenon of the control system's output. These problems will challenge the smooth operation of trains and railway energy conservation.

**Keywords:** Virtual coupling · MPC · Vehicle-to-vehicle communication

## 1 Introduction

In recent years, the urban rail transit industry has developed rapidly. However, due to the continuous growth of passenger flow in some megacities, such as Beijing and Shanghai, data show that the average daily passenger flow growth in these cities is significantly greater than the growth of line networks [1]. This undoubtedly puts forward higher requirements for the operation efficiency of urban rail transit. In addition, with the rapid development of vehicle-to-vehicle communication technology and autonomous train positioning technology, a new generation of more advanced and efficient virtual coupling train operation control technology has become a research hotspot at present.

In 2000, the concept of virtual coupling was first proposed in Europe, and it was assumed that trains could run together in a virtual connection through communication


Y. Qin et al. (Eds.): EITRT 2023, LNEE 1137, pp. 182–191, 2024.
https://doi.org/10.1007/978-981-99-9311-6_20

in the field of train control. However, the initial idea did not carry out research on how to realize it [2]. The concept comes from highway traffic, where a vehicle keeps a safe distance from the vehicle in front of it, well below the braking distance required to go to a complete stop, and the driver braking according to the brake lights of the vehicle in front. Under virtual coupling, trains connect with adjacent trains in the section through wireless communication and transmit their speed, location, and other information to each other. Combined with the reference signals of ground equipment and vehicle-to-vehicle communication data, the trains in the virtual coupling team are closely tracked and run, and the spacing of trains is significantly reduced. The spacing is not restricted by the absolute braking distance, which further improves transportation efficiency (Fig. 1).

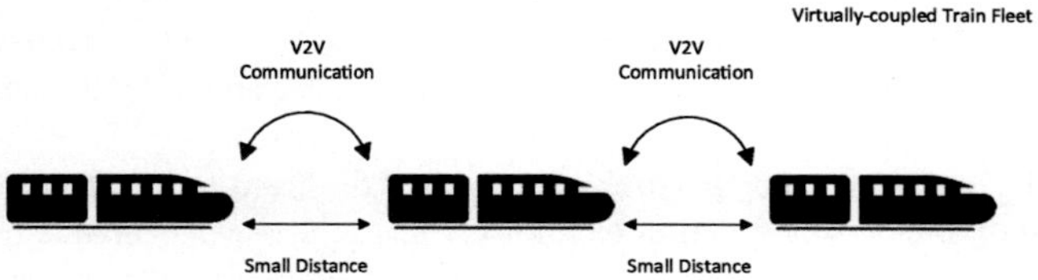


**Fig. 1.** Diagram of virtual coupling

In order to further study the virtual coupling technology based on vehicle-to-vehicle communication. The paper proposes a train kinematic model, deduces the train state space model, and uses the model predictive control method to model and simulate the virtual coupling and system control. In addition, this paper simulates the running effect of virtual coupling through the experimental simulation of the hysteresis caused by communication delay and analyzes the impact of communication delay.

## 2 Related Work

In 2015, the EU fully launched Shift2Rail, which explicitly listed virtual coupling technology as an innovation project and carried out a series of studies along the set technical route. The above research focuses on the shortening degree of train distance, but the research on its performance index is little.

At the same time, researchers have also done much research on the small distance tracking control method in virtual coupling. Jesus F's team designed a distributed model predictive control model for virtual coupling [3]. They summarize the advantages of virtual coupling by comparing virtual coupling technology with traditional moving block technology but do not consider the communication delay. Xiaolin Luo et al. optimized model predictive control with variable predictive step size [4], which significantly shortened the running distance of trains. However, problems caused by changes in communication quality during train tracks still need to be solved. In addition, in order to overcome the problem that traditional control algorithms fall into local optimal solutions due to various uncertainties, some scholars introduce artificial intelligence algorithms into the field of virtual coupling control. Chai et al. use a fusion algorithm of model prediction algorithm and Long Short Term Memory Networks [5]. Wang et al. proposed a multi-agent cooperative control algorithm [6]; when realizing the virtual coupling of trains, the artificial potential field method and reinforcement learning algorithm are combined to realize the tracking of small distances between trains. Since various uncertain errors and disturbances in the process of running, changes in train operation mode, track parameters, communication delay, and so on will have an impact on train control, the solution of virtual coupling train operation control has a higher anti-interference requirement to restrain various uncertain factors. Wu et al. enhanced the system's applicability to time-varying environments by using the adaptive parameter method [7].

In the process of realizing virtual coupling technology, the delay and faults of train communication significantly affect the safe distance between trains. When evaluating the reliability of traditional train operation control systems, for example, in Chinese Train Control System Level 3 (CCTS-3), GSM-R, the communication transmission channel between RBC and onboard equipment, has a profound impact on the safety of train operation [8].

In order to ensure the synchronous operation of the front and rear train units in the virtual coupling, the communication technology has higher performance requirements. Although researchers have done much research in the field of virtual coupling technology, the research analysis of communication delay in virtual coupling operation needs to be included. How will communication delay affect the coordinated process of trains in the virtual coupling state, and to ensure the response speed of the train control system, how much communication delay should be kept within are topics worth further discussion.

# 3 Modeling

## 3.1 Train Dynamics Model

Currently, for the train operation scene of urban rail transit, the simulation modeling is mainly based on a single-point dynamics model. The longitudinal motion process can be described as follows:

$$\begin{cases} \dot{s} = v \\ \dot{v} = a \end{cases} \tag{1}$$

In the formula, $s$ represents the train's position relative to the reference point. $v$ represents the speed of the train. $a$ represents the acceleration of the train. The calculation method of acceleration can also be expressed as follow:

$$\dot{a} = -\tau^{-1}a + \tau^{-1}a_{de} \tag{2}$$

In the formula, $a_{de}$ represents the desired acceleration determined by the train controller. $\tau$ represents a delay factor due to the fact that the train controller does not execute and produce the desired power immediately after the traction or braking command is given.

As can be seen from Fig. 2:

$$e_s = \Delta s - d_{de} - l \tag{3}$$

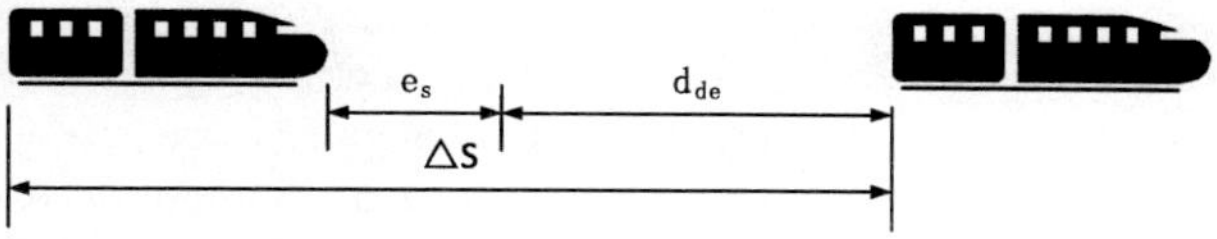


**Fig. 2.** Diagram of train tracking

In the formula, $e_s$ represents the workshop distance error of front and rear trains. $\Delta s$ represents the distance between the front and rear trains. $d_{de}$ represents the desired distance between the front and rear trains. $l$ represents the length of the train.

In addition, $e_v$ is be used to represent the speed difference between the front and rear trains, the formula can be expressed as:

$$e_v = v_{i-1} - v_i \tag{4}$$

The model considers the execution lag of train controllers. To design the model prediction controller, the train dynamics model is replaced with a discrete state equation.

$$\begin{cases} x(k+1) = Ax(k) + B_1u(k) + B_2w(k) \\ y(k) = Cx(k) \end{cases} \tag{5}$$

In the formula, $x(k) = [e_x(k), e_v(k), a(k)]^T$. Each column of the matrix respectively represents the actual workshop distance error of the front and rear trains, the speed difference between the front and rear trains, and the train acceleration. $y(k) = e_s(k)$ as output. $u(k)$ represents the control input of the system. The acceleration of the train in front is going to be $w(k)$. By linearizing the train operation process, the state matrix of A, B, and C can be obtained:

$$\begin{cases} A = \begin{bmatrix} 0 & 1 & 0 \\ 0 & 0 & -1 \\ 0 & 0 & -\frac{1}{\tau} \end{bmatrix} \\ B_1 = \begin{bmatrix} 0 \\ 0 \\ \frac{1}{\tau} \end{bmatrix} \\ B_2 = \begin{bmatrix} 0 \\ 1 \\ 0 \end{bmatrix} \\ C = \begin{bmatrix} 1 \\ 0 \\ 0 \end{bmatrix} \end{cases} \tag{6}$$

### 3.2 Model Predictive Control

Model predictive control (MPC) is an optimal control strategy based on model and target trajectory. The control output of the system at the current time can be obtained by taking the performance indicators that affect the state of the system at a particular time in the future as the targets within each control period [9]. As a control method, the MPC method mainly consists of three essential elements: prediction model, rolling optimization, and feedback correction, which correspond to three processes: predicting the future state of the system by using the spatial state equation, solving optimization online based on the objective function, and applying the optimal value of the solution to the system control [10]. The continuous optimization strategy of model predictive control solves the optimal control result through multiple iterations, which is superior to the classical control method and is not prone to optimal local problems.

Take the prediction of the system from time $k$ for time $k+j$. The input of the predicted actuator is represented by $u(k|k)$, and the control process is realized by incremental control $\Delta u$.

$$u(k+j|k) = u(k+j-1|k) + \Delta u(k+j|k) \tag{7}$$

Meanwhile, the model state equation obtained according to the incremental control form is:

$$X(k) = Ex(k) + F\Delta U(k) + Gu(k-1) + Hw(k) \tag{8}$$

In the formula, $X(k) = \begin{pmatrix} x(k+1|k) \\ \vdots \\ x(k+N_P|k) \end{pmatrix}$, $\Delta U(k) = \begin{pmatrix} \Delta U(k|k) \\ \vdots \\ \Delta U(k+N_c-1|k) \end{pmatrix}$, $N_P$ represents the forecast interval. $N_C$ represents control interval. The matrix E, F, G, and H can be calculated by Eq. (5).

The cost function to define the system is shown as follows:

$$J = \sum_{j=0}^{N_P-1} ||y(k+j|k) - y_r(k+j)||_R^2 + \sum_{j=0}^{N_C-1} ||\Delta u(k+j|k)||_W^2 \tag{9}$$

The ultimate control goal in a dynamic formation scenario is to ensure that the running interval of all trains in formation is equal to the desired ideal interval to ensure safety. The control objective designed by the controller can be expressed as the error value of the speed and position in each prediction time domain. In order to reduce the significant change of train output control quantity in the formation process and ensure the smooth running of trains, the control increment is also taken as the control objective. Then the dynamic optimization problem of train formation is converted into quadratic programming in the standard form:

$$J = \frac{1}{2}\Delta U^T(k)G\Delta U(k) + f^T\Delta U(k) + C \tag{10}$$

In the formula, $C$ represents a constant, $C = E^T(k)\overline{R}E(k)$. Using this formula can get the final cost function. Consider $\Delta u(k)$ as the incremental control input at the time $k$.

### 3.3 Communication Model

Train virtual coupling operation control is based on vehicle-to-vehicle communication technology. The following train obtains the running data of the train in front of it within a certain time, including train speed, train acceleration, etc., by means of vehicle-to-vehicle communication. By solving the quadratic programming problem, the predictive control sequence $U(k)$ of the following train at a particular time can be obtained. The train control sequence in the predictive time domain is applied to the train controller to realize the rolling optimization control of train operation. The flow chart is shown below (Fig. 3).

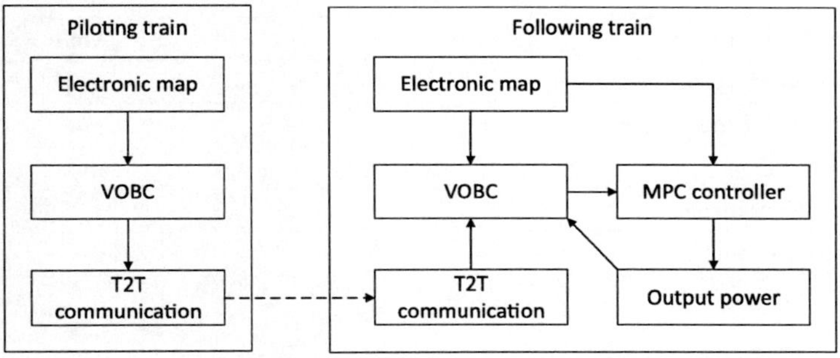


**Fig. 3.** The train control process of virtual coupling

As the critical input of the MPC control model, the acceleration information of the vehicle in front needs to be transmitted and collected through the vehicle-to-vehicle communication module. In the running scenario of virtual coupling, the quality of vehicle-to-vehicle communication will greatly influence train running control. Traditional train control systems already have stringent requirements for communications, as the train control system depends on accurate, timely, and reliable data over communication links. Since communication networks are not initially designed for high mobility environments, random packet delay, and losses are inevitable in train control systems, which could result in unnecessary traction, brake, or even emergency brake of trains, losses of line capacity, and passenger satisfaction [11].

The train dynamics model introduces the communication delay factor to supplement the train control model (Fig. 4).

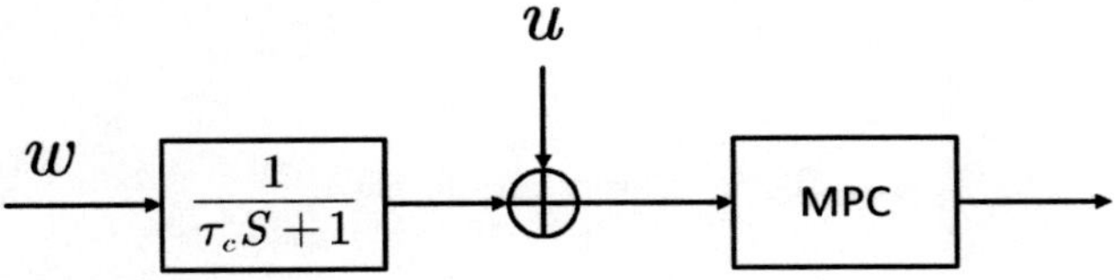


**Fig. 4.** Structural diagram of train control system with communication delay

# 4 Simulation Analysis

## 4.1 Virtual Coupling Simulation

The experiment sets up the running situation of three trains. The simulation parameters are set as follows according to the actual running status of the train:

**Table 1.** Simulation parameters

| Parameter | Numerical value | Unit |
|---|---|---|
| Train length | 120 | m |
| Limiting-velocity | 80 | km/h |
| The upper limit of acceleration | 0.9 | $m/s^2$ |
| The lower limit of acceleration | −1.25 | $m/s^2$ |
| Desired distance | 20 | m |
| Predictive time domain | 2 | s |
| Control time domain | 0.1 | s |

The simulation verifies that three trains can enter a virtual coupling coordination control state. In this state, when the pilot train's speed changes in the dynamic coupling process, the second and third trains can quickly adjust the speed according to the MPC control model. The input of the MPC solves the quadratic programming problem through a loop and sends the first column of the control sequence to the train controller as the predicted output to realize rolling control. Realize the train distance to maintain the desired train distance value. The simulation results are shown in the Figs. 5 and 6.

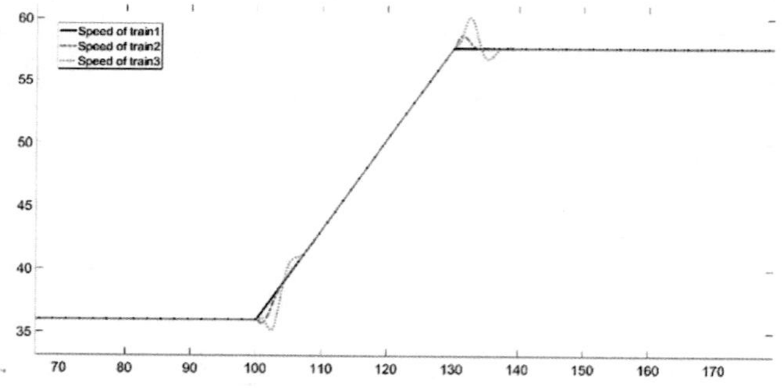


**Fig. 5.** Velocity tracking curve

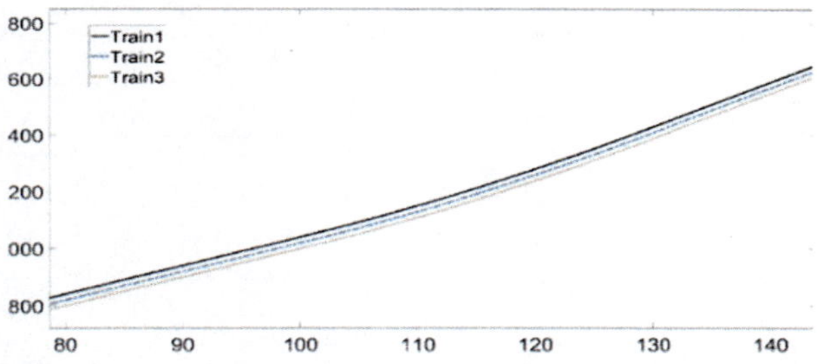


**Fig. 6.** Position error curve

## 4.2 Communication Delay Simulation

Considering the influence of communication delay on the train control system and referring to the actual performance of current communication network technology, two parameters of 20 ms and 200 ms were introduced as communication delay factors $\tau_c$ for simulation experiments. The simulation objects are two trains that have entered the virtual coupling state. The speed of the pilot train was reduced from 72 km/h to 36 km/h at the acceleration of 0.5 m/s$^2$, and the influence of different communication delay conditions on the virtual coupling operation control was analyzed. The simulation results are shown in the Figs. 7 and 8.

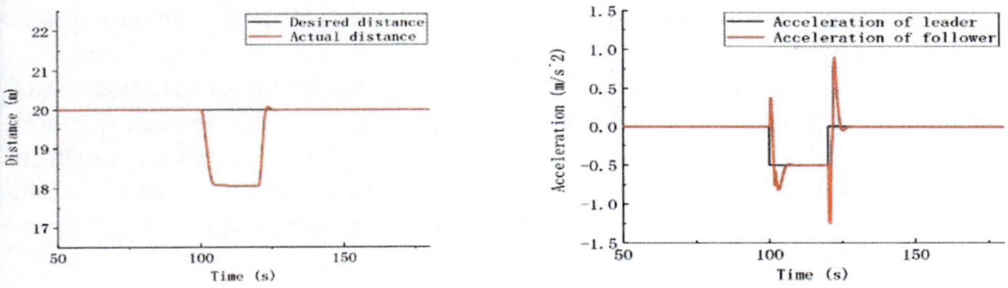


**Fig. 7.** System running status with a communication delay of 20 ms

## 4.3 Discussion

The simulation results show that the communication delay will greatly impact the following state of the train. Considerable communication delay will not only make the response of the following train lag, but also cause security risks. On the other hand, it will make the control process of the following train fluctuate. This kind of fluctuation of locomotive power output will not only cause excessive consumption of power energy but also cause wear and tear to locomotive traction and braking components and reduce their service life.

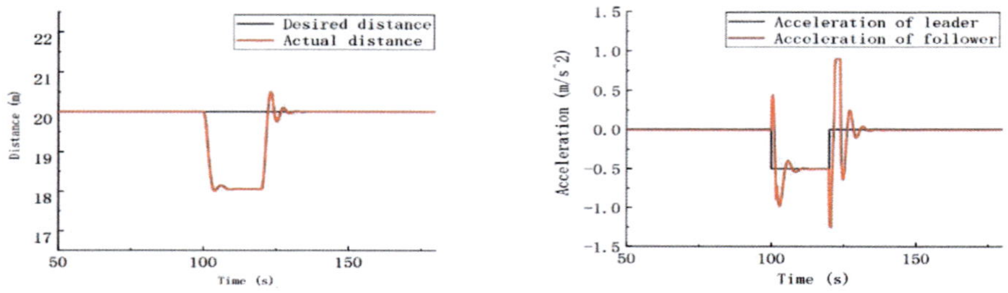


**Fig. 8.** System running status with a communication delay of 200 ms

# 5 Conclusion

In order to improve the operating efficiency of the rail transit system virtual train coupling control method, which can greatly reduce the distance between trains, has become a research hotspot at present. Since vehicle-to-vehicle communication is the key link in a virtual coupling system, the train control system will rely more on the timeliness, accuracy, and reliability of data transmission on the communication link. The experimental results show that communication delay is a very significant factor affecting the operation of virtual coupling trains. The communication delay will not only make the response time of the following train lag, but also aggravate the vibration phenomenon of the output of the control system. Therefore, a virtual coupling system has more strict requirements for communication delay. According to the experimental train dynamics model in this paper, keeping the communication delay within 50 ms and having excellent stability is suggested.

In the future, with the development of the new generation of train operation control Technology, low-delay communication technology, such as 5G (5th Generation Mobile Communication Technology), will be very integral in the field of rail transit. In the further study, the design of a virtual coupling train control system should be considered comprehensively to increase the ability of the train control system to overcome communication delay.

**Acknowledgments.** This work was sponsored by Natural Science Foundation of Shanghai (No. 22ZR1422200).

# References

1. He, B., Gu, B.: Statistics and analysis of line transfer coefficients in rapid rail transit in China's Mainland Cities – Report 8: analysis of data from China association of metros. Urban Mass Transit **23**(5), 1–5 (2020) (in Chinese)
2. Bock, U., Bikker, G.: Design and development of a future freight train concept – "virtually coupled train formations." IFAC Proc. Vol. **33**(9), 395–400 (2000)
3. Felez, J., Kim, Y., Borrelli, F.: A model predictive control approach for virtual coupling in railways. IEEE Trans. Intell. Transp. Syst. **20**(7), 2728–2739 (2019)

4. Luo, X., Tang, T., Liu, H., et al.: An Adaptive model predictive control system for virtual coupling in metros. Actuators **10**(8), 178 (2021)
5. Chai, M., Su, H., Liu, H.: Long short-term memory-based model predictive control for virtual coupling in railways. Wirel. Commun. Mob. Comput., 1–17 (2022)
6. Wang, H., Zhao, Q., Lin, S., et al.: A Reinforcement learning empowered cooperative control approach for IIoT-based virtually coupled train sets. IEEE Trans. Industr. Inf. **17**(7), 4935–4945 (2021)
7. Wu, C., Gao, S., Zhao, N., et al.: Distributed Fixed-time Cooperative Cruise Control for Train Formation. In: 2020 Chinese Automation Congress (CAC), pp. 6350–6354 (2020)
8. Shi, L., Wang, J.: Reliability analysis on the train control system in the CTCS-3 operating mode. Smart Resilient Transp. **3**(1), 25–36 (2021)
9. Mayne, D.Q., Rawlings, J.B., Rao, C.V., et al.: Constrained model predictive control: stability and optimality. Automatica **36**(6), 789–814 (2000)
10. Zhang, X., Zhao, L.: Application of model predictive control in virtual coupling train control system. Urban Mass Transit. **24**(10), 233–237 (2021) (in Chinese)
11. Bu, B., Yu, F.R., Tang, T., et al.: Performance improvements of communication-based train control (CBTC) systems with unreliable wireless networks. Wireless Netw. **20**(1), 53–71 (2014)

# Optimizing Train Platform Allocation in Busy Complex Passenger Stations with Fixed Track Utilization Scheme

Ruihua Hu[1], Qiongfang Zeng[2](✉), Yinggui Zhang[1], and Yuhang Wang[1]

[1] School of Traffic and Transportation Engineering, Central South University, Changsha 410004, China
{Ruihua,ygzhang,224211023}@csu.edu.cn

[2] School of Public Administration and Human Geography, Hunan University of Technology and Business, Changsha 410205, China
2681@hutb.edu.cn

**Abstract.** The allocation of tracks in busy complex passenger station is generally based on Fixed track utilization scheme for the station, with the aim of enhancing operational efficiency and safety at the passenger station. This paper introduces a mixed-integer programming model designed specifically to address the train platforming problem in such stations. We propose constraints that align with the fixed track utilization scheme and aim to achieve balanced track usage objectives. Additionally, an enhanced genetic algorithm is presented, incorporating improved crossover and selection methods while maintaining compliance with the fixed track utilization scheme. Taking Wuchang Station as an example, the efficiency of the model and algorithm is verified, which operates under fixed track utilization scheme. Overall, this research provides efficient approaches for the TPP in busy complex passenger stations that adhere to fixed track utilization scheme.

**Keywords:** Busy complex passenger station · Train platforming problem · Improved genetic algorithm · Fixed track utilization scheme

## 1 Introduction

With the expansion and improvement of rail transit network system scale, the contradiction between station transportation capacity and railway traffic demand is becoming more and more prominent, and the uneven utilization of complex dynamic network capacity. In order to ensure high quality passenger service, it is necessary to improve the reliability and efficiency of station capacity. Busy complex passenger station is a basic place for passenger collection and distribution, and the train platforming problem is the core part of station operation organization, and an important content of realizing network operation and online adjustment of regional rail transit in urban agglomerations.

Over the past decades, numerous scholars have devoted their efforts to developing train platforming optimization techniques that are both efficient and effective. Zwaneveld et al. [1] defined TPP as node packing problem. Sels, Cacchiniani and Zhang et al.


Y. Qin et al. (Eds.): EITRT 2023, LNEE 1137, pp. 192–201, 2024.
https://doi.org/10.1007/978-981-99-9311-6_21

[2–4] gave a broad overview of train routing problem and train platforming problem. Lusby and Janosikova et al. [5, 6] proposed the allocation method of platform in a single station; Matthew and Mojtaba et al. [7] built a optimized model based on capability, schedule and station operation plan; Corman and D 'Ariano et al. [8] combined tabu search algorithm and branch and bound algorithm. Through the design of neighborhood structure, the adjustment method of single station stock track is proposed. Valentina et al. [3, 9] optimized the train platforming problem from two aspects of train arrival time and route operation plan. Zhang et al. [10] proposed a method combining train scheduling model, stock track model and track model, designed a collaborative adjustment algorithm, and thus obtained a train operation adjustment scheme. Based on the scheduling principle, Wang et al. [11] constructed a multi-objective programming model and transformed it into a single-objective model by the minimum-maximum method. Niu et al. [12] proposed an optimization method of share allocation in intelligent high-speed railway stations from the perspective of industrial information integration. Lu et al. [13] focuses on considering real-time track reallocation problems. Zhang et al. [14] further extended the TTP problem considering locomotive operation factors.

The aforementioned research primarily focus on train rescheduling issues. Despite the considerable attention given to track allocation optimization as a research hotspot, there is a relative scarcity of studies specifically addressing the track allocation problem based on a fixed track utilization scheme in station. And the fixed track utilization scheme affects the technical operation of track in the station, then affects the integration process of the track utilization in the whole railway network. Therefore, this paper will be innovative to solve the TTP problem, the research work is challenging.

The contributions of this paper are listed as follows:

(1) A mixed integer programming model is proposed to describe TPP. This model aims to achieve a balanced occupation of tracks while considering train operation safety.
(2) The proposed improved genetic algorithm considers the fixed track utilization scheme. And the special selection operation is also designed in the algorithm to eliminate unfeasible solutions.

## 2 Ttp Model Formulation

### 2.1 Notations

### 2.2 Objectives

The optimization objective of this study in the context of the train platforming problem is to achieve a balanced utilization of tracks, aiming to mitigate train delays and prevent premature degradation of equipment caused by excessive utilization of specific tracks at the station. This optimization approach seeks to enhance the robustness of train station operations. The objective function can be formulated as follows:

$$minZ_1 = \frac{1}{n}\sum_{j=1}^{n}\left[\sum_{i=1}^{m} x_{ij}\left(t_i^{dep} - t_i^{arr}\right) - \frac{1}{n}\sum_{i=1}^{m} x_{ij}\left(t_i^{dep} - t_i^{arr}\right)\right]$$

$$\forall i \in Tr \forall j \in G \tag{1}$$

## 2.3 Constrains

(1) The uniqueness constraint of train occupation on tracks.

Each train involved in station operations can exclusively occupy only one track at the station. Once occupied, the track cannot be changed. The constraint can be expressed as follows:

$$\sum_{j=1}^{n} x_{ij} = 1 \tag{2}$$

(2) The non-overlapping constraint on track resources.

A single track can be occupied by only one train at a time, and the operational time intervals of two trains occupying the same track should not overlap. This non-overlapping constraint can be expressed as follows:

$$\left(x_{fj}t_f^{dep} - x_{sj}t_s^{arr}\right)\left(x_{fj}t_f^{arr} - x_{sj}t_s^{dep}\right) \geq 0 \forall f, s \in Tr \tag{3}$$

(3) Safety time interval constraints.

To ensure the safety of station operations, in addition to preventing conflicts in arrival and departure times of consecutive trains on the same track, it is also necessary to satisfy the constraint of track occupation time interval. This constraint is enforced through the track occupation constraint in interlocking systems. The constraint can be expressed as follows:

$$x_{fj} + x_{sj} \leq 1 \forall f, s \in Tr \tag{4}$$

$$t_f^{dep} + T_g \geq t_s^{arr} \forall f, s \in Tr \tag{5}$$

(4) Two train tracking interval constraints in the same direction.

To ensure the safety of railway operations within a specific section, the consecutive trains arriving at or leaving the station in the same direction must satisfy the constraint of train tracking interval time. This constraint ensures a minimum time gap between the arrival or departure of two successive trains. The constraint can be expressed as follows:

$$t_{i+1}^{arr} \geq t_i^{arr} + I_z \tag{6}$$

$$t_{i+1}^{dep} \geq max\left\{t_{i+1}^{arr} + T_i, t_i^{dep} + I_z\right\} \tag{7}$$

(5) Fixed track utilization scheme constraints.

The principle states that train occupancy of tracks should adhere to the fixed track utilization scheme. $m_k$ represents the total number of trains belonging to the fixed track utilization scheme of k. The constraint can be expressed as follows:

$$\sum_{i \in Tr} \sum_{j \in K} x_{ij} = m_k \tag{8}$$

## 3 GA Algorithm for TPP

The Genetic Algorithm (GA) holds a significant position among well-established artificial intelligence algorithms and has garnered extensive adoption by researchers, to address challenges related to train platforming and track allocation. Notably, GA exhibits a robust capacity for global search. In the subsequent discussion, the primary steps comprising the GA algorithm will be succinctly outlined.:

**Step 1**: Parameter initialization and chromosome encoding.

The algorithm starts by setting initial parameters: a population size of 20, crossover rate of 0.5, mutation rate of 0.15, maximum iteration count of 5000, and $T_g$ of 2 min. The algorithm reads the train timetable and modifies the arrival or departure times to integers between 0 and 300 min.

The algorithm starts with setting initial parameters, then reads the train timetable, and converts the arrival or departure time of the train into a convenient natural number time. In the TTP problem, the relationship between train operations at stations and the tracks assigned to them can be likened to the relationship between chromosomes and their genes. Therefore, this study adopts a natural number encoding scheme for chromosomes, which includes four additional sequence information besides the genetic information. Gene sorting is performed based on the order of train arrivals and departures. Figure 1 illustrates the chromosomes of the individuals involved in this process. The initial population is generated by randomly selecting from the tracks associated with the fixed track utilization scheme. This process creates an initial feasible solution, which serves as a convenient starting point for subsequent optimization.

**Step 2**: Creation of new chromosomes and fitness evaluation.

The roulette wheel selection is performed using computed and sorted fitness values of individuals. Then, chromosome crossover is performed using a single-point crossover method, and chromosome mutation is carried out using the uniform mutation technique. It is important to ensure that the fixed track utilization scheme constraints are satisfied during the mutation process. Given the possibility of large objective values, the fitness value should be set as the reciprocal of the objective function value, denoted as $1/Z_1$, tending towards 0. Additionally, it is necessary to validate whether the track utilization scheme of a chromosome individual satisfies the model constraints. When the track allocation scheme corresponding to an individual conflicts with the model constraints and becomes infeasible, we will directly assign a negative fitness value $1/Z_1 - 1$ to it. Infeasible individuals will be continually eliminated through iterations of the algorithm. Parents will be selected for breeding, and they will undergo crossover with randomly

| X1 | X2 | X3 | X4 | X5 | X6 |
|---|---|---|---|---|---|
| $t_1^{arr}$ | $t_2^{arr}$ | $t_3^{arr}$ | $t_4^{arr}$ | $t_5^{arr}$ | $t_6^{arr}$ |
| $t_1^{dep}$ | $t_2^{dep}$ | $t_3^{dep}$ | $t_4^{dep}$ | $t_5^{dep}$ | $t_6^{dep}$ |
| *from* | *from* | *from* | *from* | *from* | *from* |
| *to* | *to* | *to* | *to* | *to* | *to* |

**Fig. 1.** Illustration of individuals' chromosome

selected chromosomes to generate new offspring. After the creation of a new generation of the population, fitness evaluation will be conducted to select the best chromosome in the current iteration. If the fitness value of the new chromosome individual is better than the best chromosome found so far, it will replace the current best chromosome. Otherwise, there is a certain probability that the mutated solution will be accepted.

**Step 3**: Termination Criteria.

When the number of iterations reaches the predefined value, the algorithm will then stop and output the best solution found so far, along with its corresponding objective value.

## 4 Numerical Experiments and Results

This section demonstrates the effectiveness of the proposed model and algorithm using Wuchang Railway Station in China as an example. The layout diagram of Wuchang Station is shown in Fig. 2. The fixed track utilization scheme is presented in Table 2. For computational convenience, the arrival-departure times of each train have been converted to integers ranging from 0 to 360. The train timetable and initial track utilization scheme are provided in Table 3.

The initial parameters are set as follows: a population size of 20, crossover rate of 0.5, mutation rate of 0.15, maximum iteration count of 5000, and $T_g$ of 2 min.

To generate the initial solution, a random selection of tracks is made from the corresponding alternative track set in the fixed track utilization scheme. The improved genetic algorithm is then employed to generate an optimized solution. Figure 3 illustrates the optimized track utilization scheme obtained using the improved genetic algorithm. In the figure, red trains represent trains moving in the upward direction, blue trains represent trains moving in the downward direction, and green trains represent trains that require diversion at the originating or terminating station.

The convergence process of the algorithm is depicted in Fig. 4, which demonstrates the effectiveness of the genetic algorithm in solving the TPP problem at Wuchang Station. Over 50,000 iterations were performed within approximately 7 min, resulting in a decrease in the objective function value from over 2000 to 613. The optimized track

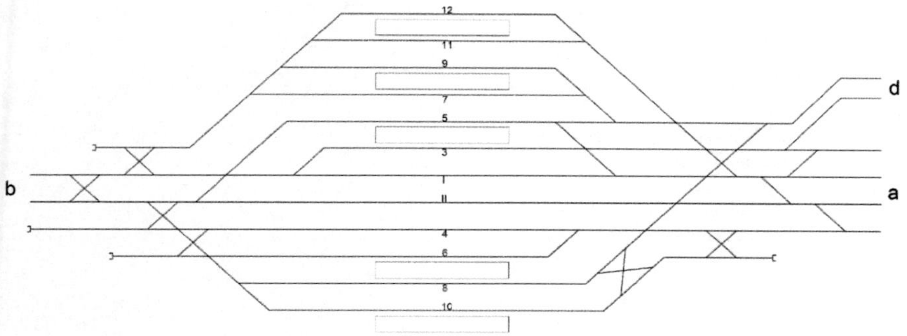


**Fig. 2.** Layout of the Wuchang station.

**Table 1.** Definitions parameters and variables

| Symbol | Description |
|---|---|
| $t_i^{arr}$ | Planned arrival time of the train *i* |
| $t_i^{dep}$ | Planned departure time of the train *i* |
| $Tr$ | Set of trains. *m* represents the number of trains occurring within a given interval |
| $G$ | Set of tracks. *n* represents the total number of tracks |
| $K$ | Set of alternative track in fixed track utilization scheme. *l* represents the total number of available tracks in a given fixed track utilization scheme |
| $T_g$ | The minimum safe operational interval time between two trains operating consecutively on the same track at the station |
| $I_z$ | The safety interval time between trains arriving or departing in the same direction at the train station |
| $x_{ij}$ | variable. = 1, If train *i* occupies track *k*, otherwise = 0 |

utilization scheme exhibits higher balance and robustness. Therefore, it can be concluded that the improved genetic algorithm designed in this study is an effective tool for addressing TPP model in complex busy railway station.

## 5 Conclusions

This paper proposes a mixed-integer programming model for the efficient utilization of tracks in busy and complex passenger stations. Additionally, a genetic algorithm based on the fixed track utilization scheme is introduced to develop a balanced track utilization plan. Firstly, the paper presents a TTP model that aims to achieve balanced track utilization while considering the constraints of the fixed track utilization scheme. Secondly, an improved TPP genetic algorithm is proposed using enhanced crossover and selection methods. A case study of Wuchang Station, it demonstrates the considerable

**Table 2.** Fixed track utilization scheme

| Up train | Down trian | Originating or terminating train |
|---|---|---|
| 3 | 3 | 3 |
| 5 | 5 | 5 |
| 6 | 7 | 6 |
| 8 | 9 | 7 |
| 10 | 11 | 8 |
| – | 13 | 9 |
| – | – | 10 |
| – | – | 11 |
| – | – | 13 |

potential of the model and algorithm in enhancing track utilization within the confines of a fixed track utilization scheme. This study emphasizes the effectiveness of the proposed model and algorithm as a rational approach for optimizing scheduling operations at passenger stations that experience high levels of complexity and activity.

In order to further advance the field, future research should focus on the development of a comprehensive model that integrates both inbound and outbound route conflicts, as well as technical operation interferences. Moreover, in the presence of disruptive factors such as track maintenance, it is imperative to pay increased attention to the adjustment of both time and track allocation to ensure the feasibility and viability of platform plans.

**Table 3.** Timetable and initial track utilization scheme

| No | arr | dep | from | to | initial track | No | arr | dep | from | to | initial track |
|---|---|---|---|---|---|---|---|---|---|---|---|
| 1 | 0 | 3 | 2 | 1 | 9 | 26 | 150 | 157 | 1 | 2 | 8 |
| 2 | 0 | 3 | 1 | 2 | 5 | 27 | 160 | 190 | 4 | 1 | 6 |
| 3 | 0 | 33 | 2 | 1 | 11 | 28 | 161 | 169 | 1 | 2 | 3 |
| 4 | 0 | 30 | 4 | 2 | 8 | 29 | 169 | 183 | 2 | 1 | 7 |
| 5 | 0 | 26 | 4 | 1 | 13 | 30 | 173 | 203 | 4 | 1 | 11 |
| 6 | 3 | 9 | 2 | 1 | 3 | 31 | 175 | 181 | 1 | 2 | 10 |
| 7 | 12 | 20 | 2 | 1 | 7 | 32 | 182 | 188 | 1 | 2 | 8 |
| 8 | 16 | 46 | 4 | 2 | 10 | 33 | 200 | 208 | 1 | 2 | 8 |
| 9 | 32 | 38 | 1 | 2 | 6 | 34 | 203 | 209 | 2 | 1 | 7 |
| 10 | 33 | 53 | 2 | 1 | 9 | 35 | 209 | 234 | 2 | 1 | 9 |
| 11 | 51 | 60 | 1 | 2 | 3 | 36 | 212 | 234 | 1 | 2 | 6 |
| 12 | 58 | 66 | 1 | 2 | 10 | 37 | 217 | 240 | 2 | 1 | 7 |
| 13 | 64 | 86 | 1 | 2 | 6 | 38 | 218 | 240 | 1 | 2 | 10 |
| 14 | 84 | 105 | 1 | 2 | 8 | 39 | 225 | 252 | 1 | 2 | 8 |
| 15 | 90 | 124 | 2 | 1 | 7 | 40 | 234 | 260 | 2 | 1 | 11 |
| 16 | 97 | 122 | 1 | 2 | 6 | 41 | 240 | 246 | 1 | 2 | 5 |
| 17 | 97 | 130 | 2 | 1 | 11 | 42 | 246 | 253 | 2 | 1 | 9 |
| 18 | 103 | 107 | 2 | 1 | 3 | 43 | 246 | 264 | 1 | 2 | 6 |
| 19 | 109 | 118 | 2 | 1 | 9 | 44 | 257 | 282 | 1 | 2 | 10 |
| 20 | 126 | 132 | 1 | 2 | 10 | 45 | 263 | 271 | 1 | 2 | 3 |
| 21 | 126 | 156 | 2 | 1 | 3 | 46 | 270 | 288 | 1 | 2 | 5 |
| 22 | 134 | 164 | 2 | 1 | 7 | 47 | 276 | 296 | 1 | 2 | 8 |
| 23 | 135 | 148 | 1 | 2 | 8 | 48 | 277 | 295 | 2 | 1 | 9 |
| 24 | 143 | 163 | 2 | 4 | 5 | 49 | 286 | 300 | 1 | 2 | 3 |
| 25 | 145 | 175 | 4 | 1 | 13 | 50 | 293 | 300 | 1 | 2 | 6 |
| - | - | - | - | - | - | 51 | 295 | 300 | 2 | 1 | 7 |

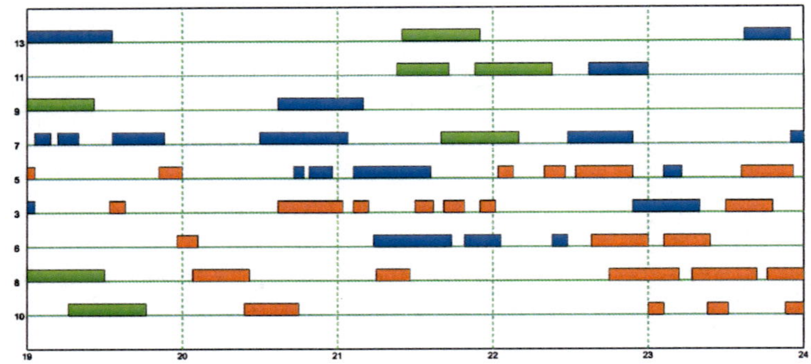


**Fig. 3.** Optimized track utilization scheme

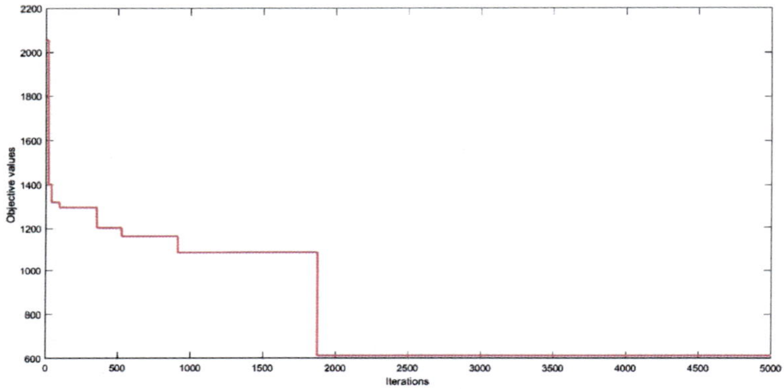


**Fig. 4.** Convergence process of GA

**Acknowledgements.** The research is supported by the National Natural Science Foundation of China (Grant No. 71971220), the Natural Science Foundation of Hunan Province, China (Grant No. 2020JJ5115, No. 2023JJ30710, No. 2022JJ31020), the Postgraduate Scicntific Research Innovation Project of Hunan Province, China (Grant No. 2023ZZTS0189).

## References

1. Zwaneveld, P.J., et al.: Routing trains through railway stations: model formulation and algorithms. Transp. Sci. **30**(3), 181–194 (1996)
2. Sels, P., Vansteenwegen, P., Dewilde, T., Cattrysse, D., Waquet, B., Joubert, A.: The train platforming problem: the infrastructure management company perspective. Transp. Res. Part B: Methodol.. **61**, 55–72 (2014)
3. Cacchiani, V., Galli, L., Toth, P.: A tutorial on non-periodic train timetabling and platforming problems. Euro J. Transp. Logist. **4**(3), 285–320 (2015)
4. Zhang, Y., Umar, A.M., An, M.: Overview of optimization models and algorithms for train platforming problem. In: Proceedings of the 4th International Conference on Electrical and

Information Technologies for Rail Transportation (EITRT) 2019: Rail Transportation System Safety and Maintenance Technologies, pp. 707–716 (2020)
5. Lusby, R.M., Larsen, J., Ehrgott, M., Ryan, D.: Railway track allocation: models and methods. OR Spectrum **33**(4), 843–883 (2011)
6. Janosikova, L., Kavicka, A., Bazant, M.: Optimal operation scheduling and platform track assignment in a passenger railway station. Journal of Rail and Rapid Transit. **228**(3), 271–284 (2014)
7. Petering, M.E., Heydar, M., Bergmann, D.R.: Mixed-integer programming for railway capacity analysis and cyclic, combined train timetabling and platforming. Transp. Sci. **50**(3), 892–909 (2016)
8. Corman, F., D'Ariano, A., Pacciarelli, D., Pranzo, M.: A tabu search algorithm for rerouting trains during rail operations. Transp. Res. Part B: Methodol. **44**(1), 175–192 (2010)
9. Cacchiani, V., Furini, F., Kidd, M.P.: Approaches to a real-world Train Timetabling Problem in a railway node. Omega **58**, 97–110 (2016)
10. Zhang, Y., Chen, Z., An, M., Umar, A.M.: An integration of train timetabling, platforming and routing-based cooperative adjustment methodology for dealing with train delay. Int. J. Software Eng. Knowl. Eng. **30**(07), 901–919 (2020)
11. Wang, H., Yue, Y.: Optimizing train-platforming plans at arrival-departure tracks of high-speed railway stations. In: 2020 IEEE 5th International Conference on Intelligent Transportation Engineering (ICITE), pp. 461–466 (2020)
12. Niu, F., Chen, Y.: Industrial information integration in track allocation optimization in high-speed train stations. J. Ind. Inf. Integr. **21**, 100193 (2014)
13. Lu, G., Ning, J., Liu, X., Nie, Y.M.: Train platforming and rescheduling with flexible interlocking mechanisms: An aggregate approach. Transp. Res. Part E: Logist. Transp. Rev. **159**, 102622 (2022)
14. Zhang, B., Zhang, Y., D'Ariano, A., Bosi, T., Lu, G., Peng, Q.: Optimal platforming, routing, and scheduling of trains and locomotives in a rail passenger station yard. Transp. Res. Part C: Emerg. Technol. **152**, 104160 (2023)

# Service Quality Evaluation for High-Speed Rail Express Transportation with EAHP

Zhongjie Sun[1], Li Wang[1,2](✉), Xiaoning Zhu[1,2], and Meiyan Chi[1]

[1] School of Traffic and Transportation, Beijing Jiaotong University, Beijing 100044, China
{22120788,liwang,xnzhu,21114066}@bjtu.edu.cn

[2] Key Laboratory of Transport Industry of Comprehensive Transportation Theory (Beijing Jiaotong University), Ministry of Transport, Beijing 100044, China

**Abstract.** To objectively evaluate the service quality of high-speed railway (HSR) express transportation and mitigate the influence of subjective factors such as evaluators' perception and preferences, this study constructs an evaluation model for HSR express transportation service quality. The model focuses on four aspects of the operation process: reservation and arrival processing, transportation, distribution, and after-sales service. Firstly, the Extenics Analytic Hierarchy Process (EAHP) is employed to determine the weights of each evaluation index in both the criterion layer and indicator layer of the evaluation system. This approach establishes the evaluation model and utilizes it to assess service quality. The results demonstrate that applying the EAHP method to evaluate the service quality of HSR express transportation effectively captures the fuzzy nature of subjective human evaluation. Furthermore, the combination of quantitative and qualitative analysis compensates for the limitations of other evaluation methods. The findings of this study provide guidance for HSR express transportation providers to identify and improve weak links in the service process. By enhancing service levels in a targeted manner, these providers can establish a stronger presence in the express market.

**Keywords:** HSR express transportation · Service quality evaluation model · EAHP

## 1 Introduction

China has established a comprehensive HSR express transportation network, covering a wide range of regions and cities. In 2013, China Railway Express Co., Ltd. Introduced HSR express transportation to meet the diverse and customized demands of society and cater to different customers' cargo transportation needs. This service has evolved into a large-scale network operation, providing convenient and fast HSR express transportation and terminal distribution services.

In 2022, the national express business witnessed a total of 110.58 billion pieces, showing a 2.1% year-on-year increase. The business revenue amounted to 1.06 trillion yuan, reflecting a 2.3% year-on-year growth, highlighting the substantial market size.


Y. Qin et al. (Eds.): EITRT 2023, LNEE 1137, pp. 202–210, 2024.
https://doi.org/10.1007/978-981-99-9311-6_22

When compared to road and air transportation [1], HSR express transportation possesses distinct advantages for transporting parcels within the 500–1000 km range, indicating a highly promising market prospect. To enhance the competitiveness of HSR express transportation in this expansive market, this study focuses on improving the service quality. By thoroughly examining the business processes of the express industry and the current organization and operation of HSR express transportation, an evaluation system is constructed. This system employs scientific evaluation methods to enhance the quality of HSR express services.

Various methods study for evaluating transportation service quality, such as hybrid Technique for Order Preference by Similarity to an Ideal Solution (TOPSIS), object-element extended model, gray comprehensive evaluation, triangular fuzzy number and fuzzy evaluation. Yang et al. [2] established an evaluation system and applied the hybrid TOPSIS model to assess the quality of railroad freight transport services by analyzing influential factors. Zhang et al. [3] proposed a service quality evaluation method based on triangular fuzzy numbers to determine weight and fuzzy evaluation matrices. Tang et al. [4] developed an object-element extended model to address evaluation indicator incompatibility under limited information conditions for railroad freight quality evaluation. Zhou et al. [5] employed the triangular fuzzy Analytic Hierarchy Process (AHP) method to determine index weights and used the gray comprehensive evaluation method to analyze and evaluate railroad freight service quality. Wei et al. [6] utilized the TOPSIS method to study the evaluation model for high-speed railway passenger service quality. However, most of these methods rely solely on numerical judgment matrices, failing to capture the subjective and fuzzy nature of evaluators' judgments and lacking objectivity in the results obtained.

The EAHP [7] is a decision analysis method that combines Extenics and AHP. It is used to tackle problems characterized by high levels of uncertainty, complexity, and vagueness. It has been applied and studied in various domains, including engineering management, economics and finance, and urban planning. In the EAHP [8], interval numbers are used instead of precise numerical values when establishing judgment matrices between two indicators. This approach reflects the fuzziness in human judgments regarding such problems while ensuring the completeness of information. The assessment of the quality of HSR express services involves an uncertain and fuzzy concept that is difficult to define using a unified standard. Therefore, it is suitable to apply the Extenics evaluation method based on the AHP.

## 2 Evaluation System

### 2.1 Business Process of HSR Express

Through the understanding and analysis of the express business, the business process of HSR express mainly has four parts: reservation and processing at the station, transportation, distribution, and after-sales, according to each part of the service content that affects the customer's feeling of service, the following specific process analysis chart was made as Fig. 1.

The process of HSR express involves customers making appointments for express services through online platforms or express outlets. Couriers collect the parcels from

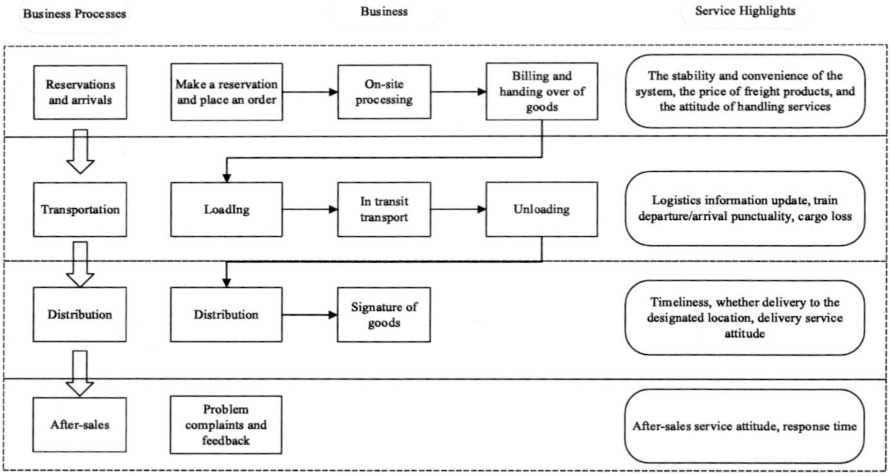


**Fig. 1.** HSR express service process and key points

customers' homes and deliver them to the HSR express station in the area. The parcels are then transported to the warehouse city by the station. In the warehouse city, the goods are packed, sealed with express bills, sorted, and packed into the collector at the HSR express base. The collector is sealed with the express face sheet and sorted before being loaded onto the HSR freight train.

The HSR freight train stopping at each distribution city along the way. The parcels are then handed over to the distribution, in collaboration with express enterprises. Finally, customers can utilize various channels such as online platforms or telephone to access related after-sales services.

### 2.2 Establishing Evaluation System

Based on the analysis of the HSR express transportation process and input from relevant personnel and experts in the field, the factors influencing customer service satisfaction can be summarized as follows: arrival service index, transportation service index, distribution service index, and after-sales service index. To ensure consistency with the evaluation index and the feasibility of the evaluation method, specific evaluation factors are identified under each major index. The overall evaluation system consists of a total of 11 evaluation factors across the four major evaluation indexes. For a detailed breakdown of the index system, please refer to Table 1.

## 3 EAHP Model

### 3.1 Build-up Judgment Matrix

In EAHP, use the extended interval number to quantitatively express the degree of pros and cons between elements of different levels, and the extended comparison matrix A is constructed by it. $A = \left[a_{ij}\right]_{nk*nk}$, $(i, j = 1, 2, \ldots, n_k)$, $a_{ij} =< a_{ij}^{-}, a_{ij}^{+} >$ is an extended

**Table 1.** HSR express service quality evaluation index system

| Target | Evaluation Indicators | Evaluation factors |
|---|---|---|
| HSR express service quality evaluation index system A | Appointment processing and arrival service indicators B1 | Stability and convenience of reservation system C1 |
| | | Acceptability of product price C2 |
| | | Service attitude of processing personnel C3 |
| | Transportation Service Indicators B2 | Real-time update of logistics information C4 |
| | | Punctuality of cargo departure and arrival C5 |
| | | Cargo loss rate C6 |
| | Distribution Service Indicators B3 | Distribution timeliness C7 |
| | | Whether the package is delivered to the specified location C8 |
| | | Service attitude of distribution staff C9 |
| | After-sales service index B4 | Service attitude of after-sales staff C10 |
| | | Timeliness of complaint acceptance C11 |

interval number, denotes the maximum and minimum evaluation value that meets the importance among the judging experts, while satisfying $a_{ii} = 1, \quad a_{ij} = 1/a_{ji}$.

### 3.2 Calculate the Eigenvectors

For the above Kth layer, the extended judgment matrix A is expressed in the form of a left-right matrix, i.e., A = A−, $A+$, the maximum eigenvalues λ− and λ+ of the left-right matrix $A-$ and $A^+$ are calculated respectively, then we have λ = [λ−, λ+] as the interval number eigenvalues of A, and the corresponding eigenvectors are x− and x+, then the weight coefficients k and m are calculated according to $A^- = [a_{ij}^-]_{nk*nk}$, $A^+ = [a_{ij}^+]_{nk*nk}$ and Eq. (1).

$$k = \sqrt{\sum_{j=1}^{n_k} \frac{1}{\sum_{i=1}^{n_k} a_{ij}^+}}, m = \sqrt{\sum_{j=1}^{n_k} \frac{1}{\sum_{i=1}^{n_k} a_{ij}^-}} \# (1) \quad (1)$$

When the derived k and m satisfy 0 ≤ k ≤ 1 ≤ m, it means that the extended judgment matrix has good consistency [9].

### 3.3 Determine the Hierarchical Single Sort Weights

For each element of Kth level, there is a weight expressed as the number of intervals $S_i^K =< \text{kx}^-, mx^+ >=< S_i^{K-}, S_i^{K+} > (i = 1, 2, ..., n_K)$.

$$V\left(S_i^K \geq S_j^K\right) = \frac{2\left(S_i^{K+} - S_j^{K-}\right)}{\left(S_i^{K+} - S_i^{K-}\right) + \left(S_j^{K+} - S_j^{K-}\right)} \tag{2}$$

For $\forall i = 1, 2, ..., n_K; i \neq j, V(S_i^K \geq S_j^K) \geq 0$, the following is taken $P_j^K = 1, P_i^K = V(S_i^K \geq S_j^K), i = 1, 2, ..., n_K; i \neq j$, then the single sort weight of the element of the Kth level with respect to an element of the K-1th level is $(P_1^K, P_2^K, ...., P_{n_K}^K)^T$.

### 3.4 Calculate the Combination Weights for the Objectives

If the combined weight ordering of the $n_{K-1}$ elements of the K-1 layer $B_j(j = 1, 2, \ldots, n_{k-1})$ with respect to the solution objective has been found as:

$$\omega^{K-1} = (\omega_1^{K-1}, \omega_2^{K-1}, ..., \omega_{n_{K-1}}^{K-1})^T$$

The $n_K$ element of the Kth layer, for the $B_j(j = 1, 2, ..., n_{K-1})$ element of the K-1th layer the hierarchical single-sort weight vector is $P_i^K = (P_{1j}^K, P_{2j}^K, ..., P_{n_K j}^K)^T (i = 1, 2, ..., n_K)$. The matrix of order $n_k * n_{k-1}$ is obtained.

$$P^k = \left(P_1^k, P_2^k, \cdots, P_{n_{k-1}}^k\right)^T = \begin{bmatrix} P_{11}^k \ P_{12}^k & \cdots P_{1n_{k-1}}^k \\ P_{21}^k \ P_{22}^k & \cdots P_{2n_{k-1}}^k \\ \vdots \quad \vdots & \vdots \quad \vdots \\ P_{n_k 1}^k \ P_{n_k 2}^k & \cdots P_{n_k n_{k-1}}^k \end{bmatrix} \tag{3}$$

K-1 layer weight vector is $\omega^{K-1} = (\omega_1^{K-1}, \omega_2^{K-1}, ..., \omega_{n_{K-1}}^{K-1})^T$, then the combined weight of the first evaluation indicator at level K for the total objective is:

$$w^K = P^k W^{k-1} = \left(w_1^k, w_2^k, \cdots, w_{n_k}^k\right)^T \tag{4}$$

## 4 HSR Express Service Quality Evaluation

### 4.1 Constructing the Criterion Layer Judgment Matrix

In EAHP, invite experts to compare the elements of the criterion layer and score according to the 1–9 scale method. The results are as follows.

$$A = \begin{bmatrix} \langle 1, 1 \rangle & \langle \frac{1}{3}, \frac{1}{2} \rangle & \langle 1, 2 \rangle & \langle 2, 5 \rangle \\ \langle 2, 3 \rangle & \langle 1, 1 \rangle & \langle 2, 5 \rangle & \langle 4, 7 \rangle \\ \langle \frac{1}{2}, 1 \rangle & \langle \frac{1}{5}, \frac{1}{2} \rangle & \langle 1, 1 \rangle & \langle 3, 4 \rangle \\ \langle \frac{1}{5}, \frac{1}{2} \rangle & \langle \frac{1}{7}, \frac{1}{4} \rangle & \langle \frac{1}{4}, \frac{1}{3} \rangle & \langle 1, 1 \rangle \end{bmatrix}$$

Calculate the largest eigenvalue of $A^-$ and $A^+$ using python: $\lambda^-$ and $\lambda^+$, the corresponding normalized eigenvectors are $x^-$ and $x^+$,

$$\lambda^- = 3.2488, \lambda^+ = 5.2097$$

$$x^- = (0.2268, 0.5055, 0.1939, 0.0738)^T, x^+ = (0.2345, 0.5078, 0.1851, 0.0727)^T$$

The values of k and m are calculated from Eq. (1) as k = 0.8973, m = 1.0964.

Satisfy $0 \le k \le 1 \le m$, the extended matrix has good consistency and passes the consistency test. Then number vector $S^k$ of the extended interval of the criterion layer in the evaluation index system is calculated, and the relative importance is solved as follows:

$$\begin{aligned} S^k &= (s_1, s_2, s_3, s_4) = < kx_i^-, mx_i^+ > (i = 1, 2, 3, 4) \\ &= (< 0.2035, 0.2571 >, < 0.4536, 0.5567 >, \\ &< 0.1740, 0.2029 >, < 0.0662, 0.0797 >) \end{aligned}$$

According to Eq. (2), the results of $V(S_i^K \ge S_j^K)$ are as follows:

$$V\left(s_1^k \ge s_4^k\right) = 5.69, V\left(s_2^k \ge s_4^k\right) = 8.4134, V\left(s_3^k \ge s_4^k\right) = 6.4481, V\left(s_4^k \ge s_4^k\right) = 1$$

The above results were normalized to obtain the weight vector for the single ranking of the criterion layers: $w = (0.2640, 0.3904, 0.2992, 0.0464)$.

## 4.2 Constructing the Indicator Layer Judgment Matrix

Establishing an extension interval number judgment matrix through expert scoring. Among them, the stability and convenience of the reservation system $C_1$, the acceptability of the product price $C_2$ and the service attitude of the processing personnel $C_3$ under the level of reservation processing and arrival service index $B_1$ have a extended interval number judgment matrix of $D_1$, and similarly, we establish $D_2$, $D_3$ and $D_4$, and the specific results are as follows:

$$D_1 = \begin{bmatrix} <1,1> & <\frac{1}{5},\frac{1}{2}> & <\frac{1}{2},1> \\ <2,5> & <1,1> & <2,4> \\ <1,2> & <\frac{1}{4},\frac{1}{2}> & <1,1> \end{bmatrix}, D_2 = \begin{bmatrix} <1,1> & <\frac{1}{3},\frac{1}{2}> & <\frac{1}{5},\frac{1}{2}> \\ <2,3> & <1,1> & <\frac{1}{3},\frac{1}{2}> \\ <2,5> & <2,3> & <1,1> \end{bmatrix}$$

$$D_3 = \begin{bmatrix} <1,1> & <2,4> & <3,5> \\ <\frac{1}{4},\frac{1}{2}> & <1,1> & <1,3> \\ <\frac{1}{5},\frac{1}{3}> & <\frac{1}{3},1> & <1,1> \end{bmatrix}, D_4 = \begin{bmatrix} <1,1> & <\frac{1}{3},1> \\ <1,3> & <1,1> \end{bmatrix}$$

The calculation process is the same. The final summary of the index weights of the HSR express service quality evaluation index layer is listed in Table 2.

Both k and m of extended intervals judgment matrix of the index layer to the criterion layer satisfy $0 \le k \le 1 \le m$, so they meet the criteria of the consistency test [10].

**Table 2.** Calculation results of relevant parameters

| | $\lambda^-$ | $\lambda^+$ | $x^-$ | $x^+$ | k | m |
|---|---|---|---|---|---|---|
| $C_1$ | 2.3692 | 3.9501 | $(0.1722, 0.5936, 0.2342)^T$ | $(0.1771, 0.6010, 0.2219)^T$ | 0.8898 | 1.1070 |
| $C_2$ | 2.5473 | 3.7545 | $(0.1373, 0.2989, 0.5638)^T$ | $(0.1536, 0.2716, 0.5747)^T$ | 0.9129 | 1.0734 |
| $C_3$ | 2.3800 | 3.9751 | $(0.6347, 0.2202, 0.1452)^T$ | $(0.5953, 0.2530, 0.1517)^T$ | 0.9073 | 1.0907 |
| $C_4$ | 1.5774 | 2.7321 | $(0.3660, 0.6340)^T$ | $(0.3660, 0.6340)^T$ | 0.8660 | 1.1180 |

## 4.3 Final Ranking Weights

Summarize the calculation results to get the comprehensive weight w = (0.0294, 0.1669, 0.0677, 0.0286, 0.1634, 0.1984, 0.2175. 0.0591, 0.0225, 0.0113, 0.0352), which are summarized as listed in Table 3.

**Table 3.** Service quality evaluation index weights

| Target Level | Criterion Level | Weight | Indicator Level | Weight | Combined Weight |
|---|---|---|---|---|---|
| A | B1 | 0.2640 | C1 | 0.1112 | 0.0294 |
| | | | C2 | 0.6324 | 0.1669 |
| | | | C3 | 0.2564 | 0.0677 |
| | B2 | 0.3904 | C4 | 0.0733 | 0.0286 |
| | | | C5 | 0.4186 | 0.1634 |
| | | | C6 | 0.5081 | 0.1984 |
| | B3 | 0.2992 | C7 | 0.7271 | 0.2175 |
| | | | C8 | 0.1976 | 0.0591 |
| | | | C9 | 0.0753 | 0.0225 |
| | B4 | 0.0464 | C10 | 0.2433 | 0.0113 |
| | | | C11 | 0.7567 | 0.0352 |
| Total | | 1.0000 | | 1.0000 | 1.0000 |

## 4.4 Further Analysis

By analyzing Table 3, it can be observed that the evaluation indexes C7, C6, C5, C2 carry significant weight within the evaluation system. This indicates that customers prioritize timeliness, price, and safety when selecting HSR express services. To establish a strong presence in the competitive express market, HSR Express should focus on enhancing service quality in these areas, as they are highly valued by customers.

In the actual application of the evaluation model, for different indicators, the evaluation indexes should be dimensionless at the end of the process, through mathematical transformation to eliminate the influence of the initial dimension of the indicators, so that the results obtained are more accurate.

## 5 Conclusion

In the background of China's HSR into a network operation, the capacity has been greatly improved, the implementation of HSR express service quality evaluation, to improve the quality of HSR logistics, to meet the needs of customers in the long-distance logistics services has a very important practical significance. This paper constructs a service quality evaluation model for the transportation logistics of HSR cargo based on EAHP and proves that the model has good reliability through empirical analysis, which provides a basis for decision-making on the direction of improving the service quality of HSR express transportation. For the healthy development of HSR express transport and HSR for the regional consumer service industry drive [8] has very important significance.

In the future, the research on the evaluation of the service quality of HSR express transportation should consider the service characteristics of different cargo categories and different HSR express transportation centers, optimize the evaluation index system, make it tend to be comprehensive and specific, and make it more suitable for the real demand.

**Acknowledgements.** This research was partially funded by the Joint Funds of the National Natural Science Foundation of China (No. U2034208).

## References

1. Cai, D., Xiao, Y., Jiang, C.: Competition between high-speed rail and airlines: considering both passenger and cargo. Transp. Policy **110**, 379–393 (2021)
2. Yang, K., Deng, C., Lin, Y.: Evaluation on railway freight services quality based on TOPSIS method. Technol. Econ. Areas Commun. **12**(6), 61–64 (2010). (in Chinese)
3. Zhang, Y., Yang, Y., Wang, J.: Study on service quality of high-speed railway express business based on triangular fuzzy number and fuzzy evaluation. Logist. Technol. **36**(5), 72–77 (2017). (in Chinese)
4. Tang, L., Zhou, L., Xu, Y., Xu, Z., Tang, H.: Research on railway logistics service quality evaluation based on object extension model. Railw. Freight Transp. **38**(12), 20–25 (2020). (in Chinese)
5. Zhou, L., Tang, L., Xu, Z., Tang, H.: Discussion on evaluation of railway freight service quality based on grey comprehensive evaluation method. Railw. Freight Transp. **39**(12), 47–53 (2021). (in Chinese)
6. Wei, B., Gong, P.: Study on comprehensive evaluation model of high-speed railway passenger service Quality. Microcomput. Appl. **36**(1), 52–54 (2020). (in Chinese)
7. Luo, D., Sun, S., Zhang, D., Wan, Y., Zhang, G., Niu, J.: Application of improved EAHP on stability evaluation of coal seam roof. Procedia Earth Planet. Sci. **3**, 384–393 (2011)

8. He, Y., Lou, R., Wang, Y., Wang, J., Fang, X.: A dual attribute weighted decision fusion system for fault classification based on an extended analytic hierarchy process. Eng. Appl. Artif. Intell. **114**, 105066 (2022)
9. Cai, W., Yang, C.: Basic theory and methodology on Extenics. Chin. Sci. Bull. **58**(13), 1190–1199 (2013). (in Chinese)
10. Jin, L., Liu, F., Ma, L., Zhang, B.: Construction and application of course teaching quality evaluation model based on EAHP. J. Hefei Norm. Univ. **39**(6), 94–98 (2021). (in Chinese)

# A Model Predictive Control Approach for Virtual-Coupling Train Operations Using Mixed-Integer Linear Programming

Junjie Wang, Bolun Zhang, and Shaofeng Lu(✉)

South Chine University of Technology, Guangzhou 511442, China
{202120160156,201910108814,lushaofeng}@scut.ecu.cn

**Abstract.** Virtual-coupling (VC) train operation is seen as an emerging technology that can effectively increase the capacity of railway transportation. Based on advanced train-to-train communication, VC allows trains to travel minimal distances without physical connections. This paper proposes a model predictive control approach based on Mixed-Integer Linear Programming (MILP) for virtually coupled trains, the nonlinear characteristics, and space-depending variables such as route gradients can be modeled based on time in the MILP model, leading to more flexibility in modeling for complex engineering scenarios. Result shows the model predictive control approach based on MILP needs about 18% solving time compared with Nonlinear Model Predictive Control with the equivalent prediction horizon. Our simulation results also demonstrate the proposed method can control the leader train to follow the reference speed and the follower train to keep an acceptable interval during the journey, which means the proposed method is effective for virtual coupled trains.

**Keywords:** Virtual Coupling · Model Predictive Control · Mixed-Integer Linear Programming

## 1 Introduction

The capacity of railway lines needs further improvement due to the demand for passenger and freight transportation. In China, passenger and freight traffic on the railways has shown a year-on-year increase. Although the capacity problem can be solved by expanding new lines based on current networks, it is usually expensive even unfeasible in available space-limited areas. Reducing the spacing or time interval between trains can also solve the capacity problem while ensuring traffic safety, which has been a common interest for the railway industries now.

Virtual coupling (VC) is considered as a more advanced method in the train control system to further reduce the spacing between trains upon moving blocks which uses absolute braking distance to calculate safety distance between trains, as shown in Fig. 1. In the conception of VC, trains can travel as a virtually coupled train set (VCTS) with relative braking distance or decouple from a VCTS autonomously. The realization of VC


Y. Qin et al. (Eds.): EITRT 2023, LNEE 1137, pp. 211–220, 2024.
https://doi.org/10.1007/978-981-99-9311-6_23

requires the support of train-to-train (T2T) communication technology [1] for exchanging train state information. European research initiative Shift2Rail has indicated that for VCTS, a cooperative train control approach is needed for trains under VC can couple as a stable formation [2].

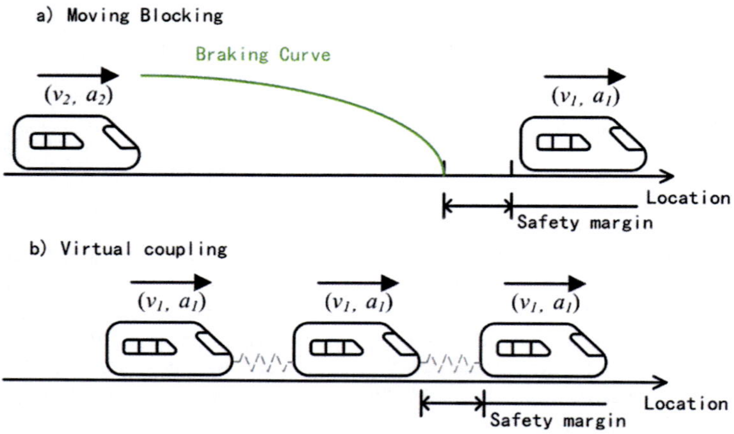


**Fig. 1.** Schematic diagram of train control systems based on moving block signaling and virtual coupling.

In the past 20 years, many train control methods focusing on VC have been proposed. Based on [3], a train-following model including multiple operation modes and train dynamics was proposed [4] and the mathematical representation of the dynamic safety margin was added in [5]. The train control method-based train following the model is rule-based; parameters in the model must be accurate enough to ensure the performance of the control method.

Feedback control has been applied in many studies concentrating on cooperative train control. Di Meo et al. [6] proposed a feedback control strategy for VCTS which considers communication time delay. Park et al. [7] proposed a robust gap controller based on sliding model control and developed a gap reference algorithm and a position error correction scheme. However, feedback control is hard to deal with control problems with constraints, such as control input limited by traction and braking characteristics.

Among the optimal control methods, model predictive control (MPC) has been widely focused on its ability to cope with hard constraints on controls and states [8]. Felez et al. [9] presented a decentralized model predictive control framework for the leader train and follower train in VCTS. Xun et al. [10] developed a multi-train cooperative control method based on self-trigger MPC to adjust headway dynamically in high-speed railways. Liu et al. [11] proposed a distributed MPC approach for VCTS in high-speed railways with consideration of individual and coupled constraints. Wu et al. [12] proposed a VCTS control method based on MPC and a coasting control strategy to improve the MPC controller.

Inspired by [9], to model nonlinear characteristics and space-depending variables flexibility and guarantee the optimality of solution [13], we propose an MPC method for virtually coupled trains in VCTS based on MILP (referred to as MMPC) in this paper. Simulation results demonstrate its effectiveness and show that it needs less solving time compared with Nonlinear MPC based on train longitudinal dynamics (referred to as NMPC).

The rest of this paper is organized as follows. Section 2 describes the train dynamics model, and Sect. 3 proposed the MPC controller based on MILP. Section 4 provides numerical experiments to illustrate the proposed method's performance. Finally, we conclude this paper in Sect. 5.

## 2 Train Dynamics Model

The train dynamics model used in this paper is simplified as a single-particle longitudinal dynamics model. The model considers the aerodynamic drag, the gravitational forces, the rolling resistance, and the traction/braking system. The dynamics model for a train can be represented as:

$$\dot{x}_i(t) = v_i(t) \tag{1a}$$

$$\dot{v}_i(t) = \frac{u_i(t) - f_i^{drag}(t) - f_i^{grad}(t)}{M} \tag{1b}$$

where $i$ is the index number of trains, $M$ is the mass of the train; $x_i(t)$ and $v_i(t)$ are the position and speed of train $i at time $t$; when $u_i(t)$ is a non-negative value it denotes the traction effort at time $t$, otherwise it denotes the opposite number of braking force; $f_i^{drag}(t)$ is the drag caused by the aerodynamic and rolling track at time $t$; $f_i^{grad}(t)$ represent the component of gravitational forces along the track at time $t$. The values of $f_i^{drag}$ and $f_i^{grad}$ can be calculated as follows:

$$f_{drag,i} = A + Bv_i(t) + Cv_i(t)^2 \tag{2}$$

$$f_{grad,i} = Mg\sin\theta(t) \tag{3}$$

where $A$, $B$, and $C$ are Davies coefficients which depend on the tracking line and type of trains; $g$ is the gravitational constant; $\theta(t)$ is the track gradient as a function of $t$.

In the practical system, there is a time lag from the controller sending the command to the command to be executed, so the traction/braking system cannot produce the corresponding traction effort or braking force immediately. The cause of the lag of the system. Therefore, a first-order damp element be introduced into the dynamics model [9] [12]:

$$\dot{u}_i(t) = \frac{1}{\tau}\left(u_i^{des}(t) - u_i(t)\right) \tag{4}$$

## 3 Model Predictive Controller Designed

Considering a VCTS consist of $n$ trains, each train runs at speed $v_i$ and has an interval distance $\Delta d_i$ after the preceding train. The VCTS can be regarded as a train set having a leader train and $n-1$ follower trains, so the control problems for VCTS can be divided into two categories, one for the leader train and the other for follower trains.

As shown in Fig. 2., we assume that each train in VCTS has a respective MPC controller that can have a Train to Train(T2T) communication with other trains and accept the speed trajectory predicted by MPC controller. The MPC controller in this paper is time-based. The Euler method is used to discretize the longitudinal train dynamics model with $\Delta t$, a constant representing the sampling time.

$$x_i(j+1) = x_i(j) + v_i(j)\Delta t \tag{5a}$$

$$v_i(j+1) = v_i(j) + \frac{u_i(j) - f_i^{drag}(j) - f_i^{grad}(j)}{M}\Delta t \tag{5b}$$

$$u_i(j+1) = u_i(j) + \frac{1}{\tau_i}\left(u_i^{des}(j) - u_i(j)\right)\Delta t \tag{5c}$$

where $*(j)$ means the corresponding state at time $j$.

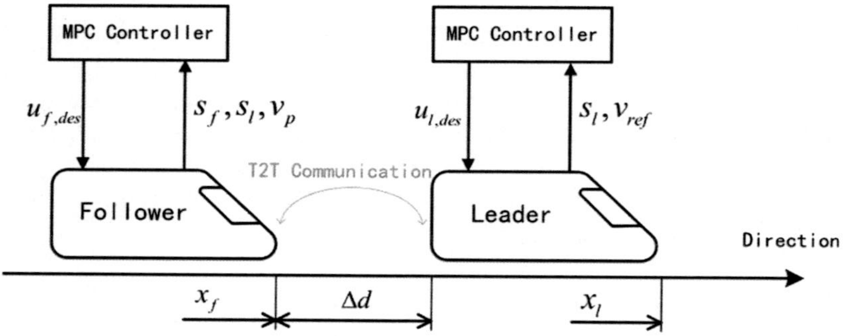


**Fig. 2.** Control architecture of MPC: $s = [x, v, u]$ represent the status of train.

To ensure the comfort of passengers during the journey, the acceleration of the train should be limited, and the speed of the train should be limited by the allowed maximum speed:

$$a_{min} \leq \frac{v_i(j+1|k) - v_i(j|k)}{\Delta t} \leq a_{max} \tag{6a}$$

$$v_i(j|k) \leq v_{max} \tag{6b}$$

where $*(j|k)$ represent the predicted state after $j$ sampling times at time $k$, $v_{max}$ is the maximum speed of the train.

Due to the traction characteristics of the traction motor, the motor power and the traction effort provided by the traction/braking system should be constrained. The traction/braking force in each segment can be calculated based on the law of conservation of energy:

$$\frac{1}{2}M\left(v_i^2(j+1) - v_i^2(j)\right) + W_i(j) = P_i(j)\Delta t \tag{7a}$$

$$W_i(j) = Mg\Delta h(j) + \left(Av_i(j) + Bv_i^2(j) + Cv_i^3(j)\right)\Delta t \tag{7b}$$

where $\Delta h(j)$ is the altitude difference in the $j^{th}$ sampling interval; $W_i(j)$ is the sum of the work done by gravity and the basic resistance; $P_i(j)$ is the traction effort/braking power of the motor in sampling time, which should be constrained as:

$$P_{min} \leq P_i(k) \leq P_{max} \tag{8a}$$

$$F_{min} \leq u_i(j|k), u_i^{des}(j|k) \leq F_{max} \tag{8b}$$

where $P_{min}$ and $P_{max}$ are the minimum and maximum motor power, $F_{min}$ and $F_{max}$ are the minimum and maximum traction/braking force, respectively.

In this time-based controller, track gradient θ, and altitude variation along the line in each sampling interval $\Delta h(j)$ is space-based. There are nonlinear terms such as $v_i^2(j)$ and $v_i^3(j)$ in Eq. (7a) and (7b). To calculate the space-based variables and linearize nonlinear terms, a series of SOS2 variables $\alpha_j$ with length $L$ and $\beta_j$ with length $L\prime$ are introduced to use the Piecewise-linear (PWL) approach. SOS2 variables is an ordered variable set in which two adjacent variables can be nonzero at most. We set the sum of $\alpha_j$ and $\beta_j$ are set as 1 to apply the PWL approach:

$$\sum\nolimits_{l=1}^{L}\alpha_{j,k} = 1, \sum\nolimits_{l=1}^{L'}\beta_{j,k} = 1 \tag{9}$$

Arrays $V_{PWL}$, $X_{PWL}$, $Y_{PWL}$ and $\theta_{PWL}$ are defined as PWL points for speed, location, altitude, and track gradient respectively. All of them are incremental. The number of elements in the set $V_{PWL}$ is $L$ and the others are $L\prime$. $v_i(j|k)$ Can be express using $V_{PWL}$ and $\alpha_j$ as:

$$v_i(j|k) = \sum\nolimits_{l=1}^{L} V_{PWL,l}\alpha_{j,l} \tag{10}$$

For the nonlinear terms $v_i^2(j)$ and $v_i^3(j|k)$ which are dependent on $v_i(j|k)$ can be approximated by:

$$v_i^2(j|k) = \sum\nolimits_{l=1}^{L} V_{PWL,l}^2\alpha_{j,l}, v_i^3(j|k) = \sum\nolimits_{l=1}^{L} V_{PWL,l}^3\alpha_{j,l} \tag{11}$$

This method also can apply to space-based variables. $x_i(j|k)$ Can be express using $X_{PWL}$ and $\beta_j$ as:

$$x_i(j|k) = \sum\nolimits_{l=1}^{L'} X_{PWL,l}\beta_{j,l}. \tag{12}$$

And then, $\Delta h(j|k)$ and $\theta(j|k)$ which depend on the value of $x_i(j|k)$ can be approximated by:

$$\Delta \mathrm{h}(\mathrm{j}|\mathrm{k}) = \sum_{l=1}^{L'} Y_{PWL,l}\beta_{j+1,l} - \sum_{l=0}^{L} Y_{PWL,l}\beta_{j,l} \tag{13}$$

$$\theta(j|k) = \sum_{l=1}^{L'} \theta_{PWL,l}\beta_{j,l}. \tag{14}$$

The aim of the leader train controller is to track the reference speed trajectory $v_i^{ref}$ in each prediction horizon $N_p$, there are no other trains before the leader train so we can only consider the longitudinal dynamics constraints. The optimization can be formulated as follows:

$$\min_{u_i^{des}(k)} \sum_{j=1}^{N_p/\Delta t} \left(v_i(j|k) - v_i^{ref}(j|k)\right)^2 \tag{15a}$$

$$\mathrm{s}.t.(5) - (14) \tag{15b}$$

$$x_i(0|k) = x_i(k), v_i(0|k) = v_i(k), u_i(0|k) = u_i(k) \tag{15c}$$

where $x_i(k)$, $v_i(k)$, and $u_i(k)$ are the detected parameters in time $k$. The recursive feasibility of the controller can be guaranteed because all constraints from the leader controller are subject to train dynamics constraints.

The aim of follower trains is to track the preceding train and keep the desired distance $d_{des}$ as close as possible under the safe premise, and we need to consider the impact of the preceding train. Therefore, the optimization problem for follower trains in each predictive horizon can be formulated as:

$$\min_{u_i^{des}(k)} c_v \sum_{j=1}^{N_p/\Delta t} \left(\frac{v_i(j|k) - v_{i-1}(j|k)}{v_{max}}\right)^2 + c_d \sum_{j=1}^{N_p/\Delta t} \left(1 - \frac{\mathrm{d_i}(j|k)}{\mathrm{d_{des}}}\right)^2 \tag{16a}$$

$$\mathrm{s}.t.(5) - (14) \tag{16b}$$

$$x_i(0|k) = x_i(k), v_i(0|k) = v_i(k), u_i(0|k) = u_i(k) \tag{16c}$$

$$d_i(j|k) = x_{l,i}(j|k) - x_i(j|k) - L \tag{16d}$$

$$d_{min} \le d_i(j|k) \tag{16e}$$

$$d_{min} \le d_i\left(N_p/\Delta t|k\right) + \frac{v_{i-1}^2\left(N_p/\Delta t|k\right)}{2a_{max}} - \frac{v_i^2\left(N_p/\Delta t|k\right)}{2a_{max}} \tag{16f}$$

where $c_v$ is the penalty weighting of the speed difference between follower and leader, $c_d$ is the penalty weighting of the deviation between interval distance and desired distance $d_{des}$. The value of $d_{des}$ should be larger than $d_{min}$ for safety reasons. (16e) make sure the interval between trains is never closer than minimal safe distance and (16f) is a terminal constraint to ensure the MPC controller is recursively feasible [9].

The optimization problem of the leader and follower train is modeled as MILP. After solving the solution in each horizon, we can get the corresponding control sequence

$$u_{i,des} = \left[u_{i,des}(0|k), u_{i,des}(1|k), \ldots, u_{i,des}\left(N_p/\Delta t - 1|k\right)\right]$$

and only the first command $u_{i,des}(0|k)$ is applied as input at time $k$. Then the optimization problem will be updated and solved again at the next sampling time to get a new input until the coupling is terminated.

## 4 Results and Discussions

The simulation defined a VCTS consisting of two trains with the same vehicle parameters driving on a single line without switches and crossing. Unless stated, we assume the model is infallible and two trains run in ideal conditions. The data of trains and lines are based on the line from Expo Center to Wenquan East, an inter-station section of Qingdao Metro Line 11 with a total length of 2707 m. The value of parameters is listed in Table 1.

**Table 1.** Dynamic model and MPC parameters

| Parameter | Value | Parameter | Value |
|---|---|---|---|
| $M$ (kg) | 144000 | $P_{min}$(kW) | −2520 |
| $A$(N) | 3001.6 | $F_{max}$(kN) | 230.81 |
| $B(N/(\text{km/h}))$ | 20.16 | $F_{min}$(kN) | −230.81 |
| $C\left(N/(\text{km/h})^2\right)$ | 0.69692 | $v_{max}$(m/s) | 35 |
| $L$(m) | 68 | $\tau$(s) | 0.25 |
| $d_{min}$(m) | 3.5 | $\Delta t$(s) | 0.5 |
| $d_{des}$(m) | 5 | $N_p$(s) | 15 |
| $P_{max}$(kW) | 2520 | $g\left(\text{m/s}^2\right)$ | 9.8 |

(1) Case 1 - Comparative study between MMPC and NMPC: To verify the effectiveness and computational performance of the MMPC, we compared it with NMPC without considering track gradients. We use the position of the leader train as a zero reference, and the initial location of the follower train is 5 m behind the leader train. The maximum solving time of NMPC is set as 2 s to avoid the optimization problem in a single horizon being solved for a long time. In case 1 $a_{max} = 1.2\text{m/s}^2$, $a_{min} = -1.2\text{m/s}^2$, and $c_v = c_d = 0.5$.

The speed (location) error at each sampling time is defined as the speed (location) from the train minus the reference speed (location). Table 2. Shows the average solving

**Table 2.** Average solving times of NMPC and MMPC for single horizon

| Algorithm | Controller type | CPU times(s) |
|---|---|---|
| NMPC | Leader controller | 1.089 |
| | Follower controller | 1.184 |
| MMPC | Leader controller | 0.195 |
| | Follower controller | 0.214 |

times of two kinds of MPC for a single horizon, results shows the solving time needed by MMPC is about 18% of NMPC needed. The performance difference between MMPC and NMPC is demonstrated in Fig. 3 and Fig. 4. In Fig. 3, we can observe that for leader train, NMPC has less speed error than MMPC most of the time. However, NMPC has a more prominent location error during traveling because of the speed error caused by NMPC from 2 s to 13 s. It can be seen from Fig. 4 that MMPC causes a more significant speed error whose value is 0.13 m/s at 40 s, which is affected by the approximation quality of PWL. No matter which MPC controller the follower adopts, the interval is always larger than the safety distance. The maximum location error caused by MMPC is 5.1 m, however, a 2% error is acceptable for desire interval is 5 m. The simulation result above verified the MMPC can control the train driving under VC effectively, and it can be solved in sampling time.

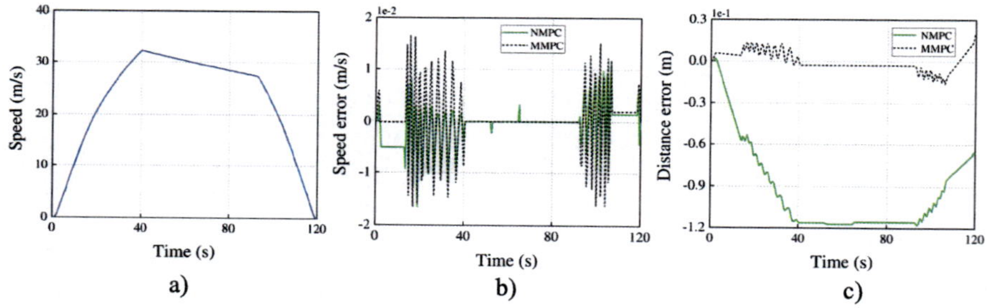


**Fig. 3.** Simulation results of the leader train in case 1: (a) reference speed trajectory for the leader train; (b) the speed error of leader train compared with reference; (c) the location error of leader train compared with reference

(2) Case 1 - Considering the control force error of follower: To verify the ability to deal with control force disturbance of the proposed MPC method, in case 2, the initial location of the follower is 80 m, and we consider the control force of the follower with a uniform distribution (-1440,1440) whose unit is N. In case 2 $a_{max} = 1.6\text{m/s}^2$ and $a_{min} = -1.6\text{m/s}^2$, $c_v = c_d = 0.5$. Figure 5 shows leader train can track the reference trajectory with a small error in each sampling time, and the follower train can reach the next station simultaneously. Because the initial interval distance is greater than $d_{des}$, the follower train will accelerate at the first eight seconds to reduce the interval distance and

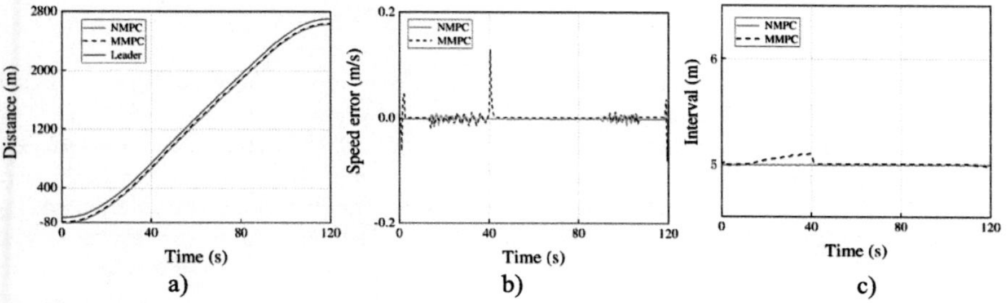


**Fig. 4.** Simulation results of follower train in case 1: (a) Reference trajectory and follower trajectory under control; (b) the speed error between the follower MPC and reference; (c) the interval distance under the control of MMPC and NMPC

then brake to run at the same speed as the leader. From 10 s after departure, the interval distance is kept about $d_{des}$ under the effect of control force error and greater than $d_{min}$. The simulation results as discussed above can verify the ability to deal with control force disturbance of the proposed MPC method.

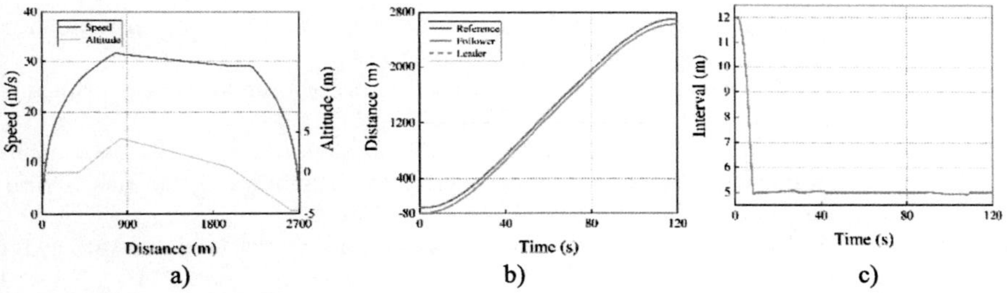


**Fig. 5.** Simulation results of follower train in case 1: (a) reference trajectory and follower trajectory under control; (b) the speed error between the follower MPC and reference; (c) the interval distance under the control of MMPC and NMPC

## 5 Conclusion and Future Work

This paper proposes an MPC method based on MILP for virtually coupled trains in VCTS. Sets of SOS2 variables are used to PWL the nonlinear terms in the constraints of the MPC controller. The space-depend variables can also be modeled with real-world data from an inter-station section of Qingdao Metro Line 11. The simulation results verify that MMPC needs less solving time than NMPC, and the simulation also demonstrates the effectiveness and robustness of MMPC.

In the future, we will consider the uncertainty of speed and location caused by sensor error and environment. The effect of time delay from T2T communication also will be addressed. Making MPC controllers with robust performance in uncertain environments will be considered in the future.

## References

1. Shift2Rail: Multi-annual action plan (2020), https://rail-research.europa.eu/wp-content/uploads/2020/09/MAAP-Part-A-and-B.pdf
2. Aoun, J., Quaglietta, E., Goverde, R., Pachl, J., Scheidt, M., Redfern, B., Blumen-feld, M.: Application roadmap for the introduction of virtual coupling (2020)
3. Quaglietta, E.: Analysis of platooning train operations under v2v communication-based signaling: Fundamental modelling and capacity impacts of virtual coupling. In: Proceeings of the 98th Transportation Research Board Annual Meeting. Trans-portation Research Board (TRB) (2019)
4. Quaglietta, E., Wang, M., Goverde, R.M.: A multi-state train-following model for the analysis of virtual coupling railway operations. J. Rail Transp. Plan. Manag. **15**, 100195 (2020)
5. Quaglietta, E., Spartalis, P., Wang, M., Goverde, R.M., van Koningsbruggen, P.: Modelling and analysis of virtual coupling with dynamic safety margin considering risk factors in railway operations. J. Rail Transp. Plan. Manag. **22**, 100313 (2022)
6. Di Meo, C., Di Vaio, M., Flammini, F., Nardone, R., Santini, S., Vittorini, V.: Ertms/etcs virtual coupling: proof of concept and numerical analysis. IEEE Trans. Intell. Transp. Syst. **21**(6), 2545–2556 (2019)
7. Park, J., Lee, B.H., Eun, Y.: Virtual coupling of railway vehicles: gap reference for merge and separation, robust control, and position measurement. IEEE Trans. Intell. Transp. Syst. (2020)
8. Mayne, D.Q., Rawlings, J.B., Rao, C.V., Scokaert, P.O.: Constrained model predictive control: Stability and optimality. Automatica **36**(6), 789–814 (2000)
9. Felez, J., Kim, Y., Borrelli, F.: A model predictive control approach for virtual coupling in railways. IEEE Trans. Intell. Transp. Syst. **20**(7), 2728–2739 (2019)
10. Xun, J., Yin, J., Liu, R., Liu, F., Zhou, Y., Tang, T.: Cooperative control of high-speed trains for headway regulation: a self-triggered model predictive control based approach. Transp. Res. Part C: Emerg. Technolo. **102**, 106–120 (2019)
11. Liu, Y., Liu, R., Wei, C., Xun, J., Tang, T.: Distributed model predictive control strategy for constrained high-speed virtually coupled train set. IEEE Trans. Veh. Technol. **71**(1), 171–183 (2021)
12. Wu, Z., Gao, C., Tang, T.: A virtually coupled metro train platoon control approach based on model predictive control. IEEE Access **9**, 56354–56363 (2021)
13. Zhang, B., et al.: Impact of on-board hybrid energy storage devices on energy-saving operation for electric trainsin dc railway systems. Batteries **8**(10), 167 (2022)

# Developing Typical Disturbance Scenario Library for High-Speed Railways

Yuxin Chen[1], Xiaoning Zhu[1,2](✉), Li Wang[1], and Zhiyuan Yao[1]

[1] School of Traffic and Transportation, Beijing Jiaotong University, Beijing 100044, China
{21120794,xnzhu,liwang,zhiyuan}@bjtu.edu.cn

[2] Key Laboratory of Transport Industry of Comprehensive Transportation Theory, Ministry of Transport, Beijing Jiaotong University, Beijing 100044, China

**Abstract.** As one of the modern transportation modes, high-speed railway plays an important role in boosting economic and social development. However, the operation of high-speed railways is susceptible to interference from various factors such as external and internal ones, and researchers have designed methodologies and approaches to minimize the impacts on high-speed railway operation of these disturbances. To comprehensively test the methodologies that cope with these unexpected events, this study analyzes the factors that could influence the normal operation of high-speed railways. Based on the classification of disturbances, a disturbance scenario library software is developed that can record different types of disturbances and can be used as input for testing and validating the railway rescheduling approaches designed by railway practitioners or railway researchers.

**Keywords:** High-speed railway · Railway disturbances · Scenario library

## 1 Introduction

As one of the modern transportation modes, high-speed railway plays important roles in boosting economic and social development, improving people's living standards, and has become an important part of China's passenger transportation infrastructure. High-speed railway gains its popularity by its reliability, high level of speed, safety and environmental friendliness.

However, the operation of high-speed railways is susceptible to interference from various factors. On the one hand, the operation of high-speed railways requires highly precise equipment and technical support such as train control systems, rolling stock performances and so on. Once equipment malfunctions or technical problems occur, it may affect the scheduled operation of the railway. On the other hand, the environment in which high-speed railways operate can also have an impact on railway operations, such as weather, natural disasters, maintenances, etc. In addition, some adverse social factors, such as terrorist attacks, theft, smoking on board, etc., may also have a negative impact on the safety and operation of high-speed railways.

Since disturbances can have significant impacts on railway operation safety and punctuality, researchers have paid much attention to reschedule train operations such that the


Y. Qin et al. (Eds.): EITRT 2023, LNEE 1137, pp. 221–227, 2024.
https://doi.org/10.1007/978-981-99-9311-6_24

impacts can be minimized [1–3], especially with consideration of passenger perspectives [4–6]. However, it is crucial to test the validity of their rescheduling approaches via trial and error. Therefore, it is necessary to build a typical disturbance scene library for high-speed railways, considering various factors that could have a wide impact on the operation and management of high-speed railways, and serves as a testbed for various rescheduling approaches and algorithms. Hence, this study considers the typical disturbance types of railway operations by analyzing historical data, and examined several typical disturbance scenes, based on which a comprehensive dataset is created, and a disturbance data management software is developed.

## 2 Causes of Disturbances of Railway Operations

### 2.1 External Factors

(1) **Weather factors**

Weather factors are one of the main disturbances in high-speed railway operation. For example, extreme weather such as heavy rain, blizzards, typhoons, etc. [7]. Can cause signal system failures, track landslides, and even ice and snow cover. When facing ice and snow weather, various factors such as snow accumulation, wind speed, and intense snowfall may cause difficulties in snow removal on the track, icing on the overhead lines, etc., resulting in abnormal train operation or extensive delays, and inconvenience to passengers' travel.

(2) **Topological factors**

High-speed railways are usually built along mountain ranges, rivers, and other terrains. The geological conditions of the areas passing along the line are variable, and the operation of trains is easily affected by the terrain. Therefore, terrain factors such as earthquakes, mudslides, landslides, and floods have great suddenness [8]. Once they occur, they will cause varying degrees of damage to infrastructure, seriously affecting the operation of high-speed railways. In addition, the impact of terrain factors such as underground pipeline explosions and ground subsidence on high-speed railways can be considered.

(3) **Emergency factors**

Emergency factors are one of the most unpredictable disturbances in high-speed railway operation. For example, incidents such as fires and terrorist attacks can have an impact on high-speed railway operations. Therefore, it is possible to simulate the operation of high-speed railways in various emergency situations, in order to develop corresponding emergency plans and response measures.

## 2.2 Internal Factors

### (1) Human factors

Human factors are also one of the common disturbances in high-speed railway operation. Human factors mainly include operator factors and non-operator factors. The operator factor mainly refers to the factors caused by operational errors, insufficient dispatching skills, and improper handling of emergencies by high-speed dispatchers [9]. These factors cause chaos in on-site operations and interfere with the normal operation of trains. Non-operational personnel factors refer to factors caused by passengers, including sudden injuries and illnesses during transportation, low-quality passengers obstructing high-speed train departure, passengers or other personnel crossing the track, passengers smoking and triggering smoke alarms, resulting in the train being unable to depart or operate normally. Excessive passenger traffic, passenger violations, and poor management can all affect the operation of high-speed railways.

### (2) Technical factors

Technical factors also have a significant impact on the operation of high-speed railways. Mainly reflected in equipment failures, which include a series of faults such as power, signals, maintaining and repairing of rolling stocks and tracks. They can be divided into fixed equipment failures and mobile equipment failures. Fixed equipment failure mainly includes signal failure (centralized dispatching system, track circuit failure, and so on [10]), power supply equipment failure, line equipment failure (bridge, tunnel, track failure), dispatching system failure (ATP failure); Mobile equipment failures mainly include EMU failures, on-board equipment failures (failures of train control on-board equipment, etc.). Technical factors such as onboard signal system failures, train control failures, and track deviations can all cause interruptions in the operation of high-speed railways.

### (3) Passenger flow factor

The passenger flow factor is also one of the common disturbances in high-speed railway operation. The passenger flow factors include large delays in trains, resulting in a large number of passengers being stranded, sudden large passenger flow during holidays or large-scale events, etc. The large number of passengers gathering at stations causes congestion, affects the boarding and getting-off time of passengers, and leads to delayed departure of trains [11]. In addition, the large number of passengers gathering can easily cause other safety issues. These all require real-time communication and coordination among dispatchers, station staff, and train attendants, in order to handle them correctly and promptly, and minimize the impact on train operation.

In summary, constructing a typical disturbance scene library based on high-speed railways requires consideration of various factors mentioned above, and obtaining data and validating models through various methods such as laboratory experiments, simulation, and on-site testing. This will help to conduct more in-depth research on the operation and management of high-speed railways, and at the same time, these disturbance scenario data can be used to develop early warning systems and intelligent control systems, improving the safety and operational efficiency of high-speed railways.

## 3 Developing Typical Disturbance Scene Library for High-Speed Railways

Based on the analysis of the typical high-speed railway disturbance scenarios mentioned above, the structure of the library of typical high-speed railway disturbance scenarios is constructed as shown in Fig. 1.

By analyzing the historical disturbance data of Chinese high-speed railways, a disturbance scenario library software is built with C# codes with ECharts as inserted data visualizer, and it records various types of disturbance scenarios with specific information such as date, occurring time, railway line, fault type, specific fault and the recovery time needed to handle the event. An example of track faults of the software is shown in Fig. 2. Besides, the software is also able to analyze the data in terms of frequencies of different types of faults, and the distribution of recovery times needed to cope with corresponding events as shown in Fig. 3.

The software reads input historical disturbance data by csv file format, and supports input of data by interfaces. The library analyzes its data in real-time and calculate the characteristics of data, which will be transferred to ECharts tool box to draw visualization diagrams.

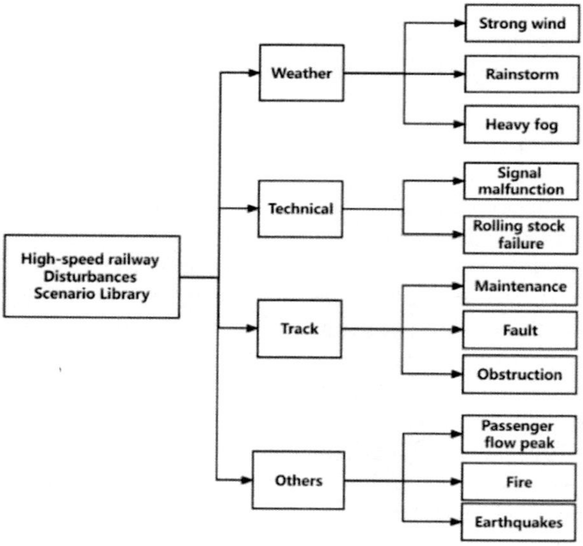


**Fig. 1.** Structure of the disturbance scene library for high-speed railway.

The software can be used by railway dispatching training program in terms of testing their capability in coping with various and comprehensive types of disturbance events, and can also be used by researchers to test the capability and robustness of their rescheduling models and algorithms.

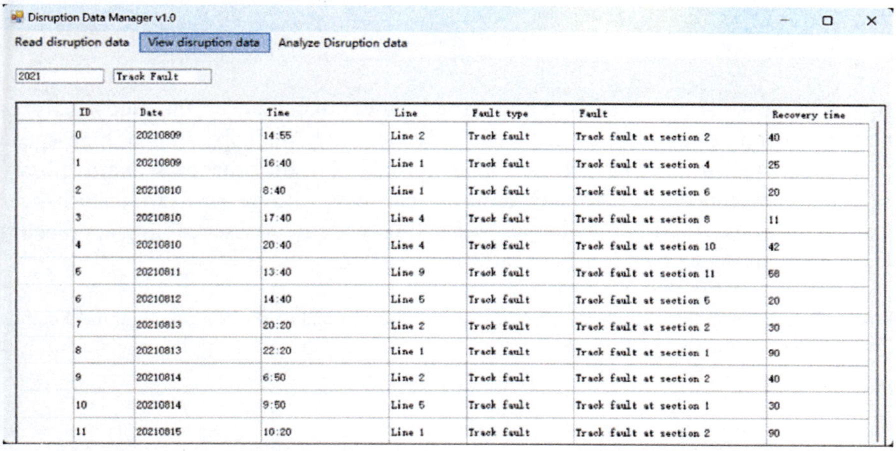


| ID | Date | Time | Line | Fault type | Fault | Recovery time |
|---|---|---|---|---|---|---|
| 0 | 20210809 | 14:55 | Line 2 | Track fault | Track fault at section 2 | 40 |
| 1 | 20210809 | 16:40 | Line 1 | Track fault | Track fault at section 4 | 25 |
| 2 | 20210810 | 8:40 | Line 1 | Track fault | Track fault at section 6 | 20 |
| 3 | 20210810 | 17:40 | Line 4 | Track fault | Track fault at section 8 | 11 |
| 4 | 20210810 | 20:40 | Line 4 | Track fault | Track fault at section 10 | 42 |
| 5 | 20210811 | 13:40 | Line 9 | Track fault | Track fault at section 11 | 58 |
| 6 | 20210812 | 14:40 | Line 5 | Track fault | Track fault at section 5 | 20 |
| 7 | 20210813 | 20:20 | Line 2 | Track fault | Track fault at section 2 | 30 |
| 8 | 20210813 | 22:20 | Line 1 | Track fault | Track fault at section 1 | 90 |
| 9 | 20210814 | 6:50 | Line 2 | Track fault | Track fault at section 2 | 40 |
| 10 | 20210814 | 9:50 | Line 5 | Track fault | Track fault at section 1 | 30 |
| 11 | 20210815 | 10:20 | Line 1 | Track fault | Track fault at section 2 | 90 |

**Fig. 2.** Demonstration of disturbance scenario data.

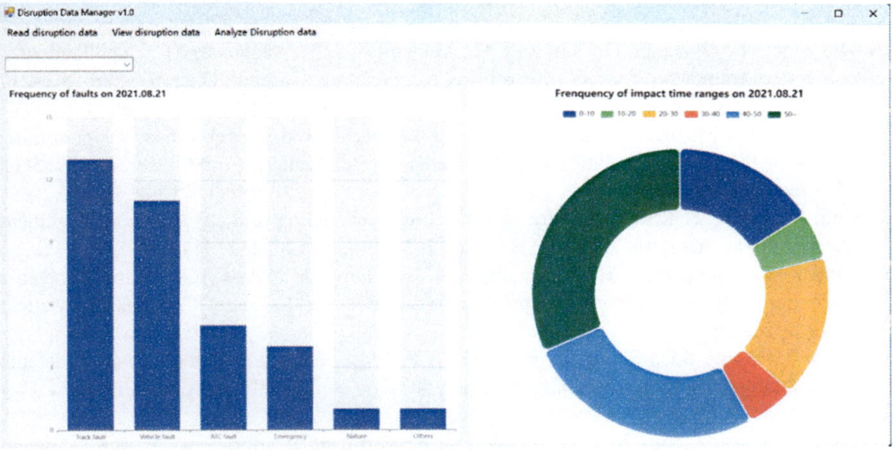


**Fig. 3.** Demonstration of fault frequencies and distribution of recovery times.

The construction and operation of high-speed railways require strict management and monitoring to ensure safe, timely, and efficient operation. The above scenario library established is a part of the disturbance scenario library that may exist during high-speed railway operation and management, and specific disturbance scenarios need to be supplemented and modified according to actual situations and needs. Meanwhile, the above scenarios are not the source of high-speed railway disturbances, but rather factors that may affect the normal operation of high-speed railways. In the future, in order to better respond to these disturbance scenarios, it is necessary to develop scientific and reasonable response strategies and emergency plans to ensure the normal, safe, and efficient operation of high-speed railways.

## 4 Conclusion

In view of the usual occurrences of high-speed railway disturbances, this study analyzes the internal and external factors that could influence the normal operation of high-speed railways. Based on the classification of disturbances, a disturbance scenario library software is developed with C# and ECharts that can record different types of disturbances and can be used as input for testing and validating the railway rescheduling approaches designed by railway practitioners or railway researchers.

**Acknowledgement.** This study is supported by the Joint Funds of the National Natural Science Foundation of China (No.U2034208).

## References

1. Veelenturf, L.P., Kidd, M.P., Cacchiani, V., Kroon, L.G., Toth, P.: A railway timetable rescheduling approach for handling large-scale disruptions. Transp. Sci. **50**(3), 841–862 (2016)
2. Dollevoet, T., Huisman, D., Kroon, L.G., Veelenturf, L.P., Wagenaar, J.C.: Application of an iterative framework for real-time railway rescheduling. Comput. Oper. Res. **78**, 203–217 (2017)
3. Fischetti, M., Monaci, M.: Using a general-purpose Mixed-Integer Linear Programming solver for the practical solution of real-time train rescheduling. Eur. J. Oper. Res. **263**(1), 258–264 (2017)
4. Zhu, Y., Goverde, R.M.P.: Integrated timetable rescheduling and passenger reassignment during railway disruptions. Transp. Res. Part B-Methodol. **140**, 282–314 (2020)
5. Binder, S., Maknoon, M.Y., Azadeh, S.S., Bierlaire, M.: Passenger-centric timetable rescheduling: a user equilibrium approach. Transp. Res. Part C-Emerg. Technol. **132**, Article 103368 (2021)
6. Zhan, S., Wong, S.C., Shang, P., Peng, Q., Xie, J., Lo, S.M.: Integrated railway timetable rescheduling and dynamic passenger routing during a complete blockage. Transp. Res. Part B-Methodol. **143**, 86–123 (2021)
7. Ling, X., Peng, Y., Sun, S., Li, P., Wang, P.: Uncovering correlation between train delay and train exposure to bad weather. Phys. A-Stat. Mech. Its Appl. **512**, 1152–1159 (2018)
8. Yin, L., Wang, Y.: Network Characteristics and Vulnerability Analysis of Chinese Railway Network under Earthquake Disasters. Isprs Int. J. Geo-Inf. **9**(12), Article 697 (2020)
9. Lei, S., Guo, Z., Tan, X., Chen, X., Li, C., Zou, J., Cao, S., Feng, G.: Cognitive abilities predict safety performance: a study examining high-speed railway dispatchers. J. Adv. Transp. **2021**, Article 5538320. (2021)

10. Shin, D., Chae, E., Lee, J.-S.: A Study on Reliability demonstration for railway signaling equipment. J. Korean Soc. Railw. **20**(4), 458–465 (2017)
11. Besinovic, N., Wang, Y., Zhu, S., Quaglietta, E., Tang, T., Goverde, R.M.P.: A Matheuristic for the Integrated disruption management of traffic, passengers and stations in urban railway lines. IEEE Trans. Intell. Transp. Syst. **23**(8), 10380–10394 (2022)

# Design of Track Utilization Plan of High-Speed Railway Passenger Station Based on Offline Sorting

Yinggui Zhang[1], Yuhang Wang[1], Min An[2(✉)], Ruihua Hu[1], and Juan Wang[3]

[1] School of Traffic and Transportation Engineering, Central South University, Changsha 410004, China
{ygzhang,224211023,ruihua}@csu.edu.cn

[2] School of Science, Engineering and Environment, University of Salford, Manchester M5 4WT, UK
M.An@Salford.ac.uk

[3] School of Logistics and Transportation, Central South University of Forestry and Technology, Changsha 410004, China
wangjuan@csuft.edu.cn

**Abstract.** Track utilization is an important part of the high-speed railway passenger station operating organization. On account of the idea of off-line sorting, taking the design of the high-speed railway passenger station track utilization plan as the research object, taking into account the constraints of track occupation, route occupation, safety time interval, etc., taking the best balance of track occupation as the optimization objective function, the parallel machine off-line sorting model that does not allow interrupt processing is constructed for the design of the high-speed railway passenger station track utilization plan, and the improved MLS algorithm based on the distribution rules is designed to solve that. In the end, based on the background of a class plan of a Railway Station, based on the proposed method, MATLAB is used to program to verify the rationality and feasibility of the constructed model and algorithm. The results of the comparative analysis of the plan show that the proposed model and algorithm can formulate the track utilization plan of the high-speed railway passenger station in a short period of time, improve the balance and robustness of the plan as much as possible, and provide a strong decision-making support for the high-speed railway dispatchers to formulate and adjust the track utilization plan of the High-speed railway passenger station.

**Keywords:** High-speed railway passenger station · Track utilization · Off-line sorting · The improved MLS algorithm

## 1 Introduction

Trains' arrival and departure at the station in strict accordance with the track utilization plan. The track utilization plan dictates the arrival and departure time of each train, which track it uses to stop, and the corresponding receiving and departure route at the station.


Y. Qin et al. (Eds.): EITRT 2023, LNEE 1137, pp. 228–236, 2024.
https://doi.org/10.1007/978-981-99-9311-6_25

Track utilization is the core of station operation organization. At present, the research on track utilization is roughly divided into two aspects: track utilization planning and rescheduling. In the aspect of track utilization planning, Abels designed an optimization solution of station track utilization planning based on Answer Set Programming (ASP) [1]. Burggraeve comprehensively considered optimizes the infrastructure occupation and the passenger robustness to formulate the track utilization plan [2]. Zhang Yinggui constructed the first kind of multi-objective window scheduling model to solve the track utilization problem [3, 4]. Zhang Qin synchronously optimizes the train working time schedule, track utilization and train working diagram, and designs different train time window modification strategies to solve [5]. From the point of view of compressing the interval between stations to improve the capacity of stations, Ma Si constructed a nonlinear model of route allocation optimization to solve the problem [6]. In the aspect of track utilization rescheduling, Ricardo constructed an integer programming model and used CPLEX to solve the problem of route and track reallocation decision [7]. Caprara constructed a directed multigraph formula to describe the railway network, and deduced the Lagrangian relaxation integer programming model [8]. Taking minimizing the deviation of train arrival time and maximizing the maintenance work completed per unit time as the optimization goal, Andrea constructed a real-time adjustment model of track utilization [9]. Considering the malpractice of insufficient flexibility of track utilization adjustment, Feng Liping designed the track utilization model based on fuzzy time knowledge reasoning algorithm and used conflict prediction algorithm to make track utilization rescheduling decision [10]. Zhan Shuguang dawning aims at the problem of track utilization rescheduling under the interference of complete congestion for a long time, and integrates the passenger transfer factors under the interference condition to reduce the impact of congestion on passengers to adjust the track utilization [11]. To summarize, the research on track utilization at home and abroad is mostly solved by the construction of integer programming model, and the sorting characteristics of station arrival and departure technical operations are rarely analyzed; at the same time, there is little theoretical research on the track utilization of train arrival and departure with special nature.

In Sect. 2, we constructed an off-line sorting model of parallel machines which do not allow interruption of processing to solve the problem of formulating track utilization plan in high-speed railway stations. In Sect. 3, we design an improved Modified List Scheduling (MLS) algorithm to solve the sorting model. In Sect. 4, we report the results of numerical experiments. The last section concludes the study.

## 2 Model

In order to facilitate the construction of the model, we make the following assumptions: (1) the station interlock is the basic route-lock-route-release; (2) the length of the train is not considered.

Let $\mathbf{J}$ be a set of all tracks contained in the station, for any track in the station $\mathbf{J} = \{j|j = 1, 2, 3..., n_{\mathbf{J}}\}$, treat the track as i harness sets, $\mathbf{J}_1 \cup \mathbf{J}_2 \cup ... \cup \mathbf{J}_i = \mathbf{J}$; let $\mathbf{M}$ be a set of trains, for arrival and departure trains, $\mathbf{M} = \{m|m = 1, 2, 3..., n_{\mathbf{M}}\}$; let $\mathbf{R}_{\mathrm{a(d)}}$ be a set of all receiving and departure routes contained in the station, $\mathrm{r_a}, \mathrm{r_d} \in \mathbf{R}_{\mathrm{a(d)}}$;

let $t_m^{arr}$ and $t_m^{dep}$ be the scheduled arrival and departure time of train m; let $t^{s}_{m,r_{a(d)}}$ and $t^{e}_{m,r_{a(d)}}$ be the time occupied by train m for the starting and ending time of the receiving and departure routes; let $T_g$ be the minimum occupancy interval of the track; let $\delta_i$ be properties of the train m; let $x_{m,j}$ and $y_{m,r_{a(d)}}$ be the decision variables in the problem, they are 0–1 variables, they represent the occupancy of track j and route $r_{a(d)}$ by train m.

## 2.1 Constraints

### 2.1.1 Track Occupancy Constraint

Each high-speed railway train can only occupy one track when performing train reception and departure operations and other operations at the station, and at the same time, at most one train can be served at the same time. In addition, it is still necessary to ensure that when two adjacent trains occupy the track one after another, it is still necessary to meet the constraint of the minimum interval between the occupation of the track. Therefore, the constraint expression is:

$$\sum_{j \in \mathbf{J}} x_{m,j} = 1, m \in \mathbf{M} \tag{1}$$

$$\sum_{j \in \mathbf{J}} (x_{m,j} t_m^{arr} - x_{m',j} t_{m'}^{dep})(x_{m,j} t_m^{dep} - x_{m',j} t_{m'}^{arr}) \geq 0, \forall m, m' \in \mathbf{M}, m \neq m' \tag{2}$$

$$\left| x_{m,j} t_m^{arr} - x_{m',j} t_{m'}^{dep} \right| \geq T_g, \forall m, m' \in \mathbf{M}, m \neq m', j \in \mathbf{J} \tag{3}$$

### 2.1.2 Path Occupancy Constraint

In order to ensure the safety of the route in the throat area of the station, the route is locked when performing train reception and departure operations. Therefore, the route occupation is unique, that is, it cannot be occupied by other trains at the same time. At the same time, there cannot be a situation in which a train occupies multiple routes. In addition, it is necessary to ensure that the train has a feasible receiving route. Therefore, the constraint expression is:

$$\sum_{r_{a(d)M} \in \mathbf{R}_{a(d)}} y_{m,r_{a(d)}} \leq 1, m \in \mathbf{M} \tag{4}$$

$$\sum_{r_{a(d)} \in \mathbf{R}_{a(d)}} (y_{m,r_{a(d)}} t^{s}_{m,r_{a(d)}} - y_{m',r_{a(d)}} t^{e}_{m',r_{a(d)}})(y_{m,r_{a(d)}} t^{e}_{m,r_{a(d)}} - y_{m',r_{a(d)}} t^{s}_{m',r_{a(d)}}) \geq 0, \quad \forall m, m' \in \mathbf{M}, m \neq m' \tag{5}$$

## 2.2 Objective Function

The high-speed railway station is mainly based on traffic operation, because the train occupies the route shortly, so we measure the advantages and disadvantages of the track utilization plan by the balance of the occupied time of the track. Considering the different types of service trains in different tracks of high-speed railway passenger stations and there are different operation requirements for different tracks. Therefore, it is more valuable to consider the balance of track occupation time from the point of view of harness than from the point of view of the station as a whole. Therefore, the occupation time balance of different harness is considered, and the occupation time balance of different harness is determined according to (6):

$$C_i = \sqrt{\frac{1}{n_{x_{m,j}}} \sum_{j \in J_i} \{ \sum_{m \in M} [(t_m^{dep} - t_m^{arr}) x_{m,j} - \frac{1}{n_{x_{m,j}}} \sum_{m \in M} (t_m^{dep} - t_m^{arr}) x_{m,j}]^2 \}} \tag{6}$$

## 2.3 Off-Line Sorting Model of Parallel Machines

To summarize, the off-line sorting model of parallel machines which do not allow interruption of processing to solve the problem of formulating track utilization plan in high-speed railway stations is as follows:

$$\mathrm{Pm} \left| \text{off - line - list}, t_m^{arr}, t_m^{dep}, \text{chain}, x_{m,j}, y_{m,r_{a(d)}}, \delta_i, \mathbf{M} \to \mathbf{J} \right| C_i$$

*s.t.*

$$\sum_{j \in J} x_{m,j} = 1, m \in \mathbf{M} \tag{7}$$

$$\sum_{j \in \mathbf{J}} (x_{m,j} t_m^{arr} - x_{m',j} t_{m'}^{dep})(x_{m,j} t_m^{dep} - x_{m',j} t_{m'}^{arr}) \geq 0, \forall m, m' \in \mathbf{M}, m \neq m' \tag{8}$$

$$\left| x_{m,j} t_m^{arr} - x_{m',j} t_{m'}^{dep} \right| \geq T_g, \forall m, m' \in \mathbf{M}, m \neq m', j \in \mathbf{J} \tag{9}$$

$$\sum_{r_{a(d)M} \in \mathbf{R}_{a(d)}} y_{m,r_{a(d)}} \leq 1, m \in \mathbf{M} \tag{10}$$

$$\sum_{r_{a(d)} \in \mathbf{R}_{a(d)}} (y_{m,r_{a(d)}} t^s_{m,r_{a(d)}} - y_{m',r_{a(d)}} t^e_{m',r_{a(d)}})(y_{m,r_{a(d)}} t^e_{m,r_{a(d)}} - y_{m',r_{a(d)}} t^s_{m',r_{a(d)}}) \geq 0, \tag{11}$$

$$\forall m, m' \in \mathbf{M}, m \neq m'$$

# 3 Solution Algorithm

## 3.1 Data Preprocessing

For a given train working time schedule, deal with the train reception and departure direction, arrival and departure time, the nature of the train, and the train stock turnaround plan; for the given station, the receiving and departure routes are represented by the track circuit section, and the basic receiving and departure route is mapped with the corresponding track.

The preprocessing result of the train data is shown in Fig. 1 below. Among them: the first represents the train number; the second represents the arrival time; the third represents the departure time; the fourth represents the receiving direction; the fifth represents the departure direction; the sixth represents the track chosen by the train.

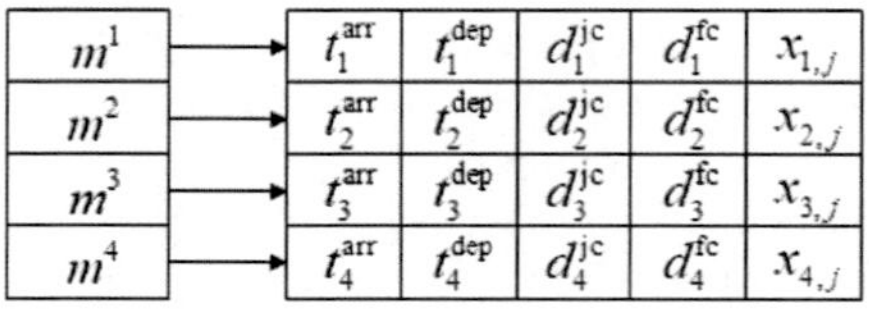


**Fig. 1.** Schematic diagram of the result of train data processing.

### 3.2 Dispatching Rules

The improved MLS algorithm is used to solve the ranking model. First of all, it is necessary to adopt different dispatching rules to modify the sequence of train arrival and departure operation, and then through the solution improvement strategy to improve the solution of the trains which are difficult to sort, in order to achieve the purpose of finding a feasible solution.

The dispatching rules are as follows:

- FIFO: the earliest arriving train chooses the track and route for receiving;
- EDD: the train that ends the occupation of the track first chooses the track and route for receiving;
- SPT: the train that occupies the shortest time on the track chooses the track and route for receiving;
- SPP: the train with special running direction preferentially chooses the track and route for receiving, such as cross-line train.

The solution improvement strategy is designed as follows: within the range of the idle time window of the track and route, the track occupied by arrival and departure operation of the adjacent trains and the route are optimized in a small range according to the time window. Special trains can be alleviated by fixed track reception treatment.

### 3.3 Algorithm Description

**Inputs**: train data after preprocessing.

**Outputs**: track utilization plan and route allocation plan.

**Step 1** Sort trains according to selected dispatching rules; pre-processes data.

**Step 2** Judge the nature of the train according to the direction of reception and departure of the train, and determines the track where it can carry out the train arrival and departure operation; if no new train needs to receive, transfer to **Step 10**.

**Step 3** According to the track that can be selected by the current train, judge the conflict constraint between the track and route occupation, select the track set without time conflict, and further select the route set according to the track set; if the train has no track and route set to choose, transfer to **Step 9**.

**Step 4** Renumber the selected track that can be used, and the numbering rule is: the track is numbered according to the minimum time occupied by the track, that is, it is sorted according to the occupied time of the track that can be used, if the two tracks are occupied by the train for the same amount of time, the track far away from the main line will be given priority in accordance with the principle of minimizing operational conflicts.

**Step 5** Judge in the selected track set, if the track set that can be selected is $\not\subset (M_1, M_2, \ldots, M_m)$. In other words, the track with the longest occupancy time and the track with the shortest occupancy time is not $\beta$ approximate (for $\beta \geq 1$, if $\frac{y}{\beta} \leq x \leq \beta y$, the two non-negative real numbers $x$ and $y$ are considered to be approximate), then the train occupies the track numbered $M_1$ and transfers to **Step 8.**

**Step 6** If $M2 + x \leq R\left(\frac{\sum Mi+x}{m}\right), R = \frac{2m-2+\beta}{m-1+\beta}$, the train occupies the track numbered $M_2$ and transfers to **Step 8.**

**Step 7** If the conditions of **Step 5** and **Step 6** are not valid, the train occupies the track numbered $M_1$ and transfers to **Step 9** and output the occupied information of the train.

**Step 8** According to the track selected by the train, update the occupied information of the track and route and transfers to **Step 2.**

**Step 9** For trains that have no route and track to choose from, they will be modified according to the solution improvement strategy and transfers to **Step 2.**

**Step 10** The track occupied information of each train, the receiving and departure route occupied by the train and the result of the target function value are output.

## 4 Case Study

Figure 2 is a station yard of a high-speed railway passenger station. The station receives and dispatches trains from four directions, with a total of 7 platforms and 15 tracks.

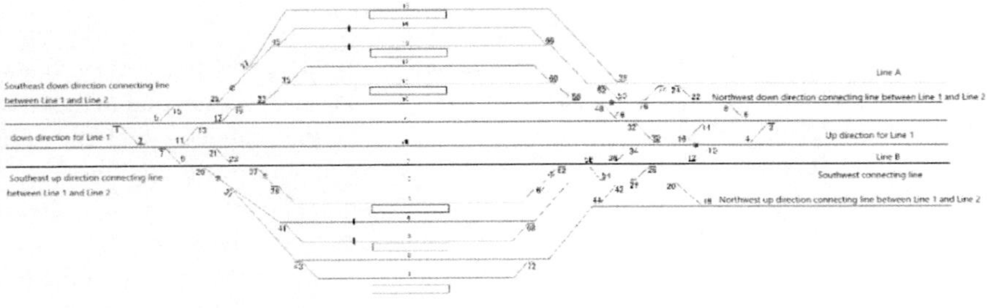


**Fig. 2.** Station yard of a high-speed railway passenger station

The standard values of the train arrival and departure operation of the station are as follows: the occupied time of receiving route is 3min, that of departure route is 4min, and the minimum interval time of track occupation is 2min.

For the preprocessed train data, different dispatching rules are used to sort the train reception and departure order, and then different track utilization plans are obtained by programming according to the improved MLS algorithm. The optimization results are as follows:

**Table 1.** Balance of track occupancy under different allocation rules

| Harness | Track | FIFO | EDD | SPT | SPP |
|---|---|---|---|---|---|
| No. 1 | 13,14,15 | 7.4 | 8.5 | 12.8 | 10.6 |
| No. 2 | 10,11,12 | 20.5 | 10.3 | 34.8 | 76.1 |
| No. 3 | 5,6,7 | 3.7 | 10.5 | 31.2 | 7.1 |
| No. 4 | 3,4 | 12.0 | 3.5 | 19 | 0.5 |
| No. 5 | 1,2 | 1.0 | 5.0 | 7 | 7 |
| Balance | – | 44.6 | 37.8 | 104.8 | 101.3 |

It can be seen from Table 1 that the occupation time balance of different harness is quite different under different dispatching rules, and the optimization of occupancy balance of different harness is improved differently in different dispatching rules, and the following conclusions can be drawn:

- In FIFO, the occupation time balance of harness 3 and 5 is better, which may be due to the less train arrival and departure operation of the harness, so it can be considered that for the stations with a small number of trains, we can use the dispatching rules of FIFO to sort the trains, and then formulate the track utilization plan, which has a good balance.
- In EDD, the harness shows a good balance as a whole, and the EDD can clear the track quickly to ensure that the train arrival and departure operation can be completed efficiently.
- In SPT, the overall balance of each harness is poor, because the sequence of priority of trains which completes the reception and departure operation first is easy to lead to the poor balance of track occupation caused by some trains connecting to the track when the technical operation time is too long, so when the train technical operation time is long, it is not recommended to use this sorting method at the station.
- In SPP, part of the harness can show a better balance, so it can be seen that giving priority to the train with special nature is of great help to the overall balance of the station track occupation, but the poor equilibrium of some harness is due to the fact that some special trains take a long time to arrive and depart the train, which has a great influence on the subsequent train selection.

## 5 Conclusions

The core of this paper lies in the construction of an off-line sorting model of parallel machines which do not allow interruption of processing to solve the problem of formulating track utilization plan in high-speed railway stations, comprehensively considering the organizational constraints of station operation, such as track and route occupation, and taking the occupation time balance of different harness as the optimization objective. At the same time, when solving the model, the dispatching rules are introduced to optimize the original data for the first time, an improved MLS algorithm is designed to solve the model, and the corresponding solution improvement strategy is designed for the trains with conflicts.

Through the analysis of the case, it can be proved that using different dispatching rules to solve train utilization problem by using the off-line sorting model can optimize the designed track utilization plan to varying degrees. When working out the track utilization plan at the station, you can refer to the following:

(1) The FIFO dispatching rules can be used in less train arrival and departure operations, which can effectively formulate the track utilization plan with high occupation time of balance.
(2) For the stations with special nature of arrival and departure trains, we can consider using the EDD dispatching rules to work out the track utilization plan.
(3) For the other two dispatching rules, the passing efficiency of passing trains and special trains can be greatly guaranteed, but it is generally not recommended to consider it when formulating the plan.

**Acknowledgment.** The research is supported by the National Natural Science Foundation of China (Grant No. 71971220), the Natural Science Foundation of Hunan Province, China (Grant No. 2023JJ30710, No. 2022JJ31020), the Fundamental Research Funds for the Central Universities (Grant No. 2023ZZTS0429).

## References

1. Abels, D., Jordi, J., Ostrowski, M.: Train scheduling with hybrid answer set programming. Theory Pract. Logic Program. **21**(3), 317–347 (2020)
2. Burggraeve, S., Vansteenwegen, P.: Robust routing and timetabling in complex railway stations. Transp. Res. Part B: Methodol. **101**, 228–244 (2017)
3. Lei, D, Zhang, Y., WANG, X.: Coupled due windows scheduling model and algorithm of carriage operations arrangements in railway technology stations. J. China Railw. Soc. **32**(04), 1–7 (2010). (in Chinese)
4. Zhang, Y., Lei, D., Tang, B.: Due windows scheduling model and algorithm of track utilization in railway passenger stations. J. China Railw. Soc. **33**(01), 1–7 (2011). (in Chinese)
5. Zhang, Q., Lusby, R., Shang, P.: A heuristic approach to integrate train timetabling, platforming, and railway network maintenance scheduling decisions. Transp. Res. Part B: Methodol. **158**, 210–238 (2022)
6. Ma, S., Ying, H.: Route allocation problem of high-speed railway station based on minimum arrival interval. J. Transp. Eng. Infor. **18**(3), 1–8 (2020). (in Chinese)
7. Ricardo, G., María, L., Luis, C.: A mixed-integer linear program for real-time train platforming management. Transp. Res. Procedia **62**, 815–823 (2022)
8. Caprara, A., Fischetti, M., Toth, P.: Modeling and solving the train timetabling problem. Oper. Res. **50**(5), 851–861 (2022)

9. Andrea, D., Meng, L., Gabriele, C.: Integrated stochastic optimization approaches for tactical scheduling of trains and railway infrastructure maintenance. Comput. Ind. Eng. **127**, 1315–1335 (2019)
10. Feng, L., Zhang, L., Gao, J.: Arrival-departure track conflict prediction based on fuzzy temporal knowledge reasoning. Railw. Transp. Econ. **44**(4), 91–98 (2022). (in Chinese)
11. Zhan, S., Wong, S., Shang, P.: Integrated railway timetable rescheduling and dynamic passenger routing during a complete blockage. Transp. Res. Part B: Methodol. **143**, 86–123 (2021)

# Research on the Evaluation Model of Urban Main Transformer DC Bias Caused by Metro Stray Current

Cheng Xinxiang[1](✉), Yu Kun[1], Ni Yanru[1], Tang Yuhang[1], Zeng Xiangjun[1], and Han Wei[2]

[1] Changsha University of Science and Technology, Changsha 410000, China
chxinee028@stu.csust.edu.cn
[2] State Grid Taiyuan Power Supply Company, Taiyuan 030000, China

**Abstract.** During the actual operation of metro locomotives, some traction current leaks and injects into the ground, forming stray currents, resulting in surface potential differences along the line and causing DC bias of the main transformer in the urban area. Reasonably evaluating the degree of influence of transformer DC bias can effectively guide the suppression of transformer DC bias. In response to the above issues, this article first established a coupling model between stray current and DC bias of the main transformer in urban substations, and further modelled the distribution of transition resistance during subway operation, which achieved dynamic calculation of stray current; By selecting indicators such as neutral point DC, excitation current DC component, and harmonic distortion deviation, effective evaluation of the degree of DC bias influence on the main transformer along the subway line was achieved. The simulation results show that as the ground potential difference increases, the DC flow at the neutral point of the transformer increases. Along with the degree of hysteresis loop distortion in the iron core increases, the DC component of the excitation current and the deviation of harmonic distortion rate gradually increase. It is show that DC bias of the transformer becomes more severe.

**Keywords:** Metro · Stray current · Transformer DC bias · Evaluation model

## 1 Introduction

With the rapid growth of China's economy and the continuous expansion of urban scale, how to alleviate the urban traffic pressure caused by residents' travel has become a hot issue. In this context, the metro system has been vigorously developed due to its advantages of safety, environmental protection, and convenient travel. However, metro locomotives can have adverse effects on the urban AC power grid along the line during actual operation, endangering the safe and stable operation of the urban AC power grid. The metro DC power supply system generally adopts 1500V or 750V. The traction current flows through the contact network to supply power to the metro locomotive, and the steel rail is used as a return conductor. Due to the inability of the steel rail


Y. Qin et al. (Eds.): EITRT 2023, LNEE 1137, pp. 237–245, 2024.
https://doi.org/10.1007/978-981-99-9311-6_26

to achieve complete insulation to the ground, some traction current will leak into the ground, forming metro stray current [1]. The stray current of the metro propagates in the ground, causing fluctuations in the surface potential along the line [1], resulting in a DC potential difference between different substations along the metro line, causing DC power to inject into the neutral grounding main transformer of the substation, causing DC bias of the transformer [2]. When a transformer experiences DC bias, its winding current generates a DC component. Under the combined action of AC and DC magnetic flux, the iron core will enter a saturation state in advance, leading to increased hysteresis and contraction [3]. As a result, transformer vibration will also intensify, causing problems such as noise and temperature rise in the transformer [4], and even leading to malignant power outages [5–7]. Therefore, it is of great significance to study the impact of metro stray current on transformer DC bias, select corresponding evaluation indicators for transformer DC bias, and establish corresponding evaluation models.

At present, research on metro stray current has achieved certain results, such as dynamic modeling, prevention and control. Shi Yuntao et al. established a simulation model for stray current distribution based on the PSCAD/EMTDC simulation platform, and studied the distribution pattern and influencing factors of subway stray current in urban power grids [8]. Wu Guoxing et al. established a static distribution model for metro stray current, a multi-layer block soil model in Shenzhen, and a simulation model for Shenzhen AC power grid, they calculated and analyzed the ground potential changes in AC power grid substations caused by stray current and the distribution characteristics of transformer neutral DC current [9]. Reference [10] conducted a joint simulation of a DC traction power supply return system and a stray current diffusion model to study the interference level of stray current. In addition, further research is needed to evaluate the degree of impact of metro stray current on transformer DC bias.

The paper establishes a coupling model between stray current and DC bias of the main transformer in urban substations, and achieves dynamic calculation of stray current. And by selecting indicators such as neutral point DC, excitation current DC component, and harmonic distortion deviation, effective evaluation of the degree of DC bias influence on the main transformer along the metro line was achieved.

## 2 Modeling and Analysis of DC Bias of Transformers Along the Line Caused by Metro Stray Current

### 2.1 Coupling Model of Metro Stray Current and Magnetic Bias of Main Transformer in Urban Substation

During the actual operation of the metro, due to the incomplete insulation of the steel rails to the ground, some traction current leaks and injects into the ground to form stray current, causing changes in the ground potential along the track, resulting in a DC potential difference. The main transformers of substations at different locations along the line are affected by potential differences $\Delta U_{DC}$, neutral point bias DC is generated, causing transformer DC bias phenomenon, as shown in Fig. 1.

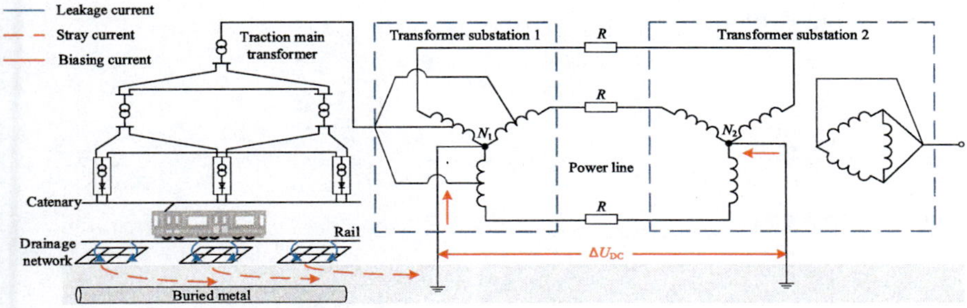


**Fig. 1.** Schematic diagram of DC bias of the main transformer along the line caused by metro stray current.

## 2.2 Analysis of the Distribution Characteristics of Stray Current in Urban Areas

Transition resistance is an important parameter that reflects the distribution characteristics of stray current and the impact of metro on the ground potential along the line. Due to the possibility of traction current leaking into the ground at any position of the power supply section steel rail, the load of the running train will also change with location and time, and the soil environment along the metro line is different, resulting in non-uniform distribution of transition resistance, as shown in Fig. 2. Therefore, ideal assumptions are used for model derivation. Assuming that the longitudinal resistance of the steel rail, drainage network, etc. is uniformly distributed, and the transition resistance between the steel rail and the earth, drainage network, etc. is uniformly distributed, as well as the parameters of the traction substations at both ends are the same when powered by two terminals.

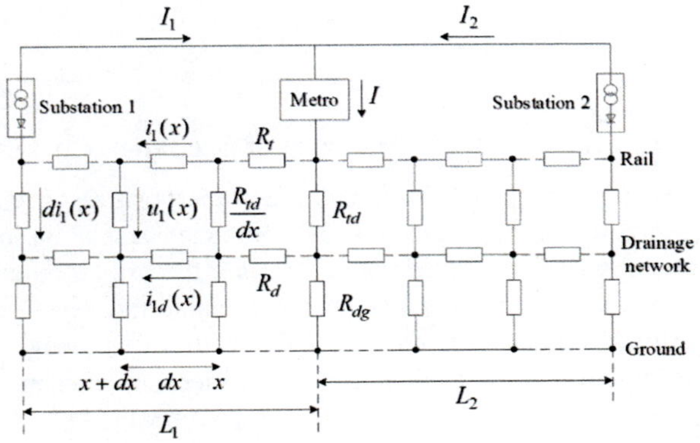


**Fig. 2.** Equivalent model of transition resistance under dual terminal power supply mode.

At the same time, the differential element method and Kirchhoff's law are used to calculate and derive the analytical formula for metro stray current. Figuring $\alpha^2 = (R_t + R_d)/R_{td}$, by writing the circuit equations in the above figure, the expressions for track voltage and stray current leakage can be obtained separately.

(1) When $x \in [0,L_1]$:

$$i_{s1}(x) = \frac{I_1 R_t}{R_t + R_d}\left(th\frac{\alpha L_1}{2} \cdot sh\alpha x\right) + \frac{I_1 R_t}{R_t + R_d}(1 - ch\alpha x) \tag{1}$$

$$u_{s1}(x) = \frac{R_t\sqrt{R_{td} \cdot (R_t + R_d)}}{R_t + R_d} I_1 \cdot sh\alpha x - \frac{R_t\sqrt{R_{td} \cdot (R_t + R_d)}}{R_t + R_d} I_1 \cdot th\frac{\alpha L_1}{2} \cdot ch\alpha x \tag{2}$$

(2) When $x \in [L_1,L_2]$:

$$i_{s2}(x) = \frac{I_2 R_t}{R_t + R_d}\left(th\frac{\alpha L_2}{2} \cdot sh\alpha(L - x)\right) + \frac{I_2 R_t}{R_t + R_d}(1 - ch\alpha(L - x)) \tag{3}$$

$$u_{s2}(x) = \frac{R_t\sqrt{R_{td} \cdot (R_t + R_d)}}{R_t + R_d} I_2 \cdot sh\alpha(L - x) - \frac{R_t\sqrt{R_{td} \cdot (R_t + R_d)}}{R_t + R_d} I_2 \cdot th\frac{\alpha L_2}{2} \cdot ch\alpha(L - x) \tag{4}$$

# 3 Evaluation Model for DC Bias of Main Transformers in Substations Along the Metro Line

## 3.1 Simulation Model for Hysteresis Characteristics of Transformers

The hysteresis characteristics of transformer cores are closely related to the operating performance of transformers under DC bias. The main model used to describe the hysteresis characteristics of iron cores is the Jiles Atherton (J-A) hysteresis model, which has the advantages of less parameter description and clear physical meaning. Therefore, it is often used to describe the hysteresis loop characteristic curve of transformers, as shown in Eq. (5).

$$\frac{dM}{dH} = \frac{c\frac{dM_{an}}{dH_e} + \frac{M_{an}-M}{\frac{\delta k}{\mu_0} - \frac{\alpha(M_{an}-M)}{1-c}}}{1 - \alpha c\frac{dM_{an}}{dM_e}} \tag{5}$$

## 3.2 Numerical Calculation of DC Component of Excitation Current

When a transformer experiences DC bias, under the superposition of AC and DC magnetic flux, the excitation current generates a DC component, resulting in an increase in harmonic content and waveform distortion. Therefore, the excitation current of transformers can be used as an important indicator to evaluate the degree of influence of DC bias on transformers. For a periodic signal $f(t)$ with a frequency of $F = 1/T$, the DC component is the 0th harmonic content of its Fourier decomposition, which is $a_0$, as shown in Eq. (6). As shown in Eq. (7), the expression for the DC component of the excitation current can be further obtained.

$$\int_0^T a_0 dt = \int_0^T f(t)dt - \int_0^T \sum_{n=1}^{+\infty} [a_n \cos n\omega t + b_n \sin n\omega t]dt \tag{6}$$

$$I_{DC} = \frac{\int_0^T i(t)dt}{T} \tag{7}$$

### 3.3 Waveform Distortion Rate Offset

After the excitation current generates a DC component, the even harmonic content increases sharply, and the waveform undergoes serious distortion. The total harmonic distortion rate (*THD*) of the excitation current represents the degree of waveform distortion of the excitation current, and its value is expressed as the square root of the ratio of the effective values of all harmonic components of a specific order N to the effective values of the fundamental component. As the amplitude of the neutral point direct flow injection increases, the even harmonic content increases, and the degree of waveform distortion further increases, the expression of waveform distortion rate offset $\Delta THD$ is shown in Eq. (8).

$$\Delta THD = \frac{\sqrt{\sum_{n=2}^{N}\left(\frac{I_n}{I_1}\right)^2} - \sqrt{\sum_{n=2}^{N}\left(\frac{I_{n'}}{I_{n'}}\right)^2}}{\sqrt{\sum_{n=2}^{N}\left(\frac{I_{n'}}{I_{n'}}\right)^2}} \cdot 100\% \tag{8}$$

### 3.4 Evaluation Indicators for the Impact of Transformer DC Bias

The State Grid Corporation of China stipulates that the maximum direct current withstand by the neutral point of transformers with voltage levels of 110 kV and above is 3A [11]. As shown in Table 1, establish evaluation indicators for transformer DC bias under the influence of metro stray current. Among them, levels C, B, A, and S correspond to the degree of weak, moderate, strong, and severe DC bias influence on the transformer.

**Table 1.** Evaluation indicators for the impact of transformer DC bias.

| Neutral point DC | Excitation current DC rate | | $\Delta THD$ | Extent |
|---|---|---|---|---|
| <1 A | <0.05 A | | <10% | C |
| 1–2 A | 0.05–0.10 | | 10–100% | B |
| 2–3 A | 0.10–1 A | | 100–200% | A |
| >3 A | >1 A | | >200% | S |

## 4 Simulation Verification and Case Analysis

Set various parameters of substations 1 and 2 to be the same, while considering the actual operation of metro locomotives, simulate the surface potential difference along the line using an adjustable DC voltage source $\Delta U_{DC}$. The parameters of the transmission line along the metro and the main transformer parameters of the urban substation are shown in Table 2.

**Table 2.** Transmission line parameters along the metro and main transformer parameters of urban substations.

| Simulation parameters | Value |
|---|---|
| Line length/kM | 25 |
| Rated capacity/MVA | 63 |
| Load impedance/Ω | 350 |

## 4.1 Simulation Results of Transformer DC Bias Under Different Ground Potential Differences

Based on the actual operation of metro locomotives, consider the distribution of surface potential along the line under the operation of multiple sections and trains [12]. Simulate the DC potential difference by setting the numerical values of different adjustable DC voltage sources.

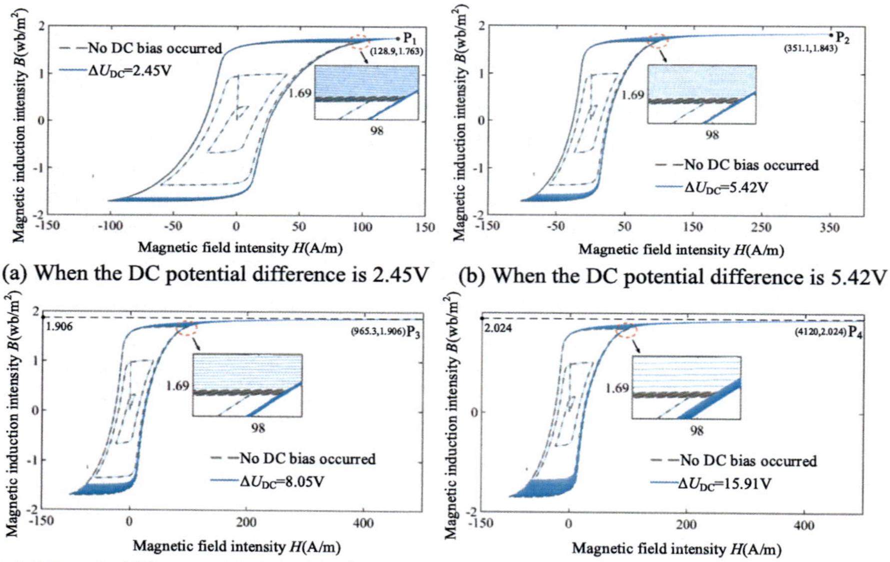


(a) When the DC potential difference is 2.45V (b) When the DC potential difference is 5.42V

(c) When the DC potential difference is 8.05V (d) When the DC potential difference is 15.91V

**Fig. 3.** Schematic diagram of transformer core hysteresis loop under different ground potential differences.

The simulation results of the transformer core hysteresis loop are shown in Fig. 3. As the ground potential difference occurs along the metro line, the main transformer in the urban area experiences DC bias, and the core hysteresis loop undergoes distortion. Due to the generation of the DC component, compared to the unbiased state, the working point of the iron core moves to the nonlinear region. As the magnetic field intensity *H* increases, the magnetic induction intensity $B$ no longer changes. As shown in Fig. 3(a), when the local potential difference is 2.45V, the magnetic induction intensity value approaches 1.763wb/m$^2$, and as the ground potential difference increases, the maximum magnetic induction intensity value continuously increases, and the degree of hysteresis loop distortion becomes more severe.

Further analysis is conducted on the waveform and data characteristics of transformer excitation current under different ground potential differences, as shown in Fig. 4.

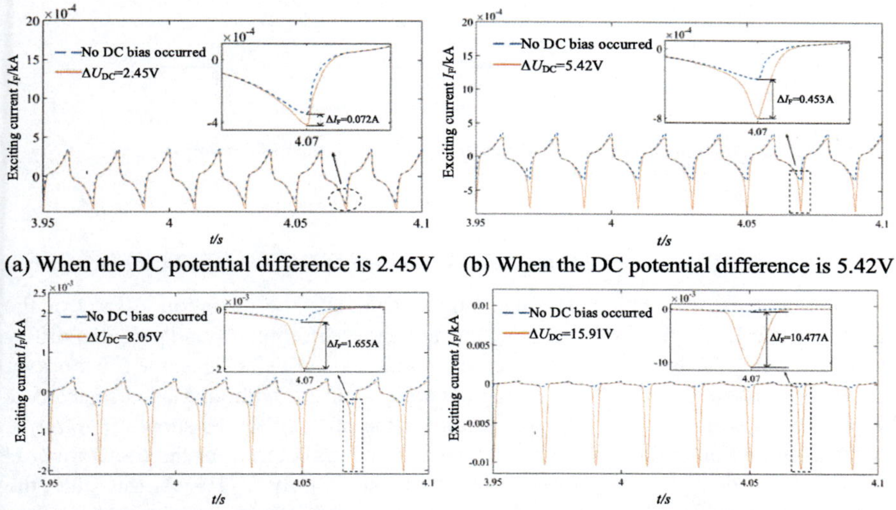


(a) When the DC potential difference is 2.45V (b) When the DC potential difference is 5.42V

(c) When the DC potential difference is 8.05V (d) When the DC potential difference is 15.91V

**Fig. 4.** Schematic diagram of transformer excitation current waveform under different ground potential differences.

The transformer core has hysteresis characteristics, and the excitation current has a non-linear relationship with the main magnetic flux, which results in the excitation current not being a standard sine wave. As shown in Fig. 4(c), when the transformer experiences DC bias, under the combined action of AC and DC magnetic flux, the peak amplitude of the negative half cycle waveform of the excitation current sharply increases, increasing by 1.655A compared to the unbiased state. From this, it can be concluded that when the DC potential difference reaches over 10V, the degree of DC bias in the transformer is more severe.

Perform Fast Fourier Transform (FFT) on the excitation current to obtain the even harmonic content amplitude of the excitation current under different ground potential differences, as shown in Fig. 5. The ground potential difference is 0, which means that when the transformer is in normal operation, the excitation current only contains odd harmonics, and the even harmonics value is 0. As the ground potential difference continues to increase, the even harmonic content of the excitation current gradually increases, and the amplitude of the second harmonic is the largest. From the above analysis, it can be concluded that by analyzing the relevant characteristics of transformer excitation current, the degree of influence of DC bias on the transformer can be further evaluated.

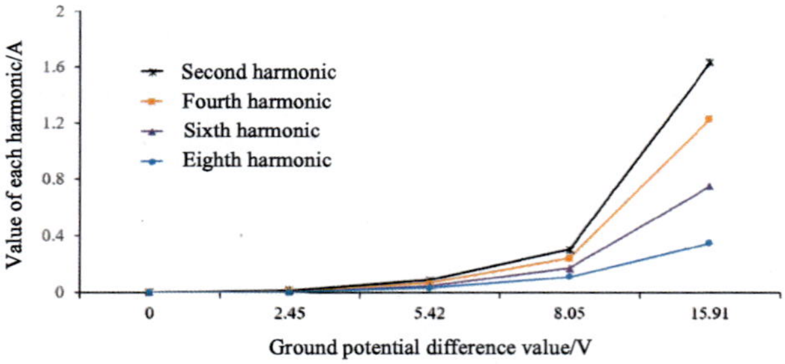


**Fig. 5.** Comparison diagram of even harmonic amplitude of excitation current.

### 4.2 Evaluation Results of the Degree of Influence of DC Bias on Transformers

By combining Eq. (8) with Fig. 3 and Fig. 4, calculate the deviation $\Delta THD$ of the harmonic distortion rate of the excitation current under different ground potential differences, in order to further evaluate the degree of influence of DC bias on the transformer. The total harmonic distortion rate of the excitation current without bias is selected as the reference harmonic distortion rate, with a value of 45.57%. As shown in Table 3, when the ground potential difference is 2.45V, the neutral current of the transformer is 0.54A, and the deviation of harmonic distortion rate is only 5.71%. At this time, the transformer is weakly affected by DC bias, and the evaluation level is C (weak); When the ground potential difference increases to 15.91V, the neutral current of the transformer increases by 9.8A, and the deviation of harmonic distortion rate is as high as 239.57%. The transformer is most affected by DC bias, and the evaluation level is S (severe).

**Table 3.** Evaluation results of transformer dc bias caused by stray current in subway.

| $\Delta U_{DC}$/V | Neutral point DC/A | Excitation current DC rate/A | $\Delta THD$/% | Extent |
|---|---|---|---|---|
| 2.45 | 0.54 | 0.02 | 5.71 | C |
| 5.42 | 1.50 | 0.09 | 88.85 | B |
| 8.05 | 2.91 | 0.29 | 198.90 | A |
| 15.91 | 10.34 | 1.43 | 239.57 | S |

## 5 Conclusion

In response to the issue of transformer DC bias caused by metro stray current along the line, the paper proposes a transformer DC bias evaluation model based on ground potential coupling. A coupling model between stray current and DC bias of the main transformer in urban substations was established, and the distribution model of transition resistance during actual subway operation was analyzed, achieving numerical calculation of stray current. On this basis, an evaluation model for the degree of DC bias influence on transformers was established. The research results indicate that as the ground potential difference increases, the impact of DC bias on the transformer continues to increase. When the ground potential difference reaches 15.91V, the transformer is evaluated as severely affected by DC bias.

**Acknowledgments.** This research was supported by the Science and Technology Project of the State Grid Corporation of China: Research on the coupling influence and suppression method of grid DC bias caused by metro stray current (5100-202199276A-0-0-00).

## References

1. Ogunsola, A., Sandrolini, L., Mariscotti, A.: Evaluation of stray current from A DC-electrified railway with Integrated electric–electromechanical modeling and traffic simulation. IEEE Trans. Ind. Appl. **51**(6), 5431–5441(2015)
2. Li, X., Chu, F., Shi, S.: Influence of rail transit on neutral current and vibration of 220 kV substation along the line. Trans. China Electrotech. Soc. **36**(S2), 423–429+437 (2021) (in Chinese)
3. Lin, S., Huangfu, Y., Zhou, Q., et al.: Evaluation and analysis model of stray current in the metro depot. IEEE Trans. Transp. Electrification **7**(7), 1780–1794 (2021)
4. Li, H., Cui, X., Lu, T., et al.: Electric circuit and magnetic circuit combined model of DC biased power transformer. Proc. CSEE **29**(27), 119–125 (2009). (in Chinese)
5. Wang, Z., Li, M., Xuan, M., et al.: Temperature Rise Test and Simulation of Single-phase Four-column Transformer under DC-bias. Trans. China Electrotech. Soc. **36**(05), 1006–1013 (2021). (in Chinese)
6. Zhang, S., Wang, K., Li, J, et al.: Tests of DC bias key performances of power transformer with single-phase three-limb core under no-load and rated-load conditions. Proc. CSEE **39**(14), 4334–4344+4281(2019) (in Chinese)
7. Chen, J., Wang, D., Cheng, S., et al.: Influence of DC-biased magnetic induction on magnetic property of silicon steel. IEEE Trans. Magn. **55**(5) (2019)
8. Wang, A., Lin, S., Hu, Z., et al.: Evaluation model of DC current distribution in AC power systems caused by stray current of DC metro systems. IEEE Trans. Power Deliv. **36**(1), 114–123 (2021)
9. Shi, Y., Zhao, L., Lin, S., et al.: Analysis of distribution of metro stray current in urban power grid and its influencing factor. Power Syst. Technol. **45**(05), 1951–1957 (2021). (in Chinese)
10. Wu, G., Xiao, L., Zhang, F., et al.: Simulation analysis of distribution characteristics of urban rail transit stray current in power system. South. Power Syst. Technol. **13**(10), 39–43+61 (2019) (in Chinese)
11. Liu, W., Zhou, L., Tang, Y., et al.: Co-simulated model of DC traction power supply return system and stray current diffusion. Trans. China Electrotech. Soc., 1–11(2023) (in Chinese)
12. Ma, S., Rong, Z., Lin, X., et al.: Study on the global optimal configuration of DC bias equipment considering the cooperation of multiple devices. Proc. CSEE **11**(06), 85–91 (2014). (in Chinese)
13. Ni, Y., Zeng, X., Yu, K., et al.: Modeling and influencing factors analysis of dynamic potential distribution caused by metro stray current. Proc. CSEE, 1–14 (2022) (in Chinese)

# Research on the Design of High-Speed Railway Express Space-Time Network

Qi Zhou[1], Li Wang[1,2](✉), and Xiaoning Zhu[1,2]

[1] School of Traffic and Transportation, Beijing Jiaotong University, Beijing 100044, China
{22121000,wang.li,xnzhu}@bjtu.edu.cn

[2] Key Laboratory of Transport Industry of Comprehensive Transportation Theory (Beijing Jiaotong University), Ministry of Transport, Beijing 100044, China

**Abstract.** Aiming at the "multi-hub-multi-distribution-spoke" multimodal network layout and specific transportation needs in the operation of high-speed rail (HSR) express transportation, this paper builds a space-time network with the coupling of physical network layer and task layer based on space-time network theory, aims at reducing the operating cost of HSR express transportation network and focuses on the transportation capacity and cargo timeliness of HSR. This paper constructs a space-time network design model of HSR express cargo considering "rail-rail" horizontal transfer mode and "highway-rail" vertical collection and distribution mode. In this paper, 20 transport demand cases are designed and solved by using CPLEX in Python to verify the validity of the model and provide an effective solution to the actual HSR express cargo hub-and-spoke network design problem.

**Keywords:** High-speed railway express · Hub-and-spoke network · Space-time network

## 1 Introduction

A hub-and-spoke network is a network structure consisting of large hub yards and demand points as spokes. Domestic and foreign research on the design issues of high-speed railroad hub-and-spoke networks is less compared to that under road, waterway, or air transport modes. High-speed railway (HSR) express transportation is a new type of express transportation. The actual share of HSR express transport is less than 1%. Therefore, scholars focus on how to effectively use the HSR network to provide HSR express transport services and improve the "highway-rail" and "rail-rail" context of network logistics operation services, operations departments.

In this paper, a space-time network with coupling of physical network and task layers is built. This research builds an objective function from the perspective of minimizing the operation cost of HSR express transportation network, focusing on the constraints of high-speed rail transportation capacity and cargo timeliness, and build a network that considers the "rail-rail" horizontal transit mode and "rail-road" vertical collection and distribution mode.


Y. Qin et al. (Eds.): EITRT 2023, LNEE 1137, pp. 246–254, 2024.
https://doi.org/10.1007/978-981-99-9311-6_27

## 2 Literature Review

In order to highlight the main contributions of this research, we now include and summarize some representative papers on hub-and-spoke network and HSR network design, as shown in Table 1.

**Table 1.** Review of the existing literature

| Paper | Transportation Mode | Model | | | Solution |
|---|---|---|---|---|---|
| | | Type | Objtective fuction | Constraint | |
| Ni and Shi [1] | Highway | MILP | Min cost | Time | SA |
| Lv [2] | highway, Railway, air, ship | MILP | Min cost | Time | Tabu search |
| Zhang [3] | Highway, air | MILP | Min cost | Time | Benders |
| Zhou et al. [4] | Highway, railway | MILP | Service level/min cost | time & capacity | Tabu search |
| Wang et al. [5] | Highway, railway | MILP | Min (cost/time) | time | Memetic algorithm |
| Delgado et al. [6] | Highway, air | MILP | Max revenue | Time & capacity | Nested Decomposition |
| Chen et al. [7] | High-speed railway | MILP | Min cost | Time & capacity | Gurobi |
| Bi et al. [8] | High-speed railway | MILP | Service level/min cost | | CPLEX |
| Shen [9] | High-speed railway, air | MILP | Min cost | Distance | GA & PSO algorithm |
| Li [10] | High-speed railway | MILP + space-time Network | Min cost | time & capacity | Benders & column-pool based approximation |
| Yan [11] | High-speed railway | MILP | Min cost | Time & capacity | LR |
| This paper | Highway, high-speed railway | MIP + space-time Network | Min cost | Time & capacity | CPLEX |

This research studies the design of HSR express network in the context of "highway-rail" and "rail-rail", analyzes the process of intermodal transportation between highway and HSR in the spoke network, constructs the objective function from the perspective of minimizing the total cost, considers the storage, reload and transportation process of cargo, customer's preference of arrival time, and the time of transportation. In addition, this study combines the space-time network theory to build the space-time network model of HSR express cargo in the context of "highway-rail" and "rail-rail", and explore the methods that can reduce the operation cost, so as to provide a practical solution to the problem. To provide an effective solution to the problem of designing HSR express network.

## 3 Modeling Work

### 3.1 Study Area

In this paper, a HSR express network in the context of "highway-rail" and "rail-rail" is constructed, and its transport operation mode is shown in Fig. 1. How to arrange the flow direction of each cargo in this network, to meet the requirements of HSR transportation capacity and cargo timeliness, and to reduce the network operation cost and reduce network structure adjustment has become a key issue in the design of HSR express network in the context of "highway-rail" and "rail-rail".

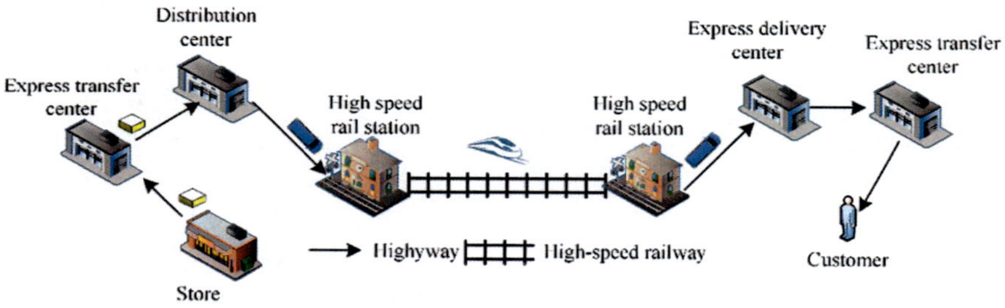


**Fig. 1.** HSR express operation mode

### 3.2 Space-Time Network Construction

The space-time network of HSR express transportation in this paper is represented as $G(V, A)$, and the time is discretized to 1 min as a time unit. Which set the actual express transit center, distribution center and HSR station in the network as the physical node set $N^{ph}$, the link set between physical nodes as $L^{ph}$, set virtual points $N^{vir}$ according to the goods, the highway virtual point of goods $k$ set as $i^k = i^{ph} \times 100 + k$, the HSR station virtual point of goods k set as $i^k = i^{ph} \times 1000 + k$, resulting in the space-time network Point set $N = N^{ph} \cup N^{vir}$. Let $L^{ph}$ be the set of links between physical nodes, $(i^{ph}, j^{ph}) \in L^{ph}$.

$V$ is the set of points for space-time network work, each vertex contains spatial node $i$ and time $t$, which can be denoted as $(i, t)$, where $i \in N$. $A_k$ denotes the set of cargo k space-time network arcs, including storage arc $A_k^{store}$, dressing arc $A_k^{reload}$ and transport arc $A_k^{trans}$, $A_k = A_k^{store} \cup A_k^{reload} \cup A_k^{trans}$, where $k$ denotes the specific shipment number.

The set of links between physical nodes $L^{ph}$, belongs to the physical network layer and $A_k$ belongs to the task layer. Figure 2 shows the space-time network coupled with the physical network layer and the task layer, and Fig. 3 shows a space-time network diagram of a HSR express with seven actual physical nodes.

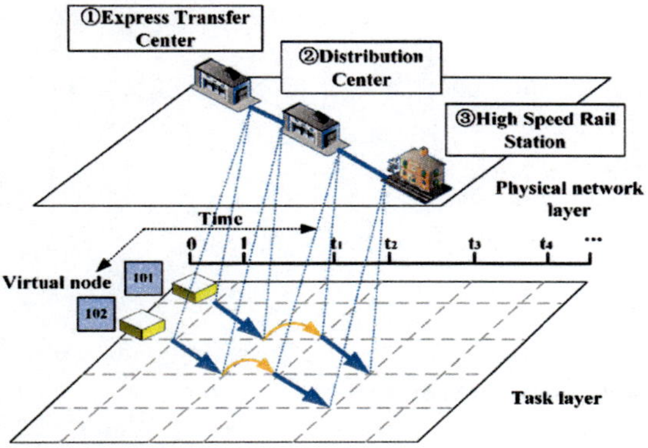


**Fig. 2.** Connection between physical network layer and task layer

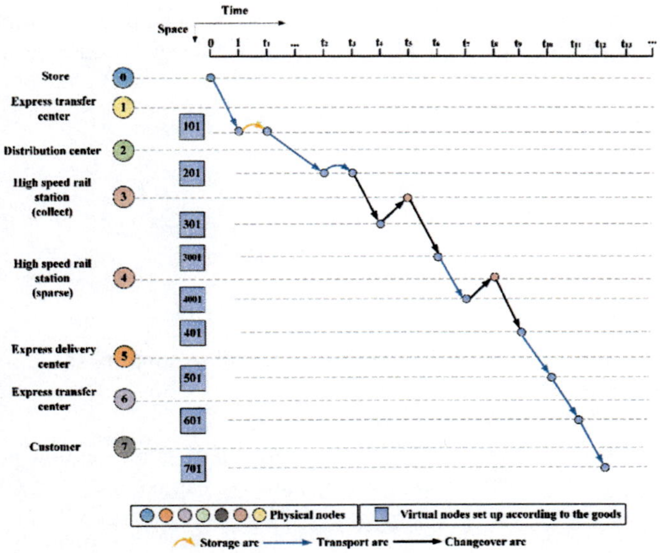


**Fig.3.** Graph of the space-time arc of the cargo operation

### 3.3 Model Establishment

The basic assumptions of this paper are as follows: distribution centers, hubs and HSR stations have the operating conditions and capacity to match freight demand; the participation of high-speed passenger trains in freight transport will not affect high-speed passenger travel; the efficiency of loading and unloading is stable; the number of supply points and the supply quantity, the number of demand points and the demand quantity are determined.

This paper aims to minimize the cost of delivering all goods to the customer. Equation (1) demotes the total cost:

$$Z = \min \sum_{k \in K} \sum_{(i,j,t,t') \in A_k} c(i,j,t,t') y^k(i,j,t,t') \tag{1}$$

where,

$K$: denotes the set of goods, denotes the set of goods, k is the number of goods, is the number of goods, $k = 1, 2, \ldots, K$,
$(i, j, t, t^{'})$: denote the arcs in the space-time network,$(i, j, t, t^{'}) \in A_k$,
$c(i, j, t, t^{'})$: The cost of the arc $\left(i, j, t, t^{'}\right)$,
$y^k(i, j, t, t^{'})$: 0-1 variable, if cargo $k$ is assigned to arc $\left(i, j, t, t^{'}\right) = 1$; otherwise, 0.

This paper adopts the space-time network research method to construct an integer programming model for HSR express operation, which needs to consider the following constraints.

(1) Space-time network flow balance constraint

$$\sum_{g_k \in \delta_v^-} y_{g_k}^k - \sum_{g'_k \in \delta_v^+} y_{g'_k}^k = \begin{cases} 1 \; n_v = o_k, t_v \in [T_{el}, T_{ll}] \\ -1 \; n_v = d_k, t_v \in [T_{ea}, T_{la}] \\ 0 \; otherwise \end{cases}, \forall k \in K \tag{2}$$

where,

$g_k$: denotes the arc in the spacetime network, and to simplify the representation,, $g_k = (i, j, t, t\prime)$,
$y_{g_k}^k$: The simplified representation of $y^k(i, j, t, t\prime)$,
$\delta_v^-, \delta_v^+$: denotes the set of arcs at inflow point $v$, the set of arcs at outflow point $v$,
$o_k, d_k$: virtual pickup points, virtual delivery points of cargo $k$, $o_k, d_k \in N^{vir}$,

$n_v, t_v$: the nodes, time corresponding to the space-time network vertices $v$,

$[T_{el}, T_{ll}]$: The earliest departure time, the latest departure time,

$[T_{ea}, T_{la}]$: The earliest arrival time, latest arrival time.

(2) Commitment to delivery time constraint

Given the maximum committed delivery time of the goods, ensure that the transit time of the goods is always less than the given delivery time limit. The constraint is represented as Eq. (3):

$$\sum_{t\in T} t\prime y^k\left(i^k, d_k, t, t\prime\right) - \sum_{\tau\in T} \tau\prime y^k\left(o_k, j^k, \tau, \tau\prime\right) \le T_k, \forall\left(i^k, d_k, t, t\prime\right), \left(o_k, j^k, \tau, \tau\prime\right) \in A_k, \forall k \in K \tag{3}$$

where,

$T_k$: denoting the committed arrival time frame of cargo *k*,

$i^k, j^k$: $i^k, j^k$ denotes the virtual nodes of cargo, $i^k, j^k \in N^{vir}$,

(3) Transportation capacity limitation

For any physical network connection $(i^{ph}, j^{ph})$, the corresponding transportation arc carries no more freight than the transportation capacity limit of the physical network connection $(i^{ph}, j^{ph})$. The constraint is represented as Eq. (4):

$$\sum_{k\in K} \sum_{(i^k, j^k, t, \tau)\in A_k^{trans} | i^k = i^{ph}, j^k = j^{ph}} D^k y^k\left(i^k, j^k, t, \tau\right) \le Cap\left(i^{ph}, j^{ph}\right), \forall\left(i^{ph}, j^{ph}\right) \in L^{ph} \tag{4}$$

where,

$D^k$: denotes the volume of cargo $k$,

$Cap\left(i^{ph}, j^{ph}\right)$: physical network link $(i^{ph}, j^{ph})$ transport capacity limit.

The model constructed in this paper is a integer programming model, which can be solved by calling CPLEX using Python.

## 4 Case Study

### 4.1 Data Preparation

This case study contains 20 goods, and there is a one-to-one correspondence between suppliers and customers, i.e., the goods in supplier 1 are sent to customer 1. Considering the whole process of transporting goods from suppliers to customers, the example part of this paper constructs a cargo transportation network with 12 physical nodes. The transport time and transport cost are marked on the network link, such as (62,41) represents the transport time and transport cost from transshipment center 100 to distribution center 300 are 62 min and 41 yuan respectively, as shown in Fig. 4.

There are four types of HSR express products, same-day delivery, next-morning delivery, next-day delivery, and the-day-after-tomorrow delivery, which account for about 8%, 6%, 76%, and 10% of the total HSR express transportation respectively [12]. In this paper, according to the proportion of the four types, 20 cargo demands are classified and 2 same-day, 2 next-morning, 14 next-day and 2 the-day-after-tomorrow

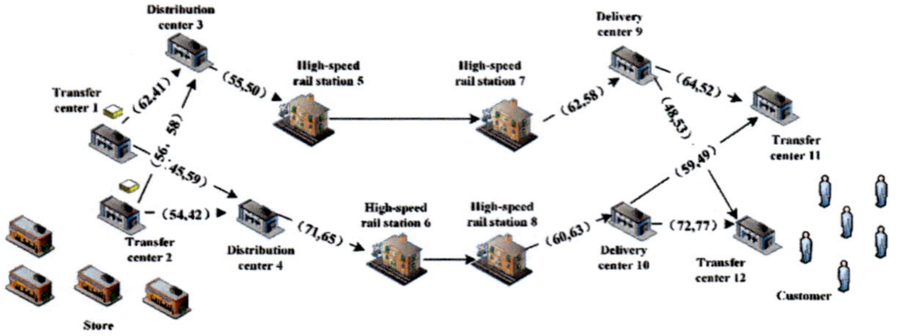


**Fig.4.** Physical network in this example

demands are set. The product time requirements for the four types of HSR express are: 12h, 18h, 24h and 30 h.

According to the research results of Yan [11], the data needed in the following calculation cases were collected, as shown in Table 2:

**Table 2.** Parameters and values

| Parameter | Numerical value |
|---|---|
| Maximum transport capacity in passenger train piggyback mode | 1300 kg |
| Highway-high-speed rail station changeover time | 10 min |
| Highway-highway changeover time | 5 min |
| Cost of loading and unloading at highway-high-speed railway station | 10 yuan |
| High-speed railway station storage cost | 5 yuan |
| Storage cost at distribution center | 3 yuan |
| Transfer center storage cost | 2 yuan |

## 4.2 Results

After using Python to call CPLEX to solve the model, the results meet the requirements of high-speed railway transportation capacity and cargo timeliness. Figure 5 shows 20 cargo transportation time diagram, solve the total cost of 7441 yuan, thus can prove that the constructed model is effective.

There are four types of HSR express service products, same-day, next-morning, next-day, and the-day-after-tomorrow. Table 3 shows the cost of different cargo. The average cost of the four types of products after passing the solution is 413.5 yuan, 374 yuan, 368.2 yuan and 355 yuan, and the model can also provide reference suggestions for pricing different products of HSR express in the future.

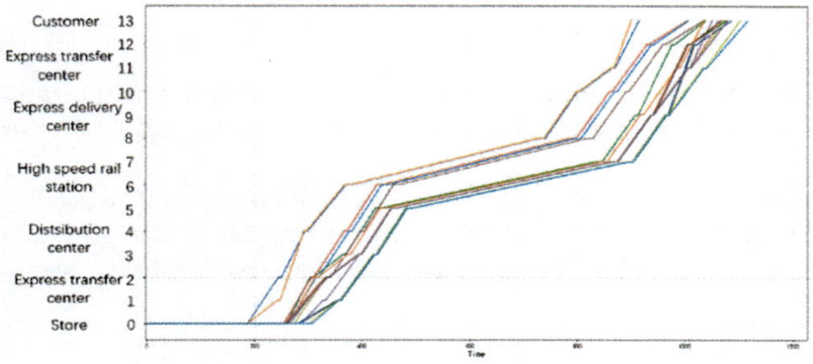


**Fig.5.** Space-time diagram of cargo transportation

**Table 3.** Different cargo and their cost

| Cargo number | Cost/yuan | Cargo number | Cost/yuan | Cargo number | Cost/yuan | Cargo number | Cost/yuan |
|---|---|---|---|---|---|---|---|
| 1 | 414 | 6 | 365 | 11 | 340 | 16 | 377 |
| 2 | 413 | 7 | 369 | 12 | 379 | 17 | 374 |
| 3 | 371 | 8 | 363 | 13 | 372 | 18 | 369 |
| 4 | 377 | 9 | 382 | 14 | 370 | 19 | 355 |
| 5 | 364 | 10 | 364 | 15 | 368 | 20 | 355 |

# 5 Conclusions

In order to reduce the operating cost of HSR express transportation network, the space-time network design model of HSR express cargo network considering "rail-rail" horizontal transfer mode and "rail-highway" vertical collection and distribution mode is constructed with the constraint requirement of cargo timeliness. The study case in this paper contains 20 demands. This research solves the model by using Python to call CPLEX. The results verify the validity of the model and providing decision basis for pricing of different products of HSR express transport.

More practical factors will be considered in the future. For example, the space-time network model should be enriched by combining four modes: passenger train piggyback mode, mixed passenger and freight, confirmation train, and special HSR express train. In addition, for the case that the running time of CPLEX solution is too long, the algorithm of reduced calculation scale can be used to solve the cases and reduce the solution time.

**Acknowledgements.** This work is supported by the Joint Funds of the National Natural Science Foundation of China (No. U2034208).

## References


1. Ni, L., Shi, F.: A hub siting and distribution optimization method for multi-distribution express spoke networks: theory and practice of system engineering, 32(02), 441–448. (2012). (in Chinese)
2. Lv, Y.: Research on Optimization of Beijing-Tianjin-Hebei Regional Logistics Network Based on Spoke-Based Network. Beijing Jiaotong University, China (2016). (in Chinese)
3. Zhang, H.: Multi-modal Hub Location and Network Design. Beijing Jiaotong University, China (2022). (in Chinese)
4. Zhou, J., Jjiang, Y., Shen, Y., Pantelous Athanasios, A., Liu, Y., Huang, C., Mei, L.: Intermodal hub-and-spoke logistic network design with differentiated services: the case of China railway express. Inf. Sci. **612**, 796–815 (2022)
5. Wang, R., Yang, K., Yang, L., Gao, Z.: Modeling and optimization of a road–rail intermodal transport system under uncertain information. Eng. Appl. Artif. Intell. **72**, 423–436 (2018)
6. Delgado, F., Trincado, R., Pagnoncelli, B.K.: A multistage stochastic programming model for the network air cargo allocation under capacity uncertainty. Transp. Res. Part E: Logist. Transp. Rev. **131**, 292–307 (2019)
7. Chen, X., Zhou, P., Lang, M., Yu, J., Li, S.: Optimization of freight flow distribution and organization mode of HSR express transport under demand changes. Transp. Syst. Eng. Inf. **22**(05), 174–186 (2022). (in Chinese)
8. Bi, M., He, S., Xu, W.: Express delivery with high-speed railway: definitely feasible or just a publicity stun. Transp. Res. Part A: Policy Pract. **120**, 165–187 (2019)
9. Hao, S.: Research on the Location of HSR Express Hub Based on the Value of Freight Tim. Southwest Jiaotong University (2021). (in Chinese)
10. Li, S., Zhu, X., Shang, P., Li, T., Liu, W.: Optimizing a shared freight and passenger high-speed railway system: a multi-commodity flow formulation with Benders decomposition solution approach. Transp. Res. Part B: Methodol. **172**, 1–31 (2023)
11. Yan, L.: Collaborative study of HSR express transportation organization. Bei-jing Jiaotong University, China (2022). (in Chinese)
12. Wu, Y.: Research on the operation mode of high-speed railroad freight trains. Southwest Jiaotong University, China (2017). (in Chinese)

# Optimization Strategy of Arrival and Departure Tracks Distribution in China Railway Highspeed Express Based on Lateral Transshipment Between Trains

Hanchen Zuo[1], Li Wang[1,2(✉)], and Xiaoning Zhu[1,2]

[1] School of Traffic and Transportation, Beijing Jiaotong University, Beijing 100044, China
{zuohanchen,wang.li,xnzhu}@bjtu.edu.cn

[2] Key Laboratory of Transport Industry of Comprehensive Transportation Theory (Beijing Jiaotong University), Ministry of Transport, Beijing 100044, China

**Abstract.** Railway highspeed rail express transportation has become a current hotspot, and its operation will inevitably have an impact on the current highspeed rail passenger transport organization. Therefore, this article takes highspeed railway stations as the research object, fully considering the resources of the section, throat area, arrival and departure tracks, and the lateral transshipment demand of China railway highspeed express, constructs a space-time network for the station. With the shortest delay time of highspeed rail passenger trains as the optimization goal, a 0–1 integer programming model is established. A case study is designed and solved using CPLEX to verify the effectiveness of the model and achieve a reasonable distribution of stations to arrival and departure tracks.

**Keywords:** Railway Transportation · Lateral Transshipment · China Railway Highspeed Express · Arrival and Departure Tracks Distribution

## 1 Introduction

With changes in consumer attitudes and the vigorous development of e-commerce, there is a growing demand for rapid freight transportation services. General Secretary Xi Jinping proposed at the Eighth Meeting of the Central Committee for Financial and Economic Affairs that efforts should be made to strengthen the construction of highspeed rail freight transportation capacity and accelerate the formation of a safe and efficient logistics network that connects domestic and international markets. In 2020, a total of 980 highspeed railway routes launched highspeed express delivery services, with an annual freight volume of 44,000 tons, which was nearly 10 times the growth rate of railway freight transportation. From the demand perspective of express delivery transportation, highspeed rail can meet the needs of high-efficiency express transportation for medium to long distances ranging from 600 km to 2,400 km. In 2021, the parcel delivery volume in China exceeded 100 billion, and it is estimated that the parcel delivery volume will exceed 180 billion by 2025. Among them, more than 40% of cross-city clusters and major economic zones with transportation distances exceeding 600 km, while high-efficiency products account for 5%.


Y. Qin et al. (Eds.): EITRT 2023, LNEE 1137, pp. 255–263, 2024.
https://doi.org/10.1007/978-981-99-9311-6_28

Compared with the boarding and alighting of passengers on highspeed trains, the loading and unloading of goods on highspeed freight trains requires additional requirements for the distribution of arrival and departure tracks resources. As China's current highspeed railway organization is still mainly focused on passenger transportation, it is necessary to consider the distribution of arrival and departure tracks resources, throat area resources, and section occupation in order to develop a feasible plan for arrival and departure tracks resource distribution, which can reduce the impact on the operation of highspeed railway passenger trains and is of great research significance for fully utilizing China's highspeed railway network.

In response to the need for lateral transshipment of goods on highspeed rail freight trains, experts in related fields have conducted the following research on how to develop arrival and departure tracks resource distribution plans.

In terms of the organization of highspeed rail freight transportation, Xu et al. considered the economic feasibility of hub construction and transformation as well as the current operating conditions of the existing highspeed rail network, who abstracted the optimization model of the highspeed rail freight transportation network into a mixed-integer second-order cone programming problem and proposed an outer approximation algorithm to solve it, which helps to select more stable hub locations [1]. Zhou et al. conducted a survey on the distribution center network layout of express delivery companies and designed an evaluation system for the layout of highspeed rail freight bases, constructed a highspeed rail freight network layout model based on maximum coverage theory and designed a solution algorithm, which has some reference significance for the construction of specialized highspeed rail freight base networks [2]. Li et al. constructed a node planning index system for the highspeed rail freight network with multiple indicators, considered geographical distance factors, and used k-means clustering algorithm to optimize the initial plan, obtaining the first and second-level node planning schemes for the highspeed rail freight network [3]. Dejax et al. compared and analyzed the differences between road and rail transportation in highspeed rail freight transportation, conducted technical analysis on long-distance transportation of highspeed rail freight on highspeed railways, and finally established an organization model for highspeed rail freight transportation by adopting a multi-network integration transport mode and considering timeliness [4]. Jennifer et al. pointed out how to reduce congestion on US highways by establishing a highspeed rail freight network and designed a service network model for highspeed rail freight transportation [5].

In terms of arrival and departure tracks distribution, Lian et al. proposed a delay-based adjustment model for highspeed railway arrival and departure tracks utilization plans to guide the adjustment of arrival and departure tracks utilization plans under the circumstance of train delays, so as to minimize the impact of delays on all trains [6]. Ren et al. proposed a multi-objective optimization model and an improved fast non-dominated sorting genetic algorithm based on time-periods for arrival and departure tracks utilization planning and optimization in highspeed rail transportation, which effectively improved the quality of passenger service, equipment utilization rate of stations, and robustness of arrival and departure tracks utilization plans [7]. Liu et al. used mixed-integer programming to establish a delayed train-to-line distribution model based on safety constraints, avoiding train accidents and delays spreading, and designed a simulated annealing algorithm to solve the model. The results showed that trains with shorter delays should be given priority to enter the station, while non-delayed trains should be given priority when delays are longer [8]. Chakroborty et al. designed an ant colony optimization algorithm (ACO) for the train route selection problem (TRSP), and using two French stations as examples, demonstrated the superiority of the ACO algorithm for large-scale problems [9]. Sels et al. proposed a model for providing as many trains as possible with access to stations and lines based on train schedules and idle station equipment and facilities. Through data analysis from several stations in Belgium, the optimized solution had a stronger robustness and reduced train delay times well [10].

There is currently a lack of research on optimization strategies for arrival and departure tracks distribution plans for highspeed rail freight trains using existing infrastructure. Therefore, based on the principle that China railway highspeed express trains use fixed arrival and departure tracks distribution while passenger trains have flexible arrival and departure tracks distributions, this paper constructs an arrival and departure tracks distribution model for highspeed rail freight transportation based on the lateral transshipment mode between trains, with the goal of minimizing the impact on passenger trains. The model considers constraints on arrival and departure tracks, throat areas, and sections, and uses the commercial solver CPLEX to solve the model and develops an distribution plan for arrival and departure tracks. An example is used to validate the effectiveness of the model.

## 2 Arrival and Departure Tracks Distribution Space-Time Network

### 2.1 Arrival and Departure Tracks Distribution Problem

The issue of platform distribution involves taking the train timetable as input and assigning all trains passing through and arriving at a particular station to their respective platforms in a reasonable manner, so as to avoid route conflicts while meeting safety requirements. Therefore, when considering platform distribution, it is necessary to take into account not only the constraints on the platform resources but also the constraints on the resources of the sections and station throat area.

### 2.2 Construction of Arrival and Departure Tracks Distribution Space-Time Network

To facilitate the construction of spatiotemporal networks for arrival and departure tracks distribution, stations and sections are simplified as shown in Fig. 1.

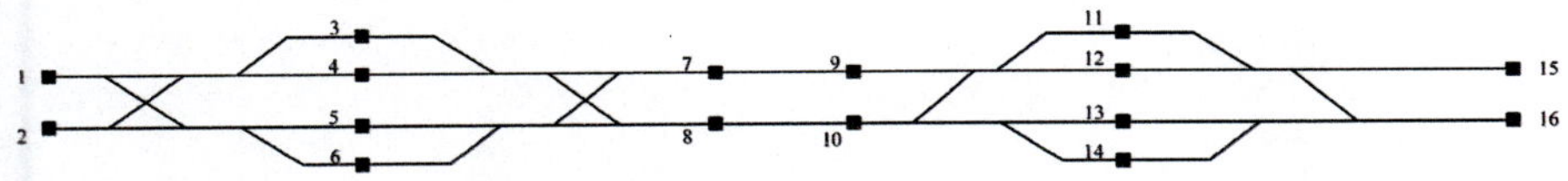


**Fig. 1.** Highspeed railway station diagram

**Table 1.** Time-space arc in station and section

| Time-space arc | Content |
|---|---|
| Section running arc | Space-time trajectory of trains at the section |
| Throat area running arc | Space-time trajectory of trains at the throat area |
| Arrival and departure tracks waiting arc | Space-time trajectory of trains waiting at the arrival and departure tracks |
| Origin waiting arc | Space-time trajectory of trains waiting at the origin point |
| Virtual running arc | Space-time trajectory of trains running during the whole period |

To describe the behavior of trains at stations and sections, the space-time arcs of trains can be abstracted into five types as shown in Table 1.

To generate a basic train timetable that minimizes the impact on highspeed rail passenger trains, given an existing timetable with conflicts, the railway authorities can make the following adjustments:

(1) Adjust the dwell time of trains on platforms.
(2) Adjust the platform distribution scheme.

# 3 Model

## 3.1 Assumption

The model assumption are as follows:

(1) The train timetable and the basic data for the railway operation are known. The conflict of train timetable is known.
(2) China railway highspeed express trains are connected to two fixed tracks on opposite sides of the same platform.
(3) The train's the minimum and maximum stop times at each station are given.
(4) The time is precise to 1 min.
(5) The train length is negligible.
(6) The throat area switch is unlocked at the same time, and the train arrives and departs at the same running time.

### 3.2 Model Construction

**Table 2.** Notation of the mathematical model.

| Sign | Content |
|---|---|
| Sets | |
| $N$ | Set of points in the physical network |
| $E$ | Set of edge in the physical network |
| $V$ | Set of points in the space-time network |
| $A$ | Set of space-time arc |
| $F$ | Set of trains |
| $F^p$ | Set of passenger trains |
| $F^w$ | Set of freight trains |
| Parameters | |
| $c_g$ | Time cost of arc $g$ |
| $\sigma_f$ | The origin points of train $f$ |
| $\gamma_f$ | The destination points of train $f$ |
| $\delta_f^+(v)$ | All space-time arcs departing from space-time point $v$ of train $f$ |
| $\delta_f^-(v)$ | All space-time arcs pointing to space-time point $v$ of train $f$ |
| $K_f$ | The set of all stations that train $f$ passes through, indexed by $k$ |
| $D_f^k$ | The minimum stopping time of train $f$ at station $k$ |
| $D_k^{fm}$ | The maximum stopping time of train $f$ at station $k$ |
| Binary variables | |
| $x_g^f$ | Whether train $f$ uses space-time arc $g$ |

The objective function is to minimize the total cost of train operation, which is (Table 2):

$$\min Z = \sum_{f \in F^w} \sum_{g \in A_f} c_g x_g^f \tag{1}$$

The space-time network flow balance constraints for a single train path are:

$$\sum_{g \in \delta_f^+(\sigma_f)} x_g^f = 1 \; \forall f \in F \tag{2}$$

$$\sum_{g \in \delta_f^+(v)} x_g^f - \sum_{g \in \delta_f^-(v)} x_g^f = 1 \; \forall f \in F \; \forall v \in V \backslash \{\sigma_f, \gamma_f\} \tag{3}$$

The minimum and maximum station dwell time constraint for trains is:

$$\left(1-\sum_{g\in A_f^{vi}} x_g^f\right)D_k^f \le \sum_{g\in A_k^{st}\cap A_f} x_g^f \le \left(1-\sum_{g\in A_f^{vi}} x_g^f\right)D_k^{fm}\ \forall f\in F\ \forall k\in K_f \tag{4}$$

In order to ensure the safety of train operations in a section, it is necessary to establish safety constraint for the section, which includes safety headway for train entry and safety headway for train departure. So incompatible arc set for train departure $\varphi_v^1$ and incompatible arc set for train arrival $\varphi_v^2$ are defined as:

$$\begin{aligned}\varphi_v^1 &= \left\{g|g\in A^{se}, o_g = n_v, 0\le \hat{o}_g - t_v \le T^d\right\}\\ \varphi_v^2 &= \left\{g|g\in A^{se}, d_g = n_v, 0\le \hat{d}_g - t_v \le T^a\right\}\end{aligned} \tag{5}$$

Each space-time arc in any incompatible arc set can only be occupied by one train, the safety headway constraints for the section are:

$$\begin{aligned}&\sum_{g\in\varphi_v^1}\sum_{g\in A_f} x_g^f = 1\ \forall v\in V^d\cup V^{se}\\ &\sum_{g\in\varphi_v^2}\sum_{g\in A_f} x_g^f = 1\ \forall v\in V^a\cup V^{se}\end{aligned} \tag{6}$$

# 4 Experiment

## 4.1 Network Construction and Parameter Assignment

The highspeed railway station is shown in Fig. 2.

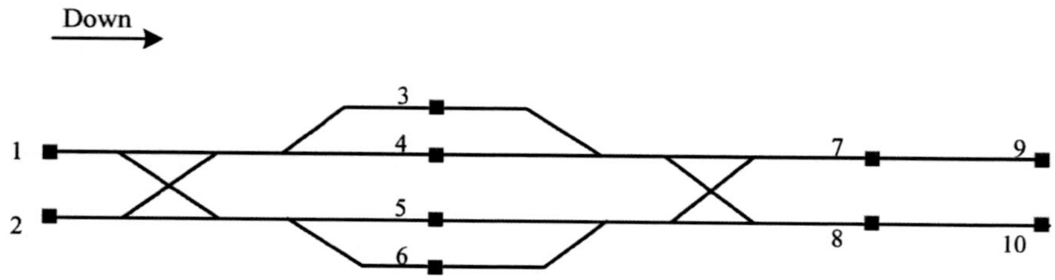


**Fig. 2.** Schematic diagram for highspeed railway station

In this figure, the station has 2 platforms and 4 tracks. There are 10 points and 18 arcs in the space-time network. The research duration is 8:00 to 9:00 and there are 7 trains being studied. The running time for each section of the main line is 4 min, while the running time for entering station is 2 min.

Assuming the departure time window at the station for the passenger train is 4 min, the minimum stopping time for the passenger train is 2 min. That is to say, the stopping time range is 2 min. And the time window of stopping time for China railway highspeed trains is 20 min to 40 min. The arrival time of the train is calculated from the time when it arrives at the throat area of the station; the departure time of the train is calculated from the time when it leaves the arrival and departure track.

For China railway highspeed express trains, they can only stop at the arrival and departure tracks located at point 3 and 4.

## 4.2 Train Timetable and Result

Taking the data from Tables 3 and 4 as input, the following station arrival and departure tracks distribution scheme with the smallest delay time of passenger trains is obtained by using CPLEX, which is shown in Table 5.

**Table 3.** Original station arrival and departure tracks distribution plan

| Train No | Type | Direction | Arrival time (min) | Track Point No |
|---|---|---|---|---|
| 1 | Passenger | Down | 8:00 | 3 |
| 2 | Passenger | Down | 8:04 | 4 |
| 3 | Passenger | Up | 8:09 | 5 |
| 4 | Passenger | Up | 8:10 | 6 |
| 5 | Passenger | Down | 8:30 | 3 |

**Table 4.** Added China railway highspeed express train arrival timetable

| Train No | Type | Direction | Arrival time (min) |
|---|---|---|---|
| 1 | Freight | Up | 8:05 |
| 2 | Freight | Up | 8:20 |

**Table 5.** Arrival and departure tracks distribution scheme

| Train No | Type | Direction | Arrival time | Track point No | Departure time | Delay time (min) |
|---|---|---|---|---|---|---|
| 1 | Passenger | Down | 08:00 | 5 | 08:07 | 1 |
| 2 | Passenger | Down | 08:04 | 6 | 08:11 | 1 |
| 3 | Passenger | Up | 08:09 | 5 | 08:13 | 0 |
| 4 | Passenger | Down | 08:10 | 4 | 08:18 | 2 |
| 5 | Passenger | Down | 08:30 | 6 | 08:32 | 0 |
| 6 | Freight | Up | 08:05 | 3 | 08:27 | – |
| 7 | Freight | Up | 08:20 | 5 | 08:42 | – |

From the results, we know that:

(1) After two China railway highspeed express trains are added, the new station arrival and departure tracks distribution scheme meets the need of safety technical operations.
(2) In the case of ensuring minimal impact on passenger trains, by adjusting to the arrival and departure line, the total delay time is only 4 min.

## 5 Conclusion

In lateral transshipment mode, considering the section, throat area, arrival and departure tracks resources, and under the condition that China railway highspeed express trains are fixed to the arrival and departure tracks as much as possible, the arrival and departure tracks distribution scheme is studied, so as to minimize the impact of passenger trains. An example is designed to verify the effectiveness of the model.

Therefore, by adjusting the distribution scheme to the arrival and departure tracks, the model reduced the impact on passenger trains when adding China railway highspeed express trains.

**Acknowledgments.** This research is supported by the Joint Funds of the National Natural Science Foundation of China (No. U2034208).

## References

1. Xu, S., Liu, K.: Hub-and-smoke Network Optimization Method of China Railway Highspeed Express Based on Distributed Robust Optimization. Chin. J. Manag. Sci., 1–11 (2023). (in Chinese)
2. Zhou, L., Ding, X., Wang, H., Wang, Y., Xu, Z.: Research on layout method of high speed railway express networks based on maximum coverage theory. Railw. Transp. Econ. **43**(12), 54–59 (2021). (in Chinese)

3. Li, Q., Sun, X., He, Z.: Research on the node planning of highspeed rail express network in China. Railw. Stand. Des. **64**(06), 32–37 (2020). (in Chinese)
4. Ceselli, A., Righini, G.: An optimization algorithm for a penalized knapsack problem. Oper. Res. Lett. **34**(4), 394–404 (2006)
5. Pazour, J.A., Meller, R.D., Pohl, L.M.: A model to design a national highspeed rail network for freight distribution. Transp. Res. Part A: Policy Pract. **44**(3), 119–135 (2010)
6. Lian, Z., Shuai, B., Xu, W.: Operation plan adjustment of arrival and departure tracks at high speed railway station in condition of train delay. Railw. Transp. Econ. **44**(02), 1–7 (2022). (in Chinese)
7. Ren, Y., Zhang, Q., Yuan, Z., Wang, T., Ding, S., Li, Z.: Optimization of train platform utilization at highspeed railway stations based on arrival and departure distribution of trains. J. Harbin Inst. Technol. **53**(08), 137–143 (2021). (in Chinese)
8. Liu, W., Zhu, X.: Distribution optimization of arrival-departure tracks for delayed trains in railway passenger station based on safety constrains. J. Saf. Sci. Technol. **13**(04), 39–45 (2017). (in Chinese)
9. Bianca, P., Marcella, S., Paola, P., Andrea, D.A., Dario, P., Joaquin, R.: Train routing selection problem: Ant colony optimization versus integer linear programming. IFAC PapersOnLine **54**(2), 167–172 (2021)
10. Sels, P., Cattrysse, D., Vansteenwegan, P.: Automated platforming & routing of trains in all Belgian railway stations. Expert Syst. Appl. **62**, 302–316 (2016)

# Design on Hub-and-Spoke Network of High-Speed Railway Express Transport

Ruijie Gao[1], Xiaoning Zhu[1,2](✉), and Li Wang[1,2]

[1] School of Traffic and Transportation, Beijing Jiaotong University, Beijing 100044, China
{22120820,xnzhu,wang.li}@bjtu.edu.cn

[2] Key Laboratory of Transport Industry of Comprehensive Transportation Theory (Beijing Jiao-Tong University), Ministry of Transport, Beijing 100044, China

**Abstract.** With the rapid development of China's economy, consumption upgrades and the booming e-commerce express delivery industry, China high-speed rail express freight transportation came into being, and at present, more than 500 cities have opened this service. This paper based on hybrid configuration hub network, taking the high-speed rail express network and stations in northern China as an example model, a two-stage algorithm is used to divide the node level in the high-speed rail express network considering the requirements of timeliness, convenience, and benefit scale in view of its imperfect transportation network. In the first stage, the network nodes are stratified by constructing a high-speed rail express network planning index system and using the system cluster analysis method; in the second stage, based on the spatial location of each node, the K-means algorithm is used to optimize the node stratification scheme, finally obtaining the high-speed rail express hub-and-spoke transportation network layout in northern China.

**Keywords:** High-speed railway express transport · Network node programming · Hub-and-spoke network

## 1 Introduction

In recent years, the construction of China's high-speed rail network has become increasingly perfect, the reform of railway marketization has achieved obvious results, and the high-speed freight sports unit with a speed of 350 km per hour has rolled off the line in Tangshan, marking a new journey in the development of China's high-speed rail freight. During the "14th Five-Year Plan" period, under the new development pattern of dual circulation, the transformation and development of China's logistics industry ushered in a new period of strategic opportunities, especially online trade forms such as e-commerce drove the exponential growth of express business, and the traditional highway-based transportation organization mode needs to be optimized and adjusted to meet the new requirements of industrial development under the new situation, and high-speed rail express cargo transportation is just in time, and China's high-speed rail express transportation will usher in a new stage of large-scale development, as shown in Figs. 1 and 2.


Y. Qin et al. (Eds.): EITRT 2023, LNEE 1137, pp. 264–272, 2024.
https://doi.org/10.1007/978-981-99-9311-6_29

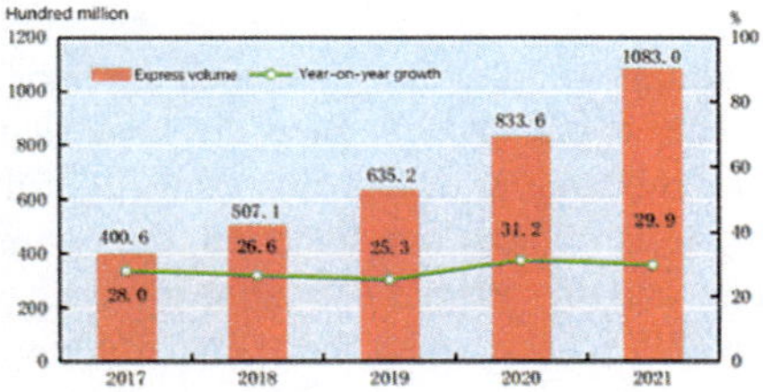


**Fig. 1.** China's express delivery business volume and its growth rate from 2017 to 2021.

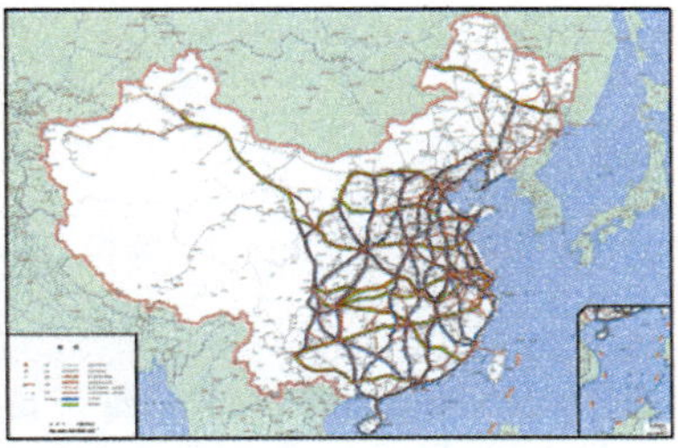

**Fig. 2.** China's medium and long-term high-speed railway network planning map.

## 2 Related Work

After studying the status quo of railway freight transportation in the United States, Don Plotkin et al. believed that only carrying out passenger business of high-speed railway could not cover its operating cost, and analyzed the feasibility of high-speed freight transportation from both technical and economic perspectives [1]. Gerhard Troche et al. demonstrated the feasibility of high-speed rail freight on the basis of analyzing the advantages of high-speed rail freight, and optimized the operation organization and management of high-speed rail freight by taking the situation of European high-speed rail freight as an example [2]. Bingguan J et al. put forward the concept of shared high-speed rail network after studying the feasibility of high-speed rail freight transportation. They believe that high-speed rail has obvious advantages in medium and long-distance freight transportation. The development of high-speed rail freight transportation can make full use of the idle capacity of railway and help the relevant departments of railway to obtain more valuable capital flow and information flow [3]. Dejax E et al. believe that the operation of high-speed freight trains can relieve the pressure of highway freight transport, and effectively reduce carbon emissions and freight costs [4].

Wang et al. established a stochastic optimization model for location selection of logistics center under uncertain environment, and built a robust optimization model for related issues in the form of regret model [5]. SUN et established a two-layer programming model to find the optimal location of logistics distribution center [6]. KLOSE et al. constructed an integer programming model based on network flow [7]. Zhou et al. designed the evaluation system of carrier city layout of high-speed rail ex-press base, and built a high-speed rail express network layout model based on maximum coverage theory [8]. Shi Lu et al. proposed the use of entropy-Topsis method for the location of high-speed rail logistics nodes [9]. Li et al. established the node planning index system of high-speed rail express network and obtained the node planning scheme of high-speed rail express network by using K-means clustering algorithm [10]. Zhang made a comprehensive decision on the optimal site selection, the optimal base scale, the optimal train operation scheme and the optimal flow allocation scheme under the condition of the high-speed railway physical network and transportation demand of a given line and station, which provided important support for the optimization of the operation organization and resource allocation of transport enterprises [11].

# 3 Principle

## 3.1 Hub-and-Spoke Network

Hub-and-spoke network refers to the organization form of retransport of goods through hub stations.

There are two types of spoke network: single hub spoke network and multi-hub spoke network. It is impossible that there is only one hub node in the high-speed rail express network of our country, so the single-hub hub-and spoke network is not considered. The multi-hub hub and spoke network is divided into single configuration hub network, multi-configuration hub network and hybrid configuration hub network as shown in Fig. 3.

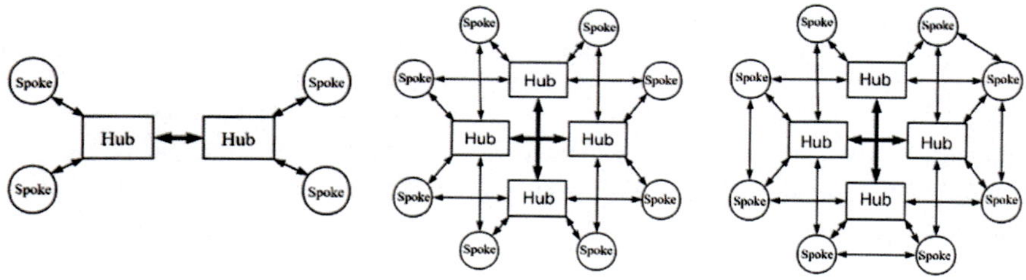


**Fig. 3.** Single, multi and hybrid configuration hub network.

Combining with the practical situation of Chinese high-speed rail express, hybrid configuration hub network is suitable for the design requirement of Chinese high-speed rail express network. Therefore, this paper studies based on this.

## 3.2 Network Node Planning

We plan to classify our high-speed rail express network nodes into the following three levels, as shown in the Fig. 4:

1. First-tier node: National hub center.
2. Second-tier node: Regional distribution center.

3. Third-tier node: Urban business site.

The key factor affecting the node hierarchy of high-speed rail express network is the express delivery volume of the city, and there are many factors affecting the express delivery volume. We used SPSS to verify the rationality of the selection. Finally, five indexes are selected to form the planning index system of high-speed rail express network, as shown in Fig. 5 and Table 1.

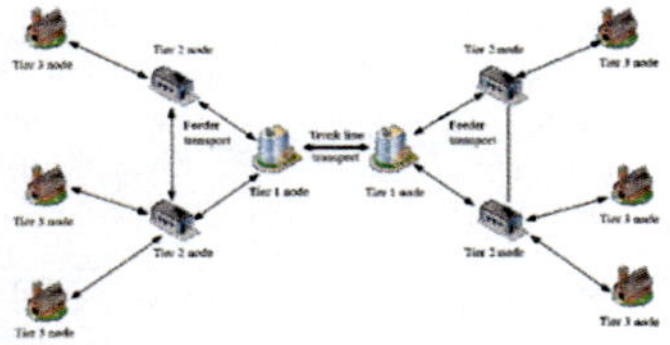

**Fig. 4.** High-speed rail express logistics node.

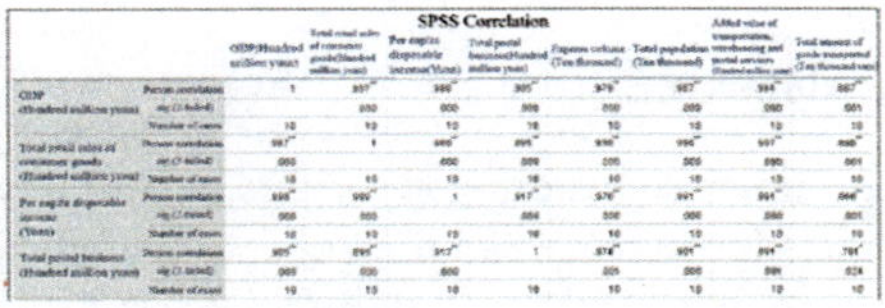

**Fig. 5.** Correlation approximation matrix.

**Table 1.** High-speed railway express freight transportation network planning index system.

| Influencing factors | Evaluation indicators |
|---|---|
| Level of economic development | GDP |
| Social consumption power | Total retail sales of consumer goods |
| Business volume of post and telecommunications industry | Express volume |
| | Total postal business |
| Conditions of transportation | Total amount of goods transported |

# 4 Methods

## 4.1 Systematic Cluster Analysis

The existing research shows that clustering analysis is an effective method to study the hierarchical planning of network nodes.

Systematic clustering is divided into the following three steps:

1. Data processing: unified standard of different statistical indicators. Range standardization is adopted in this paper.

$$x_{ij}^{'} = \frac{x_{ij} - \underset{i}{min}\{x_{ij}\}}{\underset{i}{max}\{x_{ij}\} - \underset{i}{min}\{x_{ij}\}}(i = 1, 2, ..., n; j = 1, 2, ..., n) \quad (1)$$

2. Distance calculation: Distance is the basis and basis of polymerization. Euclidean distance is used in this paper.

$$d_{ij} = \sqrt{\sum_{k=1}^{m}\left(x_{ik} - x_{jk}\right)^2}(\mathrm{i}, \mathrm{j} = 1, 2, ..., \mathrm{n}) \quad (2)$$

3. Clustering: Class mean distance method is the most widely used method with the best clustering effect. This paper adopts the class average method.

$$D_{pq}^2 = \frac{1}{n_p n_q}\sum\nolimits_{X(i)\in G_P}\sum\nolimits_{X(j)\in G_P} d_{ij}^2 \quad (3)$$

### 4.2 K-means Cluster Algorithm

K-means clustering algorithm is a clustering algorithm based on partition. The K-means clustering algorithm is introduced here, and the geographical distance between cities is used as a unified standard to measure the similarity.

The K-means algorithm uses Euclidean distance as the similarity measure and the sum of square errors as the standard function of clustering. The distance between samples and the corresponding clustering center is calculated as follows:

$$d(v_i, Z_k) = \sqrt{\left(x_i - x_j\right)^2 + \left(y_i - y_j\right)^2} \quad (4)$$

## 5 Experiment: Design of Hub-and-Spoke Transportation Network for High-Speed Railway Express in North China

### 5.1 Experimental Background

Northern China has a large population and huge potential of logistics industry. Currently, there are 53 prefecture-level cities and municipalities with high-speed railway stations in urban areas. Therefore, the 53 cities with high-speed railway service will be the candidate nodes of this experiment.

### 5.2 Basic Data Preparation

It is necessary to collect index data of 53 cities to build index system for systematic cluster analysis, and collect geographical location coordinates of main high-speed railway stations in cities for K-means cluster analysis, as shown in the Table 2.

**Table 2.** Longitude and latitude coordinates of Northern cities' rail stations (Part).

| Province | The main high-speed railway station in the city | Longitude, Latitude |
|---|---|---|
| Hebei province | Shijiazhuang railway station | 114.490829,38.016893 |
| | Baoding railway station | 115.48644,38.869177 |
| | Zhangjiakou railway station | 114.889172,40.757657 |
| | Langfang railway station | 116.715217,39.515411 |
| Shanxi province | Datongnan railway station | 113.371681,40.052174 |
| | Yangquanbei railway station | 113.444974,38.120044 |
| | Jinchengdong railway station | 112.990466,35.541883 |
| | Jinzhong railway station | 112.648974,37.692429 |
| | Yunchengbei railway station | 110.992012,35.104749 |
| | Xinzhouxi railway station | 112.699903,38.449211 |
| | Linfenxi railway station | 111.473525,36.107408 |
| Beijing | Beijingnan railway station | 116.385488,39.87128 |
| Tianjin | Tianjinnan railway station | 117.067281,39.062757 |

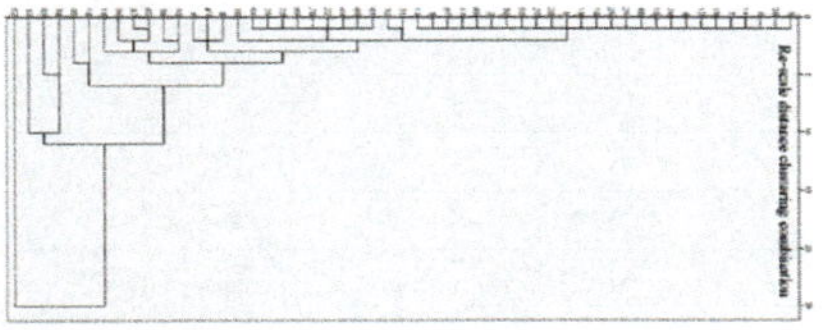

**Fig. 6.** Systematic cluster analysis icicle chart of fifty-three cities in the northern region.

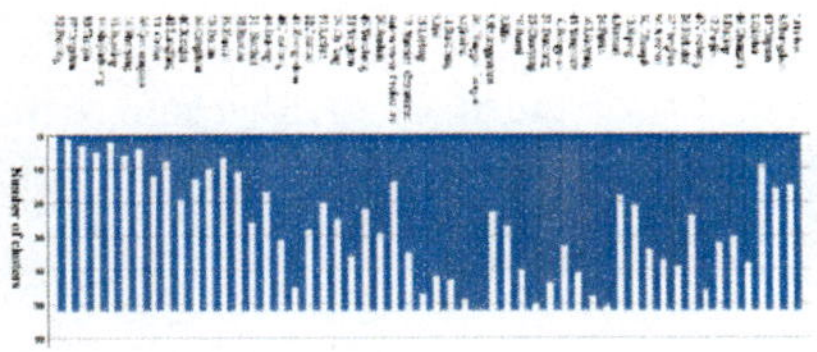

**Fig. 7.** Systematic cluster analysis genealogy map of fifty-three cities in the northern region.

## 5.3 Hierarchical Division of Nodes Based on System Cluster Analysis

SPSS software was used to perform cluster analysis on 53 nodes. The results of cluster analysis were shown in icicle chart and pedigree chart, as shown in Figs. 6 and 7.

### 5.4 Hierarchical Division of Nodes Based on K-means Cluster Algorithm

According to the existing theoretical research, the determination methods of k are human setting, root mean square method and other ways. Root mean square method is adopted in this paper, that is, $k = \sqrt{m/2}$ (m is the number of samples), where $k = 5$.

SPSS software was used to conduct K-means clustering analysis on the geographical locations of 53 cities, and the partition results were shown in Figs. 8 and 9.

**Number of cases in each cluster**

| | | |
|---|---|---|
| Clustering | 1 | 12.000 |
| | 2 | 8.000 |
| | 3 | 13.000 |
| | 4 | 7.000 |
| | 5 | 13.000 |
| Effective | | 53.000 |
| Deletion | | .000 |

**Fig. 8.** K-means cluster analysis results of fifty-three cities in the northern region.

**Cluster member**

| Case | The main high-speed railway station in the city | Cluster | Distance |
|---|---|---|---|
| 1 | Harbinxi Railway Station | 4 | 221906.521 |
| 2 | Qiqiharnan Railway Station | 2 | 234743.365 |
| 3 | Daqingxi Railway Station | 2 | 104277.321 |
| 4 | Jiamusi Railway Station | 4 | 184299.866 |
| 5 | Shuangyashan Railway Station | 4 | 205800.426 |
| 6 | Qitaihexi Railway Station | 4 | 121289.361 |
| 7 | Jixixi Railway Station | 4 | 114724.085 |
| 8 | Changchunxi Railway Station | 2 | 101561.391 |
| 9 | Jilin Railway Station | 4 | 273355.796 |
| 10 | Mudanjiang Railway Station | 2 | 79137.417 |
| 11 | Baicheng Railway Station | 2 | 89170.936 |
| 12 | Yanjixi Railway Station | 4 | 266991.456 |
| 13 | Sipingdong Railway Station | 2 | 231806.310 |
| 14 | Shenyangbei Railway Station | 3 | 116497.685 |
| 15 | Tielingxi Railway Station | 3 | 163326.701 |
| 16 | Liaoyang Railway Station | 3 | 56190.862 |
| 17 | Panjin Railway Station | 3 | 31211.146 |
| 18 | Dalianbei Railway Station | 3 | 241974.470 |
| 19 | Benxi Railway Station | 3 | 114767.046 |
| 20 | Anshanxi Railway Station | 3 | 42165.113 |
| 21 | Dandong Railway Station | 3 | 199481.099 |
| 22 | Jinzhounan Railway Station | 3 | 106391.379 |
| 23 | Yingkoudong Railway Station | 3 | 53327.494 |
| 24 | Fuxin Railway Station | 3 | 123051.851 |
| 25 | Liaoning Chaoyang Station | 3 | 176427.254 |
| 26 | Huludaobei Railway Station | 3 | 144753.102 |
| 27 | Tongliao Railway Station | 2 | 213641.630 |
| 28 | Ulanhot Railway Station | 2 | 168969.834 |
| 29 | Chifeng Railway Station | 1 | 400961.212 |
| 30 | Ulanqab Railway Station | 5 | 314660.490 |
| 31 | Hohhotdong Railway Station | 5 | 291671.999 |
| 32 | Baotou Railway Station | 5 | 329871.104 |
| 33 | Ordos Railway Station | 5 | 248365.877 |
| 34 | Shijiazhuang Railway Station | 1 | 214702.693 |
| 35 | Baoding Railway Station | 1 | 90520.201 |
| 36 | Cangzhouxi Railway Station | 1 | 102140.243 |
| 37 | Tangshanbei Railway Station | 1 | 160564.996 |
| 38 | Qinhuangdao Railway Station | 1 | 284404.010 |
| 39 | Handandong Railway Station | 1 | 329716.944 |
| 40 | Xingtaidong Railway Station | 1 | 283511.253 |
| 41 | Zhangjiakou Railway Station | 1 | 218466.587 |
| 42 | Langfang Railway Station | 1 | 42476.135 |
| 43 | Taiyuannan Railway Station | 5 | 63857.209 |
| 44 | Datongnan Railway Station | 5 | 226267.567 |
| 45 | Yangquanbei Railway Station | 5 | 113371.233 |
| 46 | Changzhidong Station | 5 | 240295.950 |
| 47 | Jinchengdong Railway Station | 5 | 309081.723 |
| 48 | Jinzhong Railway Station | 5 | 74995.836 |
| 49 | Yunchengbei Railway Station | 5 | 363873.345 |
| 50 | Xinzhouxi Railway Station | 5 | 52249.884 |
| 51 | Linfenxi Railway Station | 5 | 245935.748 |
| 52 | Beijingnan Railway Station | 1 | 75158.738 |
| 53 | Tianjinnan Railway Station | 1 | 55871.285 |

**Fig. 9.** K-means cluster membership of fifty-three cities in the northern region.

53 high-speed rail express node cities are divided into regions, and grouped scatter points are shown in the Fig. 10.

### 5.5 Node Adjustment

According to the results obtained by system clustering analysis method, combined with the results obtained by K-means clustering analysis method, the hierarchical division results of 53 high-speed rail express nodes are divided and adjusted. The final results are shown in the Fig. 11.

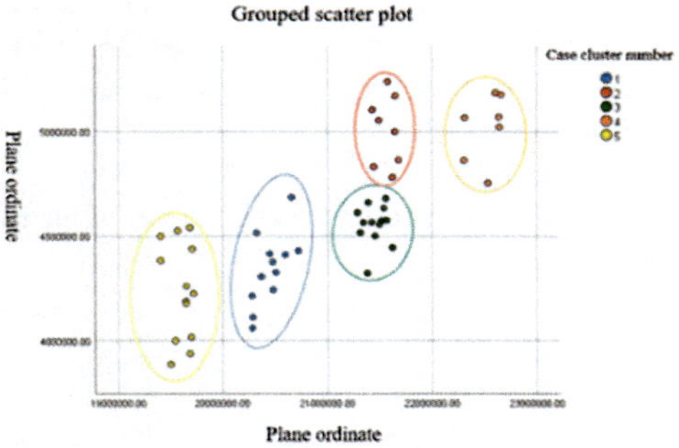


**Fig. 10.** Regional division map of fifty-three cities in the northern region.

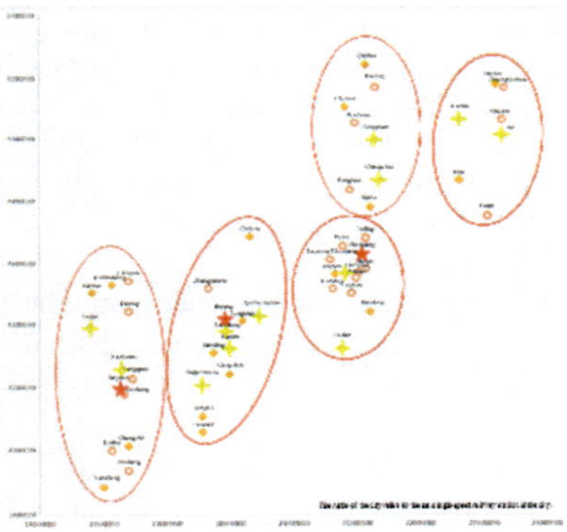

**Fig. 11.** Node level planning map of high-speed railway express freight transportation network in northern region.

# 6 Conclusion

This paper mainly achieves two achievements:

1. This paper introduces the basic theories related to the transportation network, compares and analyzes the axle and spoke transportation network mode of mixed configuration hubs which is suitable for the present situation and future development of our country. By analyzing the similarity of related evaluation indexes, this paper establishes the index system of fast freight transportation network of high-speed rail-way.
2. From the perspective of combining qualitative and quantitative methods, cluster analysis method is adopted to solve the problem by two-stage method. In particular, systematic cluster analysis method and K-means clustering algorithm are used to carry out hierarchical planning of transport nodes in the axle-spoke transport net-work model of high-speed railway fast cargo.

This paper puts forward an idea of node hierarchy planning and design for the axle-spoke transportation network of fast freight in high-speed railway. However, due to the limited capacity, there are some difficulties in the collection of relevant data, and there are still some imperfections in this study.

**Acknowledgments.** This work is supported by the Joint Funds of the National Natural Science Foundation of China (No.U2034208).

## References

1. Plotkin, D.: Carrying freight on high-speed rail lines[J]. J. Transp. Eng. **123**(3), 199–201 (1997)
2. Troche, G.: High-Speed Railway Freight[M]. KTH Railway Group (2005)
3. Jin, B., Li, Y.: A study on the feasibility of introducing high-speed freight rail system [J]. In: Koera Railway Association 2012 Spring Academic Conference (2012)
4. Dejax, P., Bookbinder, J.H.: Goods transportation by the French national railway: the measurement and marketing of reliability[J]. **25**(4), 210–225 (1991)
5. Baohua, W., Shiwei, H.E.: Robust optimization model and algorithm for logistics center location and allocation under uncertain environment[J]. J. Transp. Syst. Eng. Inf. Technol. **9**(2), 69–74 (2009). (in Chinese)
6. Sun, H., Gao, Z., Wu, J.: A bi-level programming model and solution algorithm for the location of logistics distribution centers[J]. Appl. Math. Model. **32**(4), 610–616 (2008)
7. Klose, A., Drexl, A.: Facility location models for distribution system design[J]. Eur. J. Oper. Res. **162**(1), 4–29 (2005)
8. Zhou, L., Ding, X., Wang, H., et al.: Research on layout method of high speed railway express networks based on maximum coverage theory[J]. Railw. Transp. Econ. **43**(12), 54–59 (2021). (in Chinese)
9. Lu, S.H.I., Xiaolai, M.A., Yi, C.U.I.: The location of high-speed rail logistics node based on entropy-topsis method-the case of Chengdu rail hub[J]. China Transp. Rev. **43**(5), 108–114 (2021). (in Chinese)
10. Qi, L.I., Xun, S.U.N., Zhenggang, H.E.: Research on the node planning of high-speed rail express network in China[J]. Railw. Stand. Des. **64**(6), 32–37 (2020). (in Chinese)
11. Yuzhao, Z.: Review of service network design for express freight transportation[J]. J. Transp. Syst. Eng. Inf. Technol. **21**(3), 1–12 (2021). (in Chinese)

# Research on Related Elements of Train-Cargo Matching Based on High-Speed Rail Express Transportation

Qi Zhang[1], Li Wang[1,2](✉), and Xiaoning Zhu[1,2]

[1] School of Traffic and Transportation, Beijing Jiaotong University, Beijing 100044, China
{wang.li,xnzhu}@bjtu.edu.cn

[2] Key Laboratory of Transport Industry of Comprehensive Transportation Theory (Beijing Jiaotong University), Ministry of Transport, Beijing 100044, China

**Abstract.** With the continuous development of China's high-speed rail express transportation, high-speed rail express transportation has outstanding advantages among various modes of cargo transportation, and the market is broad and highly dynamic. However, the problems such as asymmetric information of train-cargo matching and low efficiency of cargo allocation in the process of railroad transportation are becoming more and more obvious, which has drawn wide attention from government enterprises and academia. This paper summarizes the main views of scholars at home and abroad on high-speed rail express transportation delivery scheme, introduces the transportation organization modes adapted to the high-speed rail express transportation market, and analyzes the advantages and disadvantages respectively. And the factors involved in the problem of matching trains and cargoes in the context of high-speed rail express transport are listed and matched one by one to derive the mechanism of interaction between transported goods and high-speed rail express train operation scheme.

**Keywords:** High-speed rail express transportation · Train-cargo matching · Operation mode · Cargoes

## 1 Introduction

Data from the State Post Bureau shows that the total express business volume in 2022 completed 110.58 billion pieces, up 2.1% year-on-year; the total off-site business volume completed 95.77 billion pieces, up 4.0% year-on-year. The courier logistics market continued its good development trend and maintained high growth despite the larger base.

High-speed rail express transport both green, fast, punctual, efficient and other characteristics, is the ideal mode of express transport, the growing market for express transport on the further development of high-speed rail express transport has a realistic need, should be carried out in China on a large scale, the normalization of high-speed rail express transport. The market is vast and highly dynamic, and in order to improve the


Y. Qin et al. (Eds.): EITRT 2023, LNEE 1137, pp. 273–281, 2024.
https://doi.org/10.1007/978-981-99-9311-6_30

competitiveness of the railroad in the express transport market, many experts and scholars have conducted a lot of research on this aspect.

Troche [1] by giving examples of railroad transport systems in France, Sweden, Denmark and other countries, suggests that high-speed rail transport of goods has more obvious advantages in terms of timeliness and reliability compared to other modes of transport, and can achieve greater transport benefits. Hoffrichter et al. [2] summarize the transport advantages of high-speed rail freight transport and suggest that high speed rail operations should include the fast freight market. Pazour Jet al. [3] proposed a dual-optimization objective high-speed rail freight node distribution model to reduce road transportation pressure and save transportation time based on important measures to address urban public transportation congestion. Tan et al. [4–6] discussed the organization mode, operation mode and technical equipment of high-speed rail express business in Germany, France and Europe, respectively, and summarized the inspiration of high-speed rail to carry out high value-added cargo transportation, which provides experience for China's express delivery industry to use high-speed rail to deliver express shipments. Qi et al. [7] constructed a mixed integer planning model for passenger and cargo cooperative transportation scheme with passenger waiting time and transportation cost as the optimization objectives and solved the model using CPLEX to substantially increase the operating revenue while ensuring the service quality.

A number of scholars have also conducted research in the issue of train-cargo matching for high-speed rail express transportation. Jin et al. [8] constructed and solved a high-speed railroad express cargo flow alliance model based on different time-sensitive transportation products and different operational requirements to achieve the goal of accurate cargo flow matching. Chen et al. [9] carried out a comprehensive optimization of cargo delivery capacity and cargo flow allocation in the railroad network, and constructed a model based on the characteristics of traffic organization to achieve the minimization of transportation costs. Yao et al. [10] quantitatively optimized the piggyback transportation mode for passenger trains and constructed an optimization model for high-speed rail express transportation with cost minimization as the goal, and found that the load capacity and cargo arrival time estimation have important effects on the optimization effect of the scheme. Fu et al. [11] analyzed the current problems faced by China's high-speed rail express transport, proposed a flexible application strategy of combining transport organization modes with each other, and suggested transport organization modes for different high-speed rail express transport products.

## 2 High-Speed Rail Freight Train Operation Mode

At present, the domestic high-speed rail express transport organization mode includes passenger train piggyback mode, confirmation train mode, mixed mode and high-speed rail express train mode, as shown in Fig. 1. From the practice of the development of China railroad cargo transportation, high-speed rail freight train operation should try to use the existing freight facilities to achieve its own stable and rapid development without hindering the existing passenger operation. However, in addition to the special luggage and temporary cargo storage points used in the transport process, the existing stations and train equipment cannot meet the needs of large-scale cargo transportation. Therefore, combined with the development trend of high-speed railroad freight transport, based on the study of various freight transport modes, the operation mode of high-speed railroad freight transport for different scales or organization is analyzed to provide a basis for matching the transport needs of different goods with the train operation scheme at a later stage.

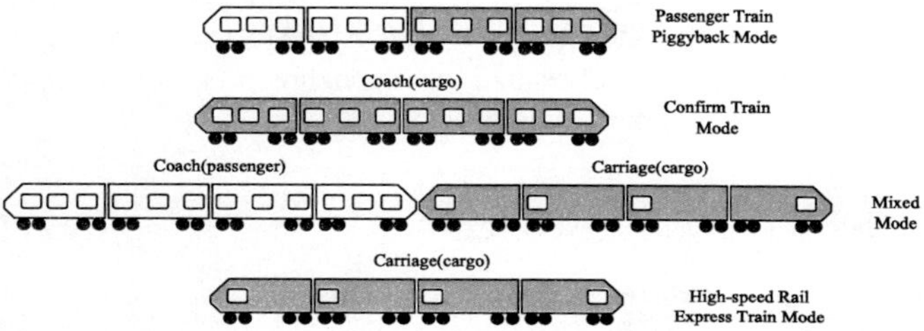


**Fig. 1.** High-speed rail freight train operation mode

## 2.1 Passenger Train Piggyback Mode

The piggyback mode makes full use of passenger trains to improve transport efficiency and flexible loading time. For cargo transportation should choose a relatively concentrated location to avoid passenger transport interference. Railroad departments adjust the ticketing system so that high-speed passenger trains have concentrated spare seats for cargo loading and unloading, reducing interference with passenger transport. It is suitable for small, light and fragmented transportation of high value-added goods.

The advantage is that it does not require additional design of special freight trains and does not have much impact on the existing passenger dispatch. And has the characteristics of high operation frequency, flexible transportation, and passenger flow cross interference is not large.

The disadvantage is that high-speed passenger trains have a short stopping time at stations along the way, and the loading and unloading of cargoes and passengers are carried out on the same platform, which has strict requirements on the loading and unloading operation time and facilities, and the existing passenger stations lack the corresponding cargo operation capacity. And limited by the capacity of the train, only part of the demand for direct transportation can be met.

## 2.2 Confirm Train Mode

The existing high-speed rail express transport mode is mainly the use of confirmation train transport, confirm train is the use of the first high-speed rail "explorer car" every morning to transport cargoes, used to check the line and equipment safety conditions, not passenger transportation services. Therefore, it can be placed in the train seats and other spare positions of cargoes, can meet the demand for medium-volume freight. The freight locomotive set can be used to complete the work of the high-speed rail confirm train, while also being able to transport cargoes.

The advantage is that it will not affect passenger transportation, and at the same time, the stopping time can be controlled more flexibly, which can effectively achieve a perfect fit with the next morning delivery, next day delivery and other express types, and at the same time, because there is no en route operation, no box inspection, handover operation, the train travels at a higher speed.

The disadvantage is that the mode is limited by the opening time and number of trips, and the volume of transported cargoes is limited, which makes it difficult to realize large-scale cargo transportation. And the interior of the carriage restricts the size of the cargoes and the location of stacking, cannot realize the mechanization of cargo operations, can only rely on human loading and unloading.

## 2.3 Mixed Mode

Mixed mode is to group two different trains in the same high-speed freight train, in order to ensure that the train in the group in the total weight, axle weight, speed and other aspects of high-speed passenger trains are basically the same, the packaging of the goods carried, the weight of certain requirements exist.

The advantages of this mode are that it has almost no effect on the operation of existing passenger trains, no need to prepare new train operation diagrams, and the cost of train transformation is low.

The disadvantage is that cargo loading and unloading and passenger loading and unloading are carried out at the same station, which has very strict requirements on cargo loading and unloading operation time and loading and unloading facilities, and also has certain requirements on the scale of cargo volume, making it difficult to maintain the balance of passenger and cargo volume.

## 2.4 High-Speed Rail Express Train Mode

The high-speed rail express train mode can meet the demand for large-volume and large-scale freight transportation, and can gain stable revenue from long-term co-operation with large manufacturing enterprises and commercial enterprises with stable sources of goods. The flexibility of its operation is strong, and it can meet the different requirements of goods for time efficiency. This mode can make full use of the idle time of passenger trains to carry out high-speed railroad moving group cargo transportation and make full use of the surplus capacity of high-speed railroad.

The advantage of this model is that the cargo trains run independently, almost no impact on the passenger operations of high-speed railway stations, cargo loading and unloading time and passenger boarding and alighting time can be inconsistent, is the form of freight products with the highest transport efficiency.

The disadvantage is that this model needs to be designed specifically for freight trains operating on high-speed rail, but the existing high-speed rail stations do not have the technical conditions to support freight, which is more difficult. At the same time in the operating chart of the laying, high-speed rail express special trains and passenger trains should be coordinated with each other to ensure normal passenger capacity.

### 2.5 Comparative Analysis of Train Operation Mode

In order to reasonably select the train operation mode of high-speed rail freight, it is necessary to analyze its operating characteristics and adaptation conditions. The above-mentioned four train operation mode of the operation time, organization mode and other related features are shown in Table 1.

**Table 1.** Comparative analysis of the characteristics of train operation mode

| Operation mode | Operation time | Organization mode | Other related features |
|---|---|---|---|
| Passenger train piggyback mode | Loading and unloading time with train stopping time | Adopt "point-to-point" transportation mode | Fix the corresponding loading point and arrange cargo stacking in a reasonable way |
| Confirm train mode | Cooperate with the time of high-speed rail confirm train | Adopt "point-to-point" transportation mode | _ |
| Mixed mode | Cargo handling to extend the dwell time at the departure station | Consistent with high-speed passenger trains | Select low occupancy, avoiding popular train lines |
| High-speed rail express train mode | Avoid high speed rail passenger and maintenance daylight hours | Selecting the central station for stopping to load and unload cargo | The number of trains is arranged according to the freight demand and station passing capacity |

## 3 Train-Cargo Matching for High-Speed Rail Express Operation Scheme

High-speed express trains are passenger trains, the use of high-speed trains carrying passengers piggyback, confirm train, mixed passenger and cargo or special freight trains to transport cargoes, the organization of which has certain similarities with passenger trains. However, in the mode of piggybacking using high-speed trains carrying passengers, both the arrangement of cargo placement on the train and the loading and unloading operation at the station need to consider the sales of passenger tickets, the attendance rate of the passenger flow, the passenger's ride and landing, and other circumstances affecting the train's capacity.

The running scheme of high-speed railroad trains is established, so the process of preparing the running scheme of high-speed railroad express transport is to match the cargoes on the trains of different trains on the basis of the determined number of high-speed trains, stopping scheme and train schedule on the existing operation chart. The conveying scheme of high-speed rail express transport means which cargoes can be transported by which train to meet the requirements of the cargo owner for the time limit and at the same time make the transport enterprise's revenue the most.

### 3.1 Analysis of the Preparation Elements of High-Speed Rail Express Operation Scheme

The specific effective conveying scheme has the following elemental attributes:

1. The train operation characteristics, including the operating range of the train and the stopping scheme en route. The choice of transport range depends on the choice of operating section, while the train running path determines the section of train operation. The key is to reasonably select the operating stops of high-speed trains according to the distribution points of cargo.
2. The properties of train, including the class, type and number of trains. The maximum running speed of trains of different grades or the number of units in different groups will affect the arrival time or capacity of trains.
3. Utilization of equipment, including train grouping content, passenger seat utilization rate, train crew, etc. These aspects will affect the cargo delivery volume, cargo transportation capacity and other transportation enterprise efficiency indicators.
4. Cargo handling plan is based on the en route stopping program and train freight capacity, considering the cargo transportation time limit requirements, and the overall arrangement of cargo loading and unloading operations at each station to maximize the comprehensive benefits.

### 3.2 Matching Relationship Between Cargo and Train Operation Scheme

This section is to be analyzed from the point of view of the matching relationship between the key elements of cargo and the elements of each part of the train operation scheme.

1. Flow rate

Cargo flow is the total amount of cargo transported, which can be matched to train operating characteristics and equipment utilization, and is the basis for selecting a train solution. At the same time, the size of the cargo flow is one of the most important factors in matching the characteristics of the train. However, in case of insufficient cargo flow, if all direct special cargo trains are still operated, it will lead to the consumption of a large amount of assembly time. In addition, the cargo needs to be loaded and unloaded at defined stops, which also involves the stopping scheme and cargo loading and unloading plan of the train in the opening section.

2. Flow Direction

OD cargo flow direction is the decisive factor for the choice of train running path, i.e. for the choice of which train for transportation service. Trains of different operation paths will have different stations along the way, so the train stopping scheme should also be compared according to the direction of cargo flow. At the same time, according to the principle of no separation of traffic flow in the cargo train grouping content, the cargo flow and traffic flow should be consistent, so the train grouping content should also consider the cargo flow direction.

3. Flow duration

The flow duration characteristics of cargoes reflect the requirements of cargo arrival time, and the requirements of different cargo categories for cargo arrival time are different.

Since the train operation scheme is closely related to each stage of transportation en route in the whole process of actual cargo transportation, the key time fac-tor of cargo transportation, i.e. the flow-time characteristics of cargo, is interlinked with the preparation and execution of the train operation scheme. The influence of the time generated at each stage of cargo transportation (i.e. cargo dwell time and travel time) on the operational scheme of the actual cargo delivery is shown in Table 2.

**Table 2.** Cargo transit time and impact on operating schemes

| Cargo transportation phase | Cargo transit time | Operating scheme related factor |
|---|---|---|
| Before shipping | Dwell time (cargo stays at the station, loading time) travel time (time for cargoes to be assembled at the starting station) | Starting station |
| In transit | Dwell time (time of cargo operation at technical stations) travel time (cargo in transit time) | Train operation characteristics |
| End of shipping | Dwell time (cargo unloading time) tTravel time (cargo evacuation time from the final destination) | Final station |

4. Properties of cargo

The properties of cargo refers to the different levels of transport demand of cargo, and the cargo of different transport demand levels should choose different levels of high-speed trains according to their respective transport demand.

To sum up, the matching relationship between cargo train operation scheme and cargo, as shown in Fig. 2.

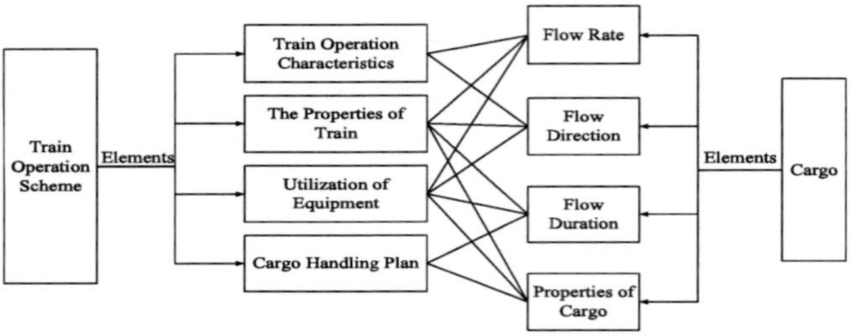


**Fig. 2.** The matching relationship between cargo train operation scheme and cargo

# 4 Conclusion

Through the analysis of the delivery scheme of high-speed rail express transportation, this paper introduces four transportation organization modes adapted to the high-speed rail express transportation market, discusses the advantages and shortcomings of each transportation mode, and makes a theoretical pavement for the optimal selection of the delivery scheme. The factors influencing the delivery scheme of high-speed rail express transportation are clarified, and the attributes of the elements that the delivery scheme should have, such as the characteristics of train operation, the properties of passenger trains and the utilization of equipment, are listed and explained.

Finally, the content of each element of cargo is refined and systematically presented and matched with the key factors of high-speed train operation scheme, and then the mechanism of interaction between transported cargo and high-speed express train operation scheme is derived.

**Acknowledgments.** This work is supported by the Joint Funds of the National Natural Science Foundation of China (No. U2034208).

# References

1. Troche, G.: High-speed rail freight: sub-report in efficient train systems for freight transport[J]. Freight Termin. (2005)
2. Hoffrichter, A., Silmon, J., Iwnicki, S.: Rail freight in 2035-traction energy analysis for high-performance freight trains[J]. Proc. Inst. Mech. Eng. Part F J. Rail Rapid Transit **226**(6), 568–574 (2012)
3. Pazour, J.A., Meller, R.D., Pohl, L.M.: A model to design a national high-speed rail network for freight distribution[J]. Transp. Res. Part A **3**, 90–98 (2015)
4. Liang, X., Tan, K.: Study on the development model of express delivery business by highspeed rail in Germany[J]. Chin. Railw. **5**, 11–15 (2014). (in Chinese)
5. Tan, K., Zhang, C.: Study on French high-speed rail freight mode[J]. Chin. Railw. **3**, 16–20 (2014). (in Chinese)
6. Tan, K., Zhang, Y.: Study on the development model of high-speed rail freight business in Europe[J]. Chin. Railw. **1**, 17–22 (2014). (in Chinese)

7. Qi, J., Zhou, H., Yang, L.: Optimization methods of combined passenger and freight transportation based on flexible train composition mode[J]. J. Transp. Syst. Eng. Inf. Technol. **22**(2), 197–205 (2022). (in Chinese)
8. Jin, W., Li, X., Zhou, L., Yu, X.: Research on optimization of high-speed railway freight transportation organization scheme based on column generation algorithm[J]. J. China Railw. Soc. **42**(9), 26–32 (2020). (in Chinese)
9. Chen, L., Lin, B., Wang, L., Wen, X., Li, J.: Integrated optimization of the maximum amount of railway freight transportation and flow distribution[J]. J. Transp. Syst. Eng. Inf. Technol. **15**(4), 45–50 (2015). (in Chinese)
10. Yao, Y., Zhang, Y.: Study on express freight transport scheme of high-speed rail in piggyback mode of passenger car[J]. J. Railw. Sci. Eng. **17**(1), 31–38 (2020). (in Chinese)
11. Fu, C., Zhang, H.: Discussion on the organization countermeasures of high-speed rail express transportation[J]. Railw. Stand. Des. **64**(5), 1–5 (2020). (in Chinese)

# Optimization of Rolling Stock Scheduling for Multi-line Suburban Railways with Minimization of Deadhead Time. A Case Study of Wenzhou Suburban Rail Network

Weiwei Song[1], Yuling Ye[1](✉), Congcong Chen[1,2], Ziyue Zhu[1], and Wentao Zhou[1]

[1] The Key Laboratory of Road and Traffic Engineering of the Ministry of Education, Tongji University, Shanghai 201804, China
yyling71@163.com

[2] School of Rail Transportation, Shandong Jiaotong University, Ji'nan 250357, China

**Abstract.** This study focuses on optimizing rolling stock utilization in multi-line suburban railway networks to minimize empty running time during operations. A mixed-integer programming model is developed, considering multiple depots, diverse train compositions, and capacity constraints. The model is applied to Wenzhou City's suburban railway network, specifically the S1 and S2 lines, using IBM ILOG CPLEX software. The results demonstrate the significance of optimizing rolling stock utilization in suburban railway networks. The implemented model effectively addresses the complexities of multi-line operation and multiple depots, improving rolling stock management. It validates the feasibility and effectiveness of minimizing empty running time in the Wenzhou City network. This research contributes to transportation planning by addressing the challenge of reducing empty running time in multi-line suburban railway systems. Optimizing rolling stock utilization enhances operational efficiency, reduces costs, and improves overall performance. Future studies can further explore additional factors and constraints to enhance optimization for diverse suburban railway networks, considering real-world operational conditions and evolving transportation demands.

**Keywords:** Suburban railways · Rolling stock scheduling · Optimization · Deadhead time · Multi-line networks · Wenzhou suburban railway network

## 1 Introduction

With urbanization on the rise in China, suburban railway development has become a priority to meet the growing transportation needs of expanding cities. Many cities have already started or are about to start operating multi-line networks for their suburban railways. In this context, it is essential to view the entire network as an integrated system, taking into account advanced operational management techniques. This approach enables unified scheduling and efficient resource sharing in terms of time and space, allowing suburban railways to optimize their operations and improve efficiency.


Y. Qin et al. (Eds.): EITRT 2023, LNEE 1137, pp. 282–291, 2024.
https://doi.org/10.1007/978-981-99-9311-6_31

Rolling stock plays a vital role in suburban railways, serving the commuting passenger flows of cities and surrounding areas. Balancing train service and the cost of rolling stock utilization is crucial, especially in multi-line operations. Optimizing rolling stock scheduling is paramount to ensuring efficient network operations. By allocating resources rationally and optimizing utilization, the challenges of managing rolling stock in suburban railways can be effectively addressed.

Ying et al. [1] propose an actor-critic deep reinforcement learning approach for metro train scheduling with limited rolling stock circulation. Zhao et al. [2] aim to minimize rolling stock coupling/decoupling and deadhead time. Li et al. [3] study the integrated optimization of urban rail transit line planning, timetabling, and rolling stock scheduling. Other influential works include [4]. Zhao et al. [5] investigate integrated planning for an oversaturated URT line, considering fairness in train scheduling and workload balance in rolling stock circulation. Wang et al. [6] address the integrated rolling stock deadhead routing and timetabling problem. Zhou et al. [7] propose a metaheuristic solution method using MATLAB. He et al. [8] develop an optimization method based on mutated dichotomy and differential evolutionary algorithm. A rolling stock utilization model is established by considering multi-depot conditions and crew scheduling [9]. An optimization model based on multi-commodity network flow is proposed, and the Lagrangian relaxation algorithm is used [10]. Wang et al. [11] propose a multi-objective rolling stock optimization model based on the MTSP model, using a greedy genetic algorithm.

Researchers have extensively studied rolling stock utilization problems both domestically and internationally. This paper fills a research gap by specifically addressing the optimization problem of rolling stock utilization in suburban railways, with a focus on minimizing deadhead time for suburban vehicles, which is essential for reducing operational costs and improving transportation efficiency.

## 2 Mathematical Model

### 2.1 Problem Description

In multi-line suburban railway operations, optimizing rolling stock scheduling involves treating rolling stock as commodities moving through a network, similar to the scheduling of buses in multiple depots. This transforms the problem into a complex multi-commodity network flow optimization.

In Fig. 1, there are seven stations labeled $S_1$ to $S_7$, with three sections identified as $A_1$, $A_2$, and $A_3$. Stations $S_1$, $S_3$, $S_5$, and $S_7$ are turn-back stations, while stations $S_2$ and $S_6$ are connected to yards $P_1$ and $P_2$. Train assignments and departure routes can be determined freely when the capacities of yards and terminal stations are not limited. Each train has three possible routes ($r_1$, $r_2$, $r_3$) to reach the starting station. These routes involve different directions and station stops, but all meet the requirements for reaching the starting station $S_4$.

The rolling stock utilization problem in this scenario aims to assign a specific number of rolling stock units from the yard to the starting station, considering yard and terminal station capacity constraints, with the objective of minimizing deadhead time and ensuring

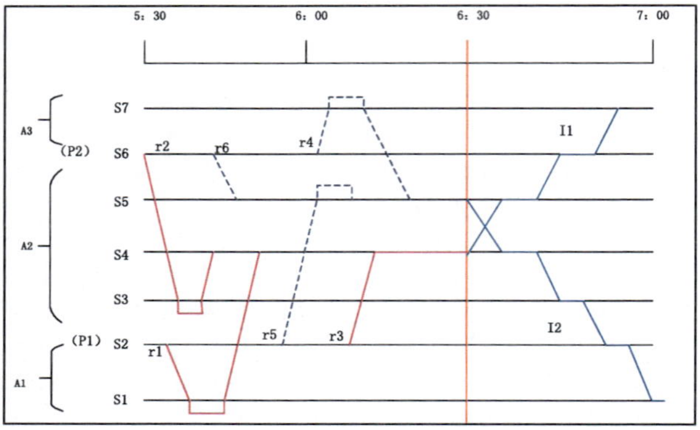


**Fig. 1.** Schematic diagram of vehicle bottom before operation hours

transportation requirements are met, taking into account various route options that impact efficiency.

## 2.2 Model Assumptions

**The train service schedules and train timetables are known.** The optimization model for rolling stock utilization in suburban railways assumes trains adhere to predetermined timetables and service schedules, operating punctually according to the schedule.

**The rolling stock types for suburban trains are uniform.** The rolling stock types operating on the railway network are uniform, following the Chinese industry standard classification of suburban trains into types A, B, C, and D, with an adequate number of rolling stock units available.

**Assumption regarding the study period.** This study primarily focuses on the process of rolling stock exiting the depot before the operational period, considering the complexity of optimizing overall operations and the reduced deadhead time when rolling stock returns to its assigned depot after the end of operation.

**Assumption regarding capacity constraints.** This study assumes that the capacities of depot facilities, railway lines, and terminal stations, as well as the maintenance regulations for rolling stock, are known constraints in the planning of rolling stock utilization in suburban railways, while excluding daily inspections, monthly inspections, and major maintenance tasks during the operational period.

## 2.3 Model Notations

Some model notations used in this paper are defined as follows, as shown in Table 1.

The following are the decision variables for the model, shown in Table 2.

**Table 1.** Notion in this model

| Notations | Definition |
|---|---|
| S | Station assembly, $s \in S$ |
| P | Yard assembly, $p \in P$ |
| K | Turn back station assembly, $k \in K$ |
| I | A collection of trains that need to leave the site empty for transportation before the operation period, $i \in I$ |
| W | Direction set, $w = 0$ up; $w = 1$ down |
| V | Collection of vehicle bottom types distinguished by the number of marshalled vehicles, $v \in V$ |
| $I_p$ | The train assembly at the bottom of the depot P is transported by the direct route, $I_p \subset I$ |
| $V_p$ | Collection of vehicle bottom types for depot P maintenance, $V_p \subset V$ |
| $P_i$ | A collection of parking lots that can take a direct path from the bottom of the vehicle, $P_i \subset P$ |
| $K_{ip}$ | The bottom of train I starts from yard P and returns to the corresponding turn back station set, $K_{ip} \subset K$ |
| $d_p$ | Station connected to the depot $p$ |
| $e_{pw}$ | The maximum number of underbody vehicles that can be sent from the parking lot $p$ to the direction w before the operation period |
| $n_p$ | The maximum number of vehicles sent by the parking lot $p$ up and down before the operation period |
| $m_{pv}$ | The maximum maintenance capacity of the parking lot $p$ to the vehicle bottom belonging to the parking lot type $v$ |
| $r_{kw}$ | Before the operation period, the turn back station $k$ turns back from the operation direction 1-W to the maximum number of underbody $w$ |
| $h_k$ | Station corresponding to turn back station $k$ |
| $a_i$ | Departure station of train $i$ |
| $b_i$ | Train $i$ running direction |
| $g_i$ | Type of underbody required by train $i$ |

### 2.4 Objective Functions

The objective is to minimize the deadhead time from the depot to the starting station in order to improve rolling stock utilization efficiency and reduce operating costs. The objective function, denoted as Z is defined as follows:

$$Z = \min \sum\nolimits_{i \in I} \sum\nolimits_{p \in P} c_{ip} \tag{1}$$

The variable $c_{ip}$ in the objective function represents the deadhead time of rolling stock unit $i$ traveling from the depot to the starting station $p$. It includes the travel time

**Table 2.** Decision variables in this model.

| Decisio variables | Definition |
|---|---|
| $x_{ip}$ | 0–1 decision variable, indicating whether the bottom of train $i$ departing from yard $p$ is a direct path |
| $y_{ipk}$ | 0–1 decision variable, indicating whether the turn back path through turn back station $k$ is adopted at the bottom of train $i$ starting from yard $p$ |
| $c_{ip}$ | Idle time from the bottom of the train to the departure station of train $i$ |
| $h_{pbi}$ | Travel time of train I on depot $p$ exit line |
| $t_{d_p a_i}$ | Time taken for train $i$ to travel from depot $p$ connecting station to departure station |
| $h_{p,1-b_i}$ | Travel time of train I on the exit line in the opposite direction of depot $p$ |
| $t_{d_p h_k}$ | Travel time of vehicle bottom from depot $p$ connection station to turn back station |
| $f_{kb_i}$ | Turn back time of train $i$ at turn back station |
| $t_{h_k a_i}$ | Time spent by train $i$ from turn back station to departure station |

along the departure path from the depot, the travel time from the depot connecting station to the starting station, and the turn back time at the terminal station. The goal is to minimize the overall deadhead time, there by optimizing the utilization of rolling stock and reducing operational inefficiencies. The equation to calculate $c_{ip}$ is given by:

$$c_{ip} = \left(f_{pb_i} + l_{d_p a_i}\right) \cdot x_{id} + \sum\nolimits_{k \in K_{ip}} \left(f_{p,1-b_i} + l_{d_p h_k} + h_{kb_i} + l_{h_k a_i}\right) \cdot y_{ip_k} \tag{2}$$

The deadhead time for rolling stock depends on the route taken. For a direct route, it includes the departure time from the depot P and travel time to the starting station. For a turn back route, it includes travel time on the opposite running line at the depot, travel time from the depot to the turn back station, turn back time at the turn back station, and travel time from the turn back l station to the starting station.

## 2.5 Constraint Conditions

**Outbound Yard Capacity Constraints.** The constraints of train headways and departure times limit the outbound capacity at the yard in each direction.

$$\sum\nolimits_{i \in I_p, b_i = w} x_{ip} + \sum\nolimits_{i \in I, b_i = 1-w} \sum\nolimits_{k \in K_{ip}} y_{ipk} \leq e_{pw} \ \forall p \in P \ \ w \in W \tag{3}$$

The total outbound capacity of each yard in both directions is also subject to constraints, expressed as follows:

$$\sum\nolimits_{i \in I_p} x_{ip} + \sum\nolimits_{i \in I} \sum\nolimits_{k \in K_{ip}} y_{ipk} \leq n_p \ \forall p \in P \tag{4}$$

**Yard maintenance capacity constraint.** The daily maintenance capacity of the yard is constrained due to limited parking tracks and maintenance resources.

$$\sum\nolimits_{i \in I_p, g_i = v} x_{ip} + \sum\nolimits_{i \in I, g_i = v} \sum\nolimits_{k \in K_{ip}} y_{ipk} \leq m_{pv} \ \forall p \in P \ \ v \in V \tag{5}$$

**Turn back capacity constraint**. The number of rolling stock units reversing at the turn back station should not exceed the maximum turn back capacity in that direction.

$$\sum_{i\in I_p,b_i=w}\sum_{p\in P,k\in K_{ip}} y_{ipk} \le r_{kw} \ \forall k\in K \ w\in W \tag{6}$$

**Unique route constraint**. To comply with the actual operational situation, each rolling stock unit can only choose one direct route or turn back route to operate at the same time.

$$\sum_{p\in P_i,g_i\in V_p} x_{ip} + \sum_{p\in P_i,g_i\in V_d}\sum_{k\in K_{ip}} y_{ipk} = 1 \ i\in I \tag{7}$$

## 2.6 Optimization Model for Suburban Railway Rolling Stock Utilization

The optimization model, with the objective of minimizing the idle time of rolling stock before the operational period, is as follows:

$$Z = \min\sum_{i\in I}\sum_{p\in P} c_{ip} \tag{8}$$

*s.t.*

$$c_{ip} = \left(f_{pb_i} + l_{d_pa_i}\right)\cdot x_{id} + \sum_{k\in K_{ip}}\left(f_{p,1-b_i} + l_{d_ph_k} + h_{kb_i} + l_{h_ka_i}\right)\cdot y_{ip_k} \tag{9}$$

$$\sum_{i\in I_p,b_i=w} x_{ip} + \sum_{i\in I,b_i=1-w}\sum_{k\in K_{ip}} y_{ipk} \le e_{pw} \ \forall p\in P \ w\in W \tag{10}$$

$$\sum_{i\in I_p,g_i=v} x_{ip} + \sum_{i\in I,g_i=v}\sum_{k\in K_{ip}} y_{ipk} \le m_{pv} \ \forall p\in P \ v\in V \tag{11}$$

$$\sum_{i\in I_p,b_i=w}\sum_{p\in P,k\in K_{ip}} y_{ipk} \le r_{kw} \ \forall k\in K \ w\in W \tag{12}$$

$$\sum_{p\in P_i,g_i\in V_p} x_{ip} + \sum_{p\in P_i,g_i\in V_d}\sum_{k\in K_{ip}} y_{ipk} = 1 \ i\in I \tag{13}$$

$$x_{ip} = \begin{cases} 1, \text{ the exit path of train } i \text{ is a direct path} \\ 0, \text{ the exit path of train } i \text{ is not a direct path} \end{cases} \ \forall i\in I \ p\in P_i \tag{14}$$

$$y_{ipk} = \begin{cases} 1, \text{ the exit path of train } i \text{ is a turn back path} \\ 0, \text{ the exit path of train } i \text{ is not a turn back path} \end{cases} \ \forall i\in I \ p\in P_i \ k\in K_{ip} \tag{15}$$

# 3 Case Study

## 3.1 Data

The suburban railway network in Wenzhou consists of two lines: S1 and S2. The S1 line is 53.5 km long, with 18 stations and suburban type D trains operating on it. The line is connected to Tongling Depot and Lingkun Depot at its respective ends. The S2 line is 63.63 km long, with 20 stations and also served by suburban type D trains. The line is connected to Xiatang Depot and Ruian Depot. Both lines intersect at Airport Station and Lingkun Station, allowing for cross-line operations using the same rolling stock and signaling system.

The S1 line and S2 line are capable of cross-line operation, utilizing the same type of rolling stock and unified signaling system (Fig. 2).

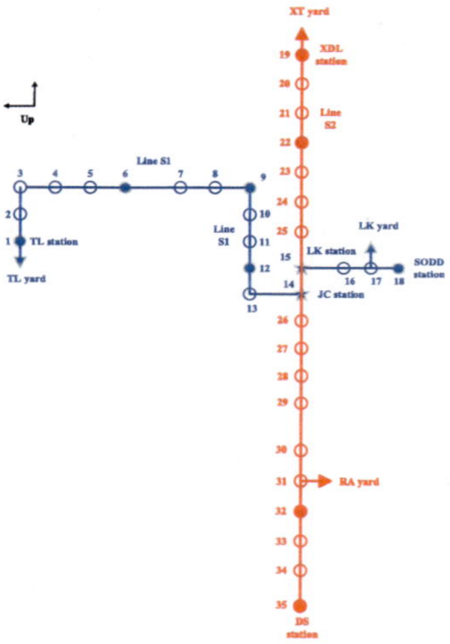


**Fig. 2.** Schematic diagram of route and yard layout

## 3.2 Turn Back and Depot Capacity

There are 9 turn-back stations in the network, including 5 up turn-back stations and 4 down turn-back stations. Each rolling stock is required to turn back in one direction at a turn-back station, and the turn-back operation should be completed at least one hour before the start of the operating period. Detailed information about the turn-back stations can be found in Table 3. The upturn and downturn exit times, as well as the exit time intervals, are uniform across all four depots. The operating hours for the entire network are from 5:30 to 23:00. Rolling stock from the four depots can undergo the exit operation up to 1 h before the start of the operating period. Detailed information about the depots can be found in Table 4.

**Table 3.** Basic information of reentry station

| Turnback station $k$ | Turnback direction $w$ | Turn back time/min | Turn back time interval/min |
|---|---|---|---|
| 1 | 1 | 3.5 | 6 |
| 6 | 0 | 5.0 | 5 |
| 9 | 0 | 5.0 | 6 |
| 12 | 1 | 5.0 | 5 |
| 18 | 0 | 3.5 | 5 |
| 19 | 1 | 3.2 | 6 |
| 22 | 0 | 5.0 | 5 |
| 32 | 0 | 5.0 | 5 |
| 35 | 0 | 3.5 | 5 |

**Table 4.** Basic information of yard

| Yard $p$ | Corresponding parking lot | Maintenance capacity | Departure time/min | Departure interval /min |
|---|---|---|---|---|
| 1 | TL yard | 32 | 3.0 | 5 |
| 2 | LK yard | 30 | 3.0 | 4 |
| 3 | XT yard | 16 | 3.0 | 4 |
| 4 | RA yard | 46 | 3.0 | 4 |

### 3.3 Rolling Stock Optimization

The optimized rolling stock utilization plan was obtained using the CPLEX software, and the specific details are presented in Tables 5 and 6. The objective function value was determined to be 225.6 min, with a total of 19 rolling stocks utilized.

From Tables 5 and 6, it can be observed that the optimized plan has resulted in a certain degree of reduction in the deadhead time for the rolling stock on both the S1 and S2 lines. Through the optimization model established in this study, the deadhead time for the S1 line was reduced to 124.9 min, saving 3.4 min compared to the original plan. Similarly, the deadhead time for the S2 line was reduced to 100.7 min, saving 7.1 min. Overall, the total reduction in deadhead time for both lines amounts to 7.1 min. It is estimated that over 20,000 min can be saved annually, which can effectively enhance the operational efficiency and reduce the operational costs of the suburban railway system.

**Table 5.** Optimized rolling stock operation information of S1 line

| Train number $i$ | Parking lot $p$ | Direction $w$ | Turn back station $k$ | Idle time /min | Reduce hours /min |
|---|---|---|---|---|---|
| 1 | 1 | 1 | – | 7.6 | – |
| 2 | 1 | 0 | 6 | 24.2 | 3.4 |
| 3 | 1 | 1 | – | 5.9 | – |
| 4 | 1 | 0 | 6 | 9.6 | – |
| 5 | 1 | 1 | – | 9.3 | – |
| 6 | 2 | 0 | 12 | 11.3 | – |
| 7 | 2 | 1 | – | 11 | – |
| 8 | 2 | 0 | 12 | 13 | – |
| 9 | 2 | 1 | – | 15 | – |
| 10 | 1 | 1 | – | 18 | – |
| Total | | | | 124.9 | 3.4 |

**Table 6.** Optimized rolling stock operation information of S2 line

| Train number $i$ | Parking lot $p$ | Direction $w$ | Turn back station $k$ | Idle time/min | Reduce hours/min |
|---|---|---|---|---|---|
| 10 | 3 | 0 | 22 | 15.4 | |
| 11 | 3 | 0 | 22 | 15.4 | |
| 12 | 3 | 1 | – | 5.3 | 6.1 |
| 13 | 3 | 0 | 22 | 8.3 | |
| 14 | 4 | 1 | – | 8.9 | 0.5 |
| 15 | 4 | 0 | 32 | 13 | |
| 16 | 4 | 1 | – | 10.7 | 0.5 |
| 17 | 4 | 0 | 35 | 10.7 | |
| 18 | 4 | 1 | – | 13 | |
| Total | | | | 100.7 | 7.1 |

## 4 Conclusion

In conclusion, this study focused on optimizing the utilization of rolling stock in suburban railways. By prioritizing the reduction of deadhead time, we developed a comprehensive model and used CPLEX software to obtain an optimized rolling stock utilization plan. The results showed significant reductions in deadhead time for both the S1 and S2 lines, resulting in a total saving of 7.1 min. This translates to substantial annual time savings and improved operational efficiency, leading to cost reductions in suburban railway operations.

The findings highlight the importance of considering rolling stock optimization in multi-line operations. By implementing advanced operational management techniques, such as resource sharing and unified scheduling, substantial improvements can be achieved in terms of reducing deadhead time and overall operational performance.

Future research can further explore additional factors such as crew scheduling, maintenance requirements, and passenger demand patterns to develop more comprehensive optimization models. Additionally, incorporating advanced algorithms and intelligent decision-making techniques can enhance the accuracy and efficiency of the optimization process.

Overall, this study provides practical insights for transportation authorities and operators to enhance the operational performance of suburban railway systems by optimizing rolling stock utilization.

## References


1. Ying, C., Chow, A.H.F., Chin. K.S.: An actor-critic deep reinforcement learning approach for metro train scheduling with rolling stock circulation under stochastic demand. Transp. Res. Part B: Methodol. **140**, 210–235 (2020)
2. Zhao, P., Li, Y., Han, B., Yang, R., Liu, Z.: Integrated optimization of rolling stock scheduling and flexible train formation based on passenger demand for an intercity high-speed railway. Sustain. **14**(9), 5650 (2022)
3. Li, C., Tang, J., Zhang, J., Zhao, Q., Wang, L., Li, J.: Integrated optimization of urban rail transit line planning, timetabling and rolling stock scheduling. PLoS ONE **18**(5), e0285932 (2023)
4. Aboueldah, T., Farag, H.: Breeding particle swarm optimization for railways rolling stock preventive maintenance scheduling. Am. J. Oper. Res. **11**, 242–251 (2021)
5. Zhao, S., Yang, H., Yinghui, W.: An integrated approach of train scheduling and rolling stock circulation with skip-stopping pattern for urban rail transit lines. Transp. Res. Part C: Emerg. Technol. **128**, 103170 (2021)
6. Wang, D., D'Ariano, A., Zhao, J. et al.: Integrated rolling stock deadhead routing and timetabling in urban rail transit lines. Eur. J. Oper. Res. **298**, 526–559 (2022)
7. Zhou, W., Oldache, M.: Integrated optimization of line planning, timetabling and rolling stock allocation for Urban railway lines. Sustain. **13**(23), 13059 (2021)
8. He, D., Zhang, L., Guo, S., et. al.: Energy-efficient train trajectory optimization based on improved differential evolution algorithm and multi-particle model. J. Clean. Prod. **128**, 127163 (2021)
9. Suhl, L., Steinzen, I., Gintner, V., et al.: A time-space network approach for the integrated vehicle-and crew-scheduling problem with multiple depots. Transp. Sci. **44**(3), 367–382 (2010)
10. Mahmoudi, M., Zhou, X.: Finding optimal solutions for vehicle routing problem with pickup and delivery services with time windows: a dynamic programming approach based on state-space-time network representations. Transp. Res. Part B: Methodol. **89**, 19–42 (2016)
11. Wang, Y., Liao, Z., Tang, T., et al.: Train scheduling and circulation planning in urban rail transit lines. Control. Eng. Pract. **61**, 112–123 (2017)

# Track Limit Detection Algorithm of Fully Automatic Train Based on Vision Sensor

Tuo Shen[1], Yuanxiang Xie[1], Lanxin Xie[1], Jinhuang Zhou[1], and Chenxin Deng[2](✉)

[1] Shanghai Key Laboratory of Rail Infrastructure Durability and System Safety, Tongji University, Shanghai 201804, China
{2233390,2033422,2133387}@tongji.edu.cn
[2] Shanghai Zegao Electrical Engineering Technology Co, Ltd., Shanghai, China
dengchenxin@shzegao.com

**Abstract.** With the benefits of a wide monitoring range, easy installation and maintenance, as well as good remarkable observability, vision sensors are gradually being applied to engineering measurement and environmental perception research in rail transportation. A sliding pane-like track limit detection algorithm based on vision sensor is proposed for the track limit detection in the field of obstacle detection for fully automatic driverless trains in rail transit. The algorithm uses the two-stage inter-frame differential thresholding method to detect the motion of video frames, and uses grayscale distribution feature extraction and adaptive Sobel operator thresholding edge detection methods to improve the accuracy and integrity of the image scene recognition and edge detection. The track limit was extracted by the tack limit search module based on the sliding pane, and Kalman filter is used to improve the accuracy and robustness of the detection results. The experimental validation results show that the algorithm has good detection performance on track limit.

**Keywords:** Rail transit · Fully automatic operation · Limit detection · Vision sensor · Obstacle detection

## 1 Introduction

Fully automatic driverless technology in rail transit field has become a new development trend. Automatic train control system (ATS), which integrates automatic train operation system (ATO), automatic train protection system (ATP) and mobile authorization system (MA +) [1] is the core of automatic trains. The fully automated driverless mode will be the main operation mode of future rail transit. By the end of 2022, China's railway mileage has reached 155,000 km, including 42,000 km of high-speed railway; the total length of China's urban rail transit lines in operation has reached 10,287.45 km, of which 15 cities had opened a total of 30 fully automatic operation system lines, with a total operational mileage of 716.83 km, and the opening of fully automatic operation lines is generally on the rise.

The intrusion of foreign objects into track limit is a major threat to the safe operation of fully automatic driverless trains. Figure 1 shows several common types of track limit


Y. Qin et al. (Eds.): EITRT 2023, LNEE 1137, pp. 292–304, 2024.
https://doi.org/10.1007/978-981-99-9311-6_32

intrusion. Rail accidents caused by track limit intrusion occur from time to time. On June 4, 2022, a train hit a mudslide derailment that intruded into the line of Guiyang-Guangzhou High-Speed Railway, resulting in 1 fatality and many injuries. On April 2, 2021, a train collided with an engineering vehicle on Taiwan Railway, causing huge casualties. Once the rail transit accident occurs, it will not only cause casualties, but also bring huge economic losses. Therefore, in order to prevent safety accidents caused by foreign objects intrusion into the boundary, ensure driving safety and reduce economic losses, many studies on the detection of intrusion into the limit of rail transit have been carried out, and the detection of rail limit is a crucial part of the process.

**Fig. 1.** Examples of track limit intrusion.

Rail transit limit is broadly defined as the dimensions of the space required for the safe operation of trains along a fixed track. The size of the space required for the safe operation of trains can be determined by building limit and vehicle limit. The track limit proposed in this study refers specifically to the two parallel track lines and the area between them in the video image, which is also the key surveillance area for track obstacle detection and track intrusion detection.

The detection accuracy of track limit will directly affect the accuracy and reliability of limit intrusion detection. In this paper, a track limit detection algorithm based on vision sensor is proposed for fully automatic trains. In the algorithm, a track limit search module based on sliding pane is designed. The dynamic and static detection of video frames is completed by two-stage inter-frame differential threshold method. Grayscale distribution feature extraction and adaptive Sobel operator threshold edge detection methods are used to improve the accuracy and integrity of image scene recognition and edge detection. Track boundary search module based on sliding pane is used to complete the track limit extraction. Kalman filter is used to improve the accuracy and robustness of detection results.

The experimental verification shows that the algorithm can track and detect the track limit area in front of the running train accurately, quickly and stably, as well as has the advantages of small memory occupation, short inference delay and light weight, which has certain practical significance to improve the safety and efficiency of train operation.

## 2 Related Work

Railway track limit detection methods according to different detection principles can be divided into two categories: contact detection and non-contact detection. Contact limit detection, such as the use of limit detection vehicles moving on the track, when a collision occurs, it means that there is limit intrusion at the place, the method is simple in principle, but the detection cost is high, the detection efficiency is low, can only be used after the track paving, and there is less application in China [2].

Non-contact limit detection methods use infrared sensors [3], ultrasonic radar [4], LIDAR [5], and laser scanners to obtain track limit locations. Huang et al. [6] developed a track mobile laser measurement and inspection system based on mobile laser scanning technology, which can meet the needs of railway and metro rail transit limit detection at the same time. Zhou et al. [7] obtained tunnel headroom 3D point cloud data by the mobile laser scanning system and achieved track line extraction from the segmented point cloud based on 3D linear fitting. Chen et al. [8] established a tunnel limit detection coordinate system based on the top surface of the rail, introduced different types of tunnel limit frames and used an improved ray algorithm to achieve limit detection for multiple tunnel types. Although the laser scanning technique can obtain the 3D spatial information of the limit, the detection results depend on the reflection of the physical entities in front of the train, which is susceptible to environmental interference and has low detection accuracy, and is more commonly used for track limit detection in tunnels.

Video surveillance, as another non-contact track limit detection method, is also widely used due to its advantages such as large monitoring area, convenient installation and maintenance, and good observable results [9]. The use of convolutional neural networks (CNN) to build feature extraction and target detection algorithms is the current research hotspot. Wang et al. [10] proposed an adaptive segmentation algorithm based on CNN, which uses the strong linear features of railway scenes and typical classes of local regions to achieve railway track limit segmentation using an adaptive feature distribution extractor. Ye et al. [11] designed an effective differential feature fusion CNN to achieve detection of track limit and intrusion obstacle. Zhang et al. [12] improved the detection rate of track limit and small obstacle by balancing the number of samples through a data enhancement strategy and fusing the features of the FPN network with the original Faster-RCNN backbone network. All the above methods require pre-collection and processing of a large number of training samples, with low algorithm timeliness and reduced algorithm reliability in the face of special scenarios with complex backgrounds. Gong et al. [13] first applied a few-shot learning algorithm to railway limit intrusion detection and developed an efficient small-sample railway limit intrusion detection algorithm using enhanced few-shot learning, where the training scale of the algorithm was reduced and the algorithm economy and performance were somewhat improved. The current neural network-based rail limit detection algorithm has limitations in practical engineering applications and deployments due to its larger model, longer time delay and higher hardware equipment requirements.

# 3 Methodology

## 3.1 Algorithm Process

The process of the track limit detection algorithm based on the sliding pane is shown in Fig. 2. Firstly, scene recognition is carried out on the input video image, including the operating status and operating environment of the train. Then the adaptive edge detection is carried out according to the scene recognition result to obtain a binarized image with track boundary features. Then track boundary candidate pixel points are searched based on the track limit search module of sliding pane to complete the track curve fitting in the image. Finally, the track curve parameters are filtered by Kalman filter and the detection results are returned.

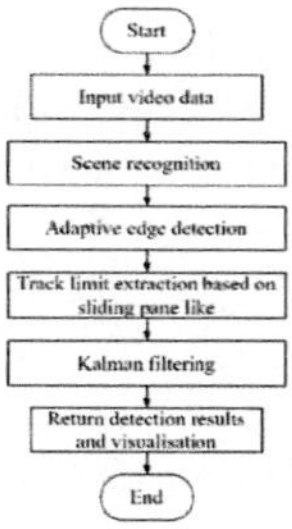


**Fig. 2.** Process of track limit detection algorithm based on sliding pane like.

## 3.2 Scene Recognition

The scene recognition consists of two steps: motion detection and scene segmentation. The image inter-frame differencing method is used to achieve the motion detection of trains, and the grayscale histogram feature is used to identify the trains operation scenes.

When a train is detected to be in a stationary state, the track limit detection results of the historical frames can be used directly to save computational resources. The train motion detection is done using inter-frame differencing, where the interval frames in a continuous video stream are differenced and the absolute value is taken to obtain a grayscale image of the same size as the original frame. The primary gray level difference threshold is set, the differential gray image is binarized and the number of changed pixels is counted, and the secondary gray level change pixel proportion threshold is set to determine whether the interval frame image has changed. Several iterations of the experiment show that when the image differential interval frames number is 3, the primary gray level differential threshold is 4, and the secondary gray level differential pixel scale threshold is 0.125, the stationary or moving state of the train can be accurately detected, and the influence of the image screen caused by the slight vibration of the camera, light changes and the movement of objects in the screen can be better overcome.

The scene segmentation strategy based on the image grayscale histogram feature is as follows:

$$c_i = \frac{H_i}{T} \tag{1}$$

$$l_i = \begin{cases} 0 c_i < r_1 \\ 1 r_1 \le c_i < r_2 \\ 2 r_2 \le c_i < r_3 \\ \vdots \\ k c_i \ge r_k \end{cases} \tag{2}$$

where, $H_i$ is the number of pixels with a grayscale value greater than or equal to the threshold W, $T$ is the total number of pixels, and $c_i$ is the ratio of the two; $l_i$ is the scene category marker of the image, and $r_k$ is the scene segmentation threshold. In this study, it is experimentally verified that when the scene is divided into two categories by setting $W = 60$ and $r_1 = 0.7$, good recognition results can be obtained and the performance of the algorithm can also be guaranteed.

### 3.3 Adaptive Edge Detection

According to the image scene recognition results, the image pre-processing is firstly performed, including image grayscale transformation, perspective transformation and calibration of the region of interest. The pre-processed image is then subjected to adaptive edge detection.

Grayscale refers to the transformation of the color image of the RGB channel into a single-channel gray image, and the conversion formula is as follows:

$$gray = 0.299 \cdot R + 0.587 \cdot G + 0.144 \cdot B \tag{3}$$

The essence of perspective transformation is to project an image onto a new viewing plane. Its general transformation formula is as follows:

$$\begin{bmatrix} X \\ Y \\ Z \end{bmatrix} = \begin{bmatrix} a_{11}\ a_{12}\ a_{13} \\ a_{21}\ a_{22}\ a_{23} \\ a_{31}\ a_{32}\ a_{33} \end{bmatrix} \begin{bmatrix} x \\ y \\ z \end{bmatrix} \tag{4}$$

$$\begin{cases} x' = \frac{X}{Z} = \frac{a_{11}x + a_{12}y + a_{13}}{a_{31}x + a_{32}y + a_{33}} \\ y' = \frac{Y}{Z} = \frac{a_{21}x + a_{22}y + a_{23}}{a_{31}x + a_{32}y + a_{33}} \\ z' = \frac{Z}{Z} = 1 \end{cases} \tag{5}$$

where, $(x, y, z)$ are the flush coordinates of the original image pixel points and $(x\prime, y\prime, z\prime)$ are the flush coordinates of the image pixel points after the perspective transformation.

The Sobel operator in the x-direction is then used to perform edge detection on the pre-processed grayscale image. A convolution kernel of size $3 \times 3$, $H_x = \begin{bmatrix} -1\ 0\ 1 \\ -2\ 0\ 2 \\ -1\ 0\ 1 \end{bmatrix}$, is selected to perform the convolution operation with the grayscale image, and the convolution result obtained for each pixel point in the image is taken as the absolute value, i.e. the gray gradient at the location of the pixel point is obtained. After obtaining the gray gradient at the location of each pixel point, the pixel point is binarized according to the parameter assignment result of scene recognition, and the final edge detection binarized image sequence is obtained.

The specific method of pixel binarization is shown in Fig. 3: set a high threshold $H_V$ and a low threshold $L_V$, when the gray gradient at the location of the pixel is greater than $H_V$ (segment A), assign a gray value of 255 to the pixel, when the gray gradient at the location of the pixel is less than $L_V$, assign a gray value of 0 to the pixel, when the gray gradient at the location of the pixel is between $H_V$ and $L_V$, , then judge whether there is a pixel with a gray value of 255 around the pixel (8 pixel points). If there is (segment B), the pixel is assigned a grayscale value of 255, if not (segment C), the pixel is assigned a grayscale value of 0.

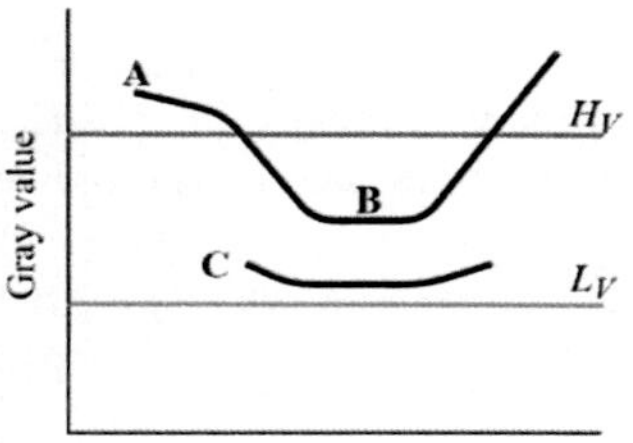


**Fig. 3.** Pixel dot binarization process.

### 3.4 Track Limit Extraction Based on Sliding Pane like

After adaptive edge detection of the original image, the resulting binarized image still has some interfering pixel points at the boundary of trackside facilities and equipment.

A sliding pane like based track boundary search module is designed to filter the track boundaries for candidate pixel points and to filter out the interfering pixels. The algorithm is divided into two sub-modules: a sliding pane search based on the current frame and a passband search based on the history frame. Both search algorithms aim to search for track boundary pixels in the edge-detected binarized image, and to fit a quadratic curve to the final track line based on least squares and the sequence of coordinates of the searched boundary points.

**Sliding Pane Search Based on the Current Frame.** The sliding pane search sub-module based on the current frame is also the initialization process for the track line extraction. The algorithm calls the sliding pane search sub-module when it does not receive the track line extraction history frame data or when the history frame track line extraction returns a failure.

(1) Manually calibrate the coordinates of the pixel point of the track boundary mid-axis at the lower edge of the image, $P_{lane\_base}$, as the base point of the track growth, and also calibrate the image parallel track spacing $d_r$.

(2) Determine the initial sliding pane position based on the calibrated track lower boundary growth base point. This is done by using the base point coordinates as the midpoint coordinates of the lower edge of the sliding pane, and determine the width $w_{win}$ and height $h_{win}$ of the pane, which are set as adjustable parameters in the algorithm design, and adjust them to obtain the best detection results.
(3) Search for white pixel points (with a grayscale value of 255) within the rectangular frame according to the initial sliding pane position coordinates determined in step (2), and record the image coordinates $P_{ij}$ of these white pixel points into the set of candidate coordinates $G$ of the track boundary, and count the number of white pixel points $N_i$ within the pane.
(4) A threshold $N_u$ is set for the number of white pixels in the pane, and $N_i$ is compared with $N_u$ to determine how the sliding pane grows towards the upper edge of the image. When $N_i \geq N_u$, the average of the horizontal coordinates of all white pixels in the pane $\overline{x}_i$ is calculated as the horizontal coordinate of the lower border of the next sliding pane, while when the number of white pixels in the pane is too small ($N_i < N_u$) to describe the growth trend of the track boundary, the horizontal coordinate of the lower border of the current pane is still used as the horizontal coordinate of the lower border of the next sliding pane. For each upward growth, the lower border of the new sliding pane that grows is the same as the vertical coordinate of the upper border of the sliding pane before the growth.
(5) Iterate, after the sliding pane grows upwards, repeat steps (3) and (4) until the upper border of the rectangular sliding pane reaches the upper edge of the image and the iteration ends.
(6) A quadratic curve $x = ay^2 + by + c$ is fitted by least squares method to a sequence of candidate coordinates $G$ at the track boundary, and the key parameters $(a, b, c)$ are obtained. (Notes: since track does not have standing points at short distances, the use of a quadratic curve to fit the track line is both practical and efficient for the algorithm).

Figure 4 shows a diagram of the sliding pane boundary pixel search based on the current frame, with the red dot at the lower boundary of the image being the calibrated track growth base point. As can be seen, the majority of interfering pixel points in the edge detection binary image are screened out, and a more accurate track line can be obtained by fitting a quadratic curve least squares to the search sequence of candidate track boundary coordinates.

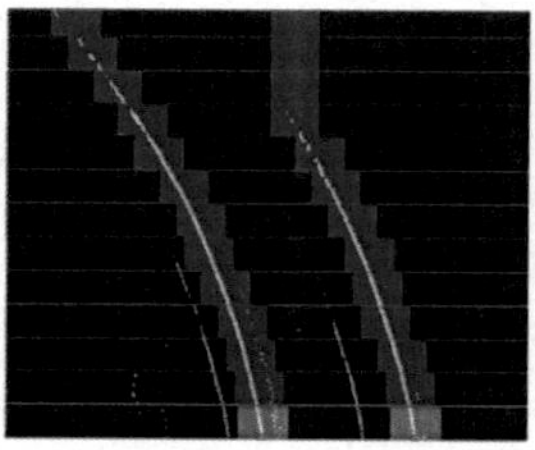

**Fig. 4.** Sliding pane track boundary pixel search schematic.

**Passband Search Based on Historical Frames.** The history frame-based passband search sub-module complements the current frame-based sliding pane like search sub-module, guarantees the effectiveness of track boundary candidate pixel point screening while also improves algorithm performance to a certain extent.

Based on the fact that the coordinates of the track lines in adjacent frames captured by the camera in a very short period of time do not change abruptly, the track boundary pixels of the current frame can be searched in the area adjacent to the track lines detected in the previous frame. When the algorithm receives the history frame data for the track line extraction, the algorithm calls the passband search sub-module.

(1) Enter the parameters $(a, b, c)$ of the last track line fit and determine the location of the last track line fit in the current frame $x = ay^2 + by + c$.
(2) The result of the track line fitted in the previous frame is translated to the left and right of the image by a certain distance $\frac{1}{2}d_m$ respectively, forming an equal width passband with the track line fitted in the previous frame as the central axis, and the width of the passband is $d_m$.
(3) Search for white pixel points covered by the passband in the edge detection binarized image and record the coordinates to generate the set of candidate coordinates $G\prime$.
(4) The quadratic curve $x = ay^2 + by + c$ is fitted using least squares to the sequence of coordinates $S\prime$ and the quadratic curve parameters $(a, b, c)$ are obtained.

Figure 5 shows a schematic of the passband search based on historical frames. As can be seen from the figure, the track boundary pixel points are basically covered and filtered out by the passband, while other interfering pixel points are eliminated.

**Fig. 5.** Passband track boundary pixel search schematic.

### 3.5 Track Boundary Smoothing Based on Kalman Filter

The Kalman filter is an optimal autoregressive data processing method [14] with good real-time performance and accuracy. The basic idea is to use the previous moment state, get the current moment estimate according to the state space model, and update the estimate of the current state by combining the current moment observation to obtain the current state quantity that satisfies the minimum mean squared error [15]. This study uses the Kalman filter to filter and smooth the fitted track curves, which can reduce the fluctuation of track line extraction from continuous video frames and improve the stability and reliability of track curve extraction.

The system dynamics of the track curve is spatially described as:

$$\begin{cases} \hat{x}_t^- = F\hat{x}_{t-1}^- + Bu_{t-1} \\ z_t = H\hat{x}_t + \varepsilon \end{cases} \quad (6)$$

The Kalman filter state update equation is:

$$\begin{cases} P_t^- = FP_{t-1}F^T + Q \\ K_t = \frac{P_t^- H^T}{HP_t^- H^T + R} \\ P_t = (I - K_t H)P_t^- | \end{cases} \quad (7)$$

The track curve prediction combined with the dynamic space equation is:

$$\hat{x}_t = \hat{x}_t^- + K_t\left(z_t - H\hat{x}_t^-\right) \quad (8)$$

where, $\hat{x}_t^-$ is the current moment forecast quantity; $\hat{x}_{t-1}$ is the previous moment state quantity; $\hat{x}_t$ is the current moment state quantity; $F$ is the system transfer matrix; $B$ is the control variable matrix; $z_t$ is the observed quantity; $H$ is the state quantity to observation transfer matrix; $\varepsilon$ and $u_{t-1}$ are Gaussian white noise; $K_t$ is the Kalman gain, also known as the state update matrix; $P_t^-$ is the predicted $\hat{x}_t^-$ covariance; $P_t$ and $P_{t-1}$ are the covariances of the state quantities $\hat{x}_t$ and $\hat{x}_{t-1}$ respectively; $Q$ and $R$ denote the covariance matrices of the predicted and measured noise; $I$ denotes the unit matrix; the state vector of the Kalman filter in this study is the track line fitting parameter $\left[a\ b\ c\right]^T$. A schematic of the Kalman filter for track limit extraction is shown in Fig. 6.

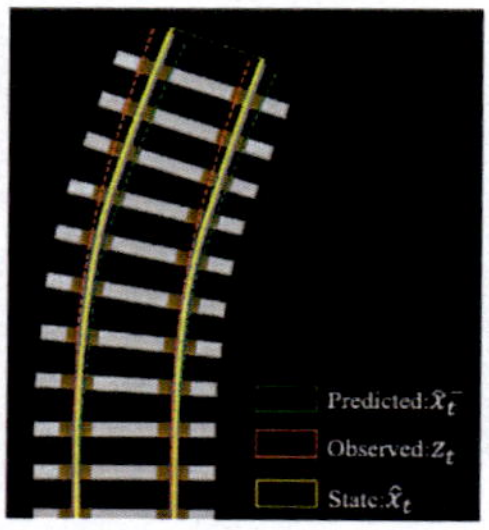


**Fig. 6.** Kalman filter for track limit extraction schematic.

# 4 Experimental Verification

## 4.1 Experimental Data

To verify the algorithm proposed in this paper, a video data of 11 min and 22 s duration was captured by the front camera during the operation of the train at ground level, elevated and tunnel, with a frame rate of 30 fps and a resolution of 1920 × 1080, between the starting station Disney and the ending station Jiading North of Shanghai Railway Line 11.

Viewing the entire video data, the train is shown in the more illuminated elevated section and the less illuminated tunnel section. The curvature of the track in the video image can be further classified as straight, left-turning and right-turning, and the illumination changes can be classified as steady, light to dark and dark to light. These special scenes are the key areas of interest for algorithm effectiveness verification.

## 4.2 Experimental Environment

The following Table 1 gives the experimental verification environment.

**Table 1.** Experimental environment

| Projects | Parameters |
|---|---|
| CPU | Inter (R) Core (TM) i9-12900k |
| GPU | GPU NVIDIA GeForce RTX 3080 |
| Memory size | 32GB |
| Operating system | Ubuntu 20.04 |
| Programming language | C++ |

## 4.3 Analysis of Experimental Results

The entire video data is fed into the algorithm for processing and the track limit detection results are shown schematically in Fig. 7. As a whole, the track limits are accurately detected in all scenes during the train operation, and the accompanying scene recognition results and algorithm detection frame rates are shown simultaneously in the upper left corner of the image. The three images in the first row show the detection results of elevated sections with sufficient illumination, which are straight, right turn and left turn respectively. The three images in the second row show the detection results of elevated sections with obvious changes in illumination, such as overpasses, trackside facilities and equipment blocking, trains entering and exiting stations, trains entering and exiting tunnels. The three images in the third row are the detection results of tunnel sections with sufficient illumination: straight, left turn and left turn. The fourth three images in the fourth row show the detection of extreme special scenes in the tunnel, such as extremely dark environments, extremely complex backgrounds, and shadow areas formed by the contrast between light and dark on the station platform. The experimental results show that the algorithm not only achieves the purpose of track limit detection, but also has good scene adaptability and is able to accurately detect track limits in various complex scenes and backgrounds of train operation.

In the track limit extraction step based on the sliding pane like, the analysis of the track line extraction "failure" distribution is shown in Fig. 8. In this experiment, the train operation scenes of video data are divided into "1" and "0". Scene 1 is mainly elevated section with sufficient light, while scene 0 is mainly tunnel section with insufficient light.

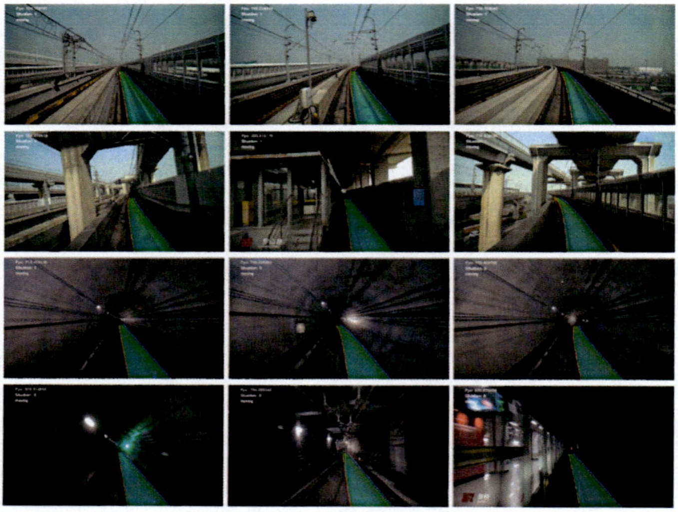

**Fig. 7.** Track limit detection results.

The total number of frames where the algorithm failed to extract the track lines was 677, accounting for 3.32% of the total number of frames, of which 452 frames were failed under the condition of 0 for scene identification and 452 frames under the condition of 1 for scene identification. The number of frames that failed to be extracted was 225, and the extraction results were significantly affected by the illumination.

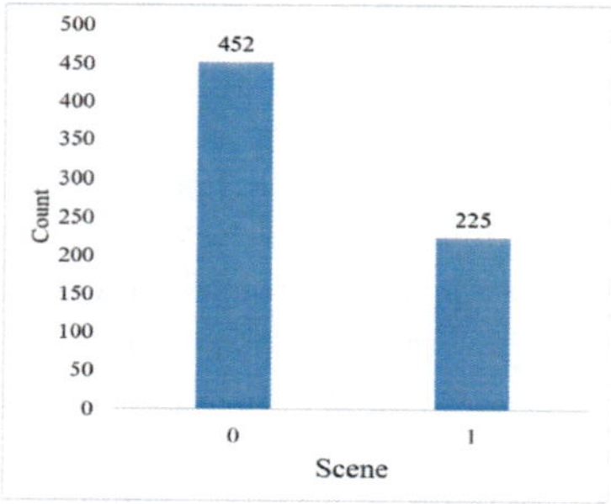


**Fig. 8.** Analysis of track extraction "failure" scenarios.

The results of the time consumption analysis of the algorithm running throughout the captured video data are shown in Figs. 9 and 10. It can be seen that the algorithm's single-frame image processing time consumption is concentrated in the range of 1ms to 2ms, with an average time consumption of about 1.35ms. The conversion of the algorithm into video frame rates is above 400 fps, much higher than the original frame rate of 30 fps for the captured video. The experiments show that the algorithm is able to meet the performance requirements of track limit detection for trains running at high speed.

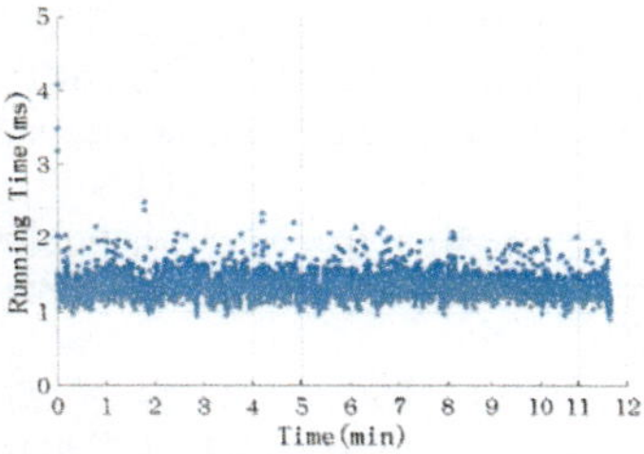


**Fig. 9.** Algorithm runtime consumption.

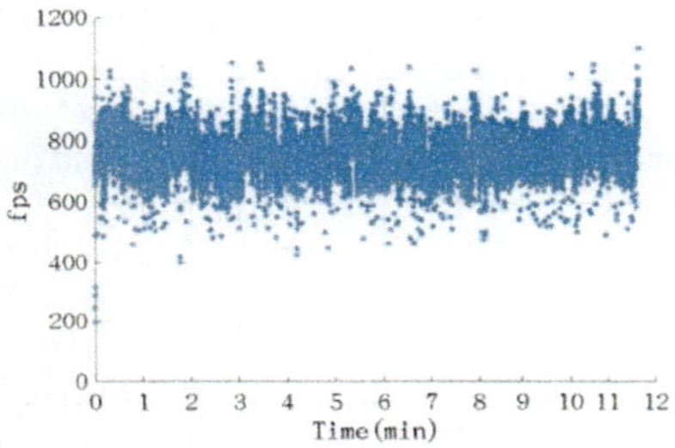


**Fig. 10.** Algorithm runtime conversion frame rate

# 5 Conclusion

In this paper, a vision sensor-based track limit detection algorithm for fully automatic trains is proposed. The algorithm uses a sliding pane like search algorithm to search for track boundary pixels in binarized images to detect track limits in front of the train. The algorithm firstly uses a two-stage inter-frame difference thresholding method to complete the motion detection of video frames, then uses grayscale distribution feature extraction and adaptive Sobel operator threshold edge detection to improve the accuracy and integrity of scene recognition and edge detection, and then uses a sliding pane like based track limit search module to complete the extraction of track limits, and finally uses Kalman filter to improve the accuracy and robustness of the detection results. The experimental validation results show that the algorithm has good track limit detection performance.

**Acknowledgments.** . This research is supported by National Key Research and Development Project of China (2022YFB4300501) and Science and Technology Commission of Shanghai Municipality (23DZ2204900).

# References

1. Song, H., Shen, T., Wang, W.: Train-centric communication-based close proximity driving train movement authority system. IEEE Intell. Transp. Syst. Mag. **10**(3), 22–34 (2018)

2. Jiang, H.: Development and application of intelligent limit detection technology for rail transportation. Railw. Constr. Technol. (02), 27–29+75 (2020). (in Chinese)
3. Kumar, P., Kumar, P.: Arduino based wireless intrusion detection using IR sensor and GSM. Int. J. Comput. Sci. Mob. Comput. **2**(5), 417–424 (2013)
4. García, J.J., Ureña, J., Hernandez, A., Mazo, M., Jiménez, J.A., Alvarez, F.J., & García, E.: Efficient multisensory barrier for obstacle detection on railways. IEEE Trans. Intell. Transp. Syst. **11**(3), 702–713 (2010)
5. Shen, T., Qian, Y., Xie, L., Yuan, T., Zeng, X., Zhang, X.: Obstacle detection algorithm of fully automatic train considering reflection intensity. J. Tongji Univ. (Nat. Sci. Ed.) **01**, 6–12 (2022). (in Chinese)
6. Huang, X., Yao, L.B., Yang, L.J.: Integrated detection of rail transit building limits based on mobile laser scanning. Mapp. Bull. **03**, 84–88 (2023). (in Chinese)
7. Zhou, Y., et al.: Railway tunnel clearance inspection method based on 3D point cloud from mobile laser scanning. Sensors **17**(9), 2055 (2017)
8. Chen, G., Han, F., Duan, X., Mou, H.: Research on the integrated automatic detection method of track structure and tunnel spatial limit based on point cloud data. Railw. Stand. Des. (12), 31–36 (2019). (in Chinese)
9. Zhan, D., Jing, D., Wu, M., Zhang, D., Yu, L., Chen, T.: An accurate and efficient vision measurement approach for railway catenary geometry parameters. IEEE Trans. Instrum. Meas. **67**(12), 2841–2853 (2018)
10. Wang, Y., Zhu, L., Yu, Z., Guo, B.: An adaptive track segmentation algorithm for a railway intrusion detection system. Sensors **19**(11), 2594 (2019)
11. Ye, T., Zhang, X., Zhang, Y., Liu, J.: Railway traffic object detection using differential feature fusion convolution neural network. IEEE Trans. Intell. Transp. Syst. **22**(3), 1375–1387 (2020)
12. Zhang, L., Shen, T., Zhang, X.: Detection of small obstacles in urban rail transit. J. Shanghai Univ. Technol. **05**, 468–473 (2021). (in Chinese)
13. Gong, X., Chen, X., Zhong, Z., Chen, W.: Enhanced few-shot learning for intrusion detection in railway video surveillance. IEEE Trans. Intell. Transp. Syst. **23**(8), 11301–11313 (2021)
14. Kalman, R.E.: A new approach to linear filtering and prediction problems (1960)
15. Yin, A., Zhu, W., Dai, Z., Ren, H.: Kalman filtered corrosion prediction with fuzzy inference and deep learning number-mode fusion. J. Electron. Meas. Instrum (2023). (in Chinese)

# Examining the Relationship Between Built Environment and Urban Rail Transit Inter-Cluster OD Passenger Flow

Lidan Chen and Xinyue Xu(✉)

State Key Laboratory of Rail Traffic Control and Safety, Beijing Jiaotong University, Beijing 100044, China
{21120790,xxy}@bjtu.edu.cn

**Abstract.** Research on the impact of built environment on Origin Destination (OD) passenger flow is essential for combining urban land planning and public transportation planning. However, the sparsity of inter-station OD makes the relationship between built environment and OD passenger flow not significant enough. To address this issue, this study proposes a clustering model based on passenger flow destination and station location. Firstly, the clustering is initialized based on the station location and station type. Then, a modular index of the road network clustering is established to optimize the clustering by considering the passenger flow destination, determine the optimal clustering, and classify it afterwards. Furthermore, considering the combination of different types of clusters, the GBRT model is used to study the relationship between different types of inter-cluster OD passenger flow and built environment. A case study in Beijing shows that the noise of the nonlinear effect of built environment on inter-cluster OD is smaller than that of the noise of the nonlinear effect of built environment on inter-station OD. This is because the aggregation of stations into clusters makes the regularity more significant, achieving better fitting results.

**Keywords:** Built environment · Inter-Cluster OD · GBRT

## 1 Introduction

Urban rail transit is favored by travelers and urban planners due to its features of large capacity and punctuality. The essential reason for the generation of urban rail transit OD passenger flow lies in the different built environments of O and D [1], so investigating the relationship between urban rail transit OD passenger flow and the built environment of O and D is crucial for integrating urban land planning with public transportation planning and achieving sustainable development of urban rail transit.

In recent years, with the development of multi-source big data technology, the impact of the built environment on OD passenger flow has been studied in various transportation modes. For example, Gan et al. [1] used the GBRT model to explore the nonlinear relationship between the built environment and OD passenger flow, and the results showed that the impact of the built environment on OD passenger flow varies heterogeneously at


Y. Qin et al. (Eds.): EITRT 2023, LNEE 1137, pp. 305–312, 2024.
https://doi.org/10.1007/978-981-99-9311-6_33

different time periods. Zhao et al. [2] used multiple regression models and multiplicative models to study the effect of the built environment on urban rail transit OD passenger flow. The results showed that the built environment at the origin and destination had heterogeneous effects on urban rail transit OD passenger flow at different times. Zhang et al. [3] used an interactive clustering method to divide Chicago's ride-hail OD into five categories and explored the spatial heterogeneity of the impact of the built environment on different types of OD flows. The results showed that estimating the different categories of OD passenger flow improved the estimation accuracy compared to a global model. Bordagaray et al. [4] divided shared bike travelers into five categories and discussed the heterogeneity of different categories based on factors. However, the research on the heterogeneity of the relationship between different types of OD passenger flow and the built environment is mainly concentrated in other transportation modes, while the research in the field of urban rail transit is not yet sufficient.

In recent years, some scholars have attempted to aggregate stations in order to improve the accuracy of OD passenger flow estimation. For example, Jia et al. [5] used a two-layer Gaussian mixture model to consider the movement trend of bicycles between stations and the geographic location information to group bicycle stations. The results showed that grouping bicycle stations improved the prediction accuracy. Jiang Yan et al. [6] considered that adjacent urban rail transit stations have similar passenger flow distribution and travel patterns, and grouped them into clusters. The results showed that station grouping can provide a basis for network operation organization arrangements. However, there is still room for improvement in the study of the relationship between OD passenger flow and the built environment in the field of urban rail transit by aggregating stations together.

Based on this, this article proposes a cluster partitioning model based on passenger flow destinations and geographic location, using the GBRT model to study the relationship between OD passenger flow and the built environment for different types of clusters. Specifically, first, the clusters are initialized by considering geographic location and station type. Then, considering the station's passenger flow destinations, a modularization index is established to optimize the routing network clusters, determine the optimal clusters, and classify them. Secondly, considering the combination of different types of clusters, the GBRT model is used to study the nonlinear relationship between OD passenger flow and the built environment among different types of clusters.

## 2 Scope of Study and Variable Description

This study mainly involves passenger flow data and built environment data. In terms of passenger flow data, this study adopts the swipe card data of Beijing rail transit system from November 1st, 2021 to November 5th, 2021 on working days. The morning peak period is from 7:00 to 10:00, and the evening peak period is from 17:00 to 20:00. As for the built environment, this study sets the influence radius of the station at 800m [7]. The built environment includes the following aspects:

(1) External transportation

External transportation characteristics includes road density (RD) and the number of buses (NB).

(2) Land use diversity

Land use diversity characteristics includes commercial and entertainment POIs (CEPOI), the number of educational POIs (EPOI), the number of large venues POIs (LVPOI), the number of hospital POIs (HPOI), the number of residential POIs (RPOI), the number of office POIs (OPOI), the number of transportation hub POIs (THPOI), the number of tourist POIs (TPOI) and mixed entropy of land use (MELU) [8]. The equation for calculating the mixed entropy of land use is:

$$E_m = \frac{-\sum_{i=1}^{N} R_i \ln(R_i)}{\ln|N|} \tag{1}$$

where $E_m$ represents the land use mix index, $R_i$ represents the proportion of the i-th type of POI quantity to the total POI quantity, and N is the number of POI types, which in this study is 8.

(3) Cluster characteristics

Cluster characteristics include the degree of clusters and the number of entrances and exits (NEE).

(4) Socio-economic factors

Socio-Economic factors include the number of office people (NOP), the number of residential people (NRP) and the office-residence balance factor (ORBF).

(5) OD impedance

OD impedance includes travel time (ODTT) and transfer times (ODTTS) for OD trips within the cluster.

# 3 Model Construction

## 3.1 Clustering and Classifying Model

**Clustering Model.** Considering that adjacent urban Rail Stations have similar OD passenger flow distribution and travel patterns [6], all railway stations are divided into clusters. The specific steps for division are as follows:

(1) Construct an initial network cluster

Divide all the stations in the network into different clusters based on the station type and location. The basic principle of clustering is that if adjacent stations on the network have the same or similar types, they will be clustered together. The initial collection of all clusters in the network is $G_0 = \{C_{01}, C_{02}, \cdots, C_{0i}, \cdots C_{0n}\}$, and $C_{0i} = \{S_{0i1}, S_{0i2}, \cdots, S_{0ij}, \cdots, S_{0im}\}$ is the set of stations contained in the cluster $C_{0i}$.

(2) Construct the set of adjacent stations for clusters

The set of adjacent stations for clusters is obtained by combining the adjacent stations of all cluster stations.

(3) Calculate the modularity index of network clusters

To measure the degree of consistency of internal destinations within the divided clusters, a modularity index is constructed. The formula for calculating the modularity index of cluster is shown as Eq. (2). The modularity index of network clusters is the average of all cluster modularity indexes.

$$Q_{0i} = \sum_{x=1}^{n}\sum_{y=1}^{m} \frac{P_{0i,0x}}{O_{0i}} - \frac{P_{0iy,0x}}{O_{0iy}} \tag{2}$$

where $P_{0i,0x}$ represents the passenger flow between cluster $C_{0i}$ and cluster $C_{0x}$, $P_{0iy,0x}$ represents the passenger flow between station $S_{0iy}$ and cluster $C_{0x}$, $O_{0i}$ represents the inflow of passengers to cluster $C_{0i}$, and $O_{0iy}$ represents the inflow of passengers to station $S_{0iy}$.

(4) Adjust and optimize clusters

Based on the existing clusters, the optimization of clusters is adjusted. The larger the index $Q_{0i}$, the more likely the cluster will be selected for adjustment. The specific adjustment method is as follows: remove stations from the set of edge points of the cluster, where stations with larger absolute values of modularity indices are more likely to be selected for removal. Then add the station to the cluster that belongs to the adjacent station set. The adjusted cluster is denoted by $G_1$. The modularity index of the adjusted cluster is denoted by $Q_1$, and the modularity change $\Delta Q = |Q_1 - Q_0|$. If $\Delta Q$ is positive, then update $Q_{\min} = Q_1$, $G_{best} = G_1$, otherwise, $Q_{\min} = Q_0$, $G_{best} = G_0$, and continue looping to adjust and update the clusters until the set number of iterations is reached.

**Cluster Classifying Model.** After dividing into clusters, cluster types are assigned based on three passenger flow indexes: residential index, office index, and deviation index.

$$I_l = \frac{\frac{q_{inm}}{q_{ins}} + \frac{q_{outn}}{q_{outs}}}{2} \tag{3}$$

$$I_w = \frac{\frac{q_{outm}}{q_{outs}} + \frac{q_{inn}}{q_{ins}}}{2} \tag{4}$$

$$I_d = \frac{I_l}{I_w} \tag{5}$$

where $I_l$,$I_w$, and $I_d$ represent the residential index, office index, and deviation index respectively, and $q_{inm}$, $q_{outn}$, $q_{outm}$, $q_{inn}$, $q_{ins}$, and $q_{outs}$ represent the morning peak entry volume, evening peak exit volume, morning peak exit volume, evening peak entry volume, all-day entry volume, and all-day exit volume respectively.

## 3.2 GBRT Model

GBRT approximates the fitting function of OD passenger flow and built environment as a linear sequence combination of multiple regression trees. The model is:

$$f_0(x) = \arg\min_{c} \sum_{i=1}^{N} L(y_i, c) \tag{6}$$

The estimate of the residual is:

$$r_{mi} = -\left[\frac{\partial L(y_i, f(x_i))}{\partial f(x_i)}\right]f(x) = f_{m-1}(x) \tag{7}$$

The weight estimation formula for the regression tree is:

$$c_{mj} = \arg\min_{c} \sum_{y_i \in R_{mj}} L(y_i, f_{m-1}(x_i + c)) \tag{8}$$

The iterative process of the overall model can be represented as follows:

$$f_m(x) = f_{m-1}(x) + \sum_{j=1}^{J} c_{mj} I(x \in R_{jm}) \tag{9}$$

The final prediction function is represented as follows:

$$F(x) = \sum_{m=1}^{M} \sum_{j=1}^{J} c_{mj} I(x \in R_{mj}) \tag{10}$$

**Feature Importance and Partial Dependence.** GBRT can measure the importance of features and display partial dependence plot. The equation is as follows:

$$I_{x_i}^2 = \frac{1}{M} \sum_{m=1}^{M} \sum_{t=1}^{T-1} d_t \tag{11}$$

$$f_s(x_s) = \frac{1}{K} \sum_{k=1}^{K} f(x_s, x_{ck}) \tag{12}$$

where $I_{x_i}^2$ denotes the importance of feature $x_i$; T is the number of nodes in each regression tree; $d_t$ is the increase in square error due to the t-th split of feature $x_i$;$f_s(x_s)$ represents the partial dependence of the fitted function; $x_s$ is the predictor under consideration; $x_{ck}$ represents the other predictors.

# 4 Case Study

## 4.1 Clustering Results and Classification Results

After the clustering is completed, the station clusters are classified according to passenger flow indexes, and the classification results are shown in Table 1.

**Table 1.** Weekday cluster classification results

| Cluster categories | Typical cluster |
|---|---|
| Residential (40.6%) | Tiantongyuan cluster, Huilongguan cluster |
| Office (40.6%) | CBD cluster, financial street cluster |
| Mixed (18.7%) | Line 2 north section cluster, Line 14 southeast cluster |

## 4.2 Comparison Analysis Between Inter-Cluster OD and Inter-Station OD

From the results in Table 2, it can be seen that the fitting error at the cluster level is better than at the station level. Aggregating stations into clusters makes the relationship between built environment properties and OD passenger flow more significant, with stronger regularity and better suited for fitting station OD passenger flow.

**Table 2.** Model fitting results comparison

| Cluster categories | Evening peak | | | |
|---|---|---|---|---|
| | Inter-cluster OD | | Inter-station OD | |
| | MAPE | $R^2$ | MAPE | $R^2$ |
| Office-residence | 0.32 | 0.66 | 0.57 | 0.45 |
| Residence-office | 0.42 | 0.54 | 0.57 | 0.44 |

## 4.3 Analysis of the Relative Importance of Built Environment on Inter-Cluster OD Passenger Flow

From the Fig. 1, during the evening rush hour in the office-residence category, the factors that are important for both inter-cluster and inter-station OD are OOPOI, DCEPOI, DMELU, DNRP, and ODTTS. The importance of DNRP is 7.8% for inter-cluster OD, which is 2.9% higher than that for inter-station OD. Meanwhile, the importance of OOPOI is 1.5% higher for inter-cluster OD than for inter-station OD.

It can be seen that there are significant differences in the relative importance of each built environment factor for inter-cluster and inter-station OD, including OCEPOI,

OEPOI, ONRP, DEPOI, and DORBF. For example, DEPOI has a relative importance of 7.2% for inter-cluster OD but only 3.5% for inter-station OD. This indicates that the results for inter-cluster OD are more reasonable. During the evening rush hour, students may go to tutoring centers after school, which should have a relatively higher importance.

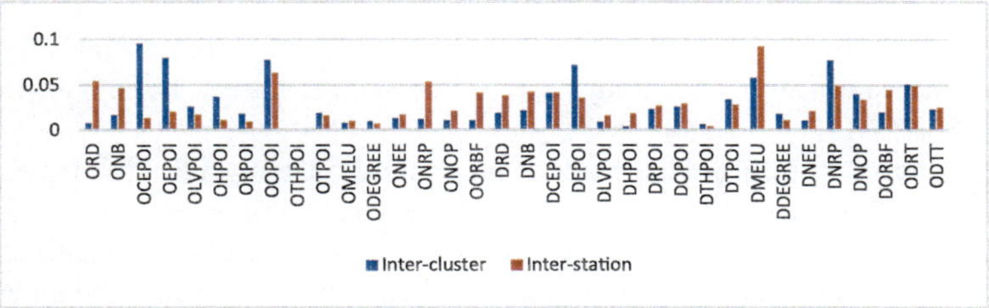


**Fig. 1.** Relative impact of built environment during evening rush hour for office-residence

## 4.4 Nonlinear Impact Analysis of Built Environment on Inter-Cluster OD Passenger Flow

Figures 2 (a) and (b) depict the nonlinear impact of residential-office morning peak OD travel time on inter-cluster and inter-station OD passenger flow. Figure 2 (a) shows that when the cluster OD travel time is less than 5 min, the cluster OD passenger flow tends to be stable, and there is a small fluctuation in the range of 5 to 40, followed by a sharp increase at 40 min, a sharp drop at 43 min, and a sharp increase at 53 min, which is more in line with the average commuting time in Beijing central city in 2021 (51 min) [10]. Figure 2 (b) shows obvious noise, with four mutation points before 50 min in OD travel time, where the OD passenger flow suddenly increases and then drops sharply to the original level. After reaching 50 min, it sharply increases, then drops again around 60 min, and starts increasing again at 100 min. Figure 2 (a) has less noise, and after exceeding 100 min, the passenger flow does not rebound, which is more reasonable and more in line with the situation of commuting passengers in Beijing.

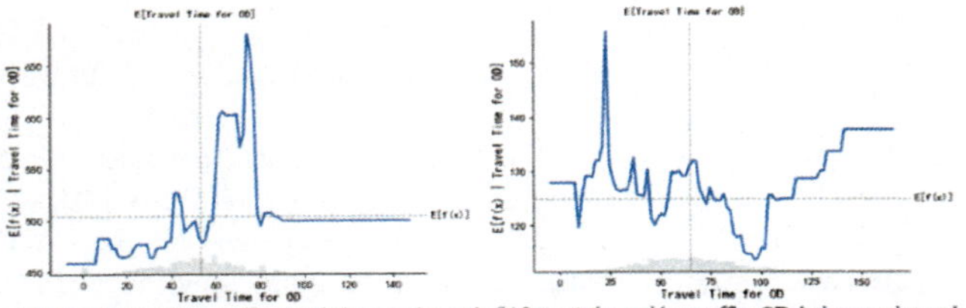


(a) Inter-cluster residence-office OD during morning peak (b) Inter-station residence-office OD during morning peak

**Fig. 2.** The effects of travel time for OD on OD passenger flow

## 5 Conclusion

This article takes urban rail transit station clusters as the unit and extracts the impact of changes in OD passenger flow of station clusters from the perspective of the built environment. Through constructing a GBRT model to analyze its nonlinear effect, the following main conclusions are obtained:

(1) The results of the GBRT model show that the fit of OD at the cluster level is better than that at the station level, indicating that the relationship between the built environment and cluster OD is more significant.
(2) The impact of the built environment on inter-station passenger flow and inter-cluster passenger flow has partial consistency, and the relative importance of the built environment on inter-cluster passenger flow is more significant. The remaining part of the impact of the built environment on inter-station passenger flow and inter-cluster passenger flow has significant differences, but the relative importance of the built environment on inter-cluster passenger flow is more reasonable.
(3) The nonlinear impact of the built environment on cluster OD passenger flow and inter-station OD passenger flow is heterogenous, and the noise of the impact of the built environment on cluster OD passenger flow is relatively smaller than that of inter-station OD passenger flow, and the overall trend is more reasonable.

## References


1. Gan, Z., Yang, M., Feng, T., Timmermans, H.J.P.: Examining the relationship between built environment and metro ridership at station-to-station level. Transp. Res. Part D: Transp. Environ. (82), 102332 (2020)
2. Zhao, J., Deng, W., Song, Y., Zhu, Y.: Analysis of Metro ridership at station level and station-to-station level in Nanjing: an approach based on direct demand models. Transportation **41**, 133–155 (2014)
3. Zhang, X., Zhao, X.: Machine learning approach for spatial modeling of ridesourcing demand. J. Transp. Geogr. (100), 103310 (2022)
4. Bordagaray, M., dell'Olio, L., Fonzone, A., Ibeas, Á.: Capturing the conditions that introduce systematic variation in bike-sharing travel behavior using data mining techniques. Transp. Res. Part C: Emerg. Technol. **71**, 231–248 (2016)
5. Jia, W., Tan, Y., Liu, L., Li, J., Zhang, H.: Hierarchical prediction based on two-level Gaussian mixture model clustering for bike-sharing system. Knowl.-Based Syst. **178**, 84–97 (2019)
6. Jiang, Y., Yang, J., Lei, Y.: Research on passenger flow forecast method during trial operation of urban transit circle line. Mod. Urban Transit **1**, 55–60 (2019). (in Chinese)
7. Du, Q., Zhou, Y., Huang, Y., Wang, Y., Bai, L.: Spatiotemporal exploration of the non-linear impacts of accessibility on metro ridership. J. Transp. Geogr. (102), 103380 (2022)
8. Chen, E., Ye, Z., Wu, H.: Nonlinear effects of built environment on intermodal transit trips considering spatial heterogeneity. Transp. Res. Part D: Transp. Environ. (90), 102677 (2021)
9. Chen, C., Feng, T., Ding, C., Yu, B., Yao, B.: Examining the spatial-temporal relationship between urban built environment and taxi ridership: results of a semi-parametric GWPR model. J. Transp. Geogr. (96), 103172 (2021)
10. Wang, X., Liu, T.: The roads one must walk down: commute and depression for Beijing's residents. Transp. Res. Part D: Transp. Environ. (109), 103316 (2022)

# Research on ATO Control Algorithm for Urban Rail Trains Considering Conversion Characteristics of Train Operating Conditions

Kaile Zhang[1,2] and Dongxiu Ou[1,2](✉)

[1] Shanghai Key Laboratory of Rail Infrastructure Durability and System Safety, Shanghai, China
{2131322,ou.dongxiu}@tongji.edu.cn
[2] School of Transportation Engineering, Tongji University, Shanghai, China

**Abstract.** Precise control of the train along the target speed curve is one of the goals of the development of urban rail transit automatic operation system. Overshoot caused by the train's working condition conversion characteristics during the actual operation of the train is an urgent need to be solved by the current train automatic operation controller as excessive overshoot will trigger emergency braking of the train. In this paper, a rigid multi-mass point model was established and the kinematic equation was constructed to describe the changing basic resistance and the additional resistance of the ramp and curve. Secondly, the characteristics of the train when changing among train operating conditions were analyzed, and the cruising operating condition of the train was decomposed into the combination of traction, coasting and braking conditions. The transmission response model of the train command acceleration and the motion model of the train in the coasting condition were given, and the conversion process between traction and braking conditions is modeled by combining these two models. Then, an early switching algorithm combined with the conventional PID controller and the sliding mode PID controller is proposed and the specific steps of the early switching algorithm for both controllers are elaborated. Finally, by simulating a specific subway line in China, it is concluded that the early switching algorithm can suppress the overshoot of the original controller without affecting arrival punctuality and stopping accuracy when the train condition is switched.

**Keywords:** Railway technology · Train operation control · Conventional PID control · Sliding mode control

## 1 Introduction

Urban rail transit is rapidly becoming the primary choice for many cities to solve their transportation problems with its advantages of high capacity, high efficiency and low pollution. However, with the rapid development of society and economy, the speed of train operation and the density of traffic is increasing, manual driving starts to show its disadvantages such as low average train speed, poor safety, inaccurate stopping, etc. In order to improve the indicators of train operation, people began to find a suitable control algorithm and apply it to the automatic train operation system to control the safe and comfortable automatic train operation [1].


Y. Qin et al. (Eds.): EITRT 2023, LNEE 1137, pp. 313–325, 2024.
https://doi.org/10.1007/978-981-99-9311-6_34

ATO system is based on the normal operation of the ATP system and relies on the information provided by the ATS system to optimize driving, improve passenger comfort, and reduce energy consumption. Once the ATO system is installed, the train can operate automatically according to the target speed curve generated by the system. Therefore, it is of practical significance to study the control algorithm of ATO system to realize automatic train operation and improve the degree of train automation.

Many scholars have conducted research on the control algorithm of ATO for trains. The train is disturbed by the basic resistance of the train itself, the additional resistance constantly changing under different slope and curvature conditions of the line, and the resistance caused by nonlinear changes in the friction coefficient due to different weather and environmental conditions during the actual operation. The different resistance mentioned above has strong nonlinearity after the combination [2, 3], which places strict requirements on the real-time and accuracy of the algorithm tracking the target curve of the train ATO controller. Conventional PID control algorithm is widely used in train ATO controller. Although it achieves satisfactory results in terms of control accuracy, but the algorithm adds and subtracts too many times in the process of speed adjustment, which is not conducive to the smooth operation of the train and also affects the ride comfort [4–7]. With the development of intelligent control theory, many intelligent control algorithms have been introduced into the ATO system. Fuzzy logic controller achieves control effects not inferior to those of skilled drivers, but the division of linguistic variables and the shape of the affiliation function in fuzzy control depend largely on expert experience and are difficult to adjust online [8, 9]. Introducing neural networks into fuzzy logic control can make fuzzy logic have online learning capability, but the convergence speed of neural networks needs to be improved and it is easy to fall into local optimum [10]. Applying iterative learning control to the ATO system can make full use of the previous useful information to optimize the current maneuvering strategy, but the realization of the high-precision tracking control effect requires sufficient iterative learning, and the real-time performance is difficult to meet the requirements [11, 12]. The adaptive train speed and position control strategy based on Lyapunov stability theory achieves high-precision tracking control effect, but its control volume changes drastically during the transition phase of operating conditions, which affects ride comfort [13]. In recent years, with the development of deep reinforcement learning algorithms, many scholars have also used deep reinforcement learning algorithms to solve train optimization control problems with positive results [14–17].

The above ATO control algorithms all use the classical first-order linear system as the control model of the train, considering only the basic resistance of the train itself, the additional resistance that changes continuously under different slope and curvature conditions of the line. However, in the actual operation of the train, in addition to the above-mentioned disturbances caused by the nonlinear drag forces caused by the synthesis of multiple drag forces, the time delay in conversion conditions caused by the conversion characteristics of the train will also affect the performance of the ATO control algorithm in speed tracking to a certain extent, resulting in a certain amount of overshoot in the automatic tracking of the train speed curve by the ATO. In some special cases, the maximum allowable overshoot may even be exceeded and trigger the emergency braking of the train, which may adversely affect the overall operation of the urban rail system.

In this paper, a train condition early switching algorithm is proposed, which can adapt to the train condition conversion characteristics, and integrate the algorithm with conventional PID control algorithm and sliding mode PID control algorithm to precisely control the starting time of train condition conversion, so that the train can reduce the overshoot during the actual operation, and make the operation curve of the train more closely match the target speed curve.

# 2 Train Operation Model

## 2.1 Train Dynamic Equation

The train will be subject to its own traction and additional resistance in the process of movement, the commonly used train kinematic models are single mass point model, multi-mass point model and rigid multi-mass point model. The single mass point model is usually used without taking into account the train length and the gradual change of additional resistance on ramps, curves and tunnels. The multi-mass point model consider the mutual interaction force between carriages, which leads to more complicated calculation and is not conducive to online calculation.

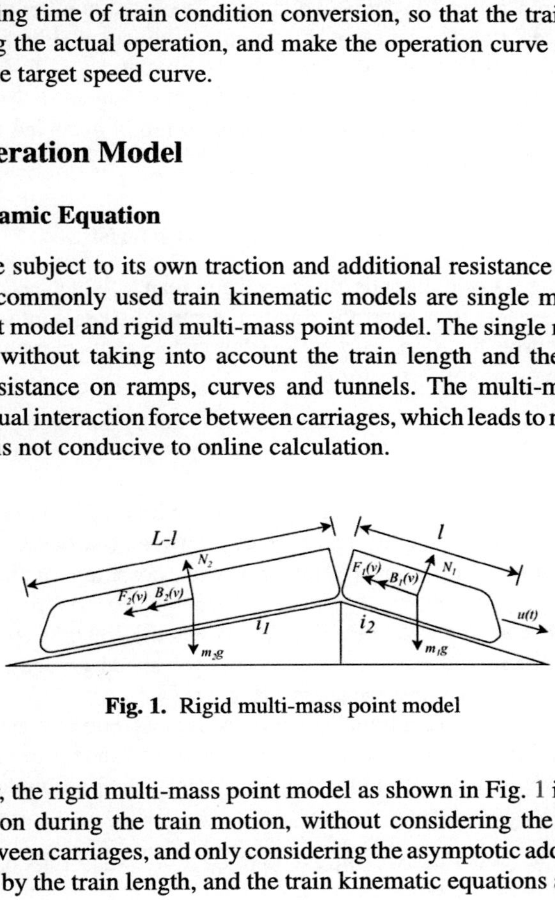


**Fig. 1.** Rigid multi-mass point model

In this paper, the rigid multi-mass point model as shown in Fig. 1 is used to analyze the force situation during the train motion, without considering the mutual coupling relationship between carriages, and only considering the asymptotic additional resistance problem caused by the train length, and the train kinematic equations as shown in Eq. 1 is construct.

$$\begin{cases} \dot{x} = v \\ M(1+\gamma)\dot{v} = u(t) - B(v) - F(v) \\ B(v) = \left(a + bv + cv^2\right)Mg \times 10^{-3} \\ F(v) = \sum_{i=1}^{n} \frac{1}{l_i}\left(10^3 \sum_{p=1}^{n_1} i_p l_p + 600 \sum_{q=1}^{n_2} \frac{l_q}{R_q} + \sum_{r=1}^{n_3} a_r l_r\right) m_i g \times 10^{-3} \end{cases} \quad (1)$$

where $x$ is the position of the train at time $t$, $v$ is the speed of the train, $M$ is the total mass of the train, $\gamma$ is the slewing mass coefficient of the train, $u(t)$ is the traction or braking force of the train, $B(v)$ is the basic resistance of the train, $F(v)$ is the additional resistance of the train which consists of ramp additional resistance, curve additional resistance and tunnel additional resistance, $a, b, c$ is the resistance coefficient of basic resistance of train, $g$ is the acceleration constant of gravity, $n$ is the number of masses in the multi-mass model, $l_i$ is the length occupied by masses, $n_1$ is the number of ramp currently occupied by masses, $i_p$ is the slope of ramp currently occupied by masses, $l_p$ is the length of ramp currently occupied by masses, $n_2$ is the number of curve currently occupied by masses, $l_q$ is the length of curve currently occupied by masses, $R_q$ is the radius of curvature of the curve currently occupied by masses, $n_3$ is the number of the tunnel currently occupied by masses, $a_r$ is the additional resistance of the tunnel currently occupied by masses, $l_r$ is the length of the tunnel currently occupied by masses, $m_i$ is the mass of the current mass point.

## 2.2 Train Operation Conditions Conversion Characteristics

In train operation, ATO system sends out control commands, which need to pass through the traction drive system to convert the traction electrical signals into traction force of electric motors to drive the train. During the whole interval operation process, the train operates under four operating conditions.

(1) Traction condition: the traction motor starts the train by continuously applying traction and makes the train speed increase continuously until it tows the train at a certain power.
(2) Cruising condition: the train uses partial traction or braking to maintain the train at a constant speed, so that the train runs smoothly between stations. Cruising condition is a combination condition of traction condition, coasting condition and braking condition.
(3) Coasting condition: the train completely unloads the traction control force, only affected by friction and other resistance, and the speed gradually decreases, the traction motor does not consume energy at this stage.
(4) Braking condition: when the train circumvents the speed limit or enters the stopping interval, the braking output power is gradually increased and the train speed is greatly reduced, and when it is close to the stopping point, the appropriate braking output command is selected to stop the train.

The process and model of switching among traction condition, coasting condition and braking condition is shown in Fig. 2.

Model 1 is the train motion model. $e^{-s\sigma}$ is the transmission time delay from the issuance of the acceleration command to the reception of this command by the train on-board controller. $1/(\tau s+1)$ is the response delay, which is the delay in the process of changing the acceleration of the train from the value of the current state to the target value corresponding to the acceleration command. $\Delta a(t)$ refers to the additional acceleration, which consists of the basic train resistance, the additional resistance of the ramp and the additional resistance of the curve. It is worth noting that the response time delay of the train is limited by the train impact rate (i.e., the rate of change of acceleration). The acceleration response delay of the train considering the maximum impact rate I is shown in Eq. 2:

$$\begin{cases} I \leq 0.75 \\ \tau = \frac{u(t)-a(t)}{I} \end{cases} \tag{2}$$

where $I$ is the train impact rate, $\tau$ is the acceleration response delay of the train, $u(t)$ is the commanded acceleration of the train at time $t$, $a(t)$ is the actual acceleration of the train at time $t$.

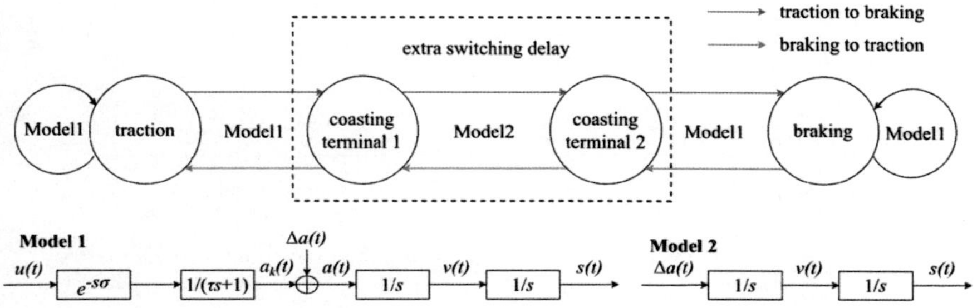


**Fig. 2.** Train operating conditions conversion characteristics

Model 2 is a train motion model with the train in the coasting condition without transmitting acceleration commands. In the coasting condition, the train does not output any traction or braking force, but only suffers from the additional resistance $\Delta a(t)$.

When the train condition is switched, if the train is to enter the braking condition from the traction condition, the traction output should be cut off from the traction condition first, so that the train enters the coasting condition from the traction condition and continues to coast for a period of time (usually about 1 s), then the train starts to output and increase the braking force, so that the train enters the braking condition from the coasting condition. Conversely, when the train is switched from braking condition to traction condition, it also needs to coast for a period of time after removing the braking force and then outputting traction. This leads to the additional time delay of the switching between train traction condition and brake condition, which is also an important reason for the overshoot of the train during the condition changeover. Table 1 summarizes the models or combinations of models involved in the switching process between traction, coasting and braking conditions.

**Table 1.** Process of train operating conditions switching

| | Traction | Coasting | Braking |
|---|---|---|---|
| Traction | Model 1 | Model 1 | Model 1-Model 2-Model 1 |
| Coasting | Model 1 | Model 2 | Model 1 |
| Braking | Model 1-Model 2-Model 1 | Model 1 | Model 1 |

Figure 3 shows a schematic diagram of the train running in the interval tracking the target speed curve.

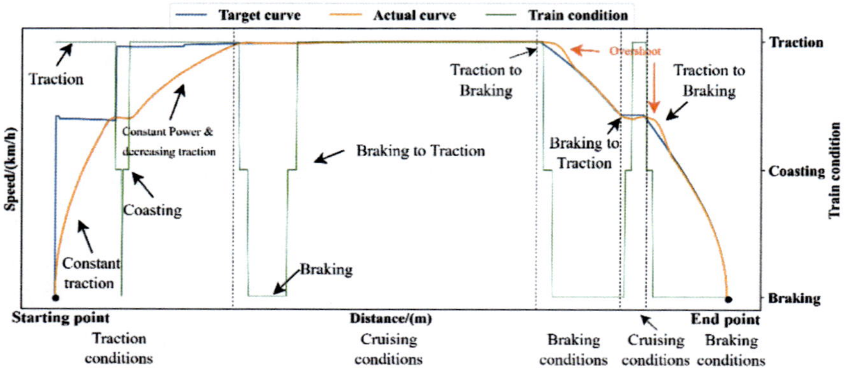


**Fig. 3.** Train operation and operating conditions switching process

### 2.3 Train Controller Design

The train ATO controller functions to continuously transmit the appropriate acceleration command to the train in each control period to make the train track the target speed curve. The current mainstream ATO control algorithms are mainly conventional PID control algorithms and sliding mode PID control algorithms. By using the current controller, the train can generally track the target curve to complete the interval operation. However, in order to ensure the safe operation of the train, the ATO controller sets a limit for the value of the actual train speed exceeding the target speed curve. If the overshoot exceeds the threshold of the maximum overshoot during the driving process, the train will automatically apply emergency braking, which will affect the operation efficiency. In order to suppress the overshoot caused by the condition switching characteristic, the train condition early switching algorithm is designed as shown in Fig. 4.

In each control period, train status information such as train position, speed, acceleration and operating conditions are first obtained, and then the current operating conditions of the train are judged.

The schematic diagram of the early switching prediction points under traction conditions is shown in Fig. 5(a). If the speed of prediction point A is higher than the target curve speed, it means the train will still overshoot if the traction condition is removed immediately, so the acceleration value corresponding to the target curve of prediction point needs to be transmitted to the train control unit as the command acceleration immediately.

The schematic diagram of the working condition conversion limit under the braking condition as shown in Fig. 5(b). If the prediction point B is higher than the target curve speed, it means that the train is very close to the parking brake phase and the train does

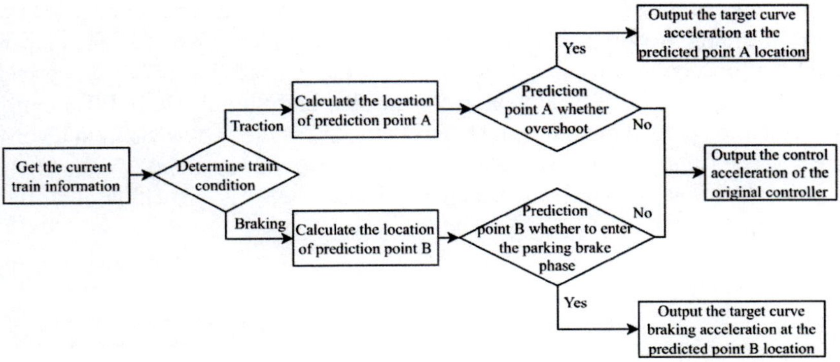


**Fig. 4.** Flow chart of early switching algorithm

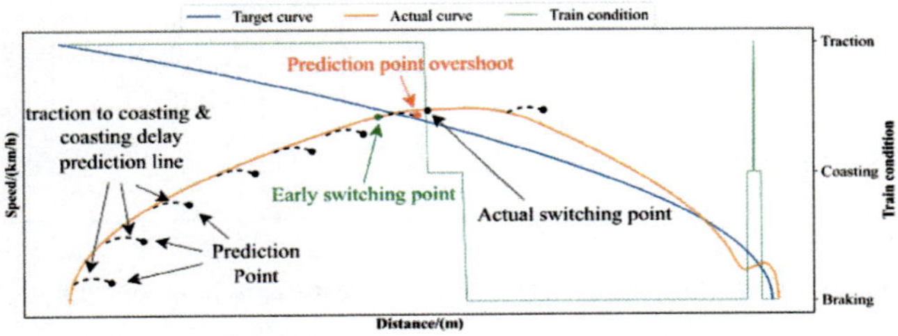


(a) Early Switching Strategy Under Traction Conditions

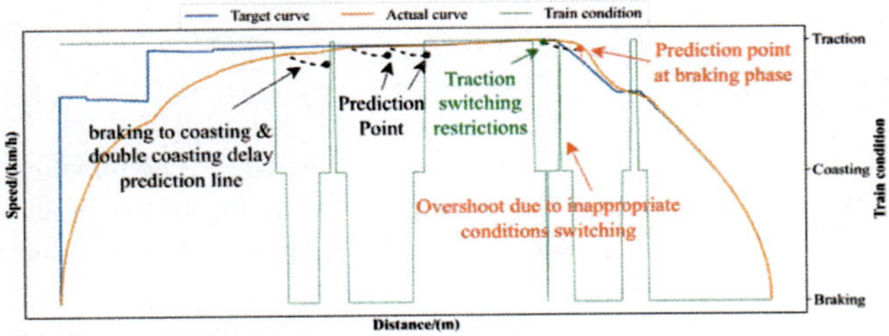


(b) Restricted Conditions Switching Strategy Under Braking Conditions

**Fig. 5.** Principle of early switching algorithm

not have sufficient distance to switch back to the braking condition from the traction condition with the early switching algorithm in the real-time traction condition, resulting in train speed overshoot. At this time, the braking acceleration value of the target curve at the prediction point B can be used as the commanded acceleration to the train control unit, so that the train maintains the gradually increasing braking acceleration to fit the target curve of the upcoming parking brake phase.

The combination of the early switching algorithm with a conventional PID controller or a sliding mode PID controller can form a train ATO controller that takes into account the conversion characteristics of train operating conditions. Since the principle and control accuracy of conventional PID controller and Sliding Mode PID controller are different, the prediction depth and control strategy of the early switching algorithm on the two controllers are different and need to be designed separately.

The conventional PID controller combined with the early switching algorithm is shown in Fig. 6.

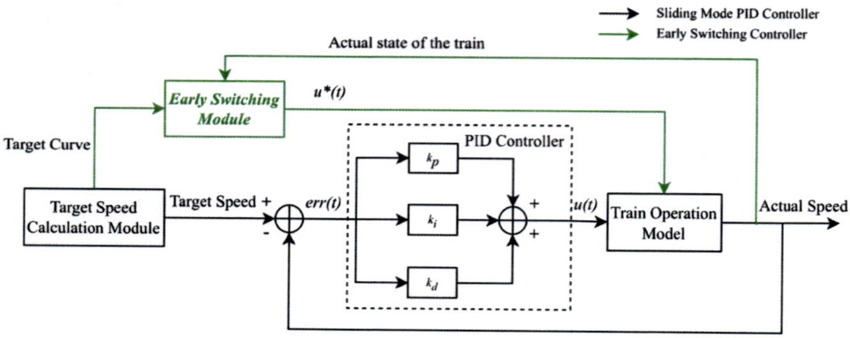


**Fig. 6.** Conventional PID controller with early switching algorithm

The calculation process of the conventional PID controller is shown in Eq. 3:

$$\begin{cases} err(t) = v^*(t) - v(t) \\ u(t) = k_p err(t) + k_i \sum err(t) + k_d(err(t) - err(t-T)) \end{cases} \quad (3)$$

where $err(t)$ is the incoming speed difference value from the PID controller, $v(t)$ is the actual speed of the train, $v^*(t)$ is the target speed of the train, $u(t)$ is the output of the PID controller, i.e.the commanded acceleration, $k_p$, $k_i$, $k_d$ are three parameters for the PID controller, $T$ is the control period.

The sliding mode PID controller is an improvement of the conventional PID controller, which can be used to force the system to follow the predetermined "sliding mode" state trajectory by changing purposely during the dynamic process according to the current state of the system, and can precisely control the train along the target speed curve by considering the changes of speed difference and acceleration difference. The sliding mode PID controller combined with the early switching algorithm is shown in Fig. 7.

The sliding mode PID control process is shown in Eq. 4:

$$\begin{cases} v_e(t) = v(t) - v_{ref}(t) \\ a_e(t) = a_{cu}(t) - a_{ref}(t) \\ a_{sl}(t) = a_{ref}(t) - k_v v_e(t) - k_a a_e(t) \\ a_{cl}(t) = a_{sl}(t) - a_{cu}(t) \\ u(t) = a_{sl}(t) + k_p a_{cl}(t) + k_i \int a_{cl}(t) + k_d \dot{a_{cl}}(t) \end{cases} \quad (4)$$

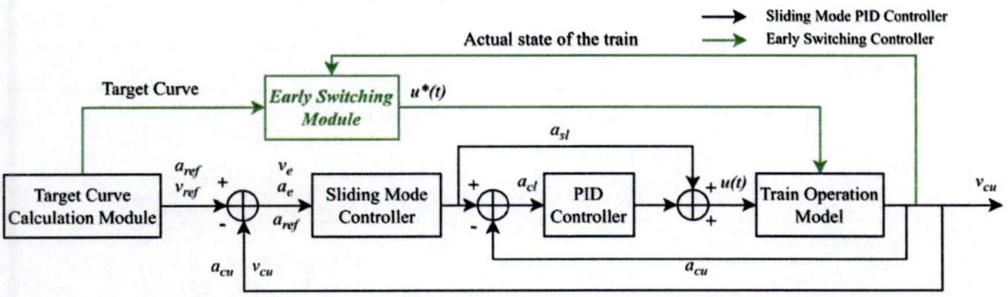


**Fig. 7.** Sliding mode PID controller with early switching algorithm

where $v_e(t)$ is the speed difference of the train, $v(t)$ is the actual speed of the train, $v_{ref}(t)$ is the target speed of the train, $a_e(t)$ is the acceleration difference of the train, $a_{cu}(t)$ is the actual acceleration of the train, $a_{ref}(t)$ s the target acceleration of the train, $a_{sl}(t)$ is the output acceleration of the sliding mode controller, $a_{cl}(t)$ is the PID controller's input acceleration difference, $u(t)$ is the commanded acceleration output from the sliding mode PID controller, $k_v$ and $k_a$ are the velocity gain parameter and acceleration gain parameter of the sliding mode controller, $k_p$, $k_i$ and $k_d$ are the three parameters of the PID controller.

## 3 Simulation and Results Analysis

In the simulation part of this paper, a subway line in China is selected to simulate the train target curve tracking effect of the conventional PID controller using the early switching algorithm and the sliding mode PID controller using the early switching algorithm, respectively. The slope and curvature information of the line is shown in Fig. 8(a), and the ATP speed limit information and ATO target speed curve are shown in Fig. 8(b).

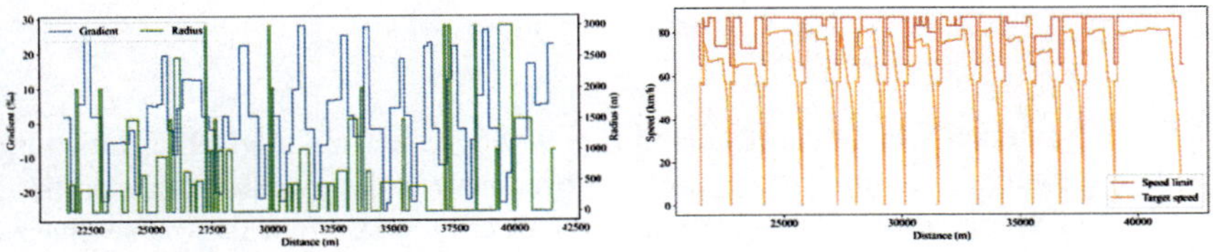


(a) Gradient and radius (b) Speed limit and target speed

**Fig. 8.** Information of the simulation rail section

The train parameters used in the simulation are shown in Table 2.

First, an interval is selected for simulation, and the results of the interval target curve tracking with the conventional PID controller and sliding mode PID controller using the early switching algorithm versus the original conventional PID controller and sliding mode PID controller without the early switching algorithm are shown in Fig. 9.

**Table 2.** Train prameters in simulation

| Train parameters | | Value |
|---|---|---|
| Train length $(m)$ | | 140.72 |
| Train quality $(kg)$ | AW0 | 22900 |
| | AW3 | 379060 |
| Maximum running speed $(km/h)$ | | 90 |
| Basic resistance $(N/KN)$ | | $2.7551 + 0.014v + 0.0075v^2$ |
| Emergency brake deceleration $(m/s^2)$ | | 1.3 |
| Coasting delay $(s)$ | | 1 |
| Impact rate $(m/s^3)$ | | $\leq 0.75$ |

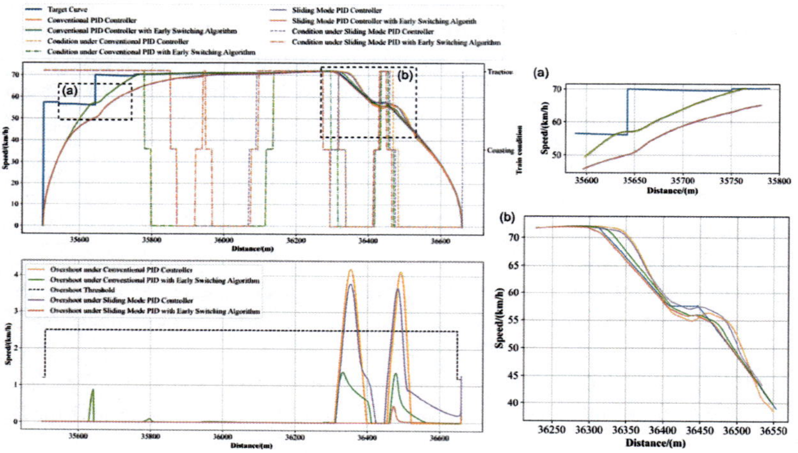


**Fig. 9.** Interval simulation results

From the interval simulation results in Fig. 9, when the train tracks the static target curve in the interval, the most likely point of overshoot is the overshoot caused by the adjustment of acceleration at the junction of the traction phase into the cruise phase as shown in the area (a), and the overshoot caused by the time delay characteristics when the train switches from traction to braking when the cruise phase enters the braking phase as shown in the area (b).

The overshoot in area (a) does not involve the coasting time delay when the condition is switched, so the overshoot is small. The overshoot occurring in area (b) is due to the combination of acceleration response time delay and coasting time delay when the train is switched from traction to braking condition, and the overshoot amount is larger.

The results show that the overshoot of the controller using the early switching algorithm is less than the maximum overshoot when the operating conditions were switched, and is significantly smaller than the overshoot generated by the original controller. At the same time, the effect of the sliding mode PID controller is better than that of the conventional PID controller.

Next, the simulation is carried out for the whole line, and the overshoot and the number of condition switching under the whole simulation section are counted, and the simulation results are shown in Fig. 10 and Table 3.

**Table 3.** Comparison of full-line simulation results under different controllers

| Controller | Maximum overshoot (km/h) | Average overshoot (km/h) | Total switching | Average stop accuracy(cm) |
|---|---|---|---|---|
| Conventional PID | 5.0 | 2.7218 | 146 | 31.4 |
| Sliding mode PID | 4.3 | 2.2295 | 121 | 5.33 |
| Conventional PID + early switching | 2.5 | 1.4107 | 138 | 11.6 |
| Sliding mode PID + early switching | 0.5 | 0.1052 | 130 | 5.22 |

The results of the full-line simulation show that the use of the early switching algorithm on the basis of the original controller can effectively suppress the overshoot of the train driving speed during the switching of the working condition. The sliding-mode PID controller with the early switching algorithm has the best suppression effect on the overshoot of the train generated during the switching of the working condition and has a higher control effect on the tracking of the target speed curve.

## 4 Conclusions

This paper firstly establishes a rigid multi-mass point model in the analysis of the train motion and control model stage, and constructs the kinematic equations under the consideration of the train subject to the changing basic resistance and the additional resistance of the ramp and curve. Next, the characteristics of the train during the condition switching are analyzed, and the cruising condition of the train is decomposed into the combination of traction, coasting and braking conditions, and the transmission response model of the train acceleration command and the motion model of the train in the coasting condition are given. Then, an early switching algorithm is proposed for the train ATO controller, and the algorithm is combined with the conventional PID controller and the sliding mode PID controller, and the specific steps of using the early switching algorithm for both controllers are elaborated. Finally, by simulating a specific subway line in China, it is concluded that using the early switching algorithm can eliminate the overshoot of the original controller during train condition switching, and it is proved that using the early switching algorithm can reduce the overshoot of the conventional PID controller and sliding mode PID controller during train condition switching to different degrees. The overshoot of the sliding mode PID controller is significantly smaller than that of

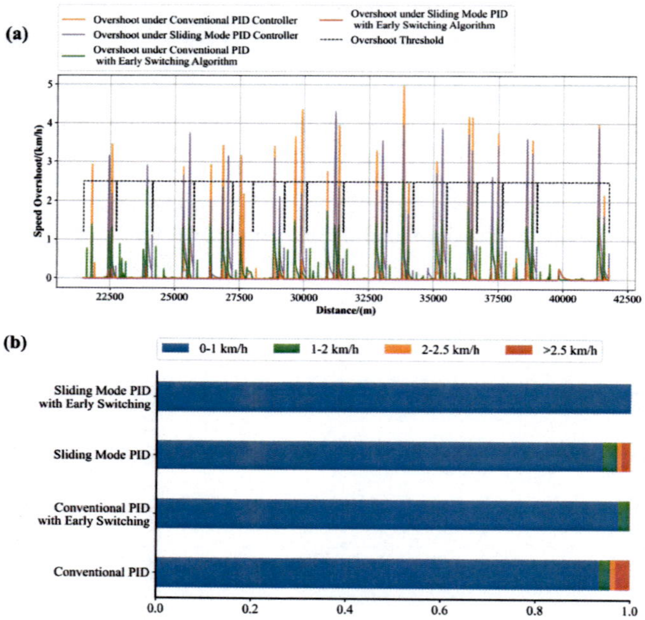


**Fig. 10.** Full-line overshoot (a) and histogram of overshoot distribution (b)

the conventional PID controller when the early switching algorithm is applied to both controllers, and the tracking effect of the target speed curve is also better.

The early switching algorithm is essentially an overshoot suppression algorithm for train condition switching, which can be applied not only to the conventional PID controller and sliding mode PID controller, but also to other ATO control algorithms as an overshoot suppression module to precisely control the train to track the target speed curve.

**Acknowledgments.** This research is supported by the National Natural Science Foundation of China (52172329), and the Research Program of Shanghai Science and Technology Committee (22ZR1422200, 21DZ2204500). The authors would like to thank the reviewers for their valuable comments and advice on this article.

## References

1. Song, Q., Song, Y.D.: Data-based fault-tolerant control of high-speed trains with traction/braking notch nonlinearities and actuator failures[J]. IEEE Trans. Neural Networks **22**(12), 2250–2261 (2011)
2. Li, D.C., Meng, G.J., Xu, R.X., Yi. M.: Adaptive and robust H_control method of high-speed train under high wind. J. RailScience **40**, 249(07):67–73 (2018). (in Chinese)

3. He, Z.Y., Xu, N.: Train autonomous driving algorithm based on adaptive iterative learning control. Transp. Syst. Eng. Information. **20**(02), 69–75 (2020). (in Chinese)
4. Wen, Y., Chenkun, Y., Zhongsheng, H.: A novel energy efficient operation strategy for a train based on modelfree adaptive predictive control. In: Proceedings of the 31st Chinese Control Conference, pp. 7286–7291 (2012)
5. Tang, T., Huang, L.: A survey of control algorithm for automatic train operation. J. China Railw. Soc. **25**(2), 98–102 (2003). (in Chinese)
6. Wang, Q.Y., Wu, P., Feng, X.Y., Zhang, Y.D.: Accurate parking algorithm based on adaptive terminal sliding mode control. J. RailScience. **38**, 220(02), 56–63 (2016). (in Chinese)
7. Hengyu, L., Hongze, X.: An integrated intelligent control algorithm for high-speed train ATO systems based on running conditions. In: 2012 Third International Conference on Digital Manufacturing & Automation, pp. 202–205 (2012)
8. Zhang, M.N., Xu, H.Z., Zhang, Y.X.: Automatic stopping algorithm for high-speed train based on adaptive fuzzy sliding mode control. Inf. Control. **44**(02), 223–229 (2015). (in Chinese)
9. Oshima, H., Yasunobu, S., Sekino, S.: Automatic train operation system based on predictive fuzzy control. In: Proceedings of IEEE International Workshop on Artificial Intelligence for Industrial Applications, pp. 485–489. Industrial Electronics Society, New York (1988)
10. Jin, Y.U., Qingquan, Q.I.A.N., Zhengyou, H.E.: Research on application of two degree fuzzy neural network in ATO of high speed train. J. China Railw. Soc. **30**(5), 52–56 (2008). (in Chinese)
11. Wang, Y., Hou, Z., Li, X.: A novel automatic train operation algorithm based on herative learning control theory. In: Proceedings of IEEE International Conforence on Service Operations, Logistics and Informatics, pp. 1766–1770. IEEE Press, New York (2008)
12. Li, Z.X., Hou, Z.S.: Adaptive iterative learning control based high-speed train operation tracking under iteration-varying parameter and measurement noise. Asian J. Control. **17**(5), 1–10 (2015)
13. Qi, S., Qing, G., Feng, L., et al.: Adaptive positing and velocity control of high speed train systems. Control. Eng. China. **17**(SO), 35–37 (2010)
14. Liu, W., Su, S., Tang, T., et al.: A DQN-based intelligent control method for heavy haul trains on long steep downhill section[J]. Transp. Res. Part C: Emerg. Technol. **129**, 103249 (2021)
15. Cao, Y., Wen, J., Hobiny, A., et al.: Parameter-varying artificial potential field control of virtual coupling system with nonlinear dynamics[J]. Fractals **30**(02), 2240099 (2022)
16. Chen, X., Guo, X., Meng, J., et al.: Research on ATO control method for Urban rail based on deep reinforcement learning[J]. IEEE Access **11**, 5919–5928 (2023)
17. Shang, M., Zhou, Y., Fujita, H.: Deep reinforcement learning with reference system to handle constraints for energy-efficient train control[J]. Inf. Sci. **570**, 708–721 (2021)

# Design Method for Facility Location and Freight Flow Allocation of China Railway Highspeed Express

Shuangsong Xu, Kanglin Liu(✉), Xinghan Chen, and Maoxiang Lang

School of Traffic and Transportation, Beijing Jiaotong University, Beijing 100089, China
{19251234,klliu,chenxinghan,mxlang}@bjtu.edu.cn

**Abstract.** China Railway Highspeed Express (CRH Express) is in the stage of mode exploration and technological innovation. This paper takes CRH Express as the research object, studies the operation and organization mode of CRH Express in the future CRH Express market, and makes a study on hub location and cargo flow distribution. This article analyzes various current CRH Express modes from multiple dimensions and proposes a "multi station-multi hub-multi mode" high-speed rail express operation mode. Then build a large-scale Mixed Integer Programming (MIP) optimization model to portray the high-speed rail express network. Based on the analysis and refinement of the CRH Express network, the optimization model is built to depict the operational performance of the CRH Express network under various transportation modes by constructing a scenario of the CRH Express hub facility location and cargo flow distribution.

**Keywords:** High-speed railway express · Freight flow prediction · Freight flow distribution

## 1 Introduction

### 1.1 A Subsection Sample

On the one hand, the express freight market is booming, and efficient, environmentally friendly, and reliable high-speed rail transportation has market potential to participate in the competition of express transportation; On the other hand, the idle capacity of the high-speed rail network has led to consecutive losses, making the railway system not only have the objective conditions to carry out express transportation but also have the need to improve business conditions through the implementation of express transportation.

Considering that the high-speed rail network places more emphasis on the organizational characteristics of passenger transportation, high-speed rail express particularly focuses on the organization of transportation modes. Xiao [1] et al. have proposed various organizational plans and analyzed and discussed them from various aspects such as station organization, operating conditions, and transportation organization. Chen [2] et al. build a freight flow distribution model based on multiple high-speed rail freight


Y. Qin et al. (Eds.): EITRT 2023, LNEE 1137, pp. 326–334, 2024.
https://doi.org/10.1007/978-981-99-9311-6_35

operation modes and obtain combinatorial optimization schemes and management suggestions. Fang [3] et al. applied revenue management theory to the study of high-speed railway express transportation in a competitive environment, enriching the theory of high-speed railway express transportation. Xu et al. [4] considered using distriutionally robust methods to optimize the high-speed rail express network under uncertain requirements.

Analyzing the technical and economic feasibility of different transportation modes in different environments involves studying the implementation forms of high-speed rail express from multiple perspectives, and new high-speed rail express operation modes still need to be developed. At present, there is relatively little research on the organization mode of high-speed rail express freight transportation, and there is also a lack of in-depth research on relevant literature, making it difficult to form a comprehensive and systematic research reserve between studies.

## 2 Problem Description

The High-speed Railway Express can be divided into three stages: collection, transportation, and distribution. The sorting and labeling of goods are carried out by the freight collection station or its supporting transfer center. After being uniformly loaded by the container, the goods are transported to the destination station through the high-speed train units. After the goods are transported to the designated station, they are unloaded, sorted, and transported by road to the customer by the distribution station or its supporting transfer center.

At present, there are three main modes of high-speed rail express high-speed rail, which are piggyback transportation using passenger sports vehicles, passenger and freight joint transportation of passenger multiple units, and freight special train transportation.

Piggyback mode mainly relies on the passenger train that has been opened, the transportation capacity is stable and guaranteed, but the transportation capacity is small, the transportation distance is limited, all types of passenger transport cars can perform piggyback transportation tasks, and the loading quality of the single car should be 0.1t.

The joint mode relies on the passenger train that has been opened, and the whole passenger car or freight car is used to be jointly linked with the passenger train, and the transportation capacity is larger, and the express transportation task can be flexibly grouped according to the freight demand and passenger transport demand, and the single car transportation volume is about 5t.

Special train mode adopts freight special train for transportation, with a single train carrying capacity of about 110t, with sufficient objective carrying capacity, but the opening of special trains requires the starting station to have the functions of cargo sports vehicle group grouping, cargo distribution, warehousing, etc., and the existing stations are mostly designed for passenger transportation, which is difficult to meet the demand.

This paper proposes a new type of "Joint-Special train" hybrid transportation mode based on a hub-and-spoke network (see Fig. 1) that adopts the joint mode to collect and distribute goods and a special train mode for transportation on the trunk line.

The hub-and-spoke network has wide applicability and high recognition, and it can effectively simulate the transportation network characteristics and economic benefits

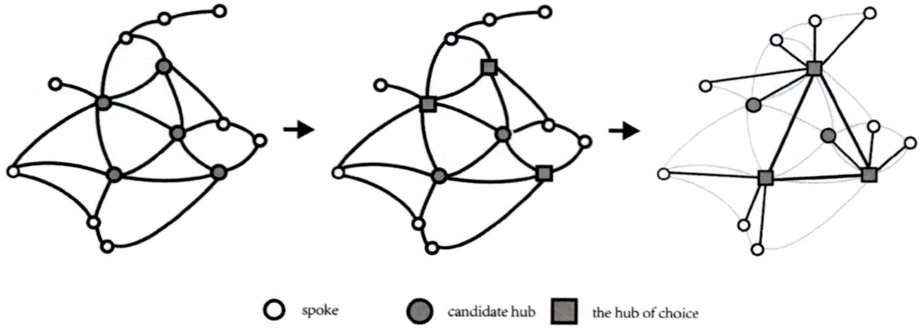


**Fig. 1.** Hub-and-spoke networks

brought by trunk line transportation [5]. The relevant models have been successfully applied to the European rail freight transport system [6], the Turkish postal system [7], and multi-modal cargo transportation [8].

Under the simultaneous operation of piggyback mode, joint mode, and special train mode, including the new hybrid transportation mode, a "multi-stop-multi-hub-multi-mode" high-speed rail express mode is proposed (see Fig. 2).

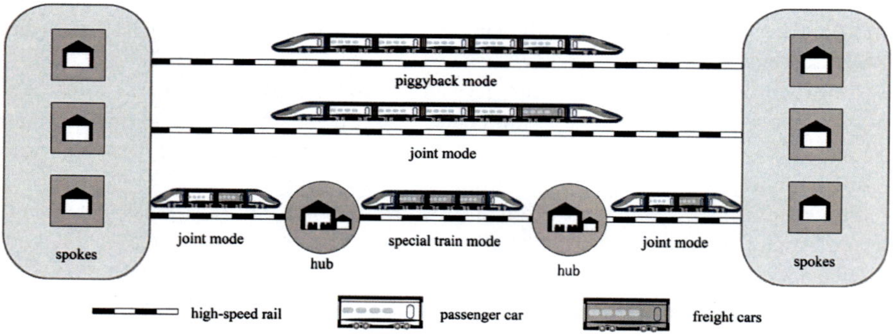


**Fig. 2.** "Multi-stop-multi-hub-multi-mode" high-speed rail express mode

## 3 Modeling Study

This paper will construct a quantitative mathematical model based on the "multi-stop-multi-hub-multi-mode" high-speed rail express cargo operation mode to carry out cargo flow allocation and hub facility location.

### 3.1 Symbol Setting

The relevant symbols are summarized in Table 1, and the vectors or matrices involved in the text are indicated in bold.

**Table 1.** Symbol setting

| Type | Symbol | Significance |
|---|---|---|
| Set | $\boldsymbol{N} = \{N\|1, 2, \ldots, N\}$ | Spoke set |
| | $\boldsymbol{H} = \{H\|1, 2, \ldots, H\}$ | Candidate hub set |
| | $\boldsymbol{G}$ | A network that contains all spoke and arcs |
| | $\boldsymbol{W}$ | Demand set |
| Parameters and other variables | $i, j \in \boldsymbol{N}$ | The spoke ordinal number |
| | $k, l \in \boldsymbol{H}$ | Candidate hub number |
| | $w_{ij} \in \boldsymbol{W}$ | Demand between spoke $i, j$, tons |
| | $arc_{a,b} \in \boldsymbol{G}$ | Directed arc between spokes $a, b$ |
| | $v^{L}_{a,b}$ | Capacity limit for piggyback mode on $arc_{a,b}$, tons |
| | $v^{R}_{a,b}$ | Capacity limit for joint mode on $arc_{a,b}$, tons |
| | $c^{L}$ | The cost of the cargo flow in piggyback mode, CNY/ton |
| | $c^{R}_{a,b}$ | The cost of the cargo flow in joint mode, CNY/ton |
| | $M$ | Infinity |
| | $c^{FT}_{k.l}$ | Inter-hub special train freight rate, CNY/ton |
| | $g$ | The number of central hubs to set up in the network |
| | $F^{L}_{a,b}$ | Piggyback mode to meet the arc capacity limits |
| | $F^{R}_{a,b}$ | Joint mode to meet the demand arc capacity limits |
| Decision variables | $X_{i,j,k,l}$ | If the cargo flow decides to use combined transportation through hubs k and l in turn, it is 1, otherwise 0 |
| | $Y_{i,j,a,b}$ | If the flow $f_{i,j}$ passes through the $arc_{a,b}$, it is 1, otherwise it is 0 |
| | $Z_k$ | 0–1 variable, select k candidate hub as the hub when it is 1, otherwise 0 |

*(continued)*

**Table 1.** (*continued*)

| Type | Symbol | Significance |
|---|---|---|
| | $w^L_{i,j}$ | The portion of the demand between stations *i* and *j* that adopts the piggyback mode |
| | $w^R_{i,j}$ | The portion of the demand between stations *i* and *j* that adopts the joint mode |
| | $w^{FT}_{i,j}$ | The portion of the demand between stations *i* and *j* that adopts the hybrid mode |
| | $s_{k,l}$ | The number of running special trains between hub $k$, $l$ |
| | $m^{FT}_{k,l}$ | Special train freight volume between hub $k$, $l$ |
| | $n^R_{a,b}$ | Integer variable, the preset number of freight cars required per group in joint mode ($n^R_{a,b} \geq, n^R_{a,b} \in \mathbf{Z}$) |

### 3.2 Model Building

**Objective function**. The piggyback mode has a weak impact on the current passenger transportation organization, which can be regarded as only generating some loading and unloading handling costs to quantify, and the $c^L$ calculation of this paper is 20 CNY/t. Because the transportation organization of the CRH Express is still immature and needs to be explored, the cost of the special train mode in this paper is set to a conservative value, taking 40 CNY/km·t [2].

$$\min \overbrace{\sum\nolimits_{arc_{a,b} \in \mathbf{G}} F^L_{a,b} c^L}^{piggyback\ mode} + \overbrace{\sum\nolimits_{arc_{a,b} \in \mathbf{G}} F^R_{a,b} c^R_{a,b}}^{jiont\ mode} + \overbrace{\sum\nolimits_{\forall k,l} m^{FT}_{k,l} c^{FT}_{k.l}}^{special\ train\ mode} \quad (1)$$

**Transportation mode selection and volume division constraints.** All requirements are met in the model. There are the following restrictions:

$$w_{i,j} \leq w^L_{i,j} + w^R_{i,j} + w^{FT}_{i,j}, \forall i, j \in \mathbf{N} \quad (2)$$

**Hub facility location constraints.** The special train must run on the passage between the two hubs, so there are the following restrictions:

$$\sum\nolimits_{k \in \mathbf{H}} Z_k = g \quad (3)$$

$$X_{i,j,k,l} \leq Z_k, X_{i,j,k,l} \leq Z_l, \forall i, j \in \mathbf{N}; \forall k, l \in \mathbf{H} \quad (4)$$

$$\sum\nolimits_{k,l\in \boldsymbol{H}} x_{i,j,k,l} = 1, \forall i,j \in \boldsymbol{N} \quad (5)$$

$$x_{i,j,k,l} = 0; i,j,k,l \in \{i \le l \le k \le j | i,j \in \boldsymbol{N}; k,l \in \boldsymbol{H}\} \quad (6)$$

$$x_{i,j,k,l} = 0; i,j,k,l \in \{j \le k \le l \le i | i,j \in \boldsymbol{N}; k,l \in \boldsymbol{H}\} \quad (7)$$

Constraint (3) limits the number of hubs to be determined; Constraints (4) make special trains only operate between selected hubs, and combined with constraints (5), they jointly limit a single demand to select only a specific set of trains that operate between two hubs; Constraints (6) and (7) limit the direction of the special train to only upward or downward when it is open.

**Hybrid mode constraints.** There are the following restrictions:

$$w_{i,j,k}^{FTR'} = \sum\nolimits_{l\in \boldsymbol{H}} w_{i,j}^{FT} x_{i,j,k,l}, \ w_{i,j,l}^{FTR''} = \sum\nolimits_{k\in \boldsymbol{H}} w_{i,j}^{FT} x_{i,j,k,l}, \forall i,j \in \boldsymbol{N}; \forall k \in \boldsymbol{H} \quad (8)$$

$$m_{k,l}^{FT} = \sum\nolimits_{i,j\in \boldsymbol{N}} x_{i,j,k,l} w_{i,j}^{FT}, \forall k,l \in \boldsymbol{H} \quad (9)$$

$$\sum\nolimits_{\forall k,l} m_{k,l}^{FT} = \sum\nolimits_{\forall i,j} w_{i,j}^{FT} \quad (10)$$

Among them, constraints (8) calculate the collecting and distributing cargo flow of special train transportation; Constraint (9) calculates the demand $m_{k,l}^{FT}$ that needs to be met to open special trains between hubs; Constraint (10) can prevent the phenomenon that the model cannot calculate the cost of the special mode.

**Line connectivity constraints.** For transport routes that carry demand $w_{i,j}$, it is necessary to ensure that their paths on the network are connected so that the goods can reach their destination smoothly. For two stations $i$ and $j$ are:

$$\sum\nolimits_{a=i} y_{i,j,a,b} = 1, \sum\nolimits_{b=i} y_{i,j,a,b} = 0, \sum\nolimits_{a=j} y_{i,j,a,b} = 0, \\ \sum\nolimits_{b=j} y_{i,j,a,b} = 1, \forall i,j \in \boldsymbol{N}; a,b \in \boldsymbol{N} \quad (11)$$

$$\sum\nolimits_{a=n} y_{i,j,a,b} = \sum\nolimits_{b=n} y_{i,j,a,b}, \forall i,j \in \boldsymbol{N}; a,b,n \in \boldsymbol{N}; a,b \notin \{i,j\} \quad (12)$$

**Volume coupling and capacity constraint constraints.** If piggyback transportation is used, the distribution of cargo flow needs to meet the capacity constraints; When the joint mode is adopted, it is necessary to additionally couple the cargo flow of the joint mode used in the process of collect and distribute cargo in the special train mode and calculate the loading rate occupied by each arc segment. Hence the following:

$$F_{a,b}^{L} = \sum\nolimits_{i,j\in \boldsymbol{N}} y_{i,j,a,b} w_{i,j}^{L} \quad (13)$$

$$F_{a,b}^{R} = \sum_{i,j\in \boldsymbol{N}} y_{i,j,a,b} w_{i,j}^{R} + \sum_{i\in \boldsymbol{N},k\in \boldsymbol{H}} y_{i,k,a,b} \sum_{c\in \boldsymbol{N}} w_{i,c,k}^{FTR''}$$

$$+\sum_{i\in N,k\in H} Y_{i,k,a,b}\sum_{c\in N} w_{c,i,k}^{FTR''}, \forall arc_{a,b}\in G \tag{14}$$

$$F_{a,b}^{L}\le v_{a,b}^{L}, F_{a,b}^{R}\le v_{a,b}^{R}n_{a,b}^{R}, \forall arc_{a,b}\in G \tag{15}$$

## 4 Case Study

We selected 9 representative express delivery business cities with moderate scale and coverage to form a research case. These cities are Beijing, Tianjin, Zhengzhou, Nanjing, Wuhan, Shanghai, Hangzhou, Chengdu, and Guangzhou, and are indicated by numbers 1–9. By investigating train schedules, we constructed a CRH Express network consisting of these nine cities (see Fig. 3).

We used Chen's [2] CRH Express market forecasting system to characterize the high-speed rail market [9], and the results are shown in Table 2. In actual operation, the capacity of the piggyback mode is affected by the density of passenger flow and the per capita baggage volume, this paper takes a more conservative estimate, it is believed that at least 50% of the capacity is available per day.

**Table 2.** Daily demand for inter-city express travel in 2023 (t)

| City | 1 | 2 | 3 | 4 | 5 | 6 | 7 | 8 | 9 |
|---|---|---|---|---|---|---|---|---|---|
| 1 | – | 35.4 | 15.8 | 24.1 | 19.7 | 58.8 | 30.6 | 20.2 | 38.9 |
| 2 | 36.8 | – | 7.9 | 11.9 | 9.9 | 31.4 | 15.5 | 10.1 | 19.1 |
| 3 | 30.6 | 14.5 | – | 21.8 | 18.1 | 50.7 | 27.7 | 20.3 | 40.1 |
| 4 | 6.7 | 3.2 | 3.1 | – | 4.5 | 14.3 | 8.2 | 4 | 8.6 |
| 5 | 26.8 | 12.9 | 12.8 | 22.3 | – | 45.7 | 27.2 | 19.6 | 42.9 |
| 6 | 29.5 | 15.2 | 13.2 | 26.5 | 16.7 | – | 49.9 | 15.4 | 37.2 |
| 7 | 15.6 | 7.7 | 7.4 | 16.7 | 10.3 | 50.9 | – | 8.6 | 21.9 |
| 8 | 41.4 | 20.1 | 21.7 | 29.6 | 29.7 | 62.5 | 34.7 | – | 54.2 |
| 9 | 55.5 | 25.8 | 29.2 | 44 | 44.9 | 105 | 60.4 | 37.3 | – |

Using the MATLAB platform, the YALMIP [10] toolkit was used as the medium of MATLAB and GUROBI to achieve the solution. The solution result is shown in Fig. 3. The meaning of the numerical value above the line segment is (upward direction channel utilization rate, downward direction channel utilization rate).

In the sharing scheme of transportation mode solved by the study, the piggyback mode, joint mode, and hybrid mode accounted for 33.72, 20.28, and 45.95%, and the total network cost was 3.5485 million CNY. The study solved the problem of considering that specialized trains only operate point-to-point between cities, and the special train transportation accounts for only 25.0%, and the total cost is 4.3226 million CNY. The result indicates that the CRH Express network using the hybrid mode has higher revenue and stronger market competitiveness.

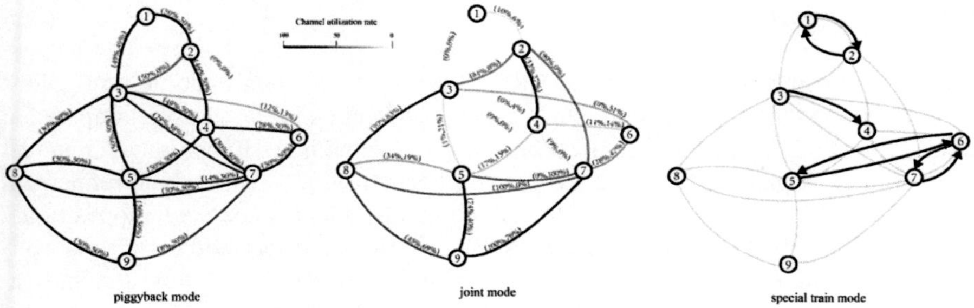


**Fig. 3.** Results of cargo flow allocation

The channel opening mode recommendation can provide decision-makers with suggestions on the reasonable operation scale of each CRH Express channel (Table 3). The decision-making suggestions for the special train opening plan are made, and the results of the suggestions are shown in Table 4. From the perspective of regional connection, suggestions can connect the large demand distribution areas and undertake the freight demand; From the perspective of operating frequency, the frequency of operation of train organizations under peak demand is low, and each railway line has sufficient capacity resources to support the operation of special trains.

**Table 3.** Channel opening mode recommendation for major O-D pairs

| CRH Express channel | | Mode | The scale of the piggyback mode | The scale of joint mode |
|---|---|---|---|---|
| Origin | Destination | | | |
| Beijing | Zhengzhou | Piggyback | 49.5% | 0.0% |
| Chengdu | Wuhan | Piggyback + joint | 50.0% | 18.7% |
| Guangzhou | Hangzhou | Piggyback + joint | 50.0% | 76.1% |

**Table 4.** Special train operation plan proposal

| Serial number | Origin | Line | Destination | Frequency |
|---|---|---|---|---|
| 1 | Beijing | Beijing → Tianjin → Nanjing → Shanghai | Shanghai | 2 |
| 2 | Shanghai | Shanghai → Nanjing → Tianjin → Beijing | Beijing | 2 |
| 3 | Wuhan | Wuhan → Nanjing → Shanghai | Shanghai | 1 |
| 4 | Shanghai | Shanghai → Nanjing → Chengdu | Chengdu | 2 |
| 5 | Chengdu | Chengdu → Nanjing → Shanghai | Shanghai | 2 |
| 6 | Zhengzhou | Zhengzhou → Nanjing → Shanghai | Shanghai | 1 |

## 5 Conclusions

This article analyzes various current CRH Express modes from multiple dimensions and proposes a "multi station-multi hub-multi mode" high-speed rail express operation mode. Then build a large-scale Mixed Integer Programming (MIP) optimization model to portray the high-speed rail express network. Based on the analysis and refinement of the CRH Express network, the optimization model is built to depict the operational performance of the CRH Express network under various transportation modes by constructing a scenario of the CRH Express hub facility location and cargo flow distribution. Finally, we get the special train operation plan proposal and the channel opening mode recommendation to provide decision support to decision makers.

**Acknowledgements.** This research is partly supported by the National Natural Science Foundation of China (grant numbers 72101021 and U2034208).

## References

1. Xiao, L., Shen, X., Yang, L., Zheng, X., Chen, S.: Research on high-speed railway freight express station operating conditions and transport organization. Railw. Stand. Des. 1–8 (2022). (in Chinese)
2. Chen, X., Zhou, P., Lang, M., Yu, X., Li, S.: High-speed railway express cargo flow allocation and operation organization optimization under varying demand. J. Transp. Syst. Eng. Inf. Technol. **22**(05), 174–186 (2022). (in Chinese)
3. Fang, Y., Zhang, X., Li, Y.: Research on multicriteria decision-making scheme of high-speed railway express product pricing and slot allocation under competitive conditions. Math. **10**(9), 1610 (2022)
4. Xu, S., Liu, K.: Hub-and-spoke network optimization method of China railway highspeed express based on distributionally robust optimization. Chin. J. Manag. Sci. 1–11 (2022). (in Chinese)
5. Fu, C., Zhu, N., Ma, S., et al.: A two-stage robust approach to integrated station location and rebalancing vehicle service design in bike-sharing systems. Eur. J. Oper. Res. **298**, 915–938 (2021)
6. Basallo-Triana, M.J., Vidal-Holguín, C.J., Bravo-Bastidas, J.J.: Planning and design of intermodal hub networks: a literature review. Comput. & Oper. Res. **136**, 105469 (2021)
7. Jeong, S.J., Lee, C.G., Bookbinder, J.H.: The European freight railway system as a hub-and-spoke network. Transp. Res. Part Policy Pract. **41**(6), 523–536 (2007)
8. Azizi, N., Chauhan, S., Salhi, S., et al.: The impact of hub failure in hub-and-spoke networks: mathematical formulations and solution techniques. Comput. & Oper. Res. **65**(JAN.), 174–188 (2014)
9. National data. https://data.stats.gov.cn/. Accessed 10 May 2023
10. YALMIP. https://yalmip.github.io/. Accessed 20 May 2023

# The Mutual Feed Response Model of Typical Disturbance and Dispatch Command in High-Speed Railway Express

Zecheng Wang[1], Li Wang[1,2(✉)], and Xiaoning Zhu[1,2]

[1] School of Traffic and Transportation, Beijing Jiaotong University, Beijing 100044, China
{21120926,liwang,xnzhu}@bjtu.edu.cn

[2] Key Laboratory of Transport Industry of Comprehensive Transportation Theory (Beijing Jiaotong University), Ministry of Transport, Beijing 100044, China

**Abstract.** As an important component of the comprehensive transportation system, high-speed railway has the characteristics of large transportation capacity, fast punctuality, efficient and convenient. In recent years, high-speed railway freight has also been booming. With the continuous increase of the scale of high-speed rail network and train operation density, the probability of various random internal or external disturbances in the high-speed rail system is also greatly increased, resulting in train delays in emergencies, and thus affecting the safe and stable operation of the entire railway system. Based on the typical disturbance library, the dispatch and command mutual feed-response model based on the spatio-temporal network is constructed to reveal the spatio-temporal evolution law of the high-speed railway operation network and realize the spatio-temporal prediction of delays under different typical emergencies, which provides theoretical and technical support for reducing the negative impact of emergencies and improving the elasticity of schedule. It has important theoretical and practical significance to ensure the smooth and safe operation of high-speed railway.

**Keywords:** High-speed rail freight · High-speed rail network · Spatio-temporal network

## 1 Introduction

### 1.1 Background

Under the support of information technology and intelligent technology, the high-speed railway dispatching and command system has comprehensively adopted information integrated management, built the whole working mode on the intelligent data transmission, and realized the collaborative work between different departments according to the integrated processing of the system [1].

Due to the joint influence of the environment, equipment, personnel and other factors, accidents are inevitable in the operation process [2]. Under the real-time transmission of data information, the whole railway can really get rid of the original manual operation


Y. Qin et al. (Eds.): EITRT 2023, LNEE 1137, pp. 335–342, 2024.
https://doi.org/10.1007/978-981-99-9311-6_36

mode in the dispatching process, and realize the effective docking of train equipment and signal personnel dispatching instructions based on signal transmission. In addition, through the construction of railway dispatching command system, each kind of data system can be guaranteed to truly reflect the current train and the transportation situation in the whole traffic line.

Meanwhile, various data information inside the car can be learned through the platform. For example, the driver, the conductor, etc. The dispatching command center will report the conditions of the car and the driving state, so that the whole system can clearly understand the data error problems existing in the current operation mode, and formulate corresponding instructions to meet the signal transmission speed and improve the quality of command.

### 1.2 Importance

Railway has become a popular transportation mode in our live. However, the operational accidents may break the normal railway operation. Therefore, the reliability, availability, maintenance and safety of railway network should be analyzed from the perspective of railway system operation [3].

The dispatching command is to conduct integrated management for daily railway transportation activities. Under the effective linkage between different departments, the rationalization and scientific organization of passenger flow, freight flow and vehicle flow can be truly realized, so as to ensure that relevant departments involved in transportation can truly realize close cooperation and collaborative processing, and realize the optimization and perfection of technical railway transportation process. Based on this, we should deeply excavate the various influencing factors of the railway dispatching command guarantee system, construct the safety indicators, and create a more complete safety guarantee system to meet the development of our railway industry.

## 2 Related Work

### 2.1 Study on Capability Failure Adjustment

Zhan et al. [4] studied the problem of real-time adjustment of train operation in the case that the capacity of a line in a certain section of a two-line high-speed railway fails due to faults, so as to reasonably arrange the sequence and time of upstream and downstream trains passing through each section from the beginning of the fault to the end of the operation period, and determine whether any train needs to be cancelled due to the reduced capacity of the line. Zhan [5] studied how to adjust train operation in real time when the capacity of a certain section of high-speed railway line is completely failed due to faults, abstracted train operation as a network composed of events and activities from a macro level, and established a mixed integer linear programming model for real-time adjustment of train operation. Zhan et al. [6] uses the optimization method of operations research for reference. By establishing a mixed integer linear programming model for train operation adjustment under interference conditions, he uses effective algorithms to quickly generate train operation adjustment schemes under interference conditions, so as to assist dispatchers in real-time scheduling decisions.

### 2.2 Study on Capability Failure Modeling

Zhan et al. [7] established the train operation adjustment optimization model considering the uncertainty of inter-train conflict, and designed the solution algorithm based on Lagrange relaxation. The above model and algorithm were verified by experiments. Wang et al. [8] studied the train operation adjustment problem under the disturbed scenario that high-speed railway trains break down in the interval, abstracted train operation on the road network as the occupation of discrete running resources from the micro level, and established a mixed integer programming model with the minimum total delay time of all trains as the target under the condition of fault duration determination. Liao et al. [9] aim at providing decision support for operational adjustment. In order to effectively reduce the external interference impact on train operation. Cheng et al. [10] analyzed the causes of high-speed railway failures are complex, and it is difficult to quantitatively and accurately describe safety evaluation. The purpose is to construct a model to ensure the safety of high-speed railway operations.

## 3 Mutual Feed Response Model Based on Space-Time Network

### 3.1 Overview of Space-Time Networks

High-speed railway network is a topology structure with a large number of nodes and connections, so complex network theory can be used for modeling. In general, the space-time network can be defined as $G = (V, E)$, where $V$ represents the set of stations in the network, and $E$ represents the connecting events between stations.

As shown in Fig. 1, different train lines form a space-time network, and each node has space-time attributes. For space-time networks, physical nodes are fixed and the connection events between physical nodes change over time. However, the existing research on the characteristics of network modeling can only be based on static connection networks. Therefore, in order to facilitate calculation and analysis, the spatio-temporal network is considered to be transformed into static weighted directed network for equivalent representation in the actual establishment of the model.

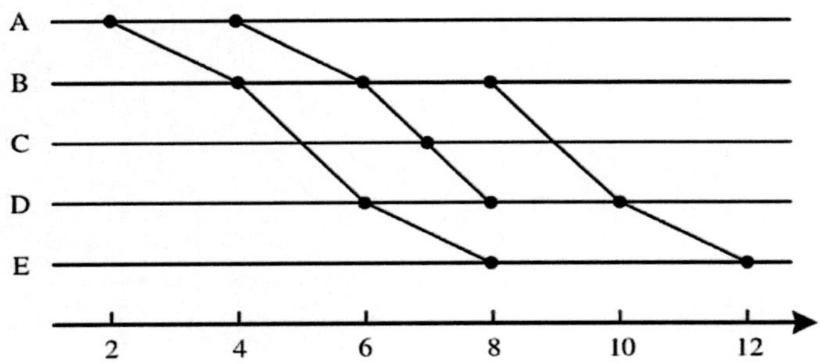


**Fig. 1.** Spatio-temporal network graph

**Initial space-time node network.** For the space-time network model, based on the train schedule information, the initial time-space node network based on the timetable is established to represent the concept of space-time. Since the schedule of a train has two arrival and departure times at each station, the freight owner of the station can choose the train until the departure time of the station, so only the departure time of a train at the station can be considered (Fig. 2).

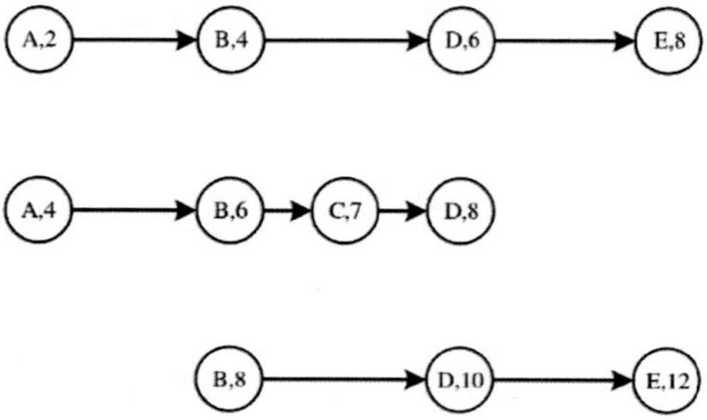


**Fig. 2.** Initial space-time node network diagram

**Set edges between all nodes in the same station.** In practice, in the space-time network established based on all train schedules, for one station, there are multiple space-time nodes in the station, and the spatial attributes of these space-time nodes are consistent, while the time attributes are inconsistent. However, in terms of cargo and time dimension, all space-time nodes at the same station are reachable in time order, that is, all these space-time nodes are connected. Therefore, according to the sequence of time, there is a connecting edge between two adjacent space-time nodes, and the weight of the corresponding edge is the difference value of the time attribute of the two space-time nodes (Fig. 3).

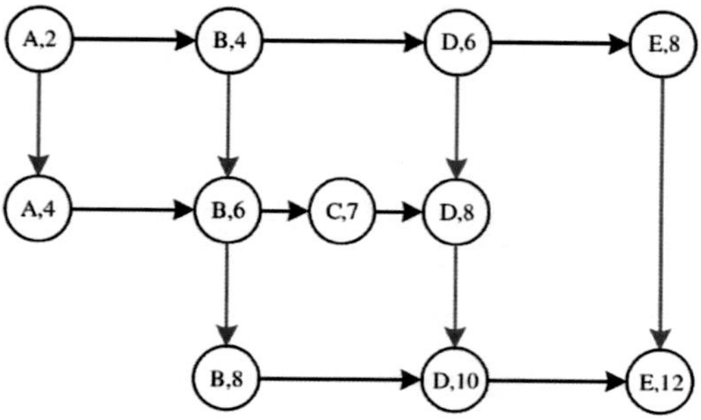


**Fig. 3.** Transition space-time node network diagram

**Transfer edges are set between different station nodes in the same city**. The operation network of high-speed railway is very complicated, and some stations are far apart from each other, so these stations can reach each other. Because there are relatively few direct trains between them, and the direct train between these stations can only be selected before a certain time (the time when the direct train leaves), after a certain time, you can only wait for the next direct train to arrive. Therefore, the high-speed railway operation network at any time can not be a fully coupled network (fully coupled network: any two nodes in the network are directly connected to the edge) (Fig. 4).

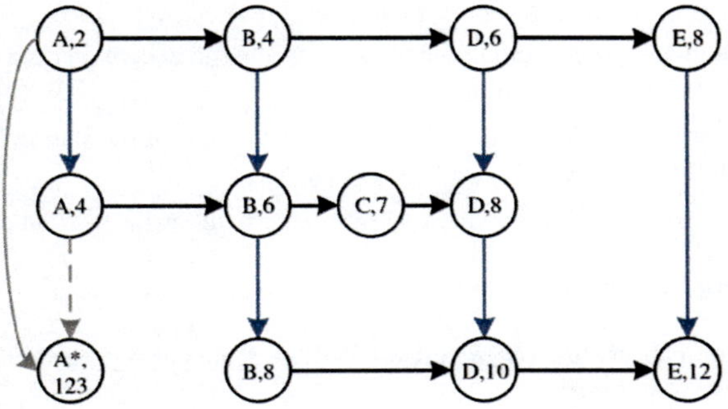


**Fig. 4.** Space-time node network diagram

In general, some indexes which can reflect the local or global characteristics of the network are selected to model and analyze the topological characteristics of the high-speed railway network. In the existing modeling and analysis research of physical network and train service network, the commonly used indicators are traditional indicators based on complex network theory, such as degree, intermediate centrality, proximity centrality, clustering coefficient, etc., which are used to study the characteristics of static network modeling.

## 3.2 Model Symbol

The parameters and decision variables of this model are shown in the following table (Table 1):

**Table 1.** Model parameter

| Symbol | Meaning |
| --- | --- |
| $f$ | The train carrying freight |
| $t_f$ | Delay time of train f |
| $l$ | The set of paths the train can use on the space-time network |
| $\hat{d}$ | The basic unit of freight carried by the train is called the demand unit |
| $g$ | Profit of a single train operation |
| $\theta_{\hat{d}f}$ | The number of shipments to demand unit $\hat{d}$ on path $l$ |
| $T$ | The total amount of time the owner is willing to accept that the goods will be delayed due to disturbance |
| $\tau_t$ | At time t, the number of trains that can be parked at the station is 0 if the disturbance occurs at time t, otherwise 1 |
| $\omega_{\hat{d},l}$ | 0–1 variables, if demand unit $\hat{d}$ choose path, the value is 1, otherwise 0 |
| $c_f^l$ | Train $f$ choose path $l$ unit costs |
| $\alpha$ | Incompatible time-space network path set that does not meet infrastructure usage requirements |
| $\lambda_f^l$ | Variable 0–1, if the path l on the space-time network is occupied by train $f$, the value is 1, otherwise 0 |
| $\eta'_{\hat{d}f'}$ | 0–1 variables, if demand unit $f$ choosing carry trains, the value is 1, otherwise 0 |
| $\zeta_f$ | The total number of trains accommodated in the station |
| $\mu_{\hat{d}ff',l,l'}$ | 0–1 variables, if demand unit $\hat{d}$ based on path $l$ and $l'$ service time interval and choose carry trains $f$, the value is 1, otherwise 0 |

## 3.3 Model Construction

**Objective function.** Taking maximization of benefits of train operation enterprises as the objective function, a train operation plan adjustment model based on space-time network is established. The calculation method of objective function is shown in the equation:

$$\max z = \sum_g \sum_{f \in F_g} \sum_{\hat{d}} \omega'_{\hat{d}f} \times \theta_{\hat{d}f} - \sum_g \sum_{f \in F_g} \sum_l c_f^l \times \lambda_f^l \tag{1}$$

**Constraint condition.** The main constraints of this model include train path selection constraint, train and demand relationship constraint, train capacity constraint and infrastructure capacity constraint.

Spatio-temporal network routing constraints:

$$\sum_l \lambda_f^l = 1, \forall f \tag{2}$$

Constraints on the relationship between train and demand:

$$\mu_{\hat{a},f,f',l,l'} \leq \sum\nolimits_{f,f'} \sum\nolimits_{l,l'} \lambda_f^l \times \lambda_{f'}^{l'} \times \omega_{\hat{a},l'}, \forall \hat{d} \tag{3}$$

$$\eta'_{\hat{a},f'} = \sum\nolimits_{f,l,l'} v_{f'}^{l'} \times \omega_{\hat{a},l'} \times \mu_{\hat{a},f,f',l,l'}, \forall \hat{a}, f' \tag{4}$$

$$\sum (t_f + t_{f'}) \leq T \tag{5}$$

Train capacity constraint:

$$\sum\nolimits_{\hat{a}} \eta'_{\hat{a},f} \leq \xi_f, \forall f \tag{6}$$

Station capacity constraint:

$$\sum\nolimits_{f \in F} \lambda_f^l \leq \tau_t, \forall t \tag{7}$$

Infrastructure capacity constraints:

$$\lambda_f^l + \lambda_{f'}^{l'} \leq 1, \forall f, f', l, l' \in \alpha \tag{8}$$

## 4 Result

In recent years, the high-speed railway with its safety and fast, high punctuality rate, comfort and other characteristics of rapid development, our high-speed railway has entered the era of network operation Ling et al. [11]. With the continuous development of high-speed railway, the train operation density is increasing, and the road network is more sensitive. When the railway system suffers from internal or external interference, it is more likely to have a chain reaction in time and space, and even lead to the collapse of the whole road network. The construction of the space-time network model of high-speed railway can not only analyze the time-varying characteristics of high-speed railway operation, but also establish the delay propagation model based on the space-time network for delay prediction, so as to predict the scope and degree of influence caused by the current emergency in time, and provide scientific data support for the adjustment of dispatchers.

## 5 Conclusion

The current high-speed railway system is complex and consists of many units. Once the system component fails or the external environment changes abruptly, the train may be delayed and spread to other trains along the train route, resulting in the delay propagation at the network level. But the space-time network model established in this paper is based on the collected schedule data and physical network data. The scale of the network is large and the computational efficiency is not high enough. Further research on the establishment of the model based on online data and the real-time evolution of the space-time network characteristics is the focus of the next research.

The reciprocal feed-back model established in this paper only considers the feedback of unidirectional disturbance, and the research on the feedback model considering larger networks and upstream and downstream bidirectional will be carried out in the future.

**Acknowledgement.** This study is supported by the Joint Funds of the National Natural Science Foundation of China (No.U2034208).

## References

1. Yang, M.: Delay Propagation Prediction for High-speed Railway Based on Spatio-temporal Network in Typical Emergent Scenarios. Beijing Jiaotong University (2022). (in Chinese)
2. Wang, J., Wang, Y., Peng, Y., Lu, J.: Examining partial proportional odds model in analyzing severity of high-speed railway accident. Smart Resilient Transp. **3**(1), 12–24 (2021)
3. Zhang, Z., Jia, L., Qin, Y.: RAMS analysis of railway network: model development and a case study in China. Smart Resilient Transp. **3**(1), 2–11 (2021)
4. Zhan, S., Zhao, J., Peng, Q.: Real-time train rescheduling on high-speed railway under partial segment blockages. J. China Railw. Soc. **38**(10), 1–13 (2016). (in Chinese)
5. Zhan, S., Zhao, J., Peng, Q.: Real-time train rescheduling on high-speed railway under complete segment blockages. J. China Railw. Soc. **37**(11), 1−9 (2015). (in Chinese)
6. Zhan, S.: Modeling and Algorithming of Real-time High Speed Train Rescheduling in Case of a Temporary Segment Blockage. Southwest Jiaotong University (2016). (in Chinese)
7. Guo J.: Train Dispatching Models with Probabilistic Conflicts on a High-speed Railway Line. Beijing Jiaotong University (2016). (in Chinese)
8. Wang, Y., Meng, L., Long, S.: An optimization model of train operation adjustment plan under the condition of train section failure in high-speed railway. China Acad. J. Electron. Publ. House **20**(1), 9–16 (2019). (in Chinese)
9. Liao, Z., Qin, Y., Wang, M.: Research on high-speed railway operation adjustment model based on priority. Smart Resilient Transp. **4**(1), 12–21 (2022)
10. Cheng, L., Wang, Y., Peng, Y.: Research on risk assessment of high-speed railway operation based on network ANP. Smart Resilient Transp. **3**(1), 37–51 (2021)
11. Ling, X., Peng, Y., Sun, S., Li, P.: Uncovering correlation between train delay and train exposure to bad weather. Phys. A-Stat. Mech. Its Appl. **512**, 1152–1159 (2018)

# Energy-Saving Optimization Study of Train Timetable Based on Regenerative Braking Technology

Zheng Yajing, Ma Zihan(✉), and Jin Wenzhou

School of Civil Engineering and Transportation, South China University of Technology, Guangzhou 510641, China
20212101037[8]@mail.scut.edu

**Abstract.** In this paper, we consider the effect of instantaneous power on the amount of regenerative braking energy recovery and divide a series of short time intervals to refine the calculation of energy consumption. Based on this, an optimization model with the objective of minimizing energy consumption is constructed, which uses the first station departure interval, inter-station travel time and station dwell time to describe the train timetable, and a running line adjustment algorithm is designed to ensure the arrival and departure intervals are reasonable. By using the hybrid genetic algorithm embedded with harmonic search, the case of a virtual line with 20 stations and 200 trips is solved and the results show that the method proposed in this paper can effectively reduce energy consumption.

**Keywords:** Regenerative braking · Train timetable · Genetic algorithm · Energy-saving optimization

## 1 Introduction

As urban rail transportation grows, the huge energy consumption of this system puts enormous pressure on the city's energy supply [1]. In order to reduce train energy consumption, scholars have explored it from many perspectives, such as train structure, track linearity, and driving strategies, etc. In addition, some scholars focus on the optimization of train operation strategies, mainly the cooperative optimization of trains based on regenerative braking technology.

Literature [2] first introduced an energy-saving method based on dwell-time and run-time control. And literature [3] proposed that increasing the overlap between capacity and consumption period can improve energy efficiency and constructed a timetable optimization model. On the basis of this, partial scholars reduced energy consumption by ensuring that the movements of the braking train and traction train were synchronized [4–6]; Liu Hongjie et al. considered the effect of fixing the first and last train departure times [7]; Gao Xin et al. added constraints on passenger travel and interchange times to the HSR timetable optimization model [8]; Ran Xinchen et al. discussed the influence of passenger flow spatial distribution on the train movement model and the overall


Y. Qin et al. (Eds.): EITRT 2023, LNEE 1137, pp. 343–357, 2024.
https://doi.org/10.1007/978-981-99-9311-6_37

energy consumption [9]; Zheng Yajing et al. added departure interval and stopping time constraints to the model to suit operational reality [10]; and Yang X constructed a two-objective model with passenger waiting time as an additional indicator and designed a binary genetic algorithm for solving [11]. Further, literature [12] proposed a "time-slot and energy cell" model to find the global optimum value of net energy consumption by selecting the running curve; literature [13] proposed that the instantaneous traction and braking power should be taken into account when calculating the amount of regenerative braking energy; and literature [14] combined both single train operation and multiple train collaboration together through the traction duration.

Existing research usually focuses on optimizing running lines by adjusting travel and dwell times but rarely considers the feasibility of train arrival and departure intervals in this process. In addition, these studies are often limited to the "single supply section" scenario, with less analysis of "multiple supply sections". Considering the above problems, this paper constructs a timetable optimization model with the objective of minimizing the amount of energy consumption, and we demonstrate the feasibility of this model through an example simulation.

## 2 Energy-Saving Optimization Model of Train Timetable

### 2.1 Problem Description and Model Assumptions

Regenerative braking is an energy recovery technology that uses the inertia of train braking to rotate the motor rotor, converting kinetic energy into electrical energy. However, the returned energy is asymmetrical in three phases, which will have some impact on the power quality of the grid (network voltage fluctuations, harmonics, negative sequence, etc.).

To solve the problem, we can supply the return energy to trains in traction under the same electric section by adjusting the train operation, which enables the immediate use of electrical energy and greatly improves the energy utilization of regenerative braking technology. Next, this paper will research the problem of energy-saving optimization of train schedules based on this idea.

Based on the actual characteristics of this problem, we make the following assumptions:

1. The utilization of regenerative braking energy occurs only between trains traveling in the same direction and under the same electric section.
2. Only the energy consumption of trains traction is considered.
3. All energy recovered from regenerative braking technology is supplied to the train for traction.

### 2.2 Notation Definition

For ease of description, we list all the involved notations in Table 1, including sets, subscripts, decision variables, auxiliary variables, etc.

**Table 1.** Notations definition

| Notations | Detailed definition |
|---|---|
| Sets: | |
| $S$ | Set of stations, $S = \{0, 1, 2 \cdots \|S\| - 1\}$ |
| $K$ | Set of trains, $K = \{0, 1, 2 \cdots \|K\| - 1\}$ |
| $M$ | Set of election sections, $M = \{0, 1, 2 \cdots \|M\| - 1\}$ |
| $S_m$ | Set of station under the same electric section m |
| Subscripts: | |
| $s$ | The index of stations, $s \in S$ |
| $k$ | The index of trains, $k \in K$ |
| $m$ | The index of election sections, $m \in M$ |
| Parameters: | |
| $[t_{\min}^{interval}, t_{\max}^{interval}]$ | The range of the departure time at the first station |
| $[t_{\min}^{travel}, t_{\max}^{travel}]$ | The range of the travel time between stations |
| $[t_{\min}^{dwell}, t_{\max}^{dwell}]$ | The standard range of the dwell time at station |
| $[\sigma_{a-a}^{\min}, \sigma_{a-a}^{\max}]$ | The standard range of the arrival-arrival intervals |
| $[\sigma_{a-d}^{\min}, \sigma_{a-d}^{\max}]$ | The standard range of the arrival-departure intervals |
| $[\sigma_{d-a}^{\min}, \sigma_{d-a}^{\max}]$ | The standard range of the departure-arrival intervals |
| $[\sigma_{d-d}^{\min}, \sigma_{d-d}^{\max}]$ | The standard range of the departure-departure intervals |
| $\xi^{traction}, \xi^{braking}$ | The duration of traction/braking |
| $t_{start}, t_{end}$ | Time to start/end service |
| Decision variables: | |
| $x_k^{up}, x_k^{down}$ | Departure interval at the first station between train $k$ and train $k + 1$ in the upward/downward direction |
| $y_{k,s}^{up}, y_{k,s}^{down}$ | Travel time of train $k$ between station $s$ and $s + 1$ in the upward/downward direction |
| $z_{k,s}^{up}, z_{k,s}^{down}$ | Dwell time of train $k$ at the station $s$ in the upward/downward direction |
| Auxiliary variables: | |
| $t_{k,s}^{up,arr}, t_{k,s}^{down,arr}$ | The moment when the train $k$ arrival at the station s in the upward/downward direction |
| $t_{k,s}^{up,dep}, t_{k,s}^{down,dep}$ | The moment when the train $k$ leaves station $s$ in the upward/downward direction |

## 2.3 Decision Variable

For an optimization problem of train timetable, the most intuitive way to present it is to solve for the arrival and departure times of each line directly, but the arrival and departure times of train $k$ at station $s$ are constrained by the arrival and departure times of all stations before station $s$, which is complicated to describe.

Thus, this paper chooses the first station departure interval, the inter-station travel time, and the station dwell time as the decision variables, and regards train arrival times and departure times as auxiliary variables. The relationship between these two types of variables is as follows:

1. For the first station, the dwell time can be considered to be 0, and the arrival time of train $k$ is equal to the arrival time of train $k$-1 plus the departure interval $k-1$.

$$\begin{cases} t_{k,0}^{up,arr} = t_{k-1,0}^{up,arr} + x_{k-1}^{up} \; \forall k \in \{K/0\} \\ t_{k,|S|-1}^{down,arr} = t_{k-1,|S|-1}^{down,arr} + x_{k-1}^{down} \; \forall k \in \{K/0\} \end{cases} \tag{1}$$

2. For any station s other than the first station, the arrival time of train $k$ is equal to its arrival time at station $s-1$ plus the dwell time and the inter-station travel time.

$$\begin{cases} t_{k,s}^{up,arr} = t_{k,s-1}^{up,arr} + y_{k,s-1}^{up} + z_{k,s-1}^{up} & \forall k \in K, s \in \{S/0\} \\ t_{k,s}^{down,arr} = t_{k,s+1}^{down,arr} + y_{k,s+1}^{down} + z_{k,s+1}^{down} & \forall k \in K, s \in \{S/|S|-1\} \end{cases} \tag{2}$$

3. For all stations, the train departure time is equal to the arrival time plus the dwell time (in particular, the dwell times of the first and last stations are 0).

$$\begin{cases} t_{k,s}^{up,dep} = t_{k,s}^{up,arr} + y_{k,s}^{up} & \forall k \in K, s \in \{S\} \\ t_{k,s}^{down,dep} = t_{k,s}^{down,arr} + y_{k,s}^{down} & \forall k \in K, s \in \{S\} \end{cases} \tag{3}$$

### 2.4 Constraint Condition

First of all, considering the practicality of service quality and driving safety, the departure intervals, travel times, and dwell times of trains are not unlimited. Therefore, there is a standard range for each decision variable.

$$\begin{cases} t_{\min}^{interval} \le x_k^{up} \le t_{\max}^{interval} \\ t_{\min}^{interval} \le x_k^{down} \le t_{\max}^{interval} \end{cases} \forall k \in K/\{|K|-1\} \tag{4}$$

$$\begin{cases} t_{\min}^{travel} \le y_{k,s}^{up} \le t_{\max}^{travel} \\ t_{\min}^{travel} \le y_{k,s}^{down} \le t_{\max}^{travle} \end{cases} \forall k \in K/\{|K|-1\}, s \in S/\{|S|-1\} \tag{5}$$

$$\begin{cases} t_{\min}^{dwell} \le z_{k,s}^{up} \le t_{\max}^{dwell} \\ t_{\min}^{dwell} \le z_{k,s}^{down} \le t_{\max}^{dwell} \end{cases} \forall k \in K/\{|K|-1\}, s \in S/\{0, |S|-1\} \tag{6}$$

Similarly, it is necessary to limit the minimum and maximum intervals between two adjacent trains. This construction can be expressed in Eqs. (7)-(10).

$\forall k \in K/\{|K|-1\}, s \in S/\{0, |S|-1\}$,

$$\begin{cases} \sigma_{a-a}^{\min}(t_{k,s}^{up,arr}) \le t_{k+1,s}^{up,arr} - t_{k,s}^{up,arr} \le \sigma_{a-a}^{\max}(t_{k,s}^{up,arr}) \\ \sigma_{a-a}^{\min}(t_{k,s}^{down,arr}) \le t_{k+1,s}^{down,arr} - t_{k,s}^{down,arr} \le \sigma_{a-a}^{\max}(t_{k,s}^{down,arr}) \end{cases} \tag{7}$$

$$\begin{cases} \sigma_{a-d}^{\min}(t_{k,s}^{up,dep}) \le t_{k+1,s}^{up,arr} - t_{k,s}^{up,dep} \le \sigma_{a-d}^{\max}(t_{k,s}^{up,dep}) \\ \sigma_{a-d}^{\min}(t_{k,s}^{down,dep}) \le t_{k+1,s}^{down,arr} - t_{k,s}^{down,dep} \le \sigma_{a-d}^{\max}(t_{k,s}^{down,dep}) \end{cases} \tag{8}$$

$$\begin{cases} \sigma_{d-a}^{\min}(t_{k,s}^{up,arr}) \le t_{k+1,s}^{up,dep} - t_{k,s}^{up,arr} \le \sigma_{d-a}^{\max}(t_{k,s}^{up,arr}) \\ \sigma_{d-a}^{\min}(t_{k,s}^{down,arr}) \le t_{k+1,s}^{down,dep} - t_{k,s}^{down,arr} \le \sigma_{d-a}^{\max}(t_{k,s}^{down,arr}) \end{cases} \tag{9}$$

$$\begin{cases} \sigma_{d-d}^{\min}(t_{k,s}^{up,dep}) \le t_{k+1,s}^{up,dep} - t_{k,s}^{up,dep} \le \sigma_{d-d}^{\max}(t_{k,s}^{up,dep}) \\ \sigma_{d-d}^{\min}(t_{k,s}^{down,dep}) \le t_{k+1,s}^{down,dep} - t_{k,s}^{down,dep} \le \sigma_{d-d}^{\max}(t_{k,s}^{down,dep}) \end{cases} \tag{10}$$

When $\sigma^{\min}$ and $\sigma^{\max}$ are the thresholds of train intervals; the subscript $'a-a'$ indicates the interval between the arrival of the previous train and the arrival of the next; the subscript $'a-d'$ indicates the interval between the arrival of the previous train and the departure of the next and so on.

For the convenience of passengers planning their journeys, the last train usually leaves at the end of its service time, which can be expressed in mathematics as:

$$\begin{cases} \sum_{k}^{|K|-1} x_k^{up} = 3600 \cdot (t_{end} - t_{start}) \\ \sum_{k}^{|K|-1} x_k^{down} = 3600 \cdot (t_{end} - t_{start}) \end{cases} \tag{11}$$

## 2.5 Objective Function

Based on the above, the objective of our optimization is to minimize the energy consumption of train travel, so we first need to quantify this indicator.

Generally, for a moment t in the operating cycle, the energy recovered by regenerative braking is the minimum value between the total braking energy and the total traction energy under the same electric section. The actual energy consumption is the energy consumption of traction minus the energy recovered by regenerative braking, so the objective function can be expressed as:

$$\min Z = E_c - E_r^{up} - E_r^{down} \tag{12}$$

When $E_c$ is the consumption of train operation; $E_r^{up}$ is the regenerative energy in the upward direction; $E_r^{down}$ is the regenerative energy in the downward direction.

And in more concrete terms:

$$E_c = \sum_{m=0}^{|M|-1} \int_0^T \sum_{k=0}^{|K|-1} \sum_{s=0}^{|S_m|-1} [e_{s,k}^{c,up}(t) + e_{s,k}^{c,down}(t)]dt \tag{13}$$

$$E_r^{up} = \sum_{m=0}^{|M|-1} \int_0^T \min(\sum_{k=0}^{|K|-1} \sum_{s=0}^{|S_m|-1} e_{s,k}^{r,up}(t), \sum_{k=0}^{|K|-1} \sum_{s=0}^{|S_m|-1} e_{s,k}^{c,up}(t))dt \tag{14}$$

$$E_r^{down} = \sum_{m=0}^{|M|-1} \int_0^T \min(\sum_{k=0}^{|K|-1} \sum_{s=0}^{|S_m|-1} e_{s,k}^{r,down}(t), \sum_{k=0}^{|K|-1} \sum_{s=0}^{|S_m|-1} e_{s,k}^{c,down}(t))dt \tag{15}$$

When $e_{s,k}^{r,up}(t)$ and $e_{s,k}^{r,down}(t)$ are the braking power of train $k$ at station $s$; $e_{s,k}^{c,up}(t)$ and $e_{s,k}^{c,down}(t)$ are the traction power of train $k$ at station $s$.

To make Eq. (12) easy to solve, the operating cycle $T$ can be discretized into $N$ equal time intervals $\Delta t$, whereupon the objective function can be transformed into Eq. (16):

$$\begin{aligned} \min Z = & \sum_{m=0}^{|M|-1} \sum_{n=1}^{N} \sum_{k=0}^{|K|-1} \sum_{s=0}^{|S_m|-1} [e_{s,k}^{c,up}(n\Delta t) + e_{s,k}^{c,down}(n\Delta t)] \cdot \Delta t \\ & - \sum_{m=0}^{|M|-1} \sum_{n=1}^{N} \min(\sum_{k=0}^{|K|-1} \sum_{s=0}^{|S_m|-1} e_{s,k}^{r,up}(n\Delta t), \sum_{k=0}^{|K|-1} \sum_{s=0}^{|S_m|-1} e_{s,k}^{c,up}(n\Delta t)) \cdot \Delta t \\ & - \sum_{m=0}^{|M|-1} \sum_{n=1}^{N} \min(\sum_{k=0}^{|K|-1} \sum_{s=0}^{|S_m|-1} e_{s,k}^{r,down}(n\Delta t), \sum_{k=0}^{|K|-1} \sum_{s=0}^{|S_m|-1} e_{s,k}^{c,down}(n\Delta t)) \cdot \Delta t \end{aligned} \tag{16}$$

That is also equaling to:

$$\begin{aligned} \max Z = & \sum_{m=0}^{|M|-1} \sum_{n=1}^{N} [\sum_{k=0}^{|K|-1} \sum_{s=0}^{|S|-1} (e_{s,k}^{c,up}(n\Delta t) + e_{s,k}^{c,down}(n\Delta t)) \\ & - \min\left(\sum_{k=0}^{|K|-1} \sum_{s=0}^{|S_m|-1} e_{s,k}^{r,up}(n\Delta t), \sum_{k=0}^{|K|-1} \sum_{s=0}^{|S_m|-1} e_{s,k}^{c,up}(n\Delta t)\right) \\ & - \min(\sum_{k=0}^{|K|-1} \sum_{s=0}^{|S_m|-1} e_{s,k}^{r,down}(n\Delta t), \sum_{k=0}^{|K|-1} \sum_{s=0}^{|S_m|-1} e_{s,k}^{c,down}(n\Delta t))] \end{aligned} \tag{17}$$

Please note that although this study involves both upward and downward directions, the two are in essence the same, only the upward direction will be explained.

As shown in Fig. 1, for a given moment $n\Delta t$, only if $n\Delta t$ is in the interval $[t_{k,s}^{up,dep}, t_{k,s}^{up,dep} + \xi^{\text{traction}}]$ will the train generate traction energy power:

$$e_{k,s}^{c,up}(n\Delta t) = \begin{cases} p(\varphi(n\Delta t), n\Delta t - t_{k,s}^{up,dep}) & t_{k,s}^{up,dep} \leq n\Delta t \leq t_{k,s}^{up,dep} + \xi^{\text{traction}} \\ 0 & else \end{cases} \tag{18}$$

When $\xi^{traction}$ is the duration of traction; $p(\varphi(n\Delta t), n\Delta t - t_{k,s}^{up,dep})$ is the value at point $n\Delta t - t_{k,s}^{up,dep}$ of the traction power curve corresponding to the full load rate $\varphi(n\Delta t)$.

Similarly, the braking capacity power $e_{s,k}^{r,up}(n\Delta t)$ can be expressed as:

$$e_{k,s}^{r,up}(n\Delta t) = \begin{cases} p(\varphi(n\Delta t), n\Delta t - (t_{k,s}^{up,arr} - \xi^{braking})) & t_{k,s}^{up,arr} - \xi^{braking} \leq n\Delta t \leq t_{k,s}^{up,arr} \\ 0 & else \end{cases} \tag{19}$$

When $\xi^{braking}$ is the duration of braking; $p(\varphi(n\Delta t), n\Delta t - (t_{k,s}^{up,arr} - \xi^{braking}))$ is the value at point $n\Delta t - (t_{k,s}^{up,arr} - \xi^{braking})$ of the braking power curve corresponding to the full load rate $\varphi(n\Delta t)$.

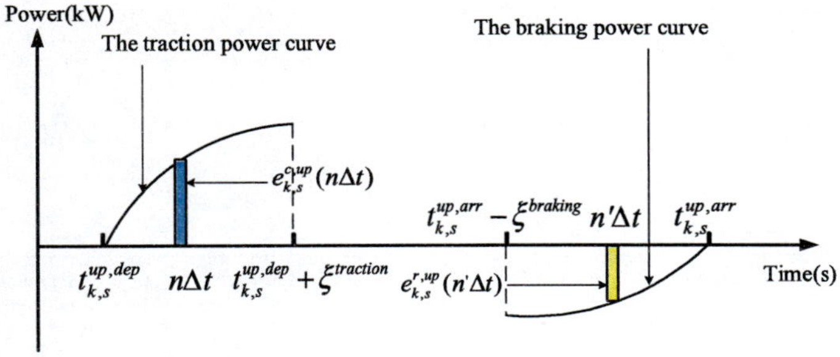


**Fig. 1.** Power curves of traction and braking

# 3 Hybrid Genetic Algorithm Embedded with Harmonic Search

The model constructed in Chap. 2 is a complex integer programming model that is hard to linearize. Due to the large size of this problem, typical computer memory can hardly support the computation of accurate algorithms, so we design a hybrid genetic algorithm embedded with harmonic search for solving.

## 3.1 Genetic Algorithm Coding Method

Use floating point coding and define a train timetable as an individual including 6 chromosomes (denoted as $r_1 \sim r_6$), of which the details are explained below:

- As shown in Fig. 2, the information saved at each point on $r_1$ and $r_4$ is a real number, which represents the departure interval of the first station in the upward and downward directions respectively.

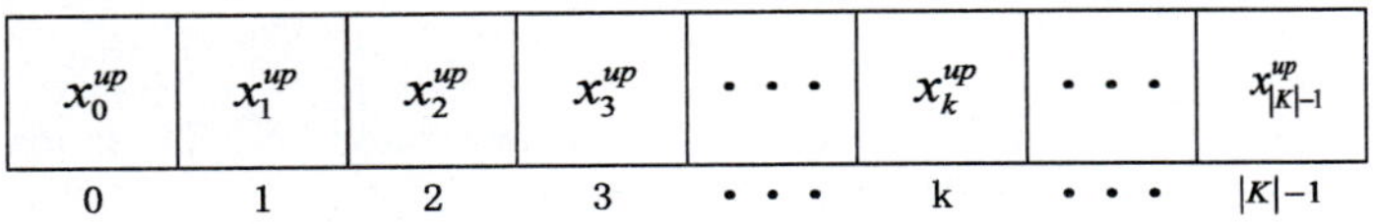


**Fig. 2.** Chromosome code of first station departure interval

- The number of points on $r_2$ and $r_5$ is $|K| \cdot (|S| - 1)$. Points 0 to $|S| - 2$ indicate the travel times of Train 0; points $|S| - 1$ to $2|S| - 3$ indicate the travel times of Train 1; more generally, points $(|S| - 1)k$ to $k(|S| - 1) + |S| - 2$ indicate the travel times of Train $k$. Further, we can divide the chromosome into k subsegments, each containing all the travel time information of a train, which are shown in Fig. 3.

- As shown in Fig. 4, the gene sequence structure of $r_3$ and $r_6$ is similar to that of $r_2$ and $r_5$, with only difference being that the length of the former subsegment is $|S| - 2$(no dwell time at first and last stations).

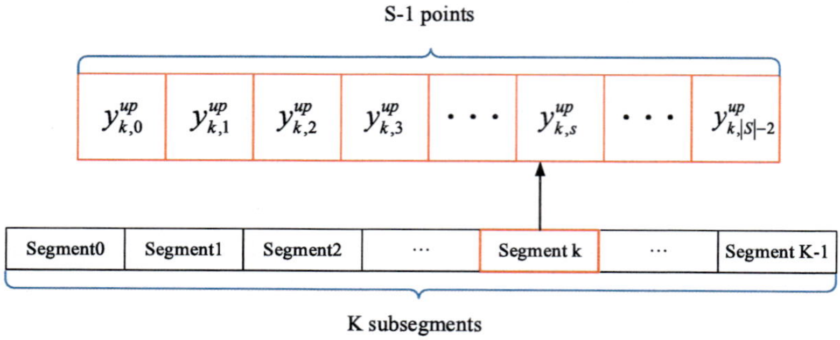


**Fig. 3.** Chromosome code of inter-station travel time

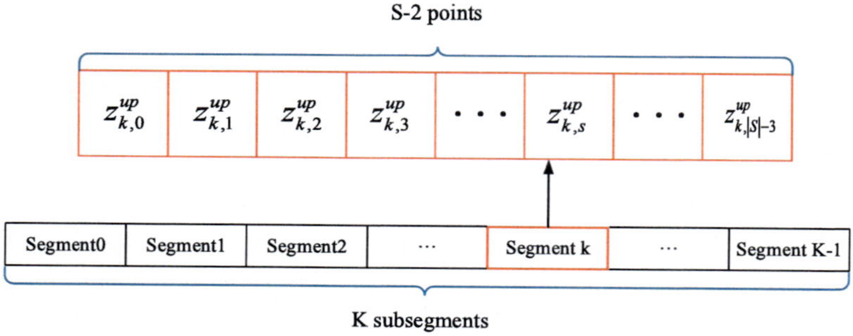


**Fig. 4.** Chromosome of station dwell time

## 3.2 Genetic Operators

1) Selection operator

This paper uses rank-based selection as the selection operator. For an individual $R$ after sorting, define the probability of it being selected as:

$$\phi(i) = \alpha(1-\alpha)^{i-1} \tag{20}$$

where $\alpha$ is the selection parameter and $i$ is the individual ranking.

According to Eq. (20) we can calculate the probability of being selected for all individuals, and then generate a 0–1 random number to complete the selection.

2) Crossover operator

For floating-point encoding, we can use the arithmetic crossover as the crossover operator. The crossover method is as follows:

$$\begin{cases} X_{1'} = \lambda X_1 + (1-\lambda)X_2 \\ X_{2'} = (1-\lambda)X_1 + \lambda X_2 \end{cases} \tag{21}$$

where $X_1$, $X_2$ are the genetic values of father and mother, and $X_1\prime$, $X_2\prime$ are the genetic values of the new offspring; $\lambda$ is a multiplier and takes a value from 0 to 1.

3) Mutation operator

This paper uses uniform mutation as the mutation operator, and the mutation value is a random integer within the range of the corresponding decision variable.

### 3.3 Population Update Based on Harmony Search

In order to increase population diversity and avoid the prematurity problem, it is necessary to perform a harmony search after each population iteration of evolution, which proceeds as follows:

Step 1: Set the parameters harmony memory choose probability *HMCR*, pitching adjust rate *PAR* and bandwidth *BW*.
Step 2: Initialize the harmony memory.
Step 3: Generate a new chromosome as follows:

For $i$ = (1 to the size of the chromosome):
If random (0,1) < *HMCR* :
The i-th gene of new chromosome $r_{new}(i) = r_x(i)$ ,when $r_x$ is randomly select from harmony memory.
If random (0,1) < *PAR* :
$r_{new}(i) = r_{new}(i) + random(-1,1) \times BW$ ;
Else:
$r_{new}(i) = LB + random(0,1) \times (UB - LB)$ ;

Step 4: Repeat step 3 until all chromosomes of the individual are generated.
Step 5: Calculate the fitness of the new individual and replace it if it is greater than the minimum value in the original harmony memory.
Step 6: Repeat step 3 to step 5 until reaching the maximum number of iterations.

### 3.4 Constraints Handling

For the constraints (4)–(6), it can be achieved by setting the corresponding range of values for the decision variables. According to Eq. (21), if the initial values are all in the standard range, the new values after the crossover will not overflow.

Obviously, the population is prone to producing individuals that violate constraints (7)–(10) during the course of their evolution. To solve this problem, we can calculate the surplus intervals and the demand intervals for each train and allocate the former to the latter to make the adjusted train timetables meet the constraints.

More specifically, this interval allocation method can be achieved by shifting the train running lines. Take the upward direction as an example, the process is shown in Figs. 5 and 6.

The adjusting algorithm is as follows:

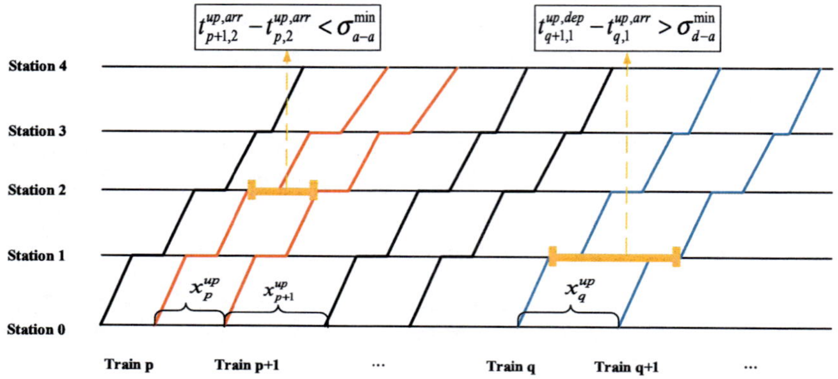


**Fig. 5.** The train timetable before adjusting

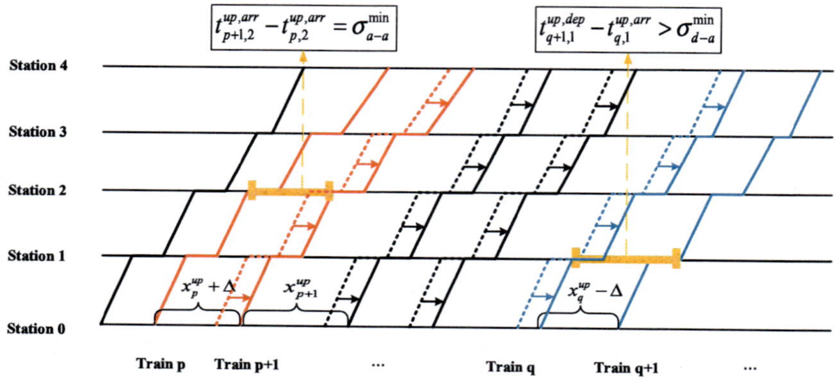


**Fig. 6.** The train timetable after adjusting

**Algorithm 1:** The adjusting of train intervals

BEGIN

Calculate arrival timetable for upward $T_{arr}^{up}$ by equation (1) and (2);

Calculate departure timetable for upward $T_{dep}^{up}$ by equation (3);

for $k = 0$ to $|K|-2$

  vector $vec$ adds $T_{arr}^{up}(k+1,0) - T_{arr}^{up}(k,0) - t_{\min}^{interval}$;

  for $s = 1$ to $|S|-1$

    $vec$ adds $T_{arr}^{up}(k+1,s) - T_{arr}^{up}(k,s) - \sigma_{a-a}^{\min}$, $T_{arr}^{up}(k+1,s) - T_{dep}^{up}(k,s) - \sigma_{a-d}^{\min}$, $T_{dep}^{up}(k+1,s) - T_{arr}^{up}(k,s) - \sigma_{d-a}^{\min}$ and $T_{dep}^{up}(k+1,s) - T_{dep}^{up}(k,s) - \sigma_{d-d}^{\min}$;

  allowance of train $k$ $A^{\min}(k) \leftarrow \min(vec)$;

  clear $vec$;

for $k = 0$ to $|K|-2$ :

  additional intervals required of train $k$ $req^{\min}(k) \leftarrow 0$;

  additional intervals provided of train $k$ $pro^{\min}(k) \leftarrow 0$;

  if $A^{\min}(k) < 0$ :

    $req^{\min}(k) \leftarrow -A^{\min}(k)$;

  else:

    $pro^{\min}(k) \leftarrow A^{\min}(k)$;

for $k_r = 0$ to $|K|-2$ :

  for $k_p = 0$ to $|K|-2$ :

    $u \leftarrow \min(req^{\min}(k_r), pro^{\min}(k_p) \cdot random(0,1))$;

    $req^{\min}(k_r) \leftarrow req^{\min}(k_r) - u$; $pro^{\min}(k_p) \leftarrow pro^{\min}(k_p) - u$;

    Interval provided by train $k_p$ for train $k_r$ $dis(k_p,k_r) \leftarrow u$;

for $k_r = 0$ to $|K|-1$ :

  for $k_p = 0$ to $|K|-2$ :

    $x_{k_r}^{up} \leftarrow x_{k_r}^{up} + dis(k_p,k_r)$; $x_{k_p}^{up} \leftarrow x_{k_p}^{up} - dis(k_p,k_r)$;

END

For constraint (11), the handling is similar to that of constraints (7)–(10). We can first calculate the departure time of the last train and then adjust all the departure intervals at the first station to make it equal to the end time of the operation.

The adjusting algorithm is as follows:

**Algorithm 2:** The adjusting of last train departure times

BWGIN

$u \leftarrow \sum_{k=0}^{|K|-2} x_k^{up} - 3600(t_{end} - t_{start})$;

while $u > 0$:

for k = 0 to $|K| - 2$:

  if $u > t_{\max}^{interval} - x_k^{up}$:

    $v \leftarrow random(0,1) \cdot (t_{\max}^{interval} - x_k^{up})$;

    $x_k^{up} \leftarrow x_k^{up} + v$, $u \leftarrow u - v$;

  else:

    $x_k^{up} \leftarrow x_k^{up} + u$, $u \leftarrow 0$;

while $u < 0$:

for k = 0 to $|K| - 2$:

  if $-u > x_k^{up} - t_{\max}^{interval}$:

    $v \leftarrow random(0,1) \cdot (x_k^{up} - t_{\max}^{interval})$

    $x_k^{up} \leftarrow x_k^{up} - v$, $u \leftarrow u + v$;

  else:

    $x_k^{up} \leftarrow x_k^{up} + u$, $u \leftarrow 0$;

END

Notice that if the original timetable meets constraint (11), the execution of Algorithm 1 does not have an effect on this, due to the fact that Algorithm 1 cannot change the position of the last train running line. In contrast, Algorithm 2 may destroy the structure of the timetable that meets constraints (7)–(10), so Algorithm 2 should be executed first before Algorithm 1 during the iteration process.

## 4 Example Analysis and Discuss

To verify the validity of the model and algorithm, a virtual line is generated for analysis. The virtual line has a total of 20 stations and 5 electric sections, and the actual service is from 6:00 a.m. to 11:00 p.m., with a total of 200 trips a day.

Considering the influence of passenger flow on train energy consumption, we divide the full load rate of the train according to the operating period, and different full load rates correspond to different traction and braking power curves.

The parameters of the genetic algorithm have a large impact on the final results. After testing, which determined in this paper are as follows: population size $cnt = 100$; number of iterations $iter_{\max} = 200$; crossover probability $p_{cross} = 0.8$; mutation probability $p_{mutate} = 0.05$; selection parameter $\alpha = 0.005$. Further, consider the following three scenarios:

1. The passenger flow changes over time and the regenerative braking technology is applied;

2. The passenger flow does not change over time (i.e., the full load rate is fixed at 60%) and the regenerative braking technology is applied;
3. The passenger flow changes over time and the regenerative braking technology is not applied (Fig. 7).

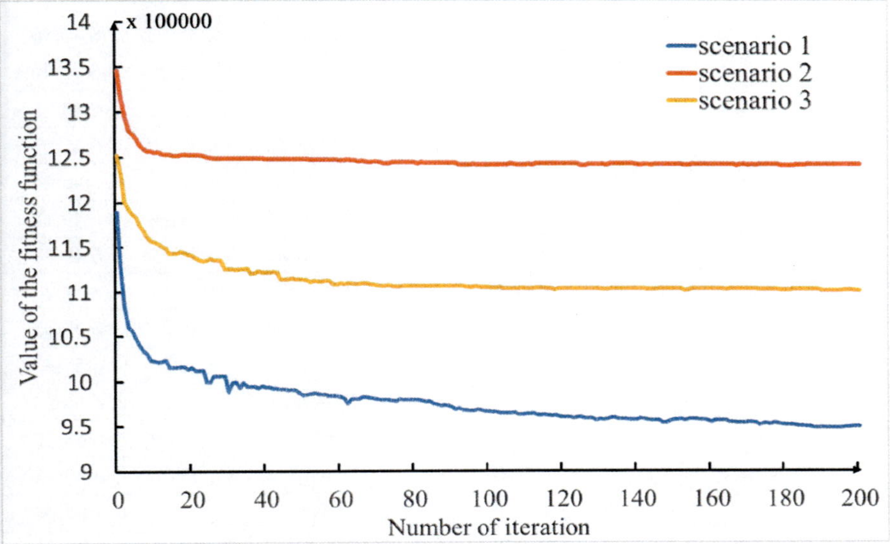


**Fig. 7.** Iterative process

Recording the optimal fitness of each iteration during the iterative process, and the calculation results are shown in Fig. 8. As a whole, the three curves have the same trend: they all decline rapidly in the first 20 iterations, and then change slightly and gradually stabilize. From a specific perspective:

- Comparing curve 1 and curve 2, it can be seen that the application of regenerative braking technology reduces energy consumption by about 14%. And the gap between the two curves grows in the process of seeking the best, which indicates that the energy returned by regenerative braking is not fixed but constantly changes with the adjustment of the timetable.
- The fitness function value of scenario 3 is greater than that of scenario 2, despite the regenerative braking technology has been applied. So there is a certain deviation between the calculation results of energy consumption under the conditions of the constant full load rate and the variable full load rate, and this fact should not be ignored.
- Based on the same algorithm, the convergence speed of curve 3 is faster than curve 1, which proves that the optimization space of the model considering the change of passenger flow is much larger than that of the model with fixed passenger flow.

**Table 2.** Part of the train schedule

| Station | | Trip 0 | Trip 1 | Station | | Trip 0 | Trip 1 |
|---|---|---|---|---|---|---|---|
| S0 | Arrive | – | – | S10 | Arrive | 6:31:53 | 6:36:06 |
| | Departure | 6:00:00 | 6:04:09 | | Departure | 6:32:39 | 6:36:50 |
| S1 | Arrive | 6:02:28 | 6:06:41 | S11 | Arrive | 6:35:09 | 6:39:25 |
| | Departure | 6:03:12 | 6:07:27 | | Departure | 6:35:55 | 6:40:10 |
| S2 | Arrive | 6:05:42 | 6:09:57 | S12 | Arrive | 6:38:28 | 6:42:40 |
| | Departure | 6:06:29 | 6:10:43 | | Departure | 6:39:11 | 6:43:25 |
| S3 | Arrive | 6:08:52 | 6:13:15 | S13 | Arrive | 6:41:40 | 6:45:56 |
| | Departure | 6:09:38 | 6:13:59 | | Departure | 6:42:24 | 6:46:44 |
| S4 | Arrive | 6:12:11 | 6:16:26 | S14 | Arrive | 6:44:54 | 6:49:12 |
| | Departure | 6:12:59 | 6:17:12 | | Departure | 6:45:39 | 6:49:58 |
| S5 | Arrive | 6:15:29 | 6:19:45 | S15 | Arrive | 6:48:12 | 6:52:23 |
| | Departure | 6:16:17 | 6:20:31 | | Departure | 6:48:57 | 6:53:05 |
| S6 | Arrive | 6:18:46 | 6:23:08 | S16 | Arrive | 6:51:28 | 6:55:36 |
| | Departure | 6:19:32 | 6:23:54 | | Departure | 6:52:13 | 6:56:23 |
| S7 | Arrive | 6:22:06 | 6:26:24 | S17 | Arrive | 6:54:39 | 6:58:52 |
| | Departure | 6:22:52 | 6:27:10 | | Departure | 6:55:26 | 6:59:38 |
| S8 | Arrive | 6:25:22 | 6:29:39 | S18 | Arrive | 6:57:55 | 7:02:04 |
| | Departure | 6:26:08 | 6:30:23 | | Departure | 6:58:43 | 7:02:53 |
| S9 | Arrive | 6:28:37 | 6:32:54 | S19 | Arrive | 7:01:09 | 7:05:23 |
| | Departure | 6:29:20 | 6:33:40 | | Departure | – | – |

Due to the large scale of the resulting output timetable (20 stations, 200 trains), only the train schedules of the first two trains in the upward direction are displayed, which can be seen in Table 2.

## 5 Conclusion

In this paper, we propose an energy-saving optimization model for train timetable based on regenerative braking, adjusting the dwell time and inter-station travel time under the condition of ensuring reasonable arrival and departure intervals. In a virtual example, we discuss three different scenarios and the results show that the proposed method can effectively optimize the energy efficiency of the train timetable.

It should be noted that only the traction and braking power curves are correlated with the train full load rate in this paper, while the effects of other factors are ignored. In future research, the calculation methods of these curves should be refined to achieve more accurate energy-saving optimization.

**Acknowledgement.** This work was supported by the National Natural Science Foundation of China (61603140).

## References

1. Su, S., Tang, T., Roberts, C.: A cooperative train control model for energy saving. IEEE Trans. Intell. Transp. Syst. **16**(2), 622–631 (2015). https://doi.org/10.1109/TITS.2014.2334061
2. Wong, K. K., Ho T. K.: Dwell-time and run-time control for DC mass rapid transit railways. IET Electr. Power Appl. **1**(6), 956–966 (2007)
3. Pena-Alcaraz, M., Fernandez, A., Cucala, A.P., et al.: Optimal underground timetable design based on power flow for maximizing the use of regenerative-braking energy. Proc. Inst. Mech. Eng. Part F J. Rail & Rapid Transit **226**(4), 397–408 (2011)
4. Qiyuan, P., Wenxin, L., Yiru, W., et al.: Research on underground train driving strategy based on regenerative braking. J. Railw. **39**(03), 7–13 (2017). (in Chinese)
5. Yang, X., Li, X., Gao, Z., et al.: A cooperative scheduling model for timetable optimization in subway systems. IEEE Trans. Intell. Transp. Syst. **14**(1), 438–447 (2013)
6. Can, L., Renzhi, W., Dewei, L., et al.: Research on energy-saving optimization of underground schedules using regenerative braking. Control Theory Appl. **36**(07), 1024–1035 (2019). (in Chinese)
7. Hongjie, L., Tao, T., Xisheng, X., et al.: An underground train schedule optimization model for regenerative braking energy utilization. J. Beijing Jiaotong Univ. **43**(01), 71–78 (2019). (in Chinese)
8. Xin, G., Yuzhao, Z., Zhipeng, H.: Optimization of high-speed railway train operating diagram based on regenerative braking energy utilization. J. Shenzhen Univ. (Sci. Technol. Ed.) 1–7 (2023). (in Chinese)
9. Xinchen, R., Shaokuan, C., Lei, C., et al.: An energy-efficient schedule optimization method for underground trains considering spatial distribution of passenger flow. Transp. Syst. Eng. Inf. **20**(3), 103–110 (2020). https://doi.org/10.16097/j.cnki.1009-6744.2020.03.016. (in Chinese)
10. Zheng, Y., Li, Y., Li, Y., Jin, W.: Energy-saving optimization of underground train operating diagram under regenerative braking conditions. J. South China Univ. Technol. (Nat. Sci. Ed.) **49**(07), 1–7(2021). (in Chinese)
11. Yang, X., Ning, B., Li, X., et al.: A two-objective timetable optimization model in subway systems. IEEE Trans. Intell. Transp. Syst. **15**(5), 1913–1921 (2014)
12. Bing, B.U., Changmin, T.E.N.G., Erchao, C.H.E.N., et al.: Research on multi-vehicle collaborative energy-saving control method for urban rail transportation. J. Railw. **40**(08), 43–51 (2018). (in Chinese)
13. Wei, R., Du, P., Yang, Y., Hu, L.: Research on regenerative braking energy utilization analysis and schedule optimization method for metro trains. J. Railw. **42**(08), 1–9 (2020). (in Chinese)
14. Feng, Y., Shaokuan, C., Xinchen, R., et al.: Research on the optimization method of energy-saving operation of urban rail transit trains considering regenerative braking energy utilization. J. Railw. **40**(02), 15–22 (2018). (in Chinese)

# Dynamics Simulation Model of Pantograph and Conductor Rail in Electrified Railway and Its Application

Jinfa Guan, Jiqin Wu(✉), Jiawei Zhang, and Zhonglin Liang

Department of Electrical Engineering, Southwest Jiaotong University, Chengdu City, Sichuan Province, China
jqwu@swjtu.edu.cn

**Abstract.** Overhead conductor rail which has become the main form of the overhead catenary, is widely used in low headroom tunnels. In order to understand the vibration behavior of pantograph and conductor rail, it is necessary to carry out the study of the interaction between pantograph and conductor rail. In this paper, we will present a dynamic simulation model of the contact force between a pantograph and a rigid catenary. The contact point change matrix with time and the calculation procedure of contact force will be given. Finally, applying the dynamic simulation model, the dynamic response and the maximum limit speed of conductor rail have been analyzed.

**Keywords:** Pantograph · Overhead Conductor Rail · Dynamic · Simulation

## 1 Introduction

Vera et al. [1] developed a new type of conductor track to improve the guiding properties of existing rails, which can achieve higher operating speeds without shortening the distance between the brackets. Montserrat Simarro et al. compared the key parametric factors in pantographs and rigid catenaries [2]. They proposed two models that can be used to calculate the wear rate of rigid contact lines and pantograph strips in subway systems [3].

The unit centralized mass module can be applied to simple chain suspension to provide new ideas for studying the dynamics of pantograph, Ambrosio et al. [4]. Through the coupling simulation platform, the parameters of the pantograph are deeply analyzed, and the influence of different pantograph parameters on the pantograph catenary contact force at the same speed is compared, and then the parameters of the pantograph are optimized to obtain the best parameters [5]. They studied the interaction between OCR and pantograph. The contact forces between 11 types of pantograph and two OCR sections are analyzed [6]. The evaluation of the dynamic performance of the pantograph model is more effective, which is more helpful for the configuration of OCRs models and the evaluation of various pantograph models [7]. Pantograph and conductor rail dynamic simulation can also refer to pantograph and cate-nary. The catenary can be used


Y. Qin et al. (Eds.): EITRT 2023, LNEE 1137, pp. 358–367, 2024.
https://doi.org/10.1007/978-981-99-9311-6_38

to establish the generalized coordinate kinetic equations by using modal superposition method [8, 9]. a pantograph–catenary system is one of the most critical chains that limit the running speed of a train [10]. Gregori et al. describe an attempt to find optimised catenary configurations in terms of current collection quality for a given train speed [11]. Su et al. combined with a variety of algorithms to optimize the parameters of the pantograph system and compare and evaluate the effects [12]. Ma [5, 13] analyzed pantograph dynamics using finite elements and gave a coupled dynamic model of pantograph and rigid catenary.

In order to obtain the dynamic characteristic parameters of pantograph at higher speed, it is necessary to study the dynamic behavior of pantograph and catenary, and a calculation method of orbital model is proposed and verified by case study.

## 2 Related Work

### 2.1 The Finite Element Model of Pantograph and Conductor Tracks

According to the contact mode and mechanism of action, the interaction between pantograph and guide rail is built, and the model diagram is as shown in Fig. 1:

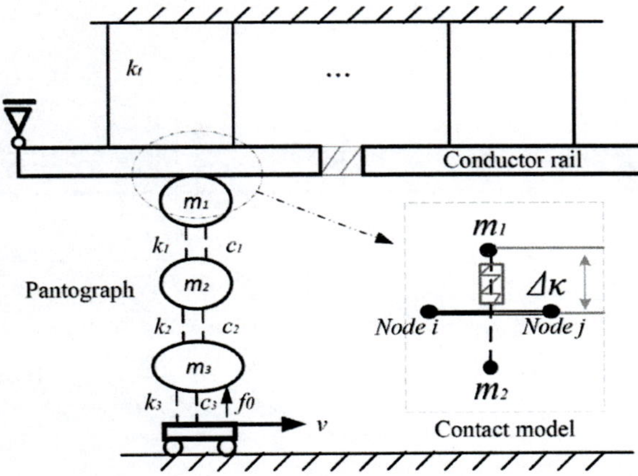


**Fig. 1.** The principle of interaction between pantograph and conductor rail

The kinetic equation of the three mass DoFs model for pantogarph is as follows:

$$m_1\ddot{y}_1(t) + c_1(\dot{y}_1(t) - \dot{y}_2(t)) + k_1(y_1(t) - y_2(t)) = -f_c(t) \tag{1}$$

$$\begin{aligned} &m_2\ddot{y}_2(t) + c_1(\dot{y}_2(t) - \dot{y}_1(t) + c_2(\dot{y}_2(t) - \dot{y}_3(t)) \\ &+k_1((\dot{y}_2(t) - \dot{y}_1(t)) + k_2(\dot{y}_2(t) - \dot{y}_3(t)) = 0 \end{aligned} \tag{2}$$

$$m_3\ddot{y}_3(t) + c_2(\dot{y}_3(t) - \dot{y}_2(t)) + c_3y_3(t) + k_2((\dot{y}_3(t) - \dot{y}_2(t)) + k_3y_3(t) = 0 \tag{3}$$

y1, y2, and y3 represent the degrees of freedom in the y direction of the three mass blocks of the panhead, upper frame, and lower arm respectively. The contact force is represented by the symbol fc.

According to the structural characteristics of the assembled beam elements, the corresponding dynamic simulation model equations are established as shown in Eq. 4:

$$M\ddot{q} + C\dot{q} + Kq = F + P_g \tag{4}$$

F is the set of contact and supporting reaction forces. Pg stands for the node's own gravity and is fixed.

The pantograph and the conductor rail interact through the contact force which is a variable associated with space and time. The coupling system needs to deal with two key issues: 1) calculating the value of the contact force; 2) modeling mathematical description of the moving contact point.

Since the conductor rail has been divided into a finite number of units, the contact model is a mass point to a beam element. The contact force is a non-node force, which needs to be converted into two equivalent nodes force. The contact force was assumed to moves to the beam element at coordinates (xf, 0), as shown in Fig. 2. The contact force would be equivalent to Eq. 5 and the calculation of the contact force in the whole coordinate system can be obtained by Eq. 6, according to one-dimensional Hermite interpolation.

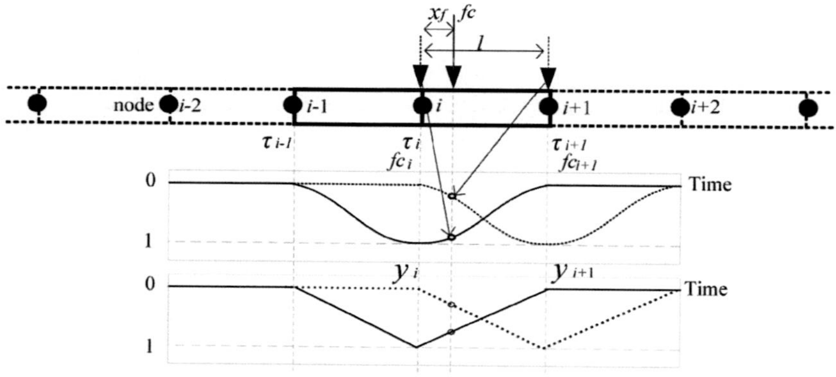


**Fig. 2.** Equivalent node force and node displacement for the contact force

$$F_c^e = f_c(t) \cdot \left[ 0 \frac{(l - x_f)^2 (l + 2x_f)}{l^3} 000 \frac{x_f (l - x_f)^2}{l^2} 0 \frac{x_f^2 (3l - 2x_f)}{l^3} 000 - \frac{x_f^2 (l - x_f)}{l^2} \right]^T \tag{5}$$

$$\overline{F}_c = f_c(t) \cdot \Gamma_{time,\, n\times m} \cdot T_{m\times 1},\ T_{m\times 1} = [0 \cdots 1 \cdots]^T \tag{6}$$

The contact force will be broken down into the component force of y and θz DOF at the two nodes of the element. Determined by the position of pantograph movement, the contact force acts on the beam element at different positions. At one moment, the contact force is only in contact with one beam element. The point of contact force changes with the integration time. $\tau i$ to $\tau i + 1$ is a time step, when the contact force appeared in element i–i + 1. The time steps includes $b\tau i - 1$ sub-steps. The contact force decreases gradually at one node, while it increases gradually at the next node. The sum of the component force is 1.

To achieve simulating the distribution of contact forces in the preceding segment and the distribution in the nodes, the cooperation between time and step size is realized in the form of a matrix of distribution integration coefficients, as shown in Fig. 3.

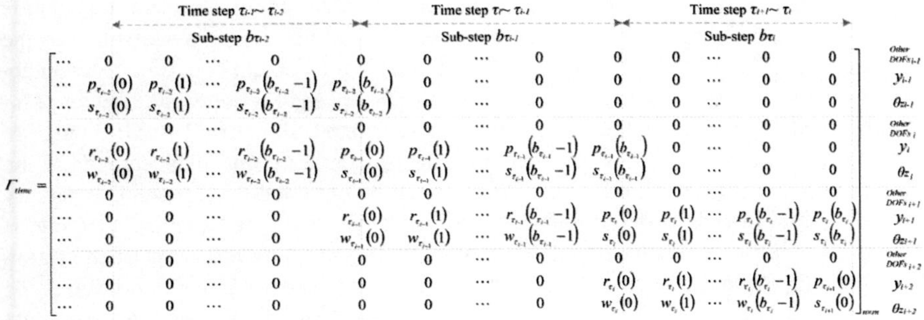

**Fig. 3.** Integral time matrix of the contact force

The Γtime matrix which is set in advance, is the n × m matrix associated with the running speed and the integration time. n is the number of DOFs. The row arrangement is consistent with the q matrix. m is the number of time steps which is includes many sub-steps in every time step. A column in the matrix reflects position of the contact force. T matrix can be used to locate the contact force and calculate the equivalent node force. Assuming that the system is running tμ minutes, then the tμ Δt row 1 is equal to 1 in the T matrix. And the rest rows is 0. The calculation of the contact force in the whole coordinate system can be obtained. The symbols in the Γtime matrix are calculated according to Eq. 7–11.

$$b_{\tau i-2} = \frac{\tau_{i-1} - \tau_{i-2}}{\Delta t}, i = 3, \ldots, t_{end} \tag{7}$$

$$p_{\tau i-2}(k) = \left(1 - \frac{k}{b_{\tau i-2}}\right)^2 \left(1 + 2\frac{k}{b_{\tau i-2}}\right), k = 0, 1, \ldots, b_{\tau i-2} \tag{8}$$

$$s_{\tau i-2}(k) = k\Delta st\left(1 - \frac{k}{b_{\tau i-2}}\right)^2 \tag{9}$$

$$w_{\tau i-2}(k) = \left(b_{\tau i-2} - k\right)\Delta t\left(\frac{k}{b_{\tau i-2}}\right)^2 \tag{10}$$

$$r_{\tau i-2}(k) = \left(\frac{k}{b_{\tau i-2}}\right)^2 \left(3 - 2\frac{k}{b_{\tau i-2}}\right) \tag{11}$$

The value of the contact force is calculated by the penalty function which is equal to the product of contact stiffness and penetration displacement, as shown in Eq. 12. Because the pantograph has been built in particle model, the contact force value is related to the y direction displacement of conductor rail. The contact stiffness depends on the material and the friction coefficient of the contact surface, while penetration displacement depends on the panhead displacement y1 and the conductor rail displacement at the contact point yc. As with the node assignment of the contact force, the contact point displacement of the conductor rail also requires node assignment. Because the beam is a line element, the displacement of the beam element can be linearly equivalent to the displacement of each node, as shown in Fig. 1. The displacement equation of a contact element is Eq. 13. As the same, a distribution coefficient matrix about the contact point displacement of the conductor rail should be constructed. Equation 14 will repalce to the Γtime matrix in Eq. 15. The calculation of the contact point displacement of the conductor rail in the whole coordinate system can be obtained by Eq. 16. The penalty function method needs to determine that whether it has an penetration displacement. If there is no penetration displacement, the contact stiffness is zero. If there is a penetration displacement, the contact stiffness is a positive constant. The judgment condition is the sign of the displacement difference Δκ between the panhead mass point and the contact point displacement of the conductor rail, as shown in Eq. 17.

$$f_c(t) = k_c \Delta k(t),\ \Delta k(t) = y_1(t) - \overline{y}_c(t) \tag{12}$$

$$y_c^e = \overline{q}_{i,j}^T \cdot \left[0\ \frac{x_f}{l} 00000 \frac{l - x_f}{l} 0000\right] \tag{13}$$

$$h_{\tau i-2}(k) = \frac{b_{\tau i-2} - k}{b_{\tau i-2}} \tag{14}$$

$$\Theta_{time,n\times m} = \Gamma_{time,n\times m|p=h,r=1-h,s=0,w=0} \tag{15}$$

$$\overline{y}_c = \overline{q}^T \cdot \Theta_{time,n\times m} \cdot T_{m\times 1} \tag{16}$$

$$k_c \begin{cases} a\ \Delta \mathrm{k} > 0 \\ 0, \Delta \mathrm{k} \le 0 \end{cases} \tag{17}$$

## 2.2 Numerical Integration Process

According to the principle of Newmark method, the numerical solution process of the dynamic equations at each time step is given, as shown in Fig. 4. There are two kinds of judgment. One is the contact judgment. The other is the convergence judgment. The contact judgment is calculated by Eq. 17. The convergence judgment is related to the difference of the displacement and speed at the current moment and the previous moment. If the difference is less than the convergence condition, the next time step will be solved. If it is not satisfied, the new value is recalculated again by using the result at the current moment. When the convergence judgment exceeds the limit value Nu times, the integration process will be run out.

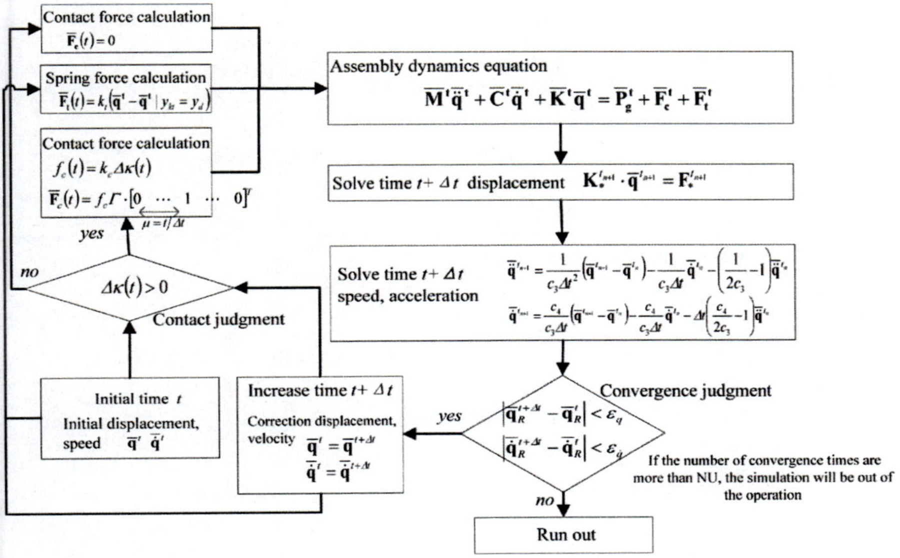


**Fig. 4.** The numerical integration process at each time step

## 3 Application of the Simulation Model

According to the standard of EN 50119 and EN 50367, the evaluation criteria for the dynamic performance of the pantograph and catenary of the rigid contact network are:

1) Maximum contact force Fmax: Fmax $\leq$ 300N (v $\leq$ 200km/h), Fmax $\leq$ 350N (v $>$ 200km/h);
2) Minimum contact force Fmin: Fmin $>$ 0;
3) Maximum standard deviation of contact force σ:0 $<$ 0.3Fm – σ, Fm is the average of contact force.

The Parameters of DSA380 pantograph and conductor rail model are as follows (Tables 1 and 2).

The contact force changes obtained by the simulation are shown in Fig. 5 below, the horizontal axis and the vertical axis correspond to the track distance and contact force respectively, and the contact force changes based on different motion conditions are obviously different (the contact force is obtained according to EN 50318).

**Table 1.** The parameters of the pantograph model

| Parameters | Value | Parameters | Value | Parameters | Value |
|---|---|---|---|---|---|
| Panhead mass(kg) | 7.12 | Panhead damp (Ns/m) | 0 | Panhead stiffness (N/m) | 9230 |
| Up-frame mass(kg) | 6 | Up-frame damp (Ns/m) | 0 | Up-frame stiffness (N/m) | 14100 |
| Down-frame mass(kg) | 5.8 | Down-frame damp(Ns/m) | 70 | Down-frame stiffness (N/m) | 0 |

**Table 2.** The parameters of the conductor rail model

| Parameters | Value | Parameters | Value |
|---|---|---|---|
| Conductor bar density(kg/m$^3$) | 2700 | Span(m) | 8 |
| Conductor bar cross section area(mm$^2$) | 2166 | Pull value(m) | 0.2 |
| Conductor bar moment of inertia(m$^4$) | 3.04e−6 | Positioning point stiffness(kN/m) | 68.6 |
| Conductor bar elastic modulus(N/m$^2$) | 7.2e10 | Positioning point error(m) | 0 or 0.002 |

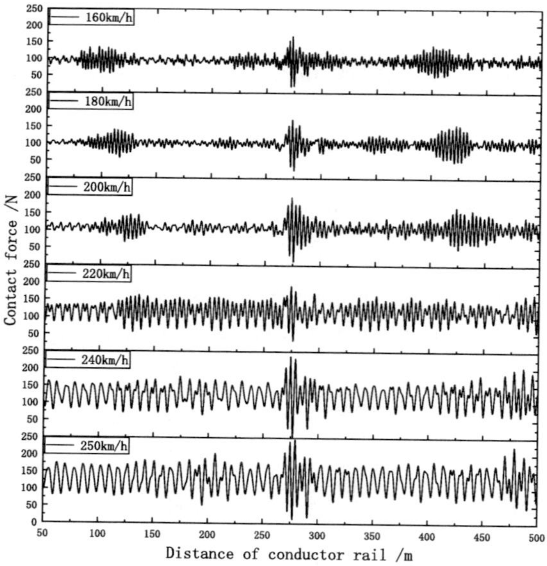


**Fig. 5.** Contact force curves by simulation

The contact force data between the pantograph and the contact rail are shown in Table 3. The data shows that all contact force indicators meet the requirements of the standard. With the increase of the pantograph operating speed, the standard deviation of contact force and the fluctuation of contact force between rails show an increasing trend.

**Table 3.** Statistical results of contact force at the speed of 160 km/h ~ 250 km/h

| Speed (km/h) | Average contact force(Fm/N) | Standard deviation of contact force( σ/ N) | Minimum contact force(N) | Maximum contact force(N) | 0.3Fm-σ(N) |
|---|---|---|---|---|---|
| 160 | 94.41 | 14.50 | 18.99 | 167.92 | 13.82 |
| 180 | 100.98 | 14.51 | 22.86 | 171.62 | 15.78 |
| 200 | 108.32 | 17.86 | 11.49 | 197.39 | 14.64 |
| 220 | 116.43 | 23.61 | 33.25 | 190.89 | 11.32 |
| 240 | 125.29 | 29.23 | 2.71 | 237.17 | 8.35 |
| 250 | 129.97 | 34.10 | 12.96 | 244.99 | 4.90 |

The simulation data in Table 3 show that with the increase of the relative motion speed of pantograph and contact rail, the average contact force and maximum contact force change trend is increasing, so the relative motion speed of 250km/h is selected for analysis (positioning point error is 2mm), the dynamic model simulation waveform is shown in Fig. 6, and the contact force statistics are shown in Table 4. When the height of the contact line of the anchor point has an error of 2mm, the contact force does not meet the standard requirements. Among them, the pantograph catenary model has good adaptability in practical applications.

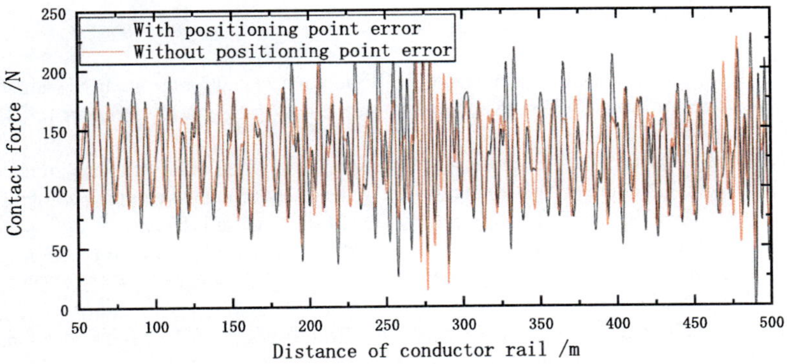


**Fig. 6.** Contact force curve under different positioning point error

**Table 4.** Statistical results of contact force under different positioning point error

| Positioning point error (m) | Average contact force(Fm/N) | Standard deviation of contact force( σ/ N) | Minimum contact force(N) | Maximum contact force(N) | 0.3Fm-σ(N) |
|---|---|---|---|---|---|
| 0.002 | 129.87 | 39.57 | 0.00 | 235.40 | −0.61 |
| 0 | 129.97 | 34.10 | 12.96 | 244.99 | 4.90 |

## 4 Conclusions

At present, the research on pantograph in the field of rigid catenary mainly focuses on dynamic modeling, and this paper proposes a pantograph modeling method (pantograph and rigid catenary) and gives specific cases to verify the correctness of the method. Dynamic simulation models, dynamic response, and maximum ultimate velocity analysis are applied. The conclusions are as follows:

1) The conductor rail system can theoretically meet the dynamic and reliable flow of pantograph-catenary in the speed range of 250km/h and below.
2) When considering the construction error of 2mm error in the height of the contact line at the positioning point, the pantograph-catenary dynamic current receiving requirement of 250km/h cannot be met, and the system and component parameters need to be further optimized.

**Acknowledgment.** This work is supported by National Natural Science Foundation of Sichuan Province in China 2022NSFSC0528, 2022NSFSC1883, 2022NSFSC1895 and Fundamental Science and Technology Innovation Project of Central Universities-2682020CX22-Friction and wear performance and wear prediction method of DC pantograph-catenary contact pair.

## References

1. Vera, C., Suarez, B., Paulin, J., Rodríguez, P.: Simulation model for the study of overhead rail current collector systems dynamics, focused on the design of a new conductor rail[J]. Veh. Syst. Dyn. **44**(8) (2006)
2. Simarro, M., Castillo, J.J., Cabrera, J.A., Postigo, S.: Evaluation of the influence of the speed, preload and span length on the contact forces in the interaction between the pantograph and the overhead conductor rail. Eng. Struct. **243**. ISSN 0141–0296 (2021)
3. Wei, X.K., Meng, H.F., He, J.H., Jia, L.M., Li, Z.G.: Wear analysis and prediction of rigid catenary contact wire and pantograph strip for railway system. Wear 442–443, 203118. ISSN 0043–1648 (2020)
4. Ambrosio, J., Pombo, J., Rauter, F., et al.: In: Bottasso, C.L. (ed.) Multibody dynamics. Springer, Dordrecht, the Netherlands (2008)
5. Guolei, M.: Research on parameter optimization of pantograph based on multibody system dynamics. J. Dalian Jiaotong University. **31**(4), 33–37 (2010)

6. Simarro, M., Postigo, S., Prado-Novoa, M., et al.: Analysis of contact forces between the pantograph and the overhead conductor rail using a validated finite element model. Eng. Struct. **225**(111265), 1–15 (2020)
7. Armbruster, K.: Modeling and dynamics of pantograph/catenary systems for high speed trains [dissertation]. Massachusetts Institute of Technology (1983)
8. Guiming, M.: Vertical coupling dynamics of pantograph and catenary [dissertation]. Southwest Jiao-tong University (2001)
9. Weihua, Z.: Research on dynamic characteristics of pantograph-catenary system [dissertation]. Southwest Jiaotong University (1989)
10. Harell, P., Drugge, L., Reijm, M.: Study of critical sections in catenary systems during multiple pantograph operation. Proc Inst Mech Eng F, J Rail Rapid Transit. **219**(4), 203–211 (2005)
11. Gregori, S., Tur, M., Nadal, E., et al.: An approach to geometric optimisation of railway catenaries. Veh. Syst. Dyn. **56**(8), 1162–1186 (2018)
12. Su, K., Zhang, J., Zhang, J., et al.: Optimisation of current collection quality of highspeed pantograph-catenary system using the combination of artificial neural network and genetic algorithm. Veh Syst Dyn. 1–26 (2022)
13. Guolei, M.: Research on pantograph system [dissertation]. Southwest Jiaotong University (2009)

# Multi-level Simulation Modeling Technology for Urban Rail Traction Power Supply System

Caiyi Chen[1], Gang Zhang[1,2(✉)], Jingjian Yang[1], and Zhigang Liu[1,2]

[1] Beijing Jiaotong University, Beijing, China
{22121432,21117023}@bjtu.edu.cnr, {gzhang,zhgliu}@bjtu.edu.cn
[2] Rail Transit Electrical Engineering Technology Research Center, Beijing, China

**Abstract.** The structure of urban rail traction (URT) power supply system is complex and includes multiple levels. Conventional modeling methods focus on a single level, neglecting the correlation between multiple levels, which cannot meet the requirements of accurate, efficient, and comprehensive analysis of the entire system. To solve this problem, this article analyzes the principle of multi-level interaction in the urban rail traction power supply system, formulates interaction strategies, and builds a URT simulation model based on multi-level modeling methods. Finally, simulations were conducted using examples to obtain simulation results that combine the macro scale of the system level with the micro scale of the station level. The results show this multi-level model has high accuracy and fast simulation speed. It can provide effective reference for the comprehensive design and operation of urban rail traction power supply systems.

**Keywords:** Urban rail traction power supply system · Multi-level modeling · Multi-time scale

## 1 Introduction

The urban rail traction power supply system provides power support for locomotives and various auxiliary electrical equipment, and is the foundation for the normal operation of urban rail transit. Due to safety constraints, modeling and simulation are currently an important means of studying the operational status of power supply systems. However, conventional modeling methods mostly only consider system level or a certain level of system level, ignoring the interaction between levels, making it difficult to achieve comprehensive analysis of the power supply system through simulation. Therefore, establishing a multi-level model of urban rail traction power supply system to achieve comprehensive observation of the power supply system is of great significance for the research and analysis of urban rail transit.

In terms of modeling methods, many researchers have built models that mostly focus on a certain aspect of system level modeling or station level modeling. When establishing system level models in reference [1–4], the main objective is to calculate the power flow, and simplify the equipment that has little impact, resulting in rough simulation results; When establishing station level models in reference [5–8], most of them model and


Y. Qin et al. (Eds.): EITRT 2023, LNEE 1137, pp. 368–376, 2024.
https://doi.org/10.1007/978-981-99-9311-6_39

simulate the key equipment inside the station, lacking the impact of the system on the equipment during actual operation. Although the current single level modeling of urban rail traction power supply systems is relatively mature, the mutual influence between the two levels has not been considered.

In summary, this article uses MATLAB and Simulink platforms to build a multi-level simulation model for urban rail traction power supply system, which can achieve comprehensive simulation of system states at multiple time scales; Design a visual interface for the urban rail traction power supply system using the MATLAB/APP Designer platform to increase the intuitiveness of the model. Finally, the feasibility of multi-level visualization modeling was verified through simulation, providing a platform for the research of urban rail traction power supply systems.

## 2 Simulation Design Framework

The goal of establishing a multi-level model for urban rail traction power supply system is to simulate the actual operating status of the system from different levels and time scales. Based on its characteristics, the overall framework of the simulation is constructed as shown in Fig. 1.

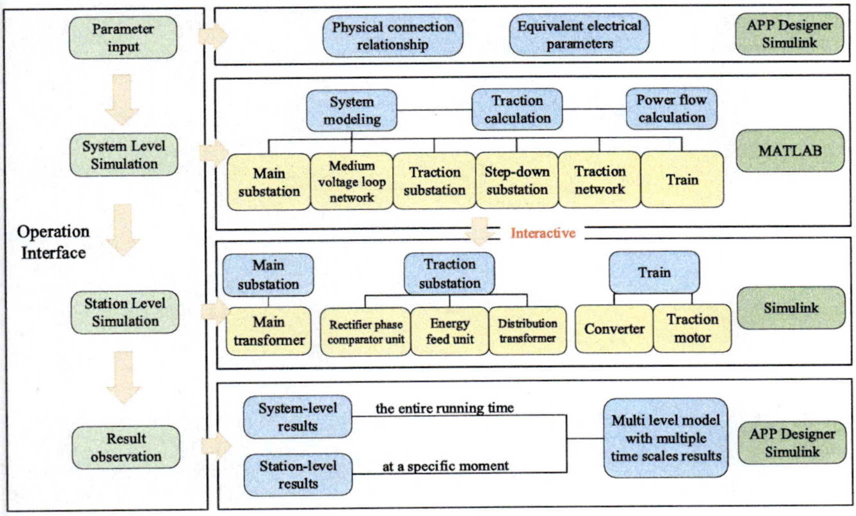


**Fig. 1.** Overall framework of simulation

The simulation is divided into two parts: a modeling and simulation module and an operation interface. The modeling and simulation module is implemented using MATLAB and Simulink, mainly used for building models and modifying parameters; The operation interface is implemented using MATLAB/APP Designer tools, which can quickly call multi-level models for simulation and observation of results.

# 3 Basic Principles of Modeling

## 3.1 System Level Modeling

The system level model includes the main substation, medium voltage loop network, traction substation, step-down substation, train, and traction network. The specific equivalent model is shown in Fig. 2. The main substation converts external power into medium voltage through the main transformer, so the core of the main substation is the transformer. Therefore, the main substation uses the commonly used T-shaped circuit of the transformer for equivalence; Medium voltage AC cables are represented by multiple π type circuits connected in series; The traction substation is divided into two parts: the AC side and the DC side, which are equivalent to two independent power sources; The step-down substation only includes the AC side, providing electricity for auxiliary equipment inside the substation; The traction network is composed of the contact network and the return rail. DC traction network uses equivalent resistance proportional to its length. The train is equivalent to a power source, and the power source value is the power affected by voltage.

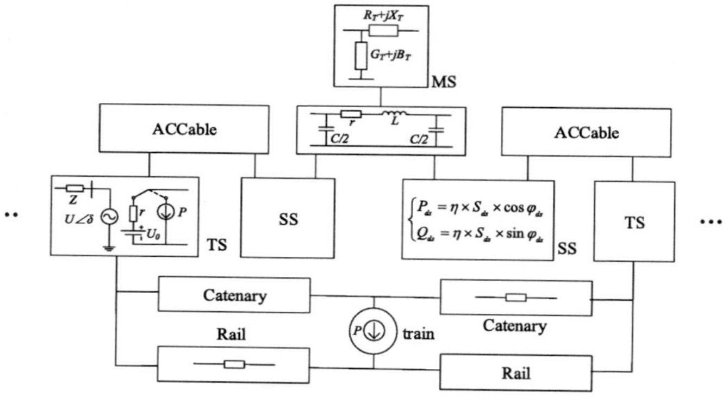


**Fig. 2.** System level model

## 3.2 Station Level Modeling

The station level mainly models the main substation, traction substation, and train. The main equipment of the main substation is the main transformer, which adopts Simulink's built-in model. This module is a linear non ideal three-phase transformer that includes linear winding leakage and linear iron core magnetization effect; The equipment of the traction substation is divided into three parts. The first part is the rectifier unit, which provides energy for train traction. The second part is the energy feedback unit, which is used to transmit the energy feedback of the train during the braking stage. The third part is the distribution transformer, which provides energy for auxiliary equipment in the station; The train receives electrical energy from the contact network, converts direct current into alternating current through an internal inverter, and drives the motor to operate. The specific model is shown in Fig. 3.

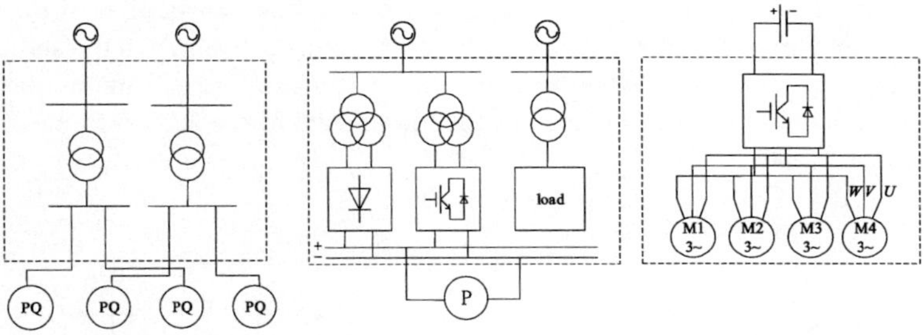


(a)Main substation (b)Traction Substation (c)Train

**Fig. 3.** Station level model

# 4 Interaction of Multi-level Models

## 4.1 Principle of Multi-level Interaction

For a single level, there are differences in state observation requirements and observation accuracy at different levels, but for the system as a whole, there are inevitable connections between different levels [9].

The modeling methods at different levels vary, and the model accuracy also varies. The station level model is a deeper and more accurate description of the system level model [10]. In system level models, various specific devices are usually simplified and equivalent based on their external characteristics, while in station level models, their operating mechanisms are finely described. For example, for train modeling, the system level only equates it to a power source, while in the station level model, the train model is driven by an inverter to drive the traction motor. So when observing the power flow operation of the urban rail traction power supply system, choose a system level model; If you need to observe the specific operating status of each device, you can choose a station level model.

Models at different levels have the characteristics of multiple time scales. The system level model simplifies each station module equivalently, and the results obtained from small-scale simulation and large-scale simulation are not significantly different. Moreover, it increases simulation time and reduces simulation efficiency; The station level model is modeled based on the actual equipment in the station, describing the operating mechanism of the equipment, which is more refined compared to the system level model. If large time scales are used for simulation, the characteristics of the actual operation process of the equipment will be ignored. Therefore, models at different levels use different time scales for simulation.

For results at different levels, multi-scale states have inherent correlations: macro scale is a rough observation result at micro scale. Therefore, when obtaining a large-scale system level state at a certain time, combined with station level models, the micro state at that time can be directly calculated without the need to calculate from the starting time, which reduces simulation time compared to directly using micro model calculations. Taking the traction substation as an example, the relationship between the system level model and the station level model is shown in Fig. 4. The three points A, B, and C in the system level model are the results of the power flow calculation of the system level model, and are directly connected to the three points a, b, and c in the station level model in physics. Therefore, when running station level simulation, the output results of the system level model can be used as the initial state of the station level model for simulation.

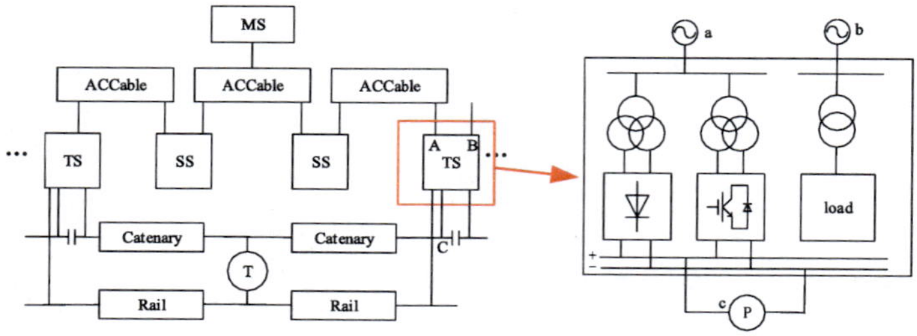


**Fig. 4.** Relationship between system level and station level

In summary, interactions at different levels are a connection between simplified models at the system level and refined models at the station level in the model; In terms of time, the connection between macro results and corresponding micro results at two time points. Therefore, the interaction between multi-level models on spatial and temporal scales is the core means to achieve effective integration between multi-level models.

## 4.2 Methods of Multi-level Interaction

Generally speaking, the key to multi-level interaction is to determine the physical interfaces of interactions between different levels. The internal simulation state of the next level is usually set as the initial value of the calculation state of the previous level. Simulation is achieved by retrieving and collaborating with corresponding data, models, and algorithms. As mentioned earlier, the interface between the system level model and the station level model is the physical connection between the two levels.

Within the same time period, the computational complexity and simulation time of microscale models are much greater than those of macroscopic scale models, making it difficult for different scale models to effectively execute and interact simultaneously through simple parallel or serial methods. Therefore, the interaction method of this multi-scale model is to select a specific time step as needed after the system level model simulation is completed, and set the corresponding time step for the station level model simulation solution. The system level simulation runs continuously, while the station level simulation runs intermittently. For system level models, due to the long simulation time step and simple model, the calculation time required for solving continuous moments can meet the actual computational efficiency. Therefore, the actual duration of simulation calculation can be kept consistent with the simulation time scale. For the station level model, the system level state value is used as the initial state of the station level, and the micro model intermittent time solution method is adopted. The simulation step size of the station level model is set according to actual needs, achieving the purpose of observing the internal equipment operation status of the station level model.

### 4.3 Simulation Verification

Due to the focus of modeling the urban rail traction power supply system in this article being on the interaction between the system level model and the station level model, simulation verification is conducted on the interaction results of the multi-layer model at different time scales. Taking Hohhot Metro Line 1 as an example, the system level calculation time step is set to seconds, and the station level calculation time step is set to 10^(−5). In order to comprehensively observe the accuracy of multi-level model interaction under different operating conditions of train operation, two typical states of train traction and energy feedback were simulated, and the third traction substation on the line was randomly selected as the station level model for simulation verification. The rated value of the voltage bus is 1.65kV; When the DC bus voltage of the traction network reaches 1700V, the energy feedback state is activated. Take 10 s from the running results to obtain two-level running results as shown in Fig. 5.

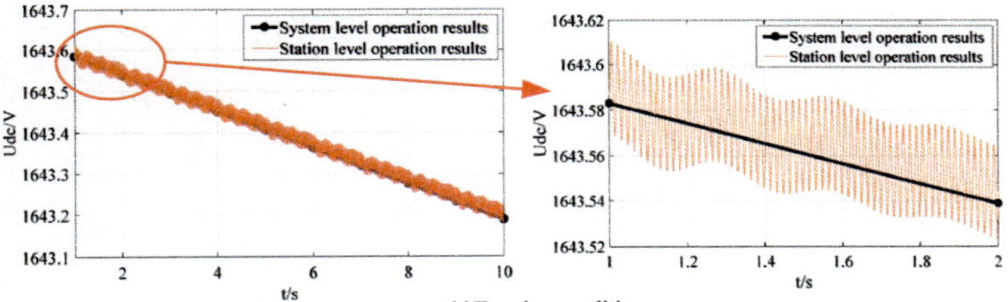


(a)Traction condition

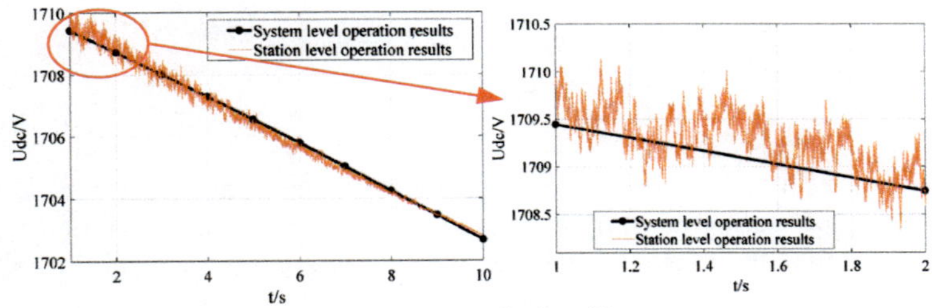


(b)Energy feedback condition

**Fig. 5.** Simulation results of multi-level models

In Fig. 5, it can be seen that under two states, the DC side voltage at the system level gradually decreases with time, which is the result of voltage droop control; The trend of operation results at the station level and system level is consistent; Compared to the data at the system level at a large time scale, the station level operation results show microscopic information on the changes in DC side voltage. In the state of energy feedback, the fluctuation range of the station level result voltage will become smaller and smaller over time, because of the dual closed-loop control strategy of energy feedback.

From the simulation results of the multi-level model in two different states, it can be seen that introducing multi-scale into the multi-level modeling of urban rail traction power supply system makes conventional modeling methods associated with physical models and time scales, ensuring the comprehensiveness and accuracy of the model while also having a faster simulation speed.

## 5 Interface Design

In order to simplify the process of building the model and make the observation results more intuitive, the corresponding model was encapsulated and an operation interface was designed.

### 5.1 Component Library

Due to the use of MATLAB language for programming the modeling of various station modules at the system level, which is not intuitive and difficult to modify parameters, the system level station modules are encapsulated in MATLAB/Simulink to establish variable connections with MATLAB programs. Users can set the parameters of the encapsulated internal equivalent circuit; At the same time, set physical interfaces for the encapsulated modules, as shown in Fig. 6, to set the physical connection relationships of each module in the system.

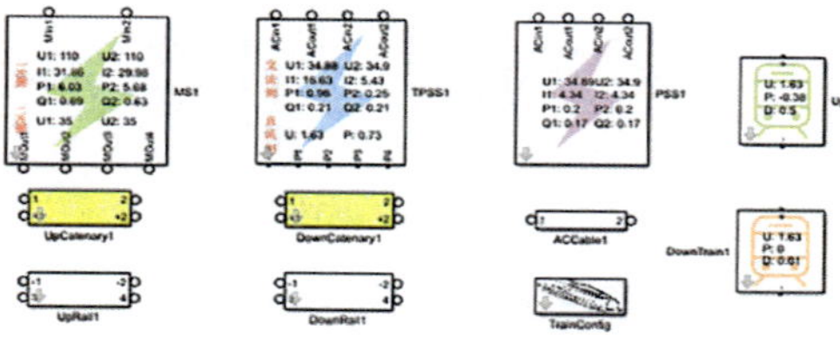


**Fig. 6.** Packaged modules

The station level model is built using the built-in equipment of the Simulink platform. The equipment of the station level model and the modules encapsulated in the system level are stored together in the newly created model library for easy calling.

### 5.2 Operation Interface Design

In order to make the operation process more intuitive, this article designs a visual operation interface based on the MATLAB/APP Designer platform, achieving functions such as calling modules and observing results. The system level operation interface mainly has two main functions: first, building a model; The second is to observe the system level results of operation, as shown in Fig. 7.

The interface operation interface at the station level has two functions, one for opening the station level model and the other for running the station level, as shown in Fig. 8. The operating results need to be observed in the Scope module of the Simulink model.

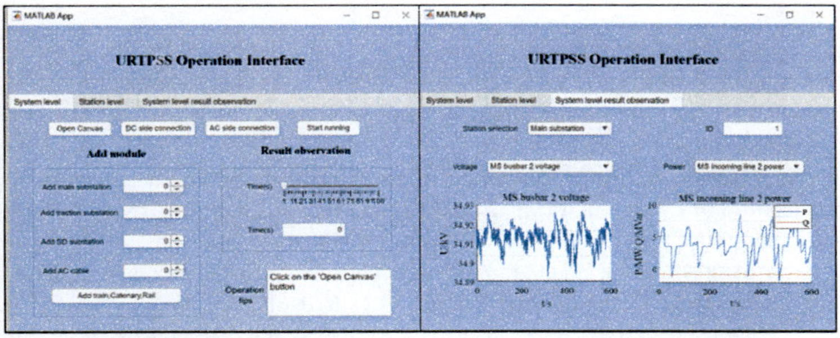


(a) Operation interface (b) Result observation interface

**Fig. 7.** System level related operation interface

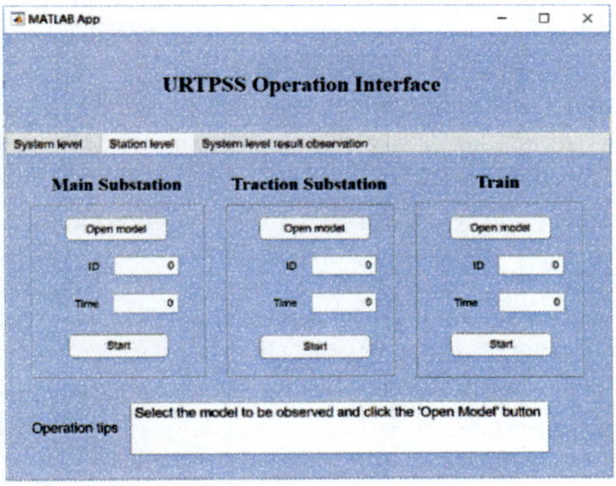


**Fig. 8.** Station level operation interface

# 6 Conclusion

In response to the problem of incomplete and inflexible simulation of conventional urban rail traction power supply system models, this paper establishes a multi-level model of the power supply system. Based on the proposed multi-level interaction principle, it ensures comprehensive and accurate simulation results while improving simulation efficiency; In order to make it more intuitive, relevant operation interfaces have been designed. The specific conclusions are as follows. Firstly, the urban rail traction power supply system consists of two levels, with different levels of modeling accuracy, and the system level model is more simplified compared to the station level model; At the same time, different levels have different time scales, and large time scales are rough values for small time scales. Secondly, this article proposes an interactive method between multi-level models, which consistently solves system level models. The station level model uses system level state values as the initial state of the station level for intermittent time solving, achieving the merging of different level models in time and space. This method greatly improves simulation efficiency.

## References

1. Liu, W., Zhang, J., Wang, H., et al.: Modified AC/DC unified power flow and energy-saving evaluation for Urban rail power supply system with energy feedback systems. IEEE Trans. Veh. Technol. **70**(10), 9898–9909 (2021)
2. Zhang, G., Hao, F., Wang, Y., et al.: Research on flexible traction power supply system and optimal control of Urban rail transit. Trans. China Electrotech. Soc. **37**(S1), 153–162 (2022). https://doi.org/10.19595/j.cnki.1000-6753.tces.L90518. (in Chinese)
3. Wang, Y., Zhang, G., Tian, Z., et al.: An online thermal deicing method for urban rail transit catenary. IEEE Trans. Transp. Electrification. **7**(2), 870–882 (2021)
4. Wang, Y., Zhang, G., Qiu, R., et al.: Distribution correction model of urban rail return system considering rail skin effect. IEEE Trans. Transp. Electrification. **7**(2), 883–891 (2021)
5. Guo, Z.: Simulation modeling of urban rail transit AC/DC hybrid power supply system based on PSCAD simulation software. Urban Mass Transit **24**(S1), 12–15+20 (2021) (in Chinese)
6. Zhou, W., Rong, S., et al.: Harmonic suppression method for urban rail traction power supply system based on braking energy feedback device. Adv. Technol. Electr. Eng. Energy Locomot. Electr. Drive. **41**(10), 20–32 (2022). (in Chinese)
7. Shi, X., Lin, F., Wang, C., et al.: Dynamic simulation method for urban rail traction power supply system based on quantized state time discretization. Locomot. Electr. Drive. **2022**(05), 151–158 (2022). https://doi.org/10.13890/j.issn.1000-128X.2022.05.022. (in Chinese)
8. Liu, W., Zhou, L., Pan, Z., et al.: Dynamic diffusion model of stray current in DC traction power supply system. IEEE Trans. Power Deliv. **38**(3), 2170–2182 (2023)
9. Zhang, G., Wang, Y., Liu, Z., et al.: Multi scale and multi physical domain modeling and simulation method for urban rail traction power supply system. Trans. China Electrotech. Soc. **37**(12), 3097–3107 (2022). (in Chinese)
10. Chen, L., Yu, Y., Gong, Z., et al.: Multi-scale modeling and simulation method of urban rail traction power supply system for digital twin. Lect. Notes Electr. Eng. **864**, 666–675 (2022)

# Embedding Group Operation Control Strategy into Heavy Haul Railway Transportation

Huang Susu[1,2], Zhang Miao[2(✉)], and Zhao Yang[2]

[1] School of Electronic and Information Engineering, Beijing Jiaotong University, Beijing 100044, China
10120289@bjtu.edu.cn

[2] Signal & Communication Research Institute, China Academy of Railway Science Co. Ltd, Beijing 100081, China
13810162613@163.com, zhyzhp@sina.com

**Abstract.** In order to improve transportation efficiency, heavy haul railways often adopt a large marshalling scheme. This involves coupling multiple locomotives to pull hundreds of carriages, resulting in a marshalling length of several thousand meters. While large marshalling improves the efficiency of heavy haul train freight transportation, it also poses challenges. For instance, it reduces the flexibility in goods transportation due to slow marshalling changes. Additionally, the longer braking distance associated with large formations requires more safety protection distance, occupying more line resources and ultimately decreasing transportation efficiency. This article examines the application of a group operation control strategy based on virtual coupling technology aimed at improving the efficiency of heavy haul railway transportation and enabling flexible organization of heavy haul trains. Compared to fixed block and moving block strategies, the train platoon under the group operation control strategy occupies smaller track segments and demonstrates higher stability. The proposed group operation strategy has the potential to achieve better conflict control in railway traffic and improve train control system efficiency. Furthermore, as physical coupling is no longer required, the marshalling process of heavy haul trains becomes more efficient, which ultimately results in improved flexibility and greater capacity.

**Keywords:** Heavy haul railway · Group operation strategy · Virtual coupling

## 1 Introduction

The heavy haul railway in China possesses unique characteristics that set it apart from other rail systems. Firstly, the complex line environment, including mountainous areas and tunnels, poses significant challenges in controlling long trains. Secondly, the railway intersects with cities, leading to substantial investments in station expansions and line supplements. Thirdly, heavy haul trains struggle to form long trains due to limitations in the braking system. Finally, there are still numerous requirements for interoperability with general speed railway lines.


Y. Qin et al. (Eds.): EITRT 2023, LNEE 1137, pp. 377–386, 2024.
https://doi.org/10.1007/978-981-99-9311-6_40

The unique characteristics of China's heavy haul railway make it incompatible with the technical routes adopted by the United States' independent road network for efficiency enhancement or Australia's approach of efficiency improvement through lengthening. Instead, China's heavy haul industry should adopt a technological approach to increase density for enhanced efficiency.

Starting from the 1970s, China began to focus on researching heavy haul technology. Relevant experts made suggestions, and in the national 'Sixth Five-Year Plan' for scientific and technological development, a project was proposed to study complete sets of technology for heavy haul railway transportation. This marked the beginning of China's breakthrough in heavy haul railway. In 2006, China's Daqin Coal Transportation Line introduced the LOCOTROL technology from American GE company to operate 20,000-ton combined trains. However, during engineering applications, these 20,000-ton long marshalling and heavy haul trains faced several development issues, including.

## 2 Numerous Operational Control Challenges

In order to enhance cargo transportation efficiency, heavy haul trains often employ large marshalling, which involves multiple locomotives pulling hundreds of wagons and can extend for several kilometers. While this improves the efficiency of heavy haul train cargo transportation, it also poses numerous operational control challenges. Long marshalling trains weighing 20,000 tons are particularly vulnerable to longitudinal impact forces, leading to hook crushing and breaking. Derailment accidents of heavy haul trains are also known to occur occasionally. Due to inadequate automation of train control system, safe operation control procedures rely heavily on driver's skills and experience, leading to high labor intensity and significant safety risks.

## 3 Inefficient Transportation

During the transportation of heavy haul trains, large-scale technical operation stations are required for marshalling and unmarshalling. Due to the long marshalling of heavy haul trains, it takes a significant amount of time to complete technical operations such as train inspection, technical inspection, coupling, and air filling. On average, it takes about 150 min for a 20,000-ton train to complete the assembly process, which negatively impacts the turnover rate of rolling stock and utilization rate of arrival-departure sidings at the station, thus limiting the flexibility of cargo transportation. Additionally, due to the long braking distance of large-scale heavy haul trains, more safety protection distances must be set aside, further reducing the overall transportation efficiency of the line.

## 4 Insufficient Transportation Capacity

Most of the existing heavy haul lines in China utilize three-display and four-display automatic block systems, while the station lines have adopted the yellow light departure method. This maximizes the capacity of train operation control system and has pushed it to its limits. However, the assembly stations at the end of each heavy haul line are constrained by throat area and the departure capacity of the arrival-departure siding, resulting in insufficient transport capacity during peak time.

In order to further enhance the transportation capacity of heavy haul railway and facilitate their digitalization, informatization, and intelligent construction, the group operation control strategy based on virtual coupling technology should be implemented. This will address practical application problems that arise due to coexisting multi-standard heavy haul railways and complex transportation conditions.

Compared to fixed block and moving block technologies, a heavy haul train platoon based on virtual coupling technology occupies fewer track segments and enhances operational efficiency. Moreover, eliminating physical coupling makes it easier to demarcate and reorganize heavy haul trains leading to increased flexibility in railway freight. This, in turn, produces more significant capacity improvement effects.

## 5 Overall Technical Scheme

In light of these circumstances, we have implemented the Chinese Train Control System-N (CTCS-N) in the heavy haul railway transportation system [1]. CTCS-N is an innovative new system that utilizes Global Navigation Satellite Systems (GNSS) for train positioning, with onboard and ground equipment working together to complete train integrity monitor [2]. Two-way IP wireless communication is employed between the trackside Radio Block Center (RBC) and onboard equipment in CTCS-N to transmit critical information such as train position, line data, speed limit commands, and movement authorization. Additionally, CTCS-N allows for virtual block or moving block, with a control system utilizing target distance continuous speed control mode to monitor train operations [3]. Figure 1 illustrates the system structure schematic diagram of CTCS-N.

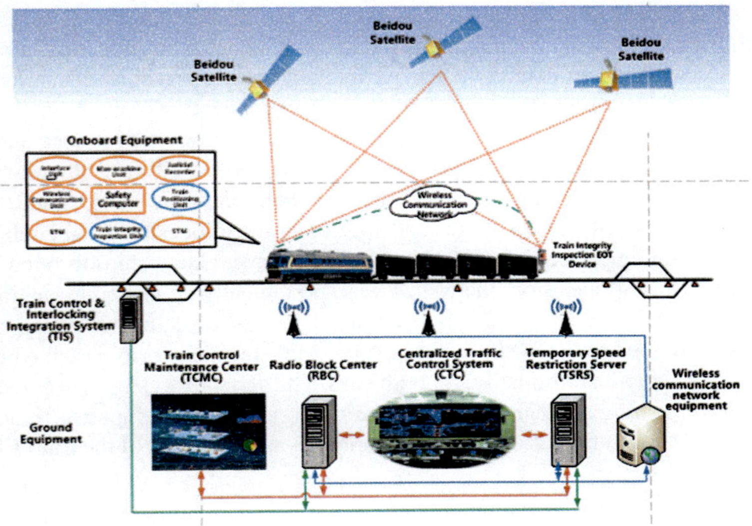


**Fig. 1.** The system structure schematic diagram of CTCS-N.

The most critical technical features of CTCS-N include utilizing GNSS-based comprehensive train positioning, conducting train integrity monitor, and employing train-ground wireless communication technology. These features enable train occupancy checks through information interaction between onboard and RBC equipment, which allows for releasing trains. Occupancy check is reliant on trackside devices, with reliable and accurate train positioning facilitating moving block function.

CTCS-N includes key technologies such as continuous train positioning based on multi-source information fusion, integrity monitor for heavy haul trains, trackside control technology, and virtual coupling technology.

## 5.1 Continuous Train Positioning Based on Multi-Source Information Fusion

The operating environment for heavy haul trains is complex and variable, with occurrences of satellite signal obstructions in long tunnels or mountainous areas that can affect train positioning results. To achieve continuous and seamless train positioning, multi-source information fusion technology must be adopted. By switching the positioning mode according to the number of visible satellites and combining multiple sensors simultaneously, system reliability and fault tolerance can be improved [4].

Integration of various measurement values from different systems enhances navigation accuracy and robustness while simplifying system construction complexity. However, an increase in the number of satellites and sensors in the multi-source integrated train positioning system also increases the probability of observation information failure. Therefore, rapid fault detection and processing technology must be studied to ensure the reliability and safety of system positioning.

## 5.2 Heavy Haul Train Integrity Monitor Technology

During the operation of heavy haul trains, it is possible for one or more wagon carriages to detach from the end and remain on the track. If not detected and addressed promptly, this can pose significant safety risks to subsequent train operations. Therefore, it is necessary to conduct integrity monitor on the overall connection of the train.

Due to the lack of wired connectivity in heavy haul trains, we propose a wireless communication-based train integrity check program. The train head and end are connected wirelessly, with the onboard ATP device obtaining train integrity information from the head. Once powered up, the head and end of the train establish a wireless connection by obtaining pairing information and master-slave relationships through configuration data. During operation, the train's head periodically collects information such as heartbeat signals and wind pressure from the end. In addition, both the head and end devices are equipped with GNSS, accelerometer, and other sensors to measure position, speed, and acceleration information for both ends. The onboard ATP system comprehensively judges the train's integrity status based on digital map and inputted train length information. This solution also enables autonomous measurement of the train's length.

## 5.3 Trackside Control Technology

The ground equipment of the CTCS-N comprises three primary components. Firstly, there is the interlocking system and interstation safety data communication network that facilitates station interlocking and directional control functions. Secondly, there is the RBC responsible for virtual automatic block or moving block functionality. Lastly, there is the Data Server Unit (DSU) which manages temporary speed limit commands, digital map data downloads, and satellite differential information transmission. Other supporting equipment includes onboard wireless communication network, location responders, and satellite differential positioning stations.

### 5.4 Virtual Coupling Technology

The IEEE 1474 standard [5] defines that trains trav-elling in the same direction on the track must be separated by a sufficient distance, as shown in Fig. 2(a). However, due to the limited braking speed caused by the adhesion between steel wheels and rails, the required train tracking interval increases significantly with speed and load. In heavy haul railways, which often deal with long marshalling and heavy loads, the tracking distance between two trains may need to reach several kilometers. The upgrade of signal systems from fixed block to moving block was aimed at maximizing track capacity within these constraints. However, with the increasing demand for capacity, the traditional emergency braking principle is now facing challenges.

The concept of Virtual Coupling emerged due to doubts regarding the train's relative position braking model, as proposed in the IEEE 1474 standard. This standard assumes that the braking point ahead is a stationary location with a speed of zero, which is not realistic [6]. In actual train tracking scenarios, the leading train is a moving body with constantly changing velocities. If two trains have the same vehicle characteristics and operational curves, they can be more closely. Communication between the trains ensures that subsequent trains will execute the same operations if the leader train begins to brake, while maintaining separation when both trains decelerate together [7, 8].

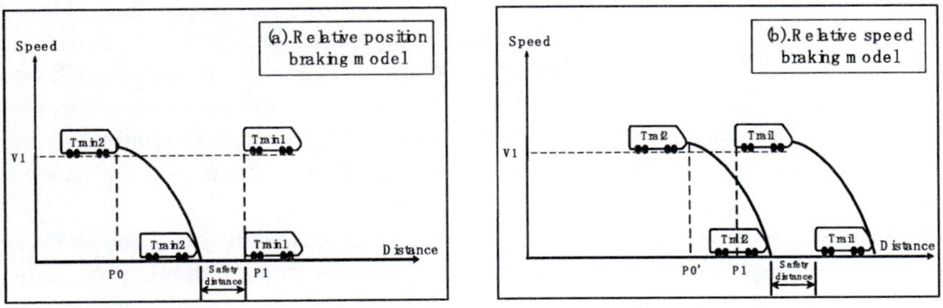


**Fig. 2.** Comparison between relative position and relative speed braking mode.

Figure 2 depicts two trains, Train1 and Train2, travelling at a speed of V1. In the relative position braking mode shown in Fig. 2(a), the minimum tracking distance is between P0 and P1 is $S_a$. . This includes Train2's braking distance($S_{follower}$) and the safety distance between the end of Train1 and the head of Train2 ($S_{safe}$), as shown in formula (1). In the relative speed braking mode shown in Fig. 2(b), the minimum tracking distance is between P0' to P1 is $S_b$. . This consists of Train2's braking distance($S_{follower}$),

, the safety distance between the end of Train1 and the head of Train2 ($S_{safe}$), the length of Train1($L_{leader}$), but subtracts Train1's braking distance($S_{leader}$), as shown in formula (2). It is apparent that the distance $S_b$ is shorter than the distance $S_a$.

$$S_a = S_{follower} + S_{safe} \tag{1}$$

$$S_b = S_{follower} + S_{safe} + L_{leader} - S_{leader} \tag{2}$$

## 6 Group Operation Control Strategy

In the early stages, heavy haul trains operated as single trains with single trips, while multiple trains obtained movement authority based on fixed block technology. Later, to increase single trip transportation capacity, multiple heavy haul trains were physi-cally linked together but still obtained movement authority based on fixed block technology. However, this method resulted in long reconnection operations, increased space occupation, and difficulty effectively distributing traction force despite improving transport capacity. The heavy haul train group operation control strategy employs wireless communication between trains to achieve virtual coupling, allowing for flexible marshalling configurations and improved overall transportation efficiency. It is an innovative strategy worthy of development [9].

As we all know, moving block operation control strategy based on relative position barking model is the most efficient block mode currently utilized in practical engineering applications. In the relative position barking mode, the maximum target braking point for a following train during tracking is the max safe rear end position the leading train, while the starting point for tracking is the max safe front end position of the following train [10, 11].

Figure 3 illustrates that in fixed block train control strategy, each section is closed off for a predetermined amount of time, and train operation is limited by the fixed block-time window. The block-time window at bottleneck points significantly impacts the efficiency of train tracking in the region. In contrast, the moving block train control strategy utilizes high capacity communication between the train and ground to create a continuous curve based on the front and rear positions of the train. This approach enables trains to operate at closer distances compared to fixed block strategy.

Figure 4 depicts the group operation control strategy based on virtual coupling technology, which allows trains to connect together and break through the limit of Train1's maximum safe rear end position and Train2's maximum safe front end position. This innovative train control strategy has the potential to reduce train tracking intervals and improve overall operational efficiency of heavy haul lines.

## 7 Typical Scenario Simulation

To assess the effectiveness of the proposed group operation strategy, an analysis and simulation were conducted for three trains operating as shown in Fig. 5. Train1 and Train3 were initially positioned on segment AB of the mainline, while Train2 was located

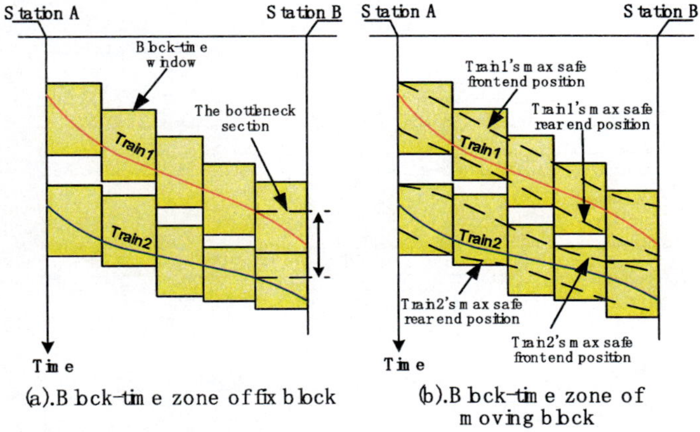


**Fig. 3.** Comparison of the block-time zone between fixed block and moving block strategies.

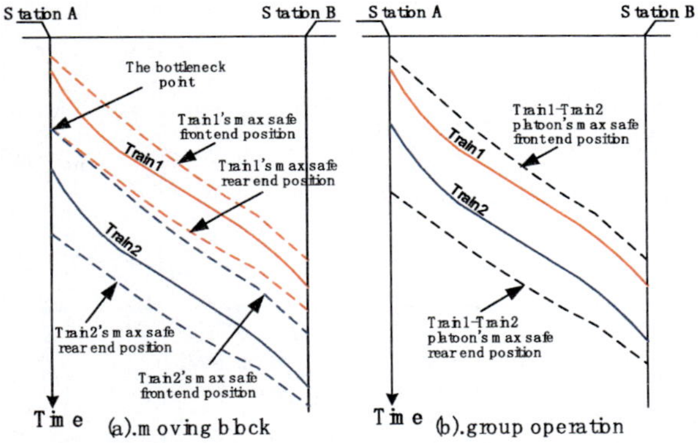


**Fig. 4.** Comparison of the block-time zone between moving block and group operation strategies.

on segment A'B' of the siding track. After passing through switch point C sequentially at a limited speed of 10 m/s, Train1, Train2, and Train3 formed a platoon designated as Train1-Train3-Train2, with a speed of 10 m/s. In Fig. 5, at time T, only Train 1 in the moving block has passed through location D, while neither Train 3 nor Train 2 has yet reached that location. However, in the group operation, all three trains have already passed through location D. Comparing the process of forming the virtual coupled platoons between moving block and group operation in Fig. 5, we can clearly observe the flexibility and efficiency of the group operation strategy.

Figures 6, 7 and 8 depict detailed information on the speeds, positions, and relative positions of the three trains during the platoon formation process under the group operation strategy.

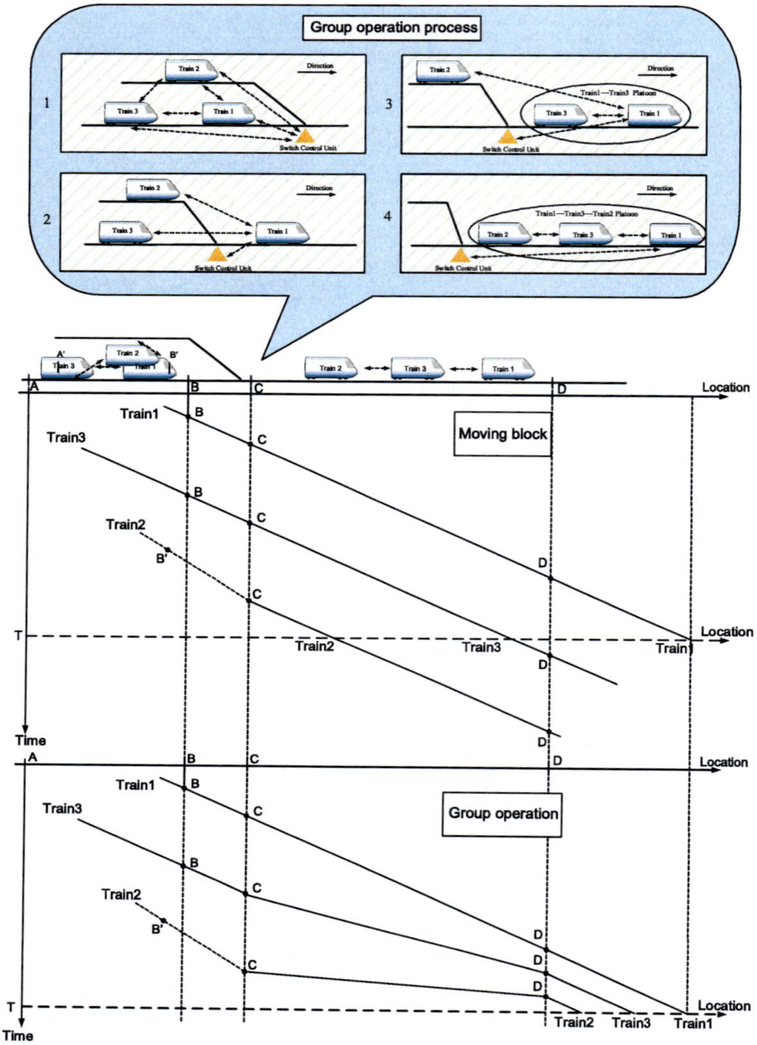


**Fig. 5.** The typical scenario in comparing between moving block and group operation strategy

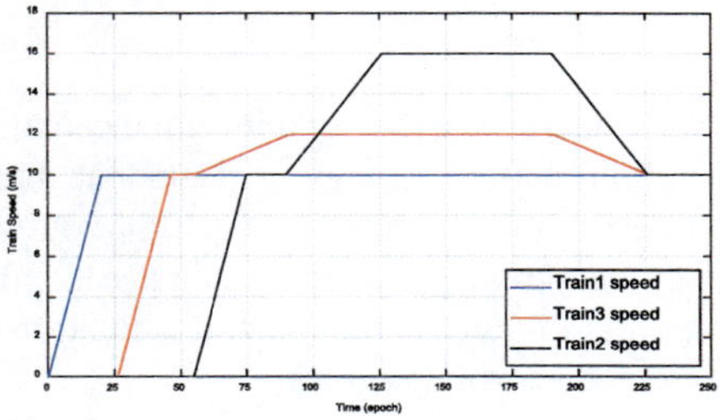


**Fig. 6.** Schematic of speed-time for the three trains.

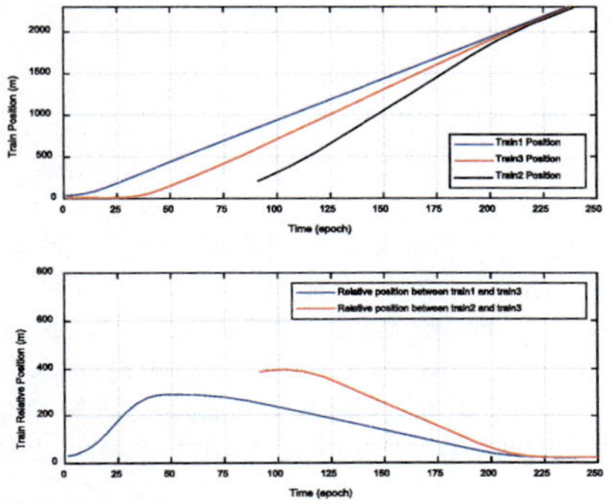

**Fig. 7.** Schematic of relative distance of three trains

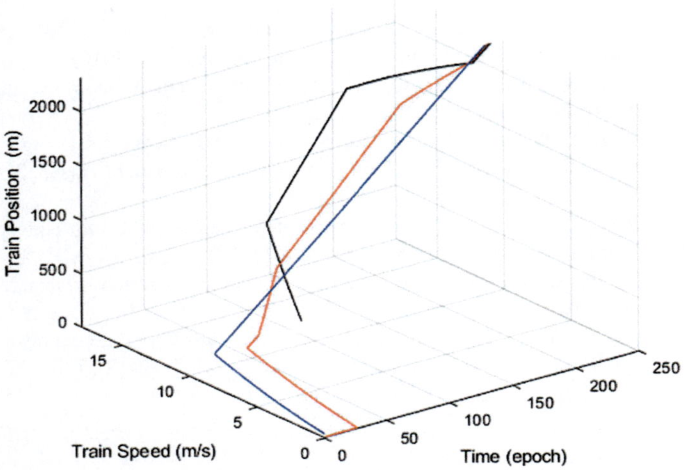


**Fig. 8.** Three-dimensional schematic of time, speed, and position for the three trains.

## 8 Conclusion

By conducting a comparative analysis of the capacity enhancement and conflict detection effects of the group operation control strategy, it is evident that the strategy utilizing virtual coupling technology is an innovative approach that outperforms moving block train control. This strategy has great potential to enhance flexibility and capacity for heavy haul railway transportation. Further testing and validation of this novel approach will be conducted in future engineering projects.

**Acknowledgments.** This work was funded by the National Natural Science Foundation of China ( No. 62120106011, 52172322), the Beijing Natural Science Foundation ( No. L211004) and the China Academy of Railway Science Foundation( No. 2022YJ159).

## References

1. Mo, Z.: Overal technical solution of CTCS-4 level train control system. Railw. Signal. & Commun. **11**(55), 34–40 (2019). (in Chinese)
2. Kai, X., Zhang, M., Gao, Y., Song, Z.: An architecture design of the train control system based on BDS. Railw. Transp. Econ. **42**(12), 77–88 (2018). (in Chinese)
3. Zhang, M., Zhao, Y., Qu, Y., Fan, K., Zhou, B.: Safety technology plan of movable block train control system of Qinghai-Tibet railway. Railw. Transp. Econ. **40**(10), 66–70 (2020). (in Chinese)
4. Zhang, M., Zhang, Q., Liu, W., Zhou, B.: A policy-based reinforcement learning algorithm for intelligent train control. J. China Railw. Soc. **42**(1), 69–75 (2020). (in Chinese)
5. IEEE.: 1474.1 IEEE standard for communication-based train control(CBTC). Perform. Funct. Requir (2004)
6. Mitchell, I., Goddard, E., Montes, F., Stanley, P., Muttram, R., et al.: ERTMS level 4, train convoys or virtual coupling. IRSE Int. Tech. Comm. **39**(219), 1–3 (2016)
7. Jia, L., Chen, X., Ma, X., Xu, Q., Yu, H., Sun, W., et al.: On autonomous transportation systems. Smart Resilient Transp. **4**(2), 66–77 (2022)
8. Wang, B., Wang, G., Wang, Y., Lou, Z., Hu, S., Ye, Y.: A K-means clustering method with feature learning for unbalanced vehicle fault diagnosis. Smart Resilient Transp. **3**(2), 162–176 (2021)
9. Bock, U., Bikker, G.: Design and development of a future freight train concept–virtually coupled train formations. IFAC Pro **33**(9), 395–400 (2000)
10. Huang, S., Feng, H.: CBTC system based on train-train communication. Uraban Mass Transit **24**(6), 188–193 (2021). (in Chinese)
11. Huang, S., Wen, T., Li, K., Wang, L., Feng, H., Cai, B.: Multi-train regional resource negotitation driven by Raft consensusu mechanism. China Saf. Sci. J. **32**(12), 110–117 (2022). (in Chinese)

# Dynamic Container Trucks Scheduling Strategy for Container Terminals Based on Station-Truck Interconnection

Xiaoyu Hu[1], Yang Chen[1], Boyu Du[1], and Li Wang[1,2](✉)

[1] School of Traffic and Transportation, Beijing Jiaotong University, Beijing 100044, China
{21251039,21251034,21251279,liwang}@bjtu.edu.cn

[2] Key Laboratory of Transport Industry of Comprehensive Transportation Theory (Beijing Jiaotong University), Ministry of Transport, Beijing 100044, China

**Abstract.** Container terminals are important nodes in the container intermodal transport network. China's rapidly growing container throughput puts forward higher requirements for the operational efficiency of container terminals. At present, due to the lack of a platform for effective communication between container terminal operators and container truck operators, a large number of container trucks arrive and leave in a centralized manner, resulting in congestion at terminal gates and the roads in the operation area, and reducing the efficiency of container terminals. Aiming at the pain points faced by the operation practice of container terminals, this project takes external container truck dynamic scheduling of container terminals as the research object, develops a station-truck interconnection APP that can realize the dynamic interaction between the two types of subjects: container terminal operators and container truck operators. A container terminal operation scheduling simulation system was built to simulate the load of operating equipment under different container truck arrival numbers, and determine the optimal container truck arrival quantity in different time periods. In the face of uncertain arrival situations such as truck in-transit failures, a dynamic adjustment algorithm for container trucks scheduling of container terminals is proposed to realize the dynamic adjustment of container truck scheduling strategy. It provides a theoretical method and platform support for the external container trucks dynamic scheduling of container terminals.

**Keywords:** Container terminal · Container truck scheduling · Station-truck interconnection · Simulation

## 1 Background

As the transit hubs of logistics and transportation network, container terminals are essential intermodal interfaces in the container transportation system. Efficient scheduling strategy is important in reducing operation costs, improving performance ability, and guaranteeing the quantity of services. As an vital patrs of scheduling strategy, container trucks scheduling problem of container terminal is a considerable aspect of studies on railway container transport organization.


Y. Qin et al. (Eds.): EITRT 2023, LNEE 1137, pp. 387–394, 2024.
https://doi.org/10.1007/978-981-99-9311-6_41

Jiang [1] treats a batch of containers as a customer and uses the multi-service-desk queuing theory to solve the resource utilization problem of a dedicated yard. The arrival and service times of containers in the queuing system are expressed in terms of ship arrival times and in-port service times, which are verified with actual port and ship arrival data in the analysis phase of the calculation. Based on the requirement of seamless connection between public and railway intermodal transport, Ye [2] focused on the interface between loading and unloading equipment and railway container vehicles to determine the optimal type and quantity of loading and unloading equipment at railway container centers; the organization of the interface between container trucks and loading and unloading equipment was also studied to determine the optimal time interval and quantity of container trucks arriving at the main container yard. Liang et al. [3] studied the layout of port stations and central stations, and provided issues that need attention in the planning and construction and operation periods of central stations according to different night flow lines. Chang et al. [4] developed a dynamic allocation model for berth allocation and terminal operation allocation based on the rolling time method, and used a hybrid parallel genetic algorithm combining parallel genetic algorithm and heuristic algorithm to solve the proposed model. Murty et al. [5] Consider the scale and complexity of these decisions and develop a decision support system that describes the various interrelated decisions made in the daily operations of a container terminal with the ultimate goal of minimising vessel berthing times, resources required to handle the workload, waiting times for customer trucks, and congestion on roads, warehouses and terminals within the terminal; and making full use of storage space. Zhu et al. [6, 7] studied the coordination and scheduling of container terminal operation equipment and proposed methods to improve related operation efficiency, and the container terminal space resource scheduling model is established.

At present, the literature related to container terminals are mostly focused on the theory of transport organisation, but further research on the dynamic scheduling of container terminal "interconnection of station and vehicle" is relatively rare. So we study on container trucks scheduling strategy based on station-truck Interconnection.

## 2 Design Principle

The main research content of this project is the simulation process design of container terminal operation scheduling. Aiming at the painful problem of low efficiency of central station collection and distribution organization and causing terminal congestion faced in practice, the dynamic scheduling of container trucks outside the container terminal is the research object, and a dynamic scheduling method of container terminal collection trucks based on interconnection of stations and vehicles is proposed, and a simulation decision platform is developed, and the specific design architecture is shown in Fig. 1.

The project is oriented to two types of subjects: container terminals and container truck operators, and the research is carried out at two aspects: operation planning and dynamic scheduling adjustment.

(i) Two types of subjects: The dynamic scheduling of container trucks outside container terminals mainly involves two types of subjects, namely container terminals and truck operators. By building an information interaction platform, the dynamic

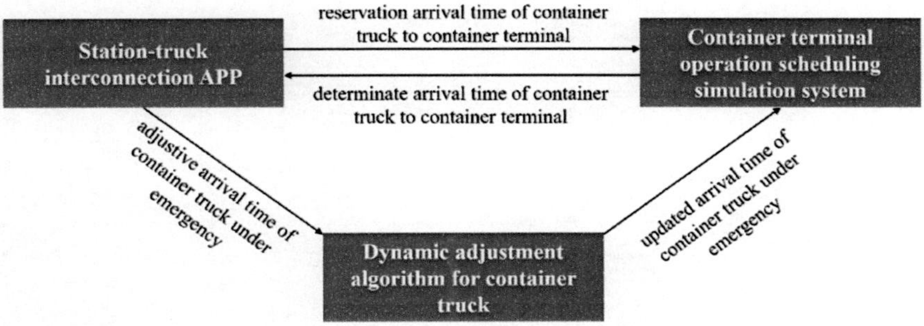


**Fig. 1.** Specific design architecture

interaction between the two types of subjects is realized, and the "station-truck interconnection" between terminals and trucks is realized.

(ii) Three aspects: Based on the information interaction between the two types of subjects, we can accurately obtain the arrival time of container trucks and the operation situation of yard stations, analyze the operation load of container terminal equipment through simulation, and determine the operation plan for each period. In view of the uncertainties such as terminal operation emergencies and truck in-transit failures, the original operation plan can be adjusted in time by using the dynamic adjustment method of container terminal operation through the timely transmission of information interaction platform, so as to realize the decision support at two levels of operation planning and operation scheduling adjustment.

## 3 Station-Truck Interconnection APP Design

The APP is designed for both truck owners and container terminal operators, providing a platform for dynamic interaction between truck users and container terminal operators, and providing a platform to support dynamic scheduling decisions for trucks (Fig. 2).

(1) For container truck users: APP mainly provides four types of services, including reservation service, en route service, arrival service and information service, etc. Among them, reservation service provides container truck users with accurate arrival time from container terminal operators according to expected operation time; en route service mainly includes route planning and reporting of unexpected faults for container truck drivers on the way to the terminal; arrival service mainly includes arrival card punching and queuing for entry after the container truck arrives at the terminal; information service mainly includes receiving one-to-one pushing of reservation The information service mainly includes receiving one-to-one information from container terminal operators on booking confirmation, booking adjustment information, and information on terminal operation released to the public, etc. [8].

(2) For container terminal operators: The APP mainly provides three types of services, including arrival time statistics of trucks at different time slots, enquiry of arrival trucks, and notification information release. Among them, the arrival time statistics for different time slots is mainly to count the appointment applications submitted by the card users at different time slots, and export the relevant appointment

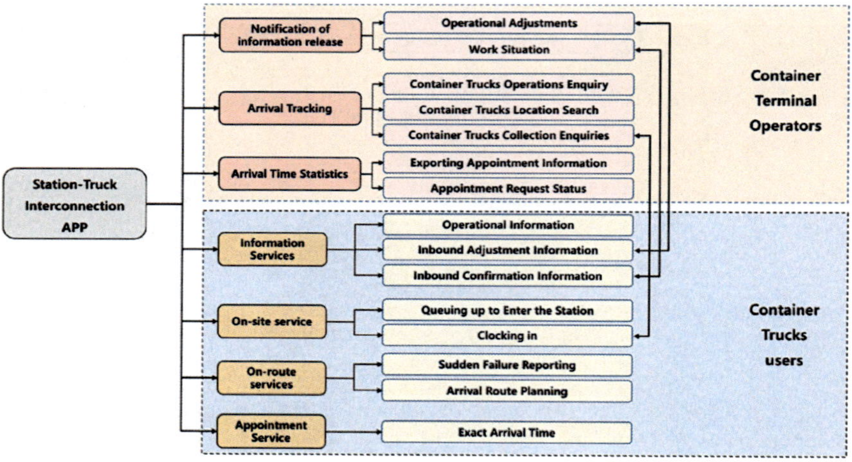


**Fig. 2.** Design ideas

arrival information to provide data support for subsequent simulation; the arrival card query is mainly to count the overall situation of cards arriving at the current time slot according to the card users' arrival card situation; the notification information release function on the one hand is to push the appointment entry information and appointment adjustment information to the card users on a one-to-one basis. On the one hand, the notification function provides one-to-one information on the booking and adjustment of the booking, and unifies the information on the operation of the terminal and the adjustment of the operation in case of emergency [9].

# 4 Container Terminal Operation Scheduling Simulation System

## 4.1 Simulation Modeling Software

FlexSim software has good displaying effect and data analysis capabilities, which can be applied to the simulation modeling of discrete events in many fields, so FlexSim was selected as a simulation modeling tool to build a simulation model for container terminal operation scheduling.

## 4.2 Mentality of Designing

According to the queuing process, queuing time, and operation organization mode and operation process of truck loading and unloading in container terminals involved in different truck queuing situations of container terminals, a simulation modeling process idea for the queuing of trucks in different container terminals is designed. The mentality of designing is shown in the Fig. 3.

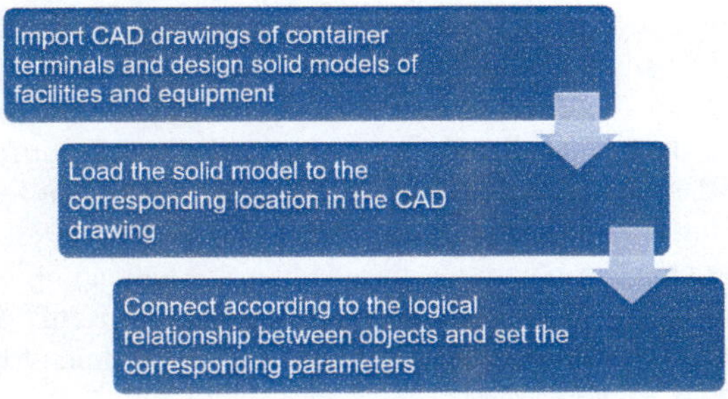


**Fig. 3.** Mentality of designing

### 4.3 Modeling Steps

The specific steps of simulation modeling are as follows:

(i) Layout and model design. Import CAD drawings of container terminals, design solid models of facilities and equipment, and load solid models into the corresponding CAD drawings.
(ii) Physical connections and parameter settings. Connections are made according to the logical relationships between objects. According to the container terminal operation scheduling logic, FlexSim is used to connect various entities. And set the parameters of truck vehicles, yard containers, truck walking, etc. [10]
(iii) Port and trigger configuration. According to the actual operation of the container terminal, clarify the trigger relationship of relevant events in the simulation model, and configure ports and triggers. Realize the truck loading and unloading process of containers entering the terminal
(iv) Simulation model adjustment and result analysis. According to the actual needs, simulate and adjust the configuration quantity of various facilities and equipment, such as the number of truck configurations, the number of yards, etc. Repeated tests are carried out based on container in-transit information and road condition information, and the simulation results are analyzed.

Through the above steps, the simulation model can make the coordinated and orderly operation between the equipment entities, plan the transportation route for the loading and unloading task of the truck, and realize the simulation of the two-way operation of the container between the terminal gate-truck-yard-loading yard.

Based on the simulation model, the arrival of trucks collected by the APP in different periods of time can be simulated in combination with the actual loading and unloading operations of container terminals to determine the impact of the arrival of different numbers of trucks on the load of related operating equipment under different time periods, so as to determine the optimal arrival number of trucks in different periods.

After determining the optimal number of trucks arriving at different times, the container terminal operator can determine the loading and unloading operation plan for different periods according to the simulation results, and then notify the specific arrival time of the relevant trucks, and notify and adjust the arrival time of the trucks that are not suitable for the arrival period.

## 5 Dynamic Adjustment Algorithm for Container Truck

On the basis of the coordinates of the operation area, the dynamic arrival of the truck is considered [6]. Under the constraints of operation duration, continuous operation movement time, pre-sequence and post-sequence operation continuation, continuous operation relationship, and operation sequence, a dynamic adjustment model of container terminal truck operation scheduling with the aim of shortest total operation time of adjusting the rear crane is constructed, and according to the characteristics of the model, with genetic algorithm as the core, a sequence optimization algorithm framework for packing operations considering the dynamic arrival of container trucks is designed. The algorithm framework is shown in the following figures (Figs. 4 and 5).

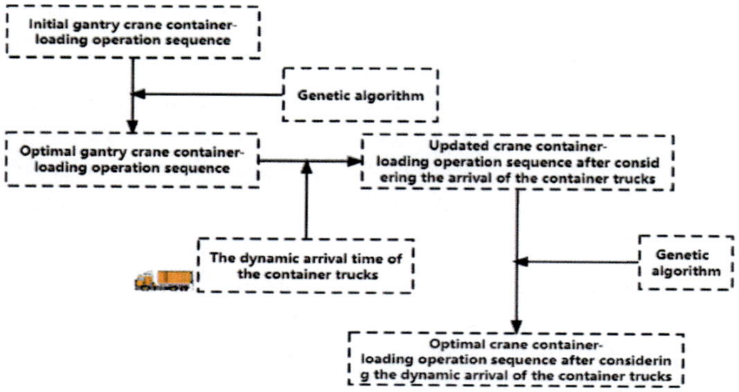


**Fig. 4.** The ideas of operation sequence optimization

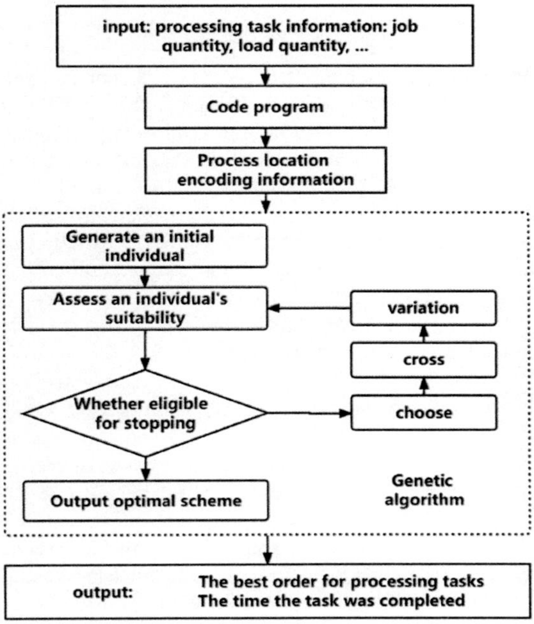


**Fig. 5.** The process of operation sequence optimization

# 6 Conclusion

With the rapid development of China's container terminals, our study can reduce congestion and queuing of road collection and transportation of collector trucks in stations. The innovative features of this work are mainly reflected in the following three aspects:

(i) To realize the "station-truck interconnection" between container terminals and trucks by building an information interaction platform for dynamic interaction between the two types of subjects: container terminals and external truck operators;
(ii) Providing decision support for both types of users from both directions (saving time of container truck drivers and minimizing container truck congestion at container terminals and balancing equipment operation load), which can meet different needs of different users;
(iii) It supports the scheduling of container truck operations from two perspectives: operation planning and operation dynamic adjustment, which is more in line with the actual operation scenarios of container terminals.

**Acknowledgments.** This research is supported by the Fundamental Research Funds for the Central Universities(2023JBZY006).

## References

1. Jiang Dongrui, F.: Space allocation problem of dedicated container yard based on queuing theory. Dalian Maritime University (Master's Thesis) (2020). (in Chinese)
2. Ye Yuanyu, F.: Research on Container Loading and Unloading Equipment Configuration and Container-Car Connection Organization Based on Public-Rail Intermodal Transportation. Beijing Jiaotong University (Master's Thesis) (2012). (in Chinese)
3. Liang Jian, F.: Study on Resource Allocation Decision of Railway Container Logistics Center Station. Southwest Jiaotong University (Master's Thesis) (2010). (in Chinese)
4. Chang, D., F., Jiang, Z., S., Yan, W., T., et al.: Integrating berth allocation and quay crane assignments. Transp. Res. Part E (2010)
5. Murty, K.G., F., Liu, J., S, Wan, Y., W. T., et al.: A decision support system for operations in a container: terminal. Decis. Support. Syst. **39**(3), 309–332 (2005)
6. Liu Wenxi, F., Zhu Xiaoning, S., Wang Li, T., et al.: Collaborative scheduling method of port equipment based on disjunctive diagram. J. China Railw. Soc. **44**(08): 14–24 (2022). (in Chinese)
7. Li, W., F., Zhu, X., S., Xie, Z., T.: Balanced scheduling model of space resources in yard of railway container central station. J. Chang'an Univ. (Nat. Sci. Ed.). **35**(S1), 86–90 (2015). (in Chinese)
8. Abdelmagid Ahmed, M., F., Gheith Mohamed, S., Eltawil Amr, T.: Scheduling external trucks appointments in container terminals to minimize cost and truck turnaround times. Logist. **6**(3), 45 (2022)
9. Li Jun, F., Yang Enze, S., Yang Guoyuan, T., et al.: Research on wireless interaction technology of railroad bag information system. Railw. Comput. Appl. **29**(05) (2020)
10. Yan Bocheng, F., Zhu Xiaoning, S., Wang Li, T., et al.: Integrated scheduling of rail-mounted gantry cranes, internal trucks and reach stackers in railway operation area of container terminal. Transp. Res. Rec.: J. Transp. Res. Board (2018)

# Continual Unsupervised Domain Adaptation for Bearing Fault Diagnosis Under Variable Working Conditions

Bojian Chen[1,2], Changqing Shen[1,2](✉), Lin Li[3], Juanjuan Shi[1,2], Weiguo Huang[1,2], and Zhongkui Zhu[1,2]

[1] School of Rail Transportation, Soochow University, Suzhou 215000, China
{cqshen,jshi091,wghuang,zhuzhongkui}@suda.edu.cn

[2] Intelligent Urban Rail Engineering Research Center of Jiangsu Province, Suzhou 215131, China

[3] Space Optoelectronic Measurement and Perception Lab, Beijing Institute of Control Engineering, Beijing 100190, China

**Abstract.** Unsupervised domain adaptation (UDA)-based fault diagnosis models have been intensively studied. UDA focuses on establishing a model that can transfer knowledge from one or more source domains with labeled data to help learn a target domain with unlabeled data. However, in the field of fault diagnosis, an obvious drawback is that UDA ignores the case of multiple target domains. In real scenarios, fault data are distributed in a continuous flow of constantly generated information due to variable working conditions. The collected unlabeled data may be in different domains, resulting in an increase in the number of target domains, which is called domain increments. Continual unsupervised domain adaptation (CUDA), which combines continual learning (CL) and UDA, is proposed to address the problem of domain increments. CL aims to create a deep learning (DL) model that can learn in dynamic environments, similar to humans. The core of CL is to prevent DL from catastrophic forgetting. In this paper, a CUDA fault diagnosis (CUDAFD) method is proposed for bearing fault diagnosis under variable working conditions. In CUDAFD, the maximum mean discrepancy is used to continually adapt the model to new domains, and the idea of prototype learning is introduced to overcome the catastrophic forgetting. Finally, CUDAFD is applied to a diagnosis case of domain increments to verify its effect.

**Keywords:** Continual learning · Unsupervised domain adaptation · Fault diagnosis · Domain increments · Catastrophic forgetting

## 1 Introduction

Fault diagnosis has become an indispensable technology in modern rail transportation systems [1, 2]. Bearings, as the key components of rail transportation, inevitably suffer from performance degradation or even failure, due to complex variable working conditions. Timely and accurate fault diagnosis is greatly significant for ensuring the safe and reliable operation of rail vehicles.


Y. Qin et al. (Eds.): EITRT 2023, LNEE 1137, pp. 395–403, 2024.
https://doi.org/10.1007/978-981-99-9311-6_42

Fault diagnosis has become more intelligent and informational in recent years due to the rapid development of modern monitoring technology. Deep learning (DL) has been introduced into the field of fault diagnosis, while research on these data-driven intelligent diagnosis methods has seen an unprecedented development. However, the simple DL method cannot meet the needs of bearing fault diagnosis under complex working conditions. DL has a limitation and a problem. The limitation is that DL requires training and testing data to meet the independent and identically distributed hypothesis. The problem is that deep neural networks suffer from catastrophic forget-ting when learning from dynamic environments. Consequently, DL cannot meet the needs of fault diagnosis under variable conditions.

Transfer learning (TL) breaks the limitation of the same distribution, with its knowledge mining and transfer ability across domains and distributions. Unsupervised domain adaptation (UDA) is one of the popular topics in TL. In fault diagnosis, UDA is used to establish a model which can transfer knowledge from one or more source domains with labeled fault data to help learn a target domain with unlabeled fault data [3]. However, an inherent problem with TL methods, such as UDA, is that they can only solve a single target domain task and cannot cope with multiple target domain tasks that continuously occur. The cause of the problem is closely related to the catastrophic forgetting of DL. Continual learning (CL) is proposed to overcome the catastrophic forgetting. CL aims to build a DL model that can learn in dynamic environments, similar to humans. Currently, several CL-based fault diagnosis methods can be used in fault diagnosis [4, 5]. However, most of these methods only focus on the increment of fault types.

In real scenarios, fault data are distributed in a continuous flow of constantly generated information due to variable working conditions. The collected unlabeled data may be in different domains, resulting in an increase in the number of target domains, which is called domain increments. Continual UDA (CUDA), which combines CL and UDA, is proposed to solve the problem of domain increments [6]. In this work, a CUDA fault diagnosis (CUDAFD) method, which is based on CUDA, is proposed for bearing fault diagnosis under variable working conditions. In CUDAFD, the maximum mean discrepancy (MMD) [7] is used to continually adapt the model to new domains, and the idea of prototype learning is introduced to construct a prototype cosine similarity classifier (PCSC) [8] to overcome the catastrophic forgetting. Finally, CUDAFD is applied to a diagnosis case of domain increments to verify its effect.

## 2 Background

### 2.1 Continual Learning

CL aims to create a DL model that can gradually extract and retain knowledge in continual training phases and transfer old knowledge to learn new knowledge, similar to human beings. The process of knowledge transfer is the fine-tuning of networks. Fine-tuning is frequently used in TL to transfer knowledge from the source domain to the target domain. However, fine-tuning cannot accumulate knowledge. Consequently, networks will lose the capability to solve the source domain tasks, which refers to the phenomenon of catastrophic forgetting.

### 2.2 Continual Unsupervised Domain Adaptation

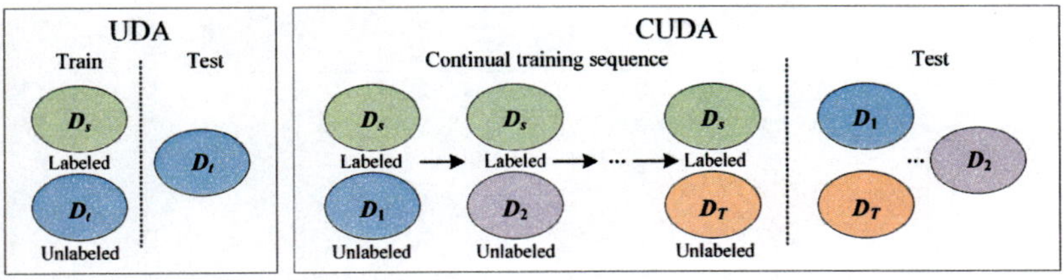


**Fig. 1.** Comparison between UDA and CUDA.

Domain adaptation is a machine learning algorithm for source domain and target domain data distribution offset. This approach is one of the TL methods. For convenience, $D_s = \{(x_i^s, y_i^s)\}_{i=1}^{n_s}$, where $y_i^s \in \{1, 2, \ldots\ldots, C\}$ represents the source domain having samples with $C$ kinds of different labels, and $D_t = \left\{\left(x_j^t\right)\right\}_{j=1}^{n_t}$ denotes the target domain with unlabeled samples. Meanwhile, $\{D_1, D_2, \ldots\ldots, D_T\}$ is used to represent a series of unlabeled target domains and each of $\{D_1, D_2, \ldots\ldots, D_T\}$ is a unlabeled target domain, similar to $D_t$.

In Fig. 1, UDA is a scenario in which the source domain $D_s$ is labeled, and the target domain $D_t$ is unlabeled. However, UDA only focuses on solving a single target domain. CUDA, which is based on UDA and CL, aims to establish a model that can continually transfer the knowledge from a single labeled source domain $D_s$ to a series of unlabeled target domains $\{D_1, D_2, \ldots\ldots, D_T\}$.

Nevertheless, UDA is not enough in the field of fault diagnosis. The complex variable working conditions result in domain increments. Hence, CUDA must be integrated into fault diagnosis.

## 3 Methodology

### 3.1 Overview of the Proposed Method

Assuming that there are a labeled source domain $D_s$ and $T$ unlabeled target domains $\{D_1, D_2, \ldots\ldots, D_T\}$, $T$ learning sessions exist. The input 1-D vibration signal training data will be transformed to frequency-domain signal by fast Fourier transform and are then reshaped into 2-D images. The model of the proposed CUDAFD contains a feature extractor $F(\cdot)$ and a PCSC classifier $P(\cdot)$. The baseline architecture of feature extractor is ResNet-32 [9]. As illustrated in Fig. 2, the specific steps of CUDAFD are as follows:

**Step 1**. The labeled source domain data $D_s = \{(x_i^s, y_i^s)\}_{i=1}^{n_s}$ is used to obtain the source domain pre-trained model $\{F^0, P^0\}$ in session 0.

**Step 2**. The PCSC classifier $P^0$ achieved in session 0 will be fixed in the following session. Fixing PCSC, based on prototype learning, can help the model overcome catastrophic forgetting.

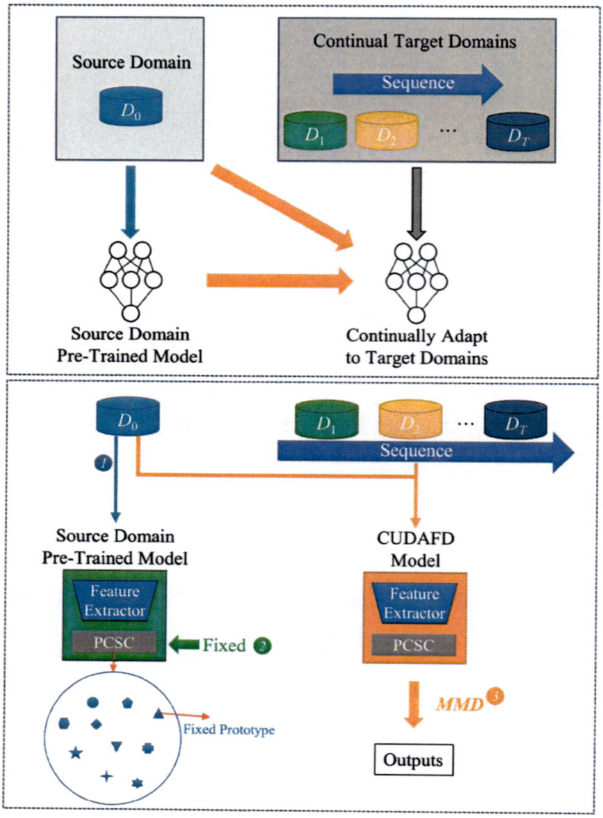


**Fig. 2.** Overview of CUDAFD.

**Step 3**. The source domain pre-trained model with fixed classifier PCSC uses MMD to narrow the distribution difference between the source and the target domain data to continuously adapt to the unlabeled target domains $\{D_1, D_2, \ldots\ldots, D_T\}$ in sequence.

The proposed CUDAFD model can continuously adapt to the new target domains without forgetting the knowledge of the domains that has been learned. The details of MMD and PCSC will be introduced in the following sections.

### 3.2 Maximum Mean Discrepancy

In order to achieve efficient and reliable domain adaptation, many UDA methods focus on reducing the distribution difference between the source and the target domains. As shown in Fig. 3, a commonly used nonparametric distance measurement method is called MMD, which can estimate the difference in marginal distribution as follows:

$$MMD(D_s, D_t) = \left\| \frac{1}{n_s} \sum_{i=1}^{n_s} ?(x_i^s) - \frac{1}{n_t} \sum_{i=1}^{n_t} ?(x_i^t) \right\|_{\mathrm{H}}^2, \tag{1}$$

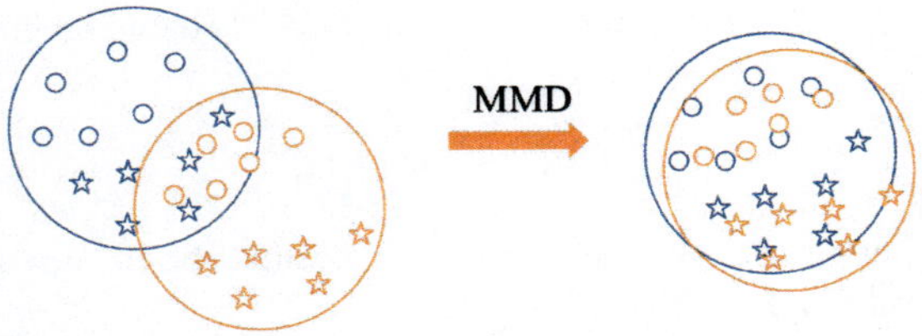


**Fig. 3.** MMD narrows the distance between the two domains.

where H denotes the reproducing kernel Hilbert space and ? is a mapping. The minimization of Eq. (1) can make the distribution of features extracted from source and target domains closer. Accordingly, the model can more accurately predict the labels of the samples from the target domains. The overall loss function used in CUDAFD includes a cross-entropy loss function and MMD as follows:

$$\text{Loss} = MMD(D_s, D_t) + \sum_{x \in D_s} \sum_{j=1}^{C} -\delta_{y_i^s = j} \cdot \log(\hat{y}_i), \tag{2}$$

where $\delta_{y_i^s=j}$ is equal to 1 when $y_i^s = j$ and 0 if otherwise, $\hat{y}_i$ is the output probability, and $D_t \in \{D_1, D_2, \ldots\ldots, D_T\}$.

In CUDAFD, the same labeled source domain will be reused to adapt to the new target domains, which means that the distribution of multiple target domains will be closer to the same source domain. The UDA methods, such as MMD, which reduces the distribution difference, can overcome catastrophic forgetting. The specific effect will be shown in the experimental part. The idea of prototype learning is introduced to further overcome catastrophic forgetting. The combination of PCSC classifier and MMD will achieve effective and continuous CUDA.

### 3.3 Prototype Cosine Similarity Classifier

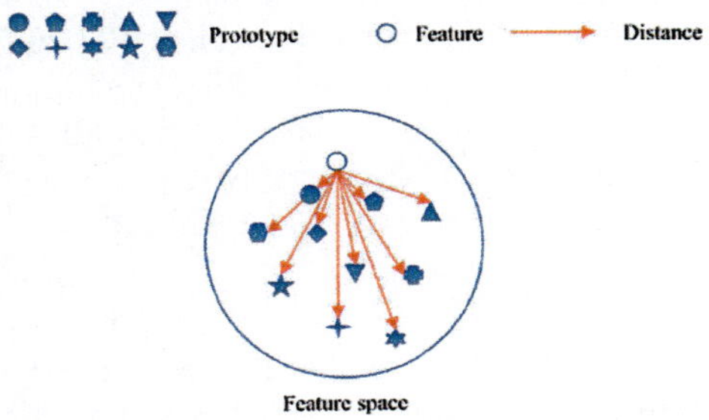


**Fig. 4.** Principle of PCSC.

The principle of PCSC is shown in Fig. 4. The PCSC is based on prototype learning, as follows:

$$\hat{y}_c = P(x) = \langle \theta_c, F(x) \rangle, \tag{3}$$

where $\theta_c$ is the prototype of fault type $c$, and $\langle \cdot, \cdot \rangle$ measures the cosine similarity between the feature $F(x)$ and prototype $\theta_c$.

In CUDAFD, fixing PCSC is equivalent to fixing the prototype of each fault type to overcome catastrophic forgetting. In this way, the new domain data will naturally move closer to the prototype from the source domain in the subsequent process of adapting to the new domain.

# 4 Experiment

## 4.1 Dataset Description

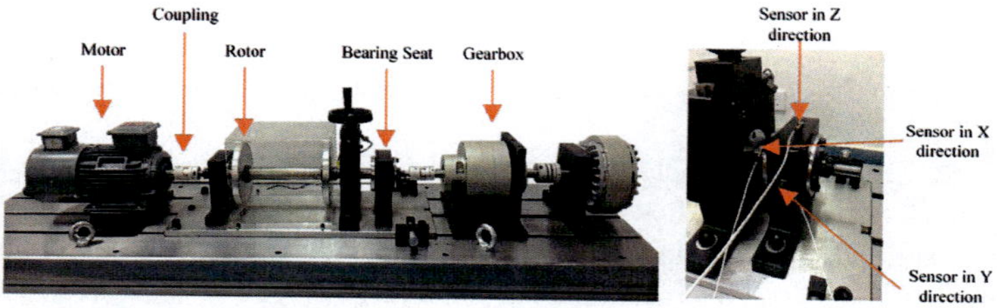


**Fig. 5.** Test rig for the collection of the datasets used.

The dataset used in this work is provided by the Shandong University of Science and Technology [10]. The left portion of Fig. 5 is the test rig. The right portion of Fig. 5 shows the three vibration sensors used to collect data. The sampling frequency is set at 25.6 kHz. This dataset contains 10 fault types and 24 working conditions combined with six speeds at 500–3000 r/min intervals of 500 r/min and four loads at 0–60 N intervals of 20 N. Select 500 r/min and 20 N as the source domain, or 500-20 for short, to simulate the working process of CUDA under complex working conditions, and the five target domains are 1000-20, 1500-40, 2000-0, 2500-60, and 3000-0. The source domain is called domain 0, and the target domains are domains 1, 2, 3, 4, and 5 for convenience.

## 4.2 Comparison with Relevant Methods

CUDAFD is compared with three related methods, namely, fine-tuning, domain adversarial neural networks (DANN) [11], and MMD, to demonstrate the effectiveness of the proposed method.

Fine-tuning is a common method in TL. In these experiments, fine-tuning will continuously fine-tune the old domain model with the new domain data in a supervised form. In fine-tuning, all domain data are labeled. The fine-tuning results can show the domain differences between the data collected under complex conditions.

DANN is a classical UDA method, which achieves effective UDA by adversarial training to ensure that the classifier cannot distinguish samples from the source and target domains. In our experiments, DANN follows the CUDA setting and continuously adapts the source domain to the new target domain.

MMD is a part of the proposed CUDAFD. This method also follows CUDA set-tings, but unlike CUDAFD, it does not use PCSC classifiers and is not fixed on classi-fiers trained from source domain data.

Figure 6 shows the confusion matrices trained by the four methods on six domains to clearly demonstrate the superiority of the proposed method. The vertical axis of the confusion matrix in Fig. 6 is the training session, and the training data of each session comes from the domain of the horizontal axis. The model obtained after each ses-sion of training is tested on all domains. The performance of the model on the un-trained domain can reflect the difference between domains, and the performance of the model on the trained domain can reflect the ability of the model to overcome catastrophic forgetting.

In Fig. 6(a), the confusion matrix of fine-tuning shows significant differences between domain 0 and domains 1–5, and the greater the differences between the working conditions, the greater the variations between domains. In addition, the differences between domains is small under high speed conditions.

In Fig. 6(b), DANN has poor ability to adapt to new domains or overcome catastrophic forgetting. Consequently, this method cannot implement CUDA. In contrast with DANN, MMD can continuously adapt to new domains, as shown in Fig. 6(c). Although the difference between domain 0 and domain 5 is large, the utilization of continuous multiple intermediate domain adaptation processes enables MMD to demonstrate good performance on domain 5 after completing training for session 5. Meanwhile, the catastrophic forgetting of MMD is not serious. Hence, MMD has the potential to implement CUDA.

CUDAFD is proposed on the basis of MMD and prototype learning. The prototype of each fault type trained from the source domain by PCSC is fixed. In combination with the ability of MMD to continuously adapt to the new domain, CUDAFD implements accurate CUDA, as shown in Fig. 6(d). In multiple training sessions, CUDAFD has satisfactory diagnostic accuracy and can better meet the diagnostic needs of actual scenarios.

## 5 Conclusion

In this paper, domain increments, which is regarded as a more challenging and meaningful diagnostic scenario, is introduced. Moreover, CUDAFD is proposed on the basis of the classical UDA method MMD to achieve continuous domain adaptation and prototype learning to overcome catastrophic forgetting. Finally, the validity and reliability of CUDAFD are verified by simulating CUDA under complex working conditions through a dataset.

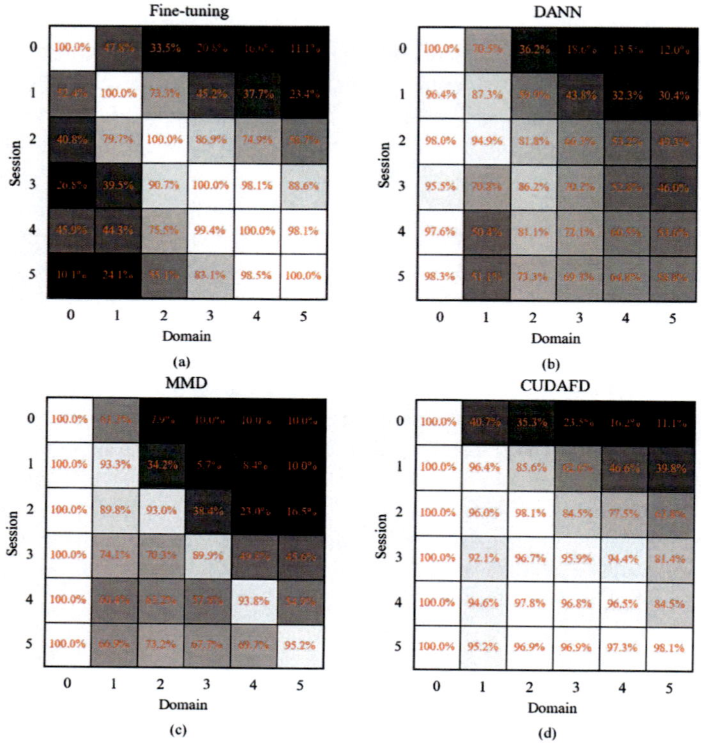


**Fig. 6.** Confusion matrix of the methods.

**Acknowledgments.** This work is financially supported by the National Natural Science Foundation of China (No. 52272440), and Soochow Science and Technology Innovation Talents Program (Nos. SYG202323, WC202168 and ZXL2022027).

## References

1. Liu, H., Wang, Y., Zhou, X., et al.: Question answering system for deterministic fault diagnosis of intelligent railway signal equipment. Smart Resilient Transp. **3**(2), 202–214 (2021)
2. Zhang, S., Su, L., Gu, J., et al.: Rotating machinery fault detection and diagnosis based on deep domain adaptation: a survey. Chin. J. Aeronaut. **36**(1), 45–74 (2023)
3. Xu, D., Li, Y., Song, Y., et al.: IFDS: An intelligent fault diagnosis system with multisource unsupervised domain adaptation for different working conditions. IEEE Trans. Instrum. Meas. **70**, 1–10 (2021)
4. Chen, B., Shen, C., Wang, D., et al.: A lifelong learning method for gearbox diagnosis with incremental fault types. IEEE Trans. Instrum. Meas. **71**, 1–10 (2022)
5. Li, J., Yue, K., Huang, R., et al.: Unsupervised continual source-free network for fault diagnosis of machines under multiple diagnostic domains. IEEE Sens. J. **23**(8), 8292–8303 (2023)

6. Chen, R., Wu, J., Peng, Y., et al.: Solving floating pollution with deep learning: A novel SSD for floating objects based on continual unsupervised domain adaptation. Eng. Appl. Artif. Intell. **120**, 105857 (2023)
7. Borgwardt, K.M., Gretton, A., Rasch, M.J., et al.: Integrating structured biological data by kernel maximum mean discrepancy. Bioinform. **22**(14), 49–57 (2006)
8. Ji, Z., Chai, X., Yu, Y., et al.: Improved prototypical networks for few-shot learning. Pattern Recogn. Lett. **140**, 81–87 (2020)
9. He, K., Zhang, X., Ren, S., Sun, J.: Deep residual learning for image recognition. In: Proceedings of the IEEE Conference on Computer Vision and Pattern Recognition (CVPR), pp. 770–778 (2016)
10. Jia, S., Wang, J., Han, B., et al.: A novel transfer learning method for fault diagnosis using maximum classifier discrepancy with marginal probability distribution adaptation. IEEE Access **8**, 71475–71485 (2020)
11. Ganin, Y., Ustinova, E., Ajakan, H., et al.: Domain-adversarial training of neural networks. J. Mach. Learn. Res. **17**(1), 2096–2030 (2016)

# Energy Minimization for Low-Latency Downlink Rate Splitting Multiple Access

Hao Tang[1,2,3], Xiaofang Sun[1,3(✉)], Li Yang[4], Guangxiang Hu[4], and Zhangdui Zhong[1,2,3]

[1] Collaborative Innovation Center of Railway Traffic Safety, Beiing Jiaotong University, Beijing 100044, China
[2] School of Electronic Information Engineering, Beijing Jiaotong University, Beijing 100044, China
[3] Beijing Engineering Research Center of High-Speed Railway Broadband Mobile Communications, Beiing Jiaotong University, Beijing 100044, China
xiaofangsun@bjtu.edu.cn
[4] Shanghai Shentie Investment Co., Ltd., Shanghai, China

**Abstract.** This paper investigates the energy minimization problem in a dual-user downlink system with strict delay constraints. To satisfy these delay constraints and balance the trade-off between block length and reliability, we use a normal approximation formula with finite blocklength code (FBC) capacity instead of the traditional Shannon formula. We consider a transmission scheme based on rate splitting multiple access (RSMA). The results show that although the objective function is not convex and does not have a closed-form expression, the RSMA problem can still be approximated to an equivalent convex function to obtain suboptimal solutions. Simulation results demonstrate that the RSMA scheme achieves significant energy savings compared to simpler non-orthogonal multiple access (NOMA) and orthogonal multiple access (OMA) schemes.

**Keywords:** Energy minimization · Finite blocklength codes · RSMA · Short packet communications · Ultra-reliable and low-latency communications

## 1 Introduction

In next-generation communication networks, the need to accommodate application scenarios such as ultra-reliable low-latency communication (uRLLC) continues to grow [1]. The system must ensure high reliability and ultra-low latency communication [2] between devices, which is a challenge given the exponential growth of various new time-critical services and data traffic [3,4]. Therefore, research into advanced multiple access transmission schemes is necessary to achieve uRLLC. Furthermore, considering the limited energy resources of most uRLLC devices, the exploration of new energy-efficient transmission methods for multi-user channels is also critical.


Y. Qin et al. (Eds.): EITRT 2023, LNEE 1137, pp. 404–414, 2024.
https://doi.org/10.1007/978-981-99-9311-6_43

The traditional Shannon formula, which is only applicable in the case of infinitely long code block lengths [5,6], does not extend to uRLLC. This limitation has motivated the investigation into short packet transmission using finite blocklength codes (FBC) [7] as a means to reduce latency. Recently, in [8], an energy efficiency analysis was conducted in a multi-user multiple-input single-output (MISO) network under short packet transmission. In [9], the focus was on studying the optimal block length and minimizing transmission energy in a hybrid non-orthogonal multiple access (NOMA) scheme. However, these studies were based on traditional multiple access technologies. To address this, this paper introduces rate-splitting multiple access (RSMA), a scheme that allows for the flexible splitting and combining of original messages and superposition transmission following certain rules. RSMA can effectively reduce the block length, making it better suited for short packet transmission. At present, [10–12] has studied the performance of RSMA in the context of users with different latency requirements, but these studies still rely on the traditional Shannon capacity which is not applicable to uRLLC. To address this, [13,14] have considered the the short packet communication performance of RSMA, taking into account factors such as power and block length and their impact on throughput. However, the aforementioned works did not consider its energy efficiency performance.

In this paper, we consider the problem of energy saving for dual user heterogeneous downlink RSMA with FBC. In particular, we solve the problem of minimizing energy subject to heterogeneous delay constraints, power constraints, and reliability constraints. As the approximate formula of FBC capacity is non-convex, the design of the formula is challenging. For this reason, we provide detailed insight and analysis on the constraint conditions and propose an iterative algorithm to solve the optimization problem.

The remainder of this paper is organized as follows. Chapter 2 describes the fundamental principles of RSMA and the proposed system modeling and optimization problems. Chapter 3 discusses equivalent transformations and approximate approximations for the original optimization problem, along with the proposed iterative algorithm. Chapter 4 presents the performance metrics of the iterative algorithm and compares RSMA with other traditional multiple access techniques such as NOMA and orthogonal multiple access (OMA). Finally, Chap. 5 concludes the paper.

## 2 System Model and Problem Formulation

### 2.1 System Model

We investigate the downlink rate splitting (RS) problem. A base station (BS) equipped with $\mathrm{N_t}$ antennas communicates with two single-antenna users. The BS sends individual messages to each user. The cell, with the BS at its center, is modeled as a disk with radius $\mathrm{R_c}$. In this network, RSMA is applied. To investigate energy-efficient beamforming design in RSMA, we assume that complete global channel state information (CSI) is fully known. CSI can be obtained by channel estimation based on [15–17].

### 2.2 RSMA Transmission Schame

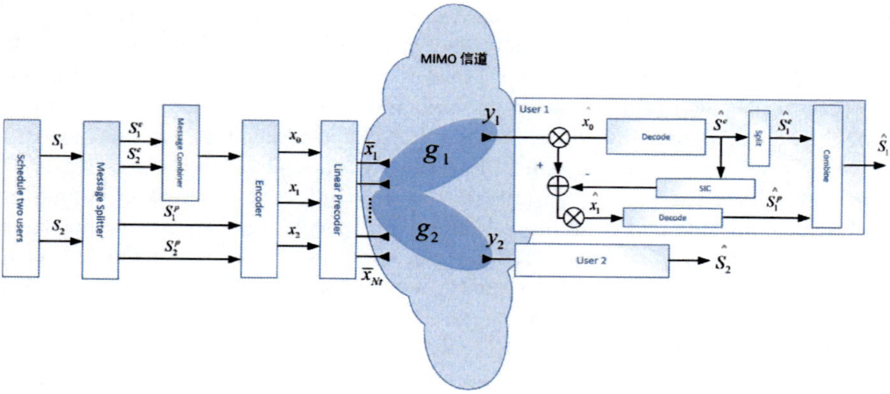


**Fig. 1.** RSMA sending and receiving process.

On the BS side, as shown in Fig. 1, the messages sent to users are represented by $S_k(k=1,2)$. Each message is divided into a common message $S_k^c$ and a private message $S_k^p$. The common messages from all users are combined and encoded together into $x_0$, which will eventually be decoded by all users. Meanwhile, each user's private message is encoded into $x_k(k=1,2)$, which will only be decoded by the user $k$. We consider that two users are constrained by the delays of $\mathrm{D}_1$ and $\mathrm{D}_2$ respectively. In our work, we consider a general case ($\mathrm{D}_1 \leqslant \mathrm{D}_2$).

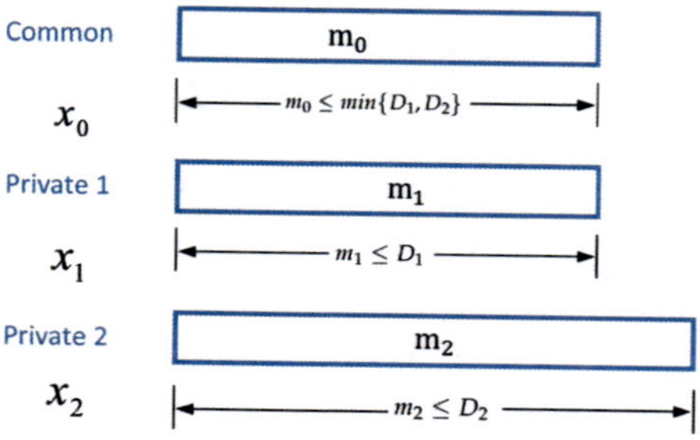


**Fig. 2.** Heterogeneous delay constraints of streams.

As shown in Fig. 2, the BS encodes the $\mathrm{N}_0^c$ bits public message into a code block of $m_0$, and the $\mathrm{N}_k^p(k=1,2)$ bits private message is encoded into a code block of $m_k(k=1,2)$. Due to the strict delay constraints imposed on both receivers, $m_j$ must be less than $\mathrm{D}_j(j=0,1,2)$. $m_k(k=1,2)$ are sent to user

$k$ in a beamforming manner respectively, and $m_0$ is superimposed on $m_k$ in a broadcast manner and sent to all users. Note that $\mathrm{N}_0^c$ and $\mathrm{N}_k^p$ are predefined.

The vector of streams is represented as $\boldsymbol{x} = [x_0, x_1, x_2]^T$. which are the unit average power coded symbols. To reduce interference, a precoding matrix $\widehat{\boldsymbol{W}} = [\boldsymbol{w}_0, \boldsymbol{w}_1, \boldsymbol{w}_2]_{\mathrm{N_t}\times 3}$ is used to give appropriate weights at each antenna. Finally, the signals are broadcast as $\boldsymbol{X} = [\bar{x}_1, \ldots, \bar{x}_{\mathrm{N}_t}]^T$. Therefore, the transmit signal is as follows:

$$\boldsymbol{X} = \widehat{\boldsymbol{W}}\boldsymbol{x} = \boldsymbol{w}_0 x_0 + \boldsymbol{w}_1 x_1 + \boldsymbol{w}_2 x_2 \tag{1}$$

The received signal of the user $k$ is as follow:

$$y_k = \tilde{\boldsymbol{g}}_k^{\mathrm{T}} \boldsymbol{X} + n_k = \tilde{\boldsymbol{g}}_k^{\mathrm{T}} \boldsymbol{w}_0 x_0 + \tilde{\boldsymbol{g}}_k^{\mathrm{T}} \boldsymbol{w}_1 x_1 + \tilde{\boldsymbol{g}}_k^{\mathrm{T}} \boldsymbol{w}_2 x_2 + n_k \tag{2}$$

where $n_k \sim \mathcal{CN}(0, \sigma_k^2)$ is complex additive white Gaussian noise (AWGN) with zero mean and variance $\sigma_k^2$. $\tilde{\boldsymbol{g}}_k$ is the channel coefficient from BS to the users. We assume that $\tilde{\boldsymbol{g}}_k$ to be fully known and $\mathrm{E}\{xx^H\} = \boldsymbol{I}$. BS's power is constrained by $\mathrm{Tr}\{\widehat{\boldsymbol{W}}\widehat{\boldsymbol{W}}^H\} \leqslant \mathrm{P_t}$, where $\mathrm{P_t}$ is the BS's maximum power.

In the RSMA, each user regards the public message as the target to receive at first, and eliminates after receiving it by successive interference cancellation (SIC). In order to cope with the delay constraints, the classic Shannon capacity is no longer applicable, so we use a normal approximation of the FBC capacity. The achievable rate of receiving common messages is given by:

$$\frac{\mathrm{N}_0^c}{m_0} = \log_2(1+\gamma_k^c) - \sqrt{\frac{1}{m_0}\left(1 - \frac{1}{(\gamma_k^c+1)^2}\right)}\frac{Q^{-1}(\epsilon_k^c)}{\ln 2}, k = 1, 2 \tag{3}$$

where $\gamma_k^c = \frac{|\tilde{\boldsymbol{g}}_k^{\mathrm{T}}\boldsymbol{w}_0|^2}{|\tilde{\boldsymbol{g}}_k^{\mathrm{T}}\boldsymbol{w}_1|^2 + |\tilde{\boldsymbol{g}}_k^{\mathrm{T}}\boldsymbol{w}_2|^2 + \sigma_k^2} = \frac{|\boldsymbol{g}_k^{\mathrm{T}}\boldsymbol{w}_0|^2}{|\boldsymbol{g}_k^{\mathrm{T}}\boldsymbol{w}_1|^2 + |\boldsymbol{g}_k^{\mathrm{T}}\boldsymbol{w}_2|^2 + 1}$.

It is noted that $\boldsymbol{g}_k = |\tilde{\boldsymbol{g}}_k|^2/\sigma_k^2$ is the normalized channel coefficient. $\gamma_k^c$ is the signal-to-interference-to-noise (SINR). $\epsilon_k^c$ is the block error probability. $Q^{-1}()$ is the derivative of the Gaussian Q-function. Here we clarify that the $x_0$ is assumed in full interference.

Then we assume that the correct decoding common message of each users. If the common message is not correctly decoded, then we consider that users cannot decode the private message correctly. Therefore, in the second step, users decode private messages. Due to remove the influence of common messgae by SIC, each user can decode its private stream $x_k$ by treating the other users' as noise. The received SINR of $x_k$ can be given by $\gamma_1^p = \frac{|\boldsymbol{g}_1^{\mathrm{T}}\boldsymbol{w}_1|^2}{|\boldsymbol{g}_1^{\mathrm{T}}\boldsymbol{w}_2|^2+1}, \gamma_2^p = \frac{|\boldsymbol{g}_2^{\mathrm{T}}\boldsymbol{w}_2|^2}{|\boldsymbol{g}_2^{\mathrm{T}}\boldsymbol{w}_1|^2+1}$.

The achievable rate of users receiving private messages is given by:

$$\frac{\mathrm{N}_k^p}{m_k} = \log_2(1+\gamma_k^p) - \sqrt{\frac{1}{m_k}\left(1 - \frac{1}{(\gamma_k^p+1)^2}\right)}\frac{Q^{-1}(\epsilon_k^p)}{\ln 2}, k = 1, 2 \tag{4}$$

### 2.3 Problem Formulation

In the considered problem, the BS jointly optimizes $\boldsymbol{W}$ and $m_j$ to minimize the energy, which is defined as $\sum_{j=0}^{3} m_j|\boldsymbol{w}_j|^2$. Here, $|\boldsymbol{w}|^2$ represents the power of the stream, and $m$ signifies the coded symbol length. The optimization problem is given by:

$$\min_{\{m,\boldsymbol{w}\}} \quad m_0|\boldsymbol{w}_0|^2 + m_1|\boldsymbol{w}_1|^2 + m_2|\boldsymbol{w}_2|^2 \tag{5a}$$

$$s.t. \quad m_j \leqslant \mathrm{D}_j,\ j = 0, 1, 2 \tag{5b}$$

$$m_0 \leqslant m_1 \tag{5c}$$

$$m_1 \leqslant m_2 \tag{5d}$$

$$|\boldsymbol{w}_0|^2 + |\boldsymbol{w}_1|^2 + |\boldsymbol{w}_2|^2 \leqslant \mathrm{P_t} \tag{5e}$$

$$\hat{m} \leqslant m_j, j = 0, 1, 2 \tag{5f}$$

$$\mathrm{F}_k^c(m_0, \gamma_k^c) = 0,\ k = 1, 2 \tag{5g}$$

$$\mathrm{F}_k^p(m_k, \gamma_k^p) = 0,\ k = 1, 2 \tag{5h}$$

$$\gamma_k^c = \frac{|\boldsymbol{g}_k^{\mathrm{T}}\boldsymbol{w}_0|^2}{|\boldsymbol{g}_k^{\mathrm{T}}\boldsymbol{w}_1|^2 + |\boldsymbol{g}_k^{\mathrm{T}}\boldsymbol{w}_2|^2 + 1}, k = 1, 2 \tag{5i}$$

$$\gamma_1^p = \frac{|\boldsymbol{g}_1^{\mathrm{T}}\boldsymbol{w}_1|^2}{|\boldsymbol{g}_1^{\mathrm{T}}\boldsymbol{w}_2|^2 + 1} \tag{5j}$$

$$\gamma_2^p = \frac{|\boldsymbol{g}_2^{\mathrm{T}}\boldsymbol{w}_2|^2}{|\boldsymbol{g}_2^{\mathrm{T}}\boldsymbol{w}_1|^2 + 1} \tag{5k}$$

where (5b), (5c), and (5d) are the latency constraints. Specifically, (5c) and (5d) imply that $m_0$ is completely interfered with by private messages $m1$ and $m2$, under the assumption that the codeword length of user 2 is greater than that of user 1. (5e) imposes a power limitation, while (5f) represents the minimum blocklength constraint for (5g) and (5h) to hold true (typically, $\hat{m} = 100$). (5i)–(5k) enforce the SINR restriction. Finally, (5g) and (5h) are the normal approximation of FBC capacity constraints with:

$$\mathrm{F}_k^c(m_0, \gamma_k^c) = \log_2(1 + \gamma_k^c) - \sqrt{\frac{1}{m_0}\left(1 - \frac{1}{(\gamma_k^c + 1)^2}\right)}\frac{Q^{-1}(\epsilon_k^c)}{\ln 2} - \frac{\mathrm{N}_0^c}{m_0} \tag{6}$$

$$\mathrm{F}_k^p(m_k, \gamma_k^p) = \log_2(1 + \gamma_k^p) - \sqrt{\frac{1}{m_k}\left(1 - \frac{1}{(\gamma_k^p + 1)^2}\right)}\frac{Q^{-1}(\epsilon_k^p)}{\ln 2} - \frac{\mathrm{N}_k^p}{m_k} \tag{7}$$

## 3 Task Scheme Efficient

### 3.1 Problem Reformulation

Given that (5) exhibits non-convexity, we adopt successive convex approximation (SCA) and semidefinite relaxation (SDR). First, inspired by SDR, we introduce auxiliary matrices $\boldsymbol{W}_j = |\boldsymbol{w}_j|^2$ and $\boldsymbol{G}_k = |\boldsymbol{g}_k|^2$. Note that when $\boldsymbol{W}_j$ and $\boldsymbol{G}_k$ are positive semidefinite matrices of rank one (5) can be rewritten as follows:

$$\min_{\{m,\boldsymbol{W}\}} \quad m_0\mathrm{Tr}(\boldsymbol{W}_0) + m_1\mathrm{Tr}(\boldsymbol{W}_1) + m_2\mathrm{Tr}(\boldsymbol{W}_2) \tag{8a}$$

$$s.t. \quad \sum_{j\in 0,1,2} \mathrm{Tr}(\boldsymbol{W}_j) \leqslant \mathrm{P_t} \tag{8b}$$

$$\gamma_k^c = \frac{\mathrm{Tr}(\boldsymbol{G}_k\boldsymbol{W}_0)}{\mathrm{Tr}(\boldsymbol{G}_k\boldsymbol{W}_1) + \mathrm{Tr}(\boldsymbol{G}_k\boldsymbol{W}_2) + 1}, k = 1, 2 \tag{8c}$$

$$\gamma_1^p = \frac{\mathrm{Tr}(\boldsymbol{G}_1\boldsymbol{W}_1)}{\mathrm{Tr}(\boldsymbol{G}_1\boldsymbol{W}_2) + 1} \tag{8d}$$

$$\gamma_2^p = \frac{\mathrm{Tr}(\boldsymbol{G}_2\boldsymbol{W}_2)}{\mathrm{Tr}(\boldsymbol{G}_2\boldsymbol{W}_1) + 1} \tag{8e}$$

$$\mathrm{Rank}(\boldsymbol{W}_j) \leqslant 1, \ j = 0, 1, 2 \tag{8f}$$

$$\boldsymbol{W}_j \succeq 0, j = 0, 1, 2 \tag{8g}$$

$$(5b)(5c)(5d)(5f)(5g)(5h)$$

The vector solution of (8) can be retrieved from the matrix solution of (5) to ensure the rank-one constraint.

### 3.2 Successive Convex Approximation

Due to the non-convexity of (8a), (8c)–(8e), (5g), and (5h), (8) is also intractable. Initially, as the objective function is non-convex with respect to $m_j$ and $\boldsymbol{W}_j$, we introduce a set of auxiliary variables $\boldsymbol{\alpha} = \{\alpha_0, \alpha_1, \alpha_2\}$ as follows:

$$\alpha_j \geqslant m_j\mathrm{Tr}(\boldsymbol{W}_j), j = 0, 1, 2 \tag{9}$$

Next, to deal with (9), we adopt SCA based on the arithmetic-geometric mean (AGM) inequality, which allows for the convexification of the upper bound of the original non-convex feasible set. Thus, (9) is replaced by:

$$\alpha_j \geqslant \frac{1}{2}\left((m_jc_j)^2 + \left(\frac{\mathrm{Tr}(\boldsymbol{W}_j)}{c_j}\right)^2\right), j = 0, 1, 2 \tag{10}$$

where $c_j$ represents fixed feasible points. To tighten the upper bounds, we iteratively update $c_j$. The update equations in the $n$-th iteration are given by: $c_j^{(n)} = \sqrt{\frac{\mathrm{Tr}(\boldsymbol{W}_j^{(n-1)})}{m_j^{(n-1)}}}, j = 0, 1, 2.$

**Proposition 1.** *The functions* $\mathrm{F}_k^c(m_0, \gamma_k^c)$ *and* $\mathrm{F}_k^p(m_k, \gamma_k^p)$*, given in* (5g) *and* (5h)*, are strictly increasing with respect to* $m_j$ *and* $\gamma_k^c, \gamma_k^p$*, respectively.*

Similarly, we introduce $\nu_k^c$ and $\nu_k^p$ to replace $\gamma_k^c$ and $\gamma_k^p$ in (5g) and (5h), respectively. Given Proposition 1, $\mathrm{F}_k^c(m_0, \nu_k^c)$ and $\mathrm{F}_k^p(m_k, \nu_k^p)$ are monotonous with respect to $m_j$. However, since $\nu_k^c = \frac{\mathrm{Tr}(\boldsymbol{G}_k\boldsymbol{W}_0)}{\mathrm{Tr}(\boldsymbol{G}_k\boldsymbol{W}_1)+\mathrm{Tr}(\boldsymbol{G}_k\boldsymbol{W}_2)+1}$, $\nu_1^p = \frac{\mathrm{Tr}(\boldsymbol{G}_1\boldsymbol{W}_1)}{\mathrm{Tr}(\boldsymbol{G}_1\boldsymbol{W}_2)+1}$, and $\nu_2^p = \frac{\mathrm{Tr}(\boldsymbol{G}_2\boldsymbol{W}_2)}{\mathrm{Tr}(\boldsymbol{G}_2\boldsymbol{W}_1)+1}$ are still non-convex, we apply the SCA as in (10). The final difficulty arises from the rank-one constraint in (8f). To address this issue, we drop (8f) to obtain a relaxed version of (11), as shown below:

$$\min_{\{m,\boldsymbol{W},\alpha,\nu\}} \quad \sum_{j=0,1,2} \alpha_j \tag{11a}$$

$$s.t. \quad f_j^1 \triangleq 2\alpha_j - ((m_j c_j)^2 + (\frac{\mathrm{Tr}(\boldsymbol{W}_j)}{c_j})^2) \geqslant 0, j=0,1,2 \tag{11b}$$

$$f_j^2 \triangleq \mathrm{D}_j - m_j \geqslant 0, \ j=0,1,2 \tag{11c}$$

$$f^3 \triangleq m_1 - m_0 \geqslant 0 \tag{11d}$$

$$f^4 \triangleq m_2 - m_1 \geqslant 0 \tag{11e}$$

$$f^5 \triangleq \mathrm{P_t} - \sum_{j=0,1,2} \mathrm{Tr}(\boldsymbol{W}_j) \geqslant 0 \tag{11f}$$

$$f_j^6 \triangleq m_j - \hat{m} \geqslant 0, j=0,1,2 \tag{11g}$$

$$f_k^7 \triangleq \mathrm{F}_k^c(m_0, \nu_k^c) \geqslant 0, \qquad k=1,2 \tag{11h}$$

$$f_k^8 \triangleq \mathrm{Tr}(\boldsymbol{G}_k\boldsymbol{W}_0) - \frac{1}{2}\left((\nu_k^c\kappa_k)^2 + \left(\frac{\mathrm{Tr}(\boldsymbol{G}_k\boldsymbol{W}_1)+\mathrm{Tr}(\boldsymbol{G}_k\boldsymbol{W}_2)+1}{\kappa_k}\right)^2\right)$$

$$\geqslant 0, k=1,2 \tag{11i}$$

$$f_k^9 \triangleq \mathrm{F}_k^p(m_k, \nu_k^p) \geqslant 0, \ k=1,2 \tag{11j}$$

$$f^{10} \triangleq \mathrm{Tr}(\boldsymbol{G}_1\boldsymbol{W}_1) - \frac{1}{2}\left((\nu_1^p\iota_1)^2 + \left(\frac{(\mathrm{Tr}(\boldsymbol{G}_1\boldsymbol{W}_2)+1)}{\iota_1}\right)^2\right) \geqslant 0 \tag{11k}$$

$$f^{11} \triangleq \mathrm{Tr}(\boldsymbol{G}_2\boldsymbol{W}_2) - \frac{1}{2}\left((\nu_2^p\iota_2)^2 + \left(\frac{\mathrm{Tr}(\boldsymbol{G}_2\boldsymbol{W}_1)+1}{\iota_2}\right)^2\right) \geqslant 0 \tag{11l}$$

$$f_j^{12} \triangleq \boldsymbol{W}_j \succeq \boldsymbol{0}, j=0,1,2 \tag{11m}$$

Now (11) is convex for any given $c_k$, $\iota_k$, $\kappa_k$, and can thus be solved efficiently using solvers of off-the-shelf graph programs, such as CVX.

It should be noted that the rank constraint is removed in (11), and the obtained solution may not satisfy this constraint. However, as proven in Theorem 1, the solution $\boldsymbol{W_k}$ obtained from (11) always satisfies the rank-one constraint.

**Theorem 1.** *The optimal solution $\boldsymbol{W}_k$ obtained in (11) is always a rank-one matrix, despite the relaxation of the rank constraint.*

The proposed algorithm based on SCA and SDR is summarized in Algorithm 1, where $\epsilon_0$ represents the tolerance of algorithm. In the iterative process, the parameters are continuously updated to tighten the upper bound, and the feasible solution $\{\boldsymbol{W}_k, m_k\}$ of the convex optimization problem is generated according to the iterative scheme.

**Algorithm 1** Algorithm to find optimal $m_k$ and $\boldsymbol{W}_k$ for problem (11)

1: **Initialize:** n←0, $v_k^{c[n-1]}$, $v_k^{p[n-1]}$, $c_j^{[n-1]}$, $\iota_k^{[n-1]}$, $\kappa_k^{[n-1]}$ and $\epsilon_0$;
2: **repeat**;
3: n←n+1;
4: Solve problem (11) using $v_k^{c[n-1]}$, $v_k^{p[n-1]}$, $c_k^{[n-1]}$, $\iota_k^{[n-1]}$, and $\kappa_k^{[n-1]}$. Denote the optimal value of the objective function as $\alpha_k$ and the optimal solutions as $v_k^{c*}$, $v_k^{p*}$, $c_k^*$, $\iota_k^*$, $\kappa_k^*$, $m_k^*$, $\boldsymbol{W}_k^*$.;
5: Update $v_k^{c[n]} \leftarrow v_k^{c*}$, $v_k^{p[n]} \leftarrow v_k^{p*}$, $c_k^{[n]} \leftarrow c_k^*$, $\iota_k^{[n]} \leftarrow \iota_k^*$, $\kappa_k^{[n]} \leftarrow \kappa_k^*$;
6: **until:**$|\sum \alpha_k^{[n]} - \sum \alpha_k^{[n-1]}| < \epsilon_0$;
7: **Output:** $m_k^*$, $\boldsymbol{W}_k^*$ ;

Finally, we discuss the connection between these optimization problems in Lemma 1.

**Lemma 1.** *: Algorithm 1 converges to a stagnation point that satisfies the Karush-Kuhn-Tucker (KKT) condition of (5).*

## 4 Simulation Results

In this section, simulation results are provided to compare the performance of RSMA with NOMA and OMA. Drawing from the discussions on URLLC in 3GPP [18,19], it is assumed that these groups are composed of packets, each with a size of 32 bytes. The fading loss is modeled as $\boldsymbol{g}_k = g_k^{-\alpha}\tilde{\boldsymbol{g}}_k$ and $\boldsymbol{h}_k = h_k^{-\alpha}\tilde{\boldsymbol{h}}_k$, where $g_k$ and $h_k$ represent the distance from the BS to the users, respectively. $\tilde{\boldsymbol{g}}_k$ and $\tilde{\boldsymbol{h}}_k$ are the Rayleigh fading coefficients, and $\alpha$ is the path loss exponent. The system bandwidth is 1 MHz. In the simulation, a BS is set up, assuming that the locations of two users are randomly distributed within a disk centered at the BS with a radius of $R_c = 500$ m. The algorithm's iteration tolerance is set to $10^{-3}$. Without loss of generality, it is assumed that all users have the same noise power density ($\sigma_1 = \sigma_2$). The energy of RSMA is solved by the CVX in MATLAB.

In Fig. 3, the energy performance is evaluated by plotting the relationship between energy consumption and the number of data packets at different bit error rates. Additionally, a comparison is made with NOMA and OMA under the same conditions. It can be observed that due to the higher spectral efficiency of non-orthogonal superimposed coding, RSMA exhibits better performance than OMA. Simultaneously, since transmitting common bits in RSMA requires more energy than private bits, RSMA achieves better performance. Furthermore, as a smaller error rate is set, the energy consumption correspondingly increases.

In Fig. 4, the relationship between the proportion of common messages to total messages and the system energy consumption in the RSMA is illustrated. The NOMA and OMA schemes are compared under the same total transmission bit condition. It can be observed that the energy consumption of the system increases with the increase of the proportion of common messages, as the common messages require substantial resources for transmission to users.

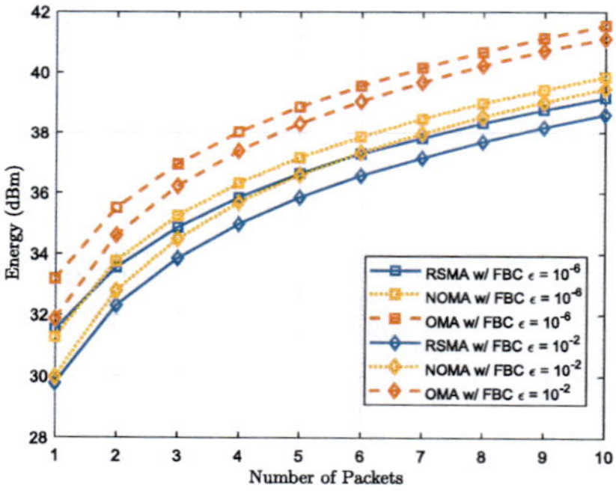


**Fig. 3.** Energy consumption of variouys schemes versus the number of packets with $P_{max} = 8dBm$

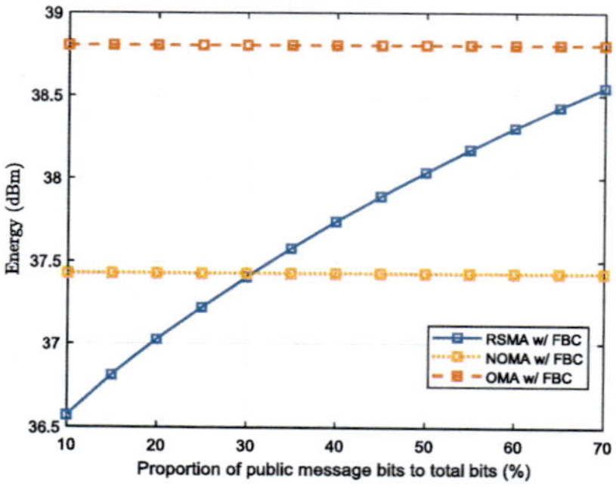


**Fig. 4.** Energy consumption of variouys schemes versus the proportion of common bits to the total number of bits

## 5 Conclusion

In this article, the energy-efficient transmission problem of RSMA under strict heterogeneous delay constraints is considered, and the normal approximation expression of FBC is used to describe the relationship between delay and reliability.

To solve this problem, a non-convex problem is constructed for the RSMA, semi-positive definite programming is conducted, and continuous convex approximation is applied to derive an equivalent convex problem from the original problem. The equivalence between the transformed convex problem and the original problem is proven. A suboptimal algorithm for this problem is proposed. The KKT conditions are used to demonstrate the convergence and stability points of the proposed algorithm on the proposed problem, and feasibility testing is finally conducted. Simulation results show that the RSMA scheme is significantly superior to both NOMA and OMA.

**Acknowledgements.** This work was supported in part by the Fundamental Research Funds for the Central Universities2022JBXT001, in part by NSFC under Grant 62101026, in part by the Project of China State Railway Group under Grant P2021G012.

## References

1. Bennis, M., Debbah, M., Poor, H.V.: Ultrareliable and low-latency wireless communication: Tail, risk, and scale. Proc. IEEE **106**(10), 1834–1853 (2018)
2. Li, Z., Uusitalo, M.A., Shariatmadari, H., Singh, B.: 5g URLLC: Design challenges and system concepts. In: 15th International Symposium on Wireless Communication Systems (ISWCS), pp. 1–6. IEEE (2018)
3. Saad, W., Bennis, M., Chen, M.: A vision of 6g wireless systems: Applications, trends, technologies, and open research problems. IEEE Netw. **34**(3), 134–142 (2020)
4. Letaief, K.B., Chen, W., Shi, Y., Zhang, J., Zhang, Y.-J.A.: The roadmap to 6g: Ai empowered wireless networks. IEEE Commun. Mag. **57**(8), 84–90 (2019)
5. Shannon, C.E.: A mathematical theory of communication. Bell Syst. Tech. J. **27**(3), 379–423 (1948)
6. Cover, T., Thomas, J.: Elements of Information Theory. Wiley (1991)
7. Durisi, G., Koch, T., Popovski, P.: Toward massive, ultrareliable, and low-latency wireless communication with short packets. Proc. IEEE **104**(9), 1711–1726 (2016)
8. Singh, K., Ku, M.-L., Flanagan, M.F.: Energy-efficient precoder design for downlink multi-user miso networks with finite blocklength codes. IEEE Trans. Green Commun. Netw. **5**(1), 160–173 (2021)
9. Xu, Y., Shen, C., Cai, D., Zhu, G.: Latency constrained non-orthogonal packets scheduling with finite blocklength codes. IEEE Trans. Veh. Technol. **69**(10), 12 312–316 (2020)
10. Yang, Z., Chen, M., Saad, W., Xu, W., Shikh-Bahaei, M.: Sum-rate maximization of uplink rate splitting multiple access (RSMS) communication. In: IEEE Global Communications Conference (GLOBECOM), pp. 1–6. IEEE (2019)
11. Tegos, S.A., Diamantoulakis, P.D., Karagiannidis, G.K.: On the performance of uplink rate-splitting multiple access. IEEE Commun. Lett. **26**(3), 523–527 (2022)
12. Yang, Z., Chen, M., Saad, W., Xu, W., Shikh-Bahaei, M.: Sum-rate maximization of uplink rate splitting multiple access (RSMA) communication. IEEE Trans. Mobile Comput. **21**(7), 2596–2609 (2022)
13. Vu, T.-H., Duong, T. Q., Khuat, T.-H., Nguyen, T. H., Kim, S.: Short-packet communication for cooperative uav-based mimo systems with rate splitting multiple access. In: 2022 IEEE Ninth International Conference on Communications and Electronics (ICCE), pp. 113–117. IEEE (2022)
14. Dhok, S., Sharma, P.K.: Rate-splitting multiple access with star RIS over spatially-correlated channels. IEEE Trans. Commun. **70**(10), 6410–6424 (2022)
15. Palleit, N., Weber, T.: Obtaining transmitter side channel state information in mimo fdd systems. In: IEEE 20th International Symposium on Personal, Indoor and Mobile Radio Communications, pp. 2439–2443. IEEE (2009)
16. Jose, J., Ashikhmin, A., Whiting, P., Vishwanath, S.: Channel estimation and linear precoding in multiuser multiple-antenna TDD systems. IEEE Trans. Veh. Technol. **60**(5), 2102–2116 (2011)

17. Thomas, T.A., Vook, F.W.: Method for obtaining full channel state information for RF beamforming. In: IEEE Global Communications Conference, pp. 3496–3500. IEEE (2014)
18. 3GPP: Summary of email discussion on the link level evaluation for LTE URLLC. TSG RAN WG1 Meeting #92, R1-1801385, Mar. 2018. Available: http://www.3gpp.org/ftp/Meetings_3GPP_SYNC/RAN1/Docs/
19. 3GPP: Discussion on PDSCH related techniques for URLLC. TSG RAN WG1 Meeting #92, R1-1801776, Mar. 2018. Available: http://www.3gpp.org/ftp/Meetings_3GPP_SYNC/RAN1/Docs/

# Spatial and Temporal Distribution of Passenger Flow on Urban Rail Transit Under Train Failure Scenarios

Xiaoke Deng[1], Yao Chen[1](✉), Xujie Feng[2], and Yun Bai[1]

[1] Key Laboratory of Transport Industry of Big Data Application Technologies for Comprehensive Transport, Beijing Jiaotong University, Beijing 100044, China
{21120802,chenyao,yunbai}@bjtu.edu.cn

[2] China Academy of Transportation Sciences, Beijing 100029, China
fengxujie@foxmail.com

**Abstract.** The deduction of passenger flow distribution is the basis of restoring normal operation order under train failure scenarios, and provides the basis for the formulation and implementation of train rescheduling and passenger flow control measures. In order to calculate passenger flow accurately and efficiently and ensure operation safety, a method for deducing the passenger flow spatial and temporal distribution is proposed to solve the above problems. Based on the route selection model, considering the changes of passengers' travel behavior after failure, the function of passengers' willingness to wait time is constructed to calculate the inbound passenger flow. Individual route choice behavior is updated according to the change of transport capacity, and reassign the route selection. The Multi-agent simulation modeling technology is used to dynamically deduce the temporal and spatial distribution of passenger flow. The empirical analysis of train faults in Hefei Metro shows that the average relative errors of the inbound and outbound stations of the fault line are 4.89% and 8.74%, respectively, which verifies the validity of the method to deduce the passenger flow distribution. In the fault scenario, compared with the normal operation scenario, the change rule of inbound traffic over time is similar but greatly reduced. The outbound passenger flow of a single station arrives at an interval, and the number of passengers in the interval is much greater than that in the same period and then slightly less than that in the same period.

**Keywords:** Urban rail transit · Spatio-temporal distribution of passenger · Path assignment · Travel behavior · Train fault

## 1 Introduction

The rapid development of urban rail transit makes it gradually shift from single line to network operation. Under the one-ticket system, there is no need to purchase tickets for transfer between different lines, but it also makes it difficult to obtain the specific passenger's route in the network. At the same time, it is more difficult to grasp the passengers'


Y. Qin et al. (Eds.): EITRT 2023, LNEE 1137, pp. 415–426, 2024.
https://doi.org/10.1007/978-981-99-9311-6_44

travel behavior with individual differences when emergencies. If the spatial and temporal distribution of passenger flow cannot be effectively deduced, the passenger flow control and the train rescheduling will be affected, which may further cause operation safety problems. Therefore, it is of great significance to study the calculation method of passenger flow distribution after faults.

The passenger flow distribution of urban rail transit is mainly realized through three steps: analyzing travel behavior, constructing passenger flow distribution model, and stimulating passenger-train interaction. Passenger behavior analysis generally adopts AFC data [1] and questionnaire survey [2] to study potential preferences and quantitatively analyze travel behavior. The passenger flow assignment model is built based on route choice behavior, which usually considers personal characteristics [3] and path attributes including on-board time, transfer convenience, route comfort, etc., and constructs a route impedance function [4, 5]. It is believed that passengers choose the route with the lowest perceived cost [6]. Based on the Logit model and its derivative model, after determining the route choice probability, the interaction between passenger flow and trains is realized by using the simulation method to obtain the passenger flow distribution [7, 8].

In view of special operation scenarios such as equipment fault, train delay and operation disruption, existing studies generally based on the idea of normal operation, take into account the duration and influence range of special scenarios [9] and the change of route impedance function [10], and re-determine the set of optional routes for passengers. However, in actual operation, emergencies are highly uncertain, and it is difficult for passengers to know the specific time when the impact of emergencies ends in advance.

Aiming at the train fault scenario and its uncertainty in the rush hour of urban rail transit, this paper analyzes the relationship between the time passengers are willing to wait for the operation to resume and the travel distance, constructs the passenger waiting time function, and explores the route choice characteristics of passengers affected by faults. The route assignment of passengers who continue to choose urban rail transit is carried out, and through the Multi-agent simulation, a more accurate calculation method of passenger flow spatial and temporal distribution is obtained.

The remainder of the paper is organized as follows: Sect. 2 analyzes the passenger behavior and presents the passenger route choice model. A simulation technology of passenger flow distribution based on Multi-agent framework is developed in Sect. 3. Section 4 takes Hefei Metro as an example to verify the model and analyzes the characteristics of passenger flow distribution; Finally, Sect. 5 covers the conclusion of the study.

## 2 Passenger Route Choice Model

### 2.1 Problem Description

When a train fault occurs in urban rail transit, the operation order will be affected to different degrees due to the different cause, severity and disposal plan of the fault. Therefore, the train fault is uncertain, and the train delay time is difficult to determine. The affected passengers cannot accurately know the fault recovery situation, and their travel behavior is strongly subjective. Through questionnaire survey, this paper analyzes the characteristics of passenger travel behavior in train fault scenarios.

In the fault scenario, urban rail transit passengers will choose to wait for a period of time to observe whether the fault can be solved in a relatively short time before deciding whether to continue the original route, change to other urban rail transit routes or turn to other modes of transportation. If the delay is longer than the passenger's acceptable waiting time, the passenger will make a new decision.

Therefore, according to the location of passengers in the research period, passengers affected by the fault are selected to calculate their possible willing waiting time. After the time is over, passengers are assigned routes and trains to complete the dynamic deduction of the spatial and temporal distribution in urban rail transit. The basic research framework is shown in Fig. 1.

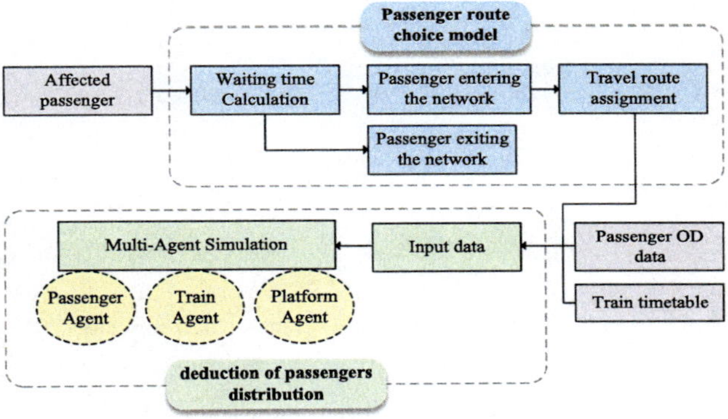


**Fig. 1.** The basic research framework

## 2.2 Passenger Behavior Characteristics in Fault Scenario

In order to explore the relationship between passengers' behavior of waiting for operation recovery and travel distance, a questionnaire was set for passenger waiting time selection under different travel distance scenarios. In order to ensure the refinement and accuracy of the questionnaire, the travel distance was selected. The average distance of most urban rail transit is between 7 and 14 km, as shown in Fig. 2. Therefore, the travel distance in the questionnaire is set to be less than 9 km, 9–16 km and more than 16 km, and the travel time is about less than 20 min, 20–40 min and more than 40 min respectively. A total of 211 valid questionnaires were collected through the social platform. The statistical results are shown in Fig. 3.

According to the survey results, more than 95% of passengers are willing to wait for some time before giving up urban rail transit. But more than 50% of passengers wait less than 15 min, and more than 97% wait 30 min or less. The relationship between waiting time and travel distance was further quantitatively analyzed, and Gaussian function was selected for fitting, as shown in Fig. 4. The selection probability distribution function $y$ of different waiting time $w$ is as follows:

$$y(w) = \begin{cases} 0.553e^{-(\frac{w-1.427}{0.960})^2} & d < 9 \\ 0.497e^{-(\frac{w-1.929}{1.019})^2} & 9 \leq d < 16 \\ 0.335e^{-(\frac{w-2.326}{1.590})^2} & d \geq 16 \end{cases} \tag{1}$$

where $w$ is the time passengers are willing to wait, the values are $1, 2, 3\ldots$, corresponding to 0 min, 1–5 min, 11–15 min.... $d$ is the route travel distance.

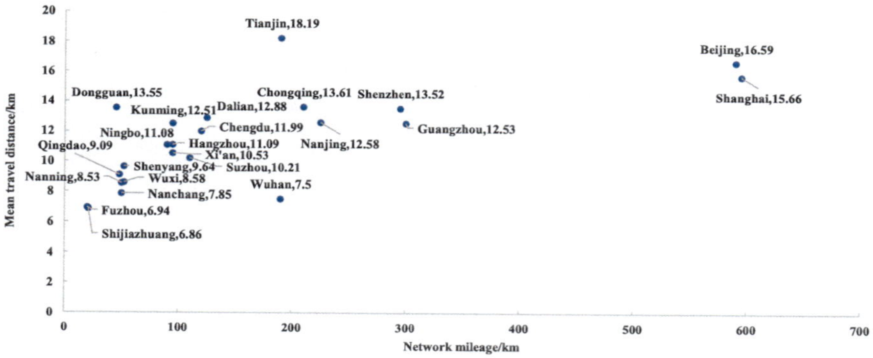


**Fig. 2.** Map of rail transit scale and average distance by city (2017)

Fig. 3.Statistical results of passenger waiting time survey

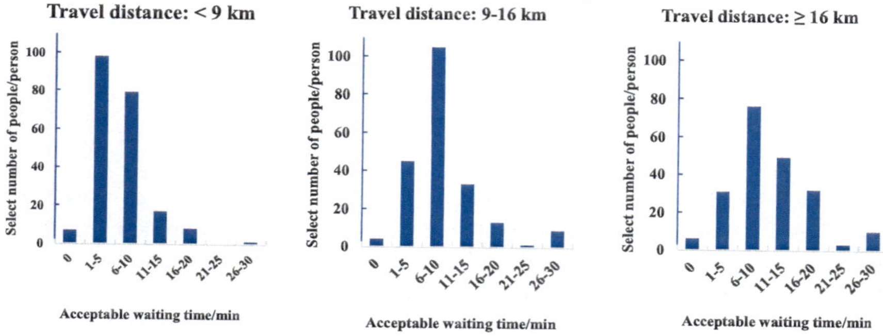


**Fig. 3.** Statistical results of passenger waiting time survey

The fitting results show that passenger waiting time is closely related to travel distance, and the waiting time increases with the increase of travel distance. At the same time, when the travel distance is short, the waiting time value is concentrated. When the travel distance is less than 9 km, more than 50% of passengers choose to wait 6–10 min. With the increase of travel distance, the values become more discrete, and the diversity and uncertainty of travel behavior increase.

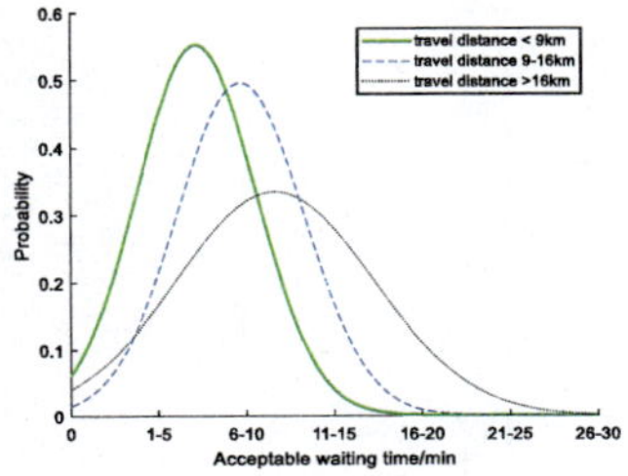


**Fig. 4.** Fitting function of passenger waiting time

Before the end of the waiting time, if a train arrives at the platform, the passenger get on the train successfully; If no train arrives at the platform during waiting time, the passenger will abandon the urban rail transit. The option $g_j^{s,u}$ for the passenger $j$ to continue using urban rail transit is as follows:

$$g_j^{s,u} = \begin{cases} 0 \ w_j^{s,u} < l_s^m \\ 1 \ w_j^{s,u} \geq l_s^m \end{cases} \quad (2)$$

where $g_j^{s,u}$ equals 0, which means that the passenger $j$ whose origin station $s$ and destination station $u$ gives up urban rail transit, otherwise the passenger continue to take. $w_j^{s,u}$ is the acceptable waiting time for passenger $j$. $l_s^m$ is the delay time of train $m$ at station $s$.

Inbound passenger number $qe_{s,u}$ is as follows:

$$qe_{s,u} = \sum_{j \in J} g_j^{s,u} \quad (3)$$

### 2.3 Passenger Route Assignment

Firstly, the physical network is transformed into a weighted and directed topology. The station is the node in the topology, and the non-transfer station $i$ on line $l$ is set as two virtual nodes $s_{i,l}^1$ and $s_{i,l}^2$; If it is a two-line transfer station $i$ for line $l$ and line $l\prime$, it is set to four virtual nodes $s_{i,l}^1$, $s_{i,l}^2$, $s_{i,l'}^1$ and $s_{i,l'}^2$. The node belonging to the same line and the same station are connected by stopping arc $(s_{i,l}^1, s_{i,l}^2)$ or $(s_{i,l}^2, s_{i,l}^1)$; The node belonging to different lines but belonging to the same station are connected by a transfer arc $(s_{i,l}^j, s_{i,l'}^j)$ or $(s_{i,l'}^j, s_{i,l}^j)$, where the value of $j$ is 1 or 2; The node at different stations are connected by running arc $(s_{i,l}^1, s_{i',l}^2)$ or $(s_{i',l}^2, s_{i,l}^1)$. Taking stop time, transfer time and section running time as the impedance of each arc, the topology structure of urban rail transit network is shown in Fig. 5.

According to the train rescheduling after the fault, K-shortest paths algorithm is used to compose the possible paths between OD into an effective path set. Then, route assignment is carried out for passengers. In the case of train fault, passengers generally expect the route to be convenient and fast. Therefore, considering travel time, transfer time and transfer frequency, the path generalized cost function $U_{s,u}^r$ is as follows:

$$U_{s,u}^r = \alpha_1 t_{run}^{s,u,r} + \alpha_2 t_{trans}^{s,u,r} + \alpha_3 \tau_{s,u}^r + \varepsilon \quad (4)$$

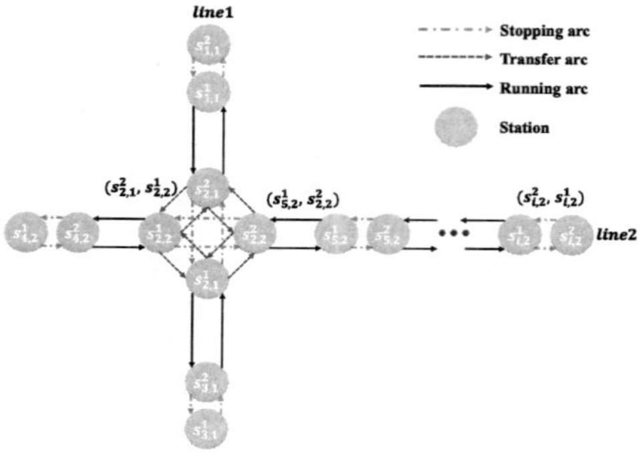


**Fig. 5.** Urban rail transit network topology

where $t_{run}^{s,u,r}$, $t_{trans}^{s,u,r}$ and $\tau_{s,u}^{r}$ are the travel time, transfer time and transfer frequency of the route $r$ from station $s$ to station $u$, respectively. $\alpha_1$, $\alpha_2$ and $\alpha_3$ are passengers' perceptual parameters respectively. $\varepsilon$ is a random term.

A passenger route choice model based on Logit model is constructed, and $\varepsilon$ obeys the Gumbel distribution, then the probability of route choice $P_{s,u}^{r}$ is as follows:

$$P_{s,u}^{r} = \frac{e^{(-\theta U_{s,u}^{r})}}{\sum_{r \in R} e^{(-\theta U_{s,u}^{r})}} \tag{5}$$

where $\theta$ is an indicator to measure passengers' familiarity with the network.

# 3 Simulation of Passenger Flow Distribution

## 3.1 Agent Attributes

Simulation is carried out based on the Multi-agent framework [11]. The system includes three types of agents, namely passengers, trains and platforms. Passenger agent attributes include the passenger's origin station $O_j$, arrival time $T_j$, destination station $D_j$, route $R_j$, and current location $W_j$. Train agent attributes include line $l$, stopping station $S_m$, arrival time $a_s^m$ and departure time $d_s^m$ at each station. Platform agent attributes include the line $l$ and trains set $M_s$.

## 3.2 Framework of Multi-Agent Simulation

Taking the activities of train agents as the center, passenger exchange between train and platform is completed based on passenger route choice behavior, so as to realize the interaction between multi-agents, as shown in Fig. 6.

Considering the train capacity limit, the number of waiting passengers $q_{wait}^{m,s,u}$, alighting passengers $q_{alight}^{m,u}$, boarding passengers $q_{aboard}^{m,s,u}$, passengers on board $q_{train}^{m,s}$ and stranding passengers $q_{strand}^{m,s,u}$ are as follows:

$$q_{wait}^{m,s,u} = \sum_{d_s^{m-1}}^{d_s^m} qe_{exit}^{s} + q_{strand}^{m-1,s,u} \tag{6}$$

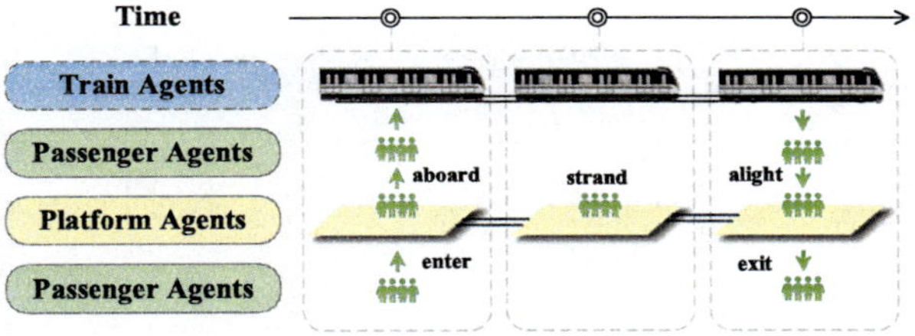


**Fig. 6.** Schematic diagram of Multi-agent simulation framework

$$q_{alight}^{m,u} = \sum_{s \in S_m^-} q_{aboard}^{m,s,u} \tag{7}$$

$$q_{aboard}^{m,s,u} = \begin{cases} (ca_m - q_{train}^{m,s-1} + q_{alight}^{m,s}) \cdot \frac{q_{wait}^{m,s,u}}{\sum_{u \in S_m^+} q_{wait}^{m,s,u}} & \sum_{u \in S_m^+} q_{wait}^{m,s,u} + q_{train}^{m,s-1} - q_{alight}^{m,s} > ca_m \\ q_{wait}^{m,s,u} & \sum_{u \in S_m^+} q_{wait}^{m,s,u} + q_{train}^{m,s-1} - q_{alight}^{m,s} \leq ca_m \end{cases} \tag{8}$$

$$q_{train}^{m,s} = q_{train}^{m,s-1} + q_{abroad}^{m,s,u} - q_{alights}^{m,s,u} \tag{9}$$

$$q_{strand}^{m,s,u} = q_{wait}^{m,s,u} - q_{abroad}^{m,s,u} \tag{10}$$

where $S_m^-$ is the station that train $m$ has passed through, and $S_m^+$ is the station that train $m$ has not passed through. $ca_m$ is the maximum number of passengers carried by train $m$. $m - 1$ is the previous train in the same line and the same direction as train $m$.

## 4 Case Study

Based on a train fault situation in Hefei Metro, this paper takes AFC data and train timetable as input to verify the validity of the model and further explore the impact of train fault on the temporal and spatial distribution of passenger flow.

### 4.1 Case Background and Parameter Value

As of January 2020, Hefei Metro network has three lines, including three transfer stations. At 6:23 a.m. on January 20, 2020, train 0102 of Line 3 ran to the Daxuechengbei - Fanhua Dadao Section (Xiangcheng Lu direction), and the train had emergency braking that could not be alleviated; Application for rescue at 6:27 a.m., train 3002 as the rescue train, 6:40 a.m. rescue train and the failure train connected completely and returned Modian depot (located at the eastern end of Line 3), the location of depot and trains are shown in Fig. 7. In the passenger route choice model, the values of parameters $\alpha_1$, $\alpha_2$, $\alpha_3$ and $\theta$ are respectively 1, 1.2640, 1.8481 and 1.8660 [12].

**Fig. 7.** Schematic diagram of the location of the depot and failure train

## 4.2 Model Validity Verification

The entrance passenger flow which the model deduced was compared with the actual entrance passenger flow of each station, as shown in Fig. 8a. The average relative error is 4.89%, and the relative error of more than 80% stations is less than 10%. The result shows that the model is effective for the analysis of passenger travel behavior under fault scenarios. The exit passenger flow which the model deduced was compared with the actual exit passenger flow of each station, as shown in Fig. 8b. The average relative error is 8.74%, and the relative error is concentrated in 0–15%. The result shows that the deduced results of the model for the temporal and spatial distribution of passenger flow are consistent with the actual situation.

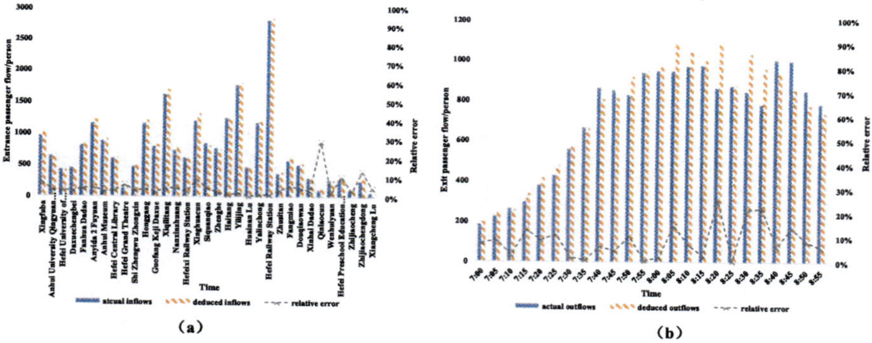


**Fig. 8.** Comparison of actual and model deduced passenger flow for Line 3

## 4.3 Passenger Flow Distribution Characteristics

**Station Passenger Flow Distribution**. The Comparison of inflows under the two scenarios is shown in Fig. 9a. The peak time of passenger flow is almost the same, and the peak time is concentrated between 7:00 a.m. and 9:00 a.m. However, the total number of inflows in the fault scenario decreased by about 22%. A single station is analyzed, taking Yilijing station with a large inbound volume as an example, as shown in Fig. 9b. The variation characteristics of station are the same as line, which increases with time from 6:10 a.m. to 7:55 a.m. and decreases with time from 7:55 a.m. to 9:00 a.m., and the inbound volume in fault scenario is less than that in normal operation scenario on the whole.

The Comparison of outflows under the two scenarios is shown in Fig. 10a. The peak period occurs between 7:35 a.m. and 8:45 a.m. under normal operation, which is basically consistent with the inflows. In the fault scenario, the outflows increases with time between 6:10 and 7:40 a.m. But at 7:40–8:45 a.m., instead of continuing to rise to a peak, it entered a flat period. Due to the increase of departure interval after train fault, inbound passengers can not board the train in a short time, and the waiting time is increased; After the resumption of operation, the train load rate increased and the outbound passenger flow resumed. A single station is analyzed, taking Shi Zhengwu Zhongxin station with a large outbound volume as an example, as shown in Fig. 10b. The passenger flow shows a gradual upward trend between 6:10 and 7:35 a.m. At this time, the passenger flow is small, the trains are not full, and the outbound volume is less affected by the delay. When the passenger flow gradually increases, the passenger flow arrival rule matches the train departure frequency, and the outflows show a trend of arriving at intervals with a period of 10–15 min, and repeatedly exceeds the peak outbound volume under normal operation. The operation department should control the passenger flow of important stations according to this phenomenon, so as to prevent the impact of unexpected large passenger flow.

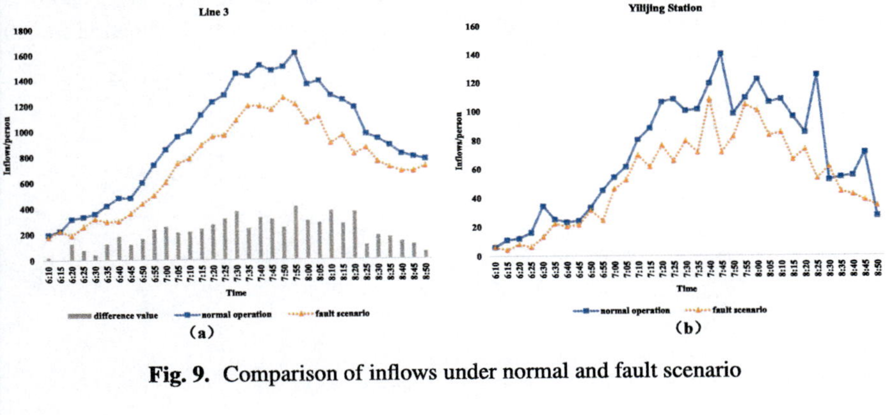


**Fig. 9.** Comparison of inflows under normal and fault scenario

**Section Passenger Flow Distribution.** Part of the train timetable of normal operation and fault scenario is shown in Fig. 11. Among them, train 0202 is the first train after the fault treatment, and train 0702 is close to its time in normal operation. Train 0902 is the first train that overlaps with the original planned operating time after the fault, and the planned operating time overlaps with train 1102 in normal operation.

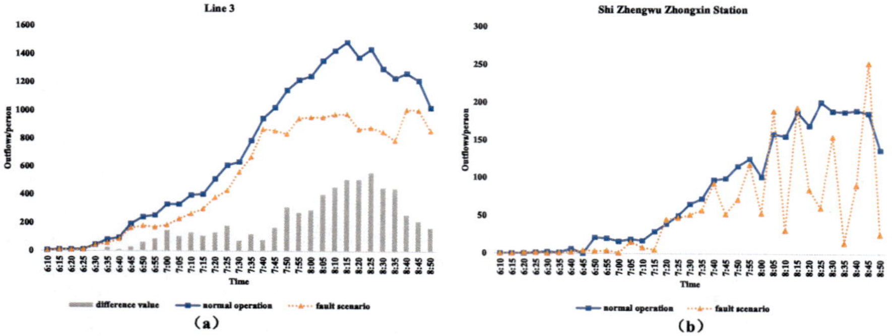


**Fig. 10.** Comparison of outflows under normal and fault scenario

The number of passengers on train 0202 and train 0702 is compared, as shown in Fig. 12a. Train 0202 needs to transport the passengers who are cleared and the passengers who arrive during the fault rescue period, so the passenger flow of train 0202 is much larger than that train 0702. If the number of passengers waiting for the train is greater than the capacity of the train, the phenomenon of passenger stranding will occur. Until the number of passengers in normal operation and fault scenario is similar, the fault has ended and normal operation will be resumed.

The number of passengers on train 0902 and train 1102 is compared, as shown in Fig. 12b. The number of passengers on train 0902 is significantly less than that on train 1102. The original plan of the departure interval of 5 min, limited by the train circulation, the train should departure as soon as possible, the trains' departure interval between train 0202 and train 0902 is shortened to 2–3 min. Because the departure interval is reduced and the previous trains has transported the affected passengers to the destination station, the number of passengers on train 0902 is reduced.

**Fig. 11.** Train timetable under normal operation and fault scenario.

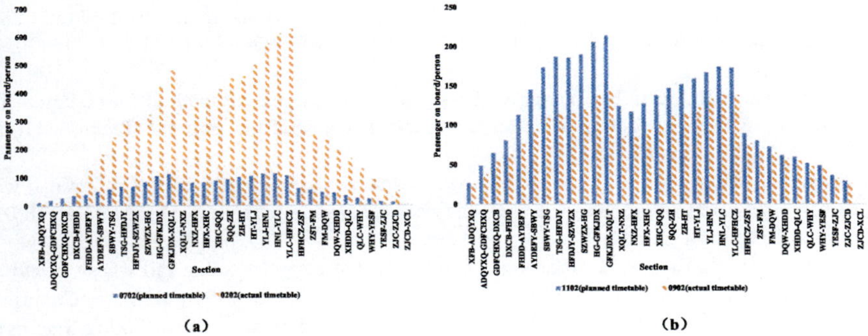


**Fig. 12.** Comparison of passengers on board between planned and actual timetable.

## 5 Conclusion

Aiming at the fault scenario of urban rail transit during peak hours, this paper proposes a dynamic calculation method of passenger flow distribution based on travel behavior analysis. The results show that: (1) The average relative error of inflows and outflows derived from the model are respectively 4.89% and 8.74%, which fully validates the travel behavior analysis is suitable for passenger flow distribution in train fault scenarios. (2) Compared with the normal operation, the rule of inbound volume in the fault scenario is similar but reduced; The outbound volume decreases, and that at a single station tends to arrive at intervals. (3) After the fault occurred, the number of passengers on board was much larger than the normal operation. With the adjustment of organization and the reduction of departure interval, the number of passengers on board after the fault was slightly smaller.

This paper guarantees the accuracy of the calculation of passenger flow distribution through the specific description of passenger travel behavior, which can provide a basis for passenger flow control after emergencies. In the further study, we can take this as the basis to further explore the train rescheduling, passenger flow control and passenger flow guidance in the emergency scenarios.

**Acknowledgments.** This research was funded by the National Natural Science Foundation of China (No. 71971016), and Beijing Natural Science Foundation (No. L221020).

## References

1. Briand, A., Come, E., Trepanier, M., Oukhellou, L.: Analyzing year-to-year changes in public transport passenger behaviour using smart card data. Transp. Res. Part C: Emerg. Technol. **79**, 274–289 (2017)
2. Yu, C., Li, H., Xu, X., Sun, Q.: Estimating left behind patterns in congested metro systems: a Bayesian model. Smart Resilient Transp. **3**(02), 149–161 (2021)
3. Raveau, S., Guo, Z., Munoz, J., Wilson, N.: A behavioural comparison of route choice on metro networks: time, transfers, crowding, topology and socio-demographics. Transp. Res. Part A: Policy Pract. **66**, 185–195 (2014)

4. Ni, S., Yang, H., Peng, Q.: Passenger flow distribution of regional multi-standard rail transit based on passenger route selection. J. Transp. Syst. Eng. Inf. Technol. **21**(01), 108–115 (2021). (in Chinese)
5. Xu, D., Gong, L., Zhu, N., Wang, X.: Passenger flow assignment method for common-line operation with multi-routing of urban rail transit. J. Transp. Syst. Eng. Inf. Technol. **21**(05), 206–213 (2021). (in Chinese)
6. Wang, G., Zhong, M., Li, J., Sun, Y.: A route dynamic assignment model of rail transit with time-varying network topology: a case study of Wuhan city. J. Transport Inform. Saf. **38**(03), 48–56 (2020). (in Chinese)
7. Lu, G., Lei, Y., Zhang, H.: Passenger flow control strategy of urban rail transit based on multi-agent simulation. J. Tongji Univ. (Nat. Sci.) **50**(08), 1189–1197 (2022). (in Chinese)
8. Lei, Y., Lu, G., Zhang, H., He, B., Fang, J.: Optimizing total passenger waiting time in an urban rail network: A passenger flow guidance strategy based on a multi-agent simulation approach. Simul. Model. Pract. Theory **117**, 1–24 (2022)
9. Zeng, L., Liu, J., Qin, Y., Wang, B., Xu, X.: Route choice behavior of passengers in urban rial transit under emergency based on AFC data. J. China Railway Soc. **41**(08), 9–18 (2019). (in Chinese)
10. Han, B., Chen, J., Zhou, W., Sun, Y.: Passenger route choice model under the condition of urban rail transit operation disruption. J. Tongji Univ. (Nat. Sci.) **51**(02), 238–246 (2023). (in Chinese)
11. Yin, H., Wu, J., Liu, Z., Yang, X., Qu, Y., Sun, H.: Optimizing the release of passenger flow guidance information in urban rail transit network via agent-based simulation. Appl. Math. Model. **72**, 337–355 (2019)
12. Lin, Z., Jiang, M., Liu, J., Si, B.: Improved Logit model and method for urban rail transit network assignment. J. Transp. Syst. Eng. Inf. Technol. **12**(06), 145–151 (2012). (in Chinese)

# Influence Analysis of the Metro Nstray Current on AC Power Grid Based on Surface Potential Gradient

Ni Yanru[1](✉), Yu Kun[1], Zeng Xiangjun[1], Cheng Xinxiang[1], Tang Yuhang[1], and Han Wei[2]

[1] Changsha University of Science and Technology, Changsha 410114, China
yanruni@qq.com, kunyu0707@163.co, chxinee028@stu.csust.edu.cn
[2] State Grid Taiyuan Power Supply Company, Taiyuan 030001, China

**Abstract.** The metro stray current can change the surface potential distribution, which will affect the grounding equipment of the power grid. In order to analyze the interference range and influence degree of metro stray current, the DC traction power supply system was equivalent to the resistance network model, so the stray current and rail potential can be calculated numerically. Based on the potential field distribution of the point current source in uniform media, by using the superposition principle, the potential distribution and gradient change of the whole line stray current at any position were realized. The research showed that the stray current is closely related to the running state of trains on the metro line. The more trains in accelerating and decelerating state, the wider the spreading range of stray current and the larger the amplitude of surface potential.

**Keywords:** Metro traction power supply system · Stray current · Earth surface potential · Potential gradient · Train operation condition

## 1 Introduction

The metro system is developing rapidly in the process of urbanization construction in China, and the interference of stray current can not be ignored in the process of ensuring the safe and reliable operation of the system [1, 2]. In the daily operation of the metro system, the leakage current are often found in the reinforced concrete structure inside the metro system and the external buried pipeline, which is called metro stray current. The stray current causes a change in the potential of the nearby surface, which affects the AC system in the area. Different potentials cause DC bias of power transformer, increase harmonic content and cause many kinds of harm to power grid equipment.

At present, the research of metro stray current mainly focuses on modeling distribution. Wuhan University established a subway simulation model and simulated and analyzed the influence of different transition resistors on rail potential [3]. Southwest Jiaotong University derived the relationship between transition resistance and stray current based on theoretical formulas, and improved the simulation model [4, 5]. Reference


Y. Qin et al. (Eds.): EITRT 2023, LNEE 1137, pp. 427–434, 2024.
https://doi.org/10.1007/978-981-99-9311-6_45

[6–8] studies the influence of rail longitudinal resistance on rail potential and stray current distribution when multiple trains run at the same time by modeling stray current in subway.

With the development of subway, the influence of stray current on the surrounding power equipment has also been studied [9–11]. The corrosive relationship of stray current to buried metal has been studied by foreign scholars. In reference [12], a stray current distribution model was established with the help of CDEGS simulation software, and parameters such as total leakage and drainage network efficiency under locomotive operation were evaluated. In reference [11], a potential distribution model was established based on Ansys finite element software, and the anodic corrosion problem of tunnel segment reinforcement was solved. Reference [13, 14] clarified that subway stray current is the main reason for DC magnetic bias of main transformer of urban power grid, and put forward corresponding solutions.

At present, the distribution of stray current in subway has been studied theoretically. However, because the stray current has the characteristics of diffusion distribution near the subway system, it is still necessary to study the impact assessment of stray current on the surrounding environment. The stray current distribution model was established in this paper. Based on the variation index of surface potential gradient, a method for evaluating the dynamic disturbance range of stray current in subway was proposed. The variation of surface potential of trains under different operating conditions was analyzed.

## 2 Metro Stray Current Distribution Model Based on Resistance Network

At present, the subway DC traction system mainly uses the running rail as the return channel of the electric train. However, in the long running process, due to the aging of rail insulation structure, metro stray current will leak to the ground. Taking the track-buried metal-earth resistance structure as an example, the grounding structure of traction power supply system is simplified as resistance network. The resistance network distribution is shown in Fig. 1. The traction substation provides current to the metro train. $I$ is the traction current of the train, A. $R_G$ is the longitudinal resistance of the rail, Ω/km. $R_P$ is the longitudinal resistance of the drainage network, Ω/km. $R_D$ is the longitudinal resistance of the earth, Ω/km. $R_{g1}$ is the rail-to-drainage transition resistance, Ω km. $R_{g2}$ is the transition resistance of the drainage net to the earth, Ω km.

For the double-ended power supply mode of subway operation, it can be regarded as two single-ended power supply models in series. So it is solved separately, that is, 0 ~ $L1$ km and $L1$ ~ $L2$ km segment.

For the x = 0 ~ $L1$ km power supply section, the differential equation is

$$\frac{d}{dx}\begin{bmatrix} I_{G1}(x) \\ I_{P1}(x) \\ U_{G1}(x) \\ U_{P1}(x) \end{bmatrix} = \begin{bmatrix} 0 & 0 & 1/R_{g1} & 0 \\ 0 & 0 & 0 & 1/R_{g2} \\ R_G & -R_P & 0 & 0 \\ R_D & (R_P + R_D) & 0 & 0 \end{bmatrix} \begin{bmatrix} I_{G1}(x) \\ I_{P1}(x) \\ U_{G1}(x) \\ U_{P1}(x) \end{bmatrix} + \begin{bmatrix} 0 \\ 0 \\ 0 \\ -I_1 \cdot R_D \end{bmatrix} \quad (1)$$

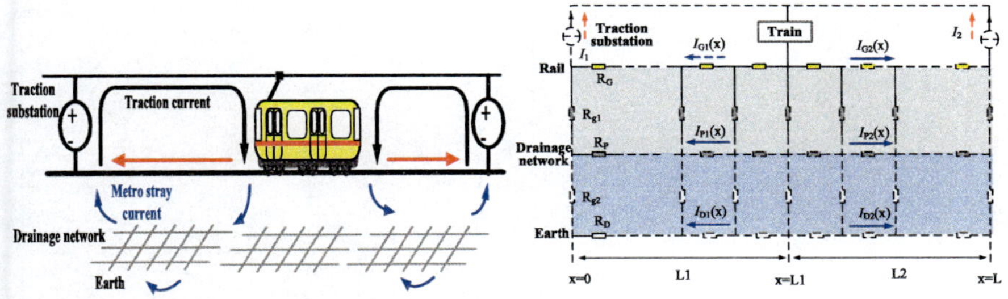


**Fig. 1.** Metro stray current distribution model

For the x = $L1 \sim L2$ km power supply section, the differential equation is

$$\frac{d}{dx}\begin{bmatrix} I_{G2}(x) \\ I_{P2}(x) \\ U_{G2}(x) \\ U_{P2}(x) \end{bmatrix} = \begin{bmatrix} 0 & 0 & -1/R_{g1} & 0 \\ 0 & 0 & 0 & -1/R_{g2} \\ -R_G & R_P & 0 & 0 \\ -R_D & -(R_P + R_D) & 0 & 0 \end{bmatrix} \begin{bmatrix} I_{G2}(x) \\ I_{P2}(x) \\ U_{G2}(x) \\ U_{P2}(x) \end{bmatrix} + \begin{bmatrix} 0 \\ 0 \\ 0 \\ I_2 \cdot R_D \end{bmatrix} \tag{2}$$

According to the boundary conditions, the differential Eqs. (1) and (2) can be solved to obtain the analytical formula of rail potential and stray current distribution in double-ended power supply mode.

$$u(x) = -IZ\frac{Rg}{Rg + Rp}th\frac{\alpha L}{2} \cdot ch\alpha x + IZ\frac{Rg}{Rg + Rp} \cdot sh\alpha x \tag{3}$$

$$i(x) = -I\frac{Rg}{Rg + Rp}th\frac{\alpha L}{2} \cdot sh\alpha x + I\frac{Rg}{Rg + Rp} \cdot ch\alpha x + I\frac{Rp}{Rg + Rp} \tag{4}$$

The total length of the power supply range is 2 km. The longitudinal resistance of the rail is 0.018 Ω/km, the longitudinal resistance of the drainage net is 0.01 Ω/km, and the transition resistance is 3Ω·km. When the train runs to 0.4 km, 1.1 km and 1.7 km respectively, the traction current is 2000A, 500A and −1600A respectively. The distribution of rail potential and stray current is shown in Fig. 2.

## 3 Calculation Method of Surface Potential Gradient Caused by Metro Stray Current

Based on the Lipschitz product differentiation simplification, the analytical formula for the distribution of the current field in a uniform medium is [15]:

$$V(x, y, z) = \frac{I\rho}{2\pi}\frac{1}{\sqrt{(z - h)^2 + x^2 + y^2}} \tag{5}$$

where, $\rho$ is the resistivity in a uniform soil medium, Ω·km; $I$ indicates the leakage current of the power supply, A; $h$ is the depth of the current source, and $x$ and $y$ are the distance between the point current source in the horizontal and vertical directions, km.

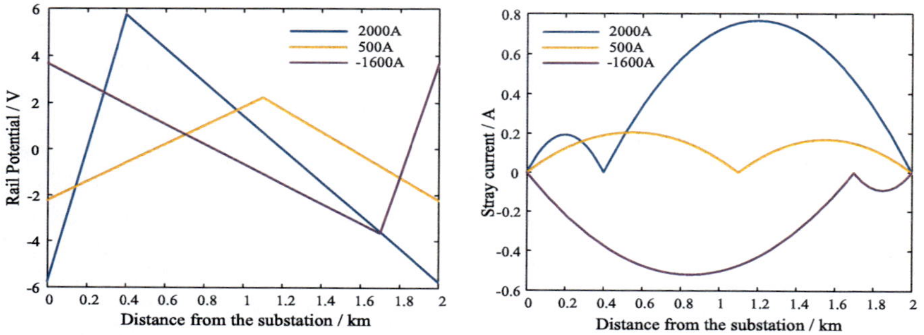


**Fig. 2.** Rail potential and stray current distribution when train runs to different positions.

As shown in Fig. 3, the stray current leaking at each point along the rail in the interval will affect the surface potential at point *i* through the current field. Therefore, it can be expressed as the integral of the potential generated at point i by all leakage currents in the interval. From this, the potential at any point on the surface caused by stray current is obtained first. Then the surface potential gradient of this point is obtained by the distribution of surface current field.

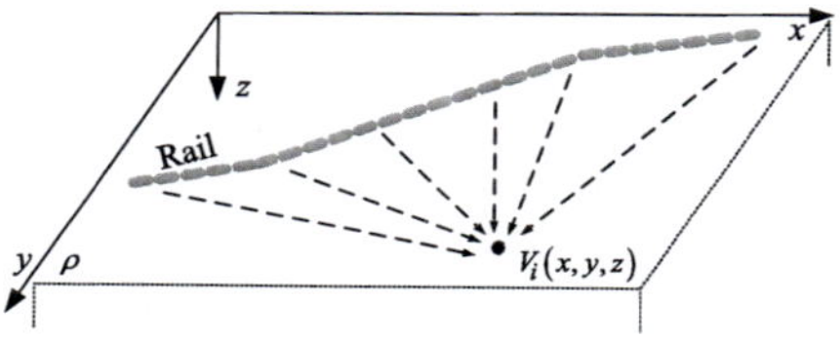


**Fig. 3.** Calculation diagram of potential distribution caused by metro stray current.

Thus, the surface potential generated by the leakage current at each point on the rail in the power supply range at any position is

$$V_i(x, y, z) = \int_0^L \frac{i(x)\rho}{2\pi} \cdot \frac{1}{\sqrt{(x-l)^2 + y^2}} dl \tag{6}$$

The surface potential gradient is further calculated as

$$gradV = \nabla Vi(x, y, z) = \frac{\partial V}{\partial x} i + \frac{\partial V}{\partial x} j + \frac{\partial V}{\partial z} k \tag{7}$$

The national standard GB 50991–2014 requires that the investigation and test of the DC traction system as the DC interference source should include the track ground potential and its distribution and the ground potential gradient near the track. As shown in Table 1, the strength of stray current in soil is judged by ground potential gradient.

**Table 1.** Relationship between ground potential gradient and stray current interference intensity

| Potential gradient (mV/m) | Stray current interference level |
|---|---|
| <0.5 | Weak |
| 0.5~5 | Average |
| >5 | Strong |

## 4 Study on the Influence of Metro Train Operating Conditions on Surface Potential Distribution

Metro trains generally go through the process of starting acceleration, inert running and deceleration braking from starting to stopping in a running section. According to the different resultant force received by the train in operation, it is divided into three states: traction state, idle state and braking state, as shown in Fig. 4. Due to the difference of locomotive traction current in different states, the stray current distribution will be different when considering the uniform distribution of rail longitudinal resistance and rail ground transition resistance in the interval. Therefore, considering the running state, the surface potential gradient will also change in different time periods, which will have different impacts on nearby power equipment and buried metal pipelines.

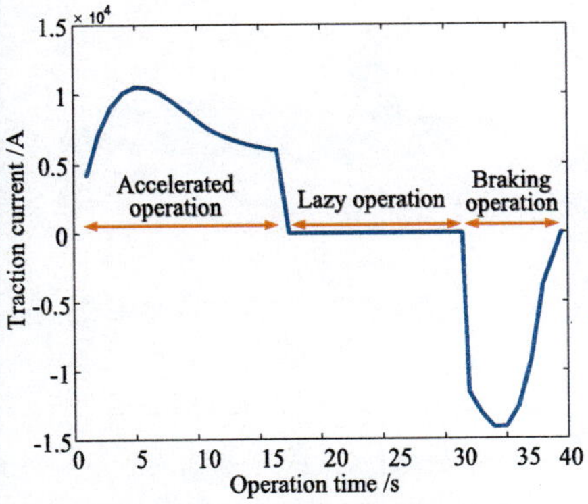


(a) Traction current changes with time

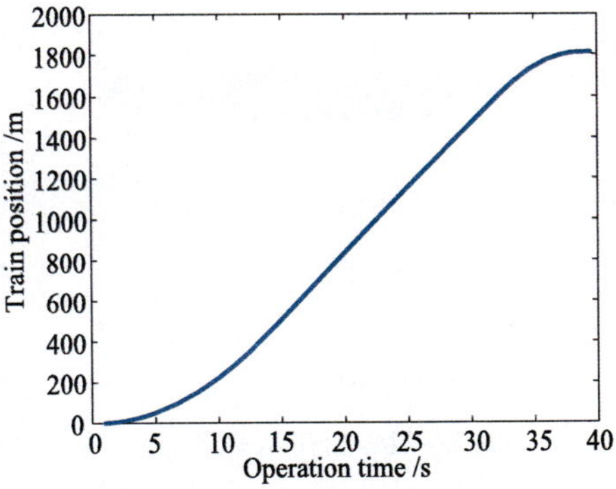


(b) Train position changes with time

**Fig. 4.** Metro train operation characteristics.

The change of ground potential caused by stray current is analyzed with a domestic metro line as an example, and the interference range and degree of stray current along the subway line are evaluated.

The city's Line 1 has a maximum operating speed of 80 km/h and a total length of 22.59 km. According to the station distance, traction strategy and train departure interval, the position of each train and the size of traction current can be obtained, and the change of ground potential and gradient can be further calculated.

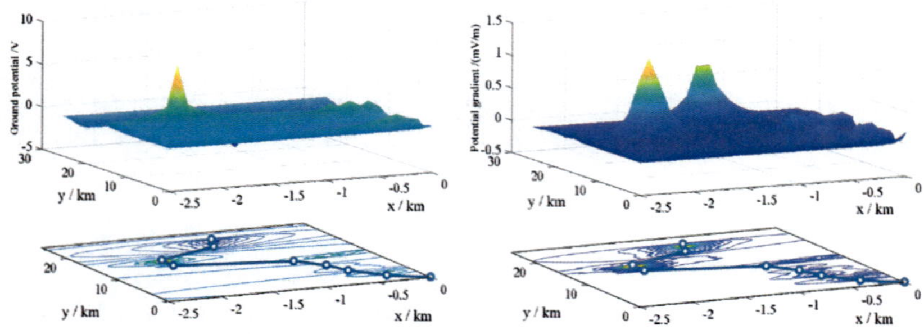


(a) Surface potential distribution (b) Potential gradient distribution

**Fig. 5.** Ground potential and gradient distribution of trains in accelerating condition.

As shown in Fig. 5, when most trains on the whole line are in the accelerating state, the ground potential is positive on the whole, and the peak value is 6.6 V at most. The equipotential lines are more dense at the train position, and with the increase of distance, the equipotential lines gradually sparse. The peak potential gradient can reach 1.2 mV/m, and within 1 km from the line, the potential gradient will be greater than 0.5 mV/m, and the interference degree will be evaluated as average.

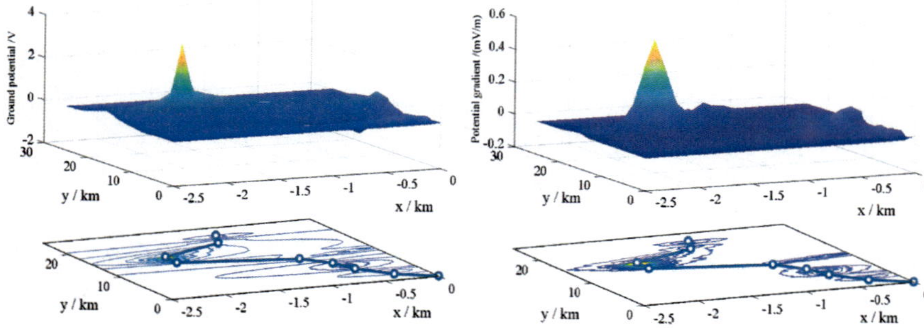


(a) Surface potential distribution (b) Potential gradient distribution

**Fig. 6.** Ground potential and gradient distribution of trains in lazy condition.

As shown in Fig. 6, when most trains on the whole line are in an inert state, the overall ground potential is basically positive. However, compared with the accelerated state, the amplitude of ground potential decreases overall, and the peak value is 3.37 V. The change of potential was relatively gentle on the whole, and the peak potential gradient is 0.56 mV/m. When the distance from the line is 1 km, the potential gradient will be less than 0.5 mV/m, and the interference degree will be evaluated as weak.

As shown in Fig. 7, when most trains on the whole line are in the braking state, the ground potential distribution is negative as a whole, and the extreme value of negative is −10 V. Based on the surface potential gradient, the impact of the entire region is similar to that of the acceleration phase, with the peak potential gradient up to 1.8 mV/m. Similarly,

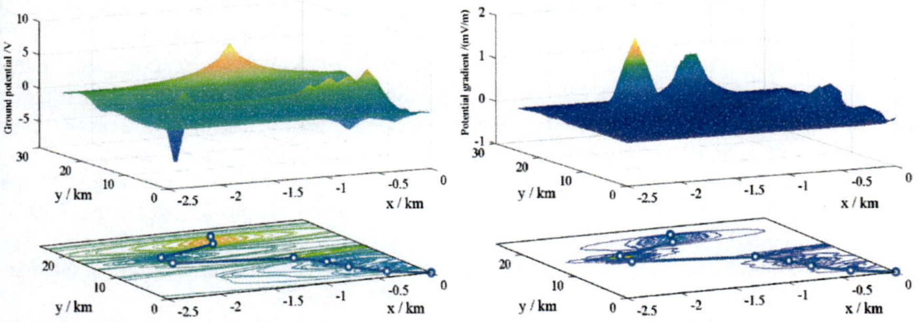


(a) Surface potential distribution (b) Potential gradient distribution

**Fig. 7.** Ground potential and gradient distribution of trains in braking conditions.

within 1 km from the line, the potential gradient will be greater than 0.5 mV/m, and the interference degree will be evaluated as average.

In summary, train operating conditions have obvious influence on the variation of surface potential gradient. Based on the change index of surface potential gradient, the influence degree of subway stray on the surrounding environment can be effectively evaluated. The research showed that the more trains in accelerating and decelerating state, the wider the stray current propagation range, the larger the surface potential amplitude, and the more obvious the gradient change.

## 5 Conclusion

Aiming at the problem of evaluating the interference of stray current to the environment along the metro line, the paper established a stray current distribution model based on the resistance network model, and studied the distribution law of rail potential and stray current under different locomotive operating conditions. The numerical calculation of ground potential and potential gradient caused by stray current was realized through the numerical integration of point current source in space electric field. The research showed that the more trains in accelerating and decelerating state, the more obvious the change of potential gradient. In order to avoid excessive rail potential and stray current, reasonable running strategies should be adjusted in time to avoid adjacent trains in the same operating conditions.

**Acknowledgments.** This research was supported by the Science and Technology Project of the State Grid Corporation of China: Research on the coupling influence and suppression method of grid DC bias caused by metro stray current (5100-202199276A-0-0-00).

## References

1. Liao, Z., Qin, Y., Wang, M.: Research on high-speed railway operation adjustment model based on priority. Smart Resilient Transp. **4**(1), 12–21 (2022)

2. Ni, Y., Zeng, X., Yu, K., et al.: A correlation analysis on transformer DC bias current caused by metro stray current. J. Electric Power Sci. Technol. **36**(06), 136–143 (2021)
3. Cai, L., Wang, J., Fan, Y., et al.: The influence of subway track on the stray current distribution of ground transition resistance. High Voltage Technol. **41**(11), 3604–3610 (2015). (in Chinese)
4. Lin, S., Wang, A., Liu, M., et al.: A multiple section model of stray current of DC metro systems. IEEE Trans. Power Delivery **36**(3), 1582–1593 (2020)
5. Zhu, F., Li, J., Zeng, H., et al.: The influence of rail-to-ground transition resistance of urban rail transit on the distribution characteristics of stray current. High Voltage Technol. **44**(8), 2738–2745 (2018). (in Chinese)
6. Lin, S., Huang, F., Zhou, Q., et al.: Evaluation and analysis model of stray current in the metro depot. IEEE Trans. Transp. Electrification **7**(7), 1780–1794 (2021)
7. Mei, J., Lin, G.: Metro stray current analysis under multi-train operation. Electrified Railway **28**(4), 68–70 (2017). (in Chinese)
8. Zhao, N.: Analysis of influencing factors of subway stray current. China Sci. Technol. Inform. **1**(3), 101–103 (2017). (in Chinese)
9. Boden, P., Evans, J.: Reduction of stray-current attack in electrochemical machining. Electrochim. Acta **16**(7), 1071–1079 (1971)
10. Zaboli, A., Vahidi, B., Yousefi, S., et al.: Evaluation and control of stray current in DC-electrified railway systems. IEEE Trans. Veh. Technol. **66**(2), 974–980 (2017)
11. Dolara, A., Foiadelli, F., Leva, S.: Stray current effects mitigation in subway tunnels. IEEE Trans. Power Delivery **27**(4), 2304–2311 (2012)
12. Charalambousc, A., Aylott, P.: Dynamic stray current evaluations on cut-and-cover sections of DC metro systems. IEEE Trans. Veh. Technol. **63**(8), 3530–3538 (2014)
13. Cai, M., Li, J., Peng, Q., et al.: Research and treatment of main transformer DC bias in urban areas. Electric. Autom. **39**(1), 7–9 (2017). (in Chinese)
14. Xue, J., Zhao, L., Feng, Y., et al.: A method for extracting the amplitude-frequency feature quantity of transformer vibration signal based on cross wavelet. High Voltage Technol. **45**(2), 505–511 (2019). (in Chinese)
15. Machczynski, W., Budnik, K., Szymenderski, J.: Assessment of DC traction stray current effects on nearby pipelines. Int. J. Comput. Math. Electr. Electron. Eng. **35**(4), 1468–1477 (2016)

# Can China's Carbon Market Pilot Cope with Uncertainty Shocks?

## Taking the Example of the COVID-19 Pandemic

Qi Wei(✉) and Kehao Wang

School of Economics and Management, Lanzhou University of Technology, Lanzhou 730050, China

weiq@lut.edu.cn, h13181989090@aliyun.com

**Abstract.** This paper introduces a novel approach by explicitly considering the COVID-19 pandemic as an analytical framework for assessing uncertainty shocks. Focusing on average trading volume and average trading prices, the study quantifies the impact of the pandemic on the carbon market in pilot regions. It compares the market performance of these regions in response to the pandemic shock while incorporating existing response measures. Relevant variables are identified and applied in a partial correlation analysis model. The findings reveal a significant correlation between the extent of the pandemic's impact on the carbon market and factors such as the proximity to the contract expiration date and the level of pandemic risk. This study provides the first empirical evidence of the effectiveness of current measures in mitigating the pandemic's impact on the carbon market. Additionally, recommendations are proposed for further optimizing response strategies.

**Keywords:** Carbon market · Uncertainty shock · COVID-19 pandemic

## 1 Introduction

In 2011, Beijing, Tianjin, Shanghai, Chongqing, Hubei, Guangdong, and Shenzhen officially launched pilot projects for carbon emission trading, hereinafter referred to as carbon trading. After a decade of development, as of June 2021, the pilot scope of the carbon emission trading market, hereinafter referred to as the carbon market, covered industries such as steel, power, and non-ferrous metals, with a cumulative transaction volume of approximately 11.4 billion yuan. However, compared to carbon markets in Europe and the United States, China's carbon market has a shorter establishment time and lower maturity, facing the problem of being "large but not strong." Uncertainty shocks can easily lead to significant losses, posing a pressing need to evaluate and determine whether the existing risk prevention mechanisms can effectively respond to uncertainty shocks and maintain the stable operation of China's carbon market [1].

Uncertainty shocks faced by the carbon market refer to disturbances that cause losses to market participants and the economy during the market operation process, mainly manifested as disruptions to transactions [2]. Based on their sources, uncertainty shocks


Y. Qin et al. (Eds.): EITRT 2023, LNEE 1137, pp. 435–447, 2024.
https://doi.org/10.1007/978-981-99-9311-6_46

can be classified into endogenous and exogenous shocks. Endogenous shocks refer to the impacts caused by operations and market rules, while exogenous shocks refer to the impacts caused by external macroeconomic factors on the carbon market. The impacts of uncertainty shocks can be divided into short-term and long-term effects. Short-term effects correspond to the micro-level elements of carbon trading and business entities, while long-term effects correspond to the macro-level aspects involving the nation, society, and market.

In order to address uncertainty shocks, carbon pilot regions have established preventive mechanisms. Taking the COVID-19 pandemic as an example of exogenous shocks, in early 2020 when the outbreak just occurred, Hubei Province repeatedly extended the closure of the market.

Scholars both domestically and internationally have conducted extensive research on carbon markets under uncertainty shocks. Regarding the impacts of uncertainty shocks on carbon markets, research has proposed various factors, such as macroeconomic indicators [3] including the Purchasing Managers' Index [4], exchange rates [5], and interest rates [6]; trading policies [7] including transaction costs [8] and allocation schemes [9]; changes induced by commodities such as natural gas [10] and oil [11]; and fluctuations arising from securities markets [4], climate [12], and other carbon markets [13]. These factors, under certain conditions, can have varying degrees of impact on carbon markets. As for how carbon markets respond to uncertainty shocks, research has suggested taking a holistic approach, actively promoting carbon initiatives within the United Nations framework, maintaining coordination between politics and markets [14], gradually exploring the linkage of carbon markets [15], clarifying the positioning of international carbon emission trading mechanisms within carbon reduction policy frameworks [16], establishing rules for stable carbon prices and financial safeguards, improving accountability and regulatory rules [17], introducing carbon floor price policies, establishing mechanisms for utilizing carbon trading revenues [18], strengthening tracking and assessment, and implementing carbon taxes [19], among others.

In summary, existing research on the impact of uncertainty shocks on carbon markets has explored changes related to commodities, trading rules, and other aspects. However, over the past three years, the COVID-19 pandemic has had extensive and sustained impacts worldwide, significantly affecting social life in China. There is a lack of research on the impact of the pandemic on carbon markets. The carbon pilot regions have continued to implement the established response measures despite the recurring outbreaks. It is therefore crucial to study and assess the effectiveness of related measures such as postponing the deadline for carbon quota compliance.

This paper takes the analysis of the COVID-19 pandemic as the carrier of uncertainty shocks. By analyzing the transaction data of major carbon pilot regions in China, it quantifies and compares the extent of the impact of the pandemic on each pilot region. It uses empirical models to analyze the effectiveness of measures such as postponing compliance deadlines and provides effective recommendations for carbon market regulatory authorities based on the research results.

## 2 Comparative Analysis

### 2.1 Data Collection and Compilation

Since September 2020, when China entered the normalized phase of COVID-19 prevention and control, Tianjin and Chongqing have not experienced large-scale outbreaks, and the carbon market has remained relatively stable. Therefore, Beijing, Shanghai, Hubei, Guangdong, and Shenzhen are selected as the research subjects, and the research data is sourced from the transaction data disclosed by the carbon exchanges in these five pilot regions.

Based on the timing of designating or canceling high-risk areas within the regions, one epidemic period and four "control periods" are selected from the five pilot regions. In terms of time duration, reference is made to relevant announcements of each carbon pilot region to understand factors such as holidays and market closures, allowing for the corresponding adjustment of the time periods. This adjustment ensures that the number of trading days in the "control periods" within each pilot region remains consistent with that of the epidemic period. Furthermore, to better analyze the impact of the pandemic on the carbon market, it is required that no large-scale outbreaks occurred during the "control periods" in each pilot region, meaning that no high-risk areas were designated. Consequently, two "control periods" are set as adjacent periods before and after the epidemic period. Considering the severe nationwide situation in 2022 and the incomplete data for 2022 in some pilot regions, which prevents the establishment of a "control period" within the same year (2022), the other two "control periods" are set as one year before and two years before the epidemic period.

Considering that the impact of uncertainty shocks on the carbon market primarily manifests in carbon prices and trading volume, the transaction data from the carbon exchanges in the five pilot regions are aggregated on a daily basis for each period, including the epidemic period and "control periods." This aggregation generates the total trading volume and total transaction value for each time period. Given the significant differences in the scale of carbon markets and the number of trading days during the research period in different pilot regions, as well as the occurrence of zero trading volume on certain trading days, the average values are used to represent the trading level of each pilot region, serving as the basis for examining the impact of the pandemic on the carbon market. The trading data for the five carbon trading pilot regions are presented in Table 1.

**Table 1.** Transaction data summary of pilot areas

| District | Trading day | Time | Volume/ton | Transaction volume/¥ | Average quantity/ton | Average price/¥ |
|---|---|---|---|---|---|---|
| Shenzhen | 20 | 2019.6.18-7.9 | 435267 | 8082942.83 | 21763.35 | 18.57 |
| | | 2020.6.18-7.20 | 117279 | 2401102.81 | 5863.95 | 20.47 |
| | | 2021.05.20-6.17 | 579526 | 7525990.22 | 28976.30 | 12.99 |
| | | 2021.06.18-7.13 | 3539560 | 46857780.29 | 176978.00 | 13.24 |
| | | 2021.07.14-8.16 | 376049 | 2289487.41 | 18802.45 | 6.09 |
| Beijing | 35 | 2018.12.23-2019.1.29 | 37910 | 2363093.2 | 1083.14 | 62.33 |
| | | 2019.12.23-2020.2.18 | 14587 | 1124932.5 | 416.77 | 77.12 |
| | | 2020.11.6-2021.12.22 | 356122 | 31415241.1 | 10174.91 | 88.21 |
| | | 2020.12.23-2021.2.8 | 59533 | 2637973.4 | 1700.94 | 44.31 |
| | | 2021.2.9-3.31 | 14348 | 388160 | 434.79 | 27.05 |
| Guangdong | 24 | 2019.5.24-6.21 | 13977054 | 249587064.7 | 582377.25 | 17.86 |
| | | 2020.5.24-6.29 | 5534696 | 116287848.5 | 230612.33 | 21.01 |
| | | 2021.4.15-5.23 | 6160795 | 226334558.1 | 256699.79 | 36.74 |
| | | 2021.5.24-6.25 | 4874097 | 185582930.8 | 203087.38 | 38.08 |
| | | 2021.6.26–7.23 | 7063853 | 300235334.8 | 294327.21 | 42.50 |
| Hubei | 20 | 2019.8.02–8.27 | 1092117 | 37857028.4 | 54605.85 | 34.66 |
| | | 2020.8.2-8.27 | 2040516 | 57545222.11 | 102025.80 | 28.20 |
| | | 2021.7.5-8.1 | 242283 | 9186831.35 | 12114.15 | 37.92 |
| | | 2021.8.2-8.27 | 128589 | 5485401.24 | 6429.45 | 42.66 |
| | | 2021.8.28–9.28 | 101874 | 4069773.12 | 5093.70 | 39.95 |
| Shanghai | 24 | 2019.8.2-2019.9.4 | 107817 | 4531692.77 | 4492.38 | 42.03 |
| | | 2020.8.2-9.4 | 491965 | 19629302.13 | 19678.60 | 39.90 |
| | | 2021.6.29-8.1 | 46478 | 1820299.76 | 1936.58 | 39.16 |
| | | 2021.8.2-9.4 | 135045 | 5468792.16 | 5626.88 | 40.50 |
| | | 2021.9.5-10.8 | 380341 | 15217340.86 | 17288.23 | 40.01 |

## 2.2 Comparative Analysis

This study assesses the extent of the impact of the pandemic on the carbon market based on the volatility of transaction data during the epidemic period. Specifically, the analysis focuses on two indicators: average trading volume and average trading price. The percentage changes for average trading volume and average trading price in the five pilot regions are calculated to two decimal places and presented in Table 2.

Based on the table, the percentage changes exhibit a wide range of distribution. The range for the average trading volume is [−93.70%, 2918.07%], while the range for the average trading price is [−49.77%, 117.44%]. The maximum value of 2918.07% occurs

**Table 2.** Summary of change rates in pilot areas

| District | Average trading volume (%) | | | | Average transaction price (%) | | | |
|---|---|---|---|---|---|---|---|---|
| | Two years ago | A year ago | Before the same year | After the same year | Two years ago | A year ago | Before the same year | After the same year |
| Beijing | 57.04 | 308.12 | −83.28 | 291.21 | −28.91 | −42.54 | −49.77 | 63.79 |
| Shenzhen | 713.19 | 2918.07 | 510.77 | 841.25 | −28.71 | −35.34 | 1.94 | 117.44 |
| Guangdong | −65.13 | −11.94 | −20.89 | −31.00 | 113.22 | 81.22 | 3.64 | −10.42 |
| Shanghai | 25.25 | −71.41 | 190.56 | −67.45 | −3.65 | 1.49 | 3.40 | 1.22 |
| Hubei | −88.23 | −93.70 | −46.93 | 26.22 | 23.06 | 51.26 | 12.50 | 6.78 |

in the average trading volume of Shenzhen during the epidemic period compared to the previous year, while the minimum value of −93.70% occurs in the average trading volume of Hubei during the epidemic period compared to the previous year. The smallest absolute value appears in the average trading price of Shanghai compared to the same period of the following year. In comparison to trading prices, uncertainty shocks have a greater impact on trading volume. The highest percentage changes in Beijing, Shenzhen, Hubei, and Shanghai are all observed in the analysis of average trading volume.

In terms of the percentage change in average trading volume, Shenzhen exhibits the highest change rate, exceeding 2900%, which is significantly higher than the other four pilot regions. Among the remaining four pilot regions, Hubei and Guangdong have relatively smaller change rates, with absolute values not exceeding 94% and 114%, respectively. Beijing and Shanghai, on the other hand, have relatively larger change rates, with the maximum absolute values surpassing 308% and 190%, respectively. Regarding the percentage change in average trading price, the values for the five pilot regions are relatively small, and the trends are similar, without significant differences. Shenzhen's trading data is the most significantly affected by the pandemic, exhibiting substantial and abnormal fluctuations.

In conclusion, the pandemic has had a certain impact on both the average trading price and average trading volume in the carbon market, with a more pronounced effect observed in the latter. Furthermore, all five pilot regions have been affected by the epidemic within the domain, with Guangdong and Hubei experiencing relatively smaller impacts, Shenzhen experiencing a relatively larger impact, and Shanghai and Beijing falling in between. The next step of this study will introduce an empirical model to examine the factors that contribute to the differences in the percentage changes in trading data among the pilot regions.

# 3 Empirical Analysis

## 3.1 Variable and Data Description

### Proximity to the Contract Deadline

The time interval between the occurrence of the epidemic in Guangdong and Shenzhen and the contract deadline is less than one month. However, compared to Shenzhen, which did not implement any intervention measures, Guangdong extended its contract deadline from the original date of June 20th to July 20th. As a result, the fluctuation of its trading data during that year was much smaller than that of Shenzhen.

The importance of the contract deadline can also be observed in the actual trading process. Taking the trading data of Guangdong between December 1, 2020, and July 20, 2021, which spanned eight months between contract deadlines, the total trading volume and total trading value were 26.18 million tons and 94,297 million yuan, respectively. The contract deadline for this period was July 20, 2021. From June 21 to July 20, the total trading volume and total trading value were 7.73 million tons and 32,808 million yuan, accounting for 29.54% and 34.79% of the entire contract period, respectively, far exceeding the average level of 12.5%, as shown in Fig. 1.

Simultaneously, studies have also indicated that postponing the contract deadline can have a significant impact on the operation of the carbon market [20]. Hence, the proximity of the epidemic's impact to the contract deadline is chosen as one of the variables for empirical analysis. This study assigns a value to this variable by measuring the number of days between the midpoint of the epidemic period (n days) and the contract deadline. By examining the relationship between the proximity to the contract deadline and the carbon market volatility, the importance of the contract deadline can be validated.

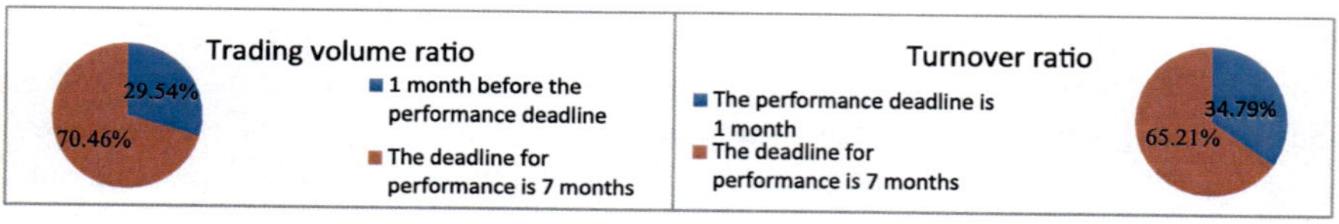


**Fig. 1.** Trading volume and proportion of trading volume on the day near the performance deadline

### Carbon Market Size

Both the Guangdong and Shenzhen pilot areas are located in economically developed provinces, and the most significant difference between the two carbon markets lies in their market sizes. Taking the total carbon quota trading volume in 2020 as an example, Guangdong and Shenzhen recorded 32.11 million tons and 1.24 million tons, respectively. However, according to the previous data, the sum of absolute value changes for Guangdong is 337.45%, with a change rate of 0.11% per 10,000 tons. For Shenzhen, the sum of absolute value changes is 5,166.71%, with a change rate of 41.67% per 10,000 tons. This indicates that as the carbon market size increases, the average trading volume change rate tends to be smaller. Furthermore, studies have suggested that the carbon market size, as a reflection of the carbon market's development, also plays a crucial role in its operation [21]. Therefore, this study introduces the carbon market size as one of the variables for empirical analysis and assigns it a value based on the total trading volume in the respective pilot areas in 2020. The carbon market sizes in the pilot areas are presented in Table 3.

**Table 3.** Scale of carbon market in pilot areas

| District | Carbon market size/tons |
|---|---|
| Beijing | 104 |
| Shenzhen | 124 |
| Guangdong | 3211 |
| Shanghai | 184 |

**Table 4.** Related data indicators

| District | Days before the deadline | Epidemic risk level | Carbon market size/ten thousand tons |
|---|---|---|---|
| Shenzhen | 0 | Medium | 124 |
| Shanghai | 44 | Medium | 184 |
| Beijing | 273 | High | 104 |
| Guangdong | 43 | High | 3211 |

**Impact of Epidemic Risk Level and Uncertainty Shock**

The severity of uncertainty shocks is significantly linked to their impact on the carbon market. Therefore, it is necessary to examine the differential effects of COVID-19 outbreaks with different severity levels on the carbon market and introduce the epidemic as one of the variables for empirical analysis. The severity level of the epidemic is classified into two categories: high risk and moderate risk, based on the presence of high-risk areas in the pilot regions. Additionally, the impact of the epidemic on the carbon market is measured using the previously mentioned change rates. This study conducts empirical analysis on the number of days to the contract deadline, epidemic risk level, and carbon market size as variables. The relevant data used for the analysis are summarized in the following Table 4.

## 3.2 Selection of Analytical Methods

As of September 1, 2022, Hubei Province has not announced the contract deadline for the previous year. Therefore, the empirical model can only use the trading data from the four carbon pilot regions, namely Beijing, Shanghai, Guangdong, and Shenzhen, from December 2018 to August 2021 as the research subjects, resulting in a total of 32 sample data points. This constitutes a small sample analysis. Additionally, considering the numerous factors that can affect trading data volatility, it is necessary to exclude the influence of other variables to accurately reflect the importance of the relevant variables during the analysis.

Correlation analysis is mainly conducted by calculating various correlation coefficients to measure the degree of association between two variables. Some scholars have introduced partial correlation models into the study of stock markets to analyze the factors affecting liquidity [22]. Therefore, in the context of correlation analysis, this study selects a partial correlation analysis model that is suitable for small sample sizes and can control the influence of other variables. Partial correlation analysis calculates partial correlation coefficients, which can measure the linear relationship between two variables while controlling for the linear effects of other variables. A higher absolute value of the partial correlation coefficient indicates a stronger linear relationship between the two variables. Based on the classification of correlation coefficient levels, under the premise of significance testing, absolute values between 0.1 and 0.3 indicate weak correlation, 0.3 and 0.5 indicate moderate correlation, and 0.5 to 1.0 indicate strong correlation.

The absolute values of the logarithmically transformed change rates of the trading data in the pilot regions are taken, and based on the Shapiro-Wilk test with a significance level of 0.596, the data is shown to follow a normal distribution. This result is significantly greater than 0.05, confirming that the sample data meets the assumptions required for the application of the partial correlation model. The results of the normality test are shown in Table 5.

**Table 5.** Normality test results

| Normal test | Shapiro-Wilk | | |
|---|---|---|---|
| | Statistics | Degree of freedom | Significance |
| Rate of change | 0.973 | 32 | 0.596 |

### 3.3 Analysis Process

To comprehensively and objectively assess the relationship between the volatility of carbon market trading data and uncertainty shocks in China, this study introduces carbon market size as a control variable to be excluded in the partial correlation analysis. The risk level of the epidemic is assigned as a binary variable (1 or 0). Using the SPSS 26.0 software, the partial correlation analysis model is employed, alternating between the number of days until the contract deadline and the risk level of the epidemic as the control variable and independent variable, respectively. The SPSS analysis results are then used to determine the impact of the two independent variables on the volatility of trading data.

### Partial Correlation Analysis with Risk Level as a Control Variable

Taking the trading data volatility in the pilot regions and the number of days until the contract deadline as variables, with carbon market size and risk level of the epidemic as control variables, the results of the partial correlation analysis are presented in Table 6. The significance of the relationship between trading data volatility and the number of days until the contract deadline in the partial correlation analysis is 0.008, which is much smaller than 0.05. This indicates a significant correlation between the two variables, with a correlation coefficient of −0.477. Controlling for the influence of carbon market size and the risk level of the epidemic, there is a negative correlation between trading data volatility and the number of days until the contract deadline.

**Table 6.** Analysis results of partial correlation model (I)

| Control variable | | | Rate of Change | Performance Deadline |
|---|---|---|---|---|
| Risk level & carbon market size | Rate of change | Correlation | 1.000 | −0.477 |
| | | Significance (double tail) | | 0.008 |
| | | Degree of freedom | 0 | 28 |
| | Performance deadline | Correlation | −0.477 | 1.000 |
| | | Significance (double tail) | 0.008 | |
| | | Degree of freedom | 28 | 0 |

### Partial Correlation Analysis with the Number of days Until the Contract Deadline as a Control Variable

Using trading data volatility in the pilot regions and the risk level of the epidemic as variables, with carbon market size and the number of days until the contract deadline as control variables, the results of the partial correlation analysis are presented in Table 7. The significance of the relationship between trading data volatility and the risk level of the epidemic in the partial correlation analysis is 0.006, which is much smaller than 0.05. This indicates a significant correlation between the two variables, with a correlation coefficient of 0.491. Controlling for the influence of the number of days until the contract deadline and carbon market size, there is a positive correlation between trading data volatility and the risk level of the epidemic.

**Table 7.** Analysis results of partial correlation model (2)

| Control variable | | Rate of change | Performance deadline | Risk level |
|---|---|---|---|---|
| Performance deadline & carbon market size | Rate of change | Correlation | 1.000 | 0.491 |
| | | Significance (double tail) | | 0.006 |
| | | Degree of freedom | 0 | 28 |
| | Risk level | Correlation | 0.491 | 1.000 |
| | | Significance (double tail) | 0.006 | |
| | | Degree of freedom | 28 | 0 |

## 3.4 Analysis Conclusion

### Postponing the Contract Deadline Can Mitigate Uncertainty Shocks

The number of days until the contract deadline shows a moderate correlation with trading data volatility in the pilot regions, which confirms the previous speculation that the contract deadline is of significant value in assessing the impact of epidemic shocks on the carbon market. Therefore, the measures taken by the pilot regions to postpone the contract deadline as a response to the epidemic are effective. The contract deadline, as an important component of the carbon trading system, can serve as an indicator for predicting the extent of epidemic impact.

Furthermore, it can be concluded that, for the current participants in the Chinese carbon market, their trading motivation is generally driven by compliance, meaning that their contractual behavior is merely aimed at fulfilling the obligations to avoid penalties from relevant authorities, rather than reflecting a strong awareness and strategic approach to carbon trading. Effective management of carbon assets within enterprises is often neglected.

### Risk Level of the Epidemic Closely Linked to Its Impact

The risk level of the epidemic also shows a moderate correlation with trading data volatility in the pilot regions. This not only confirms the significant relationship between the severity of uncertainty shocks and their impact on the carbon market, as mentioned earlier, but also suggests the feasibility of preliminary classification of the epidemic risk level based on the presence of high-risk areas within the region. Consequently, regulatory authorities in the pilot regions can conduct thorough analysis of uncertainty shocks, assess their impact levels, and determine the necessity and urgency of implementing corresponding measures.

### Significant Influence of Carbon Market Size

In the empirical model, carbon market size, as a control variable, plays a crucial role in capturing the significant correlations among other variables. Carbon market size is an important outcome of the continuous development of carbon trading and serves as a critical attribute of the carbon market. It plays a significant role in coping with uncertainty shocks and indicates that a larger carbon market size leads to greater stability and better ability to withstand the judgment of uncertainty shocks.

# 4 Recommendations

## 4.1 Reduce the Concentration of Trading and Appropriately Postpone the Contract Deadline

Relevant regulatory authorities should fully utilize existing policy tools to guide companies to engage in timely and active trading, while avoiding excessive concentration of trading activities before the contract deadline. The government can flexibly adjust the charging items and proportions based on the time remaining until the contract deadline, or provide incentives by increasing the offsetting ratio for carbon trading entities. In order to encourage diversified trading, certain restrictions on the time, type, and region of verified emission reductions can be relaxed to some extent in the existing pilot regions.

When uncertainty shocks occur within the region, carbon market regulatory authorities should carefully consider whether the carbon trading contract deadline should be postponed to cope with the shocks. It is important to determine the appropriate degree of postponement and decide on the number of days to be postponed. A flexible approach of multiple small-scale postponements can be adopted, which allows the carbon market to have enough time to mitigate the impacts of uncertainty shocks while minimizing disruptions to normal trading activities. This approach aims to maintain the stability of the carbon market and prevent significant losses to the interests of government, investors, and relevant emission control enterprises.

## 4.2 Emphasize the Analysis of Uncertainty Shocks

In addition to the delineation of high-risk areas mentioned in this paper, there are still many pieces of information that can assist carbon market regulatory authorities in assessing the severity of uncertainty shocks caused by epidemics or other factors. Relevant departments should focus on the collection and analysis of such information, ensuring timely access to relevant data on uncertainty shocks and conducting evaluations. This preparation will facilitate the judgment on whether to postpone the contract deadline and help determine the appropriate duration of the postponement.

## 4.3 Implement Multiple Measures to Expand the Carbon Market Size

The carbon market size plays a crucial role in addressing uncertainty shocks. Regulatory authorities should strengthen publicity and training efforts targeting carbon market participants, guiding them to proactively develop carbon trading strategies and actively engage in carbon trading activities. This will enhance the vitality of the carbon market and continuously expand its size, thereby increasing its resilience against uncertainty shocks.

**Acknowledgments.** This work is supported by the National Natural Science Foundation of China (71963024) and Gansu Provincial Department of Education (2021jyjbgs-08).

## References

1. Zhu, B., Tang, J., Jiang, M., Wang, P.: Research on risk simulation and regulation of carbon market based on system dynamics. Syst. Eng. Theory Practice **42**(07), 1859–1872 (2022). (in Chinese)
2. Yuan, C., Geng, C., Sun, J., Cui, H.: Vertical integration and enterprise value under uncertainty shocks: natural experimental evidence from the COVID-19 pandemic. Econ. Res. J. **22**(02), 633–652 (2022). (in Chinese)
3. Zhou, K., Li, Y.: Influencing factors and fluctuation characteristics of China's carbon emission trading price. Physica A: Stat. Mech. Appl. **524** (2019)
4. Chen, X., Liu, M., Liu, Y.: Driving factors and structural breaks in carbon trading prices: an empirical study of seven carbon trading pilots in China. Econ. Issues **11**, 29–35 (2016). (in Chinese)
5. Wang, Q., Lu, J.: Asymmetric effects of renminbi exchange rate shocks on China's carbon price: an empirical study based on the Markov switching model. J. Jilin Univ. (Social Sciences Edition) **57**(06), 95–105+205–206 (2017). (in Chinese)
6. Wang, Q., Lu, J.: Regional heterogeneity of the impact of short-term interest rate fluctuations on carbon trading prices. J. Soc. Sci. **01**, 101–110 (2018). (in Chinese)
7. Ye, S., Dai, P., Nguyen, H.T., Huynh, N.Q.A.: Is the cross-correlation of EU carbon market price with policy uncertainty really being? A multiscale multifractal perspective. J. Environ. Manage. **298** (2021)
8. Cui, L., Duan, H., Xu, J.: The impact of trading costs on the cost effectiveness of China's carbon market: a simulation analysis of domestic carbon trading pilots. Manage. Rev. **29**(06), 23–31 (2017). (in Chinese)
9. Song, Y., Liang, D., Liu, T., Song, X.: How China's current carbon trading policy affects carbon price? An investigation of the Shanghai Emission Trading Scheme pilot. J. Cleaner Prod. **181** (2018)
10. Du, Z., Liu, F.: Research on the price of regional carbon emission rights in China and its influencing factors: an empirical analysis based on the GA-BP-MIV model. Price: Theory Practice **6**, 42–45 (2018). (in Chinese)
11. Ma, Y., Wang, L., Zhang, T.: Research on the dynamic linkage among the carbon emission trading, energy and capital markets. J. Cleaner Prod. **272** (2020) (prepublish)
12. Xie, Q., Hao, J., Li, J., Zheng, X.: Carbon price prediction considering climate change: a text-based framework. Econ. Anal. Policy (2022) (prepublish)
13. Liu, Z., Ma, X., Xie, Y.: Research on the interactions among carbon market, energy market, and financial market: a comparative perspective before and after the launch of the national carbon emission trading market. Secur. Mark. Her. **06**, 36–46 (2022). (in Chinese)
14. Yu, H., Li, K.: The mixed pattern of global climate governance and China's participation. Eur. Stud. **40**(01), 64–84+6–7 (2022). (in Chinese)
15. Dang, S., Zeng, W.: Challenges and responses to the new trends in carbon trading mechanism under the Paris agreement. China Sci. Technol. Forum **01**, 181–188 (2019). (in Chinese)
16. Wang, J.: Progress, challenges, and enlightenment of international carbon emission trading mechanism construction under the Paris agreement. Environ. Prot. **49**(13), 58–62 (2021). (in Chinese)

17. Zeng, W., Jiang, L.: Development constraints and relief paths of China's carbon market mechanism under the Paris agreement. Dongyue Tribune **43**(02), 105–114+192 (2022). (in Chinese)
18. Feng, C., Yuan, X., Wang, Q., Zhou, N.: Inspiration from the UK's "Carbon Floor Price" policy for improving China's carbon pricing mechanism. Fiscal Sci. **01**, 152–160 (2022). (in Chinese)
19. Long, F., Dong, Z., Bi, F., Zhou, J., Lian, C.: Analysis of the impact and response of the EU carbon border adjustment mechanism. China Environ. Manage. **14**(02), 43–48 (2022). (in Chinese)
20. Wei, Q., Li, L.: Research on the graded progressive penalty mechanism for power generation enterprises in the carbon market. China Environ. Sci. **40**(02), 919–928 (2020). (in Chinese)
21. Xue, F., Zhou, M.: Research on the emission reduction effect of China's carbon trading market size. East China Econ. Manage. **35**(06), 11–21 (2021). (in Chinese)
22. Pan, L.: Analysis of factors affecting the liquidity of China's stock market: empirical evidence from China's a-share securities market. Financ. Account. Commun. **30**, 63–64+7 (2010). (in Chinese)

# Research on Travel Route Recommendation of Urban Rail Transit Based on Passenger Portrait

Jie He[1], Yong Qin[1,2], Xuan Sun[1,2], and Jianyuan Guo[1](✉)

[1] School of Traffic and Transportation, Beijing Jiaotong University, Beijing 100044, China
{23120804,yqin,xuansun,jyguo}@bjtu.edu.cn

[2] State Key Laboratory of Advanced Rail Autonmous Operation, Beijing Jiaotong University, Beijing 100044, China

**Abstract.** With the development of the economy and science and technology, the construction scale of urban rail transit networks is gradually deepening and improving, and it tends to the development trend of network operation. Passengers can choose more travel plans between destination stations. The diversity of passenger travel plans leads to uneven passenger flow distribution, on the other hand, which makes the operation and management of urban rail transit face great challenges. At the same time, with the rise of the concept of travel and service, passengers' requirements for travel services are gradually improving. At present, the main travel route recommendations are mainly based on the comparison of the routes themselves, and the characteristics and needs of passengers are less considered. Therefore, it is of great significance to provide passengers with efficient and accurate urban rail travel routes suitable for them. In this paper, a travel route recommendation method based on passenger portrait is proposed, and grouping design is carried out based on passenger portrait, to provide a travel plan that meets the characteristics and needs of passengers for low-frequency passengers. This method quantifies the characteristics of passengers' travel and gives a personalized travel recommendation scheme, which provides passengers with accurate operation services, improves service quality, and improves the efficiency of passenger flow organization of operation managers.

**Keywords:** Passenger portrait · Space vector model · Travel path · Recommendation

## 1 Introduction

Compared with other modes of transportation, urban rail transit has many advantages, such as large volume, high speed, and punctuality, and is an important mode of urban public transportation. In the future, urban rail transit will gradually become networked and complicated, and the complexity between network nodes will increase greatly, which will bring diversified choices to passengers and increase the difficulty of urban rail operation and management, and put forward new requirements for urban rail service level


Y. Qin et al. (Eds.): EITRT 2023, LNEE 1137, pp. 448–459, 2024.
https://doi.org/10.1007/978-981-99-9311-6_47

[1]. At the same time, with the rise of Mobility-as-a-service (MaaS), it has gradually become a trend to provide passengers with personalized and convenient travel plans centered on passengers.

In the aspect of travel route recommendation, the traditional route recommendation idea is to calculate the utility value of a travel route by establishing indicators and recommending it to passengers [2]. Zhang [3] sets up a utility function including travel time, transfer time, and transfer frequency. MNL (Multinomial Logit) model is used to calculate the utility values of all routes. Based on passenger intention survey data, Yan [4] divided passengers from the perspective of gender and whether they commuted, and used the Logit model to recommend routes to passengers. Liu [5] considered the influence of passengers' individual social and economic factors when calculating route utility, and combined it with the Logit model to make recommendations.

Passenger portrait technology is an important means to analyze the characteristics of passengers. Portrait technology was first developed by A. Cooper, the father of interactive design [6]. At present, some studies have applied passenger portrait technology to improve the quality of passenger travel services. Wu [7] constructed a portrait system of railway passengers to recommend railway trains. Yin [8] constructed the portrait system of airport passengers, and the travel mode is recommended by combining collaborative filtering algorithms. Tai [9] established the portrait of group users of air-rail passengers, and the travel scheme is recommended based on the travel planning model considering passengers' travel preferences. The application of passenger portrait technology to improve the travel experience of urban rail passengers is one of the new and effective ways to improve the service quality of urban rail. On the one hand, a passenger portrait quantifies the characteristics of passengers and facilitates the operation and management of urban rail. On the other hand, it is more personalized to recommend travel plans based on portraits.

At present, in the existing research results, more attention is paid to the influence of the characteristics of the route itself on the passenger's route choice behavior. The influence of passenger's routing preference caused by personal characteristics is less considered. By calculating the similarity between passengers and routes, the matching degree between passenger demand and route features can be better reflected. Aiming at the practical problems of urban rail transit caused by the increase in network complexity, such as uneven distribution of passenger flow on the road network, large degree of travel path congestion, and relatively poor personalized recommendation of travel path, this paper applies passenger portraits to urban rail transit to conduct travel plan recommendation research on different groups of passengers. It aims to improve the degree of refinement and individuation of urban rail transit services, improve the quality of travel, and improve the efficiency of operation organization.

## 2 Establishment of Portrait System

This paper uses multi-source data to construct the following multi-source data-driven passenger travel index system from the aspects of passenger identity information, travel behavior index, travel distribution index and passenger feeling index (Fig. 1).

According to AFC data, the passenger classification index is constructed from three aspects: urban rail dependence intensity, departure time attribute, and travel space

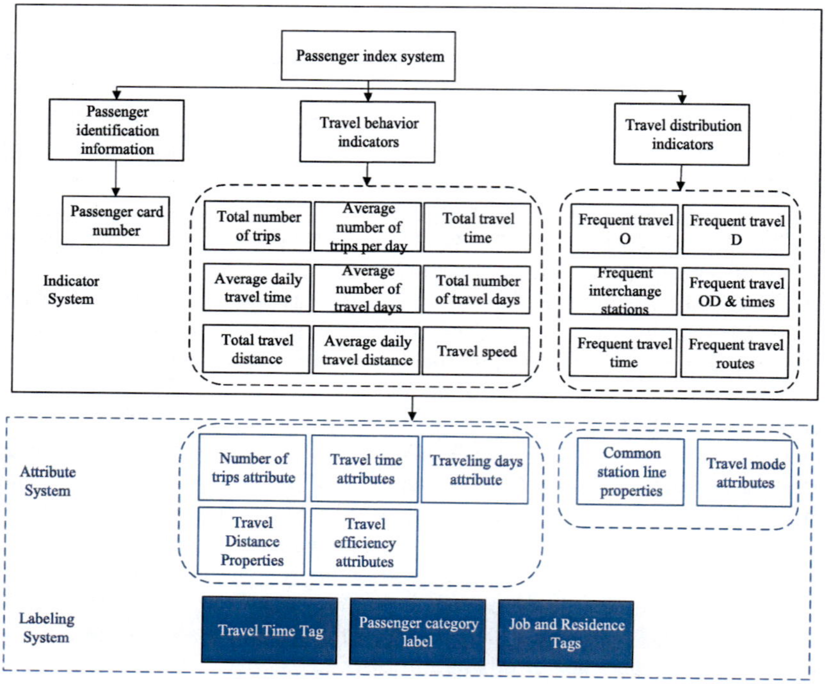


**Fig. 1.** Passenger travel index system.

attribute, and then the passenger group is divided by a k-means clustering algorithm [10, 11].

Based on the urban rail dependence strength attribute, departure time attribute, and travel space attribute, six indicators are established, which are daily average travel times, weekly average travel days, median departure time, standard deviation of departure time, OD coverage rate, and standard deviation of travel time [12] (Table 1).

**Table 1.** Passenger category label classification index.

| Variable dimension | Classification index | Unit |
|---|---|---|
| Dependency strength attribute | Number of days traveled per week | Day |
| | Average number of trips per day | Sub |
| Departure time attribute | First departure time | Minute |
| | Departure time standard deviation | / |
| Departure space attribute | OD coverage | / |
| | Length of trip standard deviation | / |

## 3 Urban Rail Travel Route Recommendation

According to relevant research, it is proved that the choice of passenger travel path has a great relationship with the characteristics of passengers' own attributes. Passengers have a high degree of familiarity with the starting and ending stations where they often travel, and a high degree of mastery of the travel time of each route under them, and vice versa. Therefore, this paper initially divides passengers into two types: familiar and unfamiliar [13]. The time limit for travel purposes is also an important factor affecting passenger route selection. A strong restriction on the duration of travel purposes means that passengers have higher requirements for travel duration, but lower requirements for comfort and transfer times, such as commuting and general study. The low limit on the duration of travel purposes means that passengers have low requirements for travel duration, but relatively high requirements for comfort and transfer times, such as traveling for entertainment and visiting relatives [14]. According to the above analysis, passengers can be classified according to their familiarity with the route of the starting and ending sites and the strength of the travel time limit, and finally, travel routes can be recommended for low-frequency passengers.

### 3.1 Space Vector Model Theory

Based on the low-frequency passenger trips obtained from the passenger target origin and destination sites combined with passenger profiles, a route recommendation scheme based on the spatial vector model is used. The basic idea is as follows: firstly, the passenger feature vector and the path feature vector under the origin and destination sites are mined through the same category label group as the target passengers, and characterized by the same dimension; secondly, the similarity between the passenger feature vector and the path feature vector is calculated; then, the top-N recommendation sequence is generated according to the similarity order. The flow of the model is as follows (Fig. 2).

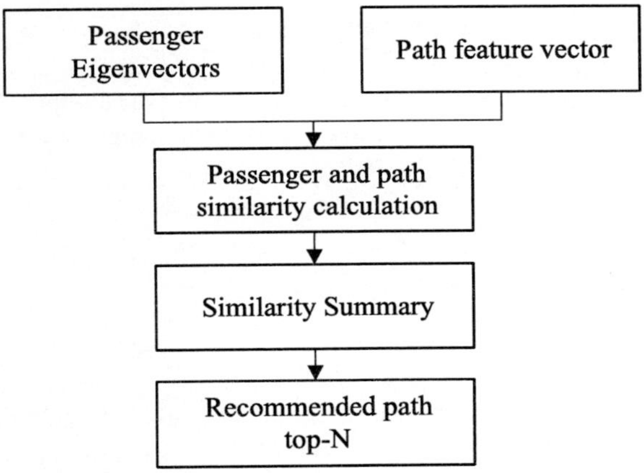


**Fig. 2.** Spatial vector model flow.

### 3.2 Feature Vector Extraction

To measure the passenger travel feature vector and the route feature vector, we mainly use the time-sharing travel record data to mine the three features, namely, the duration feature, the transfer times feature, and the congestion feature, and establish the feature vector. The calculation formula is as follows:

$$V_P = \left(V_P^{TM}, V_P^{TR}, V_P^{CR}\right) \tag{1}$$

where $V_P^{TM}$ is the duration features. $V_P^{TR}$ is the characteristics of the number of interchanges. $V_P^{CR}$ is the congestion characteristics.

(1) Duration characteristics

Duration characteristics $V_P^{TM}$ reflects the urgency of travel and the demand for travel time, by total travel time $AV_P^{TM}$ length of walk with passengers $WV_P^{TM}$ perform characterization. The calculation formula is as follows:

$$V_P^{TM} = \left(AV_P^{TM}, WV_P^{TM}\right) \tag{2}$$

$$AV_P^{TM} = \frac{1}{n}\sum_{j=1}^{n} AV_{P_{od_j}}^{TM} / avg\left(AV_{od_j}^{TM}\right) \tag{3}$$

$$WV_P^{TM} = \frac{1}{n}\sum_{j=1}^{n} WV_{P_{od_j}}^{walk\_tm_j} / \max\left(WV_{P_{od_j}}^{walk\_tm_j}\right) \tag{4}$$

where $V_P^{TM}$ is the duration characteristics. $AV_P^{TM}$ is the total trip duration, that is the average ratio of the total trip duration under all time-sharing trips to the average trip duration under OD. $WV_P^{TM}$ is the travel time, which is the average ratio of the travel time under all time-sharing trips to the maximum value of travel time under each OD.

(2) Characteristics of the number of interchanges

Characteristics of the number of interchanges $V_P^{TR}$ reflects the number of trip path interchange choices, that is the average of the number of interchanges under all time-sharing trips. The calculation formula is as follows:

$$V_P^{TR} = \frac{1}{n}\sum_{j=1}^{n} V_{P_{od_j}}^{TR} \tag{5}$$

(3) Congestion characteristics

Congestion characteristics $V_P^{CR}$ reflecting the sensitivity to congestion in travel choices. By average congestion $AV_P^{CR}$ and waiting time $WV_P^{CR}$ perform characterization. The calculation formula is as follows:

$$V_P^{CR} = \left(AV_P^{CR}, WV_P^{CR}\right) \tag{6}$$

$$AV_P^{CR} = \frac{1}{n}\sum_{j=1}^{n} AV_{P_{od_j}}^{CR} \tag{7}$$

$$WV_P^{CR} = \frac{1}{n}\sum_{j=1}^{n} WV_{P_{od_j}}^{wait\_tm_j} / \max\left(WV_{od_j}^{wait\_tm_j}\right) \tag{8}$$

where $V_P^{CR}$ is the congestion characteristics. $AV_P^{CR}$ is the Average congestion, that is the average of the trip congestion under all time-sharing trips. $WV_P^{CR}$ is the waiting time, which is the average ratio of the waiting time under all time-sharing trips to the maximum waiting time under each OD.

### 3.3 Generate Recommendations Based on Similarity

The path recommendation top-N sequence of passengers is generated based on the similarity between passenger feature vector and path feature vector. Cosine similarity is an important measure of vector distance, so this paper uses cosine similarity to calculate the similarity between passenger feature vectors and path feature vectors. Let the passenger eigenvector $V_P$, path feature vector $V_R$. Then the similarity of passengers and paths is calculated as:

$$\mathrm{sim}(V_P, V_R) = \frac{V_P^{TM} \cdot V_R^{TM} + V_P^{TR} \cdot V_R^{TR} + V_P^{CR} \cdot V_R^{CR}}{\sqrt{(V_P^{TM})^2 + (V_P^{TR})^2 + (V_P^{CR})^2} \cdot \sqrt{(V_R^{TM})^2 + (V_R^{TR})^2 + (V_R^{CR})^2}} \tag{9}$$

Based on the similarity ranking, the greater the similarity, the higher the recommendation, and then generate a list of travel route recommendations for passengers.

## 4 Case Analysis

In this chapter, a passenger is selected for case analysis. Firstly, the passenger portrait system is constructed and the passenger portrait is displayed. Secondly, the case scenario is designed for the passenger to recommend the travel plan, and the recommended results are analyzed.

### 4.1 Portrait Depiction of Passengers

In this section, passenger portrait analysis and visual display are carried out from three aspects: passenger travel index, passenger travel time and space law, and passenger label.

(1) Passenger travel indicators

Passenger travel behavior indicators mainly represent the statistical data of passengers' historical travel records, including travel days statistics, travel length statistics, travel distance statistics, travel frequency statistics, and travel speed statistics, which formally reflect the passenger behavior data and lay a foundation for subsequent characterization

**Table 2.** Passenger travel index example.

| Dimension | | Non-working days | Working day |
|---|---|---|---|
| Number of days travelled | | 6 (14.64%) | 35 (85.36%) |
| Number of trips | | 11 (13.10%) | 73 (86.90%) |
| Travel time (min) | Total | 664.000000 | 6463.000000 |
| | Average value | 60.363636 | 88.534247 |
| | Maximum value | 257.000000 | 128.000000 |
| | Minimum value | 12.000000 | 7.000000 |
| Traveling distance (m) | Total | 202225.000000 | 3.309746e+06 |
| | Mean | 18384.090909 | 4.533899e+04 |
| | Maximum value | 53841.000000 | 5.449200e+04 |
| | Minimum value | 6403.000000 | 1.759000e+03 |
| Travel speed (m/min | Average value | 423.523818 | 515.178941 |
| | Maximum value | 539.421053 | 796.608696 |
| | Minimum value | 24.914397 | 251.285714 |
| Travel time | Most frequent periods | Discrete | 5–6 |
| | Corresponding times | Discrete | 29 |

of passenger characteristics. Take the passenger with the card number 10****21 as an example. The travel indicators of this passenger from May to June 2019 are as Table 2.

From the perspective of working days and non-working days, passengers travel more on working days, less on non-working days, and the travel speed on working days is higher than that on non-working days. Passengers have obvious advantages in traveling at 5:00 to 6:00 on working days, but the non-working days are scattered and have no obvious advantages. It can be seen that passengers mostly travel by urban rail on weekdays.

(2) Temporal and spatial rules of passenger travel

Passenger travel distribution index mainly represents the temporal and spatial characteristics of passenger travel from the perspective of time and space. The combination of time and space information comprehensively characterizes the distribution characteristics of passengers' travel in many directions, so as to mine the time and space characteristics of passengers' travel and passenger labels. From the perspective of time and space, the statistics of the time and space attributes of the combination of passenger stop and time slot are as follows (Fig. 3).

Passengers usually travel from 6:00 to 8:00 in Liangxiang University Town West Station to Taiyanggong Station and from 17:00 to 19:00 in Liangxiang University Town West Station. At the same time, the statistics of working day and non-working day trips are made respectively, in which the working day trips are consistent with the overall trips, usually at 6:00 to 8:00 in Liangxiang University Town West Station to Taiyanggong Station, and at 17:00 to 19:00 in Taiyanggong Station to Liangxiang University Town

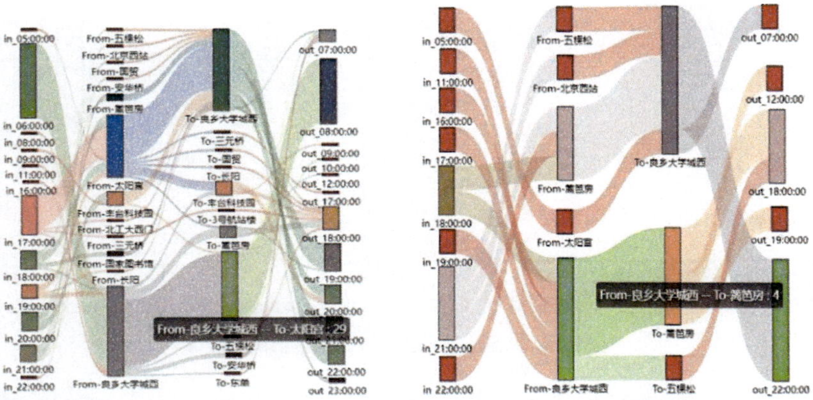


(a) Distribution of inbound stations on working days. (b) Distribution of inbound stations on non-working days.

**Fig. 3.** Time and space distribution statistics of inbound stations.

West Station. The weekend trips are scattered and the distribution regularity of OD stations is not obvious.

It can be seen that the OD of passengers on weekdays is relatively fixed, and the OD of Liangxiang University Town West Station to Taiyanggong Station and Taiyanggong Station to Liangxiang University Town West Station is frequently traveled by passengers, and the travel time is between 6:00 to 8:00 and 17:00 to 19:00. Passenger travel conforms to the characteristics of commuting.

(3) Passenger label

The passenger's category label is analyzed and verified. According to the label clustering analysis in the second chapter, the passenger's category label is a commuter, which is in line with the previous analysis of the distribution of passenger's travel time and space. By counting the passenger's first departure time on weekdays, the passenger's departure time attributes are as Fig. 4.

It can be seen that this passenger travels from 6:15 to 6:20 on weekdays, accounting for 24.7% of the total travel time distribution on weekdays, and the travel time distribution is relatively stable. The passenger's non-working day departure time distribution is discrete and has no obvious law.

According to the statistics of OD coverage and standard deviation of travel time, the stability of the passenger's departure space is as Table 3.

The passenger's working day OD coverage is relatively small, which shows that the distribution of the passenger's departure stations is relatively fixed, and the standard deviation of travel time is small, which conforms to the characteristics of commuter passengers' working day travel. On non-working days, the passenger's OD coverage and standard deviation of travel time are large, which shows that the passenger is not a commuter on non-working days.

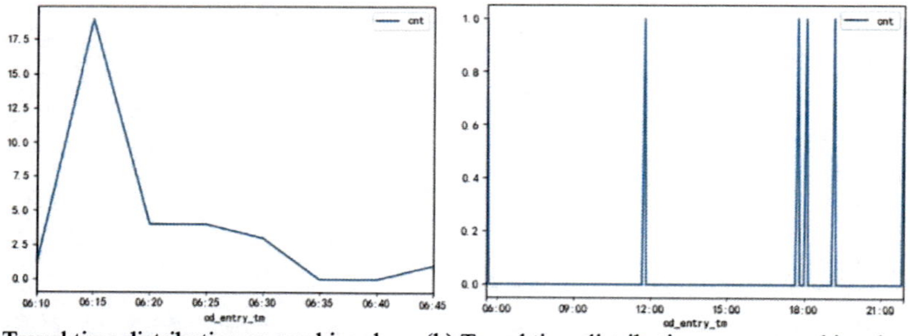


(a) Travel time distribution on working days. (b) Travel time distribution on non-working days.

**Fig. 4.** Passenger travel time distribution.

**Table 3.** Passenger travel spatial attribute statistics

| Travel space attribute | Workday | Non-workdays |
|---|---|---|
| OD coverage | 0.246575 | 0.545455 |
| Standard deviation of travel time | 27.710381 | 70.266082 |

According to the statistics of the median departure time and the standard deviation of the first departure time of the passenger, the departure time attributes are as table 4.

**Table 4.** Passenger travel time attribute statistics

| Travel time attribute | Workday | Non-workdays |
|---|---|---|
| Median time of first departure | 06:19:00 | 17:57:00 |
| Standard deviation of departure time | 0.118696 | 332.888631 |

On weekdays, the standard deviation of the passenger's departure time is small, and the departure time is mostly around 6:19, which shows that the passenger's departure time on weekdays is stable, and the standard deviation of the passenger's departure time on non-working days is large, which shows that the passenger's departure time on non-working days is unstable and there is no obvious feature in travel.

## 4.2 Travel Plan Recommendation

Design a low-frequency travel scene for this passenger, and use the space vector model to recommend the travel path for this kind of scene.

We set the case scenario as: passengers with the card number of 10****21 enter the destination from Liangxiang University Town West Station to Wukesong Station at 11:00 on July 1, 2019. Through case analysis, the number of trips made by the passenger from

Liangxiang University Town West Station to Wukesong Station was 4, and the number of trips made from Wukesong Station to Liangxiang University Town West Station was 0, accounting for 4.7% of the total passenger record, which did not exceed the OD threshold value. At the same time, the departure time is 11:00, which does not belong to the peak travel time. Therefore, it is judged that passengers travel at low frequency in this scene, and the space vector model is used to recommend the travel route.

There are three routes from the west of Liangxiang University Town West Station to Wukesong Station, numbered as No. 1, No. 2 and No. 3 respectively. Among them, the first route transfer station is Guogongzhuang Station and Military Museum Station. The second route transfer station is Guogongzhuang Station, Liuliqiao Station and Gongzhufen Station. The third route transfer station is Guogongzhuang Station, Qilizhuang Station, Xiju Station and Gongzhufen Station.

Then, we build a case model. According to the AFC data of other passengers in the same category as 10****21 from 6:00 to 6:30, the time consumption, convenience, and congestion of each travel route are counted, and the feature vectors of each travel route under the starting and ending station are obtained (Table 5).

**Table 5.** Characteristics of the route from the West of Liangxiang University Town West Station to Wukesong Station.

| | Time consumption | | Convenience | Comfort level | |
|---|---|---|---|---|---|
| | Path travel time (min) | Path travel time (min) | Path transfer times | Path full load rate | Path waiting time (min) |
| No.1 | 72.5 | 1.5 | 2 | 0.55 | 0.51 |
| No.2 | 77.9 | 2.1 | Three | 0.30 | 0.23 |
| No.3 | 84.7 | 1.8 | Four | 0.22 | 0.33 |

After normalization, the eigenvectors of each path from the west of Liangxiang University Town West Station to Wukesong Station are as Table 6.

**Table 6.** Normalized path eigenvector of Liangxiang University Town West Station to Wukesong Station.

| Path number | Normalized path feature vector |
|---|---|
| No. 1 | ((0.86, 0.48), 0.5, (0.55, 1)) |
| No. 2 | ((0.92, 1), 0.75, (0.30, 0.45)) |
| No. 3 | ((1, 0.86), 1, (0.4, 0.65)) |

According to the passenger's historical travel records, the passenger feature vector is counted from the ranking ratio of travel time, the ratio of travel time, the weighted average of transfer times, the average of section full load, and the ratio of waiting time, and the passenger feature vector is obtained as R = ((0.34, 0.56), 0, (0.36, 0.60)).

Through similarity calculation, using cosine similarity to calculate the similarity between each path feature vector and passenger feature vector, the results are as Table 7.

**Table 7.** Similarity between passenger and route feature vector

| | No. 1 | No. 2 | No. 3 |
|---|---|---|---|
| Similarity | 0.57 | 0.39 | 0.33 |

It can be seen that the maximum similarity between the passenger feature and the feature vector of Route 1 is 0.57, and the minimum similarity between the passenger feature and Route 3 is 0.33.

Finally, combined with the congestion, average travel time, and transfer times of each route, the travel routes and related information are recommended, and the recommended results of Liangxiang University Town West Station to Wukesong Station at 11: 00 are as follows (Table 8).

**Table 8.** Recommended results of Liangxiang University Town West Station to Wukesong Station.

| O-D | Path number | Recommendation order | Transfer times | Arrival time | Path congestion |
|---|---|---|---|---|---|
| LX. Univ Town West to WKS | No. 1 | 1 | 2 | 12:12 | 0.53 |
| | No. 2 | 2 | 3 | 12:17 | 0.30 |
| | No. 3 | 3 | 4 | 12:24 | 0.22 |

It can be seen that Route 1 is the most recommended in the recommendation results, with the least transfer times, constant congestion, and the shortest travel time, which is in line with the objective recommendation evaluation.

## 5 Summary and Prospect

In view of the uneven distribution of passenger flow on the road network caused by the diversity of trips, the poor personalized travel path recommendation and the relatively few studies on passenger portraits, this paper carries out a study on the travel path recommendation of urban rail transit based on passenger portraits. This paper builds a passenger portrait system for personalized travel recommendation. On the basis of mining passenger travel characteristics from a large number of data and using clustering algorithm to deeply mine passenger travel indicators, the passenger portrait system is generated. Based on passenger portraits, the group classification is carried out, and the spatial vector recommendation algorithm is used to carry out the travel path recommendation research for low-frequency passengers. The recommendation results are more in line with the characteristics and needs of passengers themselves, making the recommendation more personalized, improving the travel experience of passengers, and improving the service level and organizational efficiency of urban rail operation.

In the future, the further expansion of data can be combined with personal social attributes such as passenger age, gender, and location information, so that the establishment of urban rail transit passenger portraits will be more comprehensive and perfect. At the same time, multi-mode travel scheme recommendation research can be carried out in combination with other travel modes such as bus and bicycle, so as to enrich the form of travel scheme selection and further improve the quality of travel service. Combined with prediction method, real-time dynamic recommendation research is realized to improve the application value of passenger portrait.

## References


1. Fei, S.: Research on Passenger Flow Assignment of Urban Rail Transit Based on Passenger Familiarity with Rail Transit Network. Beijing Jiaotong University (2019). (in Chinese)
2. Li, B.: Markov models for Bayesian analysis about transit route origin–destination matrices. Transp. Res. Part B: Methodol. **43**(3), 301–310 (2009)
3. Zhang, Y., Yao, E., Yang, Y.: A route-planning algorithm based on subway passenger routes choice behavior analysis. In: ICTIS 2013: Improving Multimodal Transportation Systems-Information, Safety, and Integration, pp. 1972–1978 (2013)
4. Yan, H.: Personalized rail transit passenger travel path recommendation algorithm. Comput. Program. Skills Maintenance **22**, 87–88+91 (2015). (in Chinese)
5. Liu, W.S., Liu, X.B.: Route choice research in rail transit network based on passenger classification. J. Transp. Eng. Inform. **14**(04), 81–86+99 (2016). (in Chinese)
6. Cooper, A.: The Inmates Are Running the Asylum, p. 401. Macmillan Publishing Co., Inc. (1999)
7. Wu, Y.T.: Study on Railroad Passenger' User Profiles and Scheme Recommendation Based on Multi-Source Data. Southeast University (2020). (in Chinese)
8. Yin, X.C.: Travel Recommendation Method and Application Based on Passenger Profile and Trip Chain Model. Beijing Jiaotong University (2021). (in Chinese)
9. Tai, J.H.: Travel Planning Model and Application of Air-rail Integrated Service Based on User Portrait. Southeast University (2021). (in Chinese)
10. Liu J.W.: Research on the Effect of Time Differential Pricing of Beijing Subway on Passenger Travel Behavior. Beijing Jiaotong University (2018). (in Chinese)
11. Liu, N.N., Zhou, H.J.: Research on passenger travel elasticity of urban rail transit based on AFC data. Traffic Eng. **23**(01), 15–20+30 (2023). (in Chinese)
12. Zou, Q.R., Zhao, P., Yao, X.M.: Passenger classification for urban rail transit by mining smart card data. J. Transp. Eng. Inform. **18**(01), 223–230 (2018). (in Chinese)
13. Zhang, Z.: Research on Passenger Travel Choice and Path Generation for Urban Rail Transit Network Simulation. Beijing Jiaotong University (2015). (in Chinese)
14. Jia, F.F.: Optimization of Passenger Route Dynamic Guidance Strategy in Urban Rail Transit Network. Beijing Jiaotong University (2021). (in Chinese)

# Urban Rail Train Diagram Compilation for Unbalanced Passenger Flow

Jiaji Li[1], Jianyuan Guo[1(✉)], Yong Qin[2], and Yueyue Wang[3]

[1] School of Traffic and Transportation, Beijing Jiaotong University, Beijing 100044, China
{22120857,jyguo}@bjtu.edu.cn

[2] State Key Laboratory of Advanced Rail Autonomous Operation, Beijing Jiaotong University, Beijing 100044, China
yqin@bjtu.edu.cn

[3] Beijing Metro Network Administration Co., Ltd., Beijing 100000, China
wangyueyue@bmncc.com.cn

**Abstract.** Train diagram is an important basis for train operation and a decisive condition to ensure the safe and efficient travel of passengers. In view of the unbalanced characteristics of passenger flow in the morning commute, the train operation plan is formulated in combination with multi-routing mode, fast-slow mode and dynamic stop mode. The model is constructed to compile the train diagram with the goal of minimizing the average overload rate of the train, the average waiting time of the passengers and the travel distance of the train. The bi-level optimization algorithm is used, and the lower simulation is nested in the upper genetic algorithm to solve the model. The feasibility of the model and algorithm is verified by the actual data of Beijing Metro Line 6. The experimental results show that the mode of combining multi-routing with fast-slow train and dynamic parking reduces the average overload rate of trains and the average waiting time of passengers by 31.6% and 43.7% respectively. The method in this paper effectively improves the dynamic matching degree between the train diagram and the disequilibrium passenger flow, and meanwhile ensures the safe and efficient travel of passengers.

**Keywords:** Urban Rail Transit · Train Diagram · Disequilibrium Passenger Flow · Multi-Routing Mode · Fast-Slow Mode

## 1 Introduction

With the network development of urban rail transit, the passenger demands of urban rail transit is increasing, and the limited capacity brings about congestion. At the same time, due to the spatial characteristics of commuter passenger flow, the rail transit passenger flow shows the time imbalance of insufficient capacity in peak hours and waste of capacity in low peak hours. The passenger flow distribution of lines, sections and stations shows spatial imbalance. It is of great significance to the current urban rail transit service system to compile the operation diagram under the condition of limited capacity to meet the characteristics of disequilibrium passenger flow.


Y. Qin et al. (Eds.): EITRT 2023, LNEE 1137, pp. 460–475, 2024.
https://doi.org/10.1007/978-981-99-9311-6_48

In recent years, the research on urban rail transit operation diagram compilation for dynamic passenger flow has attracted the attention of many scholars and operators at home and abroad, which has formed a large number of research results.

Firstly, many scholars studied the key problem of train operation plan compilation. Cao [1] compiled a train operation plan based on passenger flow demand, constructed a timetable optimization model with the goal of minimizing passenger waiting time, and applied genetic harmony algorithm to solve it. Sun [2] aimed to minimize the sum of transportation costs of operating companies and generalized passenger travel costs, and solved the constructed operation plan optimization model through a simulation-based genetic algorithm. Among them, Dai et al. [3] and Duan [4] conducted in-depth research on the multi-routing mode, and Duan added the application of the fast-slow mode on the basis of it. Li et al. [5] discussed the combinatorial effect of service order and overtaking by developing four mixed integer quadratic programming timetabling models with different service order as well as overtaking conditions.

Since the problem of train diagram compilation is an NP-Hard problem, many scholars have carried out research on the solution of train diagram compilation and optimization model. Some scholars used traditional methods to solve the model. Yang [6] studied the optimization problem of the last train timetable at the level of urban rail network, and established two 0–1 integer linear programming models to characterize the accessibility based on space-time and flow levels respectively. The Lagrange multiplier dualization model constraint is introduced and the sub gradient algorithm is used to solve the problem. Su [7] proposed a special train timetable preparation method for reducing the full load rate, established a bi-level programming model of train operation plan and train diagram, and transformed it into a linear model, which was solved by CPLEX software. However, more scholars considered the superiority of heuristic algorithms and used heuristic algorithms such as genetic algorithms and large neighborhood search algorithms to solve the model. Zhu et al. [8] established a two-layer model to minimize the total cost of passengers. Considering the superiority of the heuristic algorithm, the genetic algorithm and MSA algorithm were used to solve the model to draw a one-way large routing diagram. Zhang et al. [9] established a nonlinear non-convex programming model with the goal of minimizing the total travel time of passengers, and used the adaptive large neighborhood algorithm to solve it. Zhou [10] established a timetable optimization model by adjusting the inter-station running time, residence time and departure interval of trains, aiming at minimizing passenger travel cost and enterprise operation cost, and used NSGA-II to find the optimal Pareto boundary. Shahabi [11] established a nonlinear mixed integer programming model with the goal of minimizing the total standard deviation of train departure interval, and solved it by variable neighborhood search algorithm and genetic algorithm respectively and compared the advantages of different solutions.

The existing research mainly takes the passenger flow demand as the data basis, and constructs the train diagram compilation model of dynamic passenger flow from the subway operation enterprise, the passenger or the combination of the two. At the same time, the train diagram compilation problem is integrated with the routing plan and the parking plan, so that the train diagram compilation is more systematic and applicable to the actual operation situation. The unbalanced characteristics of passenger flow put forward requirements for improving the flexibility of urban rail system's organization and meeting the needs of passengers as much as possible. However, the existing research rarely has the combination of multi-routing mode, fast-slow mode and dynamic parking mode, resulting in the failure to achieve the best matching degree between capacity and unbalanced dynamic passenger flow.

The main purpose of the paper is to meet the passenger flow demand as much as possible under the condition of limited capacity, control the train full load rate, and ensure the safe and efficient travel of passengers. Therefore, this paper analyzes the characteristics of passenger flow to formulate a reasonable operation plan considering long and short routes, fast and slow trains and flexible dwell times, establishes a train diagram preparation model, and uses a double-layer optimization algorithm to solve the problem. Compared with the previous methods, the innovation of this model is to consider multi-routing mode and fast-slow mode, and to determine the dwell time by the passenger flow. So that the matching degree between the train diagram and the passenger flow is higher.

# 2 Model of Train Diagram Compilation

## 2.1 Model Assumptions

In order to facilitate the model construction, the following assumptions are made.

(1) The passenger flow demand during the study period $T$ is known. The passenger flow of the transfer station on the line is included in the inbound passenger flow, and the passenger flow is included in the outbound passenger flow.
(2) The departure interval of the same route train issued by the same station is the same, and the departure time is only advanced or delayed when the train operation conflict occurs.
(3) When all passengers on a station exceed the upper limit of the train capacity, a random boarding method is adopted. After boarding the express train, the passengers who go to the non-stop station should get off at the starting station of the cross section, and wait for the slow train. The upper limit of train capacity is the multiplication of train capacity *cap* and full load rate $\alpha$.

## 2.2 Objective Function

In view of the research purpose of this paper, the model is constructed with the goal of minimizing the average overload rate of the train, the average waiting time of the passengers and the travel distance of the train.

After each train stops at the station, the full load rate of the train is recorded, which is used to calculate the average of the over-limit part of the full load rate limit. The calculation formula is as follows:

$$\overline{\alpha} = \frac{\sum_{n=0}^{N} \sum_{c=0}^{1} \sum_{s_l} \max(\alpha_{s_l}^{n,c} - \alpha_{limit}, 0)}{j} \tag{1}$$

where $\alpha_{s_l}^{n,c}$ is the train full load rate of the over-limit part of the train $n$ in the direction $c$ of the station $s_l$. $\alpha_{limit}$ is the limit of the train full load rate. $j$ is the number of trains which exceed the limit of the full load rate. In response to different situations such as major health problems and large passenger flow problems, the limit of the full load rate in the formula is set according to the actual situation.

For passengers who successfully board the train, the waiting time is the difference between the time when passengers get aboard the train and the time when passengers arrive at the station. For passengers who have not boarded the train until the end of the study period, the waiting time is the difference between the end of the study period and the arrival time of passengers at the station. The calculation formula is as follows:

$$\overline{p}_w = \frac{\sum_{t=0}^{T} p^{served} \times (t - t_{arrive}^{served}) + \sum_{t_{arrive}^{not\_served}=0}^{T} p^{not\_served} \times (T - t_{arrive}^{not\_served})}{\sum_{t=0}^{T}\sum_{i=0}^{S}\sum_{j=0}^{S} p_{i,j}(t)} \tag{2}$$

where $p^{served}$ and $p^{not-served}$ are the number of passengers getting on and not getting on trains respectively. $t$ is the time that the passengers get aboard the train. $t_{arrive}^{served}$ and $t_{arrive}^{not-served}$ are the arrival time of boarding and non-boarding passengers. $p_{i,j}(t)$ is the number of passengers who arrive at station $i$ and is going to station $j$ at time $t$.

In this paper, the number of kilometers traveled by train refers to the transportation cost of the enterprise, excluding the distance from the depot to the station and the turn-back distance of the train. The calculation formula is as follows:

$$m = \sum_{n=1}^{N}\sum_{t=0}^{T}\sum_{s_l} m_{s_l,s_l+1}^{n,0}(t) + \sum_{n=1}^{N}\sum_{t=0}^{T}\sum_{s_l} m_{s_l,s_l-1}^{n,1}(t) \tag{3}$$

where $m_{s_l,s_l+1}^{n,0}(t)$ and $m_{s_l,s_l-1}^{n,1}(t)$ are the running distance from station $s_l$ to station $s_l+1$ in the upward direction of train $n$ at time $t$ and the running distance from station $s_l$ to station $s_l - 1$ in the downward direction of train $n$ at time $t$.

The multi-objective is weighted and summed to obtain a single objective function. The definition of objective function is as follows:

$$\min O = \lambda_1 \times \overline{\alpha} + \lambda_2 \times \overline{p}_w + \lambda_3 \times m \tag{4}$$

where $\lambda_1$,$\lambda_2$ and $\lambda_3$ are the weights of each sub-objective, which can be manually designated by AHP method or based on experience and investigation.

## 2.3 Decision Variable

The train diagram is compiled on the basis of the train operation plan. The number of trains $v_{s_f}^l$ issued by each route, the number of express trains $k_{s_f}^l$ and the time interval $h_{s_f}^l$ of each train issued by each route are decision variables.

The number of trains and the number of express trains on each line need to meet the constraints of the number of trains.

$$0 \le \sum_{l} v_{s_f}^{l} \le v_{s_f} \tag{5}$$

$$0 \le k_{s_f}^{l} \le v_{s_f}^{l} \tag{6}$$

where $v_{s_f}$ is the upper limit of the number of trains in the depot or parking lot where the train is issued.

The time interval of each train from each route needs to meet the departure interval constraint.

$$hd_{\min} \le h_{s_f}^{l} \le hd_{\max} \tag{7}$$

where $hd_{\min}$ and $hd_{\max}$ are the minimum and maximum time intervals of trains from the station.

## 2.4 Constraint Condition

In order to ensure the safety of urban rail transit system and facilitate the model solution, the following constraints are set.

(1) Capacity constraint

After the train stops at the station, the waiting passengers should board the train under the constraint of train capacity, that is, the sum of the number of passengers aboard the train and the number of passengers on the train should less than the multiplication of the train capacity and the upper limit of the full load rate.

$$0 \le p_{s_l}^{n,c}(t) + p_{s_l}^{n,c,in}(t) \le cap \times \alpha_0 \tag{8}$$

where $p_{s_l}^{n,c}(t)$ is the number of passengers on the train $n$ after the passengers get off the train in the direction $c$ at station $s_l$ at time $t$. $p_{s_l}^{n,c,in}(t)$ is the number of people who get on the train after the train $n$ arrives at the station in the direction $c$ of station $s_l$ at time $t$.

(2) Safe driving interval constraint

For the trains in non-overtaking station and trains without overtaking behavior in overtaking station, the arrival and the departure trains need to meet the safe driving interval of the train to ensure the safety.

$$a_s^{\prime c} - d_s^c \ge h_{\min} \tag{9}$$

$$d_s^{\prime c} - d_s^c \ge h_{\min} \tag{10}$$

where $h_{\min}$ represents minimum driving interval.

(3) Arrival and departure interval constraint

In order to ensure the safety of the trains at the platform, only one train can stop at the platform, and the subsequent trains can only enter the station after the previous trains leave the station. Therefore, the departure time of the previous train $d_s^c$ and the arrival time of the subsequent train $a_s^{\prime c}$ need to meet the arrival and departure interval $\tau_{arr-dpt}$.

$$\left|a_s^{\prime c} - d_s^c\right| \geq \tau_{arr-dpt} \tag{11}$$

For trains with overtaking events, the arrival time of the subsequent overtaking express train $a_s^{\prime c}$ and the departure time of the previous slow train station $d_s^c$ should meet the arrival and departure interval (Fig. 1).

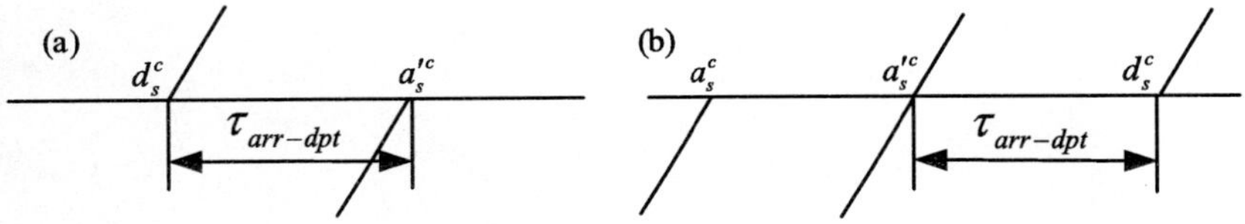


**Fig. 1.** Diagram of arrival and departure interval. **a** Arrival and departure interval without overtaking events, **b** arrival and departure interval with overtaking events.

(4) Dwell time constraint

For the train stopping at the station, the dwell time of the train is dynamically changed according to the number of passengers getting on and off the train. The sum of the time of opening and closing the door and the time of waiting for departure after closing the door is set to 20 s, and the time of each person getting on and off is set to 0.8 s.

$$w_{s_l}^{n,c}(t) = 20 + 0.8 \times (p_{s_l}^{n,c,in}(t) + p_{s_l}^{n,c,out}(t))/q \tag{12}$$

where $p_{s_l}^{n,c,out}(t)$ is the number of people getting off the train after the train $n$ arrives in the direction $c$ of station $s_l$ at time $t$.

For trains that turn back at the end of the route, the dwell time is not only affected by the passenger flow, but also by the turn-back progress of the previous turn-back train. In this paper, only one train on the turn-back line is allowed to turn back, that is, after the previous turn-back train reaches the platform, the subsequent turn-back train can enter the turn-back line and start the turn-back operation. Therefore, when all the passengers in the subsequent turn-back train get off the trains but the previous turn-back train has not completed the turn-back, the subsequent turn-back train waits on the platform and extends the dwell time.

$$w_{s_z}^c(t) = a_{s_z}^{\prime c} - a_{s_z}^{|1-c|} \tag{13}$$

(5) Turn-back time constraint

The train needs to meet the minimum turn-back time constraint to ensure the safety of the turn-back operation. The time of station-front turning-back includes five parts: getting-off time, inbound time, turn-round time, outbound time and boarding time. Since the train dynamically controls the dwell time according to the number of passengers on and off, the getting-off time and boarding time of the train at the terminal station fluctuates. So, the shortest time of return does not include the getting-off and boarding time.

$$d_{s_z}^{c} - a_{s_z}^{|1-c|} \geq h_{s_z} \tag{14}$$

(6) Studying time constraint

$$0 \leq a_s^{c,n} \leq T, n \in Trains \tag{15}$$

$$0 \leq d_s^{c,n} \leq T, n \in Trains \tag{16}$$

# 3 Bi-Level Optimization Algorithm

## 3.1 The Framework of Bi-Level Optimization Algorithm

Since compiling the train timetable is NP-hard, the traditional solution method is difficult to implement. In this paper, the genetic algorithm is used as the upper algorithm, and simulation deduction method is used in the lower algorithm to solve the train timetable compilation model. The process of bi-level optimization algorithm is shown in Fig. 2.

## 3.2 Simulation-Based Conflict Resolution

Since the train operation plan is formulated in combination with multi-routing mode, fast-slow mode and dynamic parking mode, it is necessary to make the state of the train's arrival and departure time meet the constraints through conflict resolution in the process of simulation.

(1) Non-turn-back, non-overtaking departure trains

The departure time of non-turn-back and non-overtaking departure trains is generally determined by the arrival time of the train and the dynamic dwell time of the train. However, when the safety of train operation cannot be guaranteed, the departure time is determined by the departure time of the preceding train and the safe driving interval according to formula (10). The conflict adjustment is shown in Fig. 3a.

(2) Overtaking departure trains

In this paper, the train operation plan includes fast-slow mode. The fast train can overtake the slow train at the station with side line. When the arrival time of the fast train is generated, the departure time is generated immediately. Therefore, the departure time

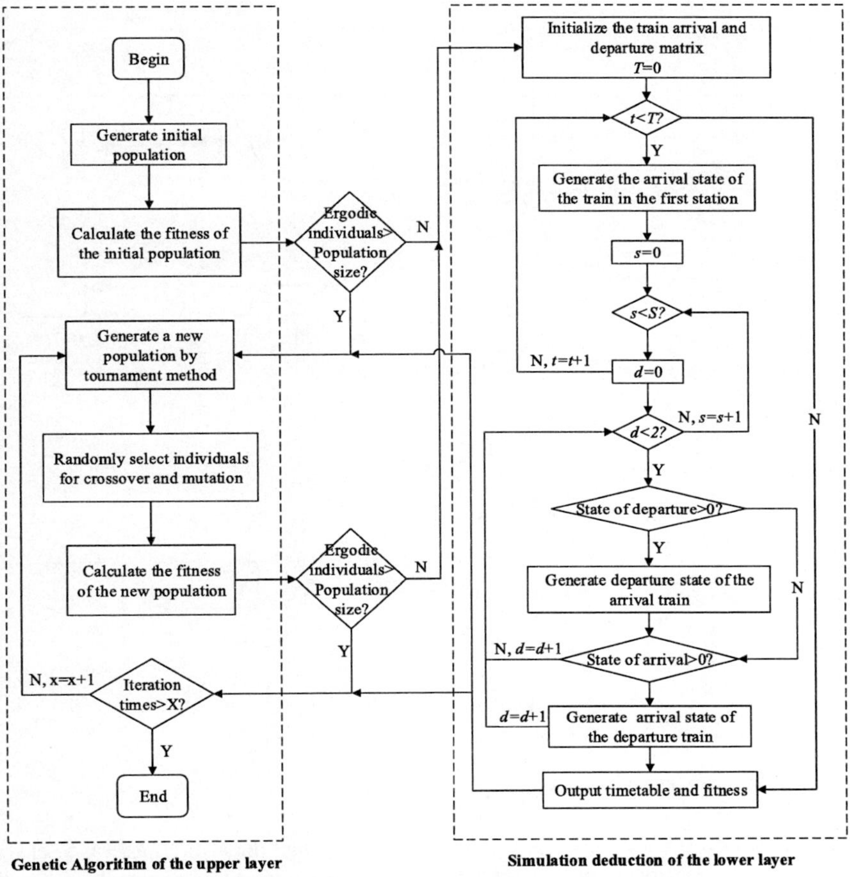


**Fig. 2.** The process of bi-level optimization algorithm

of the slow train is very likely to fail to meet the requirements of safe driving interval. In this case, the departure time of the slow train is determined by the arrival time of the express train and the time interval constraint according to formula (11). The conflict adjustment is shown in Fig. 3b.

(3) Turn-back departure trains

In order to ensure the safety of the turn-back train, this paper limits that only one train can carry out the turn-back operation on the turn-back line. When the turn-back operation of the previous train is not completed, the subsequent turn-back train takes up the platform and waits until the previous train reaches the opposite direction platform. The interaction formula is formula (13). The conflict adjustment is shown in Fig. 3c.

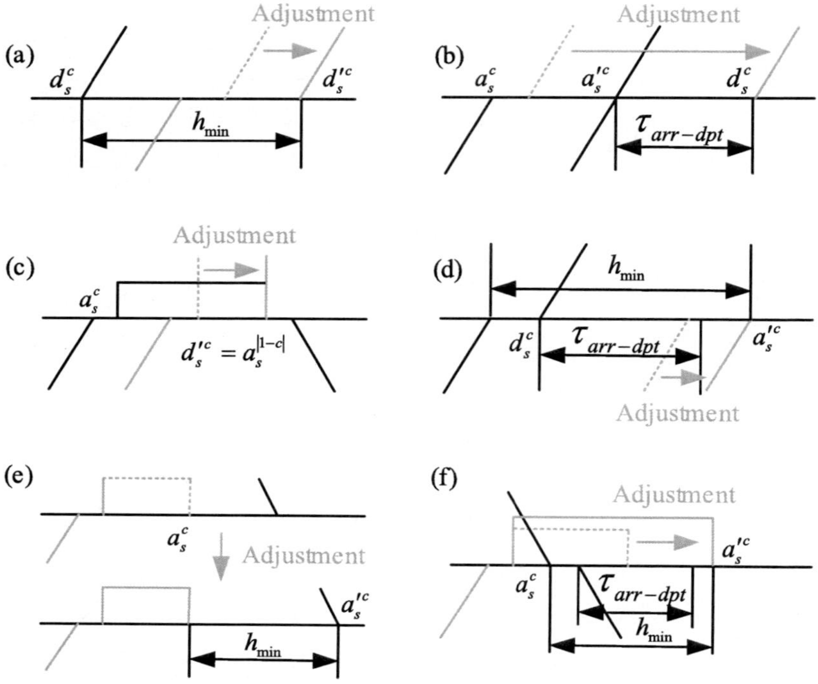


**Fig. 3.** Simulation deduction of conflicts

(4) Non-turn-back arrival trains

For the arrival trains, it is necessary to meet the arrival and departure interval to ensure the safety of the platform, and to meet the safe driving interval to ensure the safety of the line. The conflict adjustment is shown in Fig. 3f.

(5) Turn-back arrival trains

For the turn-back station in the middle of the line, it is easy to generate the case that the turn-back train and the normal running train all need to occupy the platform of the same direction. In Fig. 3e and f, the principle of 'first come, first occupied' is adopted.

If the expected arrival time of the turn-back train is earlier than the arrival time of the normal running train, the interval running time of the subsequent train is increased and the arrival time is delayed.

On the contrary, if the arrival time of the normal running train is earlier than the expected arrival time of the turn-back train, the normal running train will be the first to occupy the platform, the turn-back train will wait on the turn-back line and then entered the platform until the platform is cleared and the line is safe.

# 4 Case Analysis and Discussion

## 4.1 Data

Based on the data of a working day of Beijing Metro Line 6 in 2019, this paper verifies the validity of train diagram compilation model through simulation deduction.

As of early 2019, Beijing Metro Line 6 has opened 32 stations, numbered them from 0 to 31 in the upward direction. Among them, Jin'anqiao Station, Haidian Wuluju Station, Chegongzhuang Xi(W) Station, Chaoyang Men Station, Qingnian Lu Station, Caofang Station and Lucheng Station have the conditions to turn back, and Changying Station has overtaking conditions.

According to the data of the passenger flow, the time imbalance coefficient is calculated as shown in Fig. 4. The morning peak period is 7:00~10:00. In order to ensure the solution efficiency, the research period of this paper is 4:00~10:00 (Fig. 5).

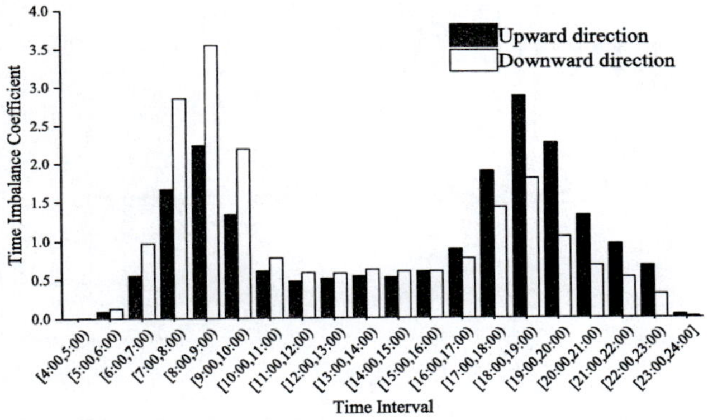


**Fig. 4.** Time imbalance coefficient of passenger flow

From the analysis of the section imbalance coefficient in Fig. 6, it can be seen that Baishiqiaonan Station to Caofang Station is suitable for the short route. Furthermore, according to the maximum section passenger flow in the morning peak, the small route from Chaoyang Men Station to Caofang Station is nested in the above route. Combined with the setting of the turn-back line of Beijing Metro Line 6, the routing scheme is determined as follows: Large-route trains run on the whole line, medium-route trains run from Chegongzhuang West Station to Caofang Station, and small-route trains run from Chaoyang Men Station to Caofang Station.

Combined with the early peak passenger flow in Fig. 6 and the side line conditions shown in Fig. 7, the express train passes through 7 stations between Hujialou Station and Wuzi Xueyuan Lu Station, and Changying station can be overtaken.

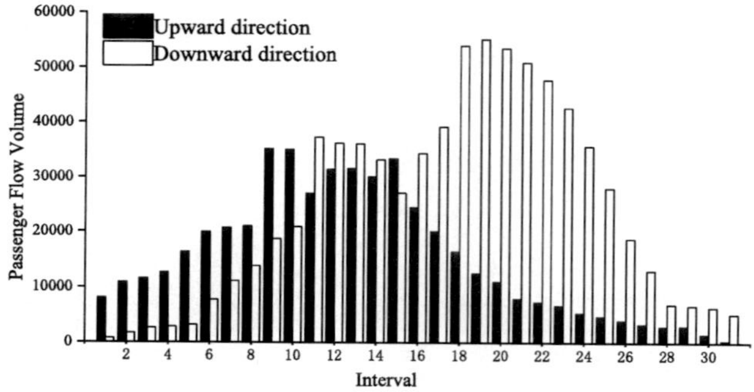


**Fig. 5.** Maximum section passenger flow in morning peak period

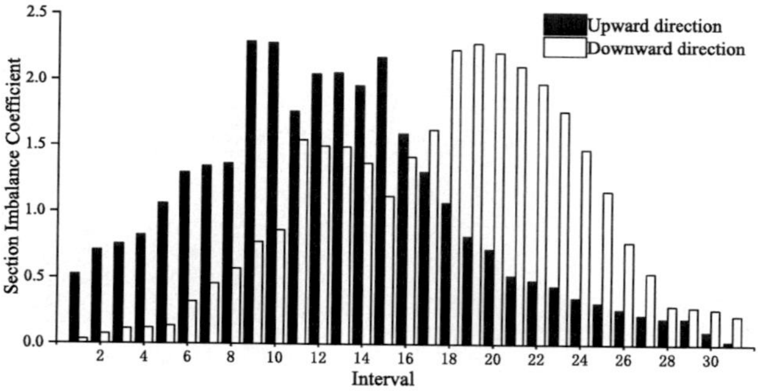


**Fig. 6.** Section imbalance coefficient of passenger flow

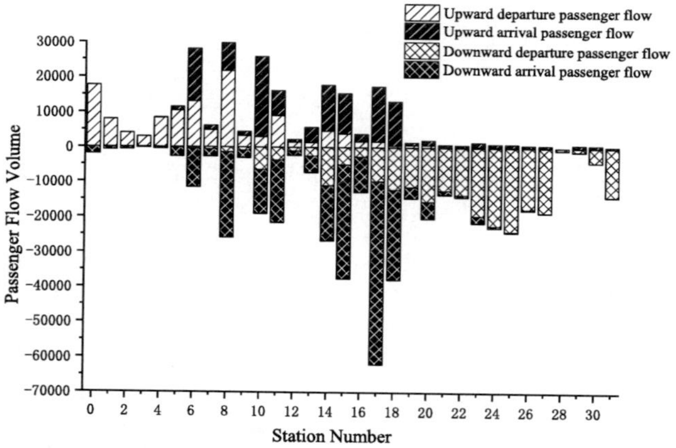


**Fig. 7.** Passenger flow OD in morning peak period

According to the actual operation diagram, the first train departure time of Dongxiaoying Depot, Wuliqiao Depot and Wulu Parking Lot is 5:00, 5:05 and 5:00 respectively.

The upper limit of the number of trains that can be routed in Dongxiaoying Depot is 16 groups, witch in Wuliqiao Depot and Wulu Parking Lot are 51 and 8 groups. After setting the average value of full load rate exceeding the limit, the average waiting time of passengers and the dimension of train running kilometers, the proportion of the objective function is 4:3:3, and the calculated weight values are 14.85, 1 and 0.000356. Other related parameters are shown in Table 1.

**Table 1.** Values of relevant parameters.

| Parameters | Values | Parameters | Values |
|---|---|---|---|
| $hd_{\min}$ | 180 | $\alpha_0$ | 1.2 |
| $hd_{\max}$ | 600 | $\alpha_{limit}$ | 0.8 |
| $h_{\min}$ | 105 | $q$ | 32 |
| $\tau_{arr-dpt}$ | 50 | $cap$ | 1980 |

### 4.2 Results

The solution process of the model and algorithm is realized by python programming. The parameters are as follows: the population size is 50, the crossover probability is 0.8, the mutation probability is 0.2, and the algorithm termination condition is that the number of iterations reaches 60 times. The computer processor is 1.6 GHz dual-core Intel Core i5, and the memory is 8G. The convergence of the algorithm is shown in Fig. 8.

Figure 9 is the operation train diagram of morning rush hour, which uses 72 sets of train. There are 13 pairs of long routing trains at 200 s departure interval in the Dongxiaoying Depot, including 7 pairs of express trains. There are 39 pairs of long route trains at 190 s departure interval, 7 pairs of medium route trains at 330 s departure interval and 2 pairs of small route trains at 450 s departure interval in the Wuliqiao Depot. Eight pairs of trains with large routing are sent out at 480 s departure intervals from the Wulu Parking Lot.

### 4.3 Analysis and Discussion

In order to verify the superiority of the model, the effect of compiling operation diagram and actual operation diagram, non-running express operation diagram and fixed dwell time operation diagram is compared in Table 2.

By analyzing the experimental results, the following conclusions can be found.

(1) Considering the passenger flow, the dynamic stop mode can better adapt to the dynamic imbalanced passenger flow, reduce the average waiting time and the full load rate of trains. So as to meet the requirements of passenger speed and comfort.

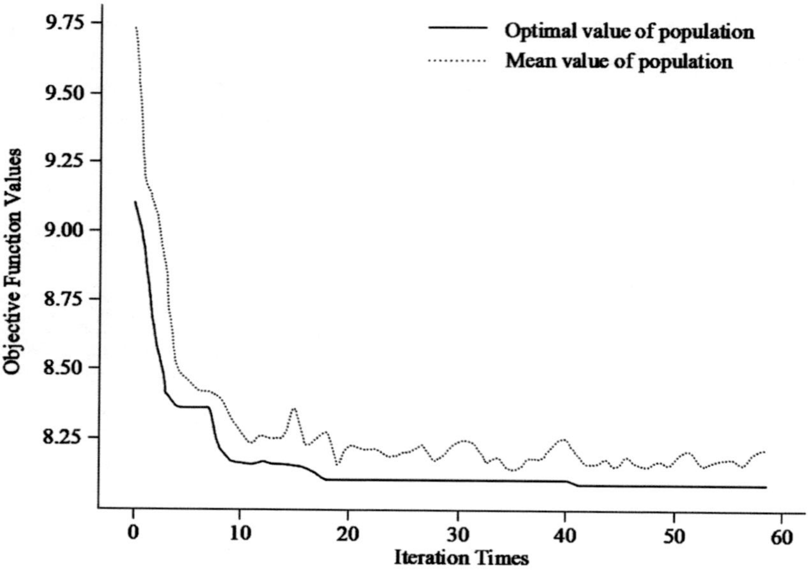


**Fig. 8.** Convergence process of genetic algorithm.

(2) By comparing the implementation effect of the diagram of dynamic stop mode with and without fast-slow mode, it is found that the diagram with fast-slow mode improves the difficulty of transportation organization, the flexibility is also improved, and the full load rate can be effectively reduced.
(3) Compared with the actual train diagram, except for the increase of train running distance, the implementation effect of the train diagram in other dimensions is better than that of the actual train diagram, which shows the important influence of passenger flow demand on the compilation of urban rail transit train diagram. Relevant departments of urban rail transit need to do a good job in passenger flow demand forecasting and survey, so as to compile a train diagram that maximizes passenger demand and improve the quality of urban rail service. At the same time, the weight of each sub-objective function in the experiment can be adjusted according to the actual situation. For example, in order to deal with major social health problems such as the epidemic, the weight of the average overload rate should be increased to reduce the overload rate as much as possible to ensure the health and safety of passengers.

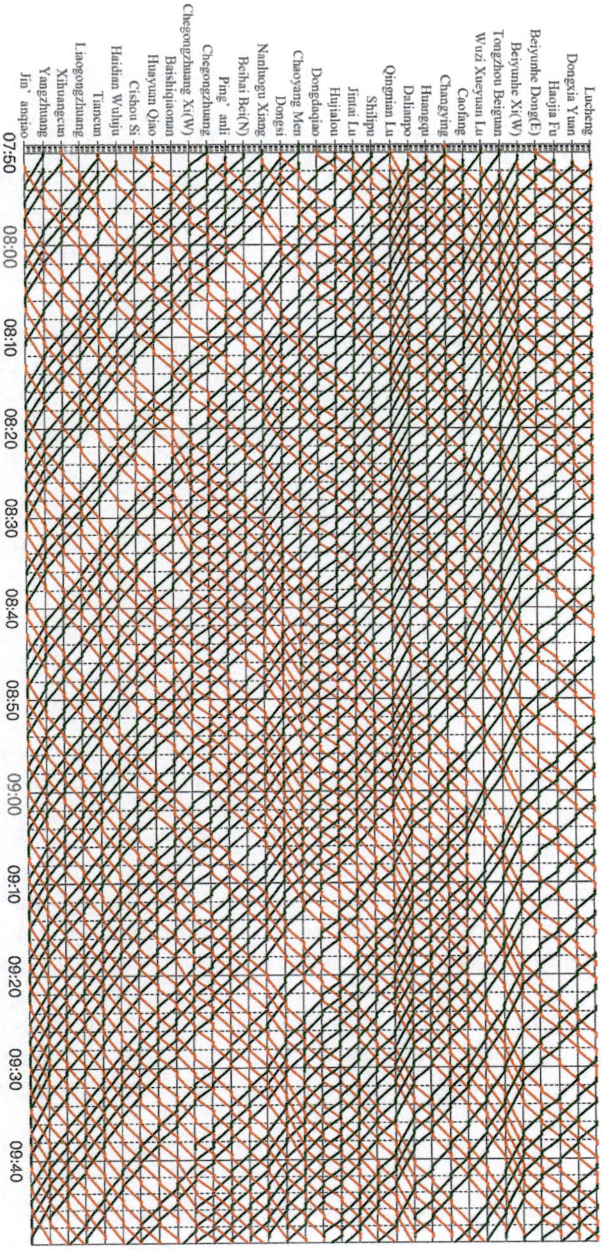


**Fig. 9.** Operation train diagram of morning rush hour

**Table 2.** Effect comparison of different train diagrams.

| | Objective function | Average overload rate | Average waiting time | Travel distance |
|---|---|---|---|---|
| Actual | 9.993 | 27.02 | 2.975 | 8442.951 |
| 50s dwell time with fast-slow mode | 9.831 | 25.50 | 3.173 | 8038.237 |
| 40s dwell time with fast-slow mode | 8.853 | 23.30 | 2.315 | 8645.064 |
| 30s dwell time with fast-slow mode | 8.379 | 21.31 | 1.909 | 9278.859 |
| Dynamic dwell time without fast-slow mode | 8.253 | 19.64 | 1.575 | 10565.231 |
| Dynamic dwell time with fast-slow mode | 8.086 | 18.49 | 1.675 | 10293.359 |

## 5 Conclusions

In order to meet the needs of passengers, take into account the efficiency of enterprises, and improve the matching degree between imbalanced passenger flow demand and capacity, this paper proposes a dynamic stop method based on passenger flow. Combined with the two operation modes of fast-slow mode and multi-routing mode, a model is proposed to minimize the average overload rate of the train, the average waiting time of the passengers and the travel distance of the train. Considering the constraints of train capacity, train operation and number of vehicles, the operation diagram of Beijing Metro Line 6 in the morning peak period is compiled. The results show that compared with the actual train diagram, the average value of train full load rate and the average waiting time of passengers are reduced by 31.6% and 43.7% respectively. On the basis of the existing research, the operation plan of the train diagram is improved, which has accumulated experience for the subway operation department to organize the transportation work.

## References


1. Cao, L., Cao, C.X.: Train operation plan and train timetable optimization of urban rail transport based on passenger demand. Shandong Sci. **33**(6), 87–95 (2020). (in Chinese)
2. Sun, M.X., Ni, S.Q., Lv, H.X.: Rail transit optimization for train operation plan under network operation. J. Transp. Eng. Inform. **18**(1), 26–33 61 (2020). (in Chinese)
3. Dai, C.J., Li, Y.Z., Zhan, Z.S., Chai, H.: Optimization of train operation scheme for urban rail transit considering dynamic passenger demand and full-length & short-turn routing modes. China Railway Sci. **39**(2), 128–136 (2018). (in Chinese)

4. Duan, L.L., Zha, W.X., Li, J., Yan, L.X.: Optimization of operation scheme for urban rail transit combining full-length and short-turn routing with express and slow train. Railway Transport Econ. **42**(5), 103–109 (2020). (in Chinese)
5. Li, D., Ding, S., Wang, Y.: Combinatorial optimization of service order and overtaking for demand-oriented timetabling in a single railway line. J. Adv. Transp. **2018**, 1–21 (2018)
6. Yang, L., Di, Z., Dessouky, M.M., Gao, Z.Y., Shi, J.G.: Collaborative optimization of last-train timetables with accessibility: A space-time network design based approach. Transp. Res. Part C: Emerg. Technol. **114**, 572–597 (2020)
7. Su, S., Liu, X., Wang, X.K., Tang, T., Cao, Y.: Determining special train timetable in epidemics for Beijing metro. J. Transp. Syst. Eng. Inform. Technol. **21**(1), 101–107 (2021)
8. Zhu, Y., Mao, B., Bai, Y., Chen, S.K.: A bi-level model for single-line rail timetable design with consideration of demand and capacity. Transp. Res. Part C: Emerg. Technol. **85**, 211–233 (2017)
9. Zhang, T., Li, D., Qiao, Y.: Comprehensive optimization of urban rail transit timetable by minimizing total travel times under time-dependent passenger demand and congested conditions. Appl. Math. Model. **58**, 421–446 (2018)
10. Zhou, Y., Bai, Y., Guo, H., Li, T., Qiu, Y., Zhang, Z.: Metro scheduling to minimize travel time and operating cost considering spatial and temporal constraints on passenger boarding. IEEE Access **8**, 114190–114210 (2020)
11. Shahabi, A., Raissi, S., Khalili-Damghani, K., Rafei, M.: Designing a resilient skip-stop schedule in rapid rail transit using a simulation-based optimization methodology. Oper. Res. 21(3):1691–1721 (2021)

# Research on Application Technology of Virtual Coupling Based on Adaptive Tracking

Hou Chengbin(✉), Liu Han, Shi Yan Hong, Zheng Chenglong, and Kang Yuan Lei

CRRC Qingdao Sifang Co., Ltd, Qingdao 266111, China
houchengbin@cqsf.com

**Abstract.** In this article, firstly, the research status of virtual coupling technology at home and abroad is summarized, and reviews the process and typical application scenarios of virtual coupling technology. Then, the minimum tracking distance model for virtual coupling operation was built, and an adaptive tracking algorithm based on error correction time was proposed to correct the speed curve of slave controlled vehicle. Finally, the application verification of the algorithm is carried out through the live vehicle test. When the speed of the virtual coupling cooperative operation reaches about 41 km/h, the tracking distance from the control vehicle is 48 m, which realizes the stable and reliable virtual coupling cooperative operation. In this article, the application technology of virtual coupling is studied, which has a good guiding significance for the scholars who are still in the stage of theoretical research.

**Keywords:** Virtual Coupling · Adaptive Tracking Algorithm · Application Scenarios

## 1 Introduction

In March 2020, the China Urban Rail Transit Smart City Rail Development Outline (2020–2035) was issued by the China Urban Rail Transit Association. In the 2025 specific goals of the intelligent technology equipment system, it is mentioned that the application of new generation communication technology and artificial intelligence technology to develop vehicles that support flexible formation and cooperative formation functions, and realize the accurate matching of transport capacity. In the construction of the intelligent train operation system, it is mentioned that the use of flexible formation and cooperative formation technology to achieve the optimal control of train cooperation, improve the operation efficiency and operation flexibility. In the construction of intelligent technology and equipment system, it is mentioned that the research of multi-train cooperative grouping technology of virtual coupling of automatic operation system.

From 2016 to 2018, CAF, a Spanish company funded by Shift2Rail, developed a virtual coupling system [1] and tested it on a tram. The two trains communicate with each other via LTE. From 2017 to 2020, TMH completed the testing of virtual coupling related technologies, and has applied the technology in a large number of locomotives. Since 2016, the Korea Railway Research Institute (KRRI) has carried out the Automatic Train


Y. Qin et al. (Eds.): EITRT 2023, LNEE 1137, pp. 476–485, 2024.
https://doi.org/10.1007/978-981-99-9311-6_49

Control System (ATCS) research program, which starts in 2021 and is expected to last until 2024, with the goal of realizing virtual train coupling. In January 2022, the company signed the project task book on environmental awareness and virtual marshalling, and will cooperate with Beijing Metro Vehicle Equipment Co., Ltd. to use autonomous virtual marshalling operation technology in the national demonstration project of Beijing Line 11.

Domestic research on virtual linking mainly focuses on the theoretical stage, and has not reached the stage of experimental verification and engineering application. On the basis of sorting out the process, scene and key technology of virtual connection, this study proposed a virtual connection control algorithm based on adaptive tracking, and finally verified the above virtual connection control algorithm through experiments.

## 2 The Technology of Virtual Coupling

Virtual coupling is a train-based operation control system based on wireless communication or vehicle-vehicle communication [2]. Any vehicle unit in a virtual grouping travels along the same route, at almost the same speed, in the same direction. Considering the communication delay, the braking characteristics of any vehicle unit should be nearly the same. Because the virtual connection with a very small tracking interval can greatly improve the operation capacity of the line or network, it is recognized as the next generation of train operation control system.

### 2.1 Description of Virtual Coupling Process

The whole process of virtual coupling mainly includes the processes of virtual coupling grouping, virtual coupling running, and virtual coupling unmarshalling, as shown in Fig. 1.

Virtual coupling grouping refers to the process in which multiple independent running units (two units are used as an example in the figure) communicate wirelessly to become a grouping after receiving the grouping instruction.

Virtual coupling operation refers to the state that after the virtual linkage grouping is completed, the grouping maintains a stable operating interval range and runs cooperatively on the same line at almost the same speed.

Virtual coupling unmarshalling refers to the process in which the virtual marshalling trains in the running state release the wireless communication lock and recover to multiple independent running units after receiving the unmarshalling instruction.

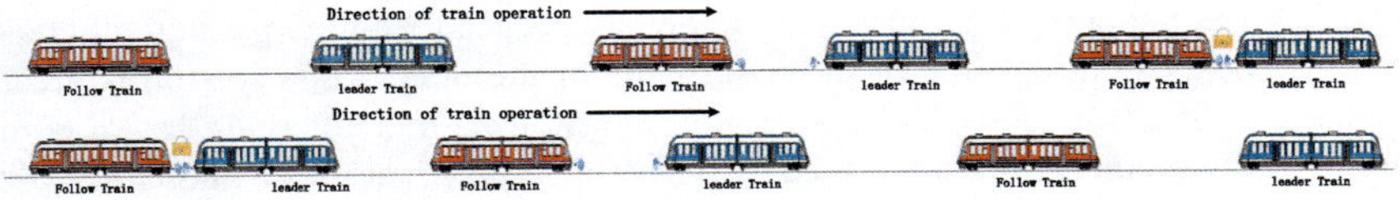


**Fig. 1.** The Whole process of virtual coupling.

## 2.2 Main Application Scenarios

Domestic and foreign scholars have carried out some theoretical studies on virtual linking scenarios, and classified major typical scenarios from the perspective of application, including high-level peak passenger flow scene (tidal passenger flow scene), Y-line scene, fast and slow car scene [2].

High-level peak passenger flow scene: During the peak passenger flow period, the train departs at the normal departure interval. During a certain passenger flow peak period, there is a large passenger flow in the upward direction, and the downward passenger flow is basically unchanged. The upward direction runs in multiple groups according to the passenger flow, which can transport the large passenger flow out in a short time. Shanghai Line 16 in 2013 for the first time in China to use online link/unlink to achieve flexible grouping, the current use is 3 + 3 link mode. The online connection time is at least 3 min, while the current average morning rush hour interval of Shanghai Line 16 is 3.5 min. In the case of parking and clearing, the entire connecting time is far more than the average morning peak traffic interval. Therefore, trains must be marshalled/dismarshalled at a specific time/place in this mode. The virtual link mode does not need to stop, the whole process only takes 1min, and can support the grouping/disassembling at any place. Recently, the 3 + 3 + 3 multi-group train coupling test has also achieved success in China. In this mode, excluding the parking and clearing time of three vehicles, the online coupling time is at least 6min, while the virtual coupling can be simultaneous, and the whole process is still 1 min. Virtual coupling in the multi-group coupling scene has more advantages (Fig. 2).

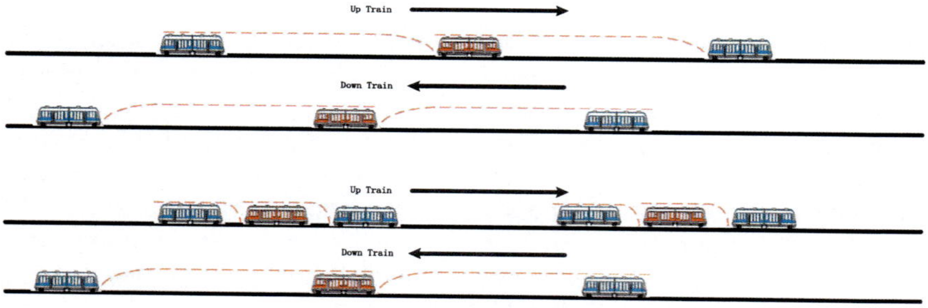


**Fig. 2.** Virtual coupling in the high-level peak passenger flow scene.

Y-line scene: The Y-line can be divided into turnout, branch line segment and common line segment, when the virtual coupling is applied to the Y-line scenario. The waiting time of the following train at the turnout is greatly reduced, and the passing efficiency is multiplied; the waiting time of the passengers of the branch line section is greatly reduced; and the advantages of the common line section are more obvious. If that virtual coupling is adopt, the vehicles produce by different vehicle factories can be linked, the models of the coupler of different vehicles are different, the coupler height of different vehicles are different, and the virtual coupling is adopted to link and combine any vehicle, There is no need to transform, no need to model (Fig. 3).

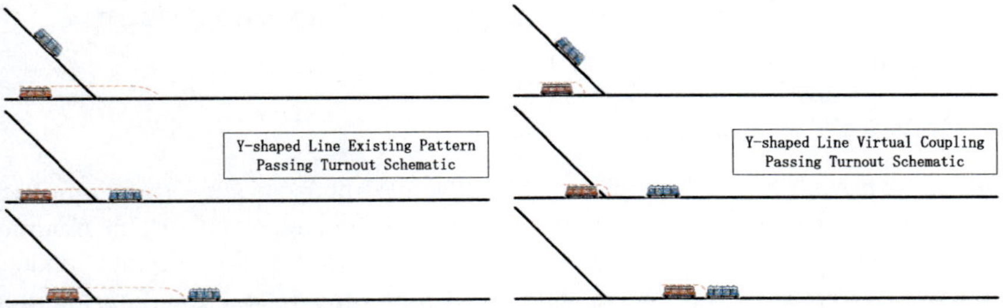


**Fig. 3.** Virtual virtual coupling in Y-line scene

Fast and Slow Train Scenario: The traditional fast and slow train adopts the mode of "big stop + stop," in which the fast train skips some stations to increase the train speed, but this mode greatly increases the waiting time of the slow train passengers and the total travel time of the system. The optimization of fast and slow trains by using virtual coupling can avoid the problem of increasing the waiting time of passengers caused by the fast train without stopping "small stop" in the traditional fast and slow trains. In that optimization mode with or without cross-over, it is necessary to form/de-group several time, and when the wired linkage mode is adopted, the time of forming/de-grouping is too long, and the scheduling requirement cannot be met at all. If a cycle of virtual marshalling/virtual unmarshalling is adopted, the maximum saving time of 6 min can be achieved, and the more marshalling/unmarshalling times the train is in operation, the virtual coupling has more advantages, and the transport capacity is obviously improved (Fig. 4).

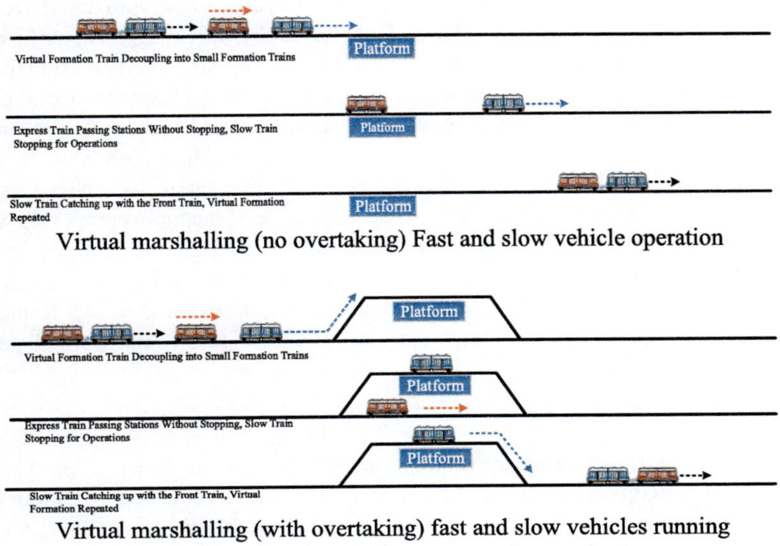


**Fig. 4.** Virtual linking application under fast and slow vehicle scenario

# 3 A Virtual Coupling Control Algorithm Based on Adaptive Tracking

## 3.1 Minimum Tracking Distance Model

Studies have proved that the key point to obtain the maximum capacity is to minimize the train safety headway while ensuring safety [3]. In some documents [4–6], the minimum interval of virtual grouping is analyzed, and the minimum tracking distance model of virtual association is established as shown in Fig. 5. The virtual coupling minimum tracking distance can be obtained when the master control vehicle performs the most favorable braking and the slave control vehicle performs the most unfavorable braking.

$$z_{vc\_min} = z_{norm} + z^{s}_{break\_max} - z^{m}_{break\_min} + z_{safe} \quad (1)$$

$$z_{normal} = t_{doc} \cdot v^{s}_{beginning} \quad (2)$$

$$z^{s}_{break\_max} = z_B + z_C + z_D + z_E + z_F \quad (3)$$

$$z^{m}_{break\_min} = (v^{m}_{beginning})^2/(2a_{ebr}) \quad (4)$$

where, $z_{vc\_min}$ representsthe minimum tracking distance of virtual connection. $z_{norm}$ Represents the distance from the control car during the braking period from the main control car to the braking period from the control car, which is calculated by formula (2). $t_{doc}$ Represents the total time of communication and other delays between the master car and the slave car. $v^{s}_{beginning}$ Represents the speed at which braking is applied from the control car. $z^{s}_{break\_max}$ Represents the most unfavorable braking distance from the control car, calculated by formula (3). $z_B$ Represents the operating distance from the reaction stage of the on-board equipment of the control vehicle. $z_C$ Represents the running distance from the power cutting stage of the control car. $z_D$ Represents the running distance from the control car before emergency braking is established. $z_E$ Represents the running distance from the control car emergency braking establishment stage. $z_F$ Represents the running distance from the emergency braking stage of the control vehicle. $z^{m}_{break\_min}$ Represents the most favorable braking distance of the main control car, which is calculated by formula (4). $v^{m}_{beginning}$ Represents the speed of the main control car when applying braking, ideally equal to the speed of applying braking from the control car. $a_{ebr}$ Represents the emergency braking rate of the main control vehicle. $z_{safe}$ Represents the distance between the two vehicles when the master and slave vehicles are all parked.

## 3.2 Adaptive Tracking Algorithm

Xun et al. Discussed the speed profile of coupled trains and added speed limit control to prevent overspeeding [7]. Park et al. designed a gap controller based on sliding mode control to manage the distance between the leading and following trains under an error correction scheme [8]. Chen et al. proposed a coordinated collision mitigation approach with model predictive control and organized a numerical analysis on emergency braking

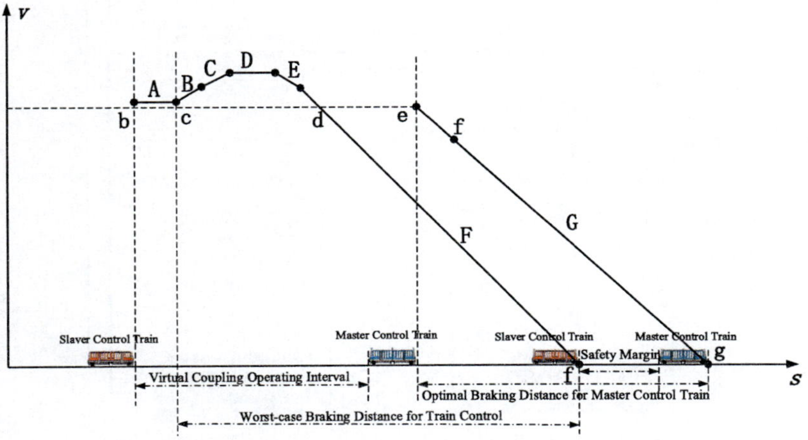


**Fig. 5.** Minimum tracking distance model of virtual coupling

of both homogeneous and heterogeneous fleets [9]. In the traditional CBTC system, the location of the ground ITP equipment is relatively fixed. The train only needs to register with the ground equipment and accept the management of the ground equipment. The ground equipment is responsible for analyzing the obstacles in front of the train operation, giving the current mobile authorized terminal point, and the operation cost is high [10, 11]. In this research, during the virtual coupling operation, the slave control car must maintain the same running state as the master control car, so the slave control car needs to track the real-time speed of the master control car. During the virtual coupling operation of the train, the master control car runs according to the speed curve generated by the ATO subsystem, and the slave control car obtains the speed curve of the master control car through wireless communication. The speed curve obtained from the control car tracking allows for collaborative operation. Figure 6 shows the schematic diagram of the virtual attached adaptive tracking algorithm.

The adaptive tracking control of slave car mainly involves ATO subsystem, wireless communication module of master car, wireless communication module of slave car, adaptive algorithm module and other on-board equipment. The ATO subsystem of the main control vehicle sends the generated speed curve to the central control unit of the main control vehicle, and the main control vehicle drives according to the speed curve. At the same time, the central control unit transmits the speed and position information of the main control vehicle to the wireless communication module of the main control vehicle, which transmits the speed and position information of the main control vehicle to the wireless communication module of the slave control vehicle through the wireless channel. The slave car wireless communication module will transmit the speed and position information of the master car to the adaptive algorithm module, and the adaptive algorithm module will also obtain the position information of the slave car in real time. The adaptive algorithm module is the core module for realizing adaptive tracking from the control car. The module optimizes the speed curve through the relative position information of the master car and slave car. The reference speed curve of the slave

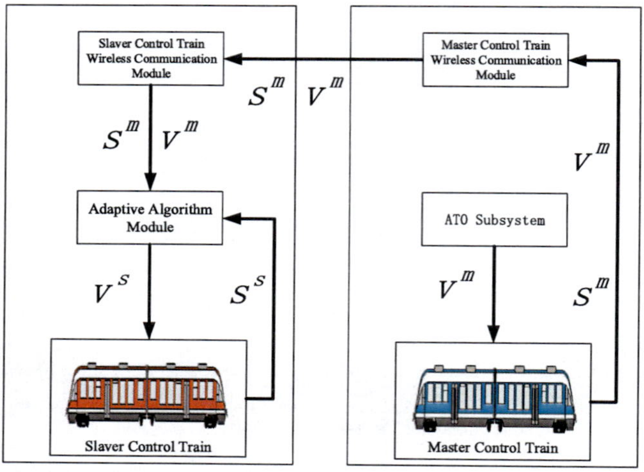


**Fig. 6.** Schematic diagram of adaptive tracking algorithm for virtual coupling

control vehicle is obtained, the adaptive tracking control of the slave control vehicle is realized, and the virtual synchronous operation between the master control vehicle and the slave control vehicle is further completed. The velocity curve is updated according to Eq. (5).

$$v_{amend} = v_m + \frac{s^s - s^m - z_m - z_{vc\_min}}{\vartheta} \tag{5}$$

where, $v_{amend}$ represents the speed correction value from the control car. $v_m$ represents the speed corresponding to the speed curve of the main control vehicle. $s^s$ represents the position coordinates of the slave control vehicle. $s^m$ represents the position coordinates of the main control vehicle. $z_m$ represents the body length of the main control vehicle. $\vartheta$ represents the error correction time, which needs to be calculated according to the communication delay, etc.

## 4 Test Verification

In order to verify the adaptive tracking virtual linking technology, a test environment was set up in the company to complete the collaborative operation of virtual linking, and the effectiveness of the adaptive tracking algorithm was verified. In the test process, the running speed and running mileage of the main control car and slave control car were collected in real time, as shown in Figs. 7, 8, 9.

As can be seen in Fig. 7, the main control vehicle and the slave control vehicle start successively, and with the increase of speed, the collaborative operation speed of 41 km/h is realized at about the 180th collection point; As can be seen from Fig. 8, the main control vehicle is always ahead of the control vehicle by a certain distance, and the distance changes from small to large to small as the speed changes. In order to better analyze the running law, Fig. 9 is obtained according to the actual mileage difference

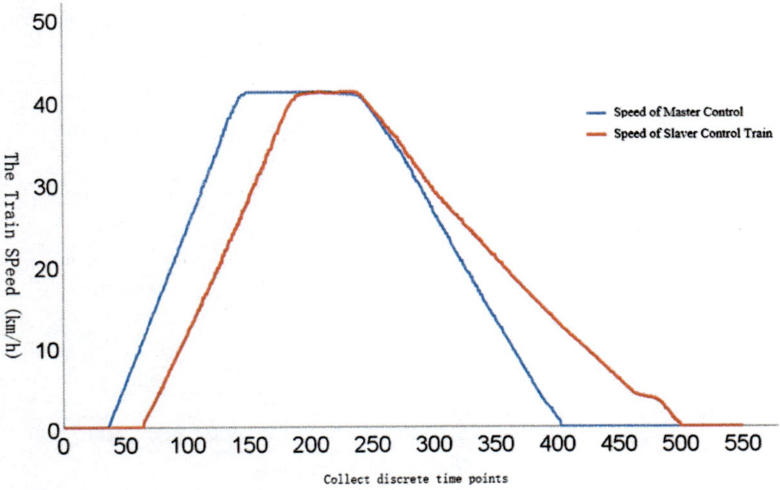


**Fig. 7.** Actual velocity curve of the main car and the secondary car

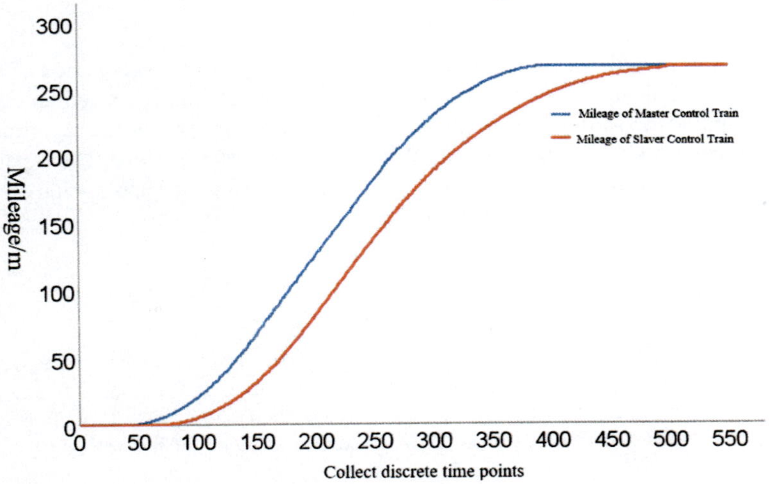


**Fig. 8.** Actual mileage accumulated curve of the main car and the secondary car

between the main control vehicle and the control vehicle. As shown in Fig. 9, when the cooperative running speed of virtual coupling reaches about 41 km/h, the mileage difference between the two vehicles is maintained at about 43 m. In the initial state, the distance between the main control vehicle and the slave control vehicle is 5 m, so the tracking distance of the slave control vehicle in the virtual coupling operation stage is 48 m, basically consistent with the designed tracking distance.

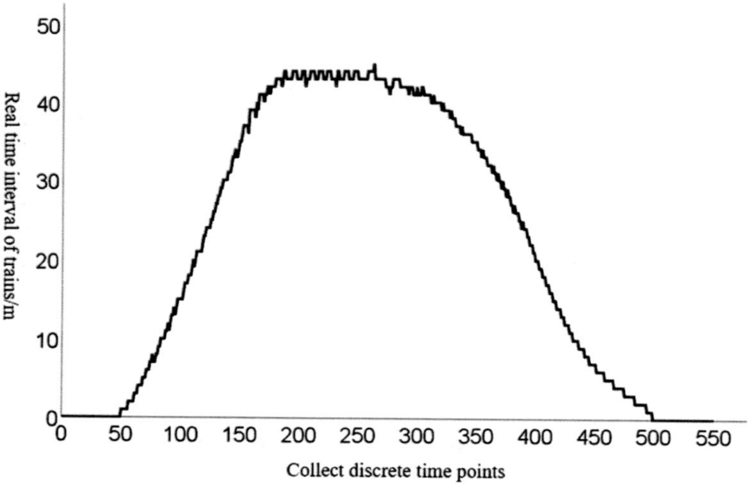


**Fig. 9.** Actual mileage difference curve of the main car and the secondary car

## 5 Conclusion

The paper first sorted out the process description and application scenarios of virtual coupling, then built the minimum tracking distance model, and proposed an adaptive tracking algorithm to correct the speed curve of the slave car. Finally, the algorithm was verified by experiments, and a good virtual coupling effect was achieved. The experiment verifies the non-interference scene, and the next step is to study the application of virtual linking technology in the interference scene.

## References

1. Wen, J.: Research on Control Method of Virtual Coupling Train Formation Based on "Dual Leader-Follower" Model. Beijing Jiaotong University (2021). (in Chinese)
2. Bai, J., Zhang, Q., Lu, F.: Skip-stop operation organization plan of virtual coupling trains in urban rail transit. Urban Rail Transit. **35**(01), 126–133 (2022). (in Chinese)
3. Flammini, F., Marrone, S., Nardone, R., Petrillo, A., Santini, S., Vittorini, V.: Towards railway virtual coupling. In: 2018 IEEE International Conference on Electrical Systems for Aircraft, Railway, Ship Propulsion and Road Vehicles & International Transportation Electrification Conference (ESARS-ITEC), pp. 1–6. IEEE (2018). (in Chinese)
4. Li, J.: Research on train control method of virtual coupling oriented to dynamic coupling and close tracking. Beijing Jiaotong University (2021). (in Chinese)
5. Xun, J., Cheng, M., Ning, B., et al.: Train tracking performance measurement under virtual coupling in subway. J. Beijing Jiaotong Univ. **43**(01), 96–103 (2019). (in Chinese)
6. Zhang, X., Zhao, L.: Application of model predictive control in virtual coupling train control system. Urban Mass Transit. **24**(10), 233–237 (2021)
7. Hao, A. Z., Yan, B. F., Niu, C. R.: A key step to virtual coupling. In: 2020 IEEE 23rd International Conference on Intelligent Transportation Systems (ITSC), pp. 1–6. IEEE (2020)

8. Di Meo, C., Di Vaio, M., Flammini, F., Nardone, R., Santini, S., Vittorini, V.: Ertms/etcs virtual coupling: proof of concept and numerical analysis. IEEE Trans. Intell. Transp. Syst. **21**(6), 2545–2556 (2019)
9. Xun, J., Chen, M., Liu, Y., Liu, F.: An overspeed protection mechanism for virtual coupling in railway. IEEE Access **8**, 187400–187410 (2020)
10. Luo, Q.P., Wu, H., Chen, L.J.: Train autonomous circumambulate system based on train to train communication Urban Mass. Transit **21**(07), 46–49 (2018)
11. Wang, J., Jin, C.M., Cai, B.G., Liu, J., Shangguan, W., Shan, H.Z.: A train control system for low density line in China. J. China Railway Soc. **37**(12), 46–52 (2015)

# Research on Collaborative Optimization of Passenger Flow Control and Train Skip-Stopping in Urban Rail Transit

Shuning Jiang[1], Jianyuan Guo[1](✉), Yong Qin[2], Qi Sun[3], and Hui Zhang[4]

[1] School of Traffic and Transportation, Beijing Jiaotong University, Beijing 100044, China
{22120842,jyguo}@bjtu.edu.cn

[2] State Key Laboratory of Advanced Rail Autonomous Operation, Beijing Jiaotong University, Beijing 100044, China
yqin@bjtu.edu.cn

[3] Beijing Metro Network Administration Co., Ltd., Beijing 100000, China
sunqi@bmncc.com.cn

[4] Sany Intelligent Mining Technology Co., Ltd, Beijing 102200, China
zhangh324@sany.com.cn

**Abstract.** Considering passenger flow will be stranded on the platform during the peak hours of urban rail transit, passenger flow congestion will be caused when the safety threshold is exceeded. A collaborative optimization model of station passenger flow control and train skip-stopping is established with the goal of minimizing the number of people exceeding the platform safety threshold and the average waiting time of passengers. A double deep Q network is used to solve it. A case study of Beijing Metro Line 13 and Changping Line is carried out with comparative analysis. In the morning peak station platform, there is a situation where the safety threshold is exceeded, and this safety problem is solved after optimization and the average waiting time of passengers is smaller than that before optimization. The case verifies the optimization effect of the model.

**Keywords:** Station Passenger Flow Control · Passenger Flow Congestion · Train Skip-Stopping · Double Deep Q Network

## 1 Introduction

During the peak hours of urban rail transit, when the passengers cannot be transported away by the train in time, there may be a situation where passengers gather on the platform because they can't board the train. When the number of passengers waiting on the platform exceeds the maximum capacity of the platform, the station will have safety problems. Therefore, according to the passenger flow and train operation, the control of the passenger flow entering the station and the adjustment of the train operation can alleviate the pressure of the platform and ensure the safety of passengers.

Currently, many scholars have proposed different control methods for passenger flow to solve the problem of too many passengers gather on the platform. Considering


Y. Qin et al. (Eds.): EITRT 2023, LNEE 1137, pp. 486–498, 2024.
https://doi.org/10.1007/978-981-99-9311-6_50

from the passenger, Wang et al. [1] establish a line level integer planning model to minimising average time of passenger delay by analysing the passenger boarding and alighting process to disperse the pressure at saturated stations. Since passenger demand is dynamic. Li [2] analyzes the characteristics of passenger flow and Wu et al. [3] introduce an equivalent time interval and use integer programming to build a model which dynamically evacuates stranded passengers. Yao et al. [4] consider OD passenger flow has volatility between different intervals, and establish a robust optimization model to make the delay time of passengers who outside the station reduce and maximizing the passenger transfer. For the perspective of passenger waiting time, Yuan et al. [5] minimise the waiting time by considering the equipment service capacity constraints with the integer programming method. Considering the internal facilities of stations, Chen et al. [6] consider multiple stations on a line and minimise the waiting time of passengers through the collaborative control optimization of station passenger flow.

The train capacity has an impact on the waiting passenger flow on the platform. Only control from the perspective of passenger flow will reduce the overall passenger flow control effect. Some scholars combined passenger flow control with train capacity adjustment to suit passenger demand.

Shang et al. [7] use commodity and network to represent passengers and trains on the line, and a model about commodity flow is constructed to obtain which station the train doesn't stop at use the Lagrangian relaxation framework for fast solution. Song [8] analyses the factors causing passenger congestion, proposes a variety of dispersal strategies, and research the control of passengers and train skip with collaborative optimization at the theoretical level. Zhang [9] using AFC data and a collaborative station flow restriction model is developed to coordinate passenger demand and train capacity and get entering passenger flow limit plan. Jiang et al. [10] propose a collaborative multi-station passenger flow method considering partial train skipping strategy to minimise the variance of station passenger detention rate and the passengers waiting time. A self-adaptive differential evolution algorithm is designed to solve it. Jiang et al. [11] made preliminary attempt to apply q-learning to passenger flow control and collaborative train skipping optimisation, and modeling to minimise the penalty of detained passengers. Q-learning algorithm is designed. With a simulation experiment on a real line in Shanghai, the model represents the validity.

Congestion causes by passenger flow retention in platform occurs frequently in the peak hours of urban rail transit network, many studies have focused on the combination of optimal train skipping and station flow control strategies. Most of the models are linear programming models, quadratic programming model and nonlinear combinatorial optimization model. The objectives often of minimising the passenger detention rate and passenger waiting time, etc. In fact, when platform passengers exceed the safe range, it has a great impact on the safety of station. When solving the model, most currently apply traditional operations research methods. To collaborative optimization of passenger flows, the non-linear combinatorial optimal passenger flow control model can more fully Optimize safety and efficiency objectives.

Therefore, optimizes the control of passengers entering the station and the train stop strategy and design a method with double deep Q-network. At the same time, pays attention to the safety of platform passengers. While controlling platform passengers below the safety threshold, passengers is consideres from the perspective of time, so that passengers can reach their destinations faster and safer.

The main contents of the remaining five chapters are as follows: in Sect. 2, the model is established to obtain strategy for the control of passengers entering the station and the train skip, and the solution algorithm is designed in Sect. 3. In Sect. 4, the model is applied and the collaborative optimization, only control the passenger flow and no control are compared and analyzed. Section 5 describes the conclusion.

# 2 Modeling

## 2.1 Model Assumptions

The model's assumptions include: trains run according to the plan operation graph; passenger demand is known and passengers will not give up urban rail transit and choose other travel modes; passengers follow the first-come-first-served principle; inbound passenger flow control strategies can be implemented at all stations.

## 2.2 Model Construction

In order to make passengers reach their destinations faster and safer, the objective is established to minimize the number of platform passengers exceeding the platform safety threshold and the average waiting time of passengers. At the same time, the model is established to obtain the control of passengers entering the station and the train stop strategy by using passenger flow demand, passenger travel randomness, train capacity and time as constraints, and the station inbound passenger flow control rate and which trains need to skip stations at stations as the decision.

1. **Model objective function**

The model aims to minimize the number of over-limit passengers and the average waiting time of passengers. The formula is as (1):

$$\min Z = \lambda \cdot \sum_{t \in T, n \in N, a \in A} t_{m,n,a} + \omega \cdot \sum_{t \in T, n \in N, a \in A} d_{m,n,a} \tag{1}$$

where,

$t_{t,n,a}$: The average waiting time of passengers at station n before train a in period t,

$d_{t,n,a}$: The number of over- threshold passengers when train a arriving at the platform n in period t.

2. **Constraints of the model**

(1) *Passenger flow control constraints*

In order to make passenger safety and avoid excessive control, the rate of control isn't too higher, and 60% is the lower limit constraint of the actual inbound passenger flow.

$$0.6D^{e}_{t,n,a} \leq F^{w}_{t,n,a} \leq D^{e}_{t,n,a} \tag{2}$$

Due to the entering passenger control, passengers cannot all enter the station, passengers waiting outside the station before train i arrives at the station n is as follows. n represents the station, a represents the arrival of the train.

$$S^{w}_{t,n,a} = D^{e}_{t,n,a} - F^{w}_{t,n,a} \tag{3}$$

where,

$D^{e}_{t,n,a}$: Number of passengers who required access to the station during period t,

$F^{w}_{t,n,a}$: Number of passengers who access the station in period t,

$S^{w}_{t,n,a}$: Number of people outside the station who wait due to control measures.

(2) *Train capacity constraints*

The capacity of the train is determined by the train capacity and train fully loaded rate. The train capacity are related to the vehicle type and the train formation. The total number of passengers on the whole line should not exceed the carrying capacity of the train.

$$\sum_{t \in T, n \in N} P^{B}_{t,n,a} \le \lambda_a \cdot C_a \tag{4}$$

where,

$P^{B}_{t,n,a}$: Number of people at station n who get on train a in period t,

$\lambda_a$: Train full load rate,

$C_a$: Train capacity.

To analyze the control from a time perspective, in time period t, the train arrives at the station, and if the train stops at the station, passengers on the train destined for this station get off and passengers from this station get on; If the train does not stop at the station, passengers cannot get on until the next train arrives. Whether passengers can all get on the train depends on the remaining train capacity, which is related to the train capacity and number of people who on the train. The difference between passengers getting on of the train and passengers getting off of the train at station 1,2,3,... n–1 represents number of people who on the train when the train arrives at the station n. n represents the station, a represents the arrival of the train. The remaining capacity calculation formula is as follows:

$$C^{r}_{t,n,a} = \lambda_a \cdot C - \left( \sum_{t \in T, n \in N_1} P^{B}_{t,n,a} - \sum_{t \in T, n \in N_1} P^{A}_{t,n,a} \right) \tag{5}$$

where,

$C^{r}_{t,n,a}$: The residual capacity,

$P^{A}_{t,n,a}$: Number of passengers get off of the train,

$P^{B}_{t,n,a}$: Number of passengers get on the train.

Use 0–1 variable $a^2_{t,n,a}$ to indicate whether the train stops. If the train doesn't skip, $a^2_{t,n,a} = 0$; if the train skips, $a^2_{t,n,a} = 1$. $a^2_{t,n,a}$ can be calculated from $a^2_{t,n}$. Where % is the remainder.

$$a^2_{t,n,a} = \begin{cases} 0 & t \in T, n \in N^0, a \in A, a^2_{t,n} = 0 \\ 1 & t \in T, n \in N^0, a \in A, i\%(a^2_{t,n} + 1) = 0, a^2_{t,n,a} \neq 0 \end{cases} \quad (6)$$

When the train doesn't skip at station n, passengers boarding train $P^B_{t,n,a}$ should not greater than the residual capacity of train $C^r_{t,n,a}$ and not less than 0; When the train skips at the station n, people boarding train $P^B_{t,n,a}$ is 0, and platform passengers continue to wait.

$$0 \leq P^B_{t,n,a} \leq (1 - a^2_{t,n,a}) \cdot C^r_{t,n,a} \quad (7)$$

If the platform passengers can't get on the train, they will be stranded on platform. After the train starts, the sum of the number of passengers stranded on the platform $P^S_{t,n,a}$ and the people get on the train $P^B_{t,n,a}$ is the number of people waiting on the platform $P^W_{t,n,a}$ before the train arrives.

$$P^w_{t,n,a} = P^B_{t,n,a} + P^S_{t,n,a+1} \quad (8)$$

(3) *Platform passenger flow constraints*

In the implementation of control measures, the passengers waiting to board on the platform are calculated, which is the sum of the passengers who have not boarded before and the passengers who enter the station. $P^w_{t,n,a}$ represents waiting people on platform in period t, $P^s_{t,n,a-1}$ represents stranded passengers and $E^w_{t,n,a}$ represents the number of people entering station before the arrival of train i-1 in period t. For the first train, $P^S_{t,n,0} = 0$.

$$P^w_{t,n,a} = P^S_{t,n,a-1} + E^w_{t,n,a} \quad (9)$$

The waiting passenger flow needs to meet the rigid and flexible constraints. The rigid constraint is the maximum number of people that can be accommodated in the physical space of the platform. The flexible constraint is that the waiting passengers of the platform need to be under the safety threshold as much as possible, and can actually exceed the safety threshold. The flexible constraint of platform capacity is expected to satisfy (11), where number of over-limit passengers before the train arrives in the period t is the difference between waiting passengers and safety threshold.

$$d_{t,n,a} \leq 0 \quad (10)$$

$$d_{t,n,a} = \max(P^w_{t.n.a} - C_n, 0) \quad (11)$$

where,

$E^w_{t,n,a}$: Number of entering station passengers,

$d_{t,n,a}$: Number of platform passengers who are greater than the platform safety threshold,

$C_n$: Platform safety threshold.

(4) *Time-constraint*

When arriving at the platform, the time of arrival of passenger is the sum of passengers' arriving time at the station, waiting time outside and the inbound travel time.

$$T_{t,n,i}^{a,p} = T_{t,n,i}^{a,s} + t_{t,n,i}^{w,s} + t_i^n \tag{12}$$

After passengers enter the station, they wait in the station. The difference between passenger boarding time and passenger arrival time represents waiting time.

$$t_{t,n,i}^{w,p} = T_{t,n,i}^{B} - T_{t,n,i}^{a,p} \tag{13}$$

The passenger boarding time of period m is the same as the arrival time of the train.

$$T_{t,n,i}^{B} = T_{a,n}^{a}, i \in I^0, a \in A \tag{14}$$

After the passengers get off, some passengers directly leave the station, and some passenger transfer. Passengers need to walk through the transfer facilities to enter another platform and the arrive time is the time when passengers get off plus the transfer time.

$$T_{t,n,i}^{t,p} = T_{t,n,i}^{A} + t_i^{n,p} \tag{15}$$

Calculating the average waiting time needs to know the overall waiting time of passengers and number of passengers. Due to the inbound control, passengers may not be able to enter the station outside the station, so they wait. In addition, passengers also need time to wait for the train on the platform. The sum of these two parts is the total waiting time of passengers. Then use $t_{t,n,a}$ represents average waiting time.

$$t_{t,n,a} = \frac{\sum\limits_{j \in J} (t_{t,n,i}^{w,s} + t_{t,n,i}^{w,p})}{D_{t,n,a}} \tag{16}$$

The passenger's entering station time and transfer time are random and described by normal distribution. The travel time of passengers entering the station is subject to the normal distribution with the mean value of $t_a$ and the variance of $t_\sigma^2$, which is recorded as $T_i^n \sim N(t_a, t_\sigma^2)$. The transfer travel time is subject to the normal distribution with the mean value of $t_b$ and the variance of $t_\delta^2$, which is recorded as $T_i^{n,q} \sim N(t_b, t_\delta^2)$. Specific expressions such as formula (17), formula (18)

$$f(t_i^n) = \frac{1}{\sqrt{2\pi}\, t_\sigma} e^{-\frac{(t_i^n - t_a)^2}{2t_\sigma{}^2}} \tag{17}$$

$$f(t_i^{n,q}) = \frac{1}{\sqrt{2\pi}\, t_\delta} e^{-\frac{(t_i^{n,q} - t_b)^2}{2t_\delta{}^2}} \tag{18}$$

where,

$T_{t,n,i}^{a,p}$: Passenger's arrive time at the platform in period t,

$T^{a,s}_{t,n,i}$: Passenger's arrive time at the station in period t,

$t^{n}_{i}$: Passenger's travel time for entering station,

$t^{w,p}_{t,n,i}$: Passenger's waiting time,

$T^{B}_{t,n,i}$: The time when passengers get on the train in period t,

$T^{a}_{a,n}$: Train arriving time at the station,

$T^{t,p}_{t,n,i}$: Transfer passenger's arrive time at the platform,

$T^{A}_{t,n,i}$: The time when passengers get off the train in period t,

$t^{n,q}_{i}$: Passenger transfer travel time.

3. **Decision variable**

The passenger flow control rate and whether train a skips at station n are decision variables where n represents the station, a represents the arrival of the train.

$$a^{1}_{t,n} = \frac{S^{w}_{t,n}}{D_{t,n}} \tag{19}$$

$$a^{2}_{t,n} = \begin{cases} 0 & \text{The train doesn't skip station} \\ 1 & \text{Skip every one trains} \\ 2 & \text{Skip every two trains} \end{cases} \tag{20}$$

The ratio of passengers who cannot enter the station due to flow restriction measures to those who need to enter the station is the control rate, which is expressed by Eq. (19). Equation (20) represents the skipping strategy of train n at the station.

## 3 Algorithm Design

Considering that the model is a nonlinear combinatorial optimization model and a non-linear multi-stage dynamic decision-making process, so the deep learning algorithm is used to solve the model.

### 3.1 Ingredients Construction of Deep Learning

Reinforcement learning mainly includes three elements: action, state and reward. The following describes the construction method of each element for the proposed large passenger flow collaborative guidance model.

1. **Action set**

Action set refers to the set of all possible decisions made by agents. The action set A includes the passenger flow control rate of station n and the skip scheme of train i. Each tuple $(a, b)$ represents an action, where a represents whether the train skips, and b represents the passenger flow control rate.

2. **State set**

A state set is a set of all possible states, commonly denoted by S. The state set contains the passenger flow that each station needs to enter. The state transition means that the agent executes action a under the current state s and interacts with the environment, and the environment generates the next moment states'.

3. **Reward function**

In reinforcement learning, rewards are used to indicate whether decisions during training are optimal or not. The optimization goal is to minimize the number of over platform passengers and the waiting time inside and outside the station. Since the maximum reward is sought during training, the original optimization target value is negative to maximize the reward.

### 3.2 Construction of Simulation Environment

In order to obtain strategy for the control of passengers entering the station and the train stop, it requires the interaction of passengers, stations and trains through the environment, to obtain rewards and realize state transfer. By inputting the basic data of passengers, stations and trains, the process of passenger arrival, getting on, getting off and transfer is simulated. The process is shown in Fig. 1.

## 4 Case Study

### 4.1 Experimental Background

Changping Line and Line 13 in Beijing are selected as the experimental lines. Based on the current situation of passenger flow aggregation and congestion caused by the transfer from Xi'erqi Station of Changping Line to Line 13 in the morning peak hour. The station conditions of Changping Line and Line 13 and the direction of large passenger flow in the morning peak are shown in Fig. 2.

The inbound passenger flow of Changping Line and Line 13 stations in the time period of 7:30–9:00 is time-sharing counted, as shown in Fig. 3. Changping, Shahe Univ. Park, Shahe, Zhuxinzhuang, Life Science Park and Xi'erqi in Changping Line and Bei Yuan, Lishui Qiao, Huoying, Huilong Guan and Longze in Line 13 have large inbound passenger flow. In order to alleviate the problem of congestion during the morning peak period, control of 11 stations of Changping Line and Line 13 in terms of inbound passenger with Co-optimization methods.

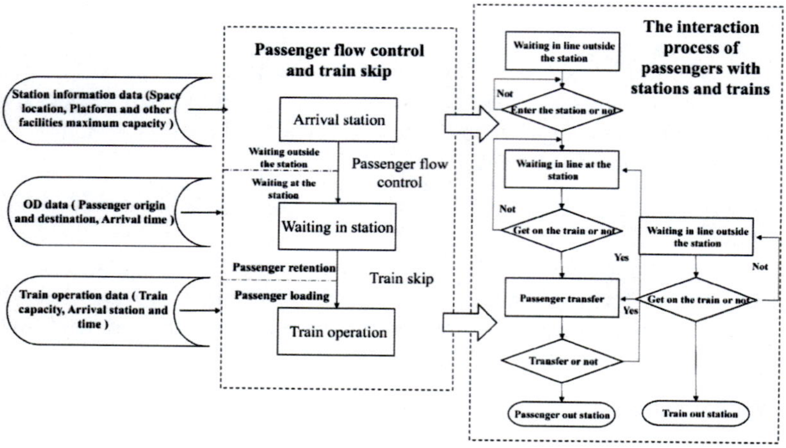


**Fig. 1.** Environment frame diagram

**Fig. 2.** Beijing Changping Line and Line 13 schematic diagram

### 4.2 Experimental Environment Construction and Parameter Setting

The experiment uses the Keras in TensorFlow2.0 to construct a double-deep Q Network and selects the relu function and the Adam optimizer, and also uses the men squared error (MSE) to calculate the loss value. 1000 generations of training are performed in the experiment and compared with the model without passenger flow control. The reinforcement learning simulation environment parameters are shown in the following Table 1.

The parameters of the reinforcement learning double-deep Q network are shown in Table 2.

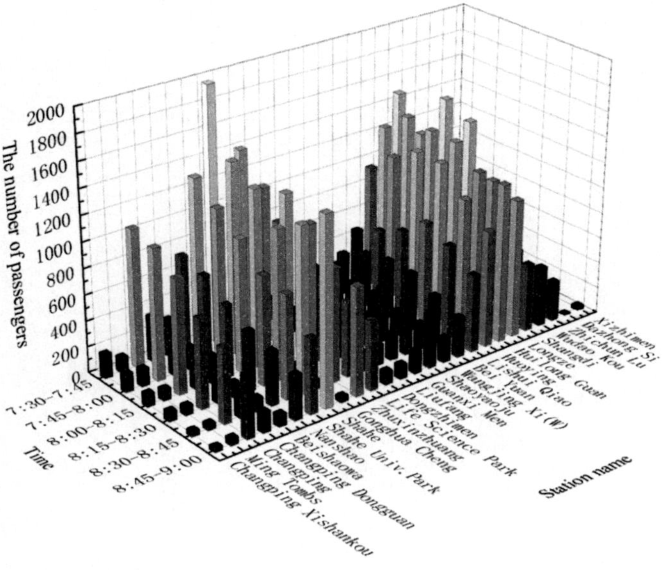


**Fig. 3.** Line passenger volume

**Table 1.** Simulation environment parameter setting

| Parameter | Meaning | Value |
|---|---|---|
| $t^c$ | Time interval | 15min |
| $\omega$ | Overrun target weight coefficient | 5 |
| $\lambda$ | Non-final period waiting time weight coefficient | 1 |
| $\lambda$ | Final period waiting time weight coefficient | 5 |
| $t_\sigma$ | Standard deviation of entering station time | 20s |
| $t_\delta$ | Standard deviation of transfer travel time | 20s |

### 4.3 Experimental Result

The training curve is shown in the Fig. 4. During the training process, due to the randomness of the experiment, there is a certain fluctuation in the reward value, but with the increase of the number of training times, the overall trend is upward and gradually stable. During training, when the reward value is higher, the objective of model is smaller. As the control strategy and train operation are adjusted, the waiting time becomes smaller, people exceeding the platform safety threshold becomes smaller, which make the platform becomes safer.

**Table 2.** Double-deep Q network parameter setting

| Parameter | Meaning | Value |
|---|---|---|
| replay_size | Training set size | 3300 |
| update_freq | Model update frequency | 660 |
| batch_size | Sample size | 330 |
| Input layer | Input layer | 56 |
| Output layer | Output layer | 14 |

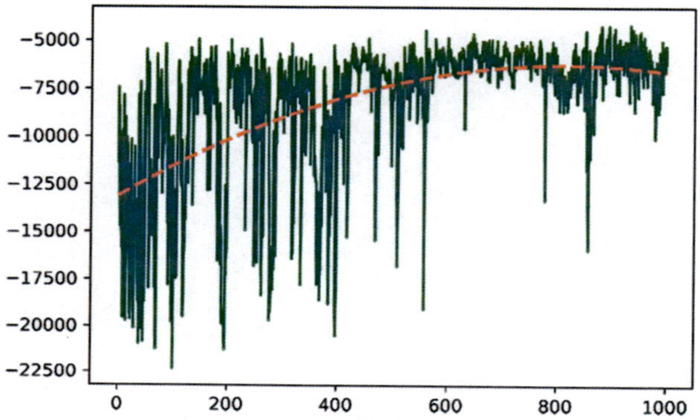


**Fig. 4.** Training curve

## 4.4 Optimal Results

The average waiting time and over-limit platform situation of Changping Line and Line 13 after optimization are shown in the Table 3.

**Table 3.** Comparative analysis

| Whether to collaborative optimization | Average waiting time | Over-limit platform situation |
|---|---|---|
| Before optimization | 134.91s | Platform passenger flow over-limit |
| After optimization | 134.56s | Platform passenger flow doesn't exceed the limit |

The table show that in the experimental station, passengers will exceed the safety capacity of the platform. After optimization, the station platform over-limit situation is solved, and the passenger waiting time is reduced, which verifies the validity of the model.

Specific control strategy is shown in Table 4.

**Table 4.** Specific control strategy of each station in different periods

| Station time | Changping | Shahe Univ.Park | Shahe | Zhuxinzhuang | Life Science Park | Xi'erqi |
|---|---|---|---|---|---|---|
| 7:30–7:45 | 0.1 | 0.1 | 0.1 | 0.1 | 0.1 | 0.1 |
| 7:45–8:00 | 0.1 | 0.1 | 0.1 | 0.1 | 0.1 | 0.2 |
| 8:00–8:15 | 0.1 | 0.1 | 0.2 | 0.1 | 0.1 | 0.2 |
| 8:15–8:30 | 0.1 | 0.1 | 0.1 | 0.1 | 0.1 | 0.2 |
| 8:30–8:45 | 0.1 | 0.1 | 0.4 | 0.1 | 0.1 | 0.3 |
| 8:45–9:00 | 0.1 | 0.6 | 0.1 | 0.1 | 0.1 | 0.2 |
| Station time | Bei Yuan | Lishui Qiao | Huoying | Huilong Guan | Longze | Xi'erqi |
| 7:30–7:45 | 0.2 | 0.2 | 0.2 | 0.2 | 0.2 | 0.1 |
| 7:45–8:00 | 0 | 0.2 | 0.2 | 0.2 | 0.2 | 0.2 |
| 8:00–8:15 | 0.2 | 0.2 | 0.4 | 0 | 0.2 | 0.2 |
| 8:15–8:30 | 0.1 | 0.1 | 0.1 | 0.2 | 0.2 | 0.2 |
| 8:30–8:45 | 0.1 | 0.1 | 0.3 | 0.2 | 0.2 | 0.3 |
| 8:45–9:00 | 0.2 | 0.2 | 0.2 | 0.2 | 0.2 | 0.2 |

## 5 Conclusion

To research the contradiction between train and station service capacity and passenger flow demand in urban rail transit network, a collaborative optimization model of passenger flow control and train skip is established with the goal of minimizing the number of over-limit platform passengers and the average waiting time of passengers. A double-deep Q network algorithm is designed to solve the problem. The model is applied to Beijing Line 13 and Changping Line. The comparative analysis shows that the number of waiting people on each station platform is less than the safety threshold and the average waiting time is reduced, which verifies the effectiveness of the model.

When collaborative optimize the passenger flow control and train skipping, the situation of double-line transfer is mainly researched. Urban rail transit is a network and the situation of multi-line transfer is more complicated and larger. The coordinated optimization of passenger flow control and train skipping station for multi-line transfer is necessary to study in the future.

## References


1. Wang, L., Yan, X., Wang, Y.: Modeling and optimization of collaborative passenger control in urban rail stations under mass passenger flow. Math. Probl. Eng. **2015**(5), 1–8 (2015)
2. Li, X.: Research on Passenger Flow Characteristics Analysis and Control Method of Urban Rail Transit. Beijing Jiaotong University, Beijing (2018). (in Chinese)
3. Wu, J., Wang, X., Yang, X., Guo, X., Yin, H., Sun, H.: Multi-station coordinated and dynamic passenger inflow control for a metro line. IET Intel. Transport Syst. **14**(9), 1068–1078 (2019)
4. Yao, X., Zhao, P., Qiao, K., Zhang, W.: Robust optimization model of peak passenger flow control scheme for urban rail transit lines. J. China Railway Soc. **43**(11), 1441–1450 (2021). (in Chinese)
5. Yuan, F., Sun, H., Kang, L., Wu, J.: Passenger flow control strategies for urban rail transit networks. Appl. Math. Model. **82**(6), 168–188 (2020)
6. Chen, Z., Gan, J., Wang, X.: Three-level passenger flow control model of urban rail transit considering safety. J. Central South Univ.: Sci. Technol. **51**(5), 1441–1450 (2020). (in Chinese)
7. Shang, P., Li, R., Liu, Z., Yang, L., Wang, Y.: Equity-oriented skip-stopping schedule optimization in an oversaturated urban rail transit network. Transp. Res. Part C: Emerg. Technol. **89**(4), 321–343 (2018)
8. Song, Y.: Metro large passenger flow organization optimization technology. Traffic Transp. **35**(6), 49–52 (2019). (in Chinese)
9. Zhang, W.: Research on Collaborative Passenger Flow Control of Urban Rail Transit Network. Beijing Jiaotong University, Beijing (2018). (in Chinese)
10. Jiang, Q., Su, J., Chen, W.: Collaborative control model of subway passenger flow considering skip-station parking. J. Railway Sci. Eng. **18**(11), 2857–2864 (2021). (in Chinese)
11. Jiang, Z., Gu, J., Fan, W., Liu, W., Zhu, B.: Q-Learning approach to coordinated optimization of passenger inflow control with train skip-stopping on a urban rail transit line. Comput. Ind. Eng. **127**(1), 1131–1142 (2018)

# Influence of Wall-Attached Antenna Distance on Channel in Tunnel Scenario

Yao Liu[1], Jianwen Ding[2,3,4,5](✉), Bin Sun[2,3,4,5], Siyu Lin[1,3], and Wei Wang[6,7]

[1] School of Electronic and Information Engineering, Beijing Jiaotong University, Beijing 100044, China
{21120095,sylin}@bjtu.edu.cn

[2] National Research Center of Railway Safety Assessment, Beijing Jiaotong University, Beijing 100044, China
{jwding,bsun}@bjtu.edu.cn

[3] Collaborative Innovation Center of Railway Traffic Safety, Beijing 100044, China

[4] Key Laboratory of Railway Industry of Broadband Mobile Information Communications, Beijing 100044, China

[5] Beijing Engineering Research Center of High-Speed Railway Broadband Mobile Communications, Beijing Jiaotong University, Beijing 100044, China

[6] State Key Laboratory of Mobile Network and Mobile Multimedia Technology, Shenzhen 518055, Guangdong, China
wang.wei183@zte.com.cn

[7] ZTE Corporation, Xian 710114, China

**Abstract.** The transformation of high-speed railway is moving toward intelligence, information technology, and digitalization. Railway 5G dedicated mobile communication (5G-R) system has become the first choice for high-speed railway communication systems and an important guarantee for the safe operation of high-speed railways. Tunnels are one of the typical railway scenarios, and the deployment of a reliable and stable 5G-R system is the guarantee for the safe operation of future intelligent high-speed railways. However, in an actual tunnel, the electric cables erected on the upper wall of the tunnel and the smooth passage of trains will limit the installation position of the antenna. Generally, the transmitting antenna is selected to be installed near the side wall. This paper takes the tunnel as the research scenario, and uses ray tracing technology to simulate the wireless channel characteristics of the transmitting antenna at different installation positions on the side wall of the tunnel in the 2.1 GHz frequency band, including received power, K factor and Root-Mean-Square Delay Spread (RMS DS), which can be used for this scenario. The construction optimization and deep coverage of the railway 5G network provide a theoretical basis to deploy a more stable and reliable 5G mobile communication system.

**Keywords:** Railway mobile communications · Tunnels · Ray-tracing · Wall-mounted antenna locations


Y. Qin et al. (Eds.): EITRT 2023, LNEE 1137, pp. 499–507, 2024.
https://doi.org/10.1007/978-981-99-9311-6_51

# 1 Introduction

In order to promote the deep integration of new generation information technology and railways, build smart railways, and enable 5G-R to better carry train control business to ensure the safety and efficiency of railway transportation, railways are developing towards intelligence. The mobile communication system is the "nerve center" for train operation safety, operation and maintenance, high-speed railway dispatch and command, and other services. 5G-R has become an important direction for the development of intelligent railways in the future [1].

However, the security of railway communication networks cannot be separated from the accurate characterization of wireless channel propagation characteristics. An accurate wireless channel model is crucial for the design and optimization of 5G-R networks, as it reflects the characteristics of channel fading in different scenarios and provides technical support for network layout, performance evaluation, and service quality of communication systems. Tunnel, as one of the typical railway scenarios, has a complex terrain, enclosed space, and significant fluctuations in wireless field strength. Its wireless channel propagation characteristics are more complex compared to other scenarios. Therefore, tunnel wireless coverage schemes have always been a focus and difficulty in the railway communication industry research.

In recent years, a large amount of research has been conducted on the wireless channel characteristics of this scenario. In [2–4] have conducted channel measurement and simulation activities for the 1.8 GHz, 5.8 GHz, and millimeter wave bands, respectively, to study channel characteristics at large and small scales such as path loss and delay spread. Due to the influence of boundary reflection on radio propagation in tunnels, the relative position of antennas plays an important role in radio wave coverage. For example, there are usually power cables and pantographs at the top in tunnels, making it difficult to install antennas. At the same time, it is necessary to consider the traffic conditions of vehicles in the tunnel, so antennas can generally only be installed near the side walls. At present, many studies only focus on the impact of antenna position on received power or path loss from the perspective of measurement and simulation for a specific tunnel scenario [5, 6]. However, few studies consider the small-scale characteristics of different antenna positions.

China Railway is applying for resources in the 2.1 GHz frequency band [7], which has the advantage of high coverage and low cost. Therefore, this article uses the CloudRT [8–10] (http://www.raytracer.cloud/) high-performance ray tracing simulation platform for dynamic simulation, with a 2.1 GHz research frequency band for dynamic simulation of tunnel scenarios. The focus is on studying the impact of the position of the wall mounted antenna on radio wave coverage, extracting channel parameters such as received power, K factor and RMS Delay Spread. This has both theoretical significance and engineering value. On the one hand, it can provide support and reference for the construction and equipment design and optimization of 5G-R tunnel scenarios; On the other hand, it can provide theoretical basis for the optimization and deep coverage of railway 5G network construction in this scenario, to ensure the deployment of stable and reliable 5G mobile communication systems in railway tunnel scenarios, and to achieve safe and stable operation of intelligent high-speed railway.

## 2 Scenario Model and Simulation Configuration

### 2.1 Scenario and Train Model Building

This article selects a typical railway tunnel scenario, uses SketchUp software to build a scenario model, and models trains in the tunnel. The key geometric parameters of the tunnel and the train are: the cross section of the tunnel is a semi-dome arch, and the length is 1000 m; The length, width and height of the train are 105 m, 2.9 m, 3.7 m, respectively. The 3D models of the tunnel cross-section and car body are shown in (a) and (b) in Fig. 1, respectively.

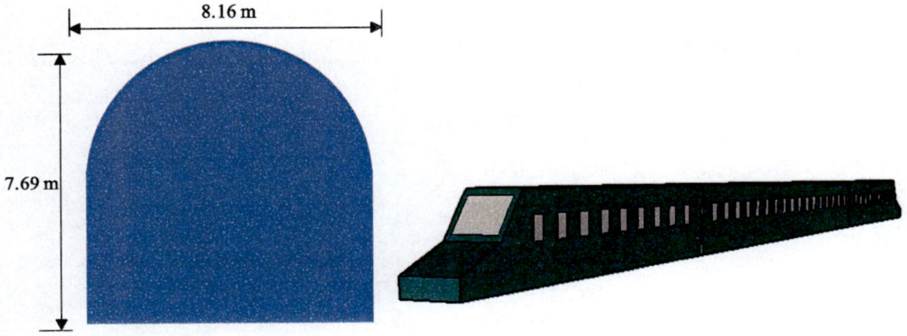


(a)Tunnel cross section (b) 3D model of the train

**Fig. 1.** Scenario of Railway Tunnel and 3D Model of Trains.

When electromagnetic waves propagate in a specific environment, phenomena such as direct radiation, reflection, scattering, diffraction, and transmission will occur. Ray tracing simulation can trace each ray between transceivers to restore the specific path of electromagnetic wave propagation. However, the scenario, object materials such as scatterers, and communication frequency bands will all have an impact on radio wave propagation methods such as backscattering. Therefore, accurate material electromagnetic parameters are needed to achieve reliable ray tracing simulation. The material parameters involved in the research model in this paper are all from the ITU-R Recommendation ITU-R P.1238-7 [11]. The specific material electromagnetic parameters are shown in Table 1.

**Table 1.** Model material parameter values.

| Material name | $\varepsilon_r$ | $\tan\delta$ | S | $\alpha$ |
|---|---|---|---|---|
| Concrete | 5.31 | 0.0958 | 0.0027 | 17.4334 |
| Metal | 1 | $10^7$ | 0.0022 | 10 |
| Glass | 6.27 | 0.014 | 0.0022 | 10 |

### 2.2 Simulation Configuration

The center frequency of this simulation is 2.1 GHz, and the bandwidth is 10 MHz. Dual-polarized antennas and vertically polarized directional antennas and omnidirectional antennas are used as transmitters and receivers respectively. The distance between the transmitter and the side wall of the tunnel is 0.03 m, 0.08 m, 0.13 m, 0.18 m, 0.23 m; The cross-section of the tunnel is located on the XoZ plane, and the apex on the left side of the tunnel bottom is placed at the origin of the coordinate system. Therefore, when the height of the transmitting antenna is 4.4 m (z = 4.4 m), the corresponding x-axis coordinate positions are 0.1 m, 0.15 m, 0.2 m, 0.25 m, and 0.3 m. The specific simulation configuration is shown in Table 2.

**Table 2.** Parameter Settings for Scenario Ray Tracking Simulation of Railway Tunnel.

| Parameter | Configuration |
|---|---|
| Frequency | 2.1 GHz |
| bandwidth | 10 MHz |
| Tx power | 46 dBm |
| Tx antenna type | "ANT/RT1721" dual polarized antenna |
| Rx antenna type | Vertically Polarized Omnidirectional Antenna |
| Rx location | On top of the first train |
| Transmission mechanism | Direct, 1~6 reflection, scattering |
| Transmitting antenna coordinates | (0 m, 0.1 m/0.15 m/0.2 m/0.25 m/0.25 m/0.3 m, 4.4 m) |
| Receive antenna coordinates | (%1–%2 m, 1~500m, 3.95 m) |

## 3 Simulation Results and Analysis

### 3.1 Received Power and Path Loss

The total field strength $\overrightarrow{E_{\text{Total}}}$ of the received signal is obtained by vector superposition of the electric field strength $\overrightarrow{E_l}$ of each received ray to calculate the signal received power, the formula can be expressed as:

$$P_r(dBm) = 10\log_{10}\left(\left|\overrightarrow{E_{\text{Total}}}\right|^2\right) \tag{1}$$

$$\overrightarrow{E_{\text{Total}}} = \sum_{i=1}^{n} \overrightarrow{E_l} \tag{2}$$

Path loss is a basic propagation characteristic for performance evaluation and system planning, and it is used to measure the radio wave propagation environment between transceivers; Shadow fading is due to the fact that radio waves are blocked by obstacles in the process of propagation, and obey the logarithmic normal distribution of zero mean. The CI model is used to express the large-scale loss model combining path loss and shadow fading, expressed as:

$$P_L(dB) = PL(d_0) + 10n\lg\left(\frac{d}{d_0}\right) + \chi_{\text{CI}}^{\sigma} \tag{3}$$

where $d_0$ is the relative reference distance, $PL(d_0)$ is the free-space path loss at the reference position, n is the path loss exponent, and $\chi_{\text{CI}}^{\sigma}$ is a Gaussian random variable with mean 0 and standard deviation $\sigma$.

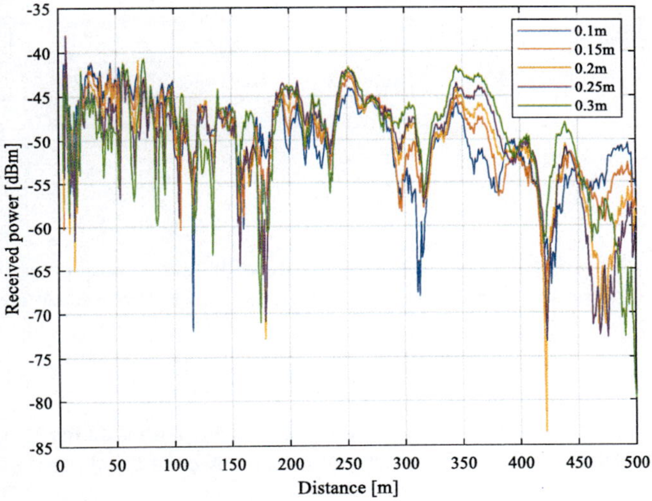


**Fig. 2.** The relationship between the received power and the distance at different distances from the transmitting antenna to the tunnel side wall.

Figure 2 shows the received power at different positions between the transmitting antenna and the tunnel side wall. It can be seen from the figure that when the distance between the transmitting and receiving antennas is relatively close (within 200 m), the influence of the different distances between the transmitting antenna and the tunnel side wall on the received power is relatively similar, but as the distance between the transmitting and receiving antennas (greater than 200 m) gradually increases, it can be clearly observed that when the transmitting antenna location is 0.3 m, there is a better coverage effect. Overall, the received power of the wireless communication system tends to increase as the position of the transmitting antenna moves away from the side wall of the tunnel. The main reason is that when the transmitting antenna gets closer to the tunnel side wall, the main beam of the transmitting antenna and the tunnel wall have higher-order reflections earlier, which reduces the received power. In order to better understand the overall impact of different distances between the transmitting antenna and the tunnel wall on the signal between the transceivers, the path loss is fitted to obtain the path loss index n and the standard deviation of shadow fading σ under different distances between the antenna and the tunnel wall, and the results are shown in Table 3. It can be found that the path loss varies between 0.87 and 0.89, which is much smaller than the loss index of free space, and the path loss index is the smallest when the position of the transmitting antenna is 0.25 m. At this time, the distance between the transmitting antenna and the side wall of the tunnel is 0.18 m.

**Table 3.** Path loss fitting and shadow fading.

| Distance between antenna and tunnel wall (m) | 0.1 m | 0.15 m | 0.2 m | 0.25 m | 0.3 m |
|---|---|---|---|---|---|
| Path loss exponent n | 0.8725 | 0.8847 | 0.8815 | 0.8808 | 0.8865 |
| $\sigma/dB$ | 10.1300 | 10.8462 | 10.4143 | 10.4720 | 10.7577 |

## 3.2 K-Factor

The Rice K factor is defined as the ratio of the path with the strongest power in the signal to the sum of the power of the remaining multipaths. The K-factor is expressed as:

$$K = 10\log_{10}\left(\frac{P_{\text{strongest}}}{\sum \text{P}_{\text{remaining}}}\right) \tag{4}$$

where K is the Rice K-factor, $P_{\text{strongest}}$ is the power of the path with the strongest signal among the multipaths, and $\text{P}_{\text{remaining}}$ represents the power of the other multipaths, and the unit is mW.

Figure 3 shows the CDF of the K factor at different distances from the transmitting antenna to the tunnel side wall, which is fitted using the normal distribution function, and the statistical parameters are shown in Table 4.

**Table 4.** Statistical parameters of rice K-factor.

| Distance between antenna and tunnel wall (m) | 0.1 | 0.15 | 0.2 | 0.25 | 0.3 |
|---|---|---|---|---|---|
| Median [dB] | 16.4778 | 16.4473 | 16.5817 | 16.6126 | 16.6675 |
| Max [dB] | 37.7697 | 47.6623 | 32.6009 | 32.5626 | 52.7021 |
| $\mu_K$ [dB] | 16.5491 | 16.7143 | 16.3974 | 16.4448 | 16.7904 |
| $\sigma_K$ [dB] | 3.1616 | 3.6469 | 3.4529 | 3.5163 | 4.2404 |

According to the analysis of the Rice K factor, more than 98% of the Rice K factors are greater than 0, which is consistent with the long and narrow structure of the tunnel, which leads to a higher order of the backscattering path received by the receiver in most cases, thereby weakening the received signal of the backscattering path. At the same time, by taking the value at the position of 50% of the Rice K factor, it can be found that the farther the distance from the tunnel side wall is, the larger the Rice K factor value is. This is because the farther away from the side wall of the tunnel, the lower the order of the backscattering path, and the stronger the dominance of the direct path.

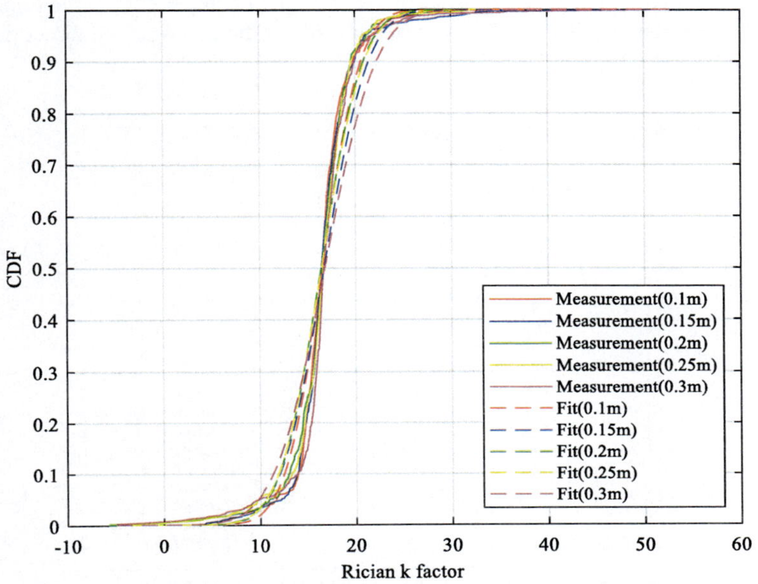


**Fig.3.** CDF of K factor at different distances from the transmitting antenna to the tunnel side wall.

### 3.3 RMS Delay Spread

The multipath effect in the tunnel scenario makes the receiver receive signals from different paths at different times, resulting in time dispersion. The root mean square delay spread can effectively characterize the dispersion degree of multipath signal energy in the time domain, and it is defined as the square root of the second-order central moment of the power delay curve. The RMS delay spread is expressed as:

$$\mathrm{RMS} = \sqrt{\frac{\sum_{n=1}^{N} \tau_n^2 \cdot P_n}{\sum_{n=1}^{N} P_n} - \left(\frac{\sum_{n=1}^{N} \tau_n \cdot P_n}{\sum_{n=1}^{N} P_n}\right)^2} \tag{5}$$

where RMS is RMS delay spread, $P_n$ and $\tau_n$ are the received power and delay of the n-th multipath respectively.

The CDF of the RMS delay spread at different distances from the transmitting antenna to the tunnel side wall is shown in Fig. 4, and the normal distribution function is used for fitting. Table 5 summarizes the relevant statistical information.

Analyzing the delay spread at different distances between the transmitting antenna and the tunnel sidewall, it can be found that the maximum value of the RMS delay spread and the value of 50% probability gradually decrease as the wall-attached antenna moves away from the tunnel sidewall. This corresponds to the characteristics obtained from the analysis of the K factor and the distribution characteristics of the received power, that is, more high-order backscattering paths will appear when the transmitting antenna is gradually closer to the tunnel side wall. On the contrary, moving away from the side wall of the tunnel will reduce the order of the multipath components in the tunnel, and the distance of the multipath will be shortened, so that the gap with the direct path will be shorter.

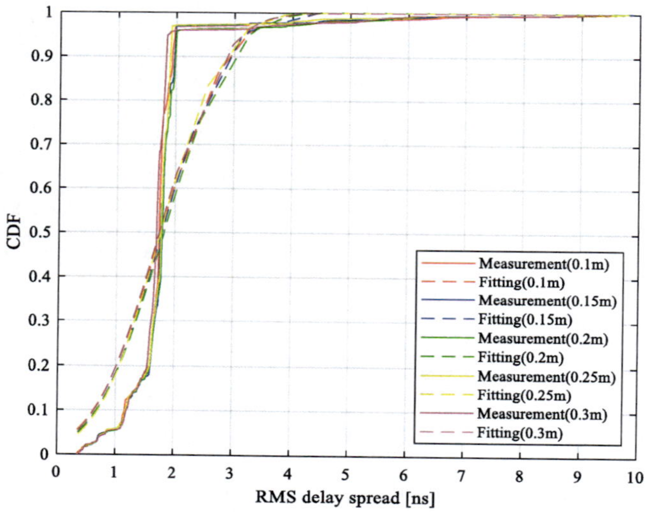


**Fig. 4.** CDF of RMS delay spread at different distances from the transmitting antenna to the tunnel side wall.

**Table 5.** Statistical parameters of delay spread.

| Distance between antenna and tunnel wall (m) | 0.1 | 0.15 | 0.2 | 0.25 | 0.3 |
|---|---|---|---|---|---|
| Median [ns] | 1.7288 | 1.7525 | 1.7763 | 1.7608 | 1.6850 |
| Max [ns] | 9.6124 | 9.8709 | 9.8169 | 9.7941 | 9.5048 |
| $\mu_{RMS}$ [ns] | 1.7555 | 1.7777 | 1.8122 | 1.7539 | 1.7299 |
| $\sigma_{RMS}$ [ns] | 0.8806 | 0.8574 | 0.8856 | 0.8194 | 0.8529 |

## 4 Conclusion

This paper uses ray tracing to simulate the wireless channel at different distances between the transmitting antenna and the tunnel side wall in the 2.1 GHz tunnel scenario, and extracts the corresponding large-scale and small-scale characteristic parameters in this scenario. The results show that the special structure of the tunnel scenario leads to the better coverage effect when the transmitting antenna is farther than 0.15 m, that is, the transmitting antenna is 0.08 m away from the side wall of the tunnel. Considering the high mobility of the train, in order to ensure the safe operation of the train, it is recommended that the installation position of the wall-mounted antenna be between 0.13 m and 0.18 m. This provides a theoretical basis for the deployment of a stable and reliable 5G-R mobile communication system in tunnel scenarios to ensure the high-speed and safe operation of trains.

**Acknowledgments.** This research was partially funded by the Fundamental Research Funds for the Central Universities under Grants 2022JBXT001, and by the National Natural Science Foundation of China under Grants 62171021, and by the Project of China State Railway Group under Grants P2021G012, and by ZTE Corporation under Grants I23L00010 and by Shanghai Shen-Tie Investment Co., Ltd. under grant W23L00450.

## References

1. Lu, Y.L., Shen, H.Y., Zhang, J.Y., et al.: Exploration and prospect of key technology of new generation mobile communication for intelligent railway. Telecommun. Sci. **39**(01), 30–41 (2023). (in Chinese)
2. Zhang, X.. et al.: Measurements and modeling of large-scale channel characteristics in subway tunnels at 1.8 and 5.8 GHz. IEEE Antennas Wireless Propag. Lett. **22**(3), 561–565 (2023)
3. Santos, A., Matos, L., Castellanos, P., Mota, V.: Channel characterization in V2I system inside a tunnel in 5.8 GHz. 2021 SBMO/IEEE MTT-S International Microwave and Optoelectronics Conference (IMOC), pp. 1–3, Fortaleza, Brazil (2021)
4. Tai, I.H.P., Lim, H.S., Diong, K.S., Alaghbari, K.A.: Effects of crowd density on radio propagation at 24 GHz in a pedestrian tunnel for 5G communications. IEEE Access **11**, 40240–40248 (2023)
5. Li, D.W.: Research on the Coverage Characteristics of Radio Waves in Confined Space. Beijing Jiaotong University (2016). (in Chinese)
6. Li, W.J.: Research on Propagation Characteristics and Positioning Method of Ultra-Broadband Electromagnetic Wave Tunnel. Wuhan University of Technology (2019). (in Chinese)
7. Duan, W., Gu, J.Y., Wen, M.W., et al.: Emerging technologies for 5G-loV networks: applications, trends and opportunities. IEEE Netw. **34**(5), 283–289 (2020)
8. Guan, K., He, D.P., Zhong, Z.D.: CloudRT: A Chinese example of open science infrastructure and services. Cult. Sci. **4**(4), 217–226 (2021)
9. He, D., Ai, B., Guan, K., et al.: The design and applications of high-performance ray-tracing simulation platform for 5G and beyond wireless communications: a tutorial. IEEE Commun. Surv. Tutor. **21**(1), 10–27 (2019)
10. CloudRT Homepage. http://cn.raytracer.cloud:9090/. Last accessed 14 June 2023
11. Series, P.: Propagation Data and Prediction Methods for the Planning of Indoor Radio Communication Systems and Radio Local Area Networks in the Frequency Range 900 MHz to 100 GHz. ITU-R Recommendation, p. 1238–7 (2012)

# Do High-Speed Railway Inhibit Earnings Management?

## Evidence from China's Capital Markets

Litian Guo[1], Zhan Wang[2], Wei Fang[2], Zhijie Liang[1(✉)], and Xiaoman Song[1]

[1] Transportation and Economic Research Institute of China Academy of Railway Sciences Co. Ltd, Beijing 100081, China
{guolitian,songxiaoman}@rails.cn, 52164300@qq.com
[2] Department China State Railway Group Co. Ltd., Beijing 100844, China

**Abstract.** The opening of the high-speed railway has made a huge difference on China's economy and society. A series of studies on its macroeconomic consequences have been carried out, but not enough attention has been paid to corporate governance. Based on the natural experimental background of the high-speed railway opening in 2008, this paper took the China's a-share listed companies from 2005 to 2021 as the research samples, and conducted an DID model to test the relationship between the high-speed railway and earnings management. It found that the opening of high-speed railway significantly inhibited the earnings management of listed companies, and the effect was stronger in prefecture-level cities than in cities at or above the sub-provincial level. This paper also discusses the internal mechanism of the effect of high-speed railway opening on earnings management, and found that the opening of high-speed railway can further affect earnings management through information transparency. This study not only enriches the research in corporate finance and other related fields, but also has practical significance for the effective supervision of regulatory authorities and rational decision-making of investors.

**Keywords:** High speed trains · Earnings management · Information transparency

## 1 Introduction

Some scholars have studied the economic consequences of the opening of high-speed rail from a macro perspective. LiuYi et al. research results show that the opening of high-speed rail has significantly promoted regional economic development [1]. Bian Yuanchao et al. found that the opening of high-speed rail can widen regional economic gap and produce polarization effect [2]. Deng Taotao et al. believe that the opening of China's high-speed rail significantly intensifies the level of urban sprawl [3]. In addition, there are also some scholars from the micro perspective to study. For example, scholar Du Xingqiang et al. pointed out that the opening of high-speed rail can significantly enhance the city's attraction to senior talents by expanding the market scale and then


Y. Qin et al. (Eds.): EITRT 2023, LNEE 1137, pp. 508–516, 2024.
https://doi.org/10.1007/978-981-99-9311-6_52

increasing the number and scale of enterprises, thus increasing the number of senior talents with doctoral degree hired by listed enterprises in the city [4]. In addition, Zhao Jing et al. found that the opening of high-speed rail reduced the risk of stock price collapse of listed companies in the region. At the same time, the paper further studies that the opening of high-speed rail has a more significant effect on reducing the risk of stock price crash when the information asymmetry is high and the external regulatory environment is weak [5]. It can be seen that the high-speed rail not only promotes inter-regional economic development but also improves the information transparency of listed companies by reducing time and other costs, thus reducing information asymmetry. From the perspective of corporate governance, the improvement of information transparency will inevitably limit the relevant illegal economic activities of enterprises.

Financial report surplus is the most commonly financial index to measure performance. There is no doubt that improving the transparency of corporate information will increase the external understanding of corporate operations and thus inhibit earnings management. This has also been confirmed in the literature [6–8]. In addition, the information transparency of listed companies in different regions is also different. The opening of high-speed rail can promote the flow of information between regions and enhance the external supervision of enterprises, thus improving the transparency of corporate information. Further, information transparency can inhibit earnings management. As mentioned above, between the opening of high-speed rail and earnings management, there may be an influence path of "opening of high-speed railway - information transparency - earnings management", that is, the opening of high-speed rail inhibits earnings management by improving information transparency. Based on the above analysis, this paper takes the opening of high-speed rail in 2008 as the natural experiment background, and adopts China's A-share listed companies from 2005 to 2021 as the research sample. The relationship between the opening of high-speed railway and earnings management is empirically tested by the method of difference-difference. The study found that the opening of high-speed rail significantly inhibited the earnings management of listed companies. Among them, the opening of high-speed rail significantly inhibited the earnings management of listed companies in prefecture-level cities, while listed companies in sub-provincial cities and above were not significant. After further testing its influence mechanism, it is found that there is a path mechanism of "opening of high-speed rail - information transparency - earnings management" between the two, that is, opening of high-speed rail inhibits earnings management by improving corporate information transparency.

## 2 Literature Review and Hypothesis

### 2.1 Opening of High-Speed Rail and Earnings Management

In recent years, geographical factors have become an important research topic in the field of corporate governance. From the perspective of distance, too far away will affect the company's personnel communication. Luo Jinhui et al.'s research confirms this point. The paper points out that independent directors are more likely to be absent from the board meeting of the company the longer the geographical distance between them and the company they work for [9]. Malloy found that a shorter spatial distance could improve the forecasting accuracy of securities analysts [10]. Li Bin pointed out that with the expansion of the geographical space distance and institutional environment distance of the parent-subsidiary company, the corporate governance level gradually declined, and the excessive investment behavior became more serious. Further research found that the distance between the parent-subsidiary company and the parent company had a significant negative impact on the corporate value [11]. Kan Jinghua pointed out that the farther the firm is from the audited unit, the worse the audit quality [12].

Why does the distance factor have such a significant impact on corporate governance? Some scholars believe that geographical factors play a more important role in economic relations than other factors because the shorter distance between market entities can facilitate information transmission [13]. Zhao Jing pointed out that the opening of high-speed rail significantly promoted the cross-regional flow of information [5]. In addition, Du Xingqiang et al. pointed out that there is a significant positive correlation between the number of high-speed rail launches in the city where the company is located and the probability of hiring auditors from other places. Further research found that the launch of high-speed rail reduces audit costs and improves audit quality of listed companies, thus increasing the probability of hiring auditors from other places [14]. Based on the foregoing, this paper proposes research hypothesis 1:

H1: Compared with listed companies that have not opened high-speed rail, the opening of high-speed rail significantly inhibits earnings management.

## 2.2 High-Speed Rail and Surplus Management: Urban Administrative Level

Nie Huihua et al. (2018) pointed out that the main advantage of cities lies in resource agglomeration, but there are obvious differences in the mode and process of resource agglomeration in Chinese and western cities. This difference is reflected in the fact that the process of urban resource aggregation in the West is spontaneous in the market, while in China, the process is affected by urban administrative levels [15]. Under China's administrative hierarchy, cities have distinct administrative divisions, such as municipalities directly under the central government, sub-provincial cities and prefecture-level cities. China's urban administrative level undertakes the dual functions of government intervention and regulation of social and economic operation, and the urban administrative level corresponds strictly to the capacity of the city in mobilizing economic resources and formulating economic development policies [16]. The source of economic and social capital tends to shift to cities with higher administrative levels [17]. As mentioned above, in cities above the sub-provincial level, external supervision, financial institutions and news media are concentrated, the time and capital cost of obtaining information are lower, and the information transparency of listed companies is high. Therefore, the opening of high-speed rail cannot significantly improve the status quo of earnings management. In contrast, in prefecture-level cities, the number of financial institutions, news media and other units stationed in such cities is small, and the market has strong information asymmetry. As a virtual information "depression", the media and regulatory resources flowing in with the opening of high-speed rail can significantly improve the information transparency of companies in such areas, and thus significantly inhibit earnings management. Therefore, based on the aforementioned research hypothesis 2, this paper proposes:

H2: The opening of high-speed rail has a stronger inhibitory effect on earnings management of listed companies in prefecture-level cities than in sub-provincial cities.

# 3 Research Design

## 3.1 The Degree of Earnings Management

In the field of earnings management, scholars at home and abroad have put forward a lot of measurement models, which has caused the academic circle to discuss the effectiveness of the measurement model of earnings management. After studying the manipulative accrual model, Huang Mei pointed out that the modified cross-section Jones model in the Chinese securities market performed better in terms of model setting and earnings management testing ability [18]. Therefore, this paper selects the modified cross-section Jones model when measuring earnings management. Specifically, we use model (1) to estimate parameters α, β1, β2, and then use formula (2) to calculate the earnings management variable DA.

$$\frac{TA_t}{A_{t-1}} = \alpha\frac{1}{A_{t-1}} + \beta_1\frac{(\Delta REV_t - \Delta REC_t)}{A_{t-1}} + \beta_2\frac{PPE_t}{A_{t-1}} + \varepsilon \tag{1}$$

$$DA = \frac{TA_t}{A_{t-1}} - \left[\alpha\frac{1}{A_{t-1}} + \beta_1\frac{(\Delta REV_t - \Delta REC_t)}{A_{t-1}} + \beta_2\frac{PPE_t}{A_{t-1}}\right] \tag{2}$$

In the above formula, $TA_t$ is the total accruals, which is equal to t year's net profit minus t year's net cash flow from operating activities; $\Delta REV_t$ Indicates the operating income in year t minus the operating income in year T-1; $\Delta REC_t$ Represents t year's accounts receivable minus t–1 year's accounts receivable; $PPE_t$ represents the original value of fixed assets at the end of t; $A_{t-1}$ indicates the total assets at the end of t–1. ε is a random disturbance term; DA stands for earnings Management.

Referring to the studies of Chen Wuchao [19] and Wu Liansheng et al. [20], this paper selects the absolute value of earnings management as the proxy variable of earnings management.

## 3.2 Multiple Regression Design

Since the opening time of high-speed rail in different cities may be different, this paper adopts multi-phase difference-difference method. Different from the ordinary differential method, in the multi-period differential model, we only pay attention to the coefficient $GT_{c,t}$, which represents the marginal effect of the opening of high-speed rail. Therefore, in order to investigate the relationship between the opening of high-speed rail and earnings management, this paper constructs a multi-stage differentially based model as follows:

$$\begin{aligned}\text{ABSDA} = {} & \beta_0 + \beta_1 GT_{c,t} + GT_{c,t} + \beta_2 SIZE + \beta_3 LEV + \beta_4 ROA \\ & + \beta_5 TOBINQ + \beta_6 BOARD + \beta_7 BIG4 + \beta_8 SOE + \beta_9 FIRST\end{aligned}$$

$$+\sum INUSTRY + \sum YEAR + \varepsilon \quad (3)$$

The research object of this paper is the relationship between the opening of HSR and earnings management, so ABSDA is the explained variable. The main explanatory variable is the opening of high-speed rail. If the city c where the listed company is located has opened high-speed rail in year t, the value is 1; otherwise, it is 0.

This paper controls a series of corporate governance and corporate characteristic variables, including: SIZE variable, which is equal to the natural logarithm of total assets; Asset-liability ratio LEV, which is equal to the total liabilities of the enterprise divided by the total assets; Return on assets (ROA) is equal to the net profit of the period divided by the total assets at the end of the period; Enterprise value variable TOBINQ, which is equal to the company's market value divided by the total assets; The variable of BOARD size, BOARD, is equal to the natural logarithm of the total number of board members; Four accounting firm variable BIG4, if the CPA is any one of PWC, Deloitte, Ernst & Young and KPMG, the value is 1, otherwise it is 0; Control the human nature variable SOE, if the company's final controller is state-owned, the value is 1, otherwise it is 0; The variable of the shareholding ratio of the largest shareholder, FIRST, is equal to the number of shares held by the largest shareholder divided by the total number of shares. Finally, the INDUSTRY dummy variable and the annual dummy variable YEAR are controlled.

In order to test hypothesis H1, in model (3), the expected β1 is less than 0, that is, after controlling other variables, the opening of high-speed rail is negatively correlated with earnings management. In addition, in order to test hypothesis H2, this paper will group the research samples according to the city administrative level, and it is expected that the group of β1 cities above the sub-provincial level is not significant, and the group of β1 prefecture-level cities is significantly negative.

### 3.3 Sample Selection and Data Sources

This paper selected listed companies in China's A-share manufacturing industry from 2005 to 2021 as research samples, combined the earnings management data of listed companies based on the registered location of listed companies, and further eliminated samples with missing variables to obtain 16,048 observed values. It covers 2,127 listed companies. In order to minimize the influence of extreme values on the research conclusion, all continuous variables were winsorize at 1% and 99%. In the research process, Excel and stata15.1 were used to complete sample data sorting, univariate analysis and multiple regression analysis.

The data sources of this paper are as follows: (1) Based on CSMAR's discretionary and non-discretionary accruals database, this paper calculates the earnings management degree of the company; (2) This paper manually collected the year data of the opening of high-speed rail in prefecture-level cities across the country; (3) The enterprise nature variable (refers to whether the enterprise is a state-owned enterprise) is taken from the CCER (Xenophon) database; (4) The remaining control variables are from the CSMAR database.

# 4 Analysis of Empirical Results

## 4.1 Univariate Analysis

In this paper, samples were grouped according to whether high-speed railway was opened, and T (WilcoxonZ) test was conducted against the difference between the mean (median) earnings management groups. First, the mean difference is 0.0736 in the group with high-speed railway service, and 0.0796 in the group without high-speed railway service. The latter is significantly higher than the former by T-test. After the value test, the median values of the two groups with and without high-speed rail were 0.0484 and 0.0514, respectively, and there was still a significant difference. Therefore, through univariate analysis, the preliminary conclusion can be drawn that there is a significant negative correlation between the opening of high-speed rail and earnings management. In other words, after the opening of high-speed rail in the place where listed companies are located, the earnings management of listed companies in the region is consistent. This is consistent with hypothesis 1. Of course, the exact relationship between the opening of high-speed rail and earnings management needs to be verified by the following more rigorous multiple regression analysis (Table 1).

**Table 1.** Univariate analysis

| Panel A: Inter-group T test was conducted on earnings management mean according to whether high-speed railway was opened | | | |
|---|---|---|---|
| Variable | Open high-speed rail | No high-speed rail service | T |
| *ABSDA* | 0.0736 | 0.0796 | −4.360*** |
| Panel B: The median earnings management value was tested by Z-test between groups according to whether the high-speed railway was opened | | | |
| Variable | Open high-speed rail | No high-speed rail service | Z |
| *ABSDA* | 0.0484 | 0.0514 | −3.604*** |

## 4.2 Multiple Regression Analysis

We then estimated model (3), and the results are shown in Table 2. The estimated coefficient of GT for the opening of high-speed rail is −0.0054, indicating that the earnings management of listed companies decreases by 0.54% for every 1% increase in the opening of high-speed rail. The statistical significance shows that the opening of high-speed rail effectively inhibits the earnings management behavior of listed companies, which supports hypothesis 1 proposed in this paper. In terms of control variables, except for the positive coefficients of TobinQ and the shareholding ratio of the largest shareholder, the remaining variables are negative. This is consistent with the conclusions of the existing literature and conforms to the theoretical expectation.

In addition, this paper also found some noteworthy corporate governance "anomalies". In terms of the Big Four accounting firms (BIG4), the Big Four accounting firms

**Table 2.** High-speed rail opening and earnings management

| Variable | Anticipatory symbol | Coefficient | T |
|---|---|---|---|
| *GT* | − | −0.0054*** | −3.35 |
| *SIZE* | − | −0.0009 | −1.24 |
| *LEV* | + | 0.0529*** | 15.77 |
| *ROA* | − | −0.0481*** | −4.29 |
| *TOBINQ* | + | 0.0085*** | 18.00 |
| *BOARD* | − | −0.0078*** | −3.10 |
| *BIG4* | − | −0.0080*** | −2.85 |
| *SOE* | − | −0.0096*** | −6.80 |
| *FIRST* | + | 0.0244*** | 5.60 |
| *CONSTANT* | / | 0.0906*** | 5.30 |
| Year and industry have been controlled | | | |
| *Observed number* | 16048 | | |
| *Adj R2* | 0.1288 | | |
| *F* | 68.76*** | | |

(BIG4) can significantly inhibit the earnings management of enterprises in cities above the sub-provincial level, but not in prefecture-level cities. In terms of estimated coefficients, the estimated coefficients of BIG4 of the prefecture-level city group and the sub-provincial city group are −0.0056 and −0.0080, respectively, indicating that the marginal inhibitory effect of the opening of high-speed rail on earnings management is lower in the prefecture-level city than in the sub-provincial city. In terms of the FIRST shareholder's shareholding ratio (FIRST), the estimated coefficient of the first major shareholder's shareholding ratio (FIRST) in the prefecture-level city group is 0.0357, which is much higher than the 0.0099 in the sub-provincial and above city group, indicating that major shareholders have more destructive effects on the corporate governance of listed companies in the prefecture-level city. In terms of company size, the company size in cities above sub-provincial level can significantly reduce the earnings management of enterprises, and the greater the company size in prefecture-level cities, the higher the earnings management degree. To sum up, the administrative levels of listed companies have different effects on corporate governance. The possible reasons for its existence are that in cities above the sub-provincial level, external supervision, financial institutions and news media are concentrated, the degree of marketization is higher, the time and capital cost of external information acquisition is lower, and the information transparency of listed companies is higher. So the Big Four accounting firms can significantly inhibit earnings management. In prefecture-level cities, the number of financial institutions and news media and other units stationed in such cities is small, the marketization level is low, and the market has strong information asymmetry, which leads to a low level of corporate governance.

## 5 Research Conclusions

In this paper, based on the natural experiment background of the opening of high-speed rail in 2008, China's A-share listed companies from 2005 to 2021 are used as research samples, and the relationship between the opening of high-speed rail and earnings management is empirically tested by using the differential differential method. It is found that the opening of high-speed railway significantly inhibits the earnings management behavior of enterprises. Among them, the opening of high-speed rail significantly inhibited earnings management in prefecture-level cities, but not in sub-provincial and above cities. After further testing the path mechanism of its influence, this paper finds the path of "opening of high-speed rail - information transparency - earnings management". In addition, the study also found that the opening of high-speed rail has not effectively inhibited the earnings management of small-scale listed companies. This study not only enriches the research of corporate finance and other fields, but also has practical significance for effective supervision of regulatory authorities and reasonable decision-making of investors. The implications of this study are as follows: First, the opening of high-speed rail not only affects the macro economy, but also affects the multi-level corporate governance. This paper finds that the opening of high-speed rail can effectively inhibit earnings management, so the regulatory authorities can focus on the supervision of listed companies in remote areas with unimpeded information; Secondly, the opening of high-speed rail can effectively improve the degree of information flow between regions and reduce various problems caused by information asymmetry; Finally, the securities regulatory authorities should improve the transparency of information, strengthen the supervision of small-scale listed companies, and reduce the information asymmetry as much as possible, so as to effectively restrain market violations.

**Acknowledgment.** This research was funded by the Science and Technology Research and Development Program of China State Railway Group Co., Ltd (2021F014).

## References

1. Liu, Y., Zhang, N., Zhou, L.: High-speed railway construction and balanced regional development: evidence from the opening of Beijing-Tianjin-Hebei high-speed Railway. J. Beijing Univ. (Philos. Soc. Sci.) **307**(3), 62–73 (2018)
2. Bian, Y., Wu, L., Bai, J.: High-speed railway opening, factor flow and regional economic gap. Financ. Trade Econ. **39**(6), 149–163 (2018)
3. Deng, T., Wang, D., Cheng, S.: The impact of high-speed rail on urban service industry agglomeration. Res. Financ. Econ. **43**(7), 119–132 (2017). (in Chinese)
4. Du, X., Peng, M.: Will the opening of high-speed rail promote the flow of senior talents in enterprises? Econ. Manage. **12**, 91–109 (2017). (in Chinese)
5. Zhao, J., Huang, J., Liu, F.: High-speed rail opening and stock price crash risk. Manage. World **1**, 157–168 (2018)
6. Yu, C.: A review on the measurement of corporate transparency—on the annual information disclosure evaluation data of Shenzhen stock exchange. Financ. Acc. Newslett. Acad. Edn. **6**, 105–106 (2008)

7. Su. X.: Empirical Study on the Relationship Between Financial Reporting Transparency, Ownership Structure and Earnings Management. Henan University (2014)
8. Yang, K.: Other comprehensive income, information transparency and earnings management. J. Central Univ. Financ. Econ. **8**, 65–73 (2016). (in Chinese)
9. Luo, J., Li, X., Lin, Z.: The influence of auditor geographical distance on the information content of client company stock price. Audit Econ. Res. **182**(04), 38–49 (2018)
10. Malloy, C.J.: The geography of equity analysis. J. Financ. **60**(2), 719–755 (2005)
11. Li, B.: Parent-subsidiary company distance, corporate governance and investment behavior. J. Shanxi Univ. Financ. Econ. **37**(4) (2015)
12. Kan, J., Wang, H.: Study on the influence of geographical relationship of accounting firms on audit quality. Audit Econ. Res. **13**(2), 74–84 (2016). (in Chinese)
13. Stephan, P.E.: Company-scientist locational links: the case of biotechnology. Am. Econ. Rev. **86**(3), 641–652 (1996)
14. Du, X., Hou, F., Lai, S.: Do transport infrastructure improvements inhibit 'geographical preferences' in auditor selection?—Empirical evidence based on natural experimental background of high-speed railway in China. Audit Res. **1**, 103–110 (2018)
15. Jiang, J., Sun, K., Nie, H.: City level, total factor productivity and resource mismatch. Manage. World **34**(3) (2018)
16. Wang, K., Chen, T.: The experience of administrative division adjustment abroad and its implications for China. World Geogr. Res. **20**(2), 57–64 (2011). (in Chinese)
17. Wasley, C.: Performance matched discretionary accrual measures. J. Account. Econ. **39**(1), 163–197 (2005)
18. Huang, M.: Empirical analysis of earnings management ability tested by manipulative accrual model. Nankai Manage. Rev. **12**(5), 136–143 (2009)
19. Chen, W.: Business cycle, industry periodicity and earnings management degree: empirical evidence from Chinese listed companies. Nankai Manage. Rev. **16**(3), 26–35 (2013)
20. Liu, H., Wang, C., Wu, L.: Allocation of decision power, earnings management and investment efficiency. Econ. Res. **8**, 93–106 (2014)

# Modeling of Spatial Transfer of Affected Passengers to and from Stations Under Delayed Urban Rail Trains


Ziyu Wang[1], Limin Jia[1], Jianyuan Guo[1](✉), and Ming Chen[2]

[1] State Key Laboratory of Advanced Rail Autonomous Operation, Beijing Jiaotong University, Beijing 100044, China
{22120790,lmjia,jyguo}@bjtu.edu.cn
[2] Beijing Metro Network Administration Co., Ltd, Beijing 100000, China
chenming@bmncc.com.cn



**Abstract.** The rapid development of urban rail transportation requires more and more emergency events such as train delays. This paper proposes a spatial transfer model for affected passengers entering and leaving stations under train delays in urban rail, aiming to provide a basis for emergency response under train delay type of emergencies. The study analyzes the influencing factors of passenger travel under train delay conditions, constructs physical network and spatio-temporal expansion network under train delay conditions, and establishes effective physical path set and spatio-temporal path set. Based on the spatio-temporal path set, the generalized cost of passenger travel under train delay is calculated, and the spatial transfer probability of affected passengers in and out of stations is obtained by Logit model, and the spatial transfer model of affected passengers in and out of stations is established by using particle swarm algorithm for parameter optimization. The prediction model was validated using the data of a train delay event on Beijing Metro Line 10.



**Keywords:** Urban Rail Transit · Train Delay · Multinomial Logit


## 1 Introduction

In recent years, urban rail transit has developed rapidly, and after long-term construction and improvement, it has become an important part of fast, convenient and safe public transportation in cities. By the end of 2022, a total of 290 urban rail transit lines were in operation in 53 cities across China, with a total operating mileage of 9,584 km, which can reach a wide range of people and provide a convenient and fast public travel mode for urban passengers. In the same year, 21 new urban rail transit lines were added, with 847 km of new operating mileage, and the passenger volume reached 19.40 billion passenger trips, with the number of inbound trips reaching 11.69 billion [1], which fully proves the necessity and importance of urban rail transit. The rapid development of urban rail transit has provided a lot of convenience for residents to travel, however, it has also brought certain hidden dangers. In recent years, a number of train delays



Y. Qin et al. (Eds.): EITRT 2023, LNEE 1137, pp. 517–529, 2024.
https://doi.org/10.1007/978-981-99-9311-6_53

have occurred in domestic urban rail transit, which has caused widespread concern and worries among the public. According to incomplete statistics, between 2017 and 2021, there were 113 delays on 22 subway lines in Beijing, including 32 train disruptions. The occurrence of these accidents not only caused casualties and economic losses, but also exposed the problems of urban rail transit safety management, including inadequate daily maintenance, lax safety management, and insufficient emergency plans. Therefore, the study of passengers' path selection behavior under train delay can provide theoretical basis for passenger flow prediction and control, emergency plan formulation and train operation adjustment.

Domestic and foreign scholars have conducted a lot of research around rail traffic passenger flow prediction under train delays. Preston et al. [2] Barron et al. [3] analyzed the impact of train operation delays on passenger travel. Hong Ling et al. [4] quantified the delayed passenger flow, detoured passenger flow and lost passenger flow of urban rail transit under the impact of unexpected events, so as to analyze the impact of unexpected events on network passenger flow. Wang et al. [5] integrated the theories of stochastic regret minimization and stochastic utility maximization to construct a travel choice model under unexpected events, and quantified the congestion propagation rate through passenger flow simulation. Sasha Liu et al. [6] considered multi-modal alternative travel options, built a travel choice model under contingencies based on MNL, and conducted a travel choice behavior survey using random sampling to extract travel choice results from it. Sihui Li [7] used questionnaire survey data to construct a travel path choice behavior model under unexpected events considering induced information based on MNL model and CL model, respectively. Yang [8] constructed a travel scheme selection model based on MNL model by constructing a set of alternative travel schemes under train delays. Most of the current domestic and international studies on the impact of urban rail train delays on analyzing passenger flow changes focus on calculating travel path selection probabilities through logit models [6–8], and rarely involve research on the change of passengers' departure and destination under train delays; in addition, most of the existing studies are based on questionnaire data, which has a strong priority interpretability, but also has the defect of subjectivity, especially Under train delay conditions, passengers' questionnaires are based on assumptions and the usability of questionnaire results is low. Based on the historical AFC data, this paper summarizes the pattern of passenger path change by analyzing passenger behavior under train delays, and then establishes a spatial transfer model of affected passengers under urban rail train delays to predict the change of passengers' departure and destination, which is important for the emergency response under train delay type emergencies.

## 2 Factors Affecting Passenger Travel Under Train Delay Conditions

Train delays in urban rail transit systems are defined as deviations from the planned schedule when trains do not arrive at their destinations at the expected time during their journey. This can have a serious impact on operational safety and passenger travel, and cause significant disturbance to urban traffic order [9]. The main influencing factors include: train malfunction, such as electrical system failure, signal system failure, etc.;

external construction operations, where the occupation of normal travel routes by construction sites and platform closures occur during construction, resulting in disruption of train scheduling; adverse weather effects, such as fog, snow, and other extreme weather can affect train travel safety; passenger factors, where a large influx of passengers onto platforms and carriages results in train terminal It takes longer time for a clearing, which leads to delays of subsequent trains; operation management, problems such as imprecise train scheduling plans and negligence of the scheduling department, which lead to inconsistent train starting and ending points at the wrong time; unexpected events, such as earthquakes, fires, security incidents, etc., can lead to serious impacts on train operations.

Train delays have become one of the frequently encountered problems in urban rail transit operations, which not only affect the normal operation of transportation, but also may have adverse effects on the public order of the city. According to the main causes and types of train delays, comprehensive and systematic measures must be taken to prevent accidents and failures, optimize the urban rail network and reduce the hazards caused by train delays. In the case of train delays, the original schedule is changed, which will affect the travel plans of passengers. If a passenger's original travel path is affected by a train delay, the passenger will judge it based on known delay information and re-decide the travel path with the goal of minimizing the total travel cost, which mainly includes the following influencing factors [10]:

(1) Travel purpose

In the case of delayed trains, some passengers need to change their original routes and re-plan their travel plans, and the purpose of travel will have an impact on the choice of route planning.

(2) Travel time

Passengers need to choose within their acceptable time frame, if time is limited, passengers will prefer to choose a shorter route and transportation mode. For example, for commuters, they may need to take a faster and more direct mode of transportation.

(3) Transportation nodes

When changing the original travel plan, passengers need to consider the properties of the transportation nodes on the new travel plan, such as where to transfer, whether to take a different mode of transportation, etc., and whether the chosen route can be reached.

(4) Travel cost

Passengers need to consider the travel cost of the selected route, such as fare, time cost, etc.

# 3 Spatio-Temporal Expansion Network Under Train Delay Conditions

## 3.1 Physical Network Model

To construct the physical network of urban rail transit [11], G(V, E) is used to represent the topology graph of the physical network of urban rail transit network, $V = \{v_i | v_1, v_2, \ldots, v_n\}$ denotes the set of nodes in graph G, and $E = \{e_i | e_1, e_2, \ldots, e_n\}$ denotes the set of nodes in graph G with to the set of arcs. The following figure shows the topology of the network, with hollow circles indicating non-interchange stations and solid circles indicating interchange stations.

For the case of train delays, the directional arc weights involved in the delayed stations of the rail transit are recalculated to increase the path relations between stations where external traffic can be connected. If a passenger prefers a path where station O or D has changed, it is defined as a spatial shift of passengers in and out of the station. As shown in the figure below, suppose the original path of passengers arriving at station N from station A through station B and station C. Since the normal travel is affected by train delays between station A and station B, the physical path achievable through external transportation between station A and station B is increased, i.e., passengers arrive at station Z through external transportation at station A, enter the inner city rail transit, and take the subway through station Z to reach their original destination station N, it is said that the passenger entry is transferred from station A to station Z (Fig. 1).

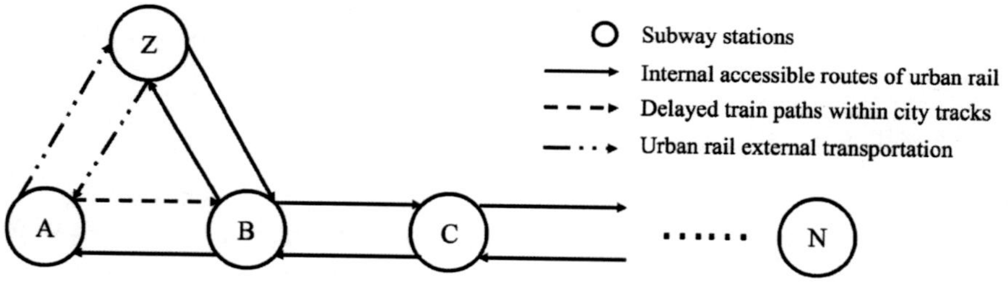


**Fig. 1.** Physical path schematic

## 3.2 Spatio-Temporal Network Construction

The set of physical paths built by the physical network model can only represent passengers' travel spatial tracks and cannot depict passengers' spatio-temporal paths. To solve this problem, a spatio-temporal expansion network needs to be constructed, which can not only represent passengers' spatial paths, but also expand on the time axis. This spatio-temporal network is built on the basis of the physical network of rail transit. The basic principle is to add temporal attributes to the network nodes by integrating information such as rail transit train operation schedules under the condition of train delays, and to represent the spatial location of passengers and subjects such as trains over time by a directed spatio-temporal arc.

The graph G(V, E) is expanded on the time axis [12] and denoted as G(V, E, L, T, F). $E = \{e_i|e_1, e_2, \ldots, e_n\}$ denotes the set of arcs in graph G, $L = \{l_i|l_1, l_2, \ldots, l_n\}$ denotes the set of subway lines, and $T = \{t_i|t_1, t_2, \ldots, t_n\}$ denotes the set of arrival and departure times of each station of different lines, and $F = \{f_i|f_1, f_2, \ldots, f_n\}$ denotes the set of trains.

(1) Spatio-temporal nodes

The spatio-temporal nodes are divided into two categories, one category is used to describe passenger behavior, including entry and exit time, entry and exit status, and node attributes, e.g. ${}_{17:35}{}^{arr}PAX^{5}_{104}$ means passenger arrived at station number 104 of line 5 at 17:35, ${}_{17:40}{}^{out}PAX^{5}_{203}$ means passenger left station number 203 of line 5 at 17:40 The first category is used to describe train-related attributes, for example, ${}_{17:35}{}^{arr}TRAIN\,1^{5}_{104}$ indicates that train No. 1 on line 5 arrived at station No. 102 at 17:35, and ${}_{17:40}{}^{out}TRAIN\,2^{5}_{203}$ indicates that train No. 2 on line 5 left station No. 203 at 17:40. Leaves the station numbered 203.

(2) Time-Space Arc

Time arc is divided into inbound arc, outbound arc, transfer arc, on-train arc and alternative transportation arc. $e(i_{t1}^{l,1}, i_{t2}^{l_f,1})$ is the inbound arc, which means passengers swipe their cards from line $l$ to enter the station at time $t1$ and walk to the station platform to take the $f$ train of line $l$ at time $t2$, where 1 means inbound and 2 means outbound; $e(i_{t1}^{l_f,1}, i_{t2}^{l,2})$ is the outbound arc, which means that the passenger arrives at the station at the moment $t1$ on the $f$ train of line $l$ and swipes his card to leave the station at the moment $t2$; $e(i_{t1}^{v_i,l_f}, i_{t2}^{v_j,l_f})$ is the inbound arc, which means that the passenger takes the $f$ train of line $l$ at the station $v_i$ at the moment $t1$ and arrives at the station $v_j$ at the moment $t2$; $e(i_{t1}^{l_f,1}, i_{t2}^{l_{f\prime},2})$ is the transfer arc, which means that the passenger leaves the $f$ train of line $l$ at moment $t1$ and transfers to the $f$ train of line $l\prime$ at moment $t2$, where 1 means get off and 2 means get on; $e(i_{t1}^{v_i,l}, i_{t2}^{v_j,l\prime})$ is the alternative traffic arc, indicating that passengers leave the $v_i$ station of line $l$ at moment $t1$ and arrive at the $v_j$ station of line $l\prime$ via external traffic at moment $t2$.

(3) Spatio-temporal paths

A spatio-temporal path is a path that contains information about the origin, destination and time, and is an ordered arrangement of spatio-temporal arcs and spatio-temporal nodes as described above. It not only depicts the spatial path of passengers, but also contains extended information on the temporal axis, which can accurately depict the travel path of passengers between different moments.

# 4 Spatial Transfer Model for Affected Passengers Entering and Leaving the Station

## 4.1 Spatio-Temporal Path Search Algorithm

To construct a spatio-temporal path collection, we first need to search for the valid paths in the physical network. In urban rail network, there will be several different physical paths between ODs, and passengers usually only consider some of these physical paths as alternatives when traveling, and these physical paths considered by passengers are the valid physical paths. The constraints defined in this paper for the physical valid paths include no duplicate stations in the valid paths, no loops in the valid paths, no duplicate lines in the paths, and the number of interchanges does not exceed the maximum threshold; the constraints for the spatio-temporal valid paths include inbound time constraints, interchange time constraints, outbound time constraints, and alternative transportation time constraints.

In this paper, we use Dijkstra algorithm to search the shortest path, search the paths of nodes in the network, construct the set of valid paths according to the neighboring relationship between inbound nodes, interchange nodes and outbound nodes, and then identify the valid spatio-temporal paths according to the historical AFC data and train schedule data.

**Step 1**. Iterate through all the AFC data, note the current data as the i-th data, and the total number of AFC data is I. Input inbound station, outbound station, inbound time and outbound time;

**Step 2**. Search the valid physical paths of the starting and ending points of this data to get the set of valid physical paths $L_{od}$;

**Step 3**. Traverse the kth valid path in $L_{od}$ and record the interchange stations of that path;

**Step 4**. Starting from the starting point O station, find the first spatio-temporal node that is larger than the passenger entry time and satisfies the minimum entry constraint, record the departure moment at the starting station, and then find the first interchange station of this path and record the arrival moment. Then find the earliest train whose spatio-temporal node of this interchange station on another line is greater than the departure time of the interchange station and satisfies the minimum interchange time, and keep finding the arrival time of the train at the end station and connect the spatio-temporal node at the end station;

**Step 5**. Starting from the end point D, find the train that arrives the latest from the end station, and search the space-time node of the train at the starting station in order to the starting station according to the method in Step 3. The running arcs between these two paths are all valid arcs of the spatio-temporal paths, and the valid arcs are connected according to the station attributes to obtain the set of all spatio-temporal paths;

**Step 6**. Judge i > I, if not, then turn Step1, otherwise the algorithm ends (Fig. 2).

## 4.2 Inbound and Outbound Station Spatial Transfer Probabilities

The generalized cost of spatio-temporal path is calculated based on the time factor, which includes: passenger on board time, transfer time, inbound time, outbound time, and external transportation time.

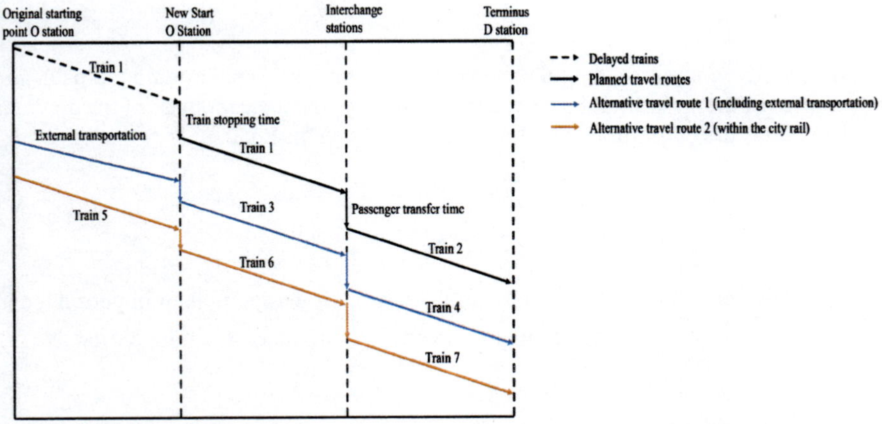


**Fig. 2.** Schematic diagram of passenger travel trajectory under delayed train conditions

(1) Passenger on board time

The time of running arc running and the time of stopping arc stopping are obtained according to the train running schedule, which is calculated by subtracting the moment of the previous node from the moment of the node after that arc. The sum of the time on all running arcs on that path gets the passenger on board time, as shown in Eq. 1.

$$T_{od,t}^{k,zc} = \sum_{e \in e_{od,t}^{k,run} \cup e_{od,t}^{k,stop}} t_2 - t_1 \tag{1}$$

where, $T_{od,t}^{k,zc}$ is the on-board time of OD for the kth space-time path in period t, $e_{od,t}^{k,run}$ is the set of running arcs of OD for the kth space-time path in period t, $e_{od,t}^{k,stop}$ is the set of stop arcs of OD for the kth spatio-temporal path in period t.

(2) Transfer time

Similarly, the passenger's interchange arc is obtained based on the passenger's spatio-temporal path, and the interchange time of that interchange arc is obtained by subtracting the moment of the previous node from the moment of the node after the interchange arc. The sum of the times on all interchange arcs on that path yields the passenger's onboard time, as shown in Eq. 2.

$$T_{od,t}^{k,tra} = \sum_{e \in e_{od,t}^{k,tra}} t_2 - t_1 \tag{2}$$

where $T_{od,t}^{k,tra}$ is the interchange time of OD for the kth space-time path in period t, and $e_{od,t}^{k,tra}$ is the set of interchange arcs of OD for the kth space-time path in period t.

(3) Inbound time

The passenger's entry arc is found by the passenger's specific spatio-temporal path, and the moment of the node after the entry arc is subtracted from the moment of the previous node, which is the entry time of that entry arc, as shown in Eq. 3.

$$T_{od,t}^{k,in} = \sum_{e \in e_{od,t}^{k,in}} t_2 - t_1 \tag{3}$$

where $T_{od,t}^{k,in}$ is the inbound time of OD for the kth spatio-temporal path in period t, and $e_{od,t}^{k,in}$ is the set of inbound arcs of OD for the kth spatio-temporal path in period t.

(4) Exit time

Similarly, after obtaining the specific spatio-temporal path of the passenger, find the outbound arc of the passenger, and subtract the moment of the previous node from the moment of the latter node of the outbound arc, which is the outbound time of the outbound arc, as shown in Eq. 4.

$$T_{od,t}^{k,out} = \sum_{e \in e_{od,t}^{k,out}} t_2 - t_1 \tag{4}$$

where $T_{od,t}^{k,out}$ is the outbound time of OD for the kth space-time path in period t, and $e_{od,t}^{k,out}$ is the set of outbound arcs of OD for the kth space-time path in time period t.

(5) External travel time

The distance between two nodes can be obtained based on the passenger's specific spatio-temporal path, which is calculated by dividing the path distance between two points by the average speed based on the average speed of the selected transportation. As passengers choose alternative stations mostly choose stations that are close to each other, cab travel is not considered; as the distance between subway stations is far for walking, walking and shared bicycle are not considered.

$$T_{od,t}^{k,v_i,v_j} = S_{v_i,v_j}/v_t \tag{5}$$

where $T_{od,t}^{k,v_i,v_j}$ is the OD travel time of external traffic between $v_i$, $v_j$ at time period t for the kth spatio-temporal path, $S_{v_i,v_j}$ is the Manhattan distance between $v_i$, $v_j$, and $v_t$ is the average speed of external traffic at time period t.

Since passengers' perceptions of onboard time, transfer time, in/outbound time and external transportation time differ, passengers are more likely to spend their time onboard than on transfer, in/outbound or external transportation, so penalty factors need to be added to transfer time, in/outbound time and external transportation time respectively. The time-path generalized cost function is shown in Eq. 6.

$$C^{k}_{od,t} = T^{k,zc}_{od,t} + \alpha T^{k,tra}_{od,t} + \beta\left(T^{k,in}_{od,t} + T^{k,out}_{od,t}\right) + \gamma T^{k,v_i,v_j}_{od,t} \quad (6)$$

where $\alpha$ is the interchange penalty coefficient, $\beta$ is the inbound and outbound penalty coefficient, and $\gamma$ is the external traffic penalty coefficient.

The Logit model is the most widely used model of non-set counting behavior because of its advantages such as easy to explain and illustrate, and fixed form of selection probability usage. Based on random utility theory, assuming that the random error terms of the physical path utility obey the same Gumbel Distribution and are independent of each other, the model equation is as follows:

$$p^{k}_{od,t} = \frac{e^{-\theta X^{k}_{od,t}}}{\sum_{k\prime\in R_{od,t}} e^{-\theta X^{k\prime}_{od,t}}} \quad (7)$$

where $p^{k}_{od,t}$ denotes the probability that the passenger chooses the kth valid path between OD pairs in period t. $\theta$ is a constant, i.e., a model parameter, which converts the generalized expense cost of physical paths into utility.

In the MNL model, the selection probability of each physical path depends on its corresponding generalized cost absolute difference. In order to solve the problem of using the generalized cost absolute difference to produce unreasonable flow allocation results, the generalized cost relative difference can be used to calculate the physical path selection probability so as to be close to the actual situation. Therefore, this paper uses the relative difference between the generalized cost and the minimum generalized cost of each physical path between ODs to construct the model.

The MNL model of physical path selection under train delay conditions constructed in this paper is as follows:

$$p^{k}_{od} = \frac{e^{-\theta \frac{X^{k}_{od,t}}{X^{min}_{od}}}}{\sum_{k\prime\in R_{od}} e^{-\theta \frac{X^{k\prime}_{od,t}}{X^{min}_{od,t}}}} \quad (8)$$

After calculating each path selection probability, the paths that do not contain external traffic are identified as the paths where no shifts occur in and out of the station. For the path containing external traffic, the inbound and outbound stations after the transfer are determined according to the location where the external traffic occurs. Specifically, if the external traffic occurs at the beginning of the path, the station that arrives is taken as the inbound station after the transfer. If the external traffic occurs at the end of the route, the station before the arrival of the route is used as the outbound station after the transfer.

## 5 Case Analysis

In this paper, the proposed algorithm and model are analyzed and validated against the background of a train delay event of more than 5 min in the Beijing urban rail transit network. On a day in 2019, a train delay event occurred in Beijing Metro Line 10, and several zones were affected, among which the delay time period was 8:12–9:37. This caused 26 trains to be stopped, 40 trains to be late, 17 trains to be cleared midway, and the maximum delay time 38min.

The experiments were conducted on the day of the delay and a week before the normal conditions of the same period AFC data, train schedule data. The AFC data from 8:00 to 10:00 was intercepted, and 704942 passenger information appearing on both days were extracted according to the card number, and 180927 passenger information entering or leaving the station on Line 10 under normal conditions were screened according to the entry and exit stations as affected passengers for the experiment. The Manhattan distance between stations near Line 10 to each station on Line 10 was crawled, and the passage time between stations was calculated based on speed to create a collection of external traffic data, resulting in 2,384 valid OD pairs.

The spatial transfer of affected passengers in and out of stations under train delays was modeled using the scheme selection model described above, and the data samples were divided into training and test sets in the ratio of 0.75:0.25, and the parameters were calibrated using a particle swarm algorithm, and the results are shown in Table 1.

**Table 1.** Parameter calibration results.

| Parameter | Parameter Value |
|---|---|
| $\alpha$ | 0.8593 |
| $\beta$ | 0.9236 |
| $\gamma$ | 0.3179 |
| $\theta$ | 0.4943 |

Taking the OD pair from station A to station F as an example, the variation of this OD passenger flow on the rail transit network is analyzed. First, the set of multi-modal alternatives between this OD pair is constructed according to the method described above. The travel options for this OD pair include: (i) taking urban rail transit from station A to station F; (ii) taking urban rail transit from station A, arriving at station C through interchange station B, and arriving at station F through external transportation; (iii) taking urban rail transit from station A, arriving at station E through interchange station D, and arriving at station F through external transportation. Second, based on the results of the travel choice model in Table 1, the proportion of each travel option being chosen is calculated, as shown in Table 2.

**Table 2.** Travel options between AF

| Scheme number | Travel options | Program attribute values | | | | Select ratio % |
|---|---|---|---|---|---|---|
| | | In-Train | Interchange | Travel | External transportation | |
| i | A-F | 11.02 | 0 | 7.48 | 0 | 38.17 |
| ii | A-B-C<br>C-F | 15.19 | 9.77 | 4.24 | 4 | 26.23 |
| iii | A-D-E<br>E-F | 32.31 | 11.45 | 6.15 | 26 | 25.60 |

According to the calculation results in Table 2, the spatial transfer of this OD passenger flow in the rail transit network can be obtained. Under this train delay, part of the path between station A and station F in the rail transit system was affected by the unexpected event. For passengers who originally planned to exit at station F, 38.17% still chose to exit at station F, 26.23% shifted to station C and 25.60 shifted to station E.

The average absolute percentage error of the model is 15.25% based on the calculated accuracy of the passenger flow prediction results obtained from the test set. In addition, the scatter plot of the predicted and true values of inter-station passenger flow during the statistical time period is compared as shown in the Fig. 3. From the figure, it can be seen that most of the inter-station passenger flow prediction values are distributed around the 45° line, which indicates that the method has a good prediction effect.

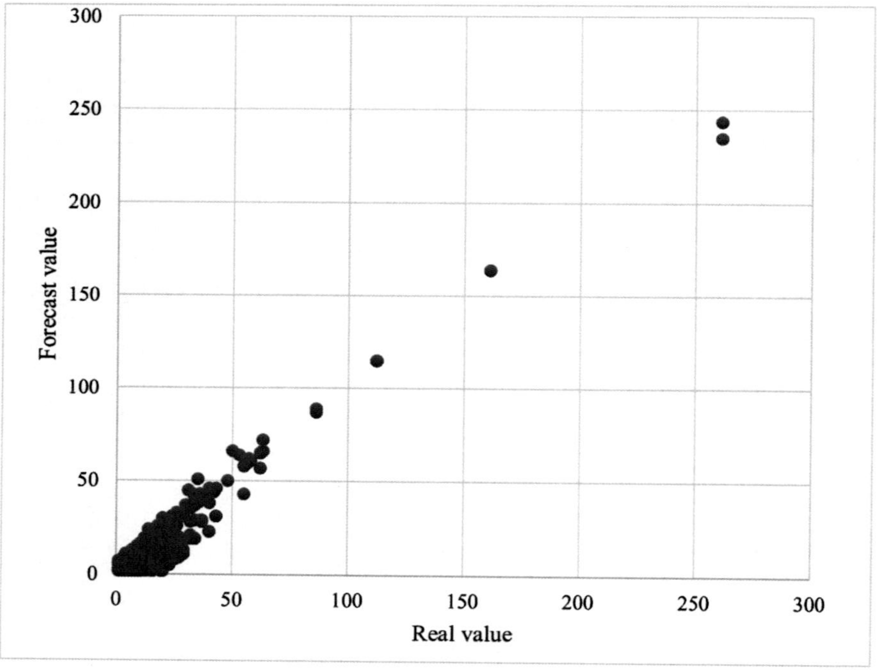


**Fig. 3.** Comparison between predicted and actual passenger flow values

## 6 Conclusion

The purpose of this paper is to explore the spatial transfer of urban rail transit passengers in and out of stations under unexpected events, construct a topological network between ODs based on actual AFC data and train running time data, and search for a collection of spatio-temporal travel paths by combining travel time constraints. In order to compensate for previous studies ignoring the influence brought by external traffic under train delays, this paper considers the physical paths formed by external traffic while constructing the internal physical paths of urban rail transit. By analyzing the travel choice behavior of rail transit passengers and the influencing factors of passenger travel under train delay conditions, we calibrate the parameters for the cost of external traffic to restore the travel decision process of passengers under train delay conditions more realistically, and further obtain the spatial transfer of passengers in and out of stations based on the calculated spatio-temporal path selection probability. However, this model does not distinguish the operation of external traffic under train delay in time, and further optimization research is needed.

## References

1. Preston, J., Wall, G., Batley, R., et al.: Impact of delays on passenger train services: evidence from great Britain. Transp. Res. Record: J. Transp. Res. Board **2117**(1), 14–23 (2009)
2. Barron, A., Melo, P.C., Cohen, J.M., et al.: A passenger-focused management approach to the measurement of train delay impacts. Transp. Res. Record: J. Transp. Res. Board **2351**(2351), 46–53 (2013)
3. Hong, L., Gao, J., Xu, R.H.: Calculation method of emergency passenger flow in urban rail network. J. Tongji Univ. (Nat. Sci.) **39**(10), 1485–1489 (2011). (in Chinese)
4. Wang, X., Yao, E., Liu, S., et al.: Simulation of metro congestion propagation based on route choice behaviors under emergency-caused delays. Appl. Sci. **9**(20), 4210 (2019)
5. Liu, S.S., Yao, E.J., Li, B.B., et al.: Forecasting passenger flow distribution between urban rail transit stations based on behavior analysis under emergent events. J. China Railway Soc. **40**(09), 22–29 (2018). (in Chinese)
6. Li, S.H.: Research on the Method of Inducing Passenger Flow in Urban Rail Transit under Emergencies. Beijing Jiaotong University (2020). (in Chinese)
7. Yang, Z.Q., Li, J.C., Shi, F.S., et al.: Modeling of passenger travel choice affected by urban rail train. Railway Comput. Appl. **29**(03), 60–64 (2020). (in Chinese)
8. Zeng, C.F., Hou, Y.F., Luo, Q., et al.: Modeling of congestion propagation of rail transit passenger flow under delay conditions. J. Shenzhen Univ. Sci. Eng. (Sci. Eng.) **36**(03), 310–316 (2019). (in Chinese)
9. Zeng, L.: Research on Passenger Flow Control Method for Urban Mass Transit under Emergency. Beijing Jiaotong University (2018). (in Chinese)
10. Deng, X.: Passenger Assignment Model in Urban Rail Transit Systems and Its Application in Emergency Scenarios. Beijing Jiaotong University (2021). (in Chinese)
11. Li, B.Y.: Prediction of Temporal and Spatial Distribution of Passenger Flow under the Condition of Urban Rail Transit Interruption. Beijing Jiaotong University (2022). (in Chinese)
12. He, Z.H., Yang, Z.X., Jiang, X.: Study on improvement of dial algorithm. J. Beijing Jiaotong Univ. **05**, 89–91 (2002). (in Chinese)

# Design of Parallel Auxiliary Converter with Different Capacity Based on Flexible Marshalling EMUs

Junbo Zhao(✉), Tao Jia, Bo Zhang, Qingwen Sun, Xinyu He, Xuan Su, Hongguo Li, and Xiang Feng

CRRC Qingdao Sifang Rolling Stock Research Institute Co, Qingdao 266031, China
zhaojunbo2018@126.com

**Abstract.** With the development and progress of high-speed railway and rail transit equipment technology, EMUs play an increasingly important role in railway transportation. In order to improve the flexibility of EMUs operation, more and more attention has been paid to the flexible marshalling technology of EMUs. Aiming at the apply problem of flexible marshalling of EMUs, this paper puts forward an auxiliary power supply system based on flexible marshalling of EMUs, introduces the power supply mode and load distribution mode. At the same time, this paper designs a parallel control strategy of auxiliary converters with different capacities based on adaptive improved droop control, which realizes the proportional allocation of active and reactive power according to the capacity of the auxiliary converter. The effectiveness and accuracy of the control strategy are verified by Matlab simulation and experiment.

**Keywords:** Flexible marshalling · Auxiliary converters · Different capacities · Improved droop control

## 1 Introduction

With the rapid development of social economy, EMUs play a more and more important role in railway transportation. The number of traditional EMU trains is usually fixed and cannot be disassembled, which restricts the railway operation more and more obviously. For example, "Hexie" and "Fuxing" EMUs of our country are mainly 8 or 16 units. In the process of operation, when the number of EMUs need to be adjusted, the method of reconnecting two EMUs is generally adopted. With the continuous development of EMUs technology, in order to improve the operation service level of high-speed railway and better meet the needs of passenger transportation, it has become the development trend of EMUs technology to adopt flexible marshalling to improve the flexibility of EMUs operation [1–4].

Flexible train marshalling technology requires flexible adjustment of train marshalling length through rapid reconnection and unmarshalling of trains [5], which puts higher requirements on auxiliary power supply system. The auxiliary converter is an


Y. Qin et al. (Eds.): EITRT 2023, LNEE 1137, pp. 530–539, 2024.
https://doi.org/10.1007/978-981-99-9311-6_54

important component of EMUs, supplying power to the three-phase AC load of the whole vehicle. Important AC loads, such as traction system cooling fans, chargers, air conditioners, etc., are either related to the safety of EMUs or closely related to the comfort of passengers, so the power supply reliability of auxiliary converters is of vital important [6–9]. The capacity of auxiliary converters in different types of flexible EMUs are also different, and each auxiliary converter needs to supply power in proportion to its own capacity. Therefore, how to quickly identify the number and type of train marshalling and complete the parallel control of auxiliary converters with different capacities is a very important topic in the process of train unmarshalling.

In the study of flexible marshalling technology, paper [1] summarizes the development context of variable marshalling EMUs at home and abroad, and qualitatively analyzes the advantages of flexible marshalling EMUs. In paper [4], based on the technical parameters of flexible EMUs at home and abroad, combined with the actual development of China's high-speed railway, the applicability of flexible marshalling EMUs in China's high-speed railway is analyzed. In paper [10], a probabilistic nonlinear programming model was established to study the passenger allocation problem of flexible marshalling EMUs, and a conclusion was drawn that flexible marshalling EMUs configuration has more economic advantages when the intensity of passenger flow demand is small. The above papers only introduces the advantages of flexible marshalling of EMUs in the application process, but does not propose an economical and applicable flexible marshalling scheme.

Many scholars have also carried out a lot of research on the parallel control strategy of different capacity auxiliary converters. In paper [11], the difference of line impedance was considered, and virtual impedance technology was adopted to achieve reactive power equalization. However, accurate line impedance information must be obtained, and the line impedance in flexible marshalling EMUs is variable. In paper [12], a method of increasing droop coefficient was proposed to achieve accurate distribution of reactive power, but too large droop coefficient would cause instability of the system. Paper [13] proposes to inject small harmonics into the output voltage to realize proportional power distribution by taking advantage of the invariance of harmonic frequency in the microgrid. However, the injection of harmonics will increase the distortion of output voltage and current and reduce the quality of electric energy.

To solve the above problems, a stable and reliable flexible marshalling auxiliary power supply system scheme of marshalling EMUs is proposed in this paper, and a parallel auxiliary converter with different capacities is designed based on this power supply system. Active power and reactive power are allocated proportionally according to capacity by using adaptive improved of droop control strategy, and the control strategy is accurate and effective.

## 2 Flexible Marshalling Auxiliary Power Supply System

As shown in Fig. 1 is the topology diagram of the flexible marshalling auxiliary power supply system proposed in this paper (taking 10 grouping as an example). The auxiliary power supply system mainly consists of the auxiliary converter, AC bus, power cabinet, loads and other components.

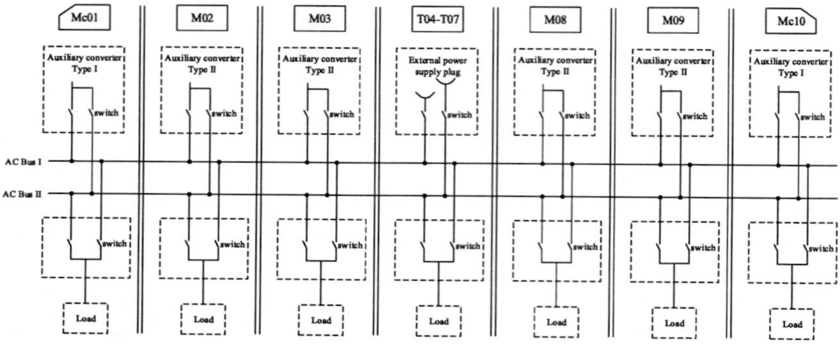


**Fig. 1.** Topology diagram of the flexible marshalling auxiliary power supply system

The flexible marshalling EMUs are mainly composed of Mc car (car number 1 and 10), Motors (car number 2, 3, 8 and 9) and trailers (car number 4–7). The auxiliary converters are installed in the Mc car and the Motor. There are two types of auxiliary converters, in which the Mc car is installed with the type I auxiliary converter, and the motor is installed with the type II auxiliary converter.

Under normal operating conditions, the auxiliary converter takes power from the DC link of the four-quadrant converter and transforms into AC380V/50Hz. When the vehicle is out of phase, the auxiliary converter converts the regenerative intermediate DC voltage of the traction converter into AC380V/50Hz, and forms two three-phase AC bus through two output contactors, namely AC bus I and AC bus II, so as to realize uninterrupted power supply to the three-phase AC bus of the vehicle. The system is three-phase four-wire system. The three-phase AC bus runs in parallel, and each phase and N line form three single-phase AC power supplies for single-phase load of the vehicle.

Considering the principle of optimal matching of circulation current and energy, the auxiliary converters of the whole vehicle are evenly divided into two groups according to the odd and even number of carriage numbers. The auxiliary converters with odd number of carriage numbers are only used for AC bus I, and the auxiliary converters with even number of carriage numbers are only used for AC bus II. The three-phase AC load is powered by the AC bus of I or II respectively, and the corresponding contactor is controlled by the train network according to the odd and even number of car numbers. When the auxiliary converter fails, the corresponding contactor is switched on or off according to the specific load of the vehicle, so as to maximize the load balance.

The auxiliary converter locates its own compartment number through the network, and in order to satisfy the free marshalling of vehicles and the free connection of the left and right ends, it can automatically select the AC bus that needs to be powered according to the compartment number. When the train turns 180°, the corresponding output contactor is closed by network control according to the head and the end of the vehicle.

The two three-phase AC bus of the whole vehicle is used to ensure the redundancy of the power supply. The auxiliary converters and AC loads are alternately hung to the two AC bus to ensure the load balance. The power supply mode and load distribution of the auxiliary power supply system is shown in Table 1.

**Table 1.** The power supply mode and load distribution of the auxiliary power supply system

| | | Mc01 | M02 | M03 | T04 | T05 | T06 | T07 | M08 | M09 | Mc10 |
|---|---|---|---|---|---|---|---|---|---|---|---|
| Auxiliary converter | | I | II | II | | | | | II | II | I |
| Load | AC Bus I | ● | | ● | | ● | | ● | | ● | |
| | AC Bus II | | ● | | ● | | ● | | ● | | ● |

## 3 Design of Parallel Auxiliary Converter with Different Capacity

Figure 2 shows the topology diagram of the auxiliary converter, which is mainly composed of pre-charging circuit, power module, LC filter, transformer, input/output contactors and other components. The main circuit devices The type I and type II auxiliary converters are the same, and the parameters of each device are different. The two types of auxiliary converters respectively take power from the DC link of traction converter and connected to the AC bus through the output contactors.

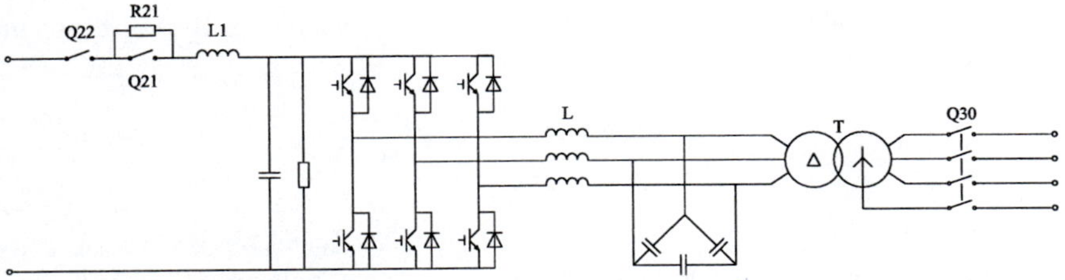


**Fig. 2.** Topology diagram of the auxiliary converter

### 3.1 Principle of Droop Control

Figure 3 shows the equivalent model of parallel inverters. $U_A$, $U_B$, $U_0$ are the output voltage of the auxiliary converter A, the output voltage of the auxiliary converter B, and the voltage of the AC parallel bus respectively. The sum of the output impedance and line impedance of the auxiliary converter is equivalent to $R_n + jX_n = Z_n \angle Z_n (n = A, B)$, where $\varphi$A and $\varphi$B are the output voltage phase of the auxiliary converter A and B respectively, and $\varphi_{Z_A}$ and $\varphi_{Z_B}$ are the phase of the output equivalent line impedance respectively [14].

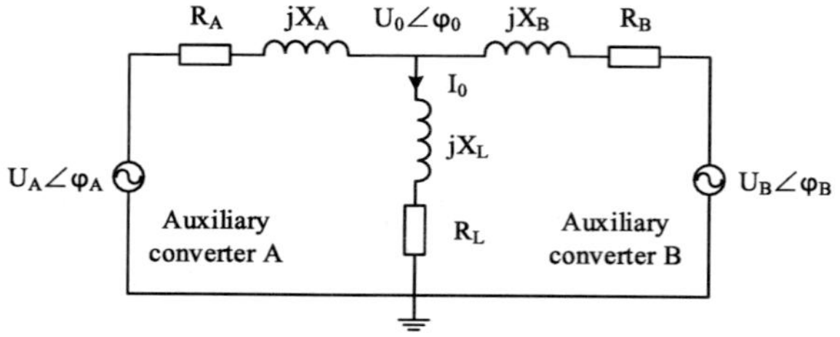


**Fig. 3.** The equivalent model of parallel inverters

According to the circuit theorem, the auxiliary converter output power can be expressed as:

$$\begin{cases} P_n = \left[U_n U_0 \cos\left(\varphi_{Z_n} - \varphi_n\right) - U_n^2 \cos\varphi_{Z_n}\right] Z_n^{-1} \\ Q_n = \left[U_n U_0 \sin\left(\varphi_{Z_n} - \varphi_n\right) - U_n^2 \sin\varphi_{Z_n}\right] Z_n^{-1} \end{cases} \quad (1)$$

Assuming that the line impedance is inductive, Eq. (1) can be simplified as:

$$\begin{cases} P_n = U_n U_0 \sin\varphi_n X_n^{-1} \approx U_0 \varphi_n X_n^{-1} \\ Q_n = \left[U_n U_0 \cos\varphi_n - U_n^2\right] X_n^{-1} \approx U_0 (U_0 - U_n) X_n^{-1} \end{cases} \quad (2)$$

It can be seen from Eq. (2) that the active power is mainly related to the phase angle difference of the output voltage of the inverter, while the reactive power is mainly related to the output voltage amplitude. Therefore, the output frequency can be changed by adjusting the active power, and the voltage phase angle difference can be controlled. The voltage amplitude is controlled by adjusting the reactive power. The corresponding droop control equation is shown in Eq. (3).

$$\begin{cases} \omega_n = \omega^* - m_n P_n \\ U_n = U^* - n_n Q_n \end{cases} \quad (3)$$

where, $\omega^*$ is no-load frequency, $U^*$ is no-load output voltage, $m_n$ and $n_n$ are active power/frequency (P-f) droop coefficient and reactive power/voltage (Q-U) droop coefficient respectively. By adjusting the frequency and amplitude of the output voltage of each inverter, it can be reduced to a new stable operating point, so as to achieve a reasonable distribution of output power. If the droop slope is equal, the output power of each inverter is equal after stability. If the droop slope is not equal, the bearing power of the large slope is small, and the bearing power of the small slope is large. Therefore, the power can be automatically adjustable through the artificial droop control.

### 3.2 Power Distribution Analysis

The control block diagram of P-f and Q-U can be obtained from Eqs. (2) and (3), as shown in Figs. 4 and 5.

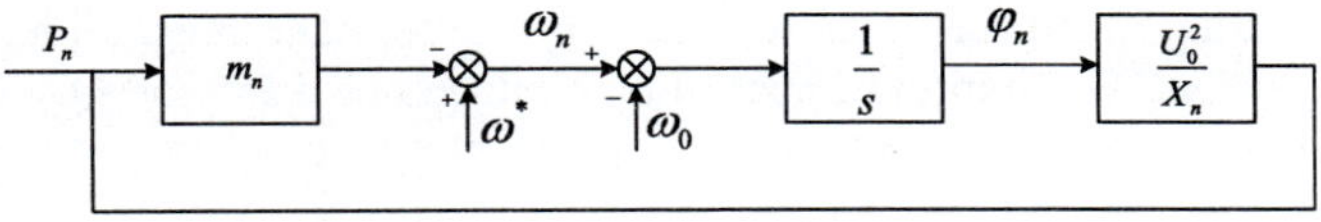


**Fig. 4.** Control block diagram of P-f

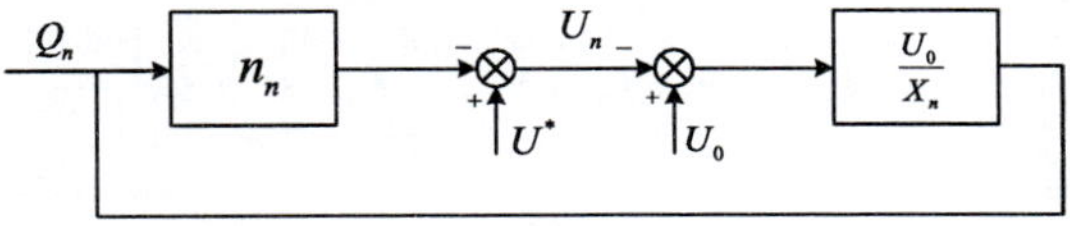


**Fig. 5.** Control block diagram of Q-U

As can be seen from Figs. 4 and 5, the transfer function of the active power and reactive power of the inverter is:

$$\begin{cases} P_n(s) = [\omega^*(s) - \omega_0(s)]\frac{1}{s}\frac{U_0^2}{X_n}\left(1 + \frac{1}{s}\frac{m_n U_0^2}{X_n}\right)^{-1} \\ Q_n(s) = (U_0 - U^*)\frac{U_0}{X_n}\left(1 + \frac{n_n U_0}{X_n}\right)^{-1} \end{cases} \tag{4}$$

By analyzing Eq. (4), it can be seen that due to the existence of integral link, the active power output of the auxiliary converter is not affected by the change of equivalent output impedance when it is in steady state. In other words, the active power output of the auxiliary converter when it is running in parallel can be distributed in proportion to its capacity through the droop control mechanism. The output reactive power is related to the equivalent connection impedance when the auxiliary converter is in steady state, and the gain of its transfer function in the entire frequency band includes the equivalent impedance. So the reactive power is subject to whether the equivalent line impedance is strictly proportional, and this proportion directly affects whether the output voltage is consistent, thus affecting the power distribution accuracy. In the auxiliary power supply system of EMUs, the equivalent line impedance is difficult to measure or calculate because of the complex AC bus connection, and the equivalent line impedance changes at any time in the flexible marshalling of EMUs. So it is difficult to achieve accurate reactive power allocation by changing the equivalent line impedance.

### 3.3 Adaptive Improved Droop Control Strategy

According to the analysis in Sect. 3.2, the transfer function of active power has an integral link, so that the active power output is not affected by the equivalent line impedance. Therefore, the integration link is also added in the transfer function of reactive power to eliminate the static error in the system and realize the proportional distribution of reactive power. At the same time, differential links are added to both active power and reactive power transfer functions to eliminate the influence on the dynamic response of the system because of the filtering algorithm in power calculation, droop coefficient, line impedance and other factors. Therefore, active power and reactive power can be adjusted according to capacity ratio by selecting appropriate integral and differential control coefficients. The improved control block diagram of P-f and Q-U is shown in Figs. 6 and 7.

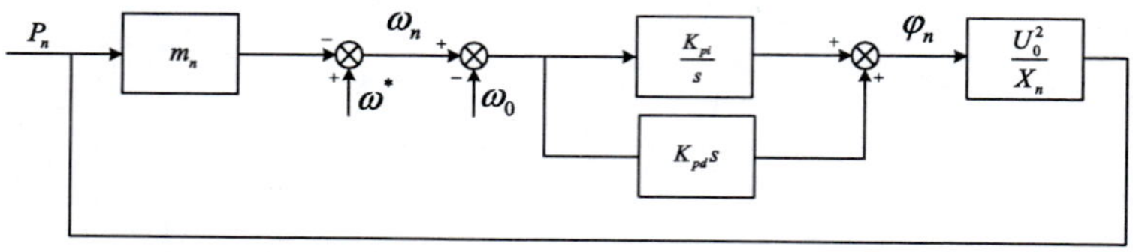


**Fig. 6.** Improved control block diagram of P-f

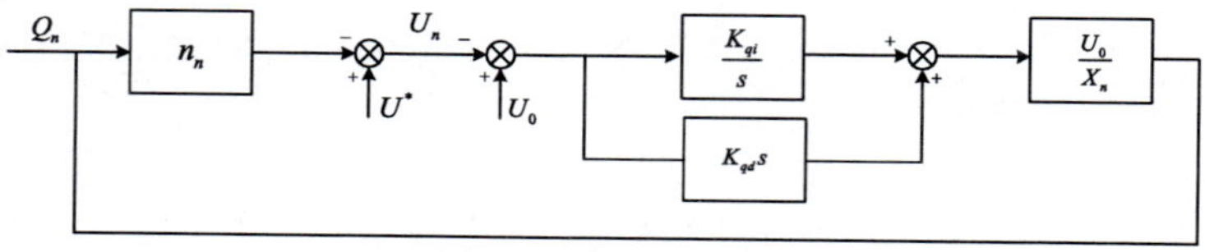


**Fig. 7.** Improved control block diagram of Q-U

As can be seen from Figs. 6 and 7, the transfer function of the improved active power and reactive power is:

$$\begin{cases} P_n(s) = \left[\omega^*(s) - \omega_0(s)\right]\left(\frac{K_{pi}}{s} + sK_{pd}\right)\frac{U_0^2}{X_n}\left(1 + \left(\frac{K_{pi}}{s} + sK_{pd}\right)\frac{m_n U_0^2}{X_n}\right)^{-1} \\ Q_n(s) = (U_0 - U^*)\left(\frac{K_{qi}}{s} + sK_{qd}\right)\frac{U_0}{X_n}\left(1 + \left(\frac{K_{qi}}{s} + sK_{qd}\right)\frac{n_n U_0}{X_n}\right)^{-1} \end{cases} \quad (5)$$

# 4 Simulation and Test Verification

## 4.1 Matlab Simulation

As shown in Fig. 8, this paper build a simulation model of different capacity auxiliary converters by using Matlab. Among them, the auxiliary converter A has a capacity of 230 kVA and the auxiliary converter B has a capacity of 100 kVA. By observing the power distribution of the two auxiliary converters under no-load, half-load and full-load conditions, the feasibility of the control strategy is verified.

As can be seen from the simulation waveforms in Figs. 9 and 10, the two auxiliary converters operate in parallel, and put in half-load and full load respectively at 2s and 3s. The two auxiliary converters adjust their own current output and gradually become stable, and both active and reactive power can be allocated in proportion to rated capacity after stabilization, which verifies the effectiveness of the proposed adaptive improved droop control strategy.

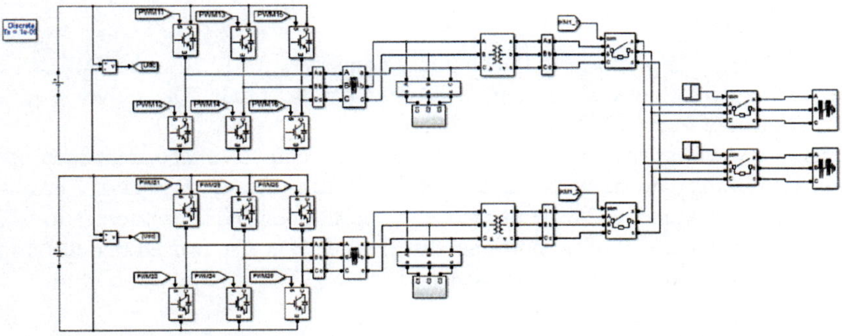

**Fig. 8.** Simulation model of different capacity auxiliary converters

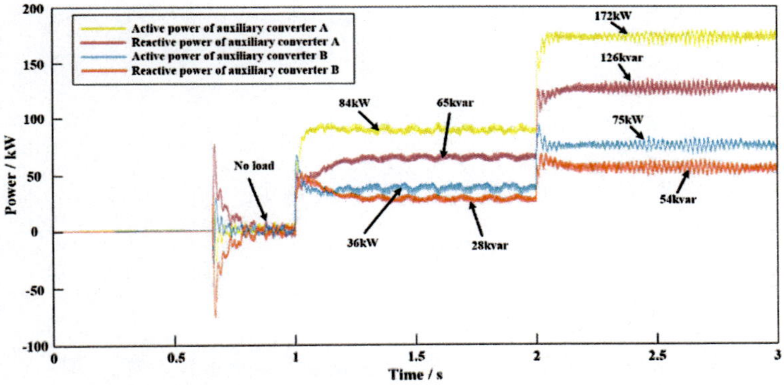


**Fig. 9.** Output power auxiliary converter

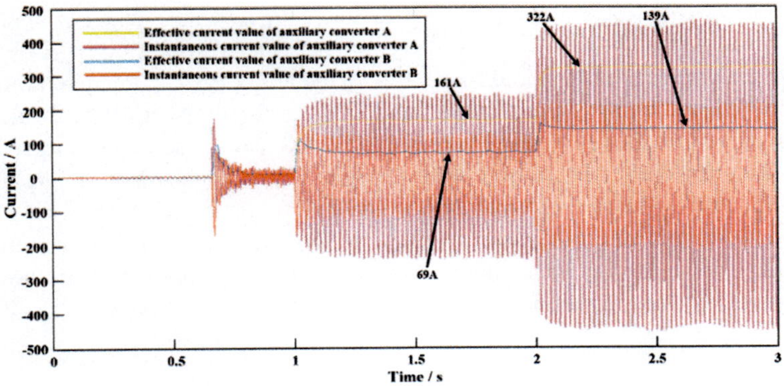


**Fig. 10.** Output current auxiliary converter

### 4.2 Test Verification

Based on the design platform of the auxiliary converters for EMUs, the 230 kVA and 100 kVA auxiliary converters are built and run in parallel mode. The active power and reactive power of load is 272 kW and 196 kvar.

The test waveform is shown in the Fig. 11. The output voltage frequency of the auxiliary converters are all 49.39 Hz, and the phase difference is very small. The output current is 353.7 A and 147.2 A respectively, and the power distribution ratio is 2.4, which achieves the effect of proportional distribution. It is also proved that the control algorithm satisfies the design requirements.

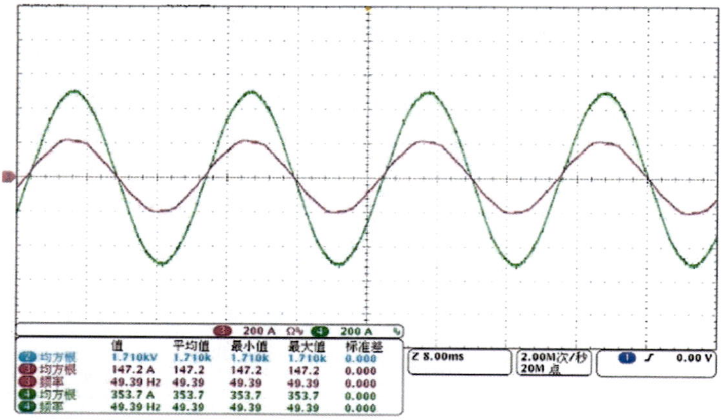


**Fig. 11.** Output waveform of auxiliary converters with different capacities

## 5 Conclusion

Based on the research background of flexible marshalling EMUs, this paper proposes an auxiliary power supply system based on flexible marshalling, in which the auxiliary converter and AC load are alternately connected to the two AC bus according to the car number, ensuring the redundancy of power supply and the balance of three-phase AC bus, and improving the flexibility of EMUs application. At the same time, an adaptive improved droop control strategy is proposed to realize the adaptive adjustment of active power and reactive power according to capacity ratio. Simulation and prototype test verify the effectiveness and reliability of the proposed control strategy.

## References

1. Gong, W., Li, B.: Analysis of the development of flexibly marshalling EMUs in China and foreign countries. Foreign Rolling Stock **52**(02), 34–37 (2015). (in Chinese)
2. Yu, Y., Han, B.: Strength of energy consumption cost for EMU based on flexible train formation. Railway Transp. Econ. **42**(S1), 144–149 (2020). (in Chinese)

3. Hu, R.: Application of flexible urban rail transit train formation technology. Urban Mass Transit. **25**(11), 143–147 (2022). (in Chinese)
4. Yang, S., Zhao, P., Kang, H., Zhang, Q.: Applicability analysis of variable configuration EMU train for high-speed railway. Railway Transp. Econ. **39**(11), 67–71 (2017). (in Chinese)
5. Zhao, C., Zhang, R.: Research on interconnection network technology of on-board PIS system based on flexible train formation. Modern Urban Transit. **10**, 18–21 (2022). (in Chinese)
6. Tang, Z., Jiang, X., Sun, X.: Topology design and application of auxiliary converter for EMU. Mach. Des. Manuf. Eng. **51**(02), 130–134 (2022). (in Chinese)
7. Zhang, B., Li, B., Zhao, J., He, X., Li, M., Sun, Q.: Research on method for grounding fault detection of AC auxiliary power supply system for multiple units. Rolling Stock **59**(3), 40–45 (2021). (in Chinese)
8. Wang, S.: Analysis on power supply mode of rail vehicle auxiliary convert. Urban Mass Transit. **22**(5), 44–46 (2019). (in Chinese)
9. Li, B., et al.: Research on network control strategy of parallel auxiliary converters of EMUs. Rolling Stock **61**(02), 30–36 (2023). (in Chinese)
10. Yan, Z., Li, X., Zhang, Q., Han, B.: Seat allocation model for high-speed railway passenger transportation based on flexible train composition. Comput. Ind. Eng. **142**, 1–12 (2020)
11. Dou, C., Zhang, Z., Yue, D., Song, M.: Improved droop control based on virtual impedance and virtual power source in low-voltage microgrid. IET Gener. Transm. Distrib. **11**(4), 1046–1054 (2017)
12. Rocabert, J., Luna, A., Blaabjerg, F.: Control of power converters in AC microgrids. IEEE Trans. Power Electron. **27**(11), 4734–4749 (2012)
13. Li, H., Yang, H., Zeng, Q, Yang, X.: Improved droop control strategy for islanded microgrid. Proc. CSU-EPSA **29**(4), 49–54 (2017). (in Chinese)
14. Hackl, C.M., Landerer, M.: Modified second-order generalized integrators with modified frequency locked loop for fast harmonics estimation of distorted single-phase signals. IEEE Trans. Power Electron. **35**(3), 3298–3309 (2020)

# Collaborative Optimization of Train Scheduling and Passenger Flow Control with Adjustable Dwell Time on a Congested Metro Line

Zhengjie Wu[1(✉)] and Kai Yang[2]

[1] School of Traffic and Transportation, Beijing Jiaotong University, Beijing 100044, China
21120925@bjtu.edu.cn
[2] School of Systems Science, Beijing Jiaotong University, Beijing 100044, China

**Abstract.** Metro lines in megacities often experience extreme congestion during peak hours, causing passengers to strand on platforms and pose serious operational risks. This paper formulates a mixed-integer nonlinear programming model (MINP) for the train scheduling and passenger flow control coordination problem with adjustable dwell time on overcrowded metro lines. It can offer a service-oriented schedule plan and implements the effective passenger flow control strategy to reduce platform congestion while maintaining transportation capacity. By employing a linearization method, this paper transformes the proposed model into a linear program that can be efficiently solved via Gurobi. Finally, numerical experiments show that the total waiting time of passengers is reduced by 40.8%, which can effectively reduce rail traffic congestion, and verify the effectiveness of the model.

**Keywords:** Passenger Flow Control · Train Timetabling · Rail Transit · Dynamic Demand

## 1 Introduction

In China, many urban metro systems have overcrowding and passenger congestion, which poses a severe danger to the operational safety and level of service [1]. A major problem in China's urban metro system is how to relieve the accumulation of people and shorten passenger wait times. First of all, based on the body of prior research, the current study concentrates on controlling passenger flow and scheduling trains to balance capacity and demand in order to reduce congestion. In terms of supply, literature mostly uses train schedule optimization to satisfy time-changing demand of passenger flow [2]. Demand-side strategies like controlling passenger flow and adopting time-sharing fares are primarily used to direct and regulate passenger movement [3].

For the train schedule optimization, in light of the significant number of passengers left stranded following a fault interruption, Gao et al. [4] developed an optimization model of the train operating diagram in stop-over mode and created an effective solution algorithm for the model. Taking the stations shared or adjacent by metro and intercity


Y. Qin et al. (Eds.): EITRT 2023, LNEE 1137, pp. 540–548, 2024.
https://doi.org/10.1007/978-981-99-9311-6_55

railway as the research object, Guo et al. [5] built a timetable model to optimize the departure time of two-way metro trains connecting intermediate stops and intercity stations under the assumption that the demand for passenger flow and the transport capacity of metro trains change over time. By carefully analyzing the running plan and schedule of urban rail transit trains, Cao et al. [6] constructed a nonlinear model to reduce waiting time of passengers on platform. They then created a genetic and harmonic algorithm to solve the model. Wang et al. [7] proposed a new deep neural network to dynamically arrange the train residence time to reduce the length of the metro voyage would increase the operating efficiency of the metro.

On the method of controlling passenger flow, to effectively monitor passenger flow and achieve accurate and reasonable control through coordinated connection among stations and allocation of inbound traffic quota, Qi et al. [8] adopted a digital method to establish a series of passenger flow control models to alleviate congestion and prevent the spread of viruses. In order to shorten the total wait times in a metro line, Meng et al. [9] examined the robust problem of controlling passenger flow under uncertain conditions. To solve the robust model, they developed a Lagrange relaxation technique. Wang [10] suggested an integer linear programming (ILP)-based optimization and coordination passenger flow control method to reduce the total waiting time of passengers at all stations during the morning peak hour. This was done to assure the safety of transfer stations. Zhang et al. [11] studied the passenger flow control problem to increase boarding passengers in light of the unknown destinations of inbound passengers. The number of passengers on all trains at every station was then dynamically calculated via a station-based dynamic restricted flow control issue that was proposed.

It is challenging to efficiently reduce rail traffic congestion with a single method of adjusting train diagrams or restricting passenger flow. Hence, we formulate an integer nonlinear programming model for the train scheduling and passenger flow control coordination problem to minimize the objective function includes passenger waiting time and train running time. Specifically, the integrated model is divided into two parts: the train diagram optimization problem where the dwell time is adjustable and the passenger flow limit problem. The coupling relationship between these two problems is stated by nonlinear constraints. In the end, a numerical example is conducted to verify the efficiency of the proposed approach.

## 2 Formulation and Linearization

The modeling technique is discussed in detail in this section. First, we list the formula's pertinent symbols and decision variables in Sect. 2.1. Section 2.2 explains the process of train operation and the behavior of passengers getting on and off the train, and a linear programming for controlling passenger flow and train diagram planning is also suggested.

### 2.1 Notations and Decision Variable

The decision variables and symbols for train diagram and passenger flow limit are defined to establish the model rigorously, as shown below.

**Set**:

$I$: trains set, $i \in I$.

$K$: metro stations set,$k \in K$.

$T$: discrete timestamps set, $t \in T$.

**Train timetable related parameters and notations**:

$t_{unit}$: the length of each timestamp.

$T^{r}_{i,k}$: the running time of train $i$ between $k$ and $k+1$.

$T^{h}_{\min}$: minimum headway between two concurrent in-service trains.

$T^{h}_{\max}$: maximum headway between two concurrent in-service trains.

$T^{dw}_{\min}$: the minimum dwell time that train $i$ stop at station $k$.

$T^{dw}_{\max}$: the maximum dwell time that train $i$ stop at station $k$.

$b_{i,k}(t)$: an indicator allowing passengers to enter the train $i$ at station $k$ with timestamp $t$.

**Passenger Flow control strategy related parameters and notations**:

$C$: maximum passenger capacity of the train.

$U$: a large positive number.

$\lambda_k(t)$: the number of people arriving at station $k$ at $t$ timestamp.

$\delta_{v,k}$: proportion of passengers whose OD is station $v$ to station $k$.

$P^{wait}_{k}(t)$: the number of people waiting when train $i$ arrives at station $k$.

$P^{in}_{i,k}$: the number of people board train when train $i$ arrives at station $k$.

$P^{on}_{i,k}$: the number of passengers on train $i$ when it runs between $k$ and $k+1$.

$P^{al}_{i,k}$: the number of people getting off when train $i$ arrives at station $k$.

**Decision variables**:

$t^{a}_{i,k}$: train $i$ arrive station $k$ at $t$ timestamp.

$t^{d}_{i,k}$: train $i$ leaves station $k$ at $t$ timestamp.

$T^{dw}_{i,k}$: the time that train $i$ dwell at station $k$.

$d_{i,k}(t)$: the variables that determines whether the train $i$ leaves the station $k$(binary variable).

$\eta_k(t)$: number of passengers allowed to use platform area of station $k$ with timestamp $t$.

## 2.2 Mathematical Model

This section describes the integrated train schedule model and the passenger flow limiting approach. We put forward some formulas which is solely applicable to schedule of train because train scheduling optimization is one of the key objectives of this work. The formula is displayed below. The discussion that follows is focused on creating systematic limitations on the train scheduling process.

We utilize binary variables $d_{i,k}(t)$ and $b_{i,k}(t)$ to represent whether train $i$ arrives at station $k$ and the loading process of train $i$ at station $k$ in addition to using variables $t_{i,k}^{a}$ and $t_{i,k}^{d}$ to describe the arrival and departure times of train $i$ at station $k$.

$$\min T_0 + \sum_{t \in T} \sum_{k \in K} P_k^{wait}(t) \cdot t_{unit} \tag{1}$$

$$s.t.$$

$$t_{i,k}^{d} - t_{i,k}^{a} = T_{i,k}^{dw}, \forall i \in I, k \in K \tag{2}$$

$$T_{\min}^{dw} \leq T_{i,k}^{dw} \leq T_{\max}^{dw}, \forall i \in I, k \in K \tag{3}$$

$$t_{i,k}^{a} - t_{i,k-1}^{d} = T_{i,k-1}^{r}, \forall i \in I, k \in K \backslash 1 \tag{4}$$

$$T_{i,k}^{h} = t_{i+1,k}^{d} - t_{i,k}^{d}, \forall i \in I \backslash |I|, k \in K \tag{5}$$

$$T_{\min}^{h} \leq T_{i,k}^{h} \leq T_{\max}^{h}, \forall i \in I \backslash |I|, k \in K \tag{6}$$

$$T_{i,k}^{h} = \sum_{t \in T} (d_{i,k}(t) - d_{i+1,k}(t)) \cdot t_{unit}, \forall i \in I \backslash |I|, k \in K \tag{7}$$

$$d_{i,k}(t) \leq d_{i+1,k}(t), \forall t \in T, i \in I, k \in K \tag{8}$$

$$b_{i,k}(t) = \begin{cases} 1 - d_{i,k}(t) & if \ \ i = 1 \\ d_{i-1,k}(t) - d_{i,k}(t) & else \ \ i \geq 2 \end{cases} \tag{9}$$

$$T_0 = \sum_{i \in I \backslash |I|} \sum_{k \in K} T_{i,k}^{h} + \sum_{i \in I} \sum_{k \in K} T_{i,k}^{dw} + \sum_{i \in I} \sum_{k \in K \backslash 1} T_{i,k-1}^{r} \tag{10}$$

$$d_{i,k}(t) \in \{0, 1\}, \forall t \in T, i \in I, k \in K \tag{11}$$

The constraints (1) objective function aims to reduce both operating time and total passenger waiting time. Constraints (2) provide the arrival times based on the interval running time $T_{i,k-1}^{r}$ and stop time $T_{i,k}^{dw}$ and the departure time is given by constraints (4). Constraints (3) represents the reasonable range of train stopping time. At the same time, the departure time $t_{1,k}^{a}$ of the first train is given in advance. The headway between two trains arriving at station $k$ in succession, $i$ and $i+1$, is specified by $T^h_{i,k}$ in constraints (5). The headway must be within a safe range, $T^h_{\min}$, $T^h_{\max}$, are guaranteed by constraints (6). We may determine the real departure time in relation to the departure interval by using constraints (7). In constraints (8), the variable $d_{i,k}(t)$ ensures the departure sequence of trains, and modes such as skipping and jumping stations are not considered here. The stopping time range of train $i$ can be obtained by constraints (9). Constraints (10) is used to calculate the total travel time of a rail line, which is expressed as the sum of train intervals at all stations, station residence times, and interval travel times.

The restrictions placed on the passenger flow control strategy are then listed as follows:

$$0 \le \eta_k(t) \le N_k^{\max}, \forall t \in T, k \in K \tag{12}$$

$$P_k^{wait}(t) = \sum_{\tau=1}^{t} \lambda_k(\tau) - \sum_{\tau=1}^{t} \eta_k(\tau), \forall t \in T, k \in K \tag{13}$$

$$P_k^{wait}(t) \ge 0, \forall t \in T, k \in K \tag{14}$$

$$P_{i,k}^{in} \le P_{\max}, \forall t \in T, k \in K \tag{15}$$

$$P_{i,k}^{in} = \sum_{t \in T} b_{i,k}(t) \cdot \eta_k(t), \forall t \in T, i \in I, k \in K \tag{16}$$

$$P_{i,k}^{on} = \sum_{r=1}^{k} P_{i,r}^{in} - \sum_{r=1}^{k} P_{i,r}^{al}, \forall i \in I, k \in K \tag{17}$$

$$P_{i,k}^{on} \le C, \forall i \in I, k \in K \tag{18}$$

$$P_{i,r}^{al} = \sum_{v=1}^{k} P_{i,v}^{in} \cdot \delta_{v,k}, \forall i \in I, k \in K \tag{19}$$

$$\sum_{t \in T} \lambda_k(t) = \sum_{i \in I} P_{i,k}^{in}, \forall k \in K \tag{20}$$

We establish a link between the train schedule and the measure of passenger flow limit by creating coupling constraints. The number of passengers who are permitted to visit the platform through the gate during the $t$ time stamp of the $k$ station is capped by constraints (12). Constraints (13) can be used to determine how many people are waiting for train $i$ at the station hall at timestamp $t$. The number of passengers waiting for passengers is determined by subtracting the total number of external arriving passengers from the total number of passengers serviced because all external arriving passengers must wait for boarding authorization. Constraints (14) ensures that passengers waiting for boarding permission in concourse is not zero. Constraints (15) Ensure that there are not too many people on board at any one time.

Additionally, we consider the passenger behavior, which includes the two different sorts of activities of getting on or off the train. The constraints (16), which links passenger demand and the train schedule closely, the $P_{i,k}^{in}$ is the number of passengers getting on the $i$ train. The number of passengers onboard are taken into account in constraints (17). All passengers must disembark by the time train $i$ arrives at the terminal, thus there are no passengers on board. Additionally, as indicated by constraints (18), the number of people in a vehicle cannot exceed the capacity of the railway. By adding the number of passengers from upstream stations to station $k$, constraints (19) calculates the number of passengers from train $i$ disembarking at station $k$. It is possible to guarantee that all passenger needs are satisfied within the taken-in time limit $T$ by applying the constraints (20).

Note that the existence of non-linear constraints can lead to difficulties in solving the model. The discussion that follows concentrates on linearizing these nonlinear restrictions in order to make the problem easier to solve.

The number of people entering the platform through the gate of station $k$ to wait for train $i$ in the $t$ timestamp is represented by the constant $\beta_{i,k}(t)$ for constraints (15). $\beta_{i,k}(t)$ satisfies the following requirements:

$$\beta_{i,k}(t) = b_{i,k}(t) \cdot \eta_k(t), \forall t \in T, i \in I, k \in K \tag{21}$$

The aforementioned constraints can be linearized to: since $b_{i,k}(t)$ is a 0–1 variable.

$$\begin{cases} 0 \leq \beta_{i,k}(t) \leq b_{i,k}(t) \cdot U \\ 0 \leq \eta_k(t) - \beta_{i,k}(t) \leq (1 - b_{i,k}(t)) \cdot U \end{cases}, \forall t \in T, i \in I, k \in K \tag{22}$$

We can infer that $\beta_{i,k}(t) = 0$ if $b_{i,k}(t) = 0$ is the case in this condition. If not, we can obtain $\beta_{i,k}(t) = \eta_k(t)$. The linearization constraints essentially guarantee that $\beta_{i,k}(t)$ takes the right value using this property. Then, the restriction (15) can be phrased as follows:

$$P_{i,k}^{in} = \sum_{t \in T} \beta_{i,k}(t), \forall i \in I, k \in K \tag{23}$$

## 3 Numerical Experiments

In this section, we showcase the performance of the proposed model on a numerical experiment. We solve the model using the Gurobi solver after specifying the necessary parameters on a Windows 10 personal computer with Intel Core i5-7200U CPU and 8G RAM. We consider a one-way metro line with stations A, B, C, and D. In the meantime, we assigned 10 service trains to the line, starting from station A and arriving at station D. For simplicity, assume that any two adjacent stations are the same distance apart. In these settings, it is assumed that the train crosses each section with a travel time of 180s. At the same time, all trains need to stop at each station for [30s, 60s]. We discretized the period into several $t = 30s$ periods at the same time and took into consideration 80 timestamps. At each station, it is expected that a single entrance facility can accommodate a maximum of 90 people at a time, and a maximum of 400 passengers can embark. The minimum and maximum headways are 120 s and 240 s, respectively, and each train can accommodate 500 passengers.

In the experiment, the total passenger demand of the A, B, C stations was 3100, 1500, 1500. Among them, A → B, A → C, A → D accounted for 50%, 30% and 20% respectively; B → C, B → D accounted for 35% and 65% respectively; C → D The number of passengers is 1500. Consider the 3,100 passengers arriving at station A as an example. They appear to have a normal distribution, as depicted in Fig. 1, with the peak hours occurring between [15, 25].

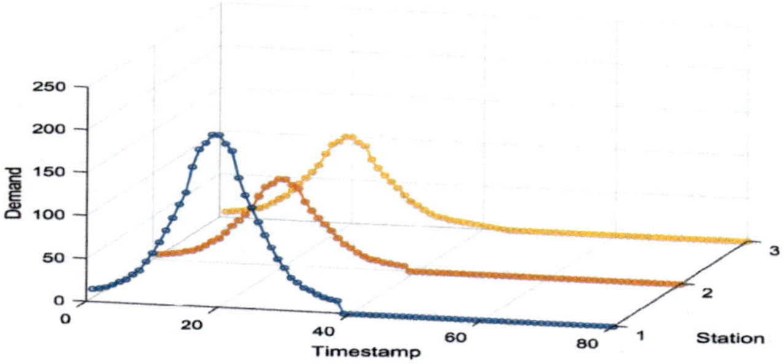


**Fig. 1.** Description of passenger demand at Station.

By contrasting the outcomes of using the proposed approach and not using the proposed method, we can determine the efficacy of the proposed method. The results are displayed in Fig. 2. The total goal value of the proposed method is calculated to be 619498.5s, where the train runs for 10320s and passengers wait for a total of 609178.5s. Passenger waiting time is reduced by 40.8% and train travel time is reduced by 4.2% compared to the results without the proposed method (passenger waiting time is 1029810s, trains running time is 10770s). Figure 2 displays the gathering of passengers at all stations at each timestamp. The blue line in Fig. 2a represents the train operating diagram, and the blue bars reflect the total number of passengers waiting on the platform. The passengers in each service train at various intervals are represented by the numbers in the rectangle. It is clear that stations A and C that without control are more crowded, whereas the proposed method keeps people inside the station hall which can prevent platform congestion (the blue bars). During peak hours and other times, it is visible that the headways of two adjacent trains vary significantly from one another. This fact demonstrates that demand-driven train moments can be accommodated by optimal train timetables that account for time-dependent passenger needs.

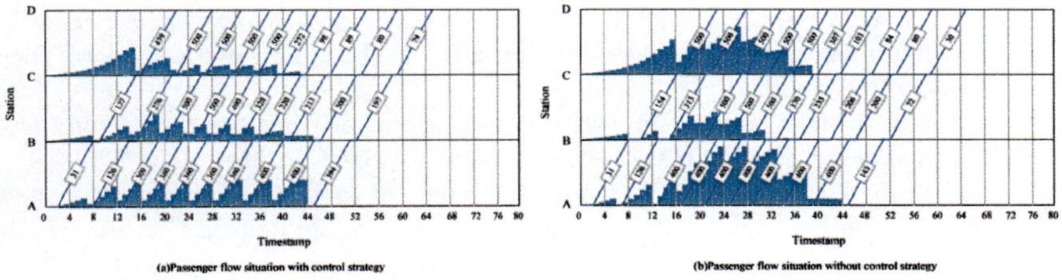


**Fig. 2.** Timetable and the number of people on the stations.

## 4 Conclusion

For overcrowded metro lines, this research examined the collaborative optimization of the schedule of trains with the passenger flow, where flexible train dwell times are taken into account. By discretizing the time, a mixed integer linear programming model is developed to mathematically represent the issue. A case study is used to analyze the suggested model's performance. The experimental results demonstrate that the suggested optimization model can avoid a buildup of passengers on the platform, and the optimized train schedule and flexible train stop time can effectively take into account the time-dependent passenger demand, allowing the metro system to run more effectively and safely.

**Acknowledgments.** This work was supported by the State Key Laboratory of Rail Traffic Control and Safety (No. RCS2022K001), Beijing Jiaotong University.

## References

1. Yao, X., Zhao, P., Qiao, K., Zhang, W.: Robust optimization model of peak passenger flow control scheme for urban rail transit lines. J. China Railway Soc. **43**(11), 29–37 (2021). (in Chinese)
2. Wang, J., Wang, Y., Peng, Y., Lu, J.: Examining partial proportional odds model in analyzing severity of high-speed railway accident. Smart Resilient Transp. **3**(1), 12–24 (2021)
3. Yuan, F., Sun, H., Kang, L., Wu, J.: Passenger flow control strategies for urban rail transit networks. Appl. Math. Model. **82**, 168–188 (2020)
4. Gao, Y., Kroon, L., Schmidt, M., Yang, L.: Rescheduling a metro line in an over-crowded situation after disruptions. Transp. Res. Part B: Methodol. **93**, 425–449 (2016)
5. Guo, H., Bai, Y., Hu, Q., Zhuang, H., Feng, X.: Optimization on metro timetable considering train capacity and passenger demand from intercity railways. Smart Resilient Transp. **3**(1), 66–77 (2021)
6. Cao, L., Cao, C.: Train planning and train timetable optimization of urban rail transit based on passenger flow demand. Shandong Sci. **33**(06), 87–95 (2020). (in Chinese)
7. Wang, Z., Pan, Z., Chen, S., et al.: Shortening passengers' travel time: a dynamic metro train scheduling approach using deep reinforcement learning. IEEE Trans. Knowl. Data Eng. **35**(5), 5282–5295 (2023)
8. Sun, Q., Sun, F., Liang, C., Yu, C., Zhang, Y.: Research on digital flow control model of urban rail transit under the situation of epidemic prevention and control. Smart Resilient Transp. **3**(1), 78–92 (2021)

9. Meng, F., Yang, L., Shi, J., Jiang, Z., Gao, Z.L.: Collaborative passenger flow control for oversaturated metro lines: a stochastic optimization method. Transportmetrica A: Transp. Sci. **18**(3), 619–658 (2021)
10. Wang, P.: An accurate collaborative programming model of passenger flow control for an oversaturated metro network. IEEE Access **9**, 141914–141926 (2021)
11. Zhang, P., Sun, H., Qu, Y., et al.: Model and algorithm of coordinated flow controlling with station-based constraints in a metro system. Transp. Res. Part E: Logistics Transp. Rev. **148**, 102274 (2021)

# Analysis of Risk Factors of Railway Trespassing Accidents Based on Multinomial Logit Model

Xu Wang[1,2], Xiaoping Ma[1,2,3(✉)], Hanqing Zhang[1,2], and Fei Chen[1,2]

[1] State Key Laboratory of Advanced Rail Autonomous Operation, Beijing Jiaotong University, Beijing 100044, China
xpma@bjtu.edu.cn

[2] School of Traffic and Transportation, Beijing Jiaotong University, Beijing 100044, China

[3] Key Laboratory of Railway Industry of Proactive Safety and Risk Control, Beijing Jiaotong University, Beijing 100044, China

**Abstract.** In order to solve the problem of imprecise alarm of foreign objects in railway encroachment, realize foreign object risk stratification and early warning. The main purpose of this paper is to analyze the trespasser factors, environmental factors and track factors associated with the severity of injuries in railway trespassing accidents from the perspective of trespasser, so as to identify features that rail monitoring needed, and propose optimization measures for accident prevention. This paper uses the Multinomial logit model to identify the factors and characteristics of trespasser injury severity based on Federal Railroad Administration accident record data. The pre-accident acts and location are the most important factors influencing the severity of the accident. Driving is the most common acts in trespassing incidents, while lying, standing and walking have a significant relationship with fatality outcomes. Combinations of some different behaviors and orbital locations can lead to polarization of accident risk with a higher risk of death, and it should be pay more attention when monitoring these features. This paper can be used to determine the relative level of risk for of each feature of illegal trespassing accidents and infer the severity of accident casualties, so as to construct an evaluation index system for foreign body characteristics. This paper helps to improve the detection and early warning accuracy of intrusion accidents by considering trespassers' acts, weather and track scene types in the framework of railway foreign object video detection, which has practical significance.

**Keywords:** Risk prevention · Railway trespassing Accidents · Multinomial Logit Model

## 1 Introduction

Railway construction is developing rapidly, and the mileage of railways is increasing, with the global high-speed rail network growing rapidly between 2000 and 2021, reaching around 58,800 km by 2021 [1]. However, railway operating scenarios are complex and diverse, and the safe operation of trains is threatened by natural disasters and foreign objects. Of these, trespassing is the most serious safety issue for the rail. According to


Y. Qin et al. (Eds.): EITRT 2023, LNEE 1137, pp. 549–560, 2024.
https://doi.org/10.1007/978-981-99-9311-6_56

FRA statistics, 94% of rail-related fatalities and injuries occur at grade crossings or as a result of trespassing [2]. With the increase of railway mileage, railways are inevitably intertwined with densely populated areas such as cities and towns, and the frequent service of trains and the movement of people increase the likelihood of collisions.

At present, in order to prevent the threat of foreign object intrusion and other emergencies to railway, each country has established railway intrusion detection system [3–5]. Although those system has played an important role in improving the safety of railway operations, it is judged only from the perspective of whether there is foreign object intrusion in the restricted area. It is difficult to prevent individual behaviors that causes casualties due to railway trespassing. It is essential to study the state of railway pedestrian features and formulate effective improvement measures and solutions.

In this regard, many scholars have studied the illegal trespass of pedestrians and the risk assessment of railway system. Based on the occurrence and countermeasures of the identified main factors affecting pedestrian intrusion, Zhang [6] comprehensively reviewed the existing research on railway intrusion risk management, and analyzed illegal intrusion behavior and characteristics, infrastructure, environment, traffic, and Influencing factors of intrusion behavior, near-miss risk analysis and active train control are proposed as the future research direction of railway intrusion. Stanchak [7] uses FRA trespass statistics to identify common characteristics of trespassers and locations where such acts occur to define risk factors that may lead to trespass incidents. Zhang [8] conducted a study of pedestrians who suffered varying degrees of injury in pedestrian trespass events, comparing the correlation between pedestrian pre-accident behaviors and collision time, location, social demographics, and other factors in railroad crossing and non-railroad crossing trespass events, and analyzed logistic regression model was established to fit various factors and injury severity. Zhang [9] proposed an artificial intelligence-based railroad trespassing detection framework, which automatically detects intrusion events through AI algorithms, distinguishes the types of violators, and sends alert text messages with important information including video clips of intrusion events to designated destinations. Railway risk assessment techniques are mainly fault tree, bow tie diagram, Monte Carlo simulation, etc. Due to the lack of reliable statistical data, these methods rely on expert knowledge and experience for estimation. Part of the study specifies some relevant risk management measures through statistical analysis of personnel factors, environmental factors, accident consequences, accident causes, and other factors of trespassing accidents on American railroads using the FRA database.

The above studies focus on the detection of trespassing events and collision frequency analysis, and are limited to post-accident risk evaluation. However, trespassers' behavior is random. It is to be noted that trespassing behavior does not necessarily cause accidents, but the consequences of train collision accidents caused by trespassing are often severe. By judging whether there are pedestrians in the track area, the railway monitoring system will over-alarm in the situation of the trespasser do not cause accident, and the effectiveness of the alarm will be reduced. Therefore, it is necessary to accurately assess the risk of trespassing accidents in real time. The level of risk is typically assessed using two parameters: accident frequency and consequence severity [10]. Accident frequency represents the number of accidents that occur within a specific time period, and consequence severity represents the number of deaths and serious injuries in accidents caused by specific causes [11]. In order to accurately assess the risk of trespassing events, it is first necessary to identify relevant risk factors of trespassing and design a reasonable comprehensive risk assessment and hierarchical early warning method.

The main research of this paper is to identify pedestrian factors, environment factors and track factors in trespassing accidents, and analyze the correlation between these factors and accident casualties through regression analysis, and infer the accident casualty probability of different combinations of factors, so as to clarify the cause of trespassing events. Focus on monitoring objects, formulate a hierarchical early warning mechanism for illegal intrusion events, and propose optimization measures for accident prevention and risk management. In this paper, these methods are used to analyze the event probability, accident consequence probability, and consequence severity of trespassing events:

1. A trespassing risk analysis framework based on foreign object characteristics is proposed. The influencing factors of trespassing accident are identified as event characteristics.
2. From the perspective of trespasser, rail environment and weather conditions, the model is constructed based on multivariate logit model.
3. Evaluate the probability of different characteristics causing different accident severity consequences, analyze the correlation between influencing factors, and quantify the relative risk.

The rest of this paper is organized as follows. Section 2 introduces the data preparation, Sect. 3 discusses the methodology used in this paper, Sect. 4 presents the results of the analysis, and Sect. 5 summarizes the conclusions and the application of this method.

## 2 Data Preparation

To analyze the relationship between pedestrian characteristics and other factors in the proposed trespassing risk analysis framework, we use the highway-rail grade crossing accident/incident data and railroad injury and illness summary data provided by the Federal Railroad Administration of the United States. These data contain the injury and illness records of the federal railroads in the United States.

The database contained rail trespass accidents from FRA over the period 2018 to 2022. We perform error check on the data and removes observations with missing or obviously invalid information from the final data Central exclusion for modeling purposes. The accident casualty data includes information such as the age of the trespasser and their actions before the accident; the highway-rail crossing casualty database includes environmental information such as the location, weather, and visibility at the time of the accident; and the cross-information of the two databases is merged using the unique identifier of the accident. A total of 1266 trespassers accidents were sampled and analyzed in this study.

# 3 Methodology

## 3.1 Trespassing Risk Analysis Framework

The risk of railroad trespassing accidents depends on factors such as the type of trespassing pedestrian, the behavior, and the environmental conditions. Different scenarios resulting from these factors can lead to railway trespass accidents. In railway trespassing accidents, pedestrians are more likely to involved in serious casualties.

In this paper, a framework is proposed to assess the degree of casualties of trespassing pedestrians as a calibrated risk. The framework is illustrated in Fig. 1. First, based on the FRA dataset, this paper analyzes track features, environmental feature, and trespasser features, then identifies the contributing factors and uses them as independent variables for regression analysis, and takes casualties as the dependent variable. Then, multinomial logit model is adopted to measure the correlation of contributing factors and casualties. By performing statistical analysis, regression estimation result analysis, and relative risk analysis on a set of trespasser factors, environmental factors, and track factors, the risk level of different trespasser scenario in causing train collisions can be predicted.

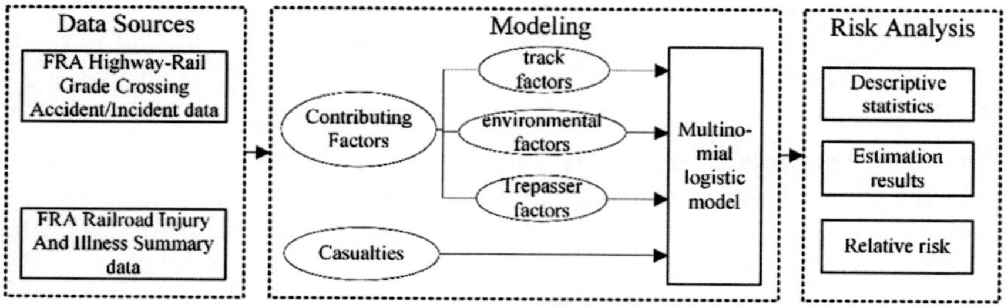


**Fig. 1.** The Framework of Trespassing Risk Analysis

## 3.2 Independent Variable: Contributing Factors

Based on the FRA accident database, the factors that affect the severity of trespassing accidents are divided into three categories: trespasser factors, environmental factors, and track factors. The Variable and descriptions of each risk factor are shown in Table 1. Trespasser factors include age and pre-accident acts. Through the descriptive statistical analysis of the data, 7 high-frequency acts were selected from 18 trespassers' pre-accident acts as key characteristics, including crossing, driving, laying, operating, riding, sitting, standing, and walking. The classification of environmental factors and track factors is based on the specifications in the FRA data file structure. Environmental factors include temperature, visibility, and weather conditions; track factors include the type of track and the location of the accident.

**Table 1.** Description of contributing factors

| Variable | Definition |
|---|---|
| Age | Age of person being reported |
| Phyact | Actions just before injury |
| Temp | Temperature in degrees Fahrenheit |
| Visibility | Visibility |
| Weather | Weather conditions |

### 3.3 Dependent Variable: Casualties

In this study, casualties' severity was modeled according to the KABCO scale [12], and the severity of casualties in trespassing is classified into three class, including "M: Minor Injured", "S: Serious Injured", and "K: Killed". Bruise or contusion, crush injury, sprain or strain, cut/laceration or abrasion and all other injuries appearing in the accident records are classified as minor injuries, dislocation, fracture (broken bone), amputation, concussion/closed head injury, internal injury is classified as serious injury.

The proportion of Casualties category is shown in Fig. 2. The frequency of Killed, Serious Injured, and minor injured were 505, 133, and 628, respectively, of the 1,266 reported incidents.

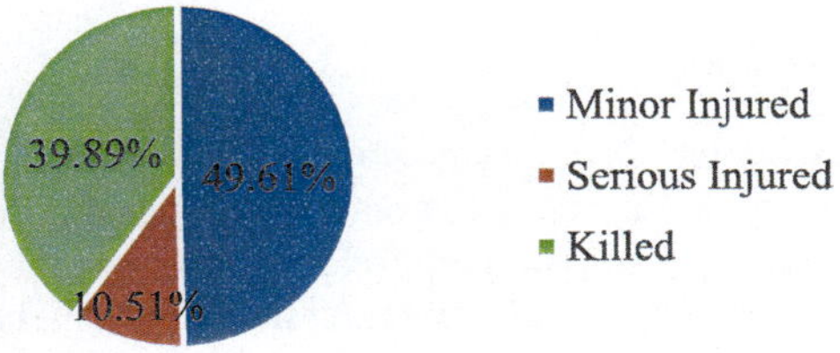


**Fig. 2.** Pie Chart of Casualties Proportion

### 3.4 Multinomial Logit Model

In order to identify the factors that had potentially contributed to trespassing accidents risk, a modeling framework called multinomial logit model [13] is adopted, which suit for multi-categorical dependent variables. This model has found success in the transportation field due to its ability to model each injury severity class as a categorical dependent variable with its unique utility function [14]. For this study, the multinomial logit model has the following function to determine the probability of casualties:

$$\ln \frac{P(Yi = m)}{P(Yi = 1)} = a_m + \sum_{k=1}^{k} \beta_{mk} X_{ik} = Z_{mi} \tag{1}$$

where Y is the category of accident consequences, the first category is the reference, then, for m = 2, …, M; *X* represents the vector of explanatory variables, including risk factors such as pre-accident acts and other; $\beta$ is a set of coefficients for *X*, $a_m$ is the intercept term.

The predicted probability of each category is calculated by the formula 2:

$$Pr(Yi = m) = \frac{\exp(Z_{mi})}{1 + \sum_{h=2}^{M} \exp(Z_{hi})} \tag{2}$$

# 4 Result

## 4.1 Descriptive Statistics Results

The descriptive statistics for each variable are shown in Table 2. According to the description of the pre-accident acts in the data, the pre-accident acts in the illegal entry accident is divided into 10 categories. In 44.31% of the accidents, the pre-accident acts of "driving" was found, and the proportion of "walking" and "riding" acts was 15.32% and 12.48%. According to Table X, intruders aged 20 to 39 have a higher percentage in the data than other groups. Among the environmental factors, the "day" with the best visibility has the same frequency of accidents as other scenarios with poor visibility. Among the types of roads, the frequency of trespassing accidents in main track is much higher than other.

## 4.2 Model Estimation Results

**Modeling the trespassing risk level.** By iteratively calculating and outputting the relative risk ratio, the effect of pre-accident acts on various Casualties can be quantitatively analyzed. Table 3 presents the results of the correlation analysis of the severity of accident consequences tested by the multivariate logit regression model, including the coefficients and the p-values, and minor injured was considered as the base outcome. The likelihood ratio statistic of the model is LR $\chi^2(25) = 282.503$, and the corresponding P value is less than 0.01, indicating that the model has a better overall fit and the model is effective.

**Analysis of Variance.** The age of trespassers was significantly correlated with the severity of accident consequences, and the coefficient of the age factor in the accident consequence of death was 0.014, indicating that with the increase of age, the severity of accidents may lead to an increase in the risk of death. Of all the pre-accident acts, lying, standing, and walking have a large impact on the fatalities of trespassers. Compared with minor injury accidents, these three acts increase the log odds of fatal accidents by 1.997, 1.381, and 1.557, respectively. For odds ratio, Fig. 3 shows the Comparison of relative risk values of pre-accident acts. Compared with crossing over, these three acts increase the risk of fatal crashes by 7.365, 3.979, and 4.745 times, respectively.

The correlation between serious injury and minor injury outcomes for trespassers in the track limit area is not significant, while the fatality outcomes for trespassers in the track limit area are significant.

**Table 2.** Descriptive Statistics of Contributing Factors

| Variable | Level of variable | Freq | Percent | Variable | Level of variable | Freq | Percent |
|---|---|---|---|---|---|---|---|
| Weather | Clear | 907 | 71.64 | Age | <10 yrs old | 34 | 2.69 |
| | Cloudy | 260 | 20.54 | | 10~19 yrs old | 100 | 7.9 |
| | Rain | 72 | 5.69 | | 20~29 yrs old | 230 | 18.17 |
| | Fog | 11 | 0.87 | | 30~39 yrs old | 237 | 18.72 |
| | Sleet | 2 | 0.16 | | 40~49 yrs old | 188 | 14.85 |
| | Snow | 14 | 1.11 | | 50~59 yrs old | 199 | 15.72 |
| phyact | Crossing over | 12 | 0.95 | | 60~69 yrs old | 145 | 11.45 |
| | Driving | 561 | 44.31 | | >70 yrs old | 133 | 10.51 |
| | Laying | 13 | 1.03 | Visibility | Dawn | 94 | 7.42 |
| | Operating | 96 | 7.58 | | Day | 626 | 49.45 |
| | Riding | 158 | 12.48 | | Dusk | 154 | 12.16 |
| | Sitting | 86 | 6.79 | | Dark | 392 | 30.96 |
| | Standing | 64 | 5.06 | Loca | Main | 1,248 | 98.58 |
| | Walking | 194 | 15.32 | | Yard | 7 | 0.55 |
| | Other | 82 | 6.48 | | Siding | 3 | 0.24 |
| Temp | <−4°F | 2 | 0.16 | | Industry | 8 | 0.63 |
| | −4~13°F | 18 | 1.42 | Locc | Track, beside | 7 | 0.55 |
| | 14~49°F | 310 | 24.49 | | Track, between | 11 | 0.87 |
| | 50~67°F | 316 | 24.96 | | Track, on | 49 | 3.87 |
| | 68~85°F | 478 | 37.76 | | In/operating vehicle | 47 | 3.71 |
| | 86~105°F | 141 | 11.14 | | On highway-rail crossing | 1,152 | 91 |
| | >105°F | 1 | 0.08 | | | | |

The log odds of trespasser that in "Track, between" having a fatal outcome versus a minor injury outcome increased by 3.4. As shown as in Fig. 4 is the comparison of relative risk values of location of the person whose injury. Trespasser are 29.187, 16.828, and 29.187 times respectively more likely to be killed at the location of "Track, on", "In/operating vehicle", and "Track, between", compared to the location of "Track, beside".

As for the environmental factors, visibility and temperature have no significant effect on the direct consequences of the accident. Additionally, the risk of death of the intruder in snow is 3.469 times higher than that of the intruder in clear weather.

Furthermore, we use the pre-accident acts as an explanatory variable and the environmental factors as an explanatory variable to analyze the correlation between the pre-accident acts and the weather factors in the railway scene. We found that there

**Table 3.** The result of multinomial logistic regression

| Natinj (killed) | Coef | St. err | p-value | Sig |
|---|---|---|---|---|
| Age | 0.014 | 0.004 | 0 | *** |
| PHYACT: base Crossing over | 0 | | | |
| Driving | −0.564 | 0.662 | – | |
| Laying | 1.997 | 1.057 | 0.059 | * |
| Operating | −0.536 | 0.697 | – | |
| Riding | 0.055 | 0.675 | – | |
| Sitting | −0.167 | 0.694 | – | |
| Standing | 1.381 | 0.722 | 0.056 | * |
| Walking | 1.557 | 0.679 | 0.022 | ** |
| Other | 1.191 | 0.718 | 0.097 | * |
| LOCA: base Main | 0 | | | |
| Yard | −15.72 | – | – | |
| Siding | 0.12 | 1.517 | – | |
| Industry | −0.842 | 1.105 | – | |
| LOCC: base Track, beside | 0 | | | |
| Track, between | 3.374 | 1.399 | 0.016 | ** |
| Track, on | 3.4 | 1.23 | 0.006 | *** |
| In/operating | 2.823 | 1.234 | 0.022 | ** |
| On highway-rail crossing | 2.933 | 1.179 | 0.013 | ** |
| Temp | 0.004 | 0.003 | – | |
| VISIBLTY: base Dawn | 0 | | | |
| Day | 0.011 | 0.264 | – | |
| Dusk | −0.25 | 0.313 | – | |
| Dark | 0.259 | 0.27 | – | |
| WEATHER: base Clear | 0 | | | |
| Cloudy | 0.171 | 0.167 | – | |
| Rain | −0.197 | 0.3 | – | |
| Fog | −0.096 | 0.743 | – | |
| Sleet | 0.675 | – | – | |
| Snow | 1.244 | 0.605 | 0.04 | ** |
| Constant | −4.159 | 1.393 | 0.003 | *** |
| Pseudo r-squared | | 0.117 | Number of obs | 1266 |
| Chi-square | | 282.503 | Prob > chi2 | 0.000 |

*** $p < 0.01$, ** $p < 0.05$, * $p < 0.1$

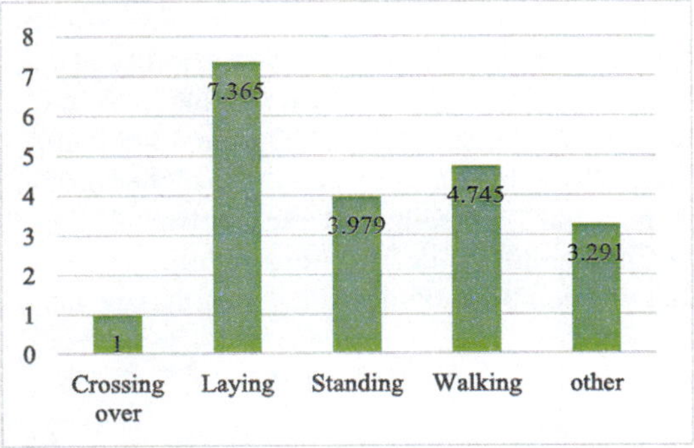


**Fig. 3.** Bar Chart of Relative Risk of Pre-Accident Acts

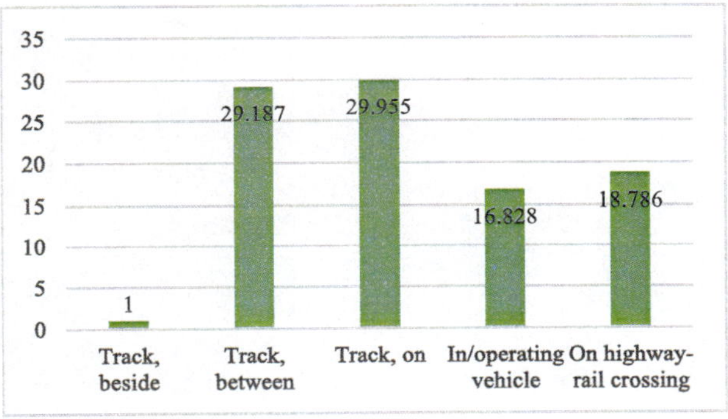


**Fig. 4.** Bar Chart of Relative Risk of Injured Location

is a correlation between operating behavior and visibility, and the coefficient is negative, which means that as visibility decreases, the probability of this behavior becomes lower. Riding is more likely to occur at dusk. Sitting is significant in cloudy and rainy environments. Lying behavior appears to be significantly dependent on temperature.

**Multi-factor Risk Prediction.** By making the probability of the pre-accident acts under specific location conditions, we can analyze the probability comparison of the two factor combinations for different accident outcomes. Based on the marginal effect analysis, if other independent variables remain unchanged, the change in a particular independent variable will induce a change in the probability of the accident. By setting the variable "locc" at five different positions, we could observe the predicted probability of minor injuries and fatalities corresponding to each behavior and draw the risk probability matrix. The probability distribution of minor injuries is shown in the Fig. 5, and the probability distribution of death is shown in the Fig. 6. The marginal effect results show that the risk of death of lying behavior at "track, on", "in/operating vehicle", "track, between", and "on highway-rail crossing" positions is greater than 80%.

If we compare the two Figs. 5 and 6, we can see the difference: If we combine the two risk levels of slight injury and death, the probability of minor injury is higher in the combination of "crossing", "driving", "operating", "riding", and "sitting" acts and the other positions except "track, beside". On the other hand, the combination of "lying", "standing", and "walking" acts resulted in a higher probability of death. The combination ("track, between", "standing") ("track, between", "walking") ("track, on", "lying") has a sum of probability of death and probability of minor injury of 1, and it can be concluded that this combination of behaviors polarizes the consequences of the accident.

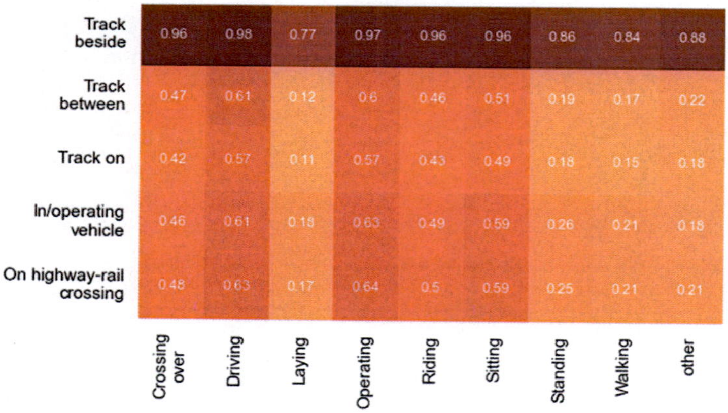


**Fig. 5.** Risk Matrix of Minor Injuries

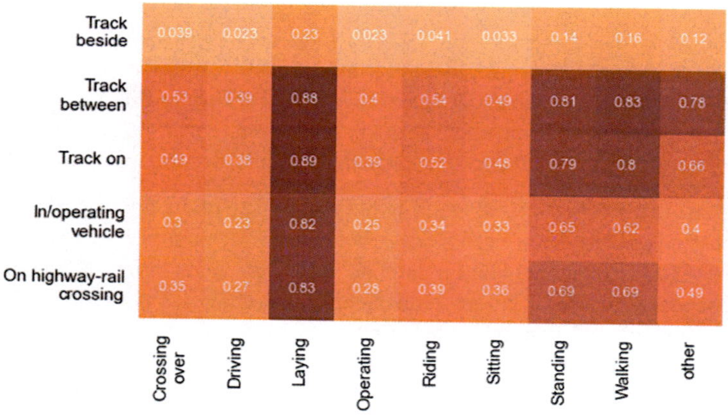


**Fig. 6.** Risk Matrix of Fatalities

### 4.3 Suggestion

By studying the risk of trespassing accidents, we can gain insight into the dynamic changes of various factors which contributing to trespassing on the changes of railway accident risks, which will help enhance current rail monitoring system, early warning and management efforts to prevent and control risks and reduce rail accidents:

1. Optimizing railroad monitoring system. The acts of the trespasserlying, standing, walking, etc.—and the location of the intruder in the railroad are the most critical factors that increase the risk of death, and these acts are easily identified automatically in the railroad monitoring system.
2. Optimizing Early Warning method. The traditional railroad monitoring system only determines whether there is a foreign object intrusion into the track area and triggers an alarm, obviously this method has a high rate of mistaken alarms. by predicting the casualty consequences of different combinations of factors, we can detect the instantaneous risk value of a specific weather, trespassing state and track condition. By determining key feature combinations, such as ("track, between", "walking") ("track, on", "lying"), etc., and dividing different risk thresholds, trespassing events that can cause serious casualties are assigned a higher alarm level to achieve more accurate and effective railway intrusion warning.
3. Improve management actions. When high-risk features are detected, more alarm measures can be taken, such as adding sound repelling devices and manual repelling in the field. For factors that are difficult for machines to detect automatically, such as age, limited visibility environments, etc., physical security measures around railways should be strengthened.

## 5 Conclusion

This study aims to solve the problem of trespassing risk analysis in the railway disaster prevention system, and a trespassing characteristic risk assessment framework is proposed to analyze the risk degree of various factors in accidents. Based on the FRA database, the risk factors related to trespassing were screened, and the modeling analysis was carried out according to these risk factor indicators and the severity of accident consequences. The results of the study show that trespassing crashes are preceded by many factors, and the combination of these factors determines the likelihood of different accident consequences.

This study reveals the correlation between the trespasser's pre-accident acts, environmental factors, and accident severity, which is useful in analyzing the risk of railway collision accidents. It can better use the existing railway detection data to extract intruder characteristics and environmental characteristics, and predict the risk of railway collision on caused by trespassers. And, it can assist railway managers to make judgments and take timely measures for railway accidents in advance, which has practical significance.

**Acknowledgements.** Supported by the Fundamental Research Funds for the Central Universities, No.2023JBZY008.

## References


1. Length of high-speed railroad network worldwide from 1964 to 2020. www.statista.com/statistics/1171776/line-kilometers-of-high-speed-railroad-worldwide. Last accessed 23 May 2023
2. Highway-Rail Grade Crossing Safety and Trespass Prevention. https://railroads.dot.gov/highway-rail-crossing-and-trespasser-programs/railroad-crossing-safety-trespass. Last accessed 30 June 2023
3. SureshKumar, M., Malar, G.P., Harinisha, N., Shanmugapriya, P.: Railway accident prevention using ultrasonic sensors. In: 2022 International Conference on Power, Energy, Control and Transmission Systems (ICPECTS), pp. 1–5. IEEE. (2022)
4. Jun, W., Jiangli, W.: Design of disaster prevention and safety monitoring system for high-speed railway. China Saf. Sci. J. **28**(S1), 39–45 (2018)
5. Arashika, T., Nakajima, S.: Enhancement of functions of Tokaido Shinkansen earthquake disaster prevention system. Japan Railway Transp. Rev. **43**(44), 64–68 (2006)
6. Zhang, Z., Casazza, A., Liu, X., Turla, T.: Railroad trespassing risk management: a literature review (2019)
7. Stanchak, K., daSilva, M.: Trespass event risk factors (No. DOT-VNTSC-FRA-14–03). United States. Federal Railroad Administration. Office of Research and Development (2014)
8. Zhang, M., Khattak, A.J., Liu, J., Clarke, D.: A comparative study of rail-pedestrian trespassing crash injury severity between highway-rail grade crossings and non-crossings. Accid. Anal. Prev.. Anal. Prev. **117**, 427–438 (2018)
9. Zhang, Z., Zaman, A., Xu, J., Liu, X.: Artificial intelligence-aided railroad trespassing detection and data analytics: methodology and a case study. Accid. Anal. Prev.. Anal. Prev. **168**, 106594 (2022)
10. An, M., Chen, Y., Baker, C.J.: A fuzzy reasoning and fuzzy-analytical hierarchy process-based approach to the process of railway risk information: a railway risk management system. Inf. Sci. **181**(18), 3946–3966 (2011)
11. Peduzzi, P., Dao, H., Herold, C., Mouton, F.: Assessing global exposure and vulnerability towards natural hazards: the disaster risk index. Nat. Hazard. **9**(4), 1149–1159 (2009)
12. KABCO Injury Classification Scale and Definitions. https://safety.fhwa.dot.gov/hsip/spm/conversion_tbl/pdfs/kabco_ctable_by_state.pdf. Last accessed 30 May 2023
13. Ying, S., Warren, F.K.: Multinomial logit models. In: SUGI 20 Conference Proceedings 1227–1234 (1995)
14. Okafor, S., Adanu, E.K., Jones, S.: Severity analysis of crashes involving in-state and out-of-state large truck drivers in Alabama: a random parameter multinomial logit model with heterogeneity in means and variances. Heliyon **8**(12), e11989 (2022)

# Design of Indoor Navigation Scheme Based on Bluetooth Low Energy

Zhaotian Wu[1,2,3], Xiaoping Ma[1,2,3](✉), Jiayin Li[1,2,3], Ruojin Wang[1,2,3], and Fei Chen[1,2,3]

[1] State Key Laboratory of Advanced Rail Autonomous Operation, Beijing, China
xpma@bjtu.edu.cn

[2] School of Traffic and Transportation, Beijing Jiaotong University, Beijing, China

[3] Key Laboratory of Railway Industry of Proactive Safety and Risk Control, Beijing, China

**Abstract.** Navigation based on mobile devices has developed rapidly in recent years, but GPS positioning technology cannot work effectively in indoor environments, so a variety of indoor wireless positioning technologies have emerged. Due to the low power consumption, convenience and cheapness of Bluetooth Low Energy (BLE) devices, Bluetooth wireless positioning has become one of the most commonly used solutions in current indoor positioning technology. In order to solve the problem of strong fluctuation of Bluetooth signal and large positioning error in the indoor navigation process, this paper uses Kalman filter to weaken the influence of random disturbance on the real Bluetooth signal, uses maximum likelihood to estimate the position coordinates of the locator, and makes full use of multiple beacon nodes to improve positioning accuracy. In order to verify the feasibility of the indoor navigation algorithm, an indoor navigation system is designed and deployed in the field. The experimental results show that the Root Mean Square Error (RMSE) of the method proposed in this paper is 2.336 m, and the RMSE in the X and Y directions is less than 3 m, which can meet the indoor positioning and navigation needs of ordinary users.

**Keywords:** Indoor navigation · System design · Bluetooth low energy · Kalman filter · Maximum likelihood method

## 1 Introduction

At present, the Global Positioning System (GPS) has long been widely used in the daily work and life of ordinary people, and it may be closely related and indispensable to people's daily life. However, due to the serious non-line-of-sight (NLOS) interference and the low penetration of GPS signals in scenes such as building buildings, underground parking lots, and tunnels [1], this technology cannot meet the needs of people. Growing demand for indoor location-based services. At the same time, with the continuous development and improvement of national modern facilities and construction, large shopping malls, office buildings and other buildings continue to increase, and people spend most of their time in indoor places such as offices, gyms, and libraries. The desire for location-based services is growing. Especially in response to emergency situations such as indoor


Y. Qin et al. (Eds.): EITRT 2023, LNEE 1137, pp. 561–569, 2024.
https://doi.org/10.1007/978-981-99-9311-6_57

fire rescue and evacuation of personnel in emergencies, the location information in the indoor environment is even more important [2] (Fig. 1).

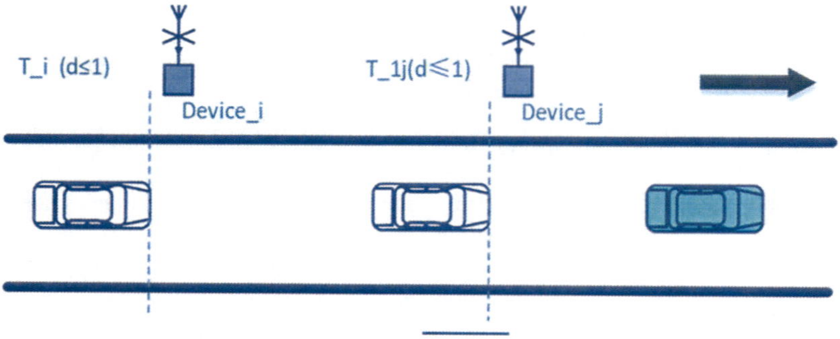


**Fig. 1.** Schematic diagram of Bluetooth beacon positioning solution

Today's indoor positioning technology and algorithms are becoming more and more diverse: RFID positioning technology has a wide transmission range and high positioning accuracy [3]. The disadvantages are that the radiation angle is large, the hardware investment is high, and the ranging range is small [4]. The accuracy of UWB positioning technology can reach cm level and the ranging range is high [5], but it needs to locate base stations, communication base stations, Devices such as wearable positioning tags are difficult to adapt to the parking background. IBeacon positioning technology is the latest short-range wireless communication technology based on the Bluetooth 4.0 protocol [6]. The advantages of iBeacon indoor positioning are easy deployment, low power consumption, and long distance, you only need a mobile phone with Bluetooth signal receiving function. The disadvantage is that it is difficult to achieve high accuracy due to signal stability and environmental factors. According to experience, when the deployment distance of modules is within 5 m or less, the accuracy is the highest. Weighing project functions, usage scenarios, software and hardware facilities, and technical advantages and disadvantages, the project team uses iBeacon as the hardware facility for indoor positioning technology.

## 2 Design of Indoor Navigation Method

### 2.1 RSSI Ranging Model and Noise Reduction Processing

In the actual environment, the iBeacon signal will continue to weaken on the propagation path. The farther the mobile phone receiver is from the Bluetooth beacon module, the weaker the Received Signal Strength Indication (RSSI) will be [7]. A large number of RSSI values at any position are collected, and the function form is obtained by function fitting:

$$RSSI = A * \ln(d) + n \tag{1}$$

The RSSI value is a time-varying signal, and there is a certain systematic error in the RSSI value of the low-power Bluetooth itself when the environment and the measurement distance are constant. At the same time, there is a certain observation error due to the RSSI value received by the mobile phone. In this project, a Kalman filter [8] is defined to denoise RSSI. The specific equation is:

(1) Prediction covariance equation: system estimation covariance at time k = system covariance at time k-1 + process noise covariance.
(2) Kalman gain equation: Kalman gain = covariance of system estimation at time k/(covariance of system estimation at time k + covariance of observation noise).
(3) Update the optimal value equation: the optimal value of the state variable at time k = predicted value of the state variable + Kalman gain * (measured value - predicted value of the state variable).
(4) Update the covariance equation: the current system covariance is passed to the next operation.

## 2.2 Indoor Positioning Algorithm Based on Maximum Likelihood Method

The positioning principle of the Maximum Likelihood Estimat (MLE) is: assuming that there is an unknown node to be measured and n known nodes in an area, when the calculated distance from the unknown node to all known nodes is known, the known node and the unknown node can be obtained [9].

Assume that the coordinates of n known nodes are $(x_1, y_1), (x_2, y_2), (x_3, y_3) \ldots (x_n, y_n)$, Their distances from unknown nodes $(x, y)$ are $d_1, d_2, d_3 \ldots d_n$, then there are the following equations:

$$\begin{cases} (x_1 - x)^2 + (y_1 - y)^2 = d_1^2 \\ \vdots \\ (x_n - x)^2 + (y_n - y)^2 = d_n^2 \end{cases} \tag{2}$$

Starting from the first equation, subtract the last equation in turn to obtain the equation system as AX = B. Among them, X, A, and B are as follows:

$$X = \begin{bmatrix} x \\ y \end{bmatrix} \tag{3}$$

$$A = \begin{bmatrix} 2(x_1 - x_n) & 2(y_1 - y_n) \\ \vdots & \vdots \\ 2(x_{n-1} - x_n) & 2(y_{n-1} - y_n) \end{bmatrix} \tag{4}$$

$$B = \begin{bmatrix} x_1^2 - x_n^2 + y_1^2 - y_n^2 - d_1^2 + d_n^2 \\ \vdots \\ x_{n-1}^2 - x_n^2 + y_{n-1}^2 - y_n^2 - d_{n-1}^2 + d_n^2 \end{bmatrix} \tag{5}$$

The standard least squares method in matrix form is used for estimation, and the coordinate solution of unknown nodes can be obtained as follows:

$$X = \left(A^T A\right)^{-1} A^T b \tag{6}$$

In this project, using the maximum likelihood estimation method for positioning has the following advantages [10]:

(1) Tolerance of measurement errors and high robustness;
(2) It is essentially a regression statistical solution, ideally able to achieve high measurement accuracy;
(3) This method has a general solution, which can be directly obtained through matrix operations, which is convenient for the realization of the program.

### 2.3 Floyd's Shortest Parking Path Algorithm Based on BRP Function

This paper hopes to be able to plan the optimal route for the user and realize the real-time navigation function when the user's destination is known. Considering the traffic conditions during the peak hours of urban roads (the time impedance is changing), we use the Bureau of Public Road road resistance function [11] to calculate the actual travel time of a certain road segment according to the relationship between the travel time and the traffic volume of the road segment. Amended, the specific formula is as follows:

$$t_i = t_{i0} \times \left(1 + \alpha\left(\frac{Q}{C}\right)^{\beta}\right) \tag{7}$$

Among them, $t_i$ is the time required to actually pass through the road section, $t_{i0}$ is the free travel time of the road section, Q is the traffic volume passing through the road section at that time, the unit is pcu/h, C is the actual traffic capacity of the road section, the unit is pcu/h, $\alpha/\beta$ is an undetermined parameter of the model, and the reference values are 0.15 and 4, respectively.

The Floyd algorithm is suitable for APSP (All Pairs Shortest Paths). The advantages of the algorithm: it can calculate the shortest distance between any two nodes and the code is simple to write [12]. The core idea is:

(1) There are two possibilities for the shortest path from node i to j:
directly from i to j;
From i through several nodes k to j.
(2) Use map(i, j) to represent the distance from node i to the shortest path j. For each node k, check that map(i, k) + map(k, j) is less than map(i, j) if map is established map(i, j) = map(i, k) + map(k, j). Traverse each k, and each update is the number except the rowk row and the column k.

## 3 Design of Indoor Navigation System

### 3.1 Demand Analysis

Based on BLE technology and WeChat applet development engine, this paper proposes a navigation system construction method in indoor scenarios. It mainly uses CAD as the base map for positioning and navigation in indoor scenes. After data preprocessing and format conversion, it is loaded into the WeChat applet platform to build and generate indoor navigation grids. Relying on the RSSI value of the signal transmitted by BLE Indoor navigation provides real-time indoor calculation coordinates, and provides shortest path planning and navigation services according to user destination (room-level) requirements. Finally, the navigation system will also conduct overall system testing and release for different application platforms (PC and mobile). The indoor navigation system construction framework combined with BLE technology and WeChat applet proposed in this paper has many characteristics such as easy development, system stability, interactivity and strong cross-platform portability, and provides a new set of framework for quickly realizing the construction of indoor navigation system. Solutions and ideas (Fig. 2).

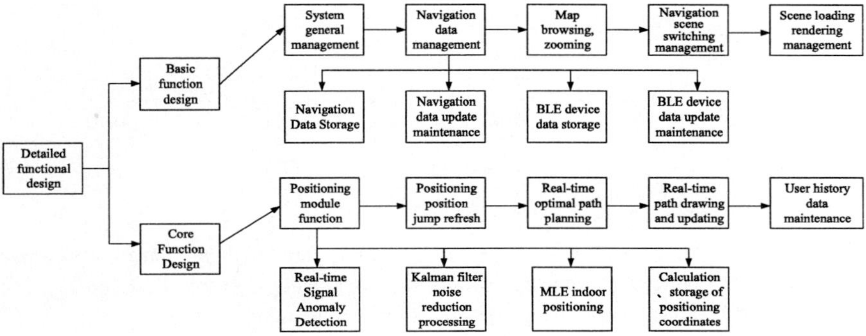


**Fig. 2.** Detailed design of system functional requirements

System functional requirements are as follows:

(1) Calculation of position information. Since this system adopts an RSSI-based indoor positioning algorithm and optimizes it, the first step is to collect the RSSI signal transmitted by the Bluetooth beacons deployed around it, upload it to the terminal software system in the format of MAC address and RSSI value, and calculate and predict it through the optimization algorithm coordinate.
(2) Real-time navigation service. The terminal software calculates its own coordinates in the building according to the received Bluetooth signal, updates the optimal path in real time according to the Floyd method, and provides navigation services for users with voice broadcast and trajectory dynamic update.
(3) Data statistics and management. The management terminal receives the real-time coordinates of the target, conducts data statistics on the location of the positioning terminal system, and automatically generates a graph of the terminal's action trajectory within a period of time.

## 3.2 System Development

In the research and development part of the system, first, based on indoor map information, BLE device information and other data, build a database that supports data writing, processing and storage. Then, realize the positioning and navigation algorithm module, realize indoor positioning based on the maximum likelihood method, and use The floyd algorithm calculates the shortest parking space path. Finally, develops the front-end navigation interactive interface, and develops the system management module that supports the operation and maintenance of the system.

The system developed in this paper mainly includes four layers, which are divided into basic layer, navigation data layer, support platform layer and application navigation layer. The specific contents are as follows:

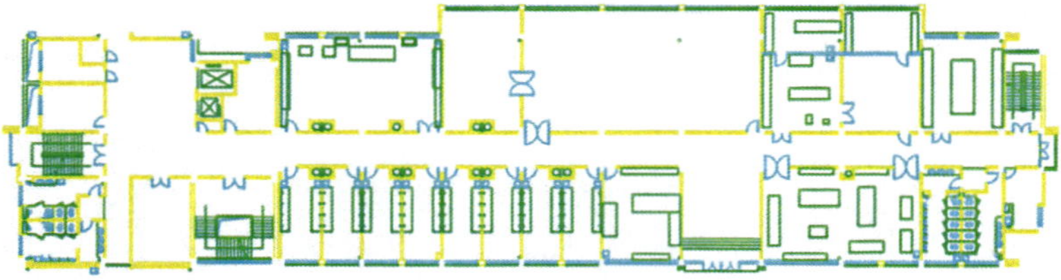

**Fig. 3.** Overall schematic diagram of the site

(1) Basic layer: The basic layer mainly includes user terminal navigation equipment, database, navigation system basic software operating system, BLE base station equipment and maintenance equipment deployed in indoor scenes.
(2) Navigation data layer: This layer mainly includes navigation base map data, coordinate data, BLE base station equipment parameter attributes and location coordinate data, user historical usage data, etc. required to provide navigation services for users.
(3) Support platform layer: In order to support the basic software platform for the smooth operation of the navigation system, this layer mainly includes WeChat applet development tools, all basic platforms involved in the system development and post-maintenance process.
(4) Application navigation layer: mainly includes real-time high-precision and high-stability positioning, quick refresh and jump of current coordinate position, real-time indoor navigation across floors in the scene, building model attribute query and other actual navigation functions required by users, etc.

### 3.3 Deployment Test

(1) Experimental software, hardware and scenarios

This paper selects the BLE (iBeacon) Smart Beacon base station produced by Zhishi Technology. This paper uses the "Bright Beacon" software, which can continuously scan and collect the MAC addresses and RSSI values of all BLE base station broadcast signals in the scene for subsequent positioning coordinates.

The experimental site is selected as the indoor corridor site on the second floor of the mechanical engineering building of Beijing Jiaotong University. The overall schematic diagram of the site is shown in Fig. 3.

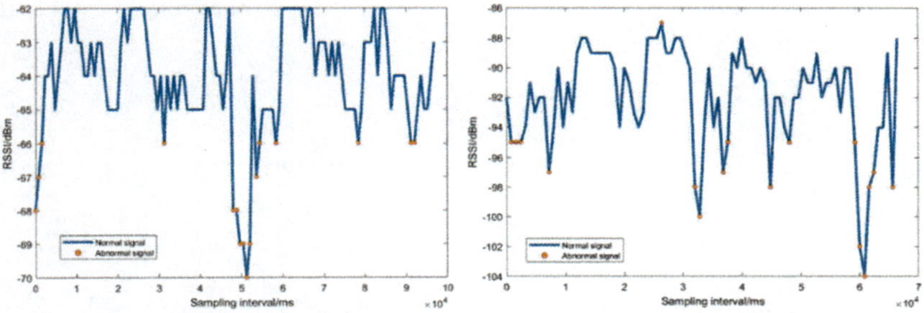


(a) 0.3 m anomaly detection (b) 9 m anomaly detection

**Fig. 4.** Abnormal detection results at distances of 0.3 m and 9 m from the signal source

(2) Ranging model fitting

First of all, in order to obtain the specific values of A and n in the trilateration ranging model, this paper takes 0.3 m as the step size, and conducts distance experiments from 0.3 m to 10.2 m. Use Kalman filter to process each signal strength value to obtain the RSSI value corresponding to the current distance. Then the ranging model is fitted under the optimization of the maximum likelihood method. The abnormal detection results at 0.3 m and 9 m away from the signal source are shown in Fig. 4, and the ranging model fitting process is shown in Fig. 5. It can be seen from Fig. 4 that the signal anomaly ratio at 9 m is higher than that at 0.3 m. After eliminating the detected anomalies on the RSSI value data of each fixed-distance point and the value, the model is fitted, and the A and n parameters in the ranging model are obtained by fitting: A = -71.58 dBm, n = 2.647.

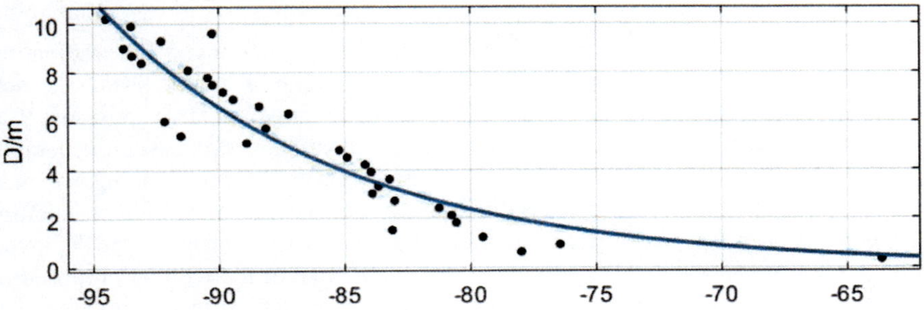


**Fig. 5.** Attenuation factor ranging model fitting

(3) Navigation test and error analysis

According to the methods mentioned in this paper, calculate the coordinates of 91 unknown points in the data collection path shown at intervals of 0.6 m. The final average plane coordinate error is 2.336 m, and the maximum plane error is 4.435 m, the maximum error and average error in the Y direction in the test site in this paper are both smaller than the error in the X direction. The comparison between the error and the reference experimental path is shown in Fig. 6.

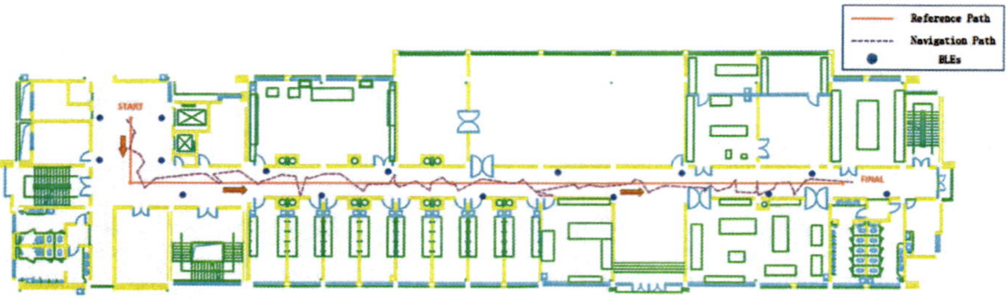

**Fig. 6.** Comparison of navigation path and reference experiment path

The RMSE of the method proposed in this paper is 2.336 m, and the RMSE in the X and Y directions is less than 3 m, which can meet the indoor positioning and navigation needs of daily general users. It has many significant advantages such as small amount of RSSI data required and quick solution. The method proposed in this paper has significantly improved the stability of positioning calculations. Among them, the errors of 91 positioning points in positioning calculations are all less than 3.5 m, 69.6% of the positioning points have an error of less than 3 m, and 79.56% of the positioning points can ensure that the error is less than 4 m. Its positioning accuracy and stability meet the needs of the application.

## 4 Conclusion

This paper proposes an appropriate solution for indoor navigation services, estimating the distance through the Bluetooth signal reception strength of mobile terminals such as mobile phones, and measuring multiple distances to solve the target position. The Kalman filter is used to reduce the impact of random disturbances on the real Bluetooth signal, and the maximum likelihood estimation method is used for positioning, which is tolerant of measurement errors and has high robustness. In order to verify the feasibility of the indoor navigation algorithm, an indoor navigation system is designed and deployed in the field. The experimental results show that the RMSE of the method proposed in this paper is 2.336 m, and the RMSE in the X and Y directions is less than 3m, which can meet the indoor positioning and navigation needs of ordinary users.

**Acknowledgment.** The present study is supported by the National Science and Technology Major Project Grant, No. 2021YFB2501001. For the writing of this paper, we thank the State Key Laboratory of Advanced Rail Autonomous Operation of Beijing Jiaotong University for their guidance and suggestions, and express our sincere gratitude to the teachers, students and staff of the research group who participated in this research.

## References


1. Yuan, Y., et al.: GPS multipath and NLOS mitigation for relative positioning in urban environments. Aerosp. Sci. Technol. **107** (2020)
2. Wong, M.O., Lee, S.: Indoor navigation and information sharing for collaborative fire emergency response with BIM and multi-user networking. Autom. Constr. **148** (2023)
3. Ji, H., et al.: Crowd Sensing: a crowd-sourcing based indoor navigation using RFID-based delay tolerant network. J. Netw. Comput. Appl. **52**, 79–89 (2015)
4. González, E., et al.: An approach based on the ifcOWL ontology to support indoor navigation. Egypt. Inf. J. **22**(1), 1–13 (2021)
5. Lin, H.-Y., Zhan, J.-R.: GNSS-denied UAV indoor navigation with UWB incorporated visual inertial odometry. Measurement **206** (2023)
6. Samuel, M., et al.: Smart indoor navigation and proximity advertising with android application using BLE technology Volume 43. Mater. Tdoay: Proc. **6**, 3799–3803 (2021)
7. Mehrabian, H., Ravanmehr, R.: Sensor fusion for indoor positioning system through improved RSSI and PDR methods. Futur. Gener. Comput. Syst. **138**, 254–269 (2023)
8. Chen, M., et al.: Multi-scene LoRa positioning algorithm based on Kalman filter and its implementation on NS3. Ad Hoc Netw. **141** (2023)
9. Shi, X., et al.: MLE-based localization and performance analysis in probabilistic LOS/NLOS environment. Neurocomputing **270**, 101–109 (2017)
10. Rajan Prabu, J., et al.: Structure of the RNA helicase MLE reveals the molecular mechanisms for uridine specificity and RNA-ATP coupling. Mol. Cell **60**(3), 487–499 (2015)
11. Pei, Q.-Y., et al.: Minimum collective dose based optimal evacuation path-planning method under nuclear accidents. Ann. Nucl. Energy **147** (2020)
12. Lyu, D., et al.: Robot path planning by leveraging the graph-encoded Floyd algorithm. Futur. Gener. Comput. Syst. **122**, 204–208 (2021)

# Research on the Optimal Coupling and Uncoupling Position Decision of Flexible Marshalling of Urban Rail Transit

O. U. Dongxiu[1,2,3,4](✉), Shaogang Dai[1,2], Xinyin Zhang[1,2], and Rei ZhG[1,5]

[1] School of Transportation Engineering, Tongji University, Shanghai 201804, China
{ou.dongxiu,reizhg}@tongji.edu.cn, dsg2002ying@163.com, zxy1998cc@163.com

[2] The Key Laboratory of Road and Traffic Engineering, Ministry of Education, Shanghai 201804, China

[3] The Key Laboratory of Railway Industry of Proactive Safety and Risk Control, Shanghai 201804, China

[4] Shanghai Railway Dispatch Research Center, Shanghai 202163, China

[5] Shanghai Zhangjiang Comprehensive National Science Center, Shanghai 201210, China

**Abstract.** The expanding scale of the rail network presents vital challenges to urban transportation operation. The demand for energy savings and improved service levels is increasing. Flexible marshalling on spatial lines can achieve a win-win situation for both operation level and passenger service. In this paper, we construct a comprehensive objective model for selecting the optimal coupling and uncoupling positions of trains under imbalanced spatial passenger flow. The model minimizes both the traction energy cost and passenger congestion penalty cost with a genetic algorithm to determine the optimal coupling and uncoupling position of trains under different operating frequencies, focusing on the entire section of the line, considering constraints such as line conditions and capacity, decision variables as the marshalling position. The optimal solution can reduce the traction energy cost by 20.08% to 27.13% and the passenger penalty congestion cost by 14.97% to 34.99%.

**Keywords:** Urban rail transit · Flexible marshalling · Coupling and uncoupling position · Multi-objective optimization

## 1 Introduction

Faced with the contradiction between the increasing demand for passenger flow and capacity supply, and the balance between operational costs and passenger service, it is necessary to make decisions on the position of coupling and uncoupling based on flexible marshalling and the spatial passenger flow. Under the "peak carbon dioxide emissions" and "carbon neutrality" policies, a low-carbon urban operation model is the primary development direction. Reducing energy consumption is the direction for energy-saving in urban rail transit. So utilizing flexible marshalling to improve the dynamic matching


Y. Qin et al. (Eds.): EITRT 2023, LNEE 1137, pp. 570–590, 2024.
https://doi.org/10.1007/978-981-99-9311-6_58

of demand on spatial lines, reduce train traction energy consumption, and optimize passenger comfort is of significance for the sustainable development of urban rail transit operations.

### 1.1 The State-of-the-Art on the Position in Flexible Marshalling

The Dutch subway and CRH3X made in China allow for flexible marshalling with multiple train types. Bi concluded that flexible marshalling is suitable for urban lines with imbalanced passenger distribution [1]. Ji considered virtual marshalling based on flexible marshalling could respond to transportation plan requirements and system safety constraints efficiently [2, 3]. The trends and applications show the technical feasibility, wide applicability and the foundation for the development of virtual marshalling.

The key to marshalling plan lies in passenger flow characteristics. Ma studied the temporal and spatial distribution characteristics of passenger flow, peak-hour maximum section passenger flow, and average passenger travel distance [4]. It shows that analyzing the spatial distribution coefficients of passenger flow plays a vital role in the construction of flexible marshalling models.

Many scholars conducted flexible marshalling plan based on the requirements of station sequencing and turnaround locations. Cadarso developed a mixed-integer programming model considering passenger and traction costs to design different marshalling scenarios for turnaround operations [5, 6]. Alejandro employed a genetic algorithm considering consumption factors to determine the location of turnaround stations [7]. Xu proposed a strategy of large-group train for major routes and small-group train for minor routes, considering the imbalanced spatial of passenger flow. The strategy aims to minimize the cost of service and operation, determining departure frequencies and turnaround locations [8].

Although flexible marshalling is becoming mature, current research focuses on designing turnaround locations and route constraints, constructing multi-objective models based on service level and energy consumption to determine turnaround stations, using genetic algorithms. There is still a lack of comprehensive methods to determine the optimal positions for flexible marshalling. This paper aims to combine flexible marshalling with routing scheme to design train marshalling plans.

### 1.2 The Contributions of This Paper

This article focuses on a subway line that based on peak-hour passenger flow and route data. The constraints are line conditions, train transportation capacity. The decision variables are the positions of train coupling and uncoupling positions. The objective is to minimize the weighted sum of traction energy costs and crowding penalty costs. A decision model is constructed to determine the optimal marshalling positions. The effectiveness of the model is then verified through case studies.

The remaining sections are as follows. In the second part present the optimal position decision model. The third part discusses the practical case and simulation. Finally, we summarize the conclusions.

## 2 Model

### 2.1 Problem Description

The spatial distribution of passenger flows on subway is often highly imbalanced. The suburb edge sections experience lower passenger flows relatively, while the central sections experience higher. If only large-group train are operated throughout the day, there will be an issue of excess capacity in suburb edge sections. To meet the passenger demand, large and small routing can be implemented. However, it introduces complexities in train operations and may inconvenience passengers during boarding and alighting. Flexible marshalling can adapt to different scenarios. It involves operating coupled trains in sections with high passenger flow and uncoupled trains in sections with lower passenger flow, which optimizes passenger comfort and traction power consumption. It is necessary to explore the optimal positions for coupling and uncoupling, considering various scenarios.

Figure 1 illustrates the operation of flexible marshalling during peak hours. The operations are divided into segmented sections, where trains couple and decouple flexibly at optimal points. The small-group train operate in the suburban edge sections, coupling with another unit train at the grouping station. In the city center sections, the coupled train operate to accommodate high passenger loads. At the uncoupling station, passengers from the coupled train transfer to the unit train, and the coupled train is decoupled into the small-group train. The decoupled unit train continue to operate in the suburban edge sections, serving passengers in those areas. The other empty unit train, after uncoupling, reverses direction via a turnaround track and couples with the unit train in the opposite direction. At this point, the passenger crowding level and traction energy consumption for different sections of the line are reasonably balanced, achieving optimal passenger service level and train energy consumption in each section.

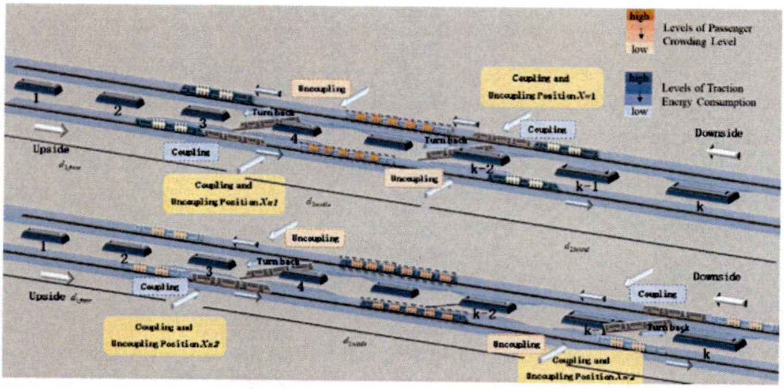


**Fig. 1.** Optimal position decision scenario for flexible coupling and uncoupling.

Improving passenger comfort and reducing waiting time are the passenger demands, while reducing train operating energy consumption is the enterprise demand. However, under the same operating time to facilitate organization and operational efficiency, we assume the frequencies of different segmented routing of the line with different-sized trains are the same. Hence, the optimization objective is to minimize the combined levels of congestion penalty and energy consumption. To determine the optimal positions, we should build constraints such as capacity and station reversing tracks. A decision model for determining the optimal marshalling positions of trains is constructed.

## 2.2 Model Formulation

The railway line is divided into separate operating sections, and the entire operating time is divided into independent time, typically on an hourly basis. The entire operating period is divided into peak hours and off-peak hours. During off-peak hours, the line operates with small-group train, while during peak hours, basic units are grouped and coupled trains are decoupled to accommodate the passenger flow within the same section. Two coupling and uncoupling positions on the line are designated as $x_u$ and $x_w$. For a metro line with a total length of $L$ and stations of $K$, we mark station 1 as the starting point and station $K$ as the terminal station. Trains operate in the direction from station 1 to station $K$, which is considered the upside direction (denoted as $m = 1$) while the opposite direction is referred to as the downside direction (denoted as $m = 2$). The set of stations is denoted as $I, I = \{1, 2, \ldots, K\}$, and the set of station intervals is denoted as $D$, $D = \{d_1, d_2, \ldots, d_K\}$. The passenger flow on the section $[i, i + 1]$ within peak hour in the $m$ direction is represented as $Q^m_{i,i+1}$, the train operating frequency as $h$, and the number of train group operating on section $[i, i + 1]$ in the $m$ direction as $X^m_{i,i+1}$. The rail transit line is represented by a schematic diagram in Fig. 2.

The operational scenario of flexible marshalling is as follows: in the upside direction, the small-group train departs from station 1 and travels to the optimal grouping position $u$, where it couples with an empty small-group train and operates as a coupled train in the central section. At the optimal grouping station $w$, the coupled train is decoupled, and the loaded small-group train continues to the terminal station while the other empty small-group train travels through the reversing line in the downside direction and couples with a loaded small-group train departing from station $K$. Then, it continues to run until arriving at the grouping position $u$ where the train is decoupled. The loaded small-group train continues to the terminal station 1, while the empty train travels through the reversing line in the upside direction and couples with a loaded train departing from station 1 to proceed. The flexible marshalling operation mode is repeated. To facilitate the organization of train operations and improve efficiency, it is assumed that the train operating frequencies are the same for each section of the segmented routing. In other words, both large-group and small-group train operate at a uniform frequency of $h$. Large-group train operate in the central section uniformly, while small-group train operate uniformly in low flow sections. Therefore, the input for the model includes the passenger flow on each section during the peak hour $\{Q^m_{i,i+1}, d^m_{i,i+1}\}$, while the output includes the optimal coupling and uncoupling position $\{x_u, x_w\}$.

**Assumptions**

There are many factors affecting the train operation plan, assumptions are needed to simplify the train operation.

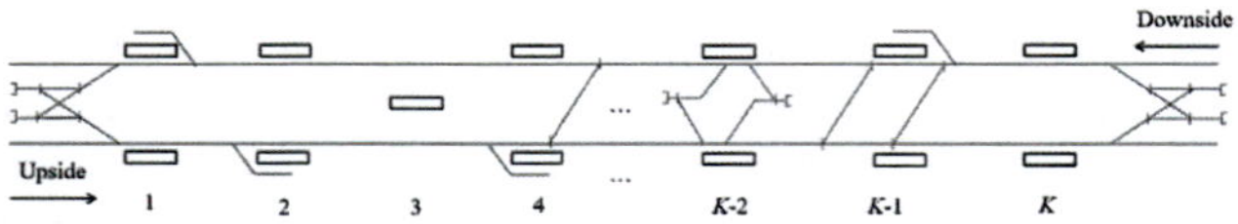


**Fig. 2.** Schematic diagram of flexible marshalling operation mode in the rail line.

(1) The peak hours of operation are selected as the study time. Large-group trains are used for high-demand passenger sections, while small-group trains are used for low-demand sections.
(2) Small-group trains are coupled to form large-group trains at the optimal coupling station and decoupled to operate as small-group trains at the optimal uncoupling station. The model does not consider the time and cost associated with train coupling, uncoupling operations, and passenger disembarking during the operational period.
(3) The capacity and maximum passenger load of the train are determined and fixed.
(4) The trains operate in a bidirectional symmetrical mode, and trains with the same basic group have the same travel speed in both directions.
(5) The train stop at every station, and express trains are not operated.
(6) The segmented routing scheme is employed, where all stations have the same turnaround time, combined with flexible marshalling for train operation.
(7) Passengers arrive at the stations uniformly, board the trains on a first-come, first-served basis, and do not have the option to choose specific train schedules. Also, passenger retention is not considered.

For a better understanding of this paper, the relevant variables and parameters are listing in Tables 1 and 2.

**Model Basics**

The multi-objective optimization model aims to optimize both passenger service level and operational efficiency. It considers the calculation and analysis of passenger crowding penalty cost and train traction energy consumption cost. The following sections will provide a detailed discussion on the optimization objectives.

$$Z_1 = \sum_{m=1}^{2}\sum_{i=1}^{K} Q_{i,i+1}^{m}\cdot\sigma_{i,i+1} = \sum \mu_{crowd} = \sum_{i=1}^{u}\mu_{crowd\ i} + \sum_{i=u}^{w}\mu_{crowd\ i} + \sum_{i=w}^{K+1}\mu_{crowd\ i}$$
$$= \sum_{m=1}^{2}\sum_{i=1}^{u} Q_{i,i+1}^{m}\cdot\sigma_{i,i+1} + \sum_{m=1}^{2}\sum_{i=u}^{w} Q_{i,i+1}^{m}\cdot\sigma_{i,i+1} + \sum_{m=1}^{2}\sum_{i=w}^{K+1} Q_{i,i+1}^{m}\cdot\sigma_{i,i+1} \quad (1)$$

Formula 1 represents the crowding level of the train, which reflects the passenger service level and indirectly indicates the impact of passenger travel quality on the marshalling strategy. This value represents the penalty value for the average standing area per passenger in the train, influenced by the passenger flow $Q$ and the unit crowding penalty price $\sigma$, as shown in Formula 1 $\sum_{m=1}^{2}\sum_{i=1}^{K} Q_{i,i+1}^{m}\cdot\sigma_{i,i+1}$. A smaller average standing area per passenger indicates a more crowded train, lower service level, and higher penalty cost [9]. However, a large standing area per passenger indicating a low passenger load

**Table 1.** The relevant variables and parameters used in the model.

| Variable | Definition |
|---|---|
| $Q_{i,i+1}$ | The passenger flow in the section [i, i + 1] |
| $X_{i,i+1}$ | The number of train group in the section [i, i + 1] |
| $C$ | The capacity of a train carriage |
| $h$ | The hourly train frequency |
| $\theta$ | The traction energy consumption coefficient |
| $E$ | The traction energy consumption |
| $n$ | Train number |
| $a, b, c$ | Train resistance coefficient |
| $v$ | Train travel speed |
| $m_{train}$ | Unit carriage mass |
| $m$ | Unit passenger mass |
| $i$ | The number of station |
| $F_0$ | Basic resistance of train operation |
| $v_c$ | Maximum sustained operating speed |
| $\Delta a$ | Residual acceleration at maximum sustained speed operation |
| $\gamma$ | Train rotation coefficient |

**Table 2.** The decision variables used in the model.

| Variable | Definition |
|---|---|
| $x_u$ | Optimal coupling and uncoupling position $u$ |
| $x_w$ | Optimal coupling and uncoupling position $w$ |

factor, can lead to excessive train capacity. To balance the energy cost and crowding cost, we introduce a crowding penalty coefficient $\sigma$ as the penalty factor. The crowding level is converted into an economic cost, and different penalties are applied to passengers within different ranges of average standing area.

Based on the 'General Technical Conditions for Subway Vehicles', the evaluation criteria for the seating capacity and standing density, as well as the classification of average standing area per person, different passenger service levels can be used to assess the crowding level [10]. This ensures that the passenger load factor remains within a scientifically and reasonably acceptable range. The crowding penalty standards for passengers on the train are shown in Table 3.

During the train operation, the number of small-group train in the suburban section before coupling is referred to as $X_{low}$, while the number of coupled train in the urban center running section after coupling is referred to as $X_{high}$. During the peak operating period in the $m$ direction, the passenger flow in the $i$ section is denoted as $Q^m_{i,i+1}$, and the station interval in the $i$ section is denoted as $d^m_{i,i+1}$. The number of train group operating in the $i$ section is $X^m_{i,i+1}$, and the train frequency is $h$. $e_s$ represents the seating capacity, and $S$ represents the standing area of a carriage. The average standing area per passenger is calculated as the ratio of the standing area to the difference between the section passenger flow and the product of the seating capacity and the number of cars. Therefore, the average standing area in the $i$ section of all trains in the $m$ direction is denoted as follow:

$$\Delta S^m_{i,i+1} = \frac{S \cdot X^m_{i,i+1} \cdot h}{Q^m_{i,i+1} - e_s \cdot X^m_{i,i+1} \cdot h} \tag{2}$$

To simplify the calculation of the train standing area, we will utilize the standard parameters from the technical specifications, including the train capacity, and the standing density per passenger. Let's consider an A-type train with 56 seats per car and a standing density of 6 passengers per square meter. The standing density of the passengers in the car is defined as the ratio between the standing area of the train and the difference between the carriage capacity $C$ and seating capacity $e_s$. This ratio should be 1/6. By applying these values, we can calculate the standing area per car S as follows:

**Table 3.** Passenger Crowding Penalty Coefficient Standards

| Service level | Above A (With seat) | B (Comfort) | C, D (With contact) | E (Overlap) | Below E (Overcrowd) | Et al |
|---|---|---|---|---|---|---|
| $\Delta s$ | $(-\infty, 0]$ | $[0.90, +\infty)$ | $[0.24, 0.90)$ | $[0.17, 0.24)$ | $[0.12, 0.17)$ | $(0, 0.12)$ |
| $\sigma$ | 0.3 | 0.2 | 0.1 | 0.4 | 0.5 | 10 |

$$S = \frac{1}{6} * (C - e_s) \tag{3}$$

$$\begin{aligned}
Z_2 &= \sum_{m=1}^{2}\sum_{i=1}^{K} F^m_{i,i+1} \cdot d^m_{i,i+1} \cdot w_e = \sum_{m=1}^{2}\sum_{i=1}^{K} \frac{P^m_{i,i+1}}{v} \cdot d^m_{i,i+1} \cdot w_e \\
&= \sum_{m=1}^{2}\sum_{i=1}^{u} E_i + \sum_{m=1}^{2}\sum_{i=u}^{w} E_i + \sum_{m=1}^{2}\sum_{i=w}^{K+1} E_i \\
&= \sum_{m=1}^{2}\sum_{i=1}^{K}\left[\frac{[(a + bv + cv^2)(X^m_{i,i+1} \cdot h^m_{i,i+1} \cdot m_{train} + Q^m_{i,i+1} \cdot m)g] \cdot v_c}{3600 \cdot v}\right. \\
&\quad \left. + \frac{[(1+\gamma)(X^m_{i,i+1} \cdot h^m_{i,i+1} \cdot m_{train} + Q^m_{i,i+1} \cdot m)\Delta a] \cdot v_c}{3600 \cdot v}\right] \cdot d^m_{i,i+1} \cdot w_e
\end{aligned}$$

$$= \left( \delta_{train} \sum_{m=1}^{2} \sum_{i=1}^{K} X_{i,i+1}^{m} \cdot h_{i,i+1}^{m} \cdot d_{i,i+1}^{m} + \delta_{passenger} \sum_{m=1}^{2} \sum_{i=1}^{K} Q_{i,i+1}^{m} \cdot d_{i,i+1}^{m} \right) \cdot w_e \tag{4}$$

$$P = \frac{[F_0 + 1 + \gamma \cdot M \cdot \Delta a] \cdot v_c}{3600} \tag{5}$$

$$F_0 = (a + bv + cv^2)Mg \tag{6}$$

$$\delta_{train} = \frac{[(a + bv + cv^2)g + 1000(1 + \gamma)\Delta a] \cdot m_{train} \cdot v_c}{3600 \cdot v} \tag{7}$$

$$\delta_{passenger} = \frac{[(a + bv + cv^2)g + 1000(1 + \gamma)\Delta a] \cdot m \cdot v_c}{3600} \tag{8}$$

Formula 4 represents the operating consumption cost, influenced by the optimal marshalling positions, determined by the traction force $F$, average traction distance $d$, and electricity price $w_e$[11]. The cost of each section is the product of the traction power cost and the electricity unit price, as shown in formula 4, $\sum_{m=1}^{2} \sum_{i=1}^{K} F_{i,i+1}^{m} \cdot d_{i,i+1}^{m} \cdot w_e$. The traction force $F_0$ is determined by the traction power $P$ and average speed $v$, as shown in formula 5 and 6. During the peak time in the $m$ direction, the passenger flow for the $i$ section is $Q_{i,i+1}^{m}$, the station distance is $d_{i,i+1}^{m}$, the group number for the $i$ section is $X_{i,i+1}^{m}$, and the train operating frequency is $h$. By using formulas 7 and 8, the parameters related to energy consumption can be transformed into variables $\delta_{train}$ and $\delta_{passenger}$.

*s.t.*

$$\frac{\max_{i=1}^{K} Q_{i,i+1}^{m}}{X_i^m \cdot h_i^m \cdot C} \le \alpha_{\max} \tag{9}$$

$$U_i \ge U_0 \tag{10}$$

$$x_u \le x_w \tag{11}$$

$$x_u x_w \in Z^+ \tag{12}$$

Formula 9 represents the constraint between capacity and passenger demand. We should ensure the alignment of capacity and demand, so it is important to ensure that the maximum cross-sectional load does not exceed the specified maximum load. Formula 10 represents the constraint on station turnaround tracks, ensuring that the selected optimal station for flexible marshalling has the necessary tracks for train departure and turnaround. The number of turnaround tracks or crossovers at the station used for marshalling should exceed the specified value that satisfies the marshalling conditions. Formula 11 represents the constraint on the optimal position for flexible marshalling, ensuring that the coupling position is earlier than uncoupling position in a single direction. Formula 12 represents the integer constraint, which requires the decision variable for positions must satisfy the constraint of being a positive integer.

## 2.3 Computational Procedure

The optimization is to minimize the cost of crowding penalties and energy consumption. There is a desire to reduce $Z_1$, involving crowding penalties decrease, which suggests operating longer sections with large-group, increasing the operational costs, while $Z_2$ suggests operating longer sections with small-group train, resulting in a decrease of service. The objectives present conflicting factors that influence each other, so we transform the multi-objective problem into a single-objective problem.

The Non-dominated Sorting Genetic Algorithm (NSGA-II) is an evolutionary algorithm that performs non-dominated sorting on obtained non-dominated solutions, addressing the multi-objective problem by obtaining a set of optimal solutions that satisfy the constraints. By employing a linear weighting method, the multi-objective problem is transformed into a single-objective problem. Then, the decision-making for the optimal locations under different frequencies is explored. The weight of $Z_1$ is $\lambda_1$, and the weight of $Z_2$ is $\lambda_2$, where $\lambda_1 + \lambda_2 = 1$. This results in the composite evaluation function $Z = \lambda_1 Z_1 + \lambda_2 Z_2$. The original model can be represented as a composite objective function, as shown in formula 13, which allows us to obtain the optimal solution $x^*$. Due to the significant differences in units and magnitudes between the functions, we normalize each function value before weighted calculation, as shown in formula 14. Formula 15 presents the normalization for the objective values of Pareto solutions, $s \in [1, 2][1, 2]$.

$$\min Z = \lambda_1 \min Z_1 + \lambda_2 \min Z_2 \tag{13}$$

$$stanZ_s^r = \frac{Z_s^r - \min Z_s^r}{\max Z_s^r - \min Z_s^r} \tag{14}$$

$$\min stanZ = \lambda_1 \min stanZ_1^r + \lambda_2 \min stanZ_2^r \tag{15}$$

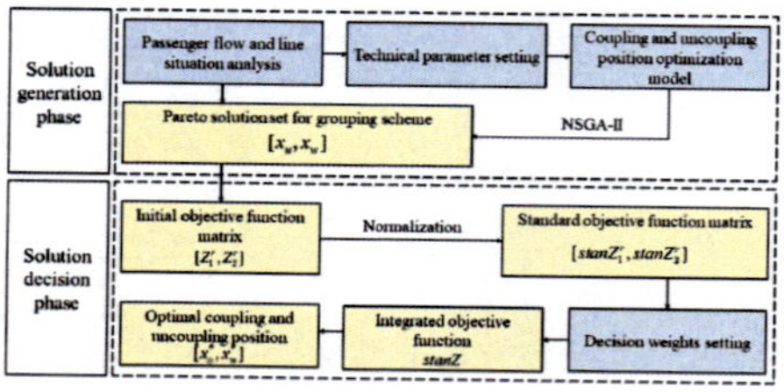


**Fig. 3.** Calculation Process for Optimal Position for Flexible Marshalling.

The calculation for determining the optimal position is shown in Fig. 3 and as follows:

Step 1: Utilize the NSGA-II to compute solutions that achieve favorable results for overall costs.
Step 2: Calculate the congestion cost and energy cost for each solution.
Step 3: Normalize the values and get the optimal solution with the minimum overall objective value.

## 3 Experimental Validation

### 3.1 Actual Engineering Line and Passenger Flow Data

Line T in a certain city exhibits significant imbalanced distribution of passenger flow, making it suitable for the proposed optimal coupling and uncoupling position model for the imbalanced spatial distribution of passenger flow, selected as the subject in this paper. Line T takes the length of 40.3 km and stations of 29. The configuration of terminal and crossover tracks at the stations is shown in Fig. 4. Trains on this line operate by reversing, coupling, and uncoupling according to the specific route conditions of each station. The stations and sections of Line T are numbered from the upstream to the downstream direction sequentially, and the inter-station distances are provided in Table 4. The line operates with Type A trains with a total length of 23 m and a width of 3 m. Each carriage has a seat capacity of 310 passengers. The design speed is 80 km per hour, while the average operating speed is 30 km per hour.

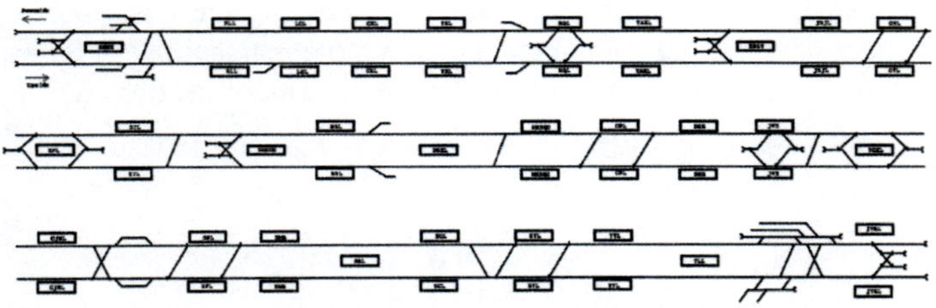

**Fig. 4.** The platform and track layout of Line T

**Table 4.** Inter-station distances of Line T

| Interval number | Interstation distance (m) | Interval number | Interstation distance (m) | Interval number | Interstation distance (m) | Interval number | Interstation distance (m) |
|---|---|---|---|---|---|---|---|
| 1 | 1.3 | 8 | 1.5 | 15 | 1.2 | 22 | 1.5 |
| 2 | 1.5 | 9 | 1.1 | 16 | 1.3 | 23 | 1.6 |
| 3 | 1 | 10 | 1.4 | 17 | 0.9 | 24 | 1.2 |
| 4 | 1.6 | 11 | 1.4 | 18 | 1.8 | 25 | 1.8 |
| 5 | 1.3 | 12 | 1.8 | 19 | 1.6 | 26 | 1 |
| 6 | 1.4 | 13 | 1.9 | 20 | 1.6 | 27 | 1.7 |
| 7 | 1 | 14 | 1.1 | 21 | 1.7 | 28 | 2.1 |

Line T operates in a single-line mode, from station S to station J in Fig. 5(a). We adopt a segmented running mode. When the train reaches the marshalling station, it will either be coupled or decoupled and continue forward or reverse direction. The middle

(a) Original Line (b) Simulated Line

**Fig. 5.** Train operating routing

segments and edge segments will operate with large-group trains and small-group trains, respectively, as depicted in Fig. 5(b).

Figure 6 shows the spatial distribution of passenger flow in different sections of Line T during the peak period. The passenger flow exhibits a trend of low at both ends and high in the middle. The upside shows highest value appearing in section 13, reaching 8,971, while the downside shows the highest value occurring in section 17, reaching 12,448. By calculating the spatial imbalance coefficient, the spatial imbalance coefficient in the upside direction is 1.57 while the downside is 1.56, indicating a significant spatial imbalance on Line T. The directional imbalance coefficient of the line is 1.08, indicating a balanced passenger flow in both upside and downside. Thus, trains can be operated in pairs. It is suggested that Line T should adopt a flexible marshalling in terms of spatial distribution, with large-group trains operating in high-demand sections during peak and small-group operating in low-demand sections.

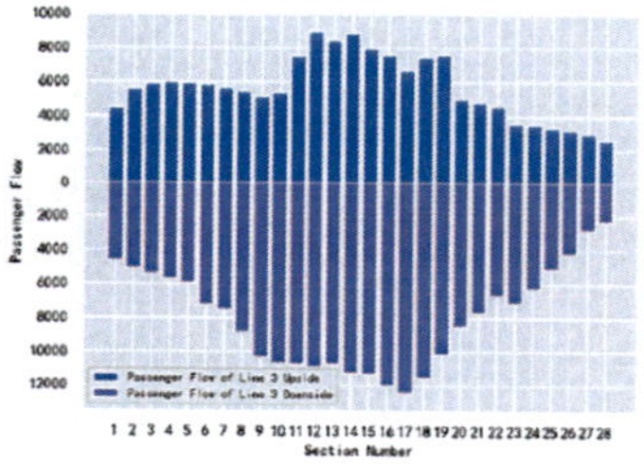


**Fig. 6.** Sectional passenger flow of Line T during the morning peak.

### 3.2 Parameter Setting and Performance Evaluation

The basic data for the optimal coupling and uncoupling decision model, which utilizes the passenger flow and line condition data of Line T is obtained. A-type vehicles are selected for operation, and the model is solved with balanced weight coefficients. Due to the operating interval during peak hours, it is recommended to keep it within the range of 2.5 min to 4 min. Therefore, three equidistant operating frequencies, namely h = 16, 20, and 24, are chosen to explore flexible coupling and uncoupling solutions. This involves operating 16, 20, and 24 train pairs on segmented routes during peak hours, aiming to achieve the optimal flexible coupling and uncoupling solutions. The values of the basic parameters are determined as shown in Table 5.

**Table 5.** Model parameters

| Parameters | Definition | Value | Unit |
|---|---|---|---|
| $L$ | Line length | 40.3 | km |
| $u$ | Number of sections | 28 | |
| $T$ | Number of time periods | 18 | |
| $C$ | Vehicle capacity | 310 | People/vehicle |
| $\alpha_{\max}$ | Maximum occupancy rate of the train | 120 | % |
| $t$ | Duration of operating time period | 1 | h |
| $I_{tra}$ | Minimum train headway | 3 | min |
| $v$ | Train operating speed | 30 | km/h |
| $v_c$ | Maximum sustainable operating speed | 80 | km/h |
| $m_{train}$ | Unit train mass | 36000 | kg |
| $m$ | Unit passenger mass | 60 | kg |
| $\gamma$ | Train turnaround coefficient | 0.06 | |
| $\Delta a$ | Residual acceleration at maximum sustainable speed | 0.05 | m/s$^2$ |
| $a, b, c$ | Train parameters | {1.9322,0.0386,0.0008} | |
| $t_{back}$ | Train turnaround time | 3 | minute |
| $t_{stop}$ | Train dwell time at stations | 0.5 | minute |
| $w_e$ | Unit electricity price | 0.7 | Yuan/Kw h |

## Parameter Values

The algorithm parameters suitable for this model were obtained by implementing the algorithm using the Python and performing debugging, as shown in Table 6.

**Table 6.** Genetic algorithm parameter values

| Parameters | Value |
|---|---|
| Termination generation | 100 |
| Initial population size | 200 |
| Cross probability | 0.9 |

**Algorithm Performance**

The performance of multi-objective optimization algorithms is evaluated using the Hypervolume (HV) indicator and the Spacing indicator [12]. HV represents the volume of the region in the objective space enclosed by the non-dominated solutions and a reference point. A larger HV value indicates better convergence and diversity. Spacing measures the standard deviation of the minimum distances between each solution and the other solutions in the converged set. A smaller Spacing value indicates a more uniform distribution of the solution set. In the following equations, $\delta$ represents Lebesgue measure, $|S|$ represents the number of non-dominated solutions, $v_i$ represents the hypervolume formed by the reference point and the i-th solution. $P$ denotes the solution set, $d_i$ represents the minimum distance of the i-th solution to the other solutions in $P$, and $\overline{d}$ represents the mean of all $d_i$ values.

$$HV = \delta\left(\bigcup_{i=1}^{|S|} v_i\right) \tag{16}$$

$$Spacing(P) = \sqrt{\frac{1}{|P|-1}\sum_{i=1}^{|P|}(\overline{d}-d_i)^2} \tag{17}$$

It shows an HV value of 0.8591, with good convergence and diversity. The Spacing value is 0.3498, indicating a uniform distribution of the set. Figure 7 shows the HV and Spacing for each generation.

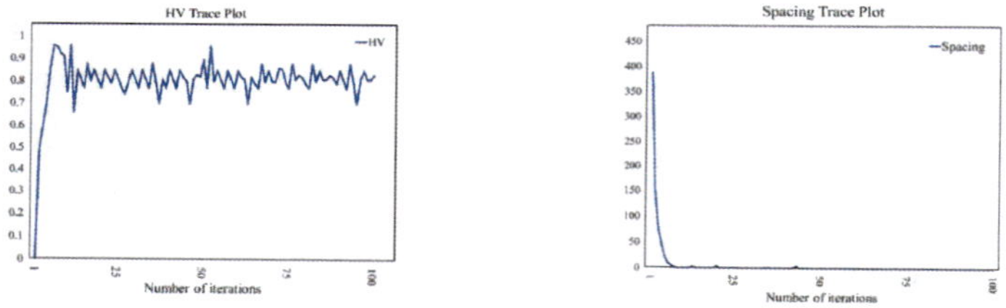


(a)HV (b)Spacing

**Fig. 7.** Model metrics

### 3.3 Simulation Results

Figure 8 shows the Pareto solution set, the consumption increases with the increment in frequency, while congestion costs show a decrease. The solutions are closer to each other, indicating smaller differences in target costs. Figure 9 presents several preferred schemes and their corresponding costs, with the red box indicating the optimal schemes. To compare the effectiveness, the costs of the fixed marshalling are showed in Table 7, with higher resource consumption. Figure 10 shows the average optimal rate, as the frequency increases, the range of sections operated with large-group trains is shortened.

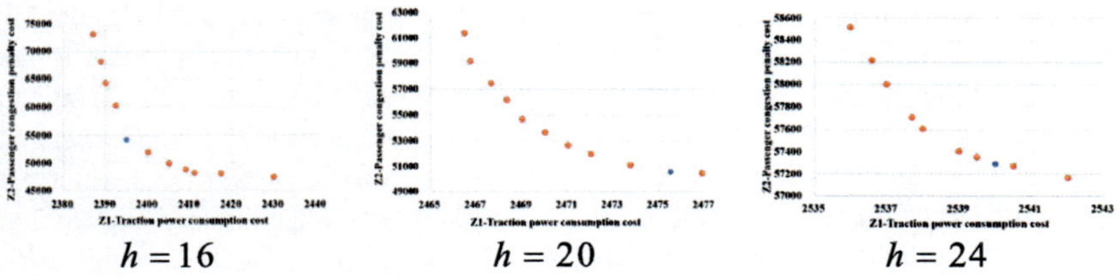


**Fig. 8.** Pareto front curve.

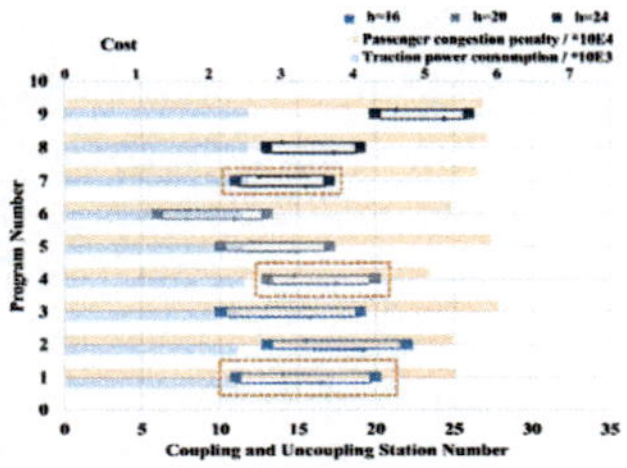


**Fig. 9.** Optimal marshalling schemes and objective costs.

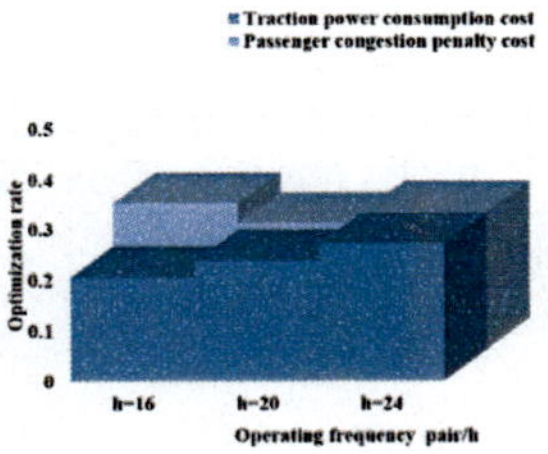


**Fig. 10.** Average optimal rate at different frequency.

**Table 7.** Objective costs of fixed marshalling operating scheme during peak hour.

| Number | Train frequency (trains per hour) | Objective function value | |
|---|---|---|---|
| | | Traction energy cost (in CNY) | Passenger congestion cost (in CNY) |
| 1 | 16 | 2882 | 72705 |
| 2 | 20 | 3049 | 67960 |
| 3 | 24 | 3218 | 73699 |

When the frequency is 16 pairs/hour, the optimal solution is achieved by flexible marshalling at stations 11 and 20. Compared to fixed marshalling, this strategy saves 20.33% and 34.39% in energy and congestion costs, showing 20.26% and 28.91% average optimization rates in energy and congestion costs of preferred solution. When the frequency is 20, the optimal solution is at 13 and 20 with savings of 23.30% and 34.52% in energy and congestion costs, showing 23.55% and 24.90% average optimization rates in energy and congestion costs of preferred solution. When the frequency is 24, the optimal solution is at 11 and 17, leading to a savings of 26.93% and 28.62% in energy and congestion costs, showing 27.05% and 27.20% average optimization rates in energy and congestion costs of preferred solution. The model demonstrates an increasing savings in energy costs with higher frequencies, optimizes congestion levels. The model exhibits good effectiveness in achieving the desired objectives.

### 3.4 Scenarios Analysis

To get better results, we provide marshalling solutions that can adapt to different scenarios. We utilize the model algorithm along with Python to find the optimal coupling and uncoupling positions based on the original passenger flow data. While keeping the remaining variables constant, we adjust factors such as flow demand, the imbalance of flow distribution, and distances between stations under design different scenarios. We conduct sensitivity analysis to analyze the impact of the scenarios and determine the optimal decision-making solutions. Under the balanced weight coefficients, where $\lambda_1 = \lambda_2 = 1/2$, we fix the variables and conduct sensitivity analysis of the model based on the optimal solution described above.

**Passenger Flow Scenarios.**

Passenger flow is a crucial factor in determining the optimal marshalling positions. We conduct an investigation on the optimal positions from the flow demand and imbalance of flow distribution.

(1) Passenger Flow Demand.

The passenger flow can be set with certain fluctuations, and the differences in marshalling schemes under different scenarios will be observed. The method for adjusting flow is as follows: Assuming equal weights for each objective and keeping other parameters constant, the initial scenario is based on the original flow in each section as $Q^r_{k,origin}$, adjustment coefficient indicating the multiplier for increasing the section flow as $\zeta_1$, $\zeta_1 \in [0.8, 2]$, and the step size is 0.1. The coefficients are numbered from 1 to 13, corresponding to different flow scenarios. The section flow is calculated as $Q^r_{k,\theta_{1,i}} = Q^r_{k,origin} \cdot \zeta_{1,i}$.

As the flow coefficient increases, the range of line segments operated with large-group train becomes longer, tending to be in areas with higher flow, leading to an increase in consumption and congestion costs. Within a unit period, higher frequency results in shorter segments for large-group, leading to higher consumption and lower congestion. As the coefficient increases, the increase in congestion costs is getting slower for solutions with higher frequencies.

The marshalling schemes under different flow scenarios are compared with the fixed marshalling scheme in Fig. 11. As the coefficient increases, the optimization in reducing congestion costs increases. Under the frequencies of 16, 20, and 24 trains per hour, it can save 21.84%–224.67%, 25.69%–253.25%, and 28.62%–309.11% of passenger congestion costs, respectively. However, the optimization in reducing consumption decreases.

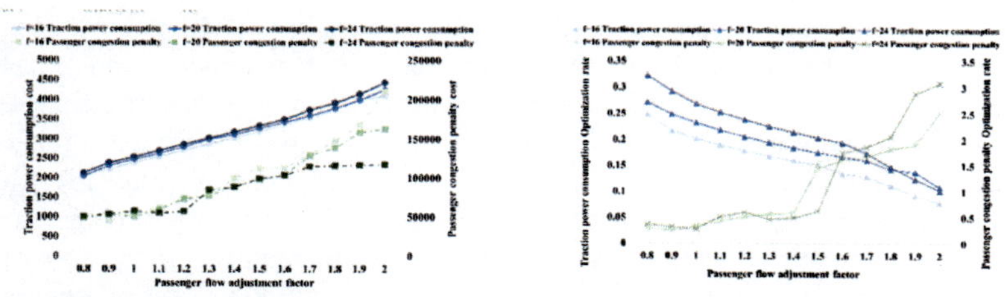


(a) The energy consumption and congestion penalty cost

(b) Optimization rate

**Fig. 11.** Objective costs and optimization rates under different passenger flow scenarios.

Under the frequencies of 16, 20, and 24, it can save 7.95%–25.97%, 10.89%–27.23%, and 10.26%–32.33% of consumption costs, respectively. The model performs better optimization in terms of congestion with higher flow, and better optimization in terms of energy consumption with lower flow. The model is more applicable at higher frequencies.

Figure 12 shows the consumption costs and congestion costs for each section under different flow scenarios, with the red dashed line showing the optimal marshalling positions. The consumption costs for large-group are higher, increasing with the flow increase, while congestion costs are lower for large-group, increasing with the flow increase. As the frequency increases, the section range operated with large-group goes shorter, and smaller-group can be operated on longer sections with the higher flow.

When the frequency is 16 trains per hour, the optimal train positions are at station 11 and 20 for the maximum flow in the range of [9958, 19917), at 10 and 20 for [19917, 22406), at 11 and 22 for [22406, 24896]. When the frequency is 20, the optimal positions are at 13 and 20 for [9958, 17427), at 11 and 19 for [17427, 22406), and at 10 and 19 for [22406, 24896]. When the frequency is 24, the optimal positions are at 11 and 17 for [9958, 21162), at 10 and 17 for [21162, 23651), and at 10 and 19 for [23651, 24896].

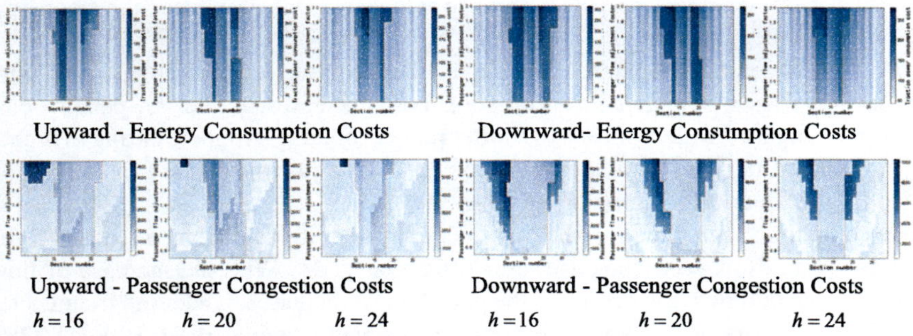


**Fig. 12.** Objective costs and optimal demand different passenger flow scenarios for each section

(2) Passenger Flow Imbalance

We set the section flow adjustment coefficient to fluctuate in a certain arithmetic progression, in order to observe the differences in marshalling plans under different passenger flow scenarios. The method of adjustment is as follows: The weight coefficients are equal and other parameters are fixed, the initial passenger flow of the line is taken as the baseline scenario. The maximum section peak passenger flow demand is denoted as $Q^r_{\max,origin}$, adjustment coefficient for the section as $\zeta_2$, $\zeta_2 \in [-6, 6]$, and the step size is 1. The maximum section peak flow is adjusted by the coefficient, resulting in the upper limit of the fluctuation range, denoted as $Q^r_{\max,\zeta_{2,i}} = Q^r_{\max,origin} + 200 \cdot \zeta_{2,i}$. The other sections are adjusted starting from the section line with the maximum and the initial, until the average flow of each remains fixed. We vary the maximum sectional flow in the order of $Q^r_{k+j,\zeta_{2,i}} = Q^r_{k+j,origin} + 200 \cdot \zeta_{2,i}/(2 \cdot j)$ and $Q^r_{k-j,\zeta_{2,i}} = Q^r_{k-j,origin} + 200 \cdot \zeta_{2,i}/(2 \cdot j)$ at each end of the line, and vary the starting stations to the maximum sectional flow in the order of $Q^r_{0+j,\zeta_{2,i}} = Q^r_{0+j,origin} - 200 \cdot \zeta_{2,i}/(2 \cdot j)$ and $Q^r_{K-j,\zeta_{2,i}} = Q^r_{K-j,origin} - 200 \cdot \zeta_{2,i}/(2 \cdot j)$. Through this adjustment, the average value and magnitude of the section peak flow remain fixed, while the spatial imbalance of the passenger flow changes.

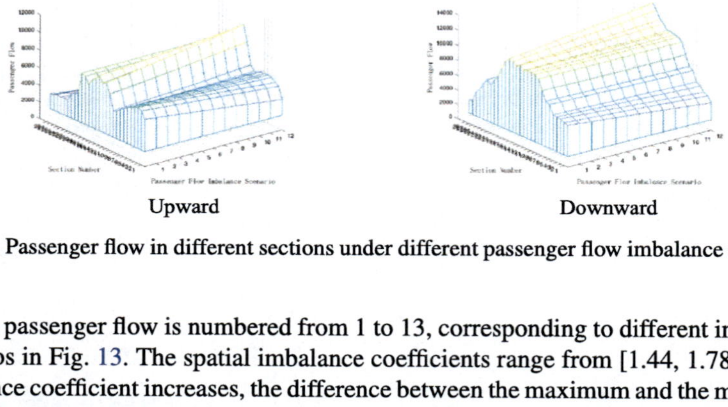

Upward Downward

**Fig. 13.** Passenger flow in different sections under different passenger flow imbalance scenarios.

The passenger flow is numbered from 1 to 13, corresponding to different imbalance scenarios in Fig. 13. The spatial imbalance coefficients range from [1.44, 1.78]. As the imbalance coefficient increases, the difference between the maximum and the mean flow in each section becomes larger.

With the increase of imbalance, the range of large-group tends to align with higher flow sections. As a result, the consumption cost decreases, while the penalty cost increases. A higher frequency leads to shorter ranges for large-group, resulting in higher consumption costs. With the increase of imbalance, the rate of increase in congestion cost for higher frequency scenarios becomes slower.

By comparing with fixed marshalling under different flow imbalance scenarios, the optimization rates of each objective are shown in Fig. 14. With the increase of flow imbalance, the optimization degree of the penalty cost increases. Under the frequencies of 16, 20, and 24 trains per hour, it is possible to achieve savings of 26.19%–41.72%, 31.07%–47.34%, and 31.96%–39.12% in congestion costs, respectively. The energy consumption costs optimization also increases, with savings of 16.84%–22.34%, 22.10%–26.85%, and 25.10%–28.47% under the above frequencies. The model demonstrates better adaptability to railway lines with higher passenger flow imbalance.

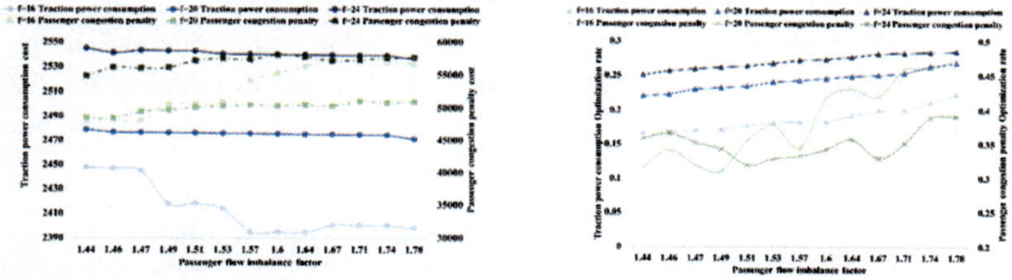

(a) The traction energy cost and congestion penalty cost

(b) Optimization Rate

**Fig. 14.** Objective costs and optimization rates in different passenger flow imbalance scenarios.

Figure 15 shows the energy consumption and crowding penalty costs in different passenger flow imbalance scenarios. The red dashed line represents the optimal marshalling position. For each section, the traction energy consumption of running larger-group is higher, and it increases with increasing passenger imbalance. As the larger group train tends to operate in higher passenger sections, the crowding costs are higher compared to the smaller group, also increase with increasing passenger flow imbalance.

When the frequency is 16 trains per hour, the optimal marshalling positions are at stations 10 and 20 for a passenger flow imbalance coefficient in the range [1.44, 1.57), at 11 and 20 for [1.57, 1.67), at 13 and 22 for [1.67, 1.78]. When the frequency is 20, the optimal positions are at 11 and 19 for [1.44, 1.51), at 13 and 20 for [1.51, 1.78), at 13 and 19 for [1.51, 1.74], at 13 and 22 for 1.78. When the frequency is 24, the optimal positions are at 10 and 17 for [1.44, 1.53), at 11 and 17 for [1.53, 1.78].

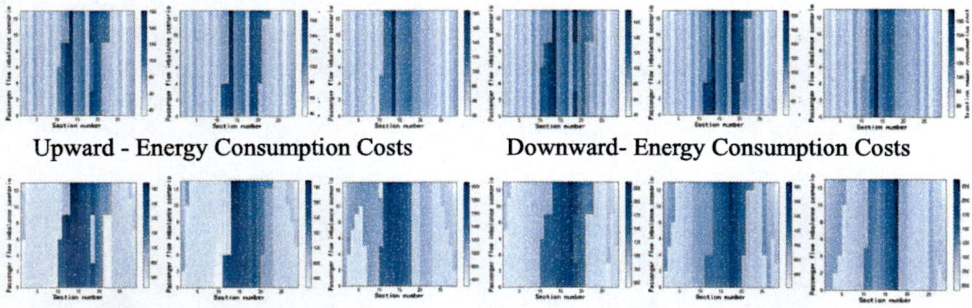

Upward - Energy Consumption Costs　　Downward- Energy Consumption Costs

Upward - Passenger Congestion Costs　　Downward - Passenger Congestion Costs

$h=16$　$h=20$　$h=24$　$h=16$　$h=20$　$h=24$

**Fig. 15.** Objective costs and optimal positions for each section in different passenger flow imbalance scenarios.

## Line Scenarios

Different line scenarios are set based on the varying distances between stations. They can be classified into urban lines with relatively uniform inter-station distances and suburban lines with longer inter-station distances and shorter inter-station distances in the center section. To explore the differences caused by different scenarios, several lines with different inter-station distances are shown in Fig. 16. These include urban lines A and C, where one side has longer inter-station distances and the other side has shorter inter-station distances, urban line B (T-Line) with relatively uniform inter-station distances, and suburban lines with longer distances in the suburb and shorter distances in the center section.

**Fig. 16.** Different line scenarios with varying inter-station distance

For suburban lines, the operating range of large-group tends to be towards the shorter inter-station distances in the center section. For urban lines, the range of large-group also tends to be towards the shorter inter-station distances section of the line. In terms of unit operational time, schemes with higher frequencies have a shorter range for operating large-group, resulting in higher energy consumption.

Figure 17 shows the consumption costs and crowding penalty costs for different line scenarios, with the red dashed lines representing the optimal marshalling positions. The higher the traction energy consumption cost for the larger the inter-station distance section. Since the range of large-group always includes sections with higher passenger demand, the passenger crowding penalty cost is also higher.

When the frequency is 16 pairs per hour, the optimal train positions are at stations 10 and 20 for the suburban line, at 13 and 22 for Line A with a relatively shorter inter-station distance on the right end of the line, at 11 and 20 for Line B with relatively inter-station distances, at 10 and 19 for Line C with a relatively shorter inter-station distance on the left end of the line. When the frequency is 20, the optimal positions are at 11 and 20 for the suburban line, at 13 and 22 for Line A, at 13 and 20 for Line B, at 10 and 17 for Line C. When the frequency is 24, the optimal positions are at 11 and 19 for the suburban line, at 13 and 19 for Line A, at 11 and 17 for Line B, at 10 and 17 for Line C.

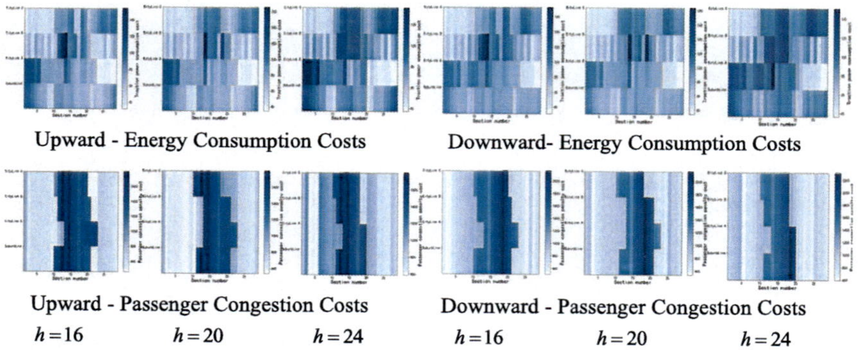


**Fig. 17.** Objective costs and optimal train formation positions for different line scenarios.

## 4 Conclusion and Future Works

This paper constructed the optimal marshalling position decision model minimizing crowding penalty and traction energy cost, using the NSGA-II and linear weighting method to get the optimal marshalling locations, and the range for trains with large and small groups. The conclusions are as follows:

(1) At an operating frequency of 16, it is suitable to adapt at stations 11 and 20. At the frequency of 20, it is suitable to adapt at stations 13 and 20. At the frequency of 24, it is suitable to adapt at stations 11 and 17. Compared to fixed marshalling, the strategy can reduce traction energy costs by 20.08%–27.13% and crowding penalty costs by 14.97%–34.99%. As the operating frequency increases, the section range for operating trains with large group becomes shorter.
(2) The model designs a combined strategy of flexible marshalling and segmented operation, the adaptability is better for lines with larger passenger spatial imbalance. The section range for operating trains with large-group increases with the growth of passenger demand. With the increase of passenger spatial imbalance, the section range for large-group trains tends to focus on sections with higher passenger flows. For different line scenarios, the section range for operating trains with large-group also tends to focus on sections with shorter inter-station distances.

This paper provides an exploration of marshalling position decision. Future research can consider integrating data-driven approaches, and introduce mature big data components to enhance the simulation.

**Acknowledgments.** This research is supported by the National Natural Science Foundation of China (52172329). The authors would like to thank the reviewer for the valuable comments and advice on this manuscript.

## References

1. Bi, Y., Liu, J., Jiang, Z.: Mixed running of different marshalling trains and its impact on URT. Urban Rail Transit Res. **9**(12), 54–47 (2006). (in Chinese)
2. Ji, Y., Ou, D., Chang, M., Ning, Z.: Research on application requirements and key technologies of train virtual formation. Urban Rail Transit Res. **25**(11), 57–61 (2022). (in Chinese)
3. Ou, D., Ning, Z., Zhang, L.: A virtual marshalling train formation structure decision method for two-line converging lines. China Patent, CN115743247A (2023). (in Chinese)
4. Ma, C., Li, P., Zhu, C.: Short-term passenger flow forecast of urban rail transit based on different time granularities. J. Chang. Univ. Natl. Sci. Edn. **40**(3), 9 (2022). (in Chinese)
5. Cadarso, L., Ángel, M.: Robust rolling stock in rapid transit networks. Comput. Oper. Res. **38**(8), 1131–1142 (2011)
6. Cadarso, L., Marin, A.: Integration of timetable planning and rolling stock in rapid transit networks. Ann. Oper. Res. **199**(Oct), 113–135 (2012)
7. Tirachini, A., Cortés, C.E., Jara-Díaz, S.R.: Optimal design and benefits of a short turning strategy for a bus corridor. Transportation **38**, 169–189 (2011)
8. Xu, D., Mao, B., Lei, L.: Optimization for train plan of full-length and short-turn routing in urban rail transit. Transp. Syst. Eng. Inf. (1) (2017). (in Chinese)

9. Chen, S., He, S., He, B.: Train service planning for urban rail transit with passenger fluctuation. Urban Rail Transit Res. **10**, 53–58 (2013). (in Chinese)
10. 20212112-T-333. General technical conditions for metro vehicles. National Standard Program, (2021). (in Chinese)
11. Zhang, Q., Li, M., Bi, H.: Energy consumption calculation of light weight high-speed train. Railw. Transp. Econ. **42**(12), 117–125 (2020). (in Chinese)
12. Paquet, L.S., Britta, S., Michael, L., Ana, C.: Computing representations using hypervolume scalarizations. Comput. Oper. Res. **137**, 105349 (2022)

# Large-Scale Language Models for PHM in Railway Systems - Potential Applications, Limitations, and Solutions

Huan Wang[1,2] and Yan-Fu Li[1,2](✉)

[1] Department of Industrial Engineering, Tsinghua University, Beijing 100084, China
liyanfu@tsinghua.edu.cn

[2] Institute for Quality and Reliability, Tsinghua University, Beijing 100084, China

**Abstract.** Prognostics and health management (PHM) technology, by monitoring the faults and degradation of railway systems, predicting the remaining useful life of equipment, and providing maintenance recommendations, can effectively improve the safety and reliability of railway systems. In recent years, large-scale language models (LLMs) like ChatGPT have achieved groundbreaking accomplishments and led a new wave of innovation in various fields. Consequently, the potential impacts of LLMs on PHM applications in railway systems are worth researching and exploring. This paper first introduces the basic principles and technical characteristics of LLMs. Subsequently, it analyzes the potential impacts of these models in PHM applications within railway systems, exploring how they can be applied in various processes of PHM, including operations management, condition monitoring, maintenance recommendations, and knowledge management, to enhance the effectiveness of PHM. Moreover, this study analyzes the limitations of LLMs in PHM applications within railway systems from a practical perspective and discusses relevant solutions accordingly. Based on these solutions, these models are expected to become more specialized and intelligent, playing a crucial role in the maintenance and management of railway systems. Finally, this paper provides an outlook on LLMs' prospects and research directions in PHM applications within railway systems.

**Keywords:** Prognostics and health management · Railway systems · Large-scale language models

## 1 Introduction

The safety and reliability of the railway system are crucial for passengers and staff, as they directly impact human life safety and have significant implications for social stability and transportation service quality [1]. Therefore, the railway system must continuously focus on and strive to enhance safety performance, ensuring that passengers can enjoy a safe and reliable travel experience, as well as a safe working environment. Prognostics and health management (PHM) involve real-time monitoring and assessment of equipment status, predicting potential failures, and taking proactive maintenance measures. This


Y. Qin et al. (Eds.): EITRT 2023, LNEE 1137, pp. 591–599, 2024.
https://doi.org/10.1007/978-981-99-9311-6_59

approach can significantly improve the safety and reliability of the railway system [2]. It not only helps reduce accident risks but also lowers maintenance costs, enhances equipment availability and reliability, and optimizes maintenance decisions.

The railway system is a complex electromechanical system that relies on the precise coordination of hundreds or even thousands of components and devices to achieve efficient cargo transportation. Therefore, conducting PHM work in the railway system requires a large number of experienced professional engineers who can perform various tasks such as analyzing and processing sensor data, inspecting and identifying equipment failures and abnormalities, as well as maintaining and replacing core equipment and components [3]. This poses significant workforce costs. Hence, the rapid development of skilled operation and maintenance personnel is crucial for reducing the workforce costs associated with railway system operation and maintenance. Furthermore, to ensure the safety of the railway system, a wide variety of sensors [4] are used to monitor the condition of various components. This results in massive sensor data that needs to be analyzed and processed. Relying solely on the manual processing of this data by engineers is impractical. Therefore, exploring highly intelligent data analysis and automated processing systems is of great significance for the operation and maintenance of the railway system.

Recently, large-scale language models (LLMs) like ChatGPT [5, 6] have achieved breakthrough accomplishments. Through training and learning on massive text data, these models possess powerful language comprehension, text generation, and conversational abilities. Language models can intelligently understand human intent and generate satisfactory responses based on contextual information and their own knowledge base. This makes them highly applicable in numerous fields [7, 8]. For example, they can be applied in search engines to help users quickly find relevant information and extract and summarize the most pertinent details. In the medical field, they can assist healthcare professionals by providing valuable medical diagnostic information. Additionally, they can assist physicians in analyzing medical data to identify disease types and severity levels, among other applications [9].

Thanks to the powerful capabilities of LLMs, such as ChatGPT [6], they have not only led a new wave of transformation in various fields but also have the potential to revolutionize the paradigm of PHM applications in the railway system. Leveraging the latest advancements in LLM technology, LLMs hold the promise of significantly reducing the cost of training professional operation and maintenance personnel in the railway system. They can also drive the development of automated railway operation and maintenance systems, enabling automated data analysis, condition monitoring, and maintenance planning [10]. For instance, LLMs can conduct in-depth analysis and mining of big data in the railway system, extracting valuable information and insights. They can identify trends, abnormal patterns, and correlations in equipment operations, providing decision-makers with data-driven decision support. LLMs can integrate maintenance knowledge and experience from the railway system to build intelligent knowledge management systems. These models can answer maintenance personnel's questions, provide real-time repair suggestions, and transform knowledge and experience into easily understandable and usable formats.

To this end, this study aims to systematically investigate the potential impact of LLMs, specifically ChatGPT, in applying PHM in the railway system. Firstly, this paper introduces LLMs' basic principles and core technical details. Then, considering the practical conditions of the railway system, it analyzes the potential applications of LMs in the railway system and explores how LLMs can be integrated into various aspects of PHM in the future to enhance the benefits of railway system PHM and improve its safety and reliability. However, existing LMs have rarely considered applications in the field of railway system PHM during their development and training processes. Therefore, if the current stage of LLMs is applied to the railway system, it inevitably has significant limitations. This study systematically analyzes these limitations and their causes and proposes targeted solutions. Finally, this paper outlines the future directions of LLMs in applying PHM in the railway system. Figure 1 shows the main content and structure of this paper. Overall, this paper analyzes the feasibility and application prospects of LMs in PHM, providing valuable references for further research.

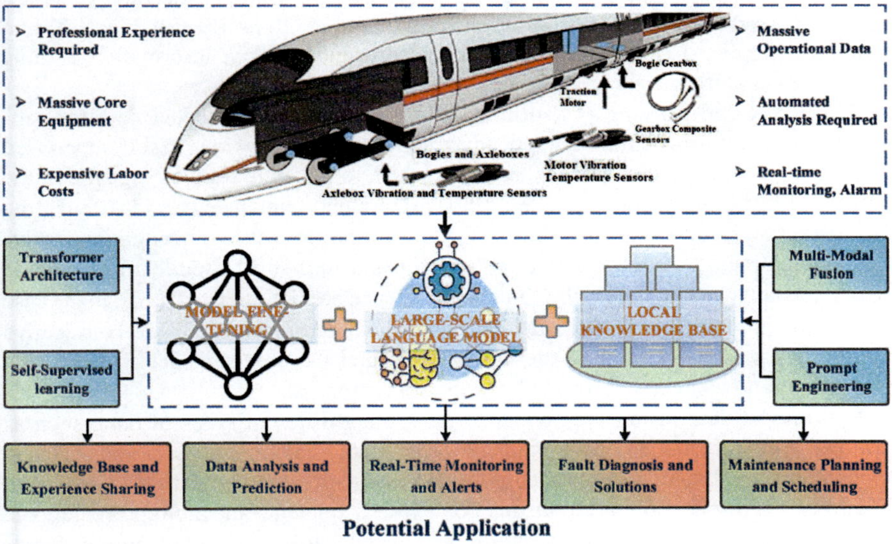


**Fig. 1.** Application potential and core components of large-scale language models in railway systems.

## 2 Large-Scale Language Models

LLMs refer to natural language processing (NLP) models with massive parameter sizes and powerful computational capabilities. These models are based on deep learning and utilize large amounts of training data and complex neural network architectures to learn the grammar, semantics, and contextual associations of language, enabling natural language understanding and generation [7]. LLMs are a recent advancement in deep learning technology, and their core components include Transformers and their variants [11, 12], self-supervised learning [13, 14], multimodal fusion [15], and prompt engineering.

LLMs typically employ Transformers [12] and their variants as their backbone network. The Transformer architecture consists of core components such as self-attention mechanism, multi-head attention (MHA) mechanism, positional encoding, and feed-forward neural networks (FFNNs).The self-attention mechanism allows the model to dynamically attend to different positions in the sequence and weight the positional information based on the relevance calculated through query, key, and value correlations, thus capturing long-range dependencies and contextual information. The MHA performs parallel computations of multiple attention heads, with each head learning different attention weights, thereby enhancing the expressive power of the model. Positional encoding embeds the positional information of the sequence into the input, capturing the sequential relationships. The FFNN serves as a nonlinear transformation, independently transforming the representation of each position. The Transformer architecture typically comprises an encoder and a decoder. The encoder is responsible for encoding the input sequence into a context-aware representation, while the decoder generates sequences based on the encoder's output and previous predictions. By utilizing the Transformer architecture, LLMs can learn richer semantic and syntactic information, leading to outstanding performance in NLP tasks.

Self-supervised learning [14] utilizes a large amount of unlabeled data for self-learning without the need for manually annotated labels. The key idea in self-supervised learning is to leverage the data's inherent structure to generate training signals. In the context of LLMs, the model can process massive amounts of unlabeled text data, enabling it to extract high-level semantic and syntactic information without needing expensive manual annotation processes. Common techniques in self-supervised learning include the masked language model (MLM) and next sentence prediction (NSP) [16]. In the case of NSP, the model is tasked with determining whether two sentences are consecutive in the original text. By training the model to predict the next sentence, it learns the associations and contextual information between sentences.

Multimodal data fusion [15] refers to integrating different types of data, such as text, images, audio, etc., to understand and generate more comprehensive and enriched language information. This approach aims to effectively combine data from different modalities to provide more accurate and contextually related semantic representations. One common approach is to use a multimodal encoder that takes input from different modalities and processes them in the model. For example, the model can simultaneously receive text and images as inputs and use a shared representation layer to capture the correlation between the two modalities. Another approach is to use multimodal attention mechanisms [17], allowing the model to weight the attention based on inputs from different modalities. The advantage of multimodal data fusion lies in leveraging the complementarity between different modalities, enhancing language models' expressive power and semantic understanding.

Prompt engineering is a key technique in improving the generation results of LLMs by designing and optimizing prompt texts. It provides an effective method to guide the model in generating outputs according to the user's intent, resulting in more accurate, consistent, and controllable outputs. When designing prompt texts, appropriate keywords need to be selected, specific instructions or constraints may be added, and it may include example inputs and outputs to guide the model in generating answers or results according to the user's intent. Through continuous iteration and optimization of prompt texts, Prompt engineering can improve the usability and adaptability of the model, providing users with a higher-quality interactive experience.

## 3 Potential Applications of LLMs in Railway Systems

PHM in the railway system refers to using various technologies to monitor, analyze, and predict the status of various core components in the railway system [18, 19]. It aims to assess the condition of railway equipment and infrastructure, predict potential failures, and take corresponding maintenance and upkeep measures to ensure the safe operation of the railway system. While LLMs are primarily known for text-based conversations, their impressive capabilities make them promising for substantial applications in the field of PHM and enhance the overall benefits of PHM. Potential applications of ChatGPT-Like LLM in railway system PHM include:

**Knowledge Base and Experience Sharing**: LLMs can serve as a knowledge base and platform for experience sharing. Through interaction with LLMs, users can query relevant knowledge, best practices, and equipment or systems case studies. LLMs can provide information on equipment operation, maintenance procedures, and fault handling to support users in making decisions in practical work scenarios.

**Data Analysis and Prediction**: LLMs can be integrated with PHM systems for data analysis and fault prediction. LLMs can converse with users to acquire relevant real-time or historical data and interact with prediction models for fault prediction and analysis. LLMs can interpret and explain the prediction results, provide insights into potential fault causes, and assist users in understanding complex model outputs.

**Real-Time Monitoring and Alerts**: LLMs can be integrated with real-time monitoring systems to periodically or instantly retrieve operating status data of equipment or systems and trigger alerts based on predefined thresholds or rules. LLMs can notify relevant users and provide suggestions and guidance when faults, abnormal conditions, or important events occur.

**Fault Diagnosis and Solutions**: LLMs can be used to interact with users, receive their descriptions of equipment or system faults, and provide possible fault diagnoses and solutions. LLMs can understand user inputs and provide information regarding fault causes, possible repair measures, and maintenance recommendations based on their language model and pre-training knowledge.

**Maintenance Planning and Scheduling**: LLMs can provide suggestions for maintenance planning and scheduling based on equipment status and historical data. Based on user-provided information, LLMs can generate optimal maintenance strategies, repair plans, or equipment replacement recommendations. This helps optimize the utilization of maintenance resources, reduce downtime, and improve system availability and performance.

# 4 Limitations and Solutions

## 4.1 Limitations of LLM in Railway Systems

Currently, the language understanding capability and knowledge reserves of LLMs like ChatGPT far surpass previous language models, and their performance in answering various types of questions is impressive. However, their inherent limitations must be considered when using them for railway systems. The major limitation of LLMs like ChatGPT lies in the lack of domain-specific expertise. This is mainly due to the limited knowledge reserve of LLM training data. The current training data for LLMs mainly come from the internet, making it impossible to fully encompass all specialized domain knowledge, particularly in railway systems where a large portion of technical manuals and maintenance details are kept confidential. This results in poor performance of LLMs when answering technical questions related to railway systems.

Additionally, LLMs also face the challenge of keeping up with the latest technological developments. For example, ChatGPT's training data only includes information until 2021, making it difficult to handle various newly emerged knowledge. In the railway domain, various technological solutions and professional content may continuously evolve and improve. If LLMs cannot quickly grasp the latest technological advancements, their practical application in real-world scenarios will be limited.

Furthermore, LLMs like ChatGPT primarily interact with users through text-based conversations, lacking visual and graphical interaction methods. This limitation restricts their application in many railway system scenarios where visual and graphical representation is essential.

## 4.2 Solutions of LLM in Railway Systems

As can be seen from the previous chapter, when applying LLMs to railway systems, the biggest limitation lies in the lack of professional technical knowledge and the unknowability of the latest technological developments. Therefore, we can resort to fine-tuning and building a local knowledge base to address this limitation. In addition, the application of large-scale visual language models can greatly expand the application scenarios of LLMs.

In the field of railway systems, LLMs like ChatGPT can provide precise and specialized solutions through a technique called fine-tuning. Fine-tuning involves further training a pre-trained language model with domain-specific data to enhance its accuracy and understanding in that particular domain. Firstly, it is necessary to collect domain-specific data related to railway systems and annotate it appropriately. This data can include fault reports, maintenance records, operation manuals, equipment specifications, and more. Next, the data is preprocessed and cleaned to transform into a format suitable for language model training. This may involve text cleaning, data labeling, and data segmentation to ensure data quality and consistency. Then, this processed data is used to fine-tune the pre-trained language model. During the fine-tuning process, the language model learns the semantics, context, and domain-specific knowledge relevant to railway systems, thereby improving its ability to express itself in that domain. Proper hyperparameter settings and optimization algorithms can be employed to accelerate the training process and improve model convergence speed. Additionally, techniques such as adjusting the data sampling strategy, incorporating data augmentation techniques, and utilizing transfer learning can enhance the robustness and generalization capability of the model. Once the fine-tuning is completed, the trained model can be used to address various problems in railway systems.

LLMs like ChatGPT can provide more accurate and practical solutions through local knowledge base enhancement. The core idea of this technique is to combine the local-specific knowledge and information of railway systems with the language model to provide more in-depth and personalized answers and suggestions. Firstly, specific knowledge and information related to the railway system domain need to be collected and organized to perform local knowledge base enhancement. Next, the local knowledge base data needs be integrated with the language model. This can be achieved by constructing a domain-specific knowledge graph or database that contains specific concepts, entities, and relationships in the railway system. The knowledge graph can be used to combine the local knowledge with the language understanding capabilities of the model, enabling the model to comprehend and answer questions related to the railway system. By combining the local knowledge base with the LLM, personalized answers and more accurate solutions to specific problems can be achieved. The model can utilize the information in the local knowledge base for real-time querying and reasoning, providing more detailed and specific answers. Additionally, the local knowledge base enhancement technique can offer more context and background information, thereby improving the quality of the model's responses.

Large-scale vision and language models [20] combine the capabilities of computer vision and natural language processing to achieve the mutual understanding of images and text, effectively addressing the limitation of LLMs in handling visual information. These models can be applied to image recognition and object detection tasks to provide support in train inspection and maintenance. By analyzing train images, they can automatically detect potential faults or damages, such as car wear, equipment malfunctions, or body damage. This helps in early problem detection and facilitates appropriate maintenance measures, thereby improving the reliability and safety of train operations. Therefore, by combining image recognition and NLP techniques, these models can play a crucial role in train safety monitoring, maintenance, and inspection, enhancing railway systems' operational efficiency and safety.

### 4.3 Future Directions of LLM in Railway Systems

Currently, there is a lack of research on the application of LLMs in the railway system. To promote the application of LLMs in the railway system, the following research directions deserve attention based on the actual situation in the railway system:

**Industrial Data Security and Privacy Protection**: PHM applications in the railway system may involve sensitive industrial data, such as train operation status, maintenance records, sensor data, etc. These data may contain trade secrets, proprietary technologies, or critical operational information. The application and research of LLMs must consider data security and protection to prevent unauthorized access, leakage, or theft. Therefore, ensuring that LLMs can operate quickly and reliably in the offline environment is necessary. Additionally, technologies such as data encryption and secure transmission, access control and permission management, data anonymization, and de-identification must be carefully considered when developing relevant LLMs applications. Furthermore, the protection of large-scale industrial data during the training and fine-tuning process of the model is also worth studying. Federated learning [21] is a feasible solution that can achieve multi-center collaborative learning and model optimization while protecting data privacy.

**Model Interpretability and Output Consistency**: Safety is paramount in the railway system, and any decision and prediction results should be reliable and trustworthy. Improving the interpretability of the model can enhance its transparency and credibility, enabling operators and decision-makers to understand the model's reasoning process, prediction basis, and conclusions. This helps improve user acceptance and system usability. In addition, output consistency refers to the stability and consistency of the prediction results produced by the model under different time and input conditions. Currently, LLM outputs exhibit high flexibility and randomness, which is unsuitable for practical industrial applications. Improper expression may lead to misunderstandings and erroneous decisions.

**Model Lightweight and Edge Device Deployment**: Existing LLMs have many parameters and computational requirements, while computing resources in the railway system are limited. Moreover, considering data security and low latency, the model often needs to be deployed offline on edge devices in the railway system. Therefore, exploring lightweight models can meet the resource constraints of edge devices, reduce latency and bandwidth requirements, and improve data privacy, security, and system reliability.

## 5 Conclusions

The significant success of LLMs like ChatGPT has the potential to have an important impact in various fields, including the railway system, driving the development of intelligent and automated processes for railway system PHM. This paper investigates the potential influence of LLMs in the railway system and explores solutions to enhance the system's safety and reliability. The paper briefly analyzes the basic principles and technical characteristics of LLMs, highlighting their powerful language understanding and text generation capabilities, which can be applied to various processes in the railway system, including operations management, condition monitoring, maintenance recommendations, and knowledge management, to improve efficiency.

However, the paper also acknowledges the limitations of current LLMs in the railway system. To address these limitations, the paper proposes solutions such as domain-specific fine-tuning, integration with local knowledge bases, and consideration of large-scale visual language models.

The paper emphasizes that LLMs have broad prospects in the railway system. With further research and development of LLM technology, it can be expected that LLMs will become more specialized and intelligent in the railway system, providing significant support for system maintenance and management, and improving reliability, safety, and efficiency.

## References


1. Li, C., Luo, S., Cole, C., Spiryagin, M.: An overview: modern techniques for railway vehicle on-board health monitoring systems. Veh. Syst. Dyn. **55**(7), 1045–1070 (2017)
2. Chen, H., Jiang, B., Ding, S.X., Huang, B.: Data-driven fault diagnosis for traction systems in high-speed trains: a survey, challenges, and perspectives. IEEE Trans. Intell. Transp. Syst. **23**(3), 1700–1716 (2022)
3. Xu, S., Chen, C., Lin, Z., Zhang, X., Dai, J., Liu, L.: Review and prospect of maintenance technology for traction system of high-speed train. Transp. Safety Environ. **3**(3), b17 (2021)
4. Bernal, E., Spiryagin, M., Cole, C.: Onboard condition monitoring sensors, systems and techniques for freight railway vehicles: a review. IEEE Sens. J. **19**(1), 4–24 (2019)
5. OpenAI: GPT-4 technical report. https://arxiv.org/abs/2303.08774
6. Ouyang, L., Wu, J., Jiang, X., Almeida, D., Wainwright, C., Mishkin, P., et al.: Training language models to follow instructions with human feedback. In: Proceedings of the NeurIPS, pp. 27730–27744 (2022)
7. Zhou, C., Li, Q., Li, C., Yu, J., Liu, Y., Wang, G., et al.: A comprehensive survey on pretrained foundation models: a history from Bert to Chatgpt. https://arxiv.org/abs/2302.09419
8. Cao, Y., Li, S., Liu, Y., Yan, Z., Dai, Y., Yu, P.S., et al.: A comprehensive survey of AI-generated content (AIGC): a history of generative AI from GAN to Chatgpt. https://arxiv.org/abs/2303.04226
9. Qin, Z., Yi, H., Lao, Q., Li, K.: Medical image understanding with pretrained vision language models: a comprehensive study. https://arxiv.org/abs/2209.15517
10. Li, Y., Wang, H., Sun, M.: ChatGPT-like large-scale foundation models for prognostics and health management: a survey and roadmaps. https://arxiv.org/abs/2305.06472
11. Lin, T., Wang, Y., Liu, X., Qiu, X.: A survey of transformers. AI Open **3**, 111–132 (2022)
12. Vaswani, A., Shazeer, N., Parmar, N., Uszkoreit, J., Jones, L., Gomez, A.N., et al.: Attention is all you need. In: Proceedings of the NeurIPS (2017)
13. Jing, L., Tian, Y.: Self-supervised visual feature learning with deep neural networks: a survey. IEEE Trans. Pattern Anal. Mach. Intell. **43**(11), 4037–4058 (2021)
14. Liu, X., Zhang, F., Hou, Z., Mian, L., Wang, Z., Zhang, J., et al.: Self-supervised learning: generative or contrastive. IEEE Trans. Knowl. Data Eng. **35**(1), 857–876 (2023)
15. Poria, S., Cambria, E., Bajpai, R., Hussain, A.: A review of affective computing: from unimodal analysis to multimodal fusion. Inf. Fusion **37**, 98–125 (2017)
16. Devlin, J., Chang, M., Lee, K., Toutanova, K.: Bert: pre-training of deep bidirectional transformers for language understanding. https://arxiv.org/abs/1810.04805
17. Nagrani, A., Yang, S., Arnab, A., Jansen, A., Schmid, C., Sun, C.: Attention bottlenecks for multimodal fusion. Adv. Neural. Inf. Process. Syst. **34**, 14200–14213 (2021)
18. Wang, H., Li, Y., Ren, J.: Machine learning for fault diagnosis of high-speed train traction systems: a review. Front. Eng. Manag. 1–17 (2023)
19. Lei, Y., Yang, B., Jiang, X., Jia, F., Li, N., Nandi, A.K.: Applications of machine learning to machine fault diagnosis: a review and roadmap. Mech. Syst. Signal Process. **138**, 106587 (2020)
20. Zhang, J., Huang, J., Jin, S., Lu, S.: Vision-language models for vision tasks: a survey. https://arxiv.org/abs/2304.00685
21. Banabilah, S., Aloqaily, M., Alsayed, E., Malik, N., Jararweh, Y.: Federated learning review: fundamentals, enabling technologies, and future applications. Inf. Process. Manag. **59**(6), 103061 (2022)

# An Adaptive Fourier Decomposition Method for Gear Fault Diagnosis of Railway Vehicle in the Non-stationary Process

Zhongshuo Hu[1], Qiang Li[1], Jinhai Wang[2(✉)], Jianwei Yang[2], and Dechen Yao[2]

[1] School of Mechanical, Electronic and Control Engineering, Beijing Jiaotong University, Beijing, China
qli3@bjtu.edu.cn

[2] School of Mechanical-Electronic and Vehicle Engineering, Beijing University of Civil Engineering and Architecture, Beijing, China
{wangjinhai,yangjianwei,yaodechen}@bucea.edu.cn

**Abstract.** The vibration signal of railway vehicle gears has prominent non-stationary characteristics, and its feature extraction is a challenging issue for fault diagnosis. To tackle this issue, we proposed an adaptive Fourier decomposition method to extract the fault frequency characteristics in non-stationary signals. This method, combined with the computational order tracking technique, can eliminate the signal's fast-changing characteristics and improve the algorithm performance obtaining intrinsic mode decomposition by taking advantage of the orthogonality, completeness, locality, and adaptability of the Fourier decomposition method (FDM). Then, a parameter setting-free AFDM method is established based on the reconstructing signal using FDM entropy and weighted kurtosis. Finally, the simulation signal and experimental signal study cases are carried out to verify its effectiveness. The results show that the proposed method's peak signal-to-noise ratio (PSNR) of the reconstructed signal is higher 23.2% than the empirical mode decomposition method (EMD), which can effectively extract the weak sideband frequency features generated by the modulation of the impact behaviour, and demodulates the weak frequency to identify the gear spalling fault of the railway vehicle.

**Keywords:** Vehicle engineering · Adaptive Fourier decomposition · Order tracking · Energy entropy

## 1 Introduction

In order to ensure the safety and reliability of urban rail transit operations, it is imperative to carry out health management for the critical components of railway vehicles. The railway vehicle gear is one of the rotating components with a high failure rate, and its fault types mainly include tooth root crack, spalling, and tooth surface wear. Dynamics research results show that gear meshing frequency contributes significantly to high-frequency vibration in railway vehicle systems; simultaneously, it has obvious


Y. Qin et al. (Eds.): EITRT 2023, LNEE 1137, pp. 600–618, 2024.
https://doi.org/10.1007/978-981-99-9311-6_60

harmonic excitation characteristics [1, 2]. A crack on the tooth root is an obvious fault type, and the sideband frequencies modulation component of shaft frequency impact is obvious on the gearbox housing [3, 4]. However, the characterization of spalling in tooth surface is weak to be observed, and it is difficult to be identified by sideband frequencies characteristics near meshing frequency. In the course of train operation, due to the changeable train operating conditions, complex operating environment and other problems, the corresponding fault frequency characteristics vary with the operating conditions, so it is difficult to identify the weak fault directly from the gearbox spectrum by using the side frequency band characteristics. Therefore, research on advanced signal processing algorithms and the extraction of weak information effectively is a challenging issue of fault diagnosis for railway vehicle gears.

Many pioneer researchers proposed various signal decomposition algorithms to extract faulty features of key components in the rotating machinery [5, 6]. At present, the development of empirical mode decomposition (EMD) tends to be mature. The method decomposes non-stationary signals into a series of intrinsic mode functions (IMFs), affecting bearing and gearbox fault diagnosis. However, this method has some problems with boundary effect and mode mixing [7, 8]. Wu et al. proposed ensemble empirical mode decomposition (EEMD), which effectively alleviates the mode mixing problem by adding white noise [9]. After that, the algorithm was combined with other algorithms to optimize the result of decomposition, and many developed algorithms as represented by complete ensemble empirical mode decomposition (CEEMD) were proposed, which were widely used for feature extraction of faulty signals. Mohammad et al. [10] proposed a method of denoising parameters based on wavelet transform to improve CEEMD. The energy level of the Combined Fault-Sensitive IMF (CFSIMF) envelope is used to describe the fault severity level. Fault Type Index (FTI) is proposed to identify the fault Type at the same time. The method has great performance on gearing fault diagnosis verified by test data. Prosvirin et al. [11] discussed the optimal selection of IMFs for EEMD, which was distinguished by the ratio of degree-of-presence (DPR) and Kullback-Leibler divergence (DPR/KLdiv). The adaptive threshold was used to select the optimal IMF, and the method showed a good performance of denoising in the analysis of experimental data, and preserved the features of frictions faults in the original signal. Wang et al. [12] proposed a Complete ensemble empirical mode decomposition with Adaptive noise (CEEMDAN). In this method, a new criterion was proposed to quantify the fault-related degree of vibration signals, it can replace the statistical parameter evaluation method in the traditional method. Finally, the signal is reconstructed based on reweighted reconstruction strategy. The ability of this method to improve the signal-to-noise ratio is verified in the simulation signals and experimental signals. Kou et al. [13] combined improved CEEMDAN with energy entropy energy Entropy and artificial fish swarm algorithm (AFSA) (ICEEMDAN). The advantage of the method is that the position of bearing defects can be determined through energy entropy, and AFSA can improve the accuracy of diagnosis model at the same time. Although EMD and other developed algorithms have excellent decomposition effects, they lack mathematical theoretical support.

In the field of fault diagnosis for a rotating machine, many signal components have their frequency range, so implementation of signal decomposition in the frequency domain is one of the main ideas for fault diagnosis. Discrete wavelet transform (DWT) decomposes signals into multiple quadrature components in the frequency domain, but the poor resolution in the high-frequency band can be its serious weakness [14]. Afterward, wavelet packet transform (WPT) and empirical wavelet transform (EWT) were proposed successively [15, 16], and there are a variety of improved methods for these algorithms [17, 18]. Merainani et al. [19] designed wavelet filter banks based on EWT. The method was proved to be very sensitive to incipient gearbox fault initiation through simulation and experimental data, and had been applied in gearbox crack fault detection. Wang et al. [20] introduced the concept of sparsity into empirical wavelet transform and proposed sparsity guided empirical wavelet transform. This method could automatically establish the Fourier segments required by empirical wavelet transform, and identify the faults of multiple railway axle bearings. The method has certain noise robustness compared with traditional methods. However, due to the computational property, the algorithms still can be easily affected by environmental noise and perform not enough robustness. The algorithm is challenging to realize feature extraction of weak failure for gearbox with a complex working environment.

Fortunately, variational mode decomposition (VMD) provides a new idea for signal analysis. The algorithm believes that the signal is composed of multiple narrowband signals with different central frequencies, the problem about signal decomposition is transformed into a constrained optimization with complete mathematical theory support [21–23]. This algorithm can reduce the problems on edge effect and mode mixing effectively, making it mainstream in the field of fault diagnosis in recent years. However, to achieve a better effect on feature extraction, many hyperparameters need to be optimized before computation [24, 25]. For example, when Wang et al. [26] applied VMD to gearbox fault diagnosis, the number of mode components k and the penalty factor a were optimized by combining envelope spectrum entropy with waveform method. Gong et al. [27] proposed tentative variational mode decomposition (TVMD) to enhance the reliability of adaptive parameter selection. Wang et al. [28] used Permutation Entropy Optimization (PEO) to accomplish selection of k adaptively.

With incredible development of machine learning in recent years, neural network was combined with signal processing by scholars [29–31]. Wang et al. [32] input the time-frequency representation results processed by dislocation into CNN network. The method increases the size of samples and solves the problem of lack of signal. After the optimization of ensemble decision approach, the method has great accuracy and stability in gearbox fault diagnosis. Bai et al. [33] optimized Markov transition field algorithm and applied it to process the frequency domain information, namely spectral Markov transfer field (SMTF) algorithm. The advantages of the algorithm are that the vibration signal spectrum can be converted into images without any parameters requiring prior knowledge, and the axlebox data with multiple working conditions and faults types can be classified accurately. Ye et al. [34] proposed a Gaussian white noise strategy (GWN-strategy) and an edge sample training strategy (EST-strategy) to process 1-dimensional vibration signals, and combined with 1D convolution network into GWN-EST-1DCNN method. The method was verified by simulation signals and measurement, and finally realized the fault detection of train suspension system. Although the theoretical result was effective, the computation speed was also reduced greatly, which lead to certain challenges for application in practical engineering.

Therefore, the Fourier mode decomposition (FDM) method is proposed, which is adaptive, a complete theory supported, and efficient; compared with the EMD, it can calculate faster and reduce the impact of modal aliasing effectively. In non-stationary signal processing, FDM has a flexible application on different fields [35]. In recent years, FDM has been optimized and applied to the fault diagnosis of key mechanical components. Huang et al. [36] combined the wavelet transform and order statistic filter based on Fourier decomposition is improved, and solved the Fourier spectrum boundary and false weight problem, bearing failure data was used to validate the method, the results show that the method can obtain more accurate weight, reflects the effective denoising performance. Deng et al. [37] proposed an adaptive bandwidth Fourier decomposition method (ABFD), which optimized the extraction of narrowband components and made improvements in terms of computational complexity and parameter optimization. Compared with traditional EMD and VMD algorithms, experimental data verification showed that the algorithm can extract single component from noise signal more accurately and realize bearing fault diagnosis. Zhou et al. [38] proposed an empirical Fourier decomposition (EFD) based on empirical wavelet transform and Fourier decomposition method. Main frequency component of Fourier decomposition process was optimized by segmentation, solved the component redundancy and modal aliasing problems, the result of time-frequency representation (TFR) and multiple decomposition are more accurate compared with traditional method. The algorithm was verified by structural health monitoring data, which showed that the algorithm has good engineering significance and practical value. Because the conditions for component decomposition can be easily satisfied in process of FDM, the over-decomposition problem is commonly found in result of FDM. Zheng et al. [39] used adaptive power spectrum search to solve the over-decomposition problem of FDM, based on which they proposed the Adaptive Power Spectrum Fourier Decomposition method (APSFDM). Compared with the current EMD, EWT and VMD methods, this method has better signal fidelity. Most of the current research work is focusing on optimizing the redundant components of FDM. However, when the signal consists of complex components or the frequency of a single intrinsic mode function changes too fast with a wide span in the frequency domain, the performance of FDM may be challenging to achieve the desired effect. Therefore, the optimization of Fourier decomposition for variable working conditions is one of the current problems to be solved.

Given this condition, we proposed an adaptive Fourier decomposition method (AFDM) to carry out this issue. In this framework, the characteristic frequency demodulation of non-stationary signals is realized by combining the computational order tracking technique (COT) and weighted kurtosis reconstruction. Based on the unique advantages of FDM, this paper introduces this method into the gear fault diagnosis of railway vehicles. Finally, the algorithm is verified by analyzing simulation data and experimental data.

This paper has the following organization: The basic theory methods about computational order tracking technique, Fourier mode decomposition, FDM entropy, and weighted kurtosis are introduced in Sect. 2. The proposed method is derived and explained in Sect. 3. It is verified by simulating data in the following Sect. 4. Analyzing results about experimental data and comparison with classical EMD method is provided in Sect. 5. Finally, the advantages of the proposed method are concluded in Sect. 6.

## 2 Basic Principle

### 2.1 Computational Order Tracking

In general, vibration signals are collected at fixed sampling frequency fs to obtain time-domain sequence signals with time interval $\Delta t$. For this kind of stationary signal, the frequency components and characteristic information can be analyzed by conventional time and frequency domain methods. However, it is unsuitable for non-stationary signals with rotating frequency variation. For this kind of problem, variable speed signal can be normalized frequency by equal angle sampling method to eliminate the amplitude and frequency interference caused by speed. By COT method, the signal with equal time sampling interval can be converted into equal Angle sampling signal to ensure that the signal has the same number of sampling points in each period. Due to acquisition equipment and installation limitations, most vibration signals are sampled at equal time intervals. Therefore, the COT method is necessary for processing non-stationary signals. The signals are resampled with same angle interval to meet the needs of time and frequency domain analysis. The basic principle is described in the following.

According to the basic characteristics of the frequency changing regularity, it is assumed that the rotation angle and time meet the following cubic polynomial relationship:

$$\theta(t) = b_0 + b_1 t + b_2 t^2 + b_3 t^3 \tag{1}$$

where $b_0$, $b_1$, $b_2$, and $b_3$ are the coefficients of the polynomial, respectively, and $t$ is the time series.

Since different rotation periods correspond to a specific rotation angle, the corresponding polynomial coefficients can be fitted when the sampling points are the same as the number of selected undetermined coefficients.

$$\begin{bmatrix} \theta_1 \\ \theta_2 \\ \theta_3 \\ \theta_4 \end{bmatrix} = \begin{bmatrix} 1 & t_1 & t_1^2 & t_1^3 \\ 1 & t_2 & t_2^2 & t_2^3 \\ 1 & t_3 & t_3^2 & t_3^3 \\ 1 & t_4 & t_4^2 & t_4^3 \end{bmatrix} \begin{bmatrix} b_1 \\ b_2 \\ b_3 \\ b_4 \end{bmatrix} \tag{2}$$

where $(\theta_1, t_1)$, $(\theta_2, t_2)$, $(\theta_3, t_3)$, and $(\theta_4, t_4)$ are four different sampling points respectively. The solution of each coefficient can be solved by matrix computation.

Furthermore, the corresponding nonlinear equation numerical algorithm can be used to calculate the time node of each angle, and the corresponding resampling value can be obtained by linear interpolation. By processing COT, the time domain signal is converted to the angular domain sequence with the same angle interval $\Delta\theta$. Thus, the stationary signal analysis technique can be used for feature extraction.

### 2.2 Fourier Mode Decomposition

FDM can obtain the Fourier instinsic band functions (FIBFs) by expanding the target signal into the Fourier series. If the signal is a finite real number sequence and meets the Dirichlet condition, the FDM can be used for decomposition. In general, for periodic signals, the Fourier series can be expanded as:

$$x_{T_0}(t) = a_0 + \frac{1}{2}\sum_{k=1}^{\infty}\left[c_k \exp(jk\omega_0 t) + c_k^* \exp(-jk\omega_0 t)\right] \tag{3}$$

where $c_k = (a_k - jb_k)$, $c_k^* = (a_k - jb_k)$, $a_k$ and $b_k$ are Fourier coefficients.

According to Eq. (3), it can be further summarized as:

$$\begin{aligned} x_{T_0}(t) &= a_0 + \frac{1}{2}\sum_{k=1}^{\infty}\left[c_k \exp(jk\omega_0 t) + c_k^* \exp(-jk\omega_0 t)\right] \\ &= a_0 + \mathrm{Re}\left[z_{T_0}(t)\right] \end{aligned} \tag{4}$$

where $z_{T_0}(t)$ is analytic function, which can be described as:

$$z_{T_0}(t) \triangleq \sum_{k=1}^{\infty} c_k \exp(j\omega_0 t) \tag{5}$$

The above function illustrates the basic property that the periodic signal can be composed of infinite trigonometric functions with a linear combination. To realize the extraction of each intrinsic mode information, the analytic Fourier intrinsic mode function can be obtained, $z_{T_0}(t)$ can be expressed as:

$$z_{T_0}(t) = \sum_{i=1}^{M} a_i(t) \exp\left[j\phi_i(t)\right] \tag{6}$$

where $a_i(t)$ is the instantaneous amplitude function, $\phi_i(t)$ is the instantaneous phase function. It can be seen from the above that the FIBFs of Fourier is an individual component after removing the mean value of the Fourier expansion signal, and Fourier decomposition of signal is realized by constructing the mathematical equation of the analytic signal. Combining Eqs. (5) and (6), we can rewrite the expression as:

$$\sum_{i=1}^{M} a_i(t) \exp(j\phi_i(t)) = \sum_{k=1}^{\infty} c_k \exp(jk\omega_0 t) \tag{7}$$

According to the different order of solving intrinsic modes, there are two kinds of FDM, the LTH algorithm calculates from low frequency to high frequency, and the HTL algorithm calculates from high frequency to low frequency [40].

In the LTH algorithm, we can get:

$$a_1(t) \exp(j\phi_1(t)) = \sum_{k=1}^{N_1} c_k \exp(jk\omega_0 t)$$

$$a_2(t)\exp(j\phi_2(t)) = \sum_{k=N_1+1}^{N_2} c_k \exp(jk\omega_0 t)$$

$$a_M(t)\exp(j\phi_M(t)) = \sum_{k=(N_{M-1}+1)}^{\infty} c_k \exp(jk\omega_0 t)$$

The above expression means the sampled signal can be expressed by $M$ Fourier intrinsic mode functions and mean components. Thus, the $i$th mode function can be constructed as:

$$a_i(t)\exp(j\phi_i(t)) = \sum_{k=N_{i-1}+1}^{N_i} c_k \exp(jk\omega_0 t) \tag{8}$$

where $i = 1, \ldots, M$, and $N_0 = 0, N_M = \infty$. These FIBFs are the real part of AFIBFs in formula (8). In order to obtain the minimum number of AFIBFs, each $i$ starts from the $(N_{i-1} + 1)$ and appends more terms until $N_i$ reaches the maximum value and satisfies $(N_{i-1} + 1) \le N_i \le \infty$ at the same time, then we can get the following function.

$$a_i(t) \ge 0,\ \omega_i(t) = \frac{d\phi_i(t)}{dt} \ge 0,\ \forall t \tag{9}$$

where $a_i(t)$ and $\omega_i(t) = 2\pi f_i(t)$ are the instantaneous amplitude (IA) and instantaneous frequency (IF).

Similar to HTL, in the LTH algorithm, we can get:

$$a_1(t)\exp(j\phi_1(t)) = \sum_{k=N_1}^{N_1} c_k \exp(jk\omega_0 t)$$

$$a_2(t)\exp(j\phi_2(t)) = \sum_{k=N_2}^{N_2} c_k \exp(jk\omega_0 t)$$

$$a_M(t)\exp(j\phi_M(t)) = \sum_{k=1)}^{(N_{M-1}-1)} c_k \exp(jk\omega_0 t)$$

Note that the lowest and highest values of the sum in Eq. (8) will also change to $k = N_i$ to $(N_{i-1} - 1)$. That is:

$$a_i(t)\exp(j\phi_i(t)) = \sum_{k=N_i}^{N_{i-1}-1} c_k \exp(jk\omega_0 t) \tag{10}$$

where $i = 1, \ldots, M$, and $N_0 = \infty, N_M = 0$. Under this situation, $i$ starts from $(N_{i-1}-1)$ and decreases gradually, finally gets to the minimum value of $N_i$, which can satisfy $1 \le N_i \le (N_{i-1} - 1)$.

Due to FDM is a kind of time-frequency analysis method based on Fourier series construction, it has the advantages of orthogonality, completeness, locality, and adaptability. It can realize the extraction of characteristic frequency effectively.

### 2.3 Fourier Mode Decomposition Entropy

Information entropy can be used to describe the average uncertainty in probabilistic systems. In practice engineering, it is often used to indicate the amount of information in an event. Mathematically, entropy is the expected value of the amount of information. If the probability distribution of each event is uniform in the system, the system has the maximum value of information entropy when each event has an equal probability, which means the uncertainty of the system reaches the maximum level. On the contrary, the information entropy of the system falls to 0 when the probability is 1 for an event.

In order to compare the advantages of the two kinds of FDM, FDM entropy is calculated to evaluate the signal. For the $M$ intrinsic modes obtained by FDM, their energies are $E_1, E_2 \ldots E_M$, they can divide signal energy in the frequency domain, and its FDM entropy can be defined as:

$$H_E = -\sum_{i=1}^{M} p_i \ln p_i \tag{11}$$

where $p_i$ is the ratio of the energy $E_i$ of the $i$th FIBF to the total energy $E$, i.e., $p_i = E_i/E$.

According to the different order of solving FIBF, there are two kinds of FDM, the LTH algorithm calculates from low frequency to high frequency, and the HTL algorithm calculates from high frequency to low frequency [35]. Due to FDM being a time-frequency analysis method based on Fourier series construction, it has the advantages of orthogonality, completeness, locality, and adaptability. It can realize the extraction of characteristic frequency effectively.

### 2.4 Weighted Kurtosis

Kurtosis is defined as the normalized fourth-order center moment, which is a classical statistic and can be used to measure the convexity of peak value for random variable sequence and identify signal features with obvious impact characteristics. In the frequency domain, gear's vibration signals with partial failures are characterized by the meshing frequency as the fundamental frequency and the sidebands modulated by shaft frequency on both sides. In the time domain, the overall level of signal amplitude increases due to the interference of noise and other vibrations, and the fault impact amplitude is often not obvious. Therefore, it is tough to extract fault features from noisy signals only by kurtosis. Therefore, the Pearson correlation coefficient was used to weight the kurtosis value, which was used to characterize the similarity between FIBFs and the original signal to avoid the wrong selection of FIBFs. Specifically, weighted kurtosis can be defined as:

$$K_W = K \cdot C \tag{12}$$

where $K$ is the kurtosis and $C$ is the Pearson correlation coefficient, respectively, and the specific formula is shown in Table 1.

where x is the original signal, $x_{FIBF}$ is FIBF, $N$ is sampling number, $\mu_x$, $\mu_{x_{IFIBF}}$ and $\sigma_x$, $\sigma_{x_{FIBF}}$ is the corresponding mean and variance value.

**Table 1.** Descriptions of the indexes.

| Indexes | Example |
|---|---|
| Kurtosis (K) | $K = \dfrac{\sum_{n=0}^{N-1} x_{FIBF}^4(n) \Big/ N}{\left[\sum_{n=0}^{N-1} x_{FIBF}^2(n) \Big/ N\right]^2}$ |
| Pearson correlation coefficient (C) | $C = \frac{1}{N-1}\sum_{n=1}^{N}\left(\frac{x_{FIBF}(n)-\mu_{xIFIBF}}{\sigma_{xFIBF}}\right)\left(\frac{x(n)-\mu_x}{\sigma_x}\right)$ |

According to the above coefficients, the appropriate FIBFs can be selected for signal reconstruction. A reconstructed signal can extract the characteristic frequency of the original signal and denoise the signal and improve the SNR.

## 3 Adaptive Fourier Mode Decomposition

The essence of FDM is extracting the intrinsic modes according to the direction of the frequency band in order. However, it is difficult to completely acquire signal features under the fast change of rotating frequency and wide range of frequency bands. Therefore, when signals are under changing working conditions, it needs to be combined with specific signal processing technique to meet the needs of FDM. The procedure is shown in Fig. 1. The steps of the adaptive Fourier decomposition method are described as follows specifically:

Step 1: Convert time-domain non-stationary vibration signal to angular stationary vibration signal by COT according to the rotational speed signal.
Step 2: FDM was used to extract transient information of the signal, and FIBFs of the LTH algorithm and the HTL algorithm were calculated.
Step 3: FDM entropy of FIBFs of FDM was calculated, and the FIBFs with the minimum entropy value were selected for reconstruction.
Step 4: Gini coefficient of each FIBF were calculated;
Step 5: The FIBFs which can describe fault characteristics were selected for signal reconstruction (the first three orders were selected in this paper).
Step 6: The Hilbert transform was implemented on the signal to construct an analytic signal.
Step 7: The spectrum of the signal was calculated. Thus the characteristic frequency was demodulated to identify the fault information.

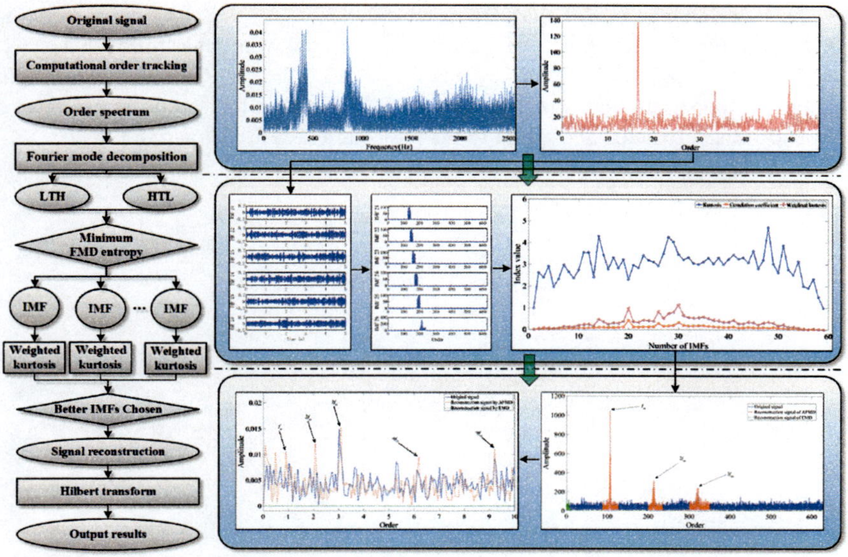


**Fig. 1.** Framework of signal processing by FDM

# 4 Simulation Signal Analysis

A simulation signal was set up to verify the feasibility of the algorithm. For gear non-stationary modulation signals with localized damage, it is obvious that the sideband frequencies are modulated by shaft frequency near the meshing frequency and its harmonics. Based on this feature, the vibration signal equation of an acceleration process can be assumed as:

$$\begin{aligned} x_s = {} & A_1 t\left[1+\sum_{i=1}^{3} B_i t \cos\left(2\pi f_o(t)t+\frac{\pi}{6}\right)\right]\cos\left(2\pi f_m(t)t+\frac{\pi}{3}\right) \\ & + A_2 t\left[1+\sum_{i=1}^{3} B_i t\cos(2\pi f_o(t)t)\right]\cos(4\pi f_m(t)t) \\ & + A_3 t\left[1+\sum_{i=1}^{3} B_i t\cos(2\pi f_o(t)t)\right]\cos(6\pi f_m(t)t)+\xi(t) \end{aligned} \tag{13}$$

where $A_1$, $A_2$, and $A_3$ are the change rates of vibration amplitude; $B_1$, $B_2$, and $B_3$ are the change rates of the amplitude of modulation frequency; $\xi(t)$ represents white noise; $f_o(t)$ is the shaft frequency; $f_M(t)$ is the meshing frequency, $f_M(t) = Zf_o(t)$. Due to the research object being a non-stationary process, the expression of shaft frequency can be described as:

$$f_o(t) = \frac{f_{o1}t}{2} \tag{14}$$

where $f_{o1}$ is the change rate of shaft frequency.

In this section, the parameters of the simulation signal were set as follows: $A_1 = 1.2$, $A_2 = 0.8$, $A_3 = 0.4$, $B_1 = 0.9$, $B_2 = 0.6$, $B_3 = 0.3$, $Z = 30$, $f_{o1} = 8$. The sampling frequency of the simulation signal was set to 5120 Hz, and the sampling time was 1s. The waveforms calculated according to Eqs. (11) and (12) are shown in Fig. 2.

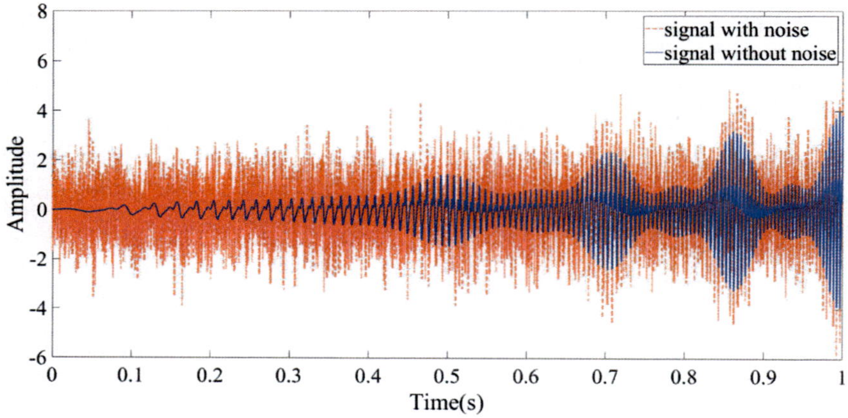


**Fig. 2.** Time-domain waveforms of simulation signal

The envelope order spectrum of the simulation signal was calculated and obtained by using AFDM is shown in Fig. 3. It can be seen that the side frequency features near the simulation signal were accurately extracted from the envelope order spectrum. The order spectrum with shaft frequency $f_o$ as the basic frequency and $2f_o$, $3f_o$, and other frequency multiple components are formed, proving that the algorithm can effectively realize frequency demodulation of non-stationary signals.

## 5 Experimental Signal Analysis

### 5.1 Experimental Description

The railway vehicle gearbox test bench, sensor layout, and test gear are shown in Fig. 4. The whole test system consists of a traction motor, torque increasing gearbox, test gearbox, and load motor. The modulus of test bevel gear was 5 mm, and the number of teeth was 17/107. The spalling failure was processed in the tooth surface of the driving gear, and its shape is like a rectangular defect 8 mm long along the direction of the contact line. In order to simulate the running process of the subway in reality, the load motor worked according to axle load and vehicle acceleration. For the gear torsional vibration system, the load torque is calculated as follows:

$$T_i = m_w r_w a \quad (15)$$

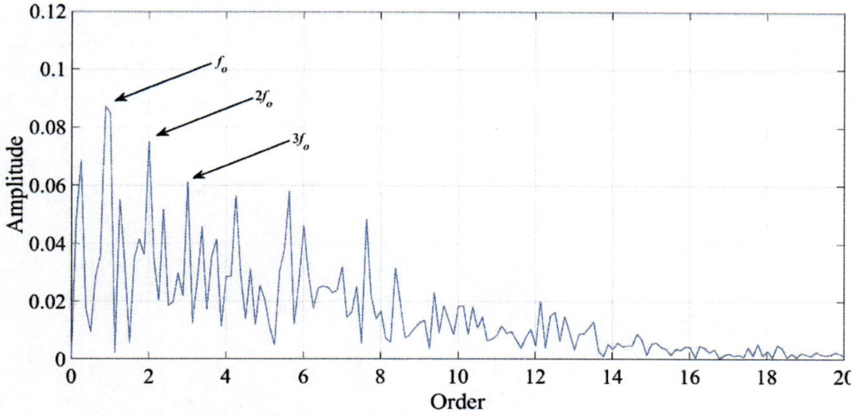


**Fig. 3.** Envelope order spectrum of the reconstructed signal

where $m_w$ is the axle load, $r_w$ is the rolling radius of the wheelset, and $a$ is the vehicle's acceleration. For the gear torsional vibration system, load torque is equal to inertial torque [41]. In this experiment, according to the traction system design for subway vehicles, loading was carried out as acceleration was 0.8 m/s$^2$ and axle load was 13t.

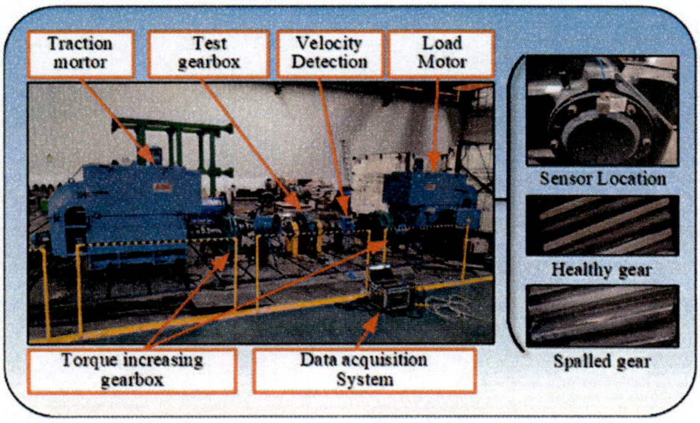


**Fig. 4.** Fault diagnosis test bench of urban railway vehicle gearbox

For ensuring the characteristic information could be collected clearly and stably, the vibration acceleration sensor is arranged on the cover of the pinion, and the spalling fault is processed on the driving gear according to actual working conditions. The sampling frequency of the equipment is set as 5120 Hz to ensure vibration information and fault features can be collected entirely.

The time-domain response and time-frequency response of vibration signals in the test are shown in Fig. 5. The transverse acceleration data was selected for analysis, which can effectively reduce the interference of shaft frequency to demodulation feature. Figure 5(a) shows the traction condition signal waveform in the time domain. Under this condition, the maximum transverse acceleration can reach 10g approximately, which is close to the experimental data, proving that the experiment can reproduce the characteristics of the actual engineering condition. Figure 5(b) shows the analysis results of transverse acceleration data in the time-frequency domain. In this map, $1/2f_m$ frequency division can be observed clearly due to the traction load and speed changes in the traction process [41].

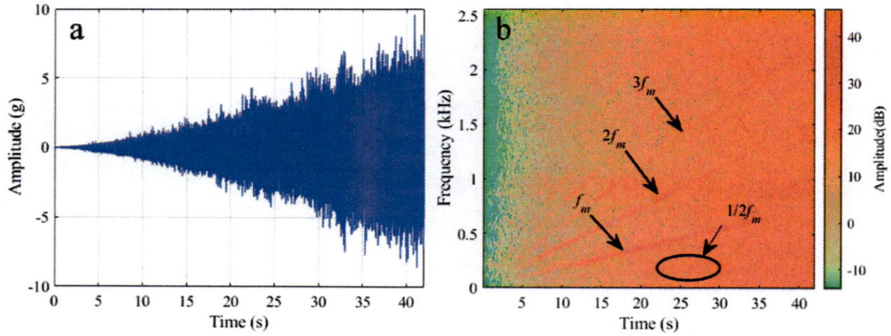


**Fig. 5.** Measured data of urban railway vehicle gearboxes under traction conditions. (a) The traction condition signal waveform in the time domain. (b) The analysis results of transverse acceleration data in the time-frequency domain.

## 5.2 Traction Condition Signal Analysis

The proposed method AFDM was used on vibration signals to obtain FIBFs. Due to space limitation, only 6 FIBFs waveforms in the time domain and order spectrum were given in Fig. 6. According to the map, any FIBF can represent the characteristic information of a particular frequency band in the order spectrum. It can be proved that the essence of FDM is a kind of band-pass filter, which divides the information into frequency bands according to certain rules.

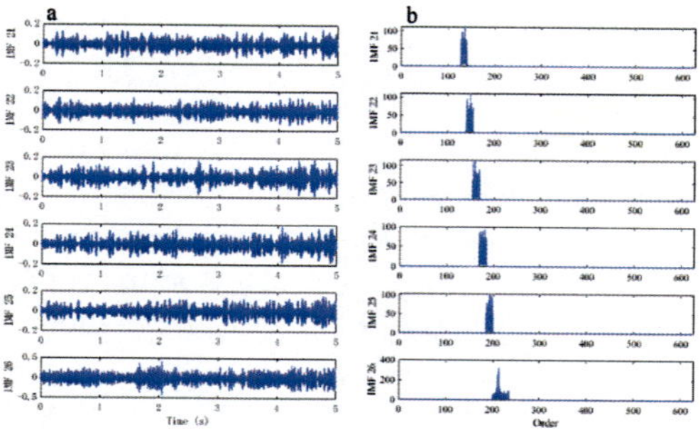


**Fig. 6.** Waveforms of intrinsic mode functions. (a) The FIBFs in the time domain. (b) The FIBFs in the frequency domain.

The weighted kurtosis (WK) of each FIBF is calculated as shown in Fig. 7. Although kurtosis is a sensitive index to a peak value of signals, abnormal kurtosis index still can be calculated due to noise interference. This abnormal value will cause serious consequences to the reconstructed signal, too many noise components will be chosen in signal reconstruction. Therefore, the correlation coefficient was used to make the kurtosis value reflect the actual situation of FIBFs, which can avoid the influence of noise and reconstruct signals accurately.

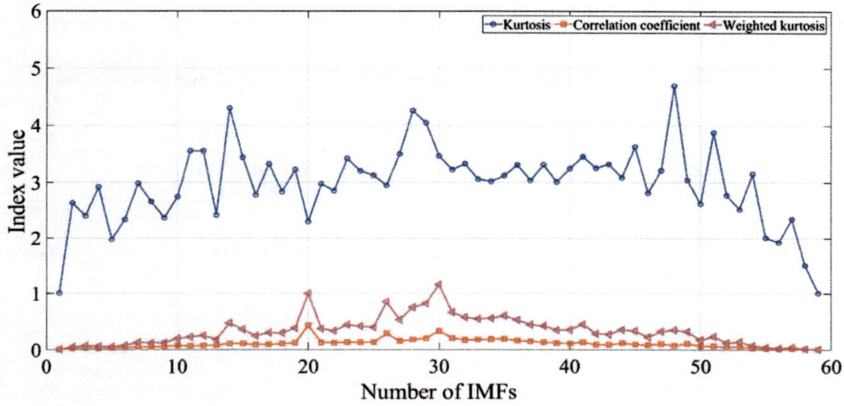


**Fig. 7.** Comparison of indicators of intrinsic mode functions

In this paper, EMD was chosen for comparison with the proposed method. The peak signal-to-noise ratio (PSNR) after decomposition with two methods is shown in Table 2. The PSNR of the proposed method is larger by 23.2% than EMD apparently, which indicates that the proposed method has a better denoising effect.

**Table 2.** T PSNR comparison of AFDM and EMD.

| | Proposed method | EMD | D-value |
|---|---|---|---|
| PSNR | 10.8301 | 8.7902 | 23.2% |

The order spectrum of the reconstructed signal for fault gear is shown in Fig. 8. The first 3 FIBFs extracted are meshing frequencies, $f_m$, $2f_m$, $3f_m$, and sideband frequencies. The order spectrum of the reconstructed signal contains the main characteristic information when a gear is in working condition, and irrelevant information such as noise in vibration signal was filtered effectively. On the contrary, the reconstructed signals obtained by EMD could not extract any components near the meshing frequency.

Figure 9 shows the envelope order spectrum of vibration signals of fault gear under traction conditions. By comparing the analysis results of the original signal and the reconstructed signal, it can be found that the shaft frequency $f_o$ and $2f_o$ are extracted

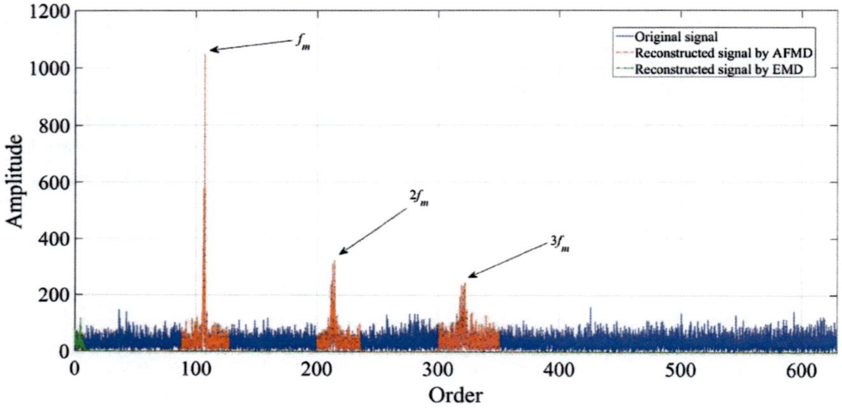


**Fig. 8.** Order spectrum of reconstruction signal for spalled gear

from the reconstructed signal, indicating that the periodic impulses caused by spalling partial failure are extracted accurately.

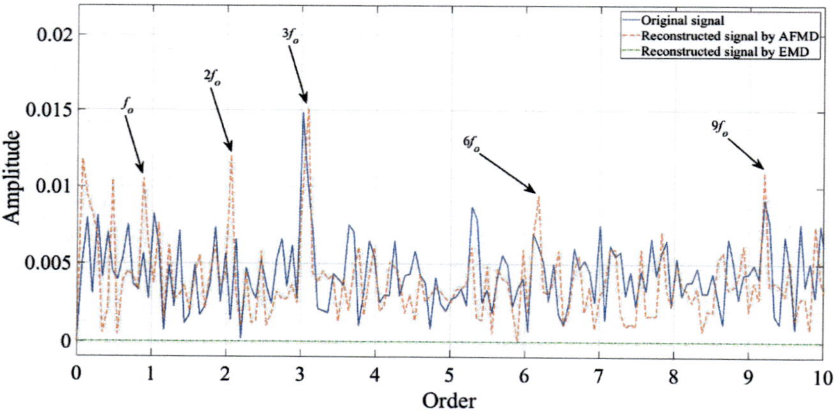


**Fig. 9.** Envelope order spectrum of vibration signal for spalled gear

In order to further verify that the AFDM can extract useful information from non-stationary signals effectively, vibration signals of the healthy gearbox are selected for comparison to test the accuracy of fault diagnosis. Envelope order spectrum analysis for both signals is shown in Fig. 10. Without the influence of periodic impulses, only multiple frequencies $3f_o$, $6f_o$, and $9f_o$ can be found in the envelope order spectrum of healthy gear, and the frequency response amplitude of faulty gearbox is more obvious than healthy gearbox can further verify that AFDM can identify the healthy state of gearbox effectively.

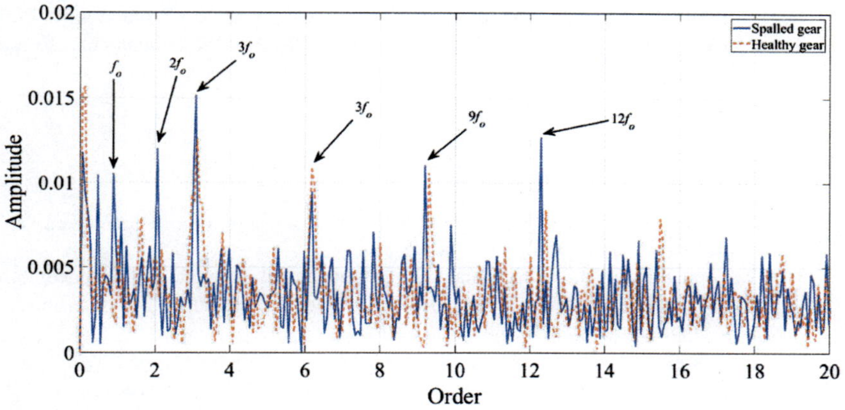


**Fig. 10.** Comparison of envelope order spectrum of spalled gears and healthy gears

## 6 Conclusions

In this paper, a novel gearbox fault diagnosis algorithm AFDM was proposed. Through the research of simulation signal and experimental signal, we can obtain the conclusions as follows.

(1) The FDM entropy was proposed. Thus, the AFDM was constructed to extract non-stationary features without any parameters to avoid problems with hyperparameters like EEMD or VMD.
(2) Weighted kurtosis can identify the impact components of FIBFs and effectively reduce noise influence. The reconstructed signal extracted by AFDM has better SNR performance, and its PSNR is 23.2% higher than the traditional EMD.
(3) AFDM can extract characteristics of meshing frequency and side frequency band from vibration signals effectively. It can also demodulate fault frequency and identify the spalling fault state in the tooth surface, providing evidence for realizing the fault diagnosis of subway gear.

Vibration information of railway vehicle gears is a non-stationary and nonlinear random signal. It is an industry challenge to extract fault features precisely under the variable working condition with fast frequency change and wide frequency band. The frequency demodulation calculated from AFDM has obvious components and a simple spectrum structure, which can be applied to the fault diagnosis of the railway vehicle gears.

**Acknowledgements.** The authors appreciate the support from the Beijing Key Laboratory of Performance Guarantee on Urban Rail Transit Vehicles for this research. This work was supported by the National Natural Science Foundation of China [51975038]; the Beijing Natural Science Foundation (3214042, KZ202010016025, L191005); the Beijing Postdoctoral Research Foundation (2021-zz-114); the Development of High-Level Teachers in Beijing Municipal Universities

(CIT&TCD201904062); the Open Research Fund Program of Beijing Key Laboratory of Performance Guarantee on Urban Rail Transit Vehicles (PGU2020K001); the Fundamental Research Funds for the Beijing Municipal Universities (X21049).

## References

1. Ren, Z.S., Xin, X., Sun, G., Wei, X.: The effect of gear meshing on the high-speed vehicle dynamics. Veh. Syst. Dyn. **59**(5), 743–764 (2020)
2. Zhang, T., Chen, Z.G., Zhai, W.M., Wang, K.Y.: Establishment and validation of a locomotive–track coupled spatial dynamics model considering dynamic effect of gear transmissions. Mech. Syst. Signal Process. **119**, 328–345 (2018)
3. Chen, Z.G., Zhai, W.M., Wang, K.Y.: Vibration feature evolution of locomotive with tooth root crack propagation of gear transmission system. Mech. Syst. Signal Process. **115**, 29–44 (2019)
4. Wang, J.H., Yang, J.W., Lin, Y.L., He, Y.P.: Analytical investigation of profile shifts on the mesh stiffness and dynamic characteristics of spur gears. Mech. Mach. Theory **167**, 104529 (2022)
5. Huang, H.P., Wei, S.Y., Chao, H.H., Hsuan, C.F., Hsu, L., Chi, S.: An investigation study on mode mixing separation in empirical mode decomposition. IEEE Access **7**, 100684–100691 (2019)
6. Zhao, X.J., Qin, Y., Fu, H.L., Jia, L.M., Zhang, X.N.: Blind source extraction based on EMD and temporal correlation for rolling element bearing fault diagnosis. Smart Resil. Transp. **3**(1), 52–65 (2021)
7. Hou, J.B., Wu, Y.X., Gong, H., Ahmad, A.S., Liu, L.: A novel intelligent method for bearing fault diagnosis based on EEMD permutation entropy and GG clustering. Appl. Sci. **10**(1), 386 (2020)
8. Ge, J.H., Niu, T.Y., Xu, D., Yin, G.B., Wang, Y.P.: A rolling bearing fault diagnosis method based on EEMD-WSST signal reconstruction and multi-scale entropy. Entropy **22**(3), 290 (2020)
9. Wu, Z.H., Huang, N.E.: Ensemble empirical mode decomposition: a noise-assisted data analysis method. Adv. Adapt. Data Anal. **1**(1), 1–41 (2011)
10. Sadegh, H.M., Esmaeilzadeh, K.S., Saleh, S.M.: Quantitative diagnosis for bearing faults by improving ensemble empirical mode decomposition. ISA Trans. **83**, 261–275 (2018)
11. Prosvirin, A.E., Islam, M.M.M., Kim, J.M.: An improved algorithm for selecting IMF components in ensemble empirical mode decomposition for domain of rub-impact fault diagnosis. IEEE Access **7**, 121728–121741 (2019)
12. Wang, L.M., Shao, Y.M.: Fault feature extraction of rotating machinery using a reweighted complete ensemble empirical mode decomposition with adaptive noise and demodulation analysis. Mech. Syst. Signal Process. **138**, 106545 (2020)
13. Kou, Z.M., Yang, F., Wu, J., Li, T.Y.: Application of ICEEMDAN energy entropy and AFSA-SVM for fault diagnosis of hoist sheave bearing. Entropy **22**(12), 1347 (2020)
14. Mallat, S.: Wavelet Tour of Signal Processing. Academic Press, San Diego (1998)
15. Akansu, A.N., Haddad, R.A.: Multiresolution Signal Decomposition: Transforms, Subbands and Wavelets. Academic Press, Boston (1992)
16. Gilles, J.: Empirical wavelet transform. IEEE Trans. Signal Process. **61**(16), 3999–4010 (2013)
17. Zheng, J.D., Pan, H.Y., Yang, S.B., Cheng, J.S.: Adaptive parameterless empirical wavelet transform based time-frequency analysis method and its application to rotor rubbing fault diagnosis. Signal Process. **130**, 305–314 (2017)

18. Hu, Y., Li, F.C., Li, H.G., Liu, C.L.: An enhanced empirical wavelet transform for noisy and non-stationary signal processing. Dig. Signal Process. **60**, 220–229 (2017)
19. Merainani, B., Benazzouz, D., Rahmoune, C.: Early detection of tooth crack damage in gearbox using empirical wavelet transform combined by Hilbert transform. J. Vib. Control **23**(10), 1623–1634 (2017)
20. Wang, D., Zhao, Y., Yi, C., Tsui, K.L., Lin, J.H.: Sparsity guided empirical wavelet transform for fault diagnosis of rolling element bearings. Mech. Syst. Signal Process. **101**, 292–308 (2018)
21. Wang, J.H., Yang, J.W., Bai, Y.L., Zhao, Y., Yao, D.C.: A comparative study of the vibration characteristics of railway vehicle axlebox bearings with inner/outer race faults. Proc. Inst. Mech. Eng. Part F: J. Rail Rapid Trans. **235**(8), 1035–1047 (2021)
22. Yang, J.W., Zhao, Y., Wang, J.H., Bai, Y.L., Liu, C.: Investigation on impact response feature of railway vehicles with wheel flat fault under variable speed conditions. J. Vib. Acoust. **142**(3), 031009.1–031009.9 (2020)
23. Dragomiretskiy, K., Zosso, D.: Variational mode decomposition. IEEE Trans. Signal Process. **62**(3), 531–544 (2014)
24. Li, Y., Cheng, G., Liu, C., Chen, X.: Study on planetary gear fault diagnosis based on variational mode decomposition and deep neural networks. Measurement **130**, 94–104 (2018)
25. Ren, H., Liu, W.Y., Shan, M.C., Wang, X.: A new wind turbine health condition monitoring method based on VMD-MPE and feature-based transfer learning. Measurement **148**, 106906 (2019)
26. Wang, J.Y., Li, J.G., Wang, H.T., Guo, L.X.: Composite fault diagnosis of gearbox based on empirical mode decomposition and improved variational mode decomposition. J. Low Freq. Noise Vib. Act. Control **40**(1), 332–346 (2020)
27. Gong, T.K., Yuan, X.H., Yuan, Y.B., Lei, X.H., Wang, X.: Application of tentative variational mode decomposition in fault feature detection of rolling element bearing. Measurement **135**, 481–492 (2019)
28. Wang, Z.J., Wang, J.Y., Du, W.H.: Research on fault diagnosis of gearbox with improved variational mode decomposition. Sensors **18**(10), 3510 (2018)
29. Cai, B.P., Huang, L., Xie, M.: Bayesian networks in fault diagnosis. IEEE Trans. Industr. Inf. **13**(5), 2227–2240 (2017)
30. Zuo, L., Zhang, L., Zhang, Z.H., Luo, X.L., Liu, Y.: A spiking neural network-based approach to bearing fault diagnosis. J. Manuf. Syst. **61**, 714–724 (2021)
31. Wang, B., Wang, G.W., Wang, Y.W., Lou, Z.Z., Hu, S.Z., Ye, Y.D.: A K-means clustering method with feature learning for unbalanced vehicle fault diagnosis. Smart Resil. Transp. **3**(2), 162–176 (2021)
32. Wang, J.H., Yang, J.W., Wang, Y.Z., Bai, Y.L., Zhang, T.L., et al.: Ensemble decision approach with dislocated time–frequency representation and pre-trained CNN for fault diagnosis of railway vehicle gearboxes under variable conditions. Int. J. Rail Transp. (2021). https://doi.org/10.1080/23248378.2021.2000897
33. Bai, Y.L., Yang, J.W., Wang, J.H., Zhao, Y., Li, Q.: Image representation of vibration signals and its application in intelligent compound fault diagnosis in railway vehicle wheelset-axlebox assemblies. Mech. Syst. Signal Process. **152**, 107421 (2021)
34. Ye, Y.G., Huang, P., Zhang, Y.X.: Deep learning-based fault diagnostic network of high-speed train secondary suspension systems for immunity to track irregularities and wheel wear. Railw. Eng. Sci. **30**(1), 96–116 (2022)
35. Singh, P., Joshi, S.D., Patney, R.K., Saha, K.: The Fourier decomposition method for nonlinear and non-stationary time series analysis. Proc. R. Soc. A-Math. Phys. Eng. Sci. **473**(2199), 20160871 (2017)

36. Huang, S., Zheng, J., Pan, H., Tong, J.Y.: Order-statistic filtering Fourier decomposition and its application to rolling bearing fault diagnosis journal of vibration and control. J. Vib. Control **11**, 107754632199759 (2021)
37. Deng, M.Q., Deng, A.D., Zhu, J., Sun, W.Q.: Adaptive bandwidth Fourier decomposition method for multi-component signal processing. IEEE Access **7**, 109776–109791 (2019)
38. Zhou, W., Feng, Z.R., Xu, Y.F., Wang, X.J., Lv, H.: Empirical Fourier decomposition: an accurate signal decomposition method for nonlinear and non-stationary time series analysis. Mech. Syst. Signal Process. **163**, 108155 (2022)
39. Zheng, J.D., Haung, S.Q., Pan, H.Y., Tong, J.Y., Wang, C.J., Liu, Q.Y.: Adaptive power spectrum Fourier decomposition method with application in fault diagnosis for rolling bearing. Measurement **183**, 109837 (2021)
40. Lian, J.J., Liu, Z., Wang, H.J., Dong, X.F.: Adaptive variational mode decomposition method for signal processing based on mode characteristic. Mech. Syst. Signal Process. **107**, 53–77 (2018)
41. Wang, J.H., Yang, J.W., Li, Q.: Quasi-static analysis of the nonlinear behavior of a railway vehicle gear system considering time-varying and stochastic excitation. Nonlinear Dyn. **93**(2), 463–485 (2018)

# Short-Term Passenger Flow Prediction for Urban Rail Based on Improved EEMD-Ensemble Learning

Yaoqin Qiao, Huijuan Zhou(✉), Xiayu Zhang, and Lufei Liu

Beijing Key Laboratory of Intelligent Control Technology for Urban Road Traffic, North China University of Technology, Beijing, China
15997774973@126.com, zhouhuijuan@ncut.edu.cn, 649303618@qq.com, 845470835@qq.com

**Abstract.** Urban rail transit passenger flow prediction encounters two primary challenges: the inability to effectively capture the variations in passenger flow patterns and the limited accuracy of prediction models. Therefore, this paper proposes three improved Ensemble Empirical Mode Decomposition (EEMD) methods to accurately capture the characteristics of passenger flow sequence data: (1) Grouped Reconstruction of IMF (Intrinsic Mode Function) Components in EEMD; (2) High-Frequency Disturbance Removal EEMD; (3) Hybrid EEMD. Simultaneously, a short-term passenger flow combination prediction model for urban rail transit stations is established. In order to enhance the prediction accuracy and robustness of the model, the base models utilized include XGBoost, LightGBM, and LSTM. Additionally, the Stacking ensemble method is employed to fuse the models together. Finally, a short-term passenger flow prediction model for urban rail transit based on improved EEMD and ensemble learning is constructed. Taking the entry passenger flow of Beijing West Station in the Beijing Subway as an example, the effectiveness of the model in predicting short-term passenger flow in urban rail transit is validated.

**Keywords:** Ensemble learning · Ensemble empirical mode decomposition · Short-term passenger flow prediction · EEMD

## 1 Introduction

Urban rail transit passenger flow prediction uses technology, data analysis, and intelligent methods to forecast indicators reflecting passenger flow characteristics.

Machine learning-based prediction methods have wider applicability, higher accuracy, and better adaptability compared to prediction methods based on statistical theories. Machine learning methods can be applied to different types of data, including nonlinear and high-dimensional data, and can automatically discover patterns and relationships in the data without prior knowledge [1]. Qiu et al. [2] used a random forest method to construct a dynamic information fusion model for predicting travel time on expressway corridors. Shi et al. [3] integrated global and local attention mechanisms


Y. Qin et al. (Eds.): EITRT 2023, LNEE 1137, pp. 619–626, 2024.
https://doi.org/10.1007/978-981-99-9311-6_61

to capture the dynamic spatiotemporal correlations among stations. They proposed a dynamic spatiotemporal neural network model equipped with attention mechanisms.

When using a single model to predict all the data, overfitting may occur. Ensemble models, by combining multiple individual models, can avoid overfitting by any single model to the data, thereby improving the generalization ability of the prediction [4]. Different datasets may have different features and patterns, and a single model may not cover all cases. Ensemble models can integrate multiple individual models to adapt to diverse datasets [5], thus improving prediction accuracy and stability. Wang et al. [6] optimized the parameters of support vector regression prediction models using the Bat Algorithm (BA). Zhu et al. [7] employed wavelet bases to analyze the passenger flow characteristics of various stations by performing wavelet transformation on the original passenger flow data. Subsequently, they constructed a short-term passenger flow prediction model known as Wavelet Transform-Autoregressive Moving Average (WT-ARMA).

To enhance prediction accuracy, researchers have explored the integration of data decomposition methods into passenger flow prediction as traditional ensemble models reached a plateau. Zhao et al. [8] utilized the advantages of the Seasonal and Trend decomposition using Loess (STL) method in preprocessing time series data, particularly in handling complex signals and reducing noise interference. They combined the STL method with Gated Recurrent Unit (GRU) to construct a model for predicting passenger flow in urban rail transit transfers. Huang et al. [9] used the Complete Ensemble Empirical Mode Decomposition with Adaptive Noise (CEEMDAN) to decompose passenger flow data and proposed a Seq2Seq model with attention mechanism. They proposed a Seq2Seq model with attention mechanism. Zeng et al. [10] proposed the utilization of an Improved Particle Swarm Optimization algorithm (IPSO) for dynamically determining the hyperparameters of the LSTM model. They further developed a subway short-term passenger flow prediction model based on CEEMDAN-IPSO-LSTM. Therefore, this study focuses on the construction of a combined model for passenger flow prediction and the use of data decomposition methods. We propose an improved EEMD-ensemble learning-based model for short-term passenger flow prediction in urban rail transit.

## 2 Model Foundation

### 2.1 Mode Decomposition

Empirical Mode Decomposition (EMD) [11] is an adaptive data analysis method suitable for nonlinear and non-stationary data. It decomposes a time series into multiple Intrinsic Mode Functions (IMFs) and a residual component using Hilbert transform. EMD effectively decomposes nonlinear time series data but has drawbacks of "end effects" and "mode mixing." To overcome these limitations, an improved method called Ensemble Empirical Mode Decomposition (EEMD) is proposed. EEMD introduces white noise sequences multiple times during the decomposition process and averages the results.

In this study, EEMD is applied to decompose urban rail transit passenger flow sequence data to uncover fluctuation patterns at different time scales within each IMF component. This serves as a foundation for passenger flow prediction work. Additionally, three improved EEMD methods are proposed.

(1) Grouped Reconstruction of IMF Components in EEMD (REEMD)

It groups and reconstructs the obtained IMF components, merging components with similar frequencies into high-frequency IMF groups and low-frequency IMF groups. The residual component $r(t)$ is treated separately. The decomposition process of REEMD can be represented as follows:

$$\begin{cases} IMF_h = \sum_{i=1}^{a} c_i \\ IMF_l = \sum_{i=a+1}^{n} c_i \\ RES = r(t) \end{cases} \tag{1}$$

In this context, $IMF_h$ represents the high-frequency IMF group, $IMF_l$ represents the low-frequency IMF group, $RES$ represents the residual component, and $a$ represents the number of IMF components in the high-frequency IMF group. $n$ represents the total number of IMF components obtained from the decomposition of the original sequence, $c_i$ represents an IMF component, and $r(t)$ represents the residual component remaining after the decomposition.

(2) High-Frequency Disturbance Removal EEMD (HFEEMD)

The HFEEMD method eliminates the highest frequency component from the IMF components extracted through EEMD decomposition of the original time series data of passenger flow. This process retains the remaining IMF components and the residual component. The decomposition process of HFEEMD can be summarized as follows:

$$x'(t) = \sum_{i=2}^{n} c_i'(t) + r(t), i = 2, ..., n \tag{2}$$

In this context, $x'(t)$ represents the original sequence of HFEEMD, and $c_i'(t)$ represents the individual IMF components obtained from the HFEEMD decomposition.

(3) Hybrid EEMD (HEEMD)

Combining the two methods, the Hybrid Ensemble Empirical Mode Decomposition (HEEMD) is proposed. In this approach, the original passenger flow data is initially decomposed using EEMD. Next, the highest frequency IMF component is eliminated, followed by a grouped reconstruction based on the similarity of frequency levels. The decomposition process of HEEMD can be represented as follows:

$$\begin{cases} IMF_h = \sum_{i=2}^{a} c_i \\ IMF_l = \sum_{i=a+1}^{n} c_i \\ RES = r(t) \end{cases} \tag{3}$$

## 2.2 Ensemble Learning

Ensemble Learning is a method in machine learning that improves the overall performance of a model by combining multiple predictive models. Given the non-linear and non-stationary nature of passenger flow in urban rail transit, accurately capturing the underlying patterns becomes challenging for a single model in the context of short-term passenger flow prediction. Hence, Ensemble Learning can be employed to enhance accuracy and robustness.

It utilizes a set of base models, such as Support Vector Machines, Decision Trees, and Neural Networks, which are then combined. To achieve optimal utility, the base models should exhibit diversity and complementary advantages. Boosting algorithm is an effective training method, wherein the weights of training samples are iteratively adjusted based on the training results to train new models, and ultimately, all the models are combined to obtain the final result. Gradient Boosting Decision Trees (GBDT) is the primary implementation algorithm, including XGBoost and LightGBM.

Stacking ensemble method takes the output results of multiple base models as new features input into the meta-model for training and prediction, thereby obtaining more accurate prediction results. The following base models will be selected in this article:

(1) LSTM: Urban rail transit passenger flow data is characterized as a typical time series data. Among neural network models, LSTM has demonstrated strong performance in predicting time series data. Hence, after decomposing the passenger flow data, LSTM is chosen as the prediction model.
(2) XGBoost: XGBoost is an improved version of the traditional GBDT algorithm. It approximates the loss function with a second-order Taylor expansion to reduce the optimization difficulty of the objective function. It introduces regularization terms to control model complexity and prevent overfitting. XGBoost demonstrates good predictive performance for urban rail transit passenger flow time series data.
(3) LightGBM: LightGBM is another implementation of the GBDT model. It improves the histogram-based algorithm by finding the best split points to divide continuous numerical features into a finite number of discrete intervals. It then computes gradient histograms for each interval to determine the optimal split points. This method enhances data aggregation efficiency and reduces algorithm complexity.

## 2.3 Overall Model Structure

Based on the HEEMD decomposition of passenger flow data, this paper constructs a prediction model based on Stacking ensemble learning. The overall structure of the model is shown in Fig. 1:

The first part involves decomposing the original passenger flow data using HEEMD into high-frequency IMF group and low-frequency IMF group, while retaining the residue component.

The second part is about partitioning the data into training and prediction sets for the high-frequency IMF group, low-frequency IMF group, and residue component.

The third part focuses on training the base models, including LSTM, XGBoost, and LightGBM. Each base model is trained independently using the same training set. The prediction results of the base models are then combined as inputs for the second layer of the Stacking ensemble model, which utilizes an LR model for the final prediction.

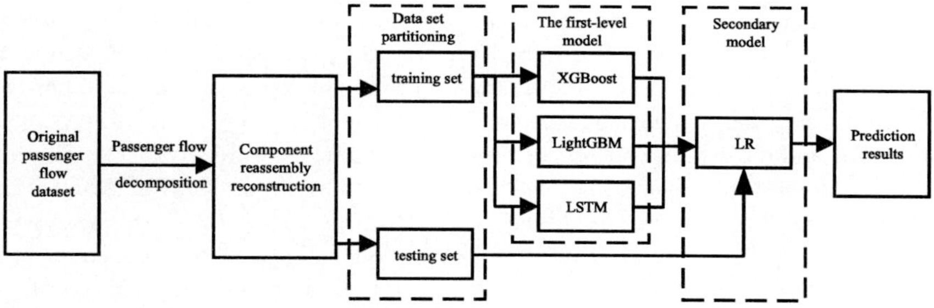


**Fig. 1.** Improved EEMD-stacking ensemble learning model structure

# 3 Case Study

## 3.1 Passenger Flow Data Processing

The experimental data in this paper are derived from the entrance passenger flow data of Beijing West Railway Station subway station from January 14, 2019, to March 29, 2019, on selected working days, totaling 20 days. The first 16 days of passenger flow data are used as the training set, while the remaining 4 days are used as the validation set. The preprocessing method applied to the raw data is the averaging of 5-min intervals to ensure the availability of continuous and complete passenger flow time series at 5-min intervals.

After applying the EEMD decomposition to the passenger flow data at Beijing West Station, 9 IMF components and 1 residual component are obtained. Each component reflects different characteristics of the passenger flow data in various dimensions. By conducting a correlation statistical analysis on each IMF component and the residual component, the summary of the correlation indicators is presented in Table 1.

**Table 1.** Statistical analysis results of IMF components and residual components

| Sequence | Pearson correlation coefficient | Kendall correlation coefficient |
|---|---|---|
| $IMF_1$ | 0.288 | 0.126 |
| $IMF_2$ | 0.597 | 0.423 |
| $IMF_3$ | 0.458 | 0.354 |
| $IMF_4$ | 0.162 | 0.129 |
| $IMF_5$ | 0.091 | 0.054 |
| $IMF_6$ | 0.087 | 0.055 |
| $IMF_7$ | 0.038 | 0.021 |
| $IMF_8$ | 0.014 | 0.013 |
| $IMF_9$ | 0.031 | 0.028 |
| RES | – | – |

Statistical analysis was conducted on the decomposed components of the passenger flow. The results revealed that $IMF_1$-$IMF_4$ displayed significantly higher Pearson Correlation Coefficients and Kendall Correlation Coefficients compared to the other IMF components and the residual component. This indicates a strong correlation between $IMF_1$-$IMF_4$ and the original passenger flow data sequence.

Based on the three proposed EEMD improvement methods, the REEMD method combines $IMF_1$-$IMF_4$ as a group of highly correlated components, $IMF_5$-$IMF_9$ as a group of weakly correlated components, and treats the residual component as a separate group. These three groups are then utilized as inputs to the prediction model. The HFEEMD method discards the highest-frequency $IMF_1$ while retaining $IMF_2$-$IMF_9$ and the residual component as separate inputs to the model. As for the HEEMD method, it eliminates the highest-frequency $IMF_1$ and combines $IMF_2$-$IMF_4$ as a group of highly correlated components, $IMF_5$-$IMF_9$ as a group of weakly correlated components, and treats the residual component as a separate group. The resulting three groups are used as inputs to the prediction model.

### 3.2 Analysis of Model Results

The proposed improved EEMD-Stacking ensemble learning prediction model is utilized to forecast the entrance passenger flow data at Beijing West Station. The predictions from each sub-model in the forecasting set are compared, and the error metrics used are Mean Absolute Error (MAE) and Mean Absolute Percentage Error (MAPE). The specific results are presented in Table 2.

**Table 2.** Prediction performance of each model

| Forecasting model | MAE | MAPE (%) |
|---|---|---|
| XGBoost | 95.28 | 23.14 |
| LightGBM | 63.98 | 16.15 |
| LSTM | 72.88 | 18.67 |
| Ensemble learning | 49.74 | 13.02 |
| HEEMD-LSTM | 39.14 | 10.42 |
| REEMD-Ensemble learning | 35.87 | 9.61 |
| HFEEMD-Ensemble learning | 39.89 | 10.69 |
| HEEMD-Ensemble learning | 30.78 | 8.55 |

Table 2 provides a detailed display of the performance of these eight models on the training and testing sets, ranked from low to high in terms of prediction accuracy as follows: XGBoost < LSTM < LightGBM < Ensemble Learning < HFEEMD-Ensemble Learning < HEEMD-LSTM < REEMD-Ensemble Learning < HEEMD-Ensemble Learning. The HEEMD-Ensemble Learning model achieves the lowest MAE and MAPE, confirming that the proposed improved EEMD-Stacking ensemble model has high prediction accuracy.

### 3.3 Model Prediction Result Comparison

(1) Comparison between Single Models and Ensemble Models

From the Table 3, it can be concluded that the ensemble learning model outperforms traditional single models in short-term passenger flow prediction in urban rail transit. This indicates that the ensemble model can leverage the advantages of individual base models while mitigating their shortcomings, resulting in a reduction in overall error.

**Table 3.** Error comparison of individual models

| Forecasting model | MAE | MAPE (%) |
|---|---|---|
| XGBoost | 95.28 | 23.14 |
| LightGBM | 63.98 | 16.15 |
| LSTM | 72.88 | 18.67 |
| Ensemble Learning | 49.74 | 13.02 |

(2) Comparison of the Utility of HEEMD Decomposition Method

From Table 4 and in combination with the above information, it is evident that the HEEMD decomposition method significantly enhances the predictive performance of both the LSTM and ensemble learning models. This demonstrates the effectiveness of passenger flow decomposition in improving the accuracy of short-term passenger flow forecasting models for urban rail transit.

**Table 4.** Error comparison of passenger flow decomposition combination models

| Forecasting model | MAE | MAPE (%) |
|---|---|---|
| HEEMD-LSTM | 39.14 | 10.42 |
| HEEMD-Ensemble learning | 30.78 | 8.55 |

## 4 Conclusion

The main conclusions drawn from this study are as follows:

(1) The three improved methods of EEMD, compared to the original EEMD model, demonstrate reduced errors and improved prediction accuracy. This validates the effectiveness of the proposed EEMD improvements in enhancing the predictive performance of passenger flow prediction.

(2) Compared to XGBoost, LightGBM, and LSTM models, the ensemble model exhibits smaller MAE and MAPE, indicating higher prediction accuracy. This verifies the predictive effectiveness of the ensemble model constructed in this study.
(3) HEEMD-Ensemble Learning outperforms all other models developed in this study. Moreover, the predictive performance of the HEEMD-Ensemble Learning model surpasses that of combining the HEEMD decomposition method with a single model.

## References

1. Yang, X., Xue, Q., Yang, X., et al.: A novel prediction model for the inbound passenger flow of urban rail transit. Inf. Sci. **566**, 347–363 (2021)
2. Qiu, B., Fan, W.: Travel time forecasting on a freeway corridor: a dynamic information fusion model based on the random forests approach. Smart Resilient Transp. **3**(2), 131–148 (2021)
3. Shi, J.Q., Li, R., Cheng, M.H., et al.: Subway passenger flow prediction based on dynamic spatiotemporal neural network model. J. Transp. Syst. Eng. Inf. Technol. **23**(02), 139–147 (2023). (in Chinese)
4. Tang, J., Liang, J., Liu, F., et al.: Multi-community passenger demand prediction at region level based on spatio-temporal graph convolutional network. Transp. Res. Part C: Emerg. Technol. **124**, 102951 (2021)
5. Wei, Y., Chen, M.C.: Forecasting the short-term metro passenger flow with empirical mode decomposition and neural networks. Transp. Res. Part C: Emerg. Technol. **21**(1), 148–162 (2012)
6. Wang, X.Y., Su, F.: Short-term passenger flow prediction model for rural tourism based on BA-SVR. Comput. Eng. Des. **39**(12), 3811–3815 (2018). (in Chinese)
7. Zhu, C.H., Sun, X.L., Li, P.K., et al.: Short-term passenger flow prediction for urban rail transit with integration of station classification and data denoising. J. Railw. Sci. Eng. **19**(08), 2182–2192 (2022). (in Chinese)
8. Zhao, J.D., Zhu, D., Liu, J.X.: Subway transfer passenger flow prediction based on time series decomposition and gated recurrent unit. J. South China Univ. Technol. (Natl. Sci. Edn.) **50**(05), 22–31 (2022). (in Chinese)
9. Huang, H., Mao, J., Lu, W., et al.: DEASeq2Seq: an attention based sequence to sequence model for short-term metro passenger flow prediction within decomposition-ensemble strategy. Transp. Res. Part C: Emerg. Technol. **146**, 103965 (2023)
10. Zeng, L., Li, Z.N., Yang, J., et al.: Research on short-term passenger flow prediction method for urban rail transit based on CEEMDAN-IPSO-LSTM. J. Railw. Sci. Eng. 1–14 [2023–05–07]. (in Chinese)
11. Zhao, X., Qin, Y., Fu, H., et al.: Blind source extraction based on EMD and temporal correlation for rolling element bearing fault diagnosis. Smart Resilient Transp. **3**(1), 52–65 (2021)

# Research on Train Delay Forecast of High-Speed Railway Based on Data Driving

Mengchen Wang[1,2], Yong Qin[2], Li Wang[1,3](✉), Xianghao Wang[1,3], and Xinyi Du[1,3]

[1] School of Traffic and Transportation, Beijing Jiaotong University, Beijing 100044, China
{21221024,wangli,xianghaow,duxinyi}@bjtu.edu.cn

[2] State Key Laboratory of Advanced Rail Autonomous Operation, Beijing Jiaotong University, Beijing 100044, China
yqin@bjtu.edu.cn

[3] Beijing Engineering Research Center of Urban Traffic Information Intelligent Sensing and Service Technologies, Beijing 100044, China

**Abstract.** China's high-speed railway exhibits high train density and small tracking intervals. Once a train is delayed, the delay easily propagates, leading to further delays. Therefore, accurate prediction of train delays can assist dispatchers in formulating timely optimization and adjustment plans, minimizing the negative impact of delay propagation, and enhancing the reliability and punctuality of train operations. This paper constructs a train delay prediction model based on the cyclic neural network (RNN) algorithm, enabling accurate prediction of train arrival delay times. The model is then validated using actual operational data from the Wuhan-Guangzhou high-speed railway, with the experimental results demonstrating high prediction accuracy.

**Keywords:** Recurrent neural network · High-speed railway · Train delay prediction

## 1 Introduction

This paper aims to address the shortcomings of existing methods by proposing a novel approach for high-speed train delay prediction using a cyclic neural network model. In recent years, China's high-speed rail has developed rapidly, and the largest high-speed rail network in the world has been built, which has become an important means of transportation for Chinese people to travel. With the breakthrough of technology and materials, the speed of high-speed rail has been continuously improved. Passengers pay more attention to the reliability of railway operation, and safety and punctuality have become important goals of railway development. In the process of train running, it will be delayed by many factors such as train equipment and external environment, which will cause serious economic losses and social losses. Therefore, the rapid and accurate prediction of high-speed train delay provides guarantee for efficient scheduling of railway transportation and reduces the large-scale passenger detention caused by delay.


Y. Qin et al. (Eds.): EITRT 2023, LNEE 1137, pp. 627–634, 2024.
https://doi.org/10.1007/978-981-99-9311-6_63

Therefore, the research on train delay prediction is of great significance with the development of related research on train delay prediction, many current theoretical studies and machine learning models have been applied in operation. Scholars at home and abroad have mainly studied from the following angles. Hu et al. put forward to comprehensively use Pearson correlation coefficient and Lasso algorithm to process high-speed train data, and use gradient lifting decision tree machine learning model to effectively predict the train delay. Tang et al. [2] proposed to use Box-Cox transformation to process train running time data, which improved the fitting effect of prediction model, and used grid search algorithm to improve the efficiency and accuracy of train running time model at the same time. Mou et al. [3] established a stochastic forest prediction model for train delay, which has more stable prediction error and higher accuracy than BP neural network model. Huang [4] used R language to establish a stochastic forest regression model of train late recovery with late recovery time as dependent variable, and achieved high accuracy. Zhang et al. [5] put forward a new prediction model based on graph neural network, which can better capture the influence of spatio-temporal characteristics and use time series fuzzy decision tree to reduce errors. Masoud Yaghini [6] and other scholars use three strategies: fast method, dynamic method and multiple combination to model Iranian railway to find the optimal method. Wen et al. [7] put forward to build a long-term and short-term memory (LSTM) prediction model, and the accuracy of LSTM model is higher than stochastic forest model and artificial intelligence network model. Zhang et al. [8] built a spatio-temporal graph convolution network for late prediction, and the prediction effect is better than the existing advanced baseline prediction. Li et al. [9] proposed to use GBRT model, and its accuracy is better than that of common models such as random forest and multiple linear regression through experiments. Chenet al. [10] proposed to use LightGBM algorithm combined with dynamic prediction to predict passenger trains from departure to delay. At the same time, compared with multiple linear regression model and XGBOOST model, the prediction effect of this model is better. From the above research, the prediction accuracy of most studies is low, and the prediction time of some studies is too long to achieve fast and accurate.

In order to make up for the shortcomings of the above methods, this paper first adopts the cyclic neural network model, then connects a completely linked Dense layer as the output, uses MSE as the loss function and Adam optimization algorithm, and considers that training the model requires a lot of time, so it is convenient to save the model.

# 2 Algorithm Design

## 2.1 Problem Description

Train delay prediction refers to the prediction of train deviation from the running plan time range. With the rise of artificial intelligence and machine learning, it has become a research hotspot to use data-driven machine learning method to predict train delay events in the field of rail transit. The steps of using machine learning method to predict train delay events are as follows:

(1) Propose the problem: In the train delay prediction problem, the model output is the train delay time, and the delay time is a continuous variable, so this problem belongs to regression problem.

(2) Feature work: Preprocess the actual train operation data by converting the original data into feature vectors. Build feature sets and divide them into training sets and test sets according to a certain proportion. The main eigenvalues are shown in Table 1.
(3) Modeling: Using the training set data to train the machine learning algorithm, the model establishes the mapping relationship between input features and output features. In the process of learning the algorithm, the internal structure and the calculation method between nodes can not be expressed by simple formulas, so it has the attribute of "black box".
(5) Evaluation model: Input the test set into the long-trained model, and use various indicators to calculate the output of the model and evaluate the prediction performance of the model.

**Table 1.** Main characteristic values

| Serial number | Characteristic name | Special description |
|---|---|---|
| 1 | preTrainArrDelays | The train ahead is late |
| 2 | actualArrInterval | Actual arrival interval of trains |
| 3 | scheduleArrInterval | Planned arrival interval of trains |
| 4 | overtakedCount | Number of times the train was overtaken |
| 5 | overtakingCount | Overtaking times of trains |
| 6 | same_track_interval | Running interval of trains on the same track |
| 7 | same_track_pre_train_dep_delay | Departure delay time of the previous train on the same track |
| 8 | pre_stationId_dep_delay | Departure delay time of train front station |
| 9 | scheduleRunningTime | Planned travel time of train |

## 2.2 RNN Based Train Delay Prediction Model

Cyclic neural network realizes this memory mechanism by using "cyclic" structure, so that it can capture long-term dependencies. In the prediction of train delay, the time data such as historical train arrival version and arrival interval are all time series data, so RNN model can capture the time dependence more accurately, and because of its memory, the model can capture the complex relationship among many factors and integrate them into the prediction, which has good application for the problem of train delay. The goal of the algorithm is to minimize the mean square error between the predicted train delay time and the actual train delay time. The main parts and steps of the model are as follows:

(1) Loss function

In this paper, the mean square error (MSE) is used as the evaluation function, and the calculation formula is as follows:

$$MSE = \frac{1}{n}\sum_{i=1}^{n}(y_i - \hat{y}_i)^2 \quad (1)$$

The smaller the error between the actual value and the predicted value, the better the prediction effect of the model.

(2) Adam Optimizer

In this paper, the Adam Optimizer is used as the optimizer for model training. The Adam Optimizer adjusts the learning rate adaptively by maintaining the first-order moment vector and the second-order moment vector of each parameter. It effectively handles noisy data and has faster convergence and better performance compared to other optimization algorithms. The update formula for the Adam optimizer is shown in formula (2).

$$\begin{cases} m_t = \beta_1 m_{t-1} + (1-\beta_1)g_t \\ v_t = \beta_2 v_{t-1} + (1-\beta_2)g_t^2 \\ \theta_t = \theta_{t-1} - \frac{\eta}{\sqrt{v_t+\epsilon}} m_t \end{cases} \quad (2)$$

$m_t$ and $v_t$ Is the first-order moment estimation and the second-order moment estimation of gradient, $\beta_1$ and $\beta_2$ Is the attenuation rate, $g_t$ is a gradient, $\theta_t$ as a parameter, during the model training, Adam optimizer updates the model parameters according to gradient descent algorithm, which makes the model output closer to the real value.

(3) Establishing RNN model

First, input the data sequence and use SimpleRNN layer. See Eq. (3) for the calculation process.

$$h_t = \sigma(W_{xh}x_t + W_{hh}h_{t-1} + b_h) \quad (3)$$

Among them, $h_t$ is the output vector of time step t, Represents the state of the network in the current time step; $W_{xh}$, $W_{hx}$, $b_h$ they are input weight matrix, hidden state weight matrix and offset vector, which need to be obtained through training and learning.

$$\sigma(x) = \frac{1}{1+e^{-x}} \quad (4)$$

Equation (4) is a sigmoid activation function, which is used for nonlinear transformation of output state.It then uses the output of SimpleRNN layer $h_t$ as the input of Dense layer.The Dense layer performs linear transformation and activation function processing through full connection layer to obtain the final predicted value $y$. The calculation formula is shown in Eq. (5).

$$y = w^T h_t + b_y \quad (5)$$

Among them, $w$ and $b_y$ they are the weight and offset of Dense layer.

(4) Train delay prediction model

The steps of establishing the prediction model of the initial delay period of train operation are as follows.

- Select the data set; This paper chooses the train operation data of Wuhan-Guangzhou Railway as the experimental data.
- Data preprocessing: Filter and delete abnormal or missing data in the experimental dataset.
- Model structure: The model consists of a SimpleRNN layer with 200 neurons and a Dense layer. The output of the SimpleRNN layer is used as the input of the Dense layer, and the final prediction result of the model is a scalar.
- Compile the model: Use the Adam optimizer as the optimization algorithm and the mean square error (MSE) as the loss function.
- Train the model: Specify parameters such as training data, validation data, batch size, and number of iterations. Start training the model using the fit function.
- Evaluate the model's performance: Observe the performance of the model by monitoring the change in the loss function during training.
- Save the model: Save the trained model as highway_model.h5, including the structure and parameters of the entire model.

# 3 Example Verification

According to the model framework and specific steps introduced above, more than 10 thousand timetables data of Wuhan-Guangzhou high-speed train are selected as the data set for model training.

(1) Parameters and training

This program adopts Tensorflow deep learning framework, and divides the experimental data set into training set and verification set according to the ratio of 8: 2 for model training. Through the loss test, in the training process of 500 Epoch, the Loss of the training set always keeps a downward trend, which shows that the fitting ability of the model is good. As shown in Fig. 1.

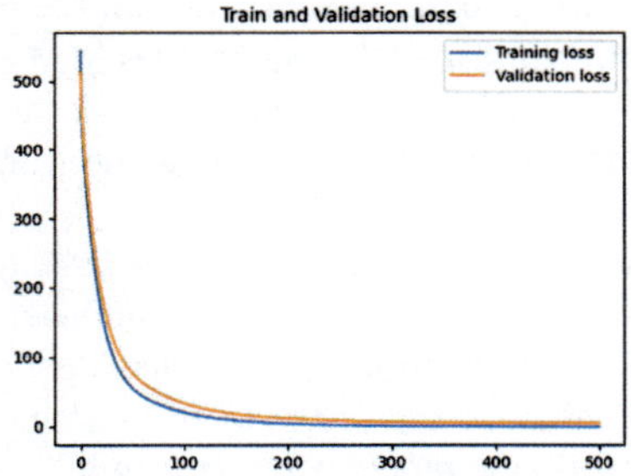


**Fig. 1.** Iterative curve of loss function

After the training is completed, the model parameters of the final optimal model are saved.

(2) Analysis of experimental results.

Load the previously saved optimal model and randomly select data for testing. The proportion of five data groups with delay times of 5–10 min, 10–30 min, 30–60 min, 60–120 min, and more than 120 min is shown in Fig. 2.

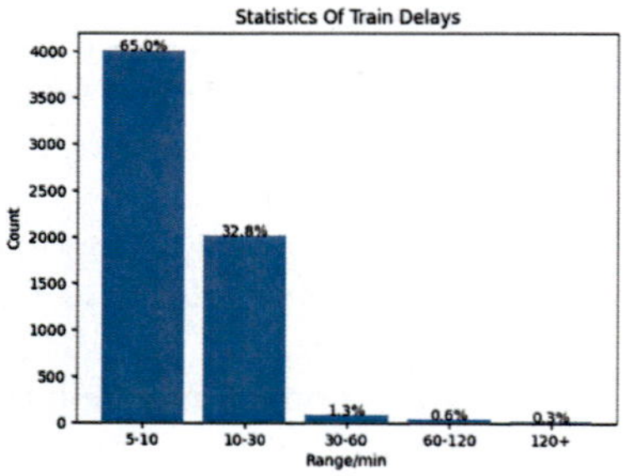


**Fig. 2.** Statistics of actual delay time distribution of trains

The test results show that when the actual delay time is within 5–10 min, the accuracy can reach more than 90% with an allowable error of 1 min. Similarly, when the delay time is within 10–30 min, the accuracy can reach more than 89% with an allowable error of 3 min. The accuracy meets the expected requirements. One data sample from each of these five test sets is randomly selected and displayed in Table 2. The analysis of the operational results is presented in Table 3.

According to the data from certain test sets in Table 3, it can be observed directly from Fig. 3 that the prediction accuracy is high and meets the expectations of the experiment.

## 4 Conclusions

Based on the historical running data of Wuhan-Guangzhou Railway, this paper constructs a model using the cyclic neural network algorithm to predict train delay events. After experimental verification, the model demonstrates high accuracy and achieves the expected results. It accurately predicts train delays, reduces dispatching workload, and improves the speed and efficiency of the train-railway network system in dealing with train delays.

However, the train delay prediction model in this paper still has some shortcomings. For instance, it classifies the reasons for delay, distinguishes different types of delay causes, studies the patterns of delay time, and predicts delay time through classification. This requires further research and analysis, considering more factors related to train operations. In future work, the model parameters will be improved to enhance the prediction accuracy for long-time train delays.

**Table 2.** Sample test suite data

| Characteristic name | The first set of data | The second set of data | The third set of data | The fourth set of data | The fifth set of data |
|---|---|---|---|---|---|
| preTrainArrDelays (min) | 0 | 15 | 18 | 96 | 192 |
| actualArrInterval (min) | 3 | 7 | 3 | 3 | 19 |
| scheduleArrInterval (min) | 5 | 5 | 10 | 6 | 11 |
| overtakedCount (min) | 1 | 0 | 3 | 0 | 15 |
| overtakingCount (min) | 0 | 0 | 1 | 2 | 0 |
| same_track_interval (min) | 3 | 29 | 3 | 3 | 19 |
| same_track_pre_ _train_dep_delay (min) | 0 | −3 | 18 | 96 | 192 |
| pre_stationId_dep_delay (min) | 9 | 17 | 39 | 75 | 162 |
| scheduleRunningTime (min) | 114 | 15 | 14 | 16 | 17 |

**Table 3.** Comparison of test set data running results

| Data group | Predict the delay time of trains (min) | Actual train delay time (min) | Error duration (min) | Error proportion (%) |
|---|---|---|---|---|
| 1 | 7.65 | 8 | −0.35 | 4.38 |
| 2 | 15.35 | 17 | −1.65 | 9.70 |
| 3 | 39.14 | 41 | −1.86 | 4.53 |
| 4 | 75.16 | 75 | 0.16 | 0.21 |
| 5 | 167.81 | 161 | 6.81 | 4.22 |

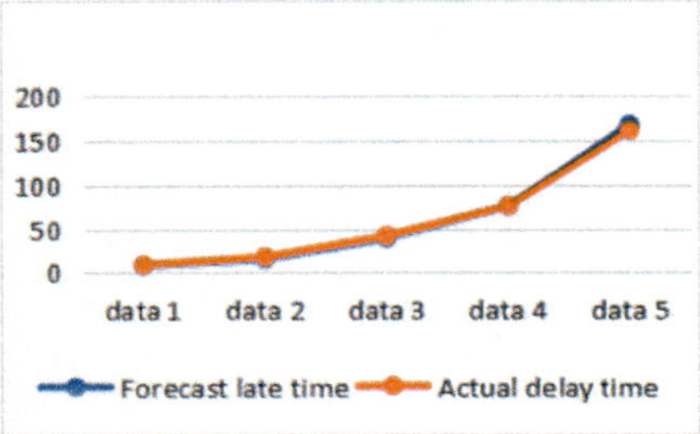


**Fig. 3.** Comparison of running results of some experimental data

**Acknowledgements.** This study is funded by the Fundamental Research Funds for the Central Universities (Grant No. 2023JBZY005).

## References

1. Rui, H., Chao, W., Mengying, Z., et al.: Machine learning model for delay prediction of high-speed trains. China Railw. **701**(11), 72–77 (2020). (in Chinese)
2. Tao, T., Jing, G.: A train running time prediction model based on domestic and foreign railway operation data. China Safety Sci. J. **32**(6), 123–130 (2022). (in Chinese)
3. Weiwei, M., Yafei, H.: Prediction model of train arrival delay based on data analysis. China Transp. Rev. **41**(06), 64–69+74 (2019). (in Chinese)
4. Ping, H., Qiyuan, P., Chao, W., Yuxiang, Y.: Random forest prediction model for Wuhan-Guangzhou HSR primary train delays recovery. J. China Railw. Soc. **40**(07), 1–9 (2018). (in Chinese)
5. Zhang, D., et al.: An interpretable station delay prediction model based on graph community neural network and time-series fuzzy decision tree. Trans. Fuzzy Syst. **31**(2), 421–433 (2023)
6. Yaghini, M., Khoshraftar, M.M., Seyedabad, M.: Railway passenger train delay prediction via neural network model. J. Adv. Transp. **47**, 355–368 (2013)
7. Wen, C., Mou, W., Huang, P., Li, Z.: A predictive model of train delays on a railway line. J. Forecast. **39**, 470–488(2020)
8. Zhang, D., Peng, Y., Zhang, Y., Wu, D., Wang, H., Zhang, H.: Train time delay prediction for high-speed train dispatching based on spatio-temporal graph convolutional network. Trans. Intell. Transp. Syst. **23**(3), 2434–2444 (2022)
9. Jin, L., Chao, W., Yuqi, D., et al.: Study on late recovery strategy of Wuhan-Guangzhou high-speed railway section based on gradient lifting regression tree. China Railw. **712**(10), 76–84 (2021)
10. Yaru, C., Hongbin, Z.: Delay prediction of Shanghai-Chengdu high-speed railway train based on LightGBM algorithm. China Transp. Rev. **44**(12), 77–83 (2022)

# Research on Factors Influencing Decision-Making Behavior for Evacuation Path Selection of Personnel Inside Subway Station

Mengmeng Yin(✉), Mengting Liu, Jingjing Wang, and Wei Zhu

Beijing Academy of Science and Technology Institute of Urban Systems Engineering, No. 16, Xizhimen South Street, Xicheng District, Beijing 100038, China
yiniashin@126.com, 13810430584@163.com

**Abstract.** Subway transportation is characterized by high passenger flow, limited space, and complex evacuation paths. In the event of a sudden incident such as a fire, flooding, or toxic gas leakage, it is crucial to study the factors that determine the selection of evacuation paths for subway station personnel. This study focuses on subway passengers in Beijing. Through a questionnaire based on cognitive mapping theory, we collected data on the emergency evacuation path selection strategies of subway participants in disaster scenarios. Based on different strategy preferences, we analyzed the determining factors of evacuation path selection behavior and proposed targeted improvements and preventive measures accordingly.

**Keywords:** Subway stations · Disaster · Personnel evacuation · Path selection · Cognitive map

## 1 Introduction

Urban subway transportation has several characteristics, such as high passenger flow, limited space, complex evacuation paths, poor ventilation, and single evacuation paths. In the event of a fire, flooding, toxic gas leakage, or other emergencies, subway station personnel may experience strong stimulation, leading to chaos and secondary incidents like stampedes, causing additional harm. This study focuses on subway passengers in Beijing. Through a questionnaire survey, we collected data on the evacuation path selection behavior of subway participants in disaster scenarios. Using cognitive mapping theory as the theoretical basis, we investigated the determining factors of evacuation path selection behavior.

The safety planning of personnel evacuation paths and signage within subway stations, as well as emergency management during unforeseen events, are essential aspects of urban rail transit construction, planning, and operation assurance. During emergency evacuations, decisions regarding path selection rely on the analysis and processing of environmental information within subway stations. Environmental information is obtained through the perception or reception of accident hazards, escape paths, and other


Y. Qin et al. (Eds.): EITRT 2023, LNEE 1137, pp. 635–644, 2024.
https://doi.org/10.1007/978-981-99-9311-6_64

environmental factors during the evacuation process. The sources of information include self-perception, information transmitted by other pedestrians, and evacuation signage. Additionally, individuals' knowledge and experience accumulated during daily activities also contribute to their understanding of the architectural spatial structure within the subway, which varies depending on individuals' familiarity with the environment. Existing evacuation studies and models often assume that subway participants have complete information about evacuation paths.

## 2 Evacuation Path Selection Strategies

The mechanisms involved in the evacuation path selection process during emergencies such as fires, flooding, and toxic gas incidents in subway stations are not yet fully understood. Downs and Stea [1] proposed four steps involved in path finding: orientation, path selection, path control, and destination recognition. By successfully constructing cognitive maps, individuals can determine their position for spatial orientation and combine it with environmental cues for path finding. Environmental cues include signage, maps, landmarks, and pathways. Pelechano and Badler [2] stated that spatial navigation problems involve decision-making, executing decisions, and information processing, requiring individuals to possess abilities such as cognitive mapping, localization, spatial exploration, and motion navigation. Wiener, Büchner and Hölscher [3] introduced a classification of path finding based on individuals' available spatial knowledge levels and external constraints. Spatial knowledge is divided into three levels: points, lines, and areas. Path finding includes unassisted and assisted methods, with aids such as signage and maps. Building upon this classification, Andresen, Chraibi and Seyfried [4] proposed path finding strategies and tools, distinguishing three types of knowledge sources.

- Internal spatial memory from previous visits, including line knowledge based on actions, and point and area knowledge represented by cognitive maps. Corresponding strategies include moving along known paths, searching for paths, and path planning.
- External sensory inputs during the path finding process, with strategies such as following maps/signage or following others.
- General experiences, such as recognizing whether the space is a transit area or a functional room.

Based on the above, this paper presents the types of emergency evacuation path selection decisions for subway station personnel, as shown in Fig. 1.

## 3 Cognitive Map Construction

A cognitive map is an internal representation of the spatial arrangement of the environment within an organism. It provides a mental representation of possible locations relevant to locating and navigating towards a specific destination, allowing us to find or plan a path to the destination. However, cognitive maps can be subjective, potentially inaccurate, oversimplified, or distorted [5]. Since humans have the ability to store topological relationships more accurately, they can successfully assist us in finding paths, especially in environments that have been visited multiple times [6, 7].

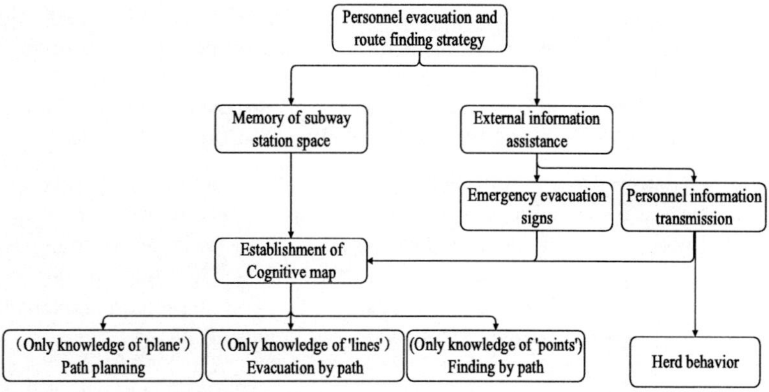


**Fig. 1.** Classification of evacuation strategies for personnel inside subway stations.

The five typical elements of a cognitive map are paths, nodes, regions, boundaries, and landmarks. Environmental image theory focuses on the different impressions of the environment formed by individuals based on their different experiences. Egocentric representations reflect the spatial relationships between individuals and the environment (self-to-object) and focus on factors such as points of direction changes, angles of direction changes, and distances between two points along the path, presenting a linear structure of paths. Allocentric representations focus on the topological relationships between significant elements or landmarks within the environment (object-to-object) and establish a global understanding, often taking the form of survey maps. These two forms of representation depend on the type of environment and individual preferences and may coexist [8].

## 4 Questionnaire Design

External influencing factors on the path selection behavior of subway station personnel during emergencies include time pressure and space pressure. Evacuation time consists of free walking time and waiting time, where the former is influenced by the distance to the exit and walking speed, and the latter is influenced by the number of people in the queue, exit width, and exit flow rate. Space pressure mainly arises from the spread and diffusion of hazardous factors such as fires, flooding, and toxic gas within the subway station. In addition to external factors, path selection also relies on the decision-maker's cognitive abilities. Based on previous research, the fundamental principles guiding path selection behavior are minimizing risk and choosing the shortest path. Risk factors include congestion and hazards, primarily influenced by external factors, and determine immediate path selection strategies. Three path selection strategies were investigated: Strategy A-following evacuation signs, Strategy B-relying on familiar paths from memory, and Strategy C-following the direction of others' evacuations.

- Strategy A is related to the construction of cognitive maps, as signage supports the egocentric construction of cognitive maps.

- Strategy B is related to memory abilities, with cognitive psychology's dual-process theory suggesting that recognition memory can be supported by two independent processes: familiarity and recollection [9–11].
- Strategy C is related to information uncertainty, as uncertainty can increase individuals' structural needs, making them more likely to conform to the crowd.

Structural needs represent the motivation to maintain structure and avoid being influenced by uncertain information. Cognitive abilities relevant to emergency evacuations include perception, attention, memory, and spatial abilities, which vary among decision-makers and can be improved through learning and training. Therefore, the questionnaire investigates the impact of individual attributes, knowledge experiences, and other factors on the preferences for these three path selection strategies (Table 1).

**Table 1.** Factors influencing the preference for evacuation path selection.

| Index | Variable | Value |
|---|---|---|
| Factor 1 | Gender | 1 = Male, 2 = Female |
| Factor 2 | Age | 1 = 18–30 years old, 2 = 30–40 years old, 3 = 40–50 years old, 4 = 50–60 years old |
| Factor 3 | Education level | 1 = Postgraduate and above, 2 = Bachelor's and Associate's degree, 3 = High school and below |
| Factor 4 | Vision | 1 = Good, 2 = Normal, 3 = Poor |
| Factor 5 | Participation in emergency training or drills | 1 = Yes, 2 = No |
| Factor 6 | Emergency knowledge level | 1 = High, 2 = Medium, 3 = Low |
| Factor 7 | Attention to signage | 1 = Always pay attention, 2 = Pay attention most of the time, 3 = Rarely pay attention |
| Strategy A | Choose according to evacuation signage instructions | 1 = Yes, 2 = No |
| Strategy B | Choose based on familiar path memory | 1 = Yes, 2 = No |
| Strategy C | Choose based on others' evacuation direction | 1 = Yes, 2 = No |

* Emergency knowledge level is evaluated based on the accuracy of answering emergency knowledge-related questions, etc.

## 5 Analysis of Influencing Factors

This study used a survey questionnaire distributed at 13 major transfer subway stations in Beijing, with a total of 511 valid responses. Analyzing the survey results, it was found that when making decisions regarding evacuation path selection, 306 respondents (59.88% of the total) preferred to follow evacuation signage instructions, 296 respondents (57.92% of the total) preferred to rely on familiar paths from memory, and 116 respondents (22.7% of the total) preferred to follow the direction of others' evacuations. The difference between the number of respondents choosing familiar paths and s-signage paths is not significant, which is consistent with the findings of Shi et al. [12]. The correlation analysis between these three path selection decisions and various influencing factors is shown in Table 2. Further analysis was conducted on factors with significant correlations ($P < 0.05$).

**Table 2.** Attributes of survey participants

| | | | | |
|---|---|---|---|---|
| Gender | Male | Female | | |
| | 46% | 54% | | |
| Age | 18–30 | 31–40 | 41–50 | 51–60 |
| | 6.3% | 41.3% | 46.0% | 6.4% |
| Education level | Postgraduate and above | Bachelor's and Associate's degree | High school and below | |
| | 3.5% | 90.2% | 6.3% | |
| Vision | Good | Normal | Poor | |
| | 36.2% | 62.4% | 1.4% | |

### 5.1 The Influence of Gender

Gender primarily influences the preference for Strategy A, as shown in Fig. 2. Women are more inclined than men to choose evacuation based on signage. This finding is consistent with the results of the VR experiment conducted by Chang and Chang [13], suggesting that women are more likely to adopt egocentric path strategies, while men are more likely to adopt allocentric strategies. In this survey, the focus was on directional guidance signage, which supports the egocentric construction of cognitive maps, as opposed to YAH (You Are Here) maps, which support allocentric strategies. Therefore, women tend to prefer Strategy A more than men do.

### 5.2 The Influence of Age

Age primarily influences the preferences for Strategy A and Strategy B. As shown in Fig. 3(a), the proportion of individuals aged 18–40 who choose evacuation based on signage is relatively low, while the proportion is higher among those aged 41–50. In Fig. 3(b), the proportion of individuals aged 31–40 who choose familiar paths for evacuation is higher, while the proportion is significantly lower among those aged 51–60. This is mainly because older adults are more likely to adopt egocentric strategies and have less memory capacity compared to younger individuals. Prominent path-finding aids such as larger signs and paint colors were helpful for the path finding of older

**Table 3.** Pearson correlation coefficient between path selection preferences and influencing factors.

| | Factor1 | Factor 2 | Factor 3 | Factor 4 | Factor 5 | Factor6 | Factor 7 | Strategy A | Strategy B | Strategy C |
|---|---|---|---|---|---|---|---|---|---|---|
| Factor 1 | 1 | −0.1905** | −0.1835** | 0.0267 | 0.2252** | 0.0112 | −0.0755 | 0.1580** | 0.0089 | −0.0061 |
| Factor 2 | | 1 | 0.1448** | 0.1372** | 0.1124* | 0.0437 | −0.0948 | 0.1114* | −0.1053* | 0.0786 |
| Factor 3 | | | 1 | −0.0060 | 0.1868** | 0.0331 | 0.2939** | 0.0206 | −0.0769 | 0.0128 |
| Factor 4 | | | | 1 | −0.1007 | −0.0613 | −0.2205** | 0.0377 | −0.0231 | 0.1095* |
| Factor 5 | | | | | 1 | −0.0576 | 0.3120** | 0.4445** | −0.2183** | 0.1499** |
| Factor 6 | | | | | | 1 | −0.0156 | −0.0927* | −0.0442 | −0.1482* |
| Factor 7 | | | | | | | 1 | −0.1230** | 0.0888* | 0.0673 |
| Strategy A | | | | | | | | 1 | −0.5924** | −0.0807 |
| Strategy B | | | | | | | | | 1 | 0.0870* |
| Strategy C | | | | | | | | | | 1 |

* $P < 0.05$, ** $P < 0.01$

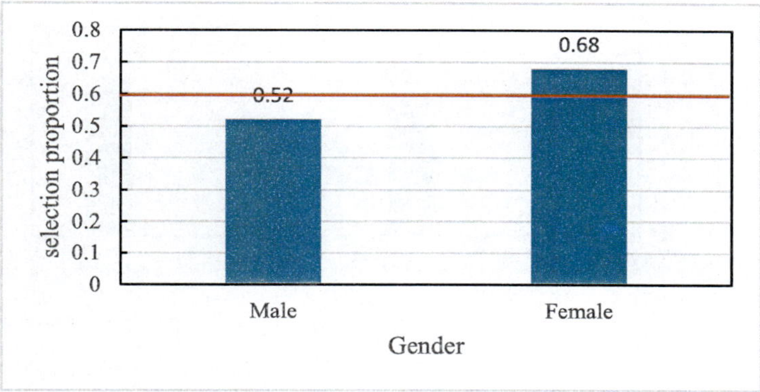


**Fig. 2.** The impact of gender on the preference of path selection strategy A.

adults. When there was insufficient path-finding aids, older adults had to rely on their limited memory and cognitive abilities, which made it more difficult for them to navigate [14]. Older adults have longer gaze durations on maps compared to younger individuals, mainly due to the decline in spatial cognition abilities [15].

### 5.3 The Influence of Educational Level

The educational level does not significantly affect the preference for the three strategies. According to the correlation analysis (see Table 3), the educational level has a significant impact on Factor 5 and Factor 7, indicating that higher educational level is associated with a higher proportion of participation in emergency training or drills and a higher level of attention to signage.

### 5.4 The Influence of Visual Acuity

Visual acuity mainly affects the preference for Strategy C. As shown in Fig. 4, there is a significant increase in the proportion of individuals with poor vision following others in the evacuation process.

### 5.5 The Influence of Emergency Knowledge Level

The emergency knowledge level mainly affects the preference for Strategies A and C. Here, the level of emergency knowledge is determined by the accuracy of the respondents' answers regarding emergency knowledge. As shown in Fig. 5, individuals with a low level of emergency knowledge has a significantly lower proportion of relying on evacuation signs or following others in the evacuation process. The higher the level of emergency knowledge, the more selective individuals are in the information search process during decision-making, focusing only on identifying and paying attention to relevant attributes.

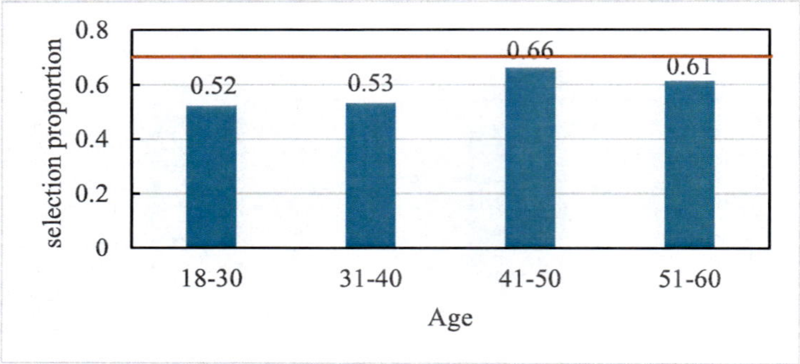


(a) Strategy A

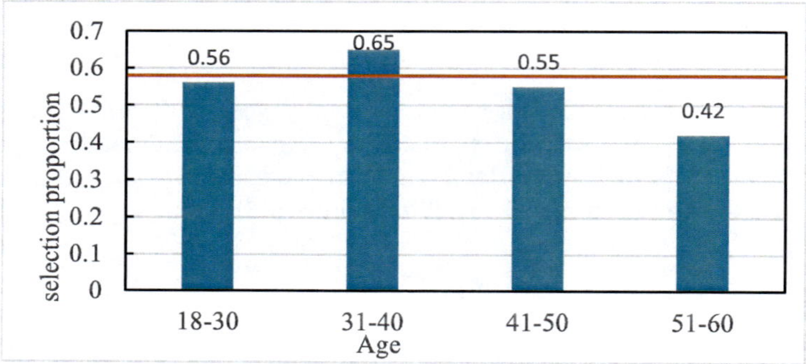


(b) Strategy B

**Fig. 3.** The influence of age on path selection strategy preferences.

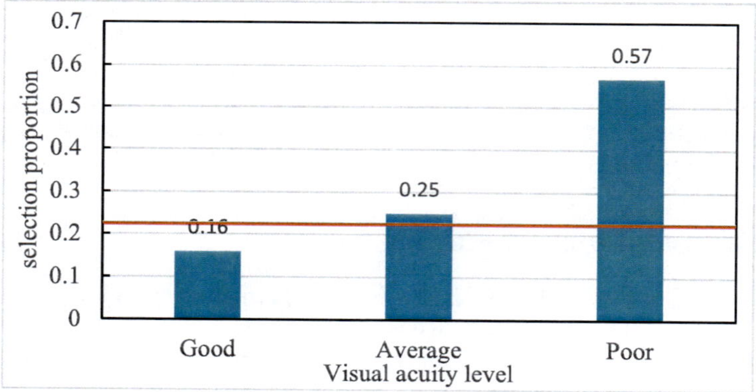


**Fig. 4.** The influence of vision level on the preference of path selection Strategy C.

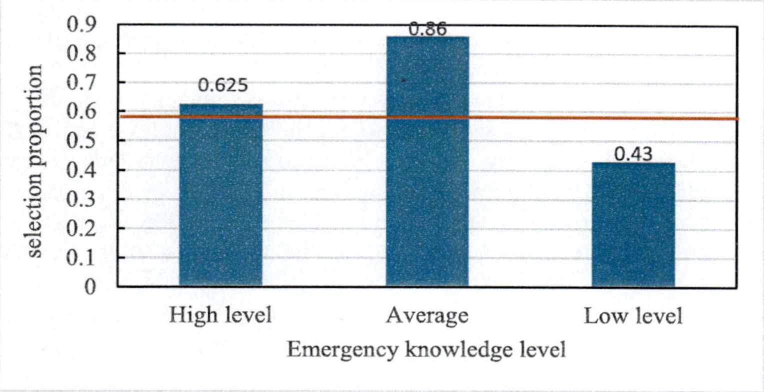


(a) Strategy A

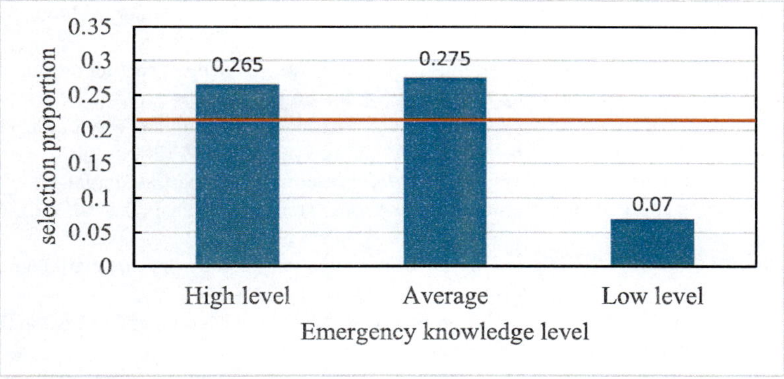


(b) Strategy C

**Fig. 5.** The influence of emergency knowledge level on export selection strategy preferences.

## 6 Conclusion

Based on the comprehensive analysis of data obtained from questionnaire surveys, the following recommendations can be made regarding emergency evacuation in the event of a sudden incident from the perspective of individual cognitive abilities of subway commuters.

- There are differences in the construction of cognitive maps based on gender and age. Females and older individuals tend to have self-centered (path-based) construction and prefer evacuating based on signage.
- The sense of uncertainty about the environment is related to age and visual acuity. Older individuals and those with poor visual acuity tend to prefer following others in evacuation. The attention to signage is influenced by visual acuity, with individuals with poor vision paying more attention to signage.

- Higher individual knowledge levels lead to more selective information search and a tendency to rely on signage and information from others during the decision-making process of selecting evacuation paths.
- The role of emergency training and evacuation drills is significant. They can improve the ability to construct a non-self-centered cognitive map, increase attention to signage, enhance safety awareness and capabilities, and improve the ability to cope with uncertainty. This helps reduce panic and uncertainty during emergencies, making it easier to construct cognitive maps and find rapid and accurate evacuation paths.
- In conclusion, these findings provide guidance and recommendations for enhancing the emergency evacuation abilities of subway passengers. This includes improving signage systems, strengthening emergency training and drills, and developing targeted evacuation strategies based on different age groups, genders, and visual acuity levels.

## References

1. Downs, R., Stea, D.: Cognitive representations. In: Image and Environment. Aldine, pp. 79–86 (1973)
2. Pelechano, N., Badler, N.I.: Modeling crowd and trained leader behavior during building evacuation. IEEE Comput. Graph. Appl. **26**(6), 80–86 (2006)
3. Wiener, J.M., Büchner, S.J., Hölscher, C.: Taxonomy of human wayfinding tasks: a knowledge-based approach. Spat. Cognit. Comput. **9**(2), 152–165 (2009)
4. Andresen, E., Chraibi, M., Seyfried, A.: A representation of partial spatial knowledge: a cognitive map approach for evacuation simulations. Transp. A: Transp. Sci. **14**(5–6), 433–467 (2018)
5. Maceachren, A.M.: Application of environmental learning theory to spatial knowledge acquisition from maps. Ann. Assoc. Am. Geogr. **82**(2), 245–274 (1992)
6. Ellard, C.: You Are Here: Why We Can Find Our Way to the Moon, but Get Lost in the Mall. Knopf Doubleday Publishing Group (2010)
7. Golledge, R.G., Jacobson, R.D., Kitchin, R., Blades, M.: Cognitive maps, spatial abilities, and human wayfinding. Geograph. Rev. Jpn. Ser. B **73**(2), 93–104 (2000)
8. Colombo, D., et al.: Egocentric and allocentric spatial reference frames in aging: a systematic review. Neurosci. Biobehav. Rev. **80**, 605–621 (2017)
9. Zejun, L., Wei, L.: The mechanism of integration promoting associative memory: familiarity and recall processing. Progr. Psychol. Sci. **30**(10), 2240–2253 (2022). (in Chinese)
10. Yonelinas, A.P.: The Nature of recollection and familiarity: a review of 30 years of research. J. Mem. Lang. **46**(3), 441–517 (2002)
11. Yonelinas, A.P.: Recognition memory ROCs for item and associative information: the contribution of recollection. Mem. Cognit. **25**(6), 747–747 (1998)
12. Ruikang, A., Jingyuan, L., Ying, M., et al.: Uncertainty and conformity: the mediating role of decisiveness and structural requirements. Psychol. Behav. Res. (2), 174–181 (2022) (in Chinese)
13. Chen, C.H., Chang, W.C., Chang, W.T.: Gender differences in relation to wayfinding strategies, navigational support design, and wayfinding task difficulty. J. Environ. Psychol. **29**(2), 220–226 (2009)
14. Lee, S., Kline, R.R.: Wayfinding study in virtual environments: the elderly vs. the younger-aged groups. Archnet-IJAR: Int. J. Archit. Res. **5**(2), 63–76 (2011)
15. Kirk Rodgers, M., Sindone III, J.A., Moffat, S.D.: The effect of age on navigation strategy. Neurobiol. Aging **33**, 202.e15–202.e22 (2012)

# Research and Simulation of Multi-objective Optimization of Urban Rail Train Automatic Driving System

Changzhi Lv[1], Dongyu Liu[1], Ke Li[2], and Xiaoming Wang[3](✉)

[1] Shandong University of Science and Technology, Qingdao 266500, China
3421644827@qq.com
[2] University of Technology MARA (UiTM), 40450 Shah Alam, Malaysia
2022843172@isiswa.uitm.edu.my
[3] Shandong University of Science and Technology, Jinan 250031, China
sdkdwxm@163.com

**Abstract.** The automatic driving system serves as the central component for efficiently managing the operation of urban rail trains, enabling train control through algorithmic means. Evolutionary algorithms are commonly employed to optimize control strategies for autonomous driving systems. This paper presents a segmented multi-objective optimization model for the train operation phase and employs a non-dominated sorting genetic algorithm with an elite retention strategy to solve the model. This approach achieves global optimization with respect to the smoothness, accuracy, punctuality, and energy efficiency of urban rail trains. The simulation results demonstrate that traction and braking operations can be optimized for stability and energy conservation, while intermediate operations can be optimized for precision, punctuality, and energy efficiency. This optimization strategy enhances the multi-objective optimization performance of the automatic driving system, accommodating the intricacies of the rail line, and offering valuable insights for enhancing the automatic driving system of urban rail trains.

**Keywords:** Autonomous driving · Multi-objective optimization · Genetic algorithm · Simulation

## 1 Introduction

To address urban traffic congestion and cater to intercity transportation needs, rail transit has emerged as a pivotal aspect of urban development. As the central control unit of urban rail transit, the automatic train driving system controls the traction and braking system of the train by coordinating with the ground signal and vehicle status. This makes it possible to replace human drivers with automatic train control [2].

Research on the automatic driving system of urban rail trains typically revolves around analyzing either specific station sections or the entire rail line. It primarily emphasizes the optimization of individual objectives and utilizes the fitting speed curve


Y. Qin et al. (Eds.): EITRT 2023, LNEE 1137, pp. 645–652, 2024.
https://doi.org/10.1007/978-981-99-9311-6_65

as a control mechanism to achieve automated train driving. This research methodology ensures comprehensive analysis coverage, fulfills the criteria for standard parking accuracy, and aligns with energy consumption indicators. Consequently, this approach has yielded favorable outcomes when widely implemented [3]. The Automatic Train Operation system (ATO) serves as an onboard control system specifically designed to achieve automated and secure operation of urban rail trains. It plays a vital role in meeting the operational requirements of train services. Given the presence of human operators in urban rail train operations, adherence to specific routes, schedules, designated passenger stops, and operational costs is necessary. The urban rail transit system sets forth four key demands for the train control system: stability, accuracy, punctuality, and energy efficiency. Consequently, designing a control system becomes a multi-objective optimization challenge. Many multi-objective optimization problems are non-convex or even non-continuous in nature, and evolutionary algorithms prove effective in addressing such problems to attain superior solutions [11]. Evolutionary algorithms encompass a range of techniques, including NSGA-II and its variations, the Differential Evolution (DE) algorithm, the Multi-Objective Evolutionary Algorithm Based on Decomposition (MOEA/D) algorithm, among others [6, 15]. The ATO control system can achieve notable performance improvements by employing evolutionary algorithms for control purposes [8, 16, 17]. These algorithms utilize distinct selection mechanisms and evolutionary mechanisms to optimize the control system. Through iterative processes, they aim to obtain a set of Pareto optimal solutions or the Pareto frontier [7, 12, 13].

By employing evolutionary algorithms and incorporating constraints related to performance indicators, it is possible to achieve multi-objective optimization design with a specific focus [10, 14]. Moreover, segmented control strategies can be implemented to cater to the control requirements at different stages of the system, ultimately leading to improved overall control outcomes [1]. In line with the control accuracy requirements of the urban rail train control system and the objective of achieving multiple performance indexes throughout the train's operation, this paper utilizes the NSGA-II algorithm. The algorithm is employed to optimize the train running control system for various stages, taking into account the distinct characteristics exhibited by the train during each stage.

## 2 Optimization Objective Analysis

### 2.1 Analysis on Automatic Driving Targets

The automatic train control system is an onboard control system designed to achieve the automated and safe operation of urban rail trains. It calculates the target speed curve based on input data and governs the execution speed curve of the train's traction and braking system. Urban rail transit imposes specific demands on train operations, which can be classified into four aspects. Firstly, passenger comfort requirements necessitate gentle acceleration and deceleration of trains to ensure a stable riding environment. Secondly, when stopping at a station, the train must align its doors with the platform doors to provide ample overlapping space, facilitating seamless passenger boarding and alighting [4, 5]. Thirdly, the train must adhere to the timetable and adhere to the specified interval running time as per the routing diagram. This ensures punctual arrival and departure at stations and enhances train density. Fourthly, train operations should aim to minimize electricity consumption and enhance operational efficiency [5, 9]. Automatic train driving plays a pivotal role in enhancing the quality and efficiency of passenger transportation.

### 2.2 Dynamics Characteristics of the Train

According to the performance of the train and the characteristics of the track line, one section is designated as the outbound speed increase (inbound deceleration) section, and the rest is the intermediate running section. In this paper, $\Delta S$ is defined as each unit step, $n\Delta S$ is the total length of each running stage, and the system uses unit step to control the acceleration of the train.

The resistance encountered in train operation includes basic resistance and additional resistance. The basic resistance is simplified by empirical formula, as shown in Eq. (1).

$$F_{JB} = A + Bv + Cv^2 \tag{1}$$

In the formula, the constants A, B, and C are related to the construction characteristics of the train. The variable $v$ represents the train speed, measured in kilometers per hour (km/h). $F_{JB}$ denotes the basic resistance, measured in newtons per kilonewton (N/kN).

According to the curve data, the relationship between the additional resistance $F_c$ and the central angle $\alpha$ and arc length $L$ of the curve section can be expressed as formula (2).

$$F_c = \frac{10.5\alpha}{L} \tag{2}$$

The additional resistance of the tunnel is calculated by empirical formula (3).

$$F_t = \alpha L_s \tag{3}$$

In the equation, $\alpha$ represents the tunnel resistance coefficient, and $L_S$ denotes the tunnel length. The traction force exerted by the train overcomes the resistance to propel the train forward. The traction force significantly exceeds the resistance, leading to the following relationship:

$$F >> F_f \text{ i.e. } F \approx F_{join} \tag{4}$$

$$W = \frac{\int F_{join} vdt + \int F_f vdt}{\xi} = \frac{\int Fvdt}{\xi} \approx \frac{\int F_{join} vdt}{\xi} \tag{5}$$

### 2.3 Algorithm Constraint

Based on the train's performance characteristics and the conditions of the rail line, a specific portion of the section is designated as the outbound speed increase segment, while the remaining portion is referred to as the intermediate operation section. In this context, the term $\Delta S$ represents the unit step length, and $n\Delta S$ denotes the total length of each running stage. The train acceleration is controlled based on the unit step length, thereby ensuring precise control over the train's movement.

Traction energy consumption can be considered as the work determined by the combined force exerted on the train. This can be modeled as follows:

$$f_1(\boldsymbol{a}) = E = \sum_{i=1}^{n} |ma_i \Delta S| \tag{6}$$

In Eq. (6), $E$ represents the energy consumption of the train during traction, m denotes the total mass of the train, $a_i$ represents the acceleration at the ith step length, and $\Delta S$ signifies the unit step length.

The stationarity index is closely related to the acceleration of the train, and the stationarity of the train is quantified by the change of traction acceleration, as shown in Eq. (7).

$$f_2(\boldsymbol{a}) = Jerk = \sum_{i=1}^{n} (a_i - a_{i-1})^2 \tag{7}$$

The time accuracy of train operation improves as the deviation between train operation and the schedule decreases. The punctuality index model is defined by Formula (8):

$$f_3(\boldsymbol{t}) = \Delta T = \sum_{i=1}^{n} |t_i - T_i| \tag{8}$$

In Formula (8), $t_i$ represents the time at which the train arrives at the end of step i, and $T_i$ represents the corresponding scheduled arrival time at the end of step i. The accuracy index is defined by Formula (9):

$$f_4(\boldsymbol{s}) = |S - n\Delta S| + \frac{1}{2}\sum_{i=1}^{k} (s_i - \Delta S_i)^2 \tag{9}$$

In Formula (9), $s_i$ represents the running distance the train to travels to the i-th stopping point, while $\Delta S_i$ represents the scheduled running distance the train travels to the i-th stopping point.

## 3 Multi-objective Optimization Modeling and Algorithm Setting

### 3.1 Multi-objective Optimization Modeling

The traction operation stage has a relatively short duration and does not involve the issue of parking accuracy. Therefore, the optimization focus primarily lies on stability and energy-saving indicators. The optimization model is represented by Eq. (10):

$$F = \min[f_1(\boldsymbol{a}), f_2(\boldsymbol{a})]^T$$
$$s.t.\begin{cases} v_n = \sqrt{\sum_{i=1}^{n} 2a_i \Delta S + v_0^2} = v_{op} \\ v_0 = 0 \\ v_i = \sqrt{\sum_{i=1}^{i} 2a_i \Delta S + v_0^2} < v_{\max} \\ 0 < a_i \le |a_{\max}| \end{cases} \tag{10}$$

During the braking operation stage, optimizing both stationarity and accuracy becomes essential due to the requirement of considering the stopping position. The corresponding optimization model is presented as Eq. (11):

$$F = \min[f_2(\boldsymbol{a}), f_3(\boldsymbol{t})]^T$$
$$s.t.\begin{cases} v_0 = v_{ad} \\ v_n = 0 \\ v_i = \sqrt{\sum_{i=1}^{i} 2a_i \Delta S + v_0^2} < v_{\max} \\ -|a_{\max}| < a_i \le 0 \end{cases} \tag{11}$$

The intermediate operation stage is divided into n unit steps ΔS, each step calculates the sum of the basic resistance and additional resistance, and is denoted as Bi. Assuming that the positive direction of the train is taken as reference and the resistance at each step remains approximately constant, Eq. (12) can be obtained.

$$F = \min[f_1(\boldsymbol{a}), f_2(\boldsymbol{a}), f_3(\boldsymbol{t}), f_4(\boldsymbol{s})]^T$$
$$s.t.\begin{cases} v_0 = v_{op} \\ v_n = v_{ad} \\ v_i \le v_{op} \\ 0 \le \Delta s \le S_{\lim} \\ \Delta t \le T_{\lim} \\ \Delta s \le \frac{S_{\lim}}{n} \end{cases} \tag{12}$$

### 3.2 Algorithm Setting

Given the characteristics of the model, the non-dominated sorting genetic algorithm with an elite retention strategy can be chosen for solving it. This algorithm is capable of preserving elite individuals more effectively, reducing computational complexity, and achieving rapid convergence towards global optimization [6].

Encoding: adopt real number encoding. In the optimization model for the traction and braking operation stages, the independent variable is the acceleration at each step. Using real number coding, individual chromosomes are represented as $A = (a_1, a_2, a_3, \ldots, a_n)$ $(a \in R^+, n = 1, 2, 3, \ldots, i)$. In the optimization model of the intermediate running stage, the independent variables are the final velocity and acceleration at each step length. Real number encoding of individual chromosomes $B = (v_1, v_2, v_3, \ldots, v_n)$ $(v \in R^+, n = 1, 2, 3, \ldots, i)$, $C = (a_1, a_2, a_3, \ldots, a_n)$ $(a \in R^+, n = 1, 2, 3, \ldots, i)$, generate initial population randomly.

The tournament method with a size of 2 is employed for the selection of offspring individuals. Mutation operation follows the principle that mutation intensity decreases with the increase of the number of iterations. And the multi-objective optimization algorithm of automatic driving system adopts the predetermined maximum number of iterations as the termination condition. The specific value can be determined based on the convergence observed in the optimization experiment.

## 4 Simulation of Multi-objective Optimization

MATLAB was employed for simulation based on the operation data provided by Jinan Rail Transit Group and the configured simulation data. The simulation process involves the design of two different parts, one is the undulating curve and the other is the elevated tunnel.

As for the undulating curve section, the train's traction stage has a length of 800 m with a step length of 100 m and a total of 8 steps. The train's braking stage has a length of 400 m with a step length of 100 m and a total of 4 steps. As for the middle running stage of the train, the parameters are set as follows: step length of 126.5 m and a total of 20 steps.

For the elevated tunnel section, the train's traction stage, the length is set to 800 m with a step length of 100 m, and there are a total of 8 steps. The train's brake stage has a length of 400 m with a step length of 100 m, and there are 4 steps in total. Regarding the middle running stage of the train, the parameters are set as follows: the step length is 131.5 m, and there are 10 steps.

The initial population is determined by considering large amounts of actual operational data, optimizing trend predictions, and the coverage of the range. The population size is set to 50, the number of evolutionary iterations is 40, the crossover probability is 0.85, and the mutation probability is 0.04. The velocity curve is visualized in Fig. 1:

For the undulating curve section, the final solution is represented by $\Delta L = 0.12$, $\Delta t = 4.27$, and $\Delta E = 17.9$. For the elevated tunnel section, the final solution is represented by $\Delta L = 0.11$, $\Delta t = 2$, and $\Delta E = 9.4$.

## 5 Conclusion

In light of the existing issues with the current automatic train driving systems, such as limited universality, single optimization goals, and challenges in fitting the speed curve, this paper aims to address these concerns. It conducts an analysis of the automatic train

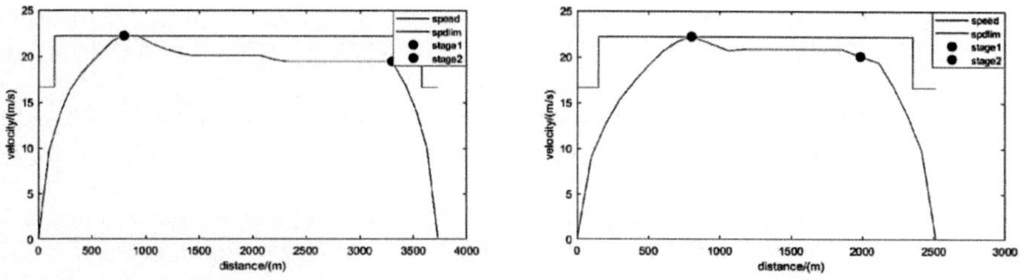


(a) Undulating curve.

(b) Elevated tunnel.

**Fig. 1.** The whole speed curve solved by MATLAB.

driving system in urban rail and integrates the principles of multi-objective optimization theory to propose a comprehensive multi-objective optimization model for the automatic driving system across different stages of train operation. The model is then solved using genetic algorithms.

By optimizing and integrating the traction operation stage, braking operation stage, and intermediate operation stage, the study achieves multi-objective optimization for the train's smoothness, accuracy, punctuality, and energy saving throughout the entire section. Through a comprehensive analysis of the simulation data, the study demonstrates significant improvements in various performance metrics. Specifically, the smoothness is enhanced by 50–60%, accuracy improves by approximately 26%, punctuality improves by around 32%, and energy saving is improved by about 4.6%. The proposed optimization strategy proves to be feasible, universal, and provides valuable insights for the research on automatic control systems for urban rail train operations.

**Acknowledgments.** This work is supported by Natural Science Foundation of Shandong Province for Innovation and Development Joint Funds (ZR2022LZH001).

## References


1. Yang, Y., Liu, H., Ma, C., et al.: Target speed control optimization of train movement for saving energy. J. Transp. Syst. Eng. Inf. Technol. **19**(01), 138–144 (2019). (in Chinese)
2. Li, Z., Hu, E.: Study of integrated intelligent completely automated and driverless train operation control system. Railw. Signal. Commun. **55**(10), 69–73+79 (2019). (in Chinese)
3. Qi, P., Wang, X., Liao, L., et al.: Study on dynamic adjust technology of inboard control curve based on vehicle braking performance. Railw. Signal. Commun. **56**(01), 9–12 (2020). (in Chinese)
4. Qi, Z., Zhu, F., Cui, Q.: Study on ATO mode accurate stopping of 01A02 train on Shanghai rail transit line 1. Electr. Locomot. Mass Trans. Veh. **04**, 138–141 (2020). (in Chinese)
5. Tang, F.L., Han, Q.J.: Analysis of energy consumption of subway vehicles and research on energy-saving measures. Locomot. Electr. Drive **43**(06), 27–31 (2020). (in Chinese)
6. Zheng, X., Ma, L.: An improved NSGA-II algorithm for multi-objective nonlinear optomization. Microelectron. Comput. **37**(07), 47–53 (2020). (in Chinese)
7. Xu, K., Yang, F., Tu, Y., et al.: Multi-objective optimization of speed profile of urban rail train base on multiple particle swarms co-evolutionary. J. China Railw. Soc. **43**(2), 95–102 (2021). (in Chinese)

8. Zhang, J., Wu, X.: Study on multi-objective optimization of high-speed railway ATO operation process based on improved MH algorithm. J. Railw. Sci. Eng. **18**(02), 334–342 (2021). (in Chinese)
9. Zhang, J., Zhu, A.: Optimization method of automatic train operation speed curve based on genetic algorithm and particle swarm optimization. J. Comput. Appl. **42**(02), 599–605 (2022). (in Chinese)
10. Li, J., Xun, J., Yin, X., et al.: Research and application of energy conservation program for train operation control of Beijing rail transit. Railw. Transp. Econ. **44**(06), 136–141 (2022). (in Chinese)
11. Deb, K., Pratap, A., Agarwal, S., Meyarivan, T.: A fast and elitist multiobjective genetic algorithm: NSGA-II. IEEE Trans. Evol. Comput. **6**(2), 182–196 (2002)
12. Jornada, D., Jorge Leon, V.: Biobjective robust optimization over the efficient set for Pareto set reduction. Eur. J. Oper. Res. **252**(2), 573–586 (2016)
13. Wang, Z., Li, H.: A novel multi-objective evolutionary algorithm based on linear programming. In: 14th International Conference on Computational Intelligence and Security (CIS), pp. 345–348. Hangzhou (2018)
14. Hou, Z., Dong, H., Gao, S., et al.: Energy-saving metro train timetable rescheduling model considering ATO profiles and dynamic passenger flow. IEEE Trans. Intell. Transp. Syst. **20**(7), 2774–2785 (2019)
15. Kochan, A., Daszczuk, W.B., Grabski, W., et al.: Formal verification of the European train control system (ETCS) for better energy efficiency using a timed and asynchronous model. Energies **16**(8), 3602 (2023)
16. Djordjević, B., Fröidh, O., Krmac, E.: Determinants of autonomous train operation adoption in rail freight: knowledge-based assessment with Delphi-ANP approach. Soft Comput. **27**(11), 7051–7069 (2023)
17. Zhong, G., Lu, Y., Chen, W., et al.: Multi-objective optimization approach of shelter location with maximum equity: an empirical study in Xin Jiekou district of Nanjing, China. Geomat. Natl. Hazards Risk **14**(1), 2165973 (2023)

# Research and Implementation of Video Intelligent Auxiliary Recognition System for Smart Stations in Urban Rail Transit

Li Youwen(✉), Li Yongzheng, and Le Linzhi

NANJING SAC Rail Traffic Engineering CO., LTD., Xinghuo Road 8, Nanjing 210032, China
{youwen-li,yongzheng-li,linzhi-le}@sac-china.com,
leiyuziqq@163.com

**Abstract.** This paper focused on the operation and management for smart stations in urban rail transit, we analyzed the key technologies of intelligent video auxiliary recognition, designed and implemented a video intelligent analysis subsystem which suitable for subway stations. This system monitors and analyzes various videos in the subway station in real-time, alerts and notifies the station operators when analyzing important information, abnormal events, and abnormal behaviors in the station, which supports typical application scenarios to assist in operation management.

**Keywords:** Smart station · Intelligent video analysis · Auxiliary recognition

## 1 Introduction

With the progress of society, video monitoring technology has been widely used in various industries. Currently, cameras are deployed in stations of urban rail transit such as subways, light rails, and intercity trains. By utilizing the security prevention, information capture, and management command functions of video monitoring, station operations can be monitored in real-time, guiding station operation management personnel to conduct inspections and handle abnormalities.

However, traditional video monitoring only serves the function of real-time image monitoring, it requires monitors to keep an eye on all camera feeds to achieve effective monitoring, which is inefficient and cannot simultaneously cover all cameras in a station. Intelligent video monitoring technology is based on digital and networked video monitoring, which is different from traditional video monitoring. Intelligent video monitoring technology utilizes artificial intelligence technology [1, 2], pattern recognition, deep learning, and other technologies to identify different objects and postures [3], thereby discovering abnormal situations in the monitoring screen and issuing sound and light alarms in the shortest time [4], this technology can effectively reduce missed and false alarms, especially in the daily operation management and passenger behavior analysis of subway smart stations, by notifying relevant personnel to detect and handle crises as soon as possible.


Y. Qin et al. (Eds.): EITRT 2023, LNEE 1137, pp. 653–661, 2024.
https://doi.org/10.1007/978-981-99-9311-6_66

## 2 System Functional Design

Based on the specific application of smart station in urban rail transit, the main functions implemented by the video intelligent auxiliary recognition system are divided into conventional basic functions and station auxiliary operation functions. The conventional basic functions include passenger and item detection, environmental anomaly detection, etc., as shown in Fig. 1; Station auxiliary operation functions including operation safety, operation efficiency, and operation management. The operation period can be divided into station opening, operation, before station closing, after station closing, emergency period, etc., as shown in Fig. 2.

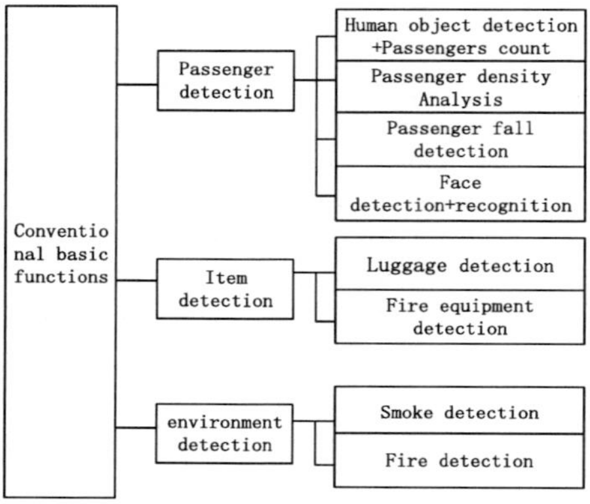


**Fig. 1.** Conventional basic functions

## 3 System Structure Design

### 3.1 System Framework

The whole smart station integration system is divided into station video monitoring subsystem, video intelligent auxiliary recognition subsystem, and smart station management platform, as shown in Fig. 3. The video intelligent auxiliary recognition subsystem, as a core subsystem, includes access module, high-performance computing module and interface module with smart station management platform.

The access module accesses real-time video data from cameras in the station video monitoring subsystem through the station video aggregation switch; The High-performance computing module uses the server GPU card to analyze the video after hard decoding by loading the video analysis algorithm trained offline, and according to the platform's scheduling of the corresponding algorithm of the camera in specific period; Through the interface module with smart station management platform, it accepts the platform's configuration and algorithm scheduling of the video intelligent auxiliary recognition subsystem, and output the video analysis results to the platform through

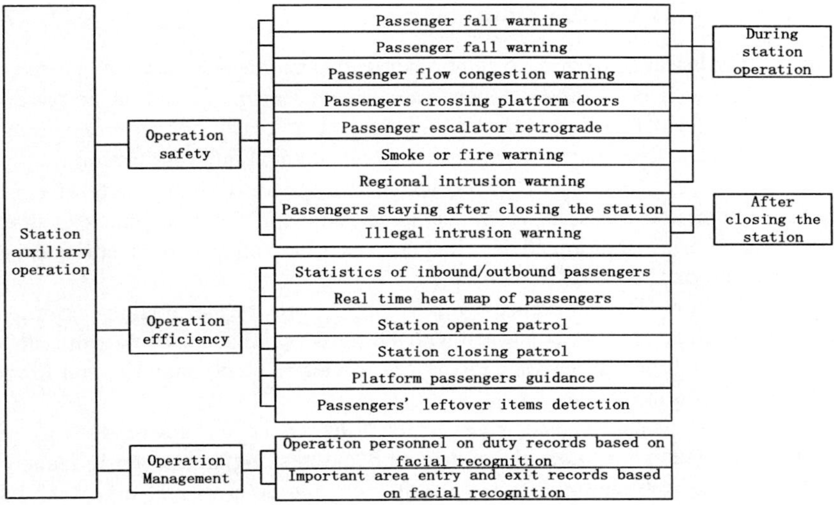


**Fig. 2.** Station auxiliary operation functions

the Kafka interface. The smart station management platform provides real-time alarms and notifies station operators of abnormal events analyzed from received videos. It also allows browsing of images of abnormal event times stored in the video intelligent auxiliary recognition subsystem, as well as 5 s videos before and after each.

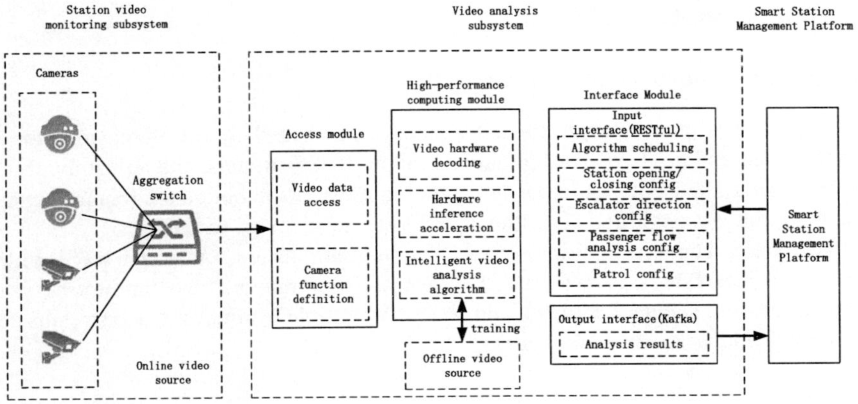


**Fig. 3.** Framework of video intelligent auxiliary recognition system

### 3.2 System Software Architecture

Using machine learning methods to comprehensively analyze videos, achieve perception of information such as passenger flow and abnormal events in the station, and based on video analysis results, achieve effective application of auxiliary operation. From the system level, the software architecture [5–7] can be divided into four layers:

The first layer is data access layer, which can access the video data of existing cameras, support the access of RTSP video streams, and the data access method of the video monitoring platform SDK. Flexible access methods can fully utilize existing resources and reduce construction costs.

The second layer is AI inference layer, which supports the use of High-performance computing accelerator cards, combined with specially optimized deep network models, to achieve efficient neural network inference, accelerate video analysis, and improve hardware utilization.

The third layer is data analysis layer, which utilizes deep learning based video analysis algorithms (such as human object detection and tracking, human posture analysis, passenger density estimation, device detection, etc.) to conduct real-time analysis of data for accessed video [8–10], including passenger flow analysis, device operation status monitoring, abnormal event analysis, etc.

The fourth layer is statistics and application layer, making full use of the result data of video analysis on the previous layer to generate data and applications conducive to auxiliary operations, including: passenger flow statistics, analysis results and visual display in the station, abnormal event alarm and interface design, data interface design with other systems, etc. The software architecture is shown in Fig. 4.

## 4 System Implementation

### 4.1 Design of Video Access Module

We read camera data through the SDK interface provided by the video monitoring manufacturer, and use GPU card to hard decoding RTSP streams, and adjust the frame rate and resolution according to the detection requirements to obtain the required image data, then image recognition algorithms processes these image data (Fig. 5).

The system supports multiple types of video data access, including RTSP video streams (H.246/H.265), as well as SDK access to common video monitoring platforms (hikvision, Dahua, Uniview), and supports multiple resolution video streams and hardware decoding methods for access.

### 4.2 Design of Video Analysis Module

The core of this system is video analysis module, which calls various intelligent analysis algorithms to detect the input video data. The intelligent analysis function module includes a passenger counting module, a passenger density analysis module, and an abnormal event detection module.

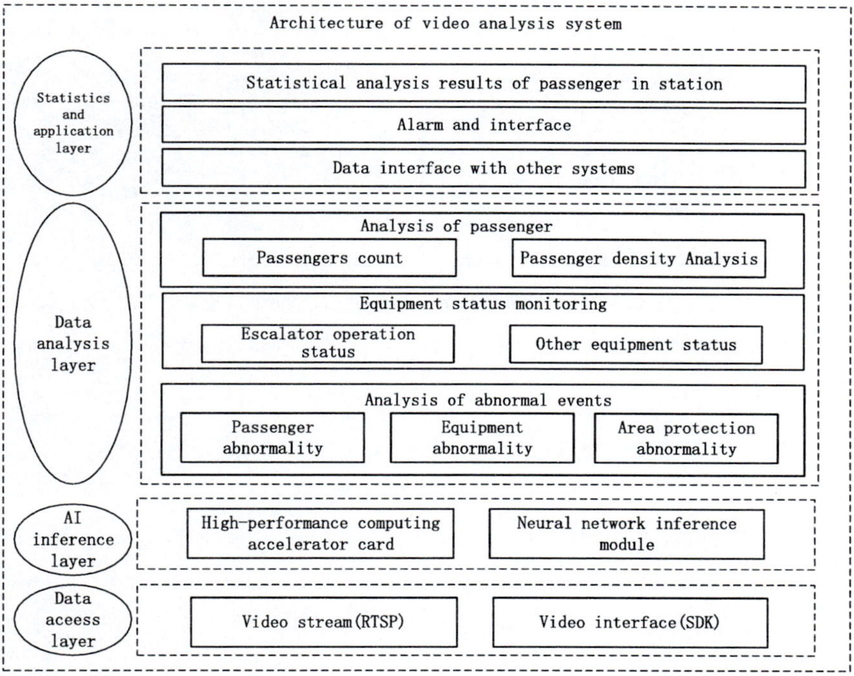


**Fig. 4.** System software architecture of video intelligent auxiliary recognition system

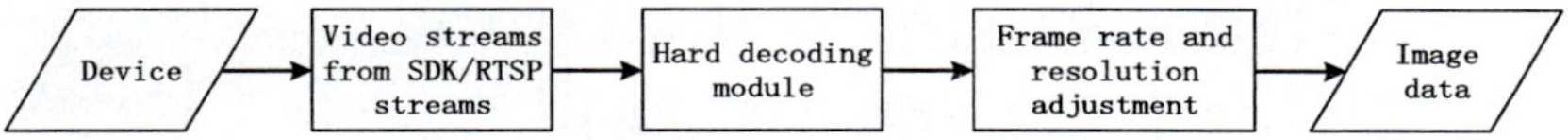


**Fig. 5.** Video access and decoding process

**Passenger counting module.** The principle of the passenger counting module need to determine a baseline in the video area and count all people passing through the line, including those entering and exiting in two directions. The baseline position is adjustable, and the number of people in two or a single direction can be selected for statistics. The counting principle is based on a deep learning based object detection method, which detects human targets, tracks multiple target human bodies, and determines the direction of target movement. When the target moves beyond the baseline, the counting value increases. The target detection model is shown in Fig. 6, and target tracking algorithm is shown in Fig. 7, also the basic function demonstration is shown in Fig. 9, which not represent the final presentation effect.

**Passenger density analysis module.** The principle of passenger density analysis need to delineate an area on the screen and estimate the density within the area. Grading passenger density and setting warning values, and when the density is too high, an alarm message will be sent. The density detection algorithm is shown in Fig. 8, and the basic function demonstration is shown in Fig. 10.

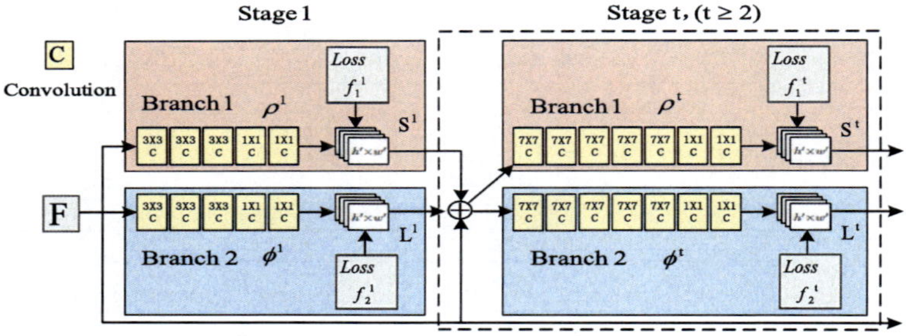


**Fig. 6.** Structure of human object detection model

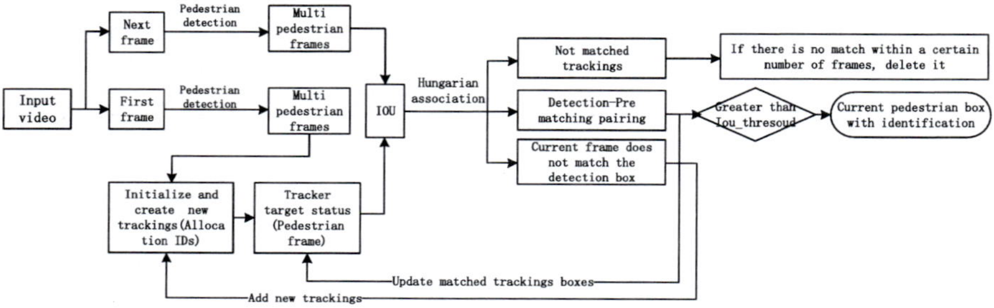


**Fig. 7.** Target tracking algorithm

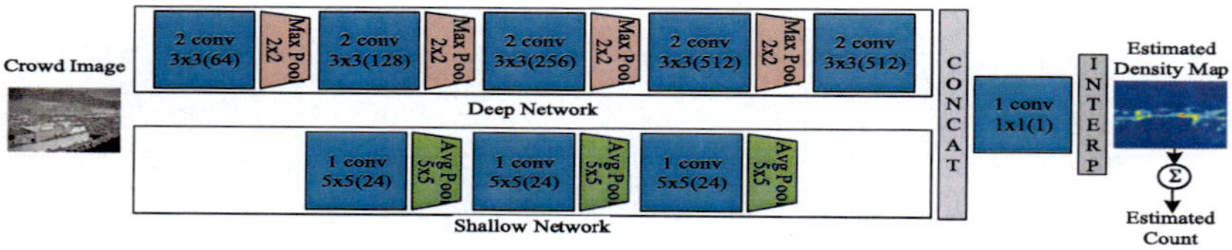


**Fig. 8.** Passenger density analysis Model

**Fig. 9.** Passenger counting demonstration

**Abnormal event detection module.** Analyze and record abnormal events in the cameras, and send alarm messages when high-risk abnormal behaviors occur.

**Fig. 10.** Passenger density demonstration

(1) Sudden Congestion Event, based on the results of passenger count and density analysis, determine the congestion situation. If the congestion situation exceeds a certain level, an alarm message will be sent. The specific implementation principles are shown in passenger counting and passenger density analysis above.
(2) Passenger fall abnormal Event, if a passenger falls and does not get up for a certain period of time, it is judged that the passenger may fall and be injured. At this time, the system prompts the monitoring operator to handle it. The detection method uses deep learning based pose detection to detect all pedestrian targets in the scene and track each target. When the pedestrian's center of gravity suddenly moves downwards and there is a significant change in the pedestrian's posture, it is judged that the pedestrian has fallen and an alarm is generated. This algorithm will not cause false positives for normal behaviors such as squatting and bending down. The algorithm principle is shown in Fig. 11.

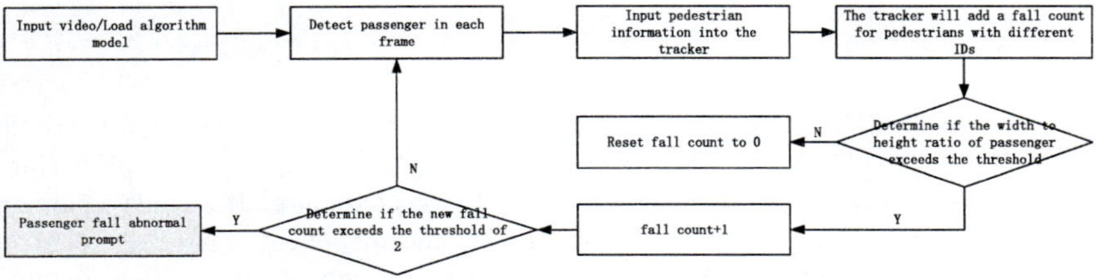


**Fig. 11.** Passenger fall abnormal detect flowchart

(3) Escalator retrograde monitoring, detect and continuously track all passengers on the escalator, determine the direction of pedestrian movement, and compare it with the set direction of the escalator. If the direction of pedestrian movement is opposite to the direction of escalator movement, it is judged as retrograde. The algorithm principle is shown in Fig. 12.
(4) Luggage leftover monitoring, collect unmanned scenes in station as background images, detect new people and objects through frame difference method, and distinguish pedestrian targets through target detection algorithm. When new objects appear in scene and there are no pedestrian targets around the object coordinates for more than N seconds, it is determined that the object is a leftover luggage and an alarm is generated. The algorithm principle is shown in Fig. 13.

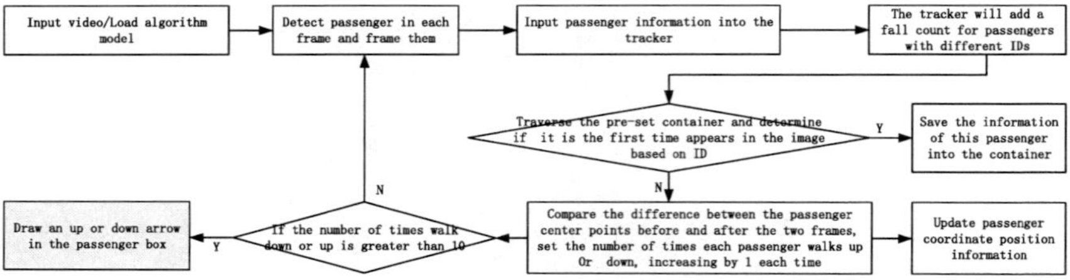


**Fig. 12.** Escalator retrograde monitoring flowchart

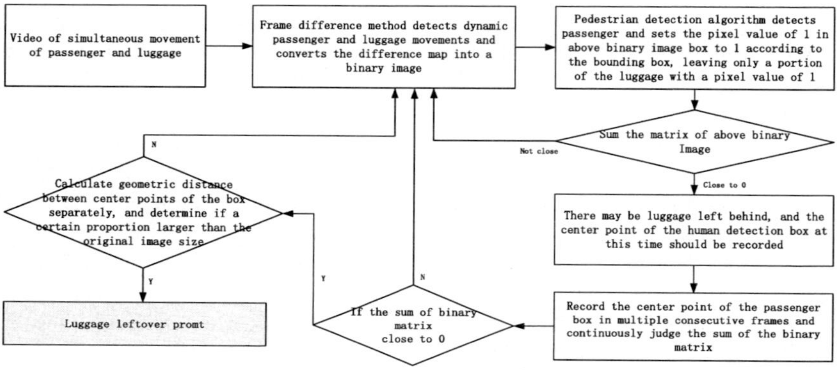


**Fig. 13.** Luggage leftover monitoring flowchart

## 5 Conclusion

Video monitoring and intelligent analysis have become necessary means for building smart stations and security systems. This paper aims to build an intelligent auxiliary recognition system while building a smart station management platform. By analyzing passengers and equipment through AI algorithms, abnormal events can be detected and reported in a short period of time, which can effectively improve the operational level of subway stations.

## References

1. Liu, S., Zhang, W., Yang, E., Liu, Y., Zhao S., Zhang, Y.: Intelligent video analysis system for railway station. In: 16th IEEE International Conference on Signal Processing (ICSP), pp. 320–324. Publisher, Beijing (2022)
2. Hidayat, F., Hidayat, F., Hamami, F., Dahlan, I.A., Supangkat, S.H., Fadillah, A., Hidayatuloh, A.: Real time video analytics based on deep learning and big data for smart station. J. Phys. Conf. Ser. **1577**(1), 012019 (2020)
3. Prati, A., Vezzani, R., Fornaciari, M., Cucchiara, R.: Intelligent Video Surveillance as a Service, pp. 1–16. Springer, Berlin, Heidelberg (2013)
4. Mabrouk, A.B., Zagrouba, E.: Abnormal behavior recognition for intelligent video surveillance systems. Expert Syst. Appl. **75**, 480–491 (2018)

5. Limna, T., Tandayya, P.: A flexible and scalable component-based system architecture for video surveillance as a service, running on infrastructure as a service. Multimed. Tools Appl. **75**(4), 1765–1791 (2016)
6. Hernández, G., Rodríguez, S., González, A., Corchado, J.M., Prieto, J.: Video analysis system using deep learning algorithms. In: Novais, P., Vercelli, G., Larriba-Pey, J.L., Herrera, F., Chamoso, P. (eds.) Ambient Intelligence – Software and Applications. ISAmI 2020. Advances in Intelligent Systems and Computing, vol. 1239, pp. 186–199. Springer, Cham (2021)
7. Chen, T., Lu, S.: Object-level motion detection from moving cameras. IEEE Trans. Circuits Syst. Video Technol. **27**(11), 2333–2343 (2017)
8. Ananthanarayanan, G.: Real-time video analytics. In: The killer app for edge computing. Computer **50**(10), 58–67 (2017)
9. Ibrahim, N., Maurya, P., Jafari, O., et al.: A survey of performance optimization in neural network-based video analytics systems. arXiv preprint arXiv:2105.14195 (2021)
10. Zaman, A., Liu, X., Zhang, Z.: Video analytics for railroad safety research: an artificial intelligence approach. Transp. Res. Record **2672**(10), 269–277 (2018)

# Establishment and Validation of a Model for Predicting Snow Drift Distance in Cold Area Tunnel

Hailong Ling[1], Chen Wang[2], Bing Wang[1(✉)], and Yingqi Li[1]

[1] College of Transportation Engineering, Xinjiang University, Urumqi 8300492, China
2388088104@qq.com
[2] China Communication Construction General Contractor Branch Company, Urumqi 8300492, China

**Abstract.** To study the snow drifting distance law and the main influencing factors of the entrance and exit section of the alpine tunnel, and to improve the safety of the tunnel entrance and exit section in the cold area, firstly, the data of wind speed, snowfall and snow drifting distance of the entrance and exit section of the alpine tunnel at high altitude were collected, and the correlation analysis of the independent variables of wind speed and snowfall on the distance of the entrance drifting snow was investigated by using the method of statistical analysis to set up a prediction model of the snow drifting distance of the tunnel in snowy days; Secondly, in order to verify the accuracy of the prediction model, FLUENT software is used to simulate the wind-blown snow scenario in the tunnel entrance section, and then the simulated values of the simulation results are analyzed in terms of the error with the calculated values calculated by the prediction model. The results show that the simulation results are in good agreement with the calculation results of the prediction model, indicating that the prediction model is able to better quantify the important characteristic of the distance of drifting snow in the tunnel entrance and exit sections. This provides a reference for further research on the evolution of the friction coefficient of tunnel pavement in cold regions, and also lays a foundation for safe driving in the entrance and exit sections of high-altitude tunnels in cold regions.

**Keywords:** Cold areas tunnels · Snow drift distance · Prediction model · Fluid simulation · Comparative analysis

## 1 Introduction

Tunnel wind-blown snow is one of the important factors considered in the planning and construction of tunnel infrastructure in high-altitude areas in cold regions [1, 2]. The impact of tunnel drifting snow on safe driving mainly includes two aspects: First, due to the "black hole effect" and "white hole effect" in the tunnel entrance and exit sections and the tunnel wind-blown snow scenarios interact with each other, resulting in reduced visibility, and in the high-altitude cold areas meet the strong winds, visibility is almost


Y. Qin et al. (Eds.): EITRT 2023, LNEE 1137, pp. 662–670, 2024.
https://doi.org/10.1007/978-981-99-9311-6_67

zero, seriously affecting the safety of high-altitude tunnel entrances and exits section. Visibility is almost zero, seriously affecting the safety of high-altitude tunnel entrance and exit section; on the other hand, due to the tunnel entrance and exit section wind-blown snow leads to uneven distribution of snow on the road surface of the entrance and exit section, the overall from the tunnel entrance to the tunnel is a decreasing distribution, the formation of road surface friction coefficient decreases gradually of ice and snow gradient pavement, uneven snow distribution of ice and snow gradient pavement seriously affects the safety of tunnel entrance and exit section.

As early as in the 1930s, there have been a lot of foreign research on tunnel wind-blown snow, the main research results are focused on tunnel wind-blown snow prevention and control of windbreaks and tunnel infrastructure [3]. Zhang Linlin [4] proposed the distribution law of snow accumulation in wind-blown snow in different roadbed sections by computer simulation method, and gave the proposed values of suitable roadbed height and slope gradient in different average wind speed areas for weakening the harm of wind-blown snow.

In summary, the current domestic and international research in the field of wind tunnel test wind-blown snow mainly focuses on the mechanism and prevention of the formation of wind-blown snow damage under the roadbed section and there is no regular research on the drifting distance of the entrance and exit section of high-altitude tunnels and its quantitative analysis [5]. In addition, the authors' team found that the friction coefficient of the ice-covered road in the urban section is relatively stable and single, while the friction coefficient of the road in the tunnel entrance and exit section is gradually decreasing or increasing. Therefore, this paper applies the statistical analysis method to characterize the prediction model of snow drifting distance in the tunnel entrance and exit section and analyzes and verifies the accuracy of the snow drifting distance prediction model through fluid simulation FLUENT simulation. It can provide a reference basis for the study of the evolution law of the friction coefficient of ice and snow gradient pavement in the above tunnel entrance section.

## 2 Establishment of Prediction Model

### 2.1 Influencing Factors of Drifting Snow Distance and Data Collection

Through the research of Tian Zhonghao [6], and Lv Xiaohui [7] on the trajectory of snow and ice particles and glass snow-covering characteristics in the train climate chamber and the wind tunnel experimental research on the wind-blown snow con-tour, it can be seen that the wind speed at the entrance of the tunnel and the amount of snowfall at the entrance of the tunnel show a strong positive correlation between the distance of snow-covering on the roadway surface at the tunnel entrance and exit section, which indicates that there is a regularity of snow and ice particles trajectory in the entrance and exit section of the tunnel under the condition of multiple-parameter coupling.

To explore the regular phenomena at the entrance and exit sections of the tunnel, four sets of experimental data were measured at the Shengli Tunnel (No. 1) in Tianshan Mountain, Xinjiang, and to increase the generality of the conclusions of this experiment, eight sets of high-altitude tunnels in the cold region around Urumqi (Gejiagou Tunnel (No. 2) and Xuelian Mountain Tunnel (No. 3)) were selected as observation objects for supplementary experiments. The main measurement parameters of the measured data in this experiment are wind speed (*V*) at the entrance of the tunnel, snowfall (*S*) at the entrance of the tunnel, temperature (*T*) and the distance of drifting snow (*L*) at the entrance and exit ends. The measured data were pre-processed by MATLAB on the experimental data obtained from the measurements, and the following Table 1 was obtained.

**Table 1.** Snowfall distance data collection under multi-parameter coupling conditions.

| Tunnel no | Hole wind speed *V* (m/s) | Snowfall *S* (mm/h) | Temperature *T* (°C) | Distance *L* (m) |
|---|---|---|---|---|
| 1 | 10.26 | 34.4 | −23 | 86.34 |
| 1 | 8.7 | 23.2 | −18 | 68.52 |
| 1 | 9.2 | 22.6 | −10 | 71.46 |
| 1 | 9.8 | 25.3 | −15 | 77.89 |
| 2 | 7.3 | 12.4 | −9 | 52.69 |
| 2 | 8.6 | 18.6 | −14 | 64.52 |
| 2 | 6.1 | 9.3 | −11 | 30.27 |
| 2 | 7.9 | 3.2 | −17 | 59.33 |
| 3 | 6.9 | 11.8 | −20 | 45.73 |
| 3 | 6.7 | 9.4 | −16 | 37.80 |
| 3 | 8.1 | 16.7 | −18 | 61.25 |
| 3 | 3.7 | 4.1 | −9 | 21.54 |

## 2.2 Data Processing and Analysis

Since this paper is to analyze the relationship between the drifting snow distance of the tunnel entrance and exit section affected by the wind speed and snowfall at the tunnel entrance, and it does not involve the order ranking of the variables and the explanatory notes of the independent variables on the dependent variables, etc.,
where *r* denotes the Pearson correlation coefficient;

$$r = \frac{\sum_{i=1}^{n}(x_i - \overline{x})(y_i - \overline{y})}{\sqrt{\sum_{i=1}^{n}((x_i - \overline{x}))^2}\sqrt{\sum_{i=1}^{n}((y_i - \overline{y}))^2}} \tag{1}$$

$x_i$ and $y_i$ denote the ith observation of the two variables, respectively;
$\overline{x}$ and $\overline{y}$ denote the mean values of the x and y variables, respectively.

Pearson correlation coefficient is chosen to quantitatively study the degree of the influence of the wind speed and snowfall at the tunnel entrance and exit section on the drifting snow distance of the tunnel entrance and exit section, which is given in the following formula:

Combined with SPSS software, analyzing the data in Table 1, it is concluded that the correlation coefficient between wind speed and snow drifting distance at the cave entrance is 0.974, and the correlation coefficient between snowfall and snow drifting distance at the cave entrance is 0.848, which both show a high degree of positive correlation, i.e., with the gradual increase of wind speed and snowfall at the cave entrance, the snow drifting distance will also show an increasing trend.

## 2.3 Predictive Model

In order to effectively characterize the snow drifting distance of the tunnel entrance and exit section under the coupled influence of wind speed, snowfall and temperature at the tunnel entrance, a prediction model under the coupled influence of wind speed, snowfall and temperature at the tunnel entrance is established based on the above experimental results in Table 1, as shown in Eq. (2).

$$L = 0.968V^2 + 0.019S^2 - 2.13V - 0.09S + 14.59 \tag{2}$$

where, $L$ is the distance of drifting snow at the hole, m;

$V$ is the wind speed at the cave entrance, m/s;

$S$ is the amount of snowfall, mm/h;

Scope of application: cave entrance temperature $T < 0$ °C.

The wind speed and snowfall at the tunnel entrance are set as independent variables, and the snow drifting distance at the entrance/exit section is the dependent variable, and the nonlinear fitting surface of the wind speed and snowfall at the tunnel entrance on the snow drifting distance at the entrance/exit section is plotted as shown in Fig. 1.

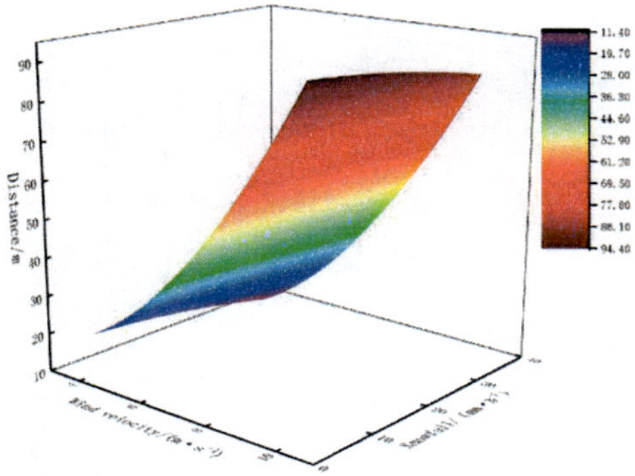


**Fig. 1.** Nonlinear fitting surface of wind speed and snowfall on drifting distance

In order to determine whether the difference between the groups of the mathematical model was significant, the coefficient of determination of its model in the ANOVA table was obtained as $R^2 = 038548$, and the probability statistic value of the model, $= 1.244 \times 10^{-8} < 0.01$, indicated a high degree of fit of the nonlinearly fitted surface.

# 3 Fluent Simulation

## 3.1 Three-Dimensional Scene Modeling

To verify the accuracy of the tunnel entrance and exit section drifting snow distance prediction model, the Xinjiang Tianshan Victory Tunnel for field examination, depicting the tunnel entrance section of the outer contour of the physical map, such as Fig. 2.

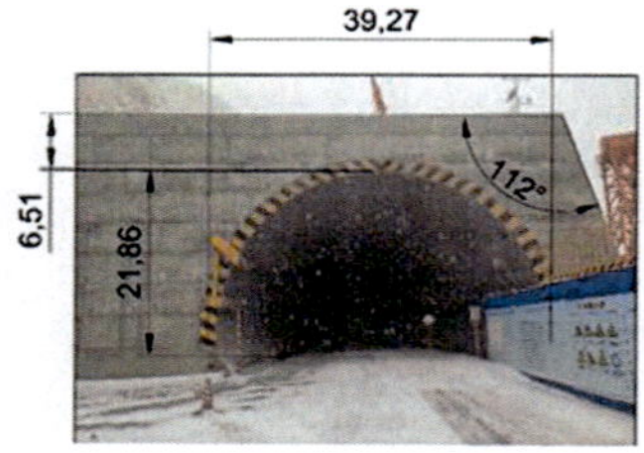


**Fig. 2.** Outline of WUYU Tianshan Shengli Tunnel

According to the Fig. 2 scale dimensions and measurements of the tunnel exit of the actual size of the tunnel exit in the three-dimensional software SOLIDWORKS to establish the 10:1 tunnel entrance section of the three-dimensional scene of the physical model, as shown in Fig. 3.

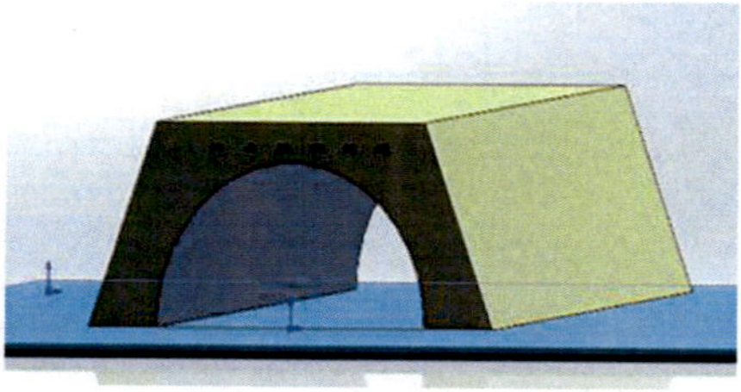

**Fig. 3.** Physical model of the tunnel entrance scene

## 3.2 Network Division and Defining Boundary Conditions

Firstly, the model information is imported into ANSYS-FLUENT, a software dedicated to flow field analysis, according to the 3D software SOLIDWORKS. Save the model file in .x_t format and load it into Geometry in WORKBENCH, use DesignModeler to process the model, and then use the Fill command to generate the computed fluid domain, i.e., the space from the inner wall of the tunnel to the tunnel pavement, and finally suppress the model of the outer wall of the tunnel, just keep the computed fluid domain and change the model attribute to Fluid. The fluid domain of the tunnel 3D stereo model is meshed using Mesh, as shown in Fig. 4, the total number of meshes is about 2,500,000, where the smallest mesh size is 0.5 m, and finally the boundary conditions such as inlet, outlet, roadway, and wall are defined.

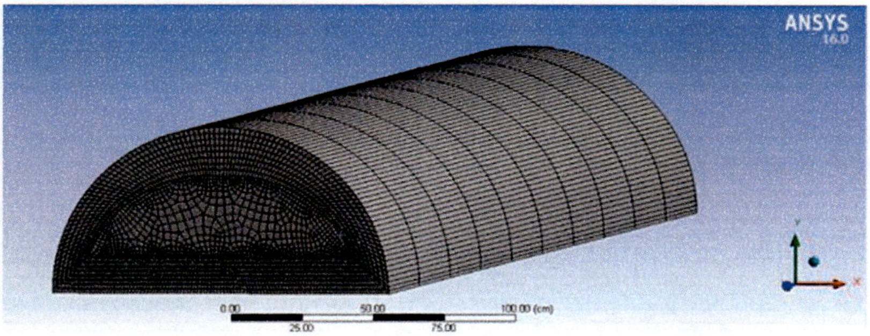


**Fig. 4.** Scene model meshing of tunnel entrance section.

### 3.3 Calculation Methods and Simulation Scenarios

The incompressible three-dimensional transient turbulence model $k-\varepsilon$ *Realizable* equation is used to solve the phenomenon of wind-snow flow in the tunnel entrance section [8]. The snow particles enter the tunnel through the tunnel entrance end face to simulate the trajectory of snow particles, and the motion of snow particles is selected from the gas-solid two-phase flow DPM model of the Eulerian-Lagrange method, and the two-way coupling of the continuous and discrete phases is ticked [9]. For the mutual coupling of pressure and velocity fields in the tunnel entrance section, the SIMPLE algorithm [10–12] is used, which can meet the motion law of snow particles in the tunnel entrance section (Table 2).

**Table 2.** Simulated working conditions.

| Working conditions | Snow grain size mm | Temperature T/°C | Hole wind speed *V* (m/s) | Snowfall (mm/h) |
|---|---|---|---|---|
| 1 | 2.4 | −5.43 | 8.326 | 27.67 |
| 2 | 2.4 | −5.43 | 14.49 | 41.58 |
| 3 | 2.4 | −5.43 | 2.864 | 17.3 |
| 4 | 2.4 | −5.43 | 6.972 | 31.73 |

In order to verify the accuracy of the prediction model of snow drifting distance in the tunnel entrance and exit section, the average value of wind speed and snowfall at the entrance of Tianshan Shengli Tunnel (Case 1), the extreme value (Case 2 for maximum value, Case 3 for minimum value) and the plural value (Case 4) are selected, and the four sets of parameters are taken as the inputs of the simulation parameters for the simulation of the working conditions. And the boundary conditions are shown in Table 3, while other settings are defaulted.

**Table 3.** Simulation scenario boundary conditions

| Name | Categories | |
|---|---|---|
| | Continuous phase | Discrete phase |
| Entrance | Velocity-intel | Escape |
| Wall | Wall | Reflect |
| Pavement | Road | Trap |
| Exit | Outflow | Escape |

## 4 Fluent Simulation

According to the simulation conditions, four simulation scenarios are set up in FLUENT for testing, Case 1 is the convergence result of the average value of the parameters such as wind speed and snowfall at the mouth of the cave, Case 2 is the convergence result of the maximum value of the parameters such as wind speed and snowfall at the mouth of the cave, Case 2 is the convergence result of the minimum value of the parameters such as wind speed and snowfall at the mouth of the cave and Case 4 is the convergence result of the multitude value of the parameters such as wind speed and snowfall at the mouth of the cave and the results of the simulation are shown in Table 4.

**Table 4.** Simulation results of snow grains under the influence of multiple parameters

| Working conditions | Snow grain size mm | Temperature T/°C | Hole wind speed *V* (m/s) | Snowfall (mm/h) | Snowdrift distance (L/m) |
|---|---|---|---|---|---|
| 1 | 2.4 | −5.43 | 8.326 | 27.67 | 68.9736 |
| 2 | 2.4 | −5.43 | 14.49 | 41.58 | 176.621 |
| 3 | 2.4 | −5.43 | 2.864 | 17.3 | 29.9664 |
| 4 | 2.4 | −5.43 | 6.972 | 31.73 | 49.7786 |

In order to verify the accuracy of the prediction model of snow drifting distance, the above four conditions of the sample parameters into the formula (2), the four conditions of snow drifting distance of the calculated value of 62.27981775 m, 166.8696814 m, 21.45853221 m, 55.65333633 m, and the four conditions of the scenario of the simulation to make comparisons, the different simulation conditions Comparison of the calculated and simulated snow drifting distance L under different simulation conditions is shown in Fig. 5.

As can be seen from Fig. 5, the deviation of the drifting snow distance L under both results is between −9.75 and 5.87 m, and the relative error is between −5.2 and 9.7%, which is within the error range of simulation, and the prediction model of the drifting snow distance it establishes can accurately predict the distance of the movement of snow particles.

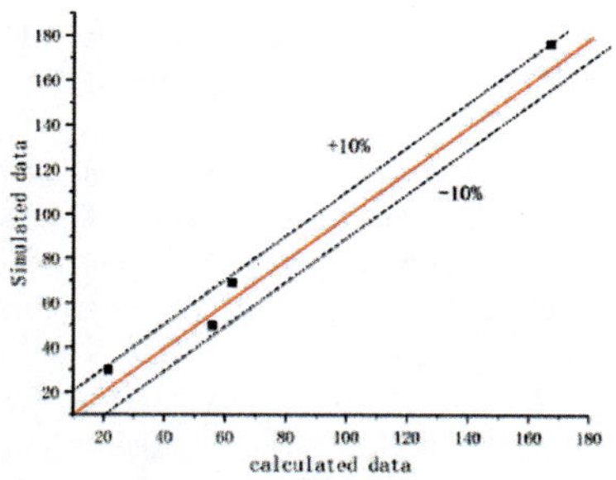


**Fig. 5.** Comparison of simulated and calculated snow drift distances under different operating conditions.

## 5 Fluent Simulation

In order to study the difficult problems of winter traffic in the entrance and exit section of the tunnel in the cold area, this paper takes the alpine high-altitude tunnel as the research object, takes the drifting snow distance on the road surface of the entrance and exit section of the tunnel as the research purpose, applies the method of statistical analysis, experimentally measures the wind speed, snowfall and drifting distance of the entrance and exit section of the tunnel in the cold area and high-altitude tunnel in the cold area and conducts correlation analysis of the collected data, applies the method of nonlinear fitting and obtains a prediction model of the drifting snow distance of the tunnel. The prediction model of snow drifting distance of the tunnel is obtained, and the accuracy of the prediction model is verified by simulating and analyzing the wind-blown snow at the tunnel entrance section through FLUENT fluid simulation. It can provide theoretical support for the safe driving and road friction coefficient of the entrance and exit section of the tunnel in the late cold area and lay the foundation for the realization of vehicle-road cooperation and intelligent tunnel construction of highway tunnels in the cold area.

**Acknowledgments.** This work was financially supported by the Major Science and Technology Special Project of Xinjiang Uygur Autonomous Region.

## References

1. Zhonglong, W.: Research on snow damage and its prevention in China. Mt. Res. (03), 22–31+65–66 (1983)
2. Zhicheng, R.: Research on snow distribution law and prevention technology of typical roadbed sections in wind-blown snow areas. Harbin Institute of Technology (2020). https://doi.org/10.27061/d.cnki.ghgdu.2020.002195.[D]
3. Zhao, G., Zuo, H., Zhang, H., et al.: Experimental study on wind tunnel simulation of snow blocking effect of snow retaining wall. Arid Zone Resources Environ. **26**(04), 107–112 (2012). https://doi.org/10.13448/j.cnki.jar.2012.04.016
4. Linlin, Z.: Research on the simulation technology of highway windblown snow and snow damage. Jilin University (2005)

5. Liu, Q., Liang, P., Gao, L., et al.: Study on the effect of slope gradient on wind-blown snow on embankment. Vib. Shock **40**(03), 87–94 (2021). https://doi.org/10.13465/j.cnki.jvs.2021.03.012
6. Zhonghao, T.I.A.N.: Research on the motion trajectory of snow and ice particles and the characteristics of glass-covered snow in train climate chamber. Qingdao University of Technology (2022). https://doi.org/10.27263/d.cnki.guide.2022.000644
7. Xiaohui, L., Ning, H., Lei, G., et al.: Wind tunnel experimental study of wind-blown snow contours. Mech. Pract. **35**(01), 20–25 (2013)
8. ANSYS fluent user's guide, release 19.0 [m].Canons burg: ANSYS, Inc. (2018)
9. Feng, G., Mingyang, L., Dongli, M., et al.: Analysis of the causes of snow and ice accumulation on bogies of urban rail trains and their prevention and control. J. Central South University (Natl. Sci. Edn.) **51**(07), 2039–2047 (2020)
10. Li, Z., Xue, C., Shi, L.: Numerical simulation analysis of the response law of wind and snow flow to railroad embankment. Railw. Stand. Des. **65**(09), 34–39 (2021). https://doi.org/10.13238/j.issn.1004-2954.202008290001
11. He, S.-Y., Feng, Y.-H., Zhou, Y.-H., et al.: Numerical simulation study on the influence law of slope gradient on snow distribution around embankment. Eng. Mech. **38**(S1), 178–183 (2021)
12. Li, G., Yu, H., Zhang, J., et al.: Wind-blown snow multiphase flow motion and its application to snow hydrology in cold regions. J. Aerodyn. **39**(03), 170–181 (2021)

# Evaluating Regional Rail Transit Safety: A Data-Driven Approach

Qing Li, Ling Liu(✉), Jun Liu, Wanqiu Zhang, and Gehui Liu

Beijing National Railway Research & Design Institute of Signal & Communication, Beijing, China
{liqing,627LL,junliu,zhangwq,liugehui}@crscd.com.cn

**Abstract.** To grasp the trend of safety conditions in the operation process of multisystem rail transit, and allocate transport capacity and maintenance resources reasonably, the managers should master accurate and comprehensive safety evaluation of regional rail transit system. A data-driven model for safety evaluation of regional rail transit system was proposed in this study. The deep autoencoder networks were employed to reduce the dimensions of the evaluation index system. The hybrid hierarchical k-means clustering method was applied to obtain the set of all possible safety status. The tree-augmented naïve Bayes algorithm was used to evaluate the overall safety. The validity and practicality of the model were verified using actual operations data from a rail transit network in regional urban agglomeration in China. A comparison with the actual situation shows that the proposed approach can evaluate the safety level of the network effectively.

**Keywords:** Regional rail transit · Safety evaluation · Data-driven approach

## 1 Introduction

With the regionalization development of urban agglomerations and the continuous extension of passenger travel distances, the scale of regional rail transit networks continues to expand, presenting a greater challenge for management in ensuring the safe operation of a network with multiple system integration. The safety evaluation of regional rail transit system (RRTS) is an important part of rail transit safety management. The application of scientific and effective methods to evaluate the safety of rail transit systems is of great significance [1]. The fundamental purpose of establishing a safety evaluation method for regional rail transit operations is to assist managers in comprehensively and accurately monitoring the status of network operation processes and risks in real time, achieving early warning and precontrol, minimizing risks, preventing problems before they occur, and ensuring operational safety.

Currently, there is a lack of research on the overall safety level evaluation of regional rail transit networks both domestically and internationally. Research on rail transit networks mainly focuses on several aspects, such as service quality, network structure, social benefits, economic benefits and environmental benefits [2, 3]. However, there is currently no comprehensive evaluation index system for the safety evaluation of regional


Y. Qin et al. (Eds.): EITRT 2023, LNEE 1137, pp. 671–681, 2024.
https://doi.org/10.1007/978-981-99-9311-6_68

or networked rail transit systems, and existing studies mainly focus on objects such as lines, stations, or equipment [4, 5]. The methods of safety evaluation are relatively limited, mainly adopting qualitative, quantitative, or combined methods, such as the analytic hierarchy process, fuzzy mathematics, neural networks, trapezoidal fuzzy mathematical theory, and system dynamics [6–12]. These methods generally have subjective assessments and poor applicability. With in-depth study of comprehensive evaluation theories, evaluation methods have become increasingly complex and diverse, with different emphases and starting points for each method. Comprehensive safety evaluation indexes are rich and have numerous quantitative indicators. The physical element method can combine qualitative and quantitative evaluations, thereby making the calculation results more accurate. The use of data-driven methods can minimize the impact of manual intervention on the calculation results, resulting in higher precision and conclusions that are more consistent with reality. Given these features, this study used a data-driven approach for a comprehensive evaluation.

In summary, this study first analyzed the main factors affecting the safety level of regional rail transit and constructed the corresponding safety evaluation index system from the perspective of "people-machine-environment-management." Subsequently, a data-driven comprehensive evaluation model for the safety status of the RRTS was established without the need to predetermine analytical mathematical models. Through machine learning algorithms, significant amounts of data on multidimensional status indexes for evaluation objects are mined to determine the regional rail transit safety index (RRTSI) and assess the overall safety status of the evaluation object as "good" or "bad."

## 2 Approach for Safety Evaluation

### 2.1 Index System for Safety Evaluation of Regional Rail Transit

The safety status of the RRTS is determined using various indicators within its evaluation scope. Therefore, the safety evaluation index system should be designed based on the overall performance of the evaluation object. The features of the RRTS, such as multi-line coupling, wide coverage, a high degree of technical system integration, and large passenger volume, make the propagation of local risks more uncertain, thereby causing more prominent impacts on the system. Its safety level is influenced by numerous dynamic risk sources, such as passenger flow, equipment conditions, and transportation resources. Accordingly, this study establishes an index system from four aspects–passenger flow, infrastructure equipment, operational environment, and management and operation–which can comprehensively reflect the attributes of risk factors, reflect the overall operation status of regional rail transit from a macroperspective, and quantify it to clarify the changes in unfavorable risk factors faced by operational safety management.

Passenger flow condition: In rail transit systems, a strong correlation between passenger flow volume and operational accidents is shown from statistical data. Similarly, the distribution state of passenger flow is a direct reflection of the safety management level of regional rail transit operations. Its representative indicators include passenger flow distribution equity, average capacity rate, and station passenger flow congestion.

Infrastructure equipment condition: Infrastructure equipment, including vehicles, signaling, power supplies, and track facility systems, is the most important component of a RRTS. Owing to the high coupling of regional systems, any system failure can significantly affect operational safety. Its representative indicators include failure rate, fault concentration, mean time between failures (MTBF) and mean time to recovery.

Operational environment condition. A solid operating environment is the foundation for the safe operation of regional rail transit. A statistical analysis of rail transit accidents found that extreme weather conditions (heavy rain, strong winds, snow, fog, lightning, etc.) account for most operational accidents that affect safety. Noise pollution is directly risky to the health of staff, thereby increasing the rate of human error. Therefore, noise should also be given attention. In addition, tracks are the foundation of train travel. Once the track geometry is determined, it cannot be changed, regardless of its quality. Poor design can become a traffic safety hazard and affect train safety and passenger comfort. Representative indicators of a good operating environment include the frequency of extreme weather events, noise levels, and track conditions.

Management and operation condition: Management and operation conditions primarily reflect the level of train dispatching and organization. It evaluates whether dispatchers command trains to operate safely and promptly according to the timetable. A low level of transport organization can cause chaos in a RRTS, leading to a decrease in the overall safety level of the system. Representative indicators of management and operation status include average train density, train punctuality rate, and service reliability.

### 2.2 Regional Rail Transit Safety Index

To comprehensively evaluate the safety status of the RRTS, the authors proposed the concept of the regional rail transit safety index (RRTSI), as defined in (1). RRTSI is a discrete quantitative indicator that measures the safety status of the RRTS, where $RRTSI \in [1, 2, \ldots, S]$; different RRTSI values represent different safety features of the RRTS.

$$RRTSI = f(q_i, q_i, \ldots, q_M) \tag{1}$$

## 3 Safety Evaluation Model

### 3.1 Process of the Model Algorithm

The algorithm process of the safety evaluation model for regional rail transit system (SEM-RRTS) is as follows, as shown in Fig. 1.

Step 1: Recognition of safety features in the RRTS. Deep autoencoder networks (DANs) can be used to reduce the dimensions of safety evaluation indexes in RRTS and reconstruct new data for safety evaluation indexes. The hybrid hierarchical k-means clustering (HHKMC) algorithm can then be used to analyze and cluster the safety status of samples, determine the value range of the RRTSI, and identify the safety features corresponding to each RRTSI value. Furthermore, the RRTSI values of the RRTS samples can be obtained in this step.

Step 2: Construction of RRTSI classifier. Based on the results from Step 1, a tree-augmented naïve Bayes (TAN) classification model is used to perform classification learning on the historical multidimensional status index data of samples with the determined RRTSI values. This helps obtain the association rules of RRTSI (i.e., the safety index classifier).

Step 3: Determination of the RRTSI. Using the safety index classifier determined in Step 2, the corresponding RRTSI value can be obtained directly based on the current multidimensional status index data of the RRTS.

Steps 1 and 2 are part of the learning stage of the model, whereas Step 3 is part of the classification stage.

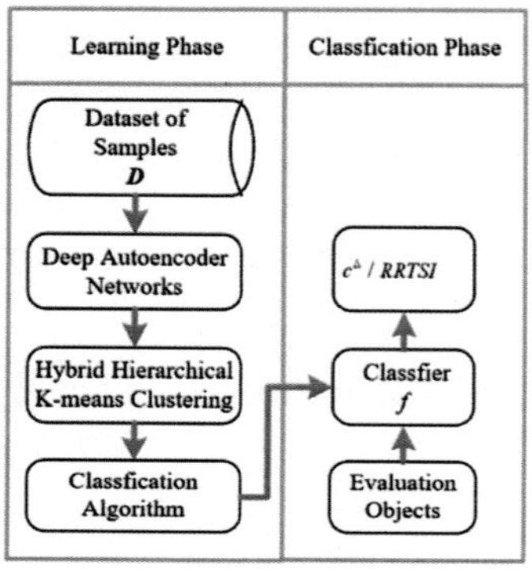


**Fig. 1.** Model algorithm flowchart

### 3.2 Safety Feature Recognition Submodel

**Deep Autoencoder Networks.** The proposed model utilizes DANs to reduce the dimensions of the safety evaluation indexes for the RRTS, compress the features of multidimensional status index data, and reconstruct new state evaluation index data. This method employs an adaptive multilayer encoding network that encodes high-dimensional data to low-dimensional data and uses a similar decoding network to reconstruct and recover low-dimensional data, as shown in Fig. 2. The neural network in the figure consists of 7 layers, where the square column in each layer represents a neuron, and the dashed lines represent connections between neurons in different layers. The network is initialized with random weights and trained by minimizing the difference between the original and reconstructed data. The gradient is calculated using the chain rule, and the weights of the network are updated. Compared with principal component analysis, DANs have better dimensionality reduction performance.

**Hybrid Hierarchical K-means Clustering.** The SEM-RRTS model is based on the multidimensional status indexes of the RRTS after dimensionality reduction. It uses HHKMC to cluster RRTS status with similar safety features into one cluster (e.g., unbalanced passenger distribution and poor infrastructure equipment status are clustered into

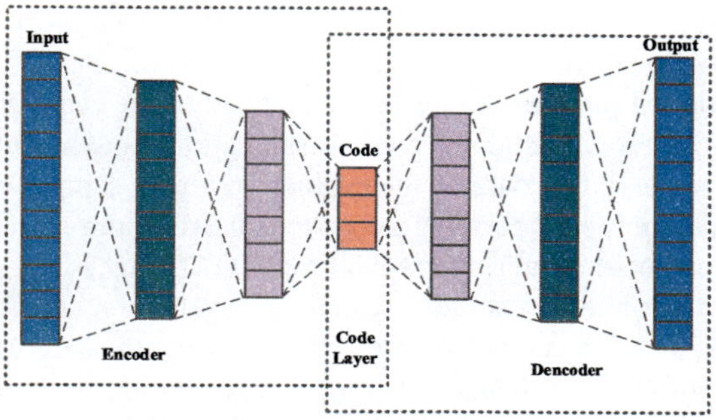


**Fig. 2.** Schematic diagram of DAN

different clusters). Evaluation objects belonging to the same cluster have the same RRTSI values. The range of the RRTSI values is determined by comparing the silhouette coefficients corresponding to different cluster numbers and selecting the cluster number that corresponds to the maximum silhouette coefficient.

The HHKMC algorithm primarily uses hierarchical clustering algorithms to perform preliminary clustering on the safety features of the RRTS and subsequently adjust the clustering results using k-means clustering algorithms. The optimal number of clusters for clustering varies for different evaluation objects and is determined based on the sample size and multidimensional status index data of the evaluation object.

### 3.3 Safety Feature Determination Submodel

**Variable Declaration.** Description of variables required to establish the model:

$B=\{q_1, q_2, \ldots, q_i, \ldots, q_M, C\}$ represents the set of variables that describe the attribute features of the evaluation object samples in classification learning, including regional rail transit safety evaluation indexes and RRTSI.

$q_i$ represents the i-th attribute variable of the evaluation object sample in classification learning, that is, the i-th state index of the evaluation object sample.

$q_1, q_2, \ldots, q_i, \ldots, q_M$ represents the M attribute variables of the evaluation object sample in classification learning, that is, the M state indexes of the evaluation object sample.

C represents the class variable of the evaluation object sample in classification learning, that is, RRTSI.

$Values(C) = \{c_1, c_2, \ldots, c_S\}$ represents the set of all possible values of class variable C in classification learning, that is, the range of RRTSI values.

S represents the number of all possible values of the RRTSI.

$c_i$ represents a specific value of class attribute C, that is, the RRTSI value of a certain evaluation object sample, $c_i \in Values(C)$.

**Model Building.** The safety status of the RRTS constantly changes during its operation. When safety status evaluation is required, managers need to quickly reevaluate RRTSI based on the latest state-detection data.

The changes in the safety status of the RRTS exhibit a "memory effect." Based on the historical multidimensional status index data of the RRTS, the set of all possible safety features obtained includes future safety features of the evaluation object. Therefore, the SEM-RRTS model adopts a TAN classification model to mine the historical multidimensional status index data of the evaluation object sample with a determined RRTSI value and determine the correlation between the multidimensional status indexes of the evaluation object and their corresponding RRTSI (that is, constructing a safety index classifier). When new status index data for an evaluation object are obtained, the corresponding RRTSI value can be obtained directly using the safety index classifier.

If there are $M$ status index data $q = (q_1, q_2, \ldots, q_M)$ of a certain evaluation object, according to Bayes' theorem, the probability that the RRTSI value of the evaluation object is $c_i$ can be expressed as

$$P(c_i|q) = \frac{P(c_i)P(q|c_i)}{P(q)} \tag{2}$$

where $P(c_i)$ represents the prior probability that the RRTSI value of the evaluation object is equal to $c_i$. $P(c_i|q)$, also known as posterior probability, reflects the influence of M status index data $q = (q_1, q_2, \ldots, q_M)$ on the probability of the RRTSI value of the evaluation object being $c_i$.

According to the Bayesian maximum a posteriori rule, the Bayesian classification learning model selects the $c^{\Delta}$ that maximizes the posterior probability $P(c_i|q)$ as the determined RRTSI value.

This study approximated the prior probability $P(c_i)$ with the probability distribution of the RRTSI value of the evaluation object being $c_i$ in the historical sample data, which is represented by ① in (3). Solving $\prod_{j=1}^{M} P\big(q_j|q_1, q_2, \ldots, q_{j-1}, c_i\big)$, which is represented by ② in (3), is essential for the Bayesian classification learning model.

$$\begin{aligned} c^{\Delta} &= \underset{c_i \in \{c_1, c_2, \ldots, c_S\}}{\arg\max} \frac{P(c_i)P(q|c_i)}{P(q)} \\ &= \underset{c_i \in \{c_1, c_2, \ldots, c_S\}}{\arg\max} P(c_i)P(q|c_i) \\ &= \underset{c_i \in \{c_1, c_2, \ldots, c_S\}}{\arg\max} P(c_i)P(q_1, q_2, \ldots, q_M|c_i) \\ &= \underset{c_i \in \{c_1, c_2, \ldots, c_S\}}{\arg\max} \underbrace{P(c_i)}_{1} \underbrace{\prod_{j=1}^{M} P\big(q_j|q_1, q_2, \ldots, q_{j-1}, c_i\big)}_{2} \end{aligned} \tag{3}$$

The naïve Bayes classifier is a simple yet efficient classification algorithm that is widely used in text classification and spam filtering tasks. Its assumption of the conditional independence between attribute variables greatly simplifies the computation process of the model. In many cases, the classification performance of naïve Bayes can surpass that of complex algorithms, such as decision trees, k-nearest neighbors, neural networks, and rule-based methods [13]. To further improve its accuracy, an extended naïve Bayes classifier can be developed by incorporating the dependencies between attribute variables into the model, which relaxes the assumption of conditional independence but requires more training samples and sophisticated computational techniques.

This study utilized the SuperParent algorithm to construct a TAN classifier learning model, which aims to explore the hidden dependency relationships among multidimensional status indices of the evaluation object, determine $\prod_{j=1}^{M} P(q_j|q_1, q_2, \ldots, q_{j-1}, c_i)$ (i.e., ② in (3)), and effectively assess the RRTSI of the evaluation object. In the SuperParent algorithm, an orphan node $q_{op}$ refers to a node that has only a class variable as its parent; a favorite child node $q_{FC}$ refers to a node that has another attribute variable as its parent in addition to the class variable; and a hyperparent node exists, which points to other nodes through arcs, namely, the parent node of the favorite child node. The specific process of the SuperParent algorithm is as follows:

Step 1: Initialize the naïve Bayes network for evaluating the multidimensional status indexes of the evaluation objects.

Step 2: Estimate the accuracy of the current classification learning model.

Step 3: Let each node $q_i \in \{q_1, q_2, \ldots, q_M\}$ be a hyperparent node and record the node $q_{SP}$ that results in the highest increase in classification learning model accuracy.

Step 4: Add an arc from s to each orphan node. If arc $(q_{SP}, q_{FC})$ provides the greatest improvement in accuracy compared to all other arcs, add the arc, and create a dependency between attributes $q_{SP}$ and $q_{FC}$. Finally, return to Step 3. Otherwise, return the current classification learning model.

Assuming that $SP(q_i)$ represents the set of parent nodes of attribute $q_i$, (3) can be further expressed as follows:

$$c^{\Delta} = \underset{c_i \in \{c_1, c_2, \ldots, c_S\}}{\arg\max} P(c_i) \prod_{j=1}^{M} P(q_j|SP(q_j)) \tag{4}$$

If $q_j$ is an orphan node, then

$$P(q_j|SP(q_j)) = P(q_j|c_i) \tag{5}$$

If the hyperparent node of $q_j$ is $q_k$, $SP(q_j) = \{q_k, C\}$, then

$$P(a_j|SP(q_j)) = P(q_j|q_k, c_i) \tag{6}$$

# 4 Empirical Analysis

## 4.1 Background Introduction

By collecting operational data on the 2020 rail transit network in a certain city in China's regional urban agglomeration and setting different operational scenarios, the safety level of the regional rail transit network was evaluated, and the proposed evaluation method was validated. This study focuses on 609 multidimensional status index data of the selected region's urban rail transit system. These data were collected during morning and evening peak hours, off-peak hours, and periods of system failure. Owing to data collection limitations, this study selected several typical indexes, denoted as $Q = (q_1, q_2, \ldots, q_7)^T$ in Table 1, from the proposed index system to verify the effectiveness of the evaluation method.

**Table 1.** Selected indexes for safety evaluation

| Safety evaluation index | Symbols |
|---|---|
| Equilibrium of passenger flow distribution | $q_1$ |
| Congestion level of passenger flow at the station (person/m$^2$) | $q_2$ |
| Failure rate (times/month) | $q_3$ |
| MTBF (min) | $q_4$ |
| Average train operation density (times/month) | $q_5$ |
| Service reliability | $q_6$ |
| Frequency of extreme weather conditions (times/month) | $q_7$ |

### 4.2 Result Analysis

Initially, the safety evaluation index data were normalized in this study. The SEM-RRTS model utilized the deep autoencoder algorithm with the R package $H_2O$ to generate a new set of status index data, denoted as $z_1$ and $z_2$. The basic structure of the DAN consists of three hidden layers with 5, 2, and 5 nodes, respectively, as shown in Fig. 3. The activation function used in the model was tanh.

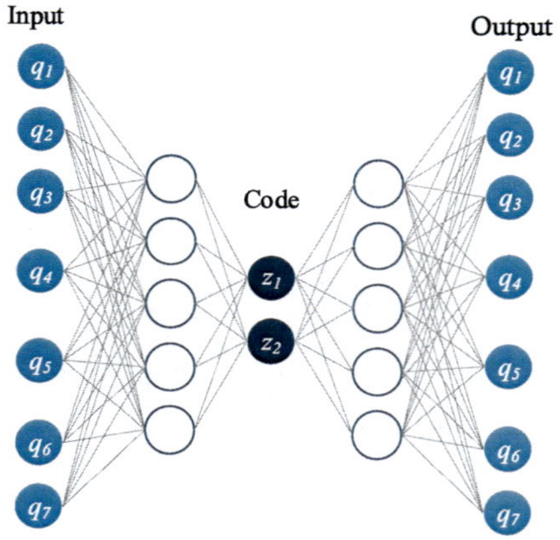


**Fig. 3.** Structure diagram of the deep autoencoder used in the example

In this example, the SEM-RRTS model utilized the mixed HHKMC algorithm with the R package FactoMineR [14]. The relationship between the number of clusters and the overall silhouette coefficient is shown in Fig. 4. By comparing the overall silhouette coefficient under different cluster numbers, the best clustering effect occurred when the number of clusters was four, indicating that the safety features of the RRTS in this example were divided into four categories, denoted as RRTSI $\in [1, 2, 3, 4]$.

The safety feature clustering results of the evaluation samples based on the new safety status attribute variables $z_1$ and $z_2$ are shown in Fig. 4, where the horizontal and vertical axes represent the status attribute variable $z_1$ and $z_2$, respectively. Evaluation samples from different safety feature categories are represented by points with different colors and shapes.

The reconstructed status attribute variables $z_1$ and $z_2$ lack corresponding physical meanings, making it difficult to directly determine which sample data with different RRTSI values have a good or bad status based on their relative sizes (i.e., their positions in Fig. 4). However, two samples are likely to exhibit similar safety features if they have similar $z_1$ and $z_2$ values. If one sample has a safety issue, the other sample may also have a similar problem.

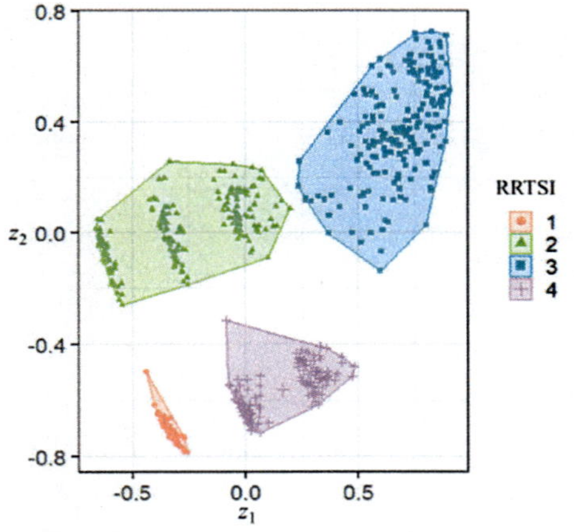


**Fig. 4.** Clustering results based on the SEM-RRTS model

The analysis results of the SEM-RRTS model were overlaid with the operational scenarios, as shown in Fig. 5. The evaluation objects with different RRTSI values are represented by different colors. Red, green, blue, and purple represent evaluation objects with RRTSI = 1, 2, 3, and 4, respectively. Different operational scenarios are represented by different numbers. The evaluation objects during off-peak hours, morning and evening peak hours, and periods of system failure are represented by "1," "2," and "3," respectively.

Table 2 shows the confusion matrix of the operational scenarios and RRTSI values of the evaluation objects.

Based on the analysis of Table 2 and Fig. 5, the evaluation objects during off-peak hours, morning and evening peak hours, and periods of system failures mainly correspond to RRTSI = 2, RRTSI = 1 and 4, and RRTSI = 3, respectively. This example clearly shows that the safety status of the evaluation objects based on RRTSI is consistent with the classification of the operational scenarios. The main difference is that RRTSI further subdivides the safety status of the evaluation objects during peak hours into two subcategories. Further analysis reveals that this result is mainly attributed to the slightly differing safety features in the morning and evening peak hours.

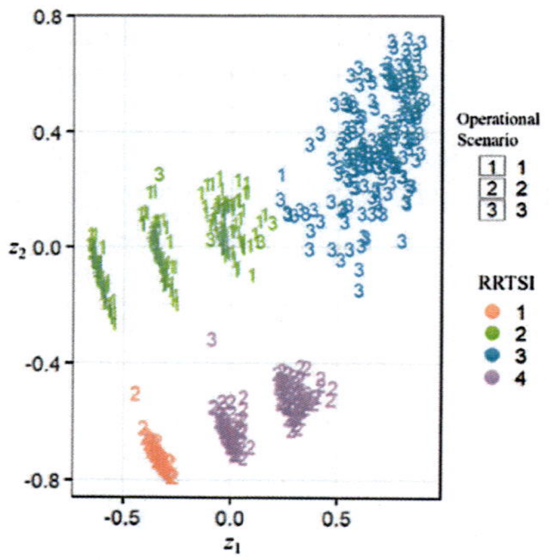


**Fig. 5.** Overlaying results of the SEM-RRTS model and operational scenario

**Table 2.** Confusion matrix of the operational scenarios and RRTSI values

| | | RRTSI | | | | Total |
|---|---|---|---|---|---|---|
| | | 1 | 2 | 3 | 4 | |
| Operational scenario | Off-peak hours | 0 | 201 | 1 | 0 | 202 |
| | Morning and evening peak hours | 62 | 0 | 0 | 141 | 203 |
| | Periods of system failure | 0 | 4 | 199 | 1 | 204 |
| Total | | 62 | 205 | 200 | 142 | 609 |

The above analysis indicates that the SEM-RRTS model can distinguish evaluation samples with significant status differences at different levels, effectively cluster and distinguish evaluation samples with similar status, and provide a clear and intuitive display of the safety status distribution.

## 5 Conclusion

This study proposed a regional rail transit safety evaluation index system from the perspective of evaluating the safety level of the RRTS. It covers four aspects enabling multiangle evaluation of safety status. Based on this, a data-driven model for evaluating the safety status of a RRTS was proposed. The model uses DANs and HHKMC algorithms to establish a safety feature recognition submodel that determines the set of all possible safety features of the evaluation objects. A TAN classification algorithm was used to establish a safety feature classification submodel, enabling the fast evaluation of the regional rail transit safety index. The effectiveness of the proposed model was validated using actual data. Compared with the actual situation, the proposed method has good applicability and can accurately evaluate the safety level of the current system.

In light of the research findings, we will attempt to use more comprehensive inspection data and methods to clarify the physical meaning of RRTSI and further improve the proposed model.

**Acknowledgments.** This work is supported by National Key R&D Program of China (2022YFB4300603).

## References

1. Chinese Urban Rail Transit Association: Overview of urban rail transit lines in mainland China in 2022. Chinese Urban Rail Transit Association, Beijing (2023). (in Chinese)
2. Yan, H., Gao, C., Elzarka, H., Mostafa, K., Tang, W.: Risk assessment for construction of urban rail transit projects. Saf. Sci. **118**, 583–594 (2019)
3. Berg, M.V.D.: Health based approach for regulations for vibrations from rail traffic. J. Acoust. Soc. Am. **135**, 2289 (2014)
4. Jiao, Z.D., Gan, C.H.: Risk assessment of crowd crushing and tramping accidents in urban rail transit station based on DEA. Appl. Mech. Mater. **730**, 327–330 (2014)
5. Xingchen, R.: Research on safety evaluation system of urban rail transit operation facilities. M.S. Thesis, Beijing Jiaotong University, Beijing, China (2012). (in Chinese)
6. An, M., Qin, Y., Jia, L.M., Chen, Y.: Aggregation of group fuzzy risk information in the railway risk decision making process. Saf. Sci. **82**, 18–28 (2016)
7. Van Weyenberge, B., Deckers, X., Caspeele, R., Merci, M.: Development of a risk assessment method for life safety in case of fire in rail tunnels. Fire Technol. **52**, 1465–1479 (2016)
8. Bunker, J.M.: Planning for transit system reliability using productive performance and risk assessment. In: Transportation Research Board 92nd Annual Meeting, Washington DC, United States (2013)
9. Peng, G., Zhao, T., Kunhong, Y., et al.: Safely risk assessment of urban rail transit operation equipment. J. Transp. Eng. Inf. **18**, 91–98 (2020). (in Chinese)
10. Qiang, Y.: Safety risk of metro operation based on ANP-SD-fuzzy evaluation research. M.S. Thesis, Lanzhou Jiaotong University, Lanzhou, China (2019). (in Chinese)
11. Zhang, J., Liu, B., Li, X., et al.: Comprehensive evaluation method for operation security situation of rail transit network. Sci. Technol. Eng. **21**(17), 7340–7347 (2021). (in Chinese)
12. Li, Q., Liu, L., Zhang, W., Liu, J., Bai, G.: Evaluating regional rail transit safety: a matter-element analysis method. In: 2020 IEEE International Conference on Intelligent Transportation, Rhodes, Greece (2020)
13. Pazzani, M.J.: Searching for dependencies in Bayesian classifiers. In: Fisher, D., Lenz, H. (eds.) Learning from Data, pp. 239–248. Springer-Verlag, New York (1996)
14. Lê, S., Josse, J., Husson, F.: FactoMineR: an R package for multivariate analysis. J. Stat. Softw. **25**, 1–18 (2008)

# Variable Channelization and Signal Coordination of Level Crossing

Yingqi Li[1], Lipeng Zhang[2], Bing Wang[1](✉), and Hailong Ling[1]

[1] School of Transportation Engineering, Xinjiang University, Urumqi 830047, China
451619037@qq.com
[2] China Communications Construction General Contracting and Operation Branch, Beijing 100120, China

**Abstract.** To solve the problem that the current intersection control form is fixed and can not adapt to the dynamic and random traffic flow. A new type of intersection dynamic control, variable channelization and traffic signal coordination control based on video traffic flow detection, is proposed. Real-time traffic flow is detected through video, and different channelization and information control schemes are adopted according to different traffic flows. Finally, taking the most common two-way four-lane lane as an example, the variable channelization and traffic signal coordination control scheme based on video traffic flow detection is designed and completed, including the channelization scheme design, the coordination and cooperation design of the main pre-signal light, the peak shift threshold design, the variable channelization and information control coordination design under different flow states and the transition design between the peak channelization scheme. And through the VISSIM simulation, from the parking times, queue length, delay three aspects of the operation benefit assessment under different flow state, verify the feasibility of the scheme, reduce the overall delay of the intersection, improve the efficiency of the intersection. It provides a new idea for the improvement of urban intersection.

**Keywords:** Urban traffic · Variable channelization · Traffic signal coordination control · Flow characteristics

## 1 Introduction

The core of traffic improvement design is scientific and technological research and engineering technology implementation. The improvement design of level crossing traffic usually adopts fixed channelization signs, markings, and corresponding timing signal control phase and timing design methods. The core data of the design is the peak traffic flow. Through continuous observation of the traffic efficiency of actual renovated level crossings in recent years, it has been found that the method of channelization and signal control coordination design based on the peak traffic flow has a significant effect on improving traffic congestion at level crossings in the morning and evening rush hours. However, the improvement in traffic flow efficiency at level crossings during peak hours


Y. Qin et al. (Eds.): EITRT 2023, LNEE 1137, pp. 682–690, 2024.
https://doi.org/10.1007/978-981-99-9311-6_69

is not significant, and there is a phenomenon of local channelization lane direction phase green light loss. The combination of traffic flow, variable channelization, and coordinated signal control can effectively alleviate this phenomenon. The so-called variable channelization and signal control coordination refers to the combination of dynamic channelization and adaptive signal control. By controlling intersection signs and markings, lane functions are changed according to traffic flow conditions to control traffic flow. At the same time, changes in signal control schemes are carried out according to the channelization method and the amount of traffic flow, achieving coordination and cooperation between channelization and signal control schemes.

The current research on variable channelization and signal control coordination mainly includes the following aspects: [1] considered the optimization control of intersection time and space allocation, established an optimization model for dynamic lane group and intersection signal control coordination, in order to reduce intersection delay and queue length. [2] proposes a simplified continuous flow intersection by assigning left turn traffic flow to shifted left turn lanes at upstream intersections. [3] designed a universal pre signal system for intersections, which can neither prohibit left turns nor change the intersection mode, improving the intersection capacity. [4] compared several channelization schemes and proposed the channelization design of turning left on the bypass. [5] studied the relevant parameters of turning left on borrowed lanes, including length of borrowed lanes, number of lanes, constraints, and signal optimization. [6] established a delay model in the case of turning left on a detour. [7, 8] respectively conducted research on the dynamic lanes of right turns and straight right turns on borrowed lanes to improve the traffic efficiency of intersections. [9] proposed that when there are short lanes at intersections, the lane length and signal phase scheme have an impact on traffic capacity.

In summary, current research mainly focuses on the exchange of straight ahead, left turn, and right turn under good road conditions, lacking consideration for the limited road conditions at intersections. In addition, there is a lack of research on solving the problem of congestion at intersections during peak hours, loss of green light time during peak hours, and low traffic efficiency. And in terms of current intersection signal control, it is mostly reflected in adaptive signal control systems. However, with the increase of traffic flow, sometimes the optimal timing method is no longer sufficient to improve the traffic efficiency of the intersection. Therefore, changes in lane functions should be considered. Based on the above issues, this article proposes a new dynamic control method for intersections. Taking the bidirectional four lane road as an example, real-time traffic flow was detected through video. Through the design of channelization schemes, the coordination and coordination of main and pre signal lights, the threshold design for video detection of changes in flat peak channelization schemes, and the coordination design of variable channelization and signal control under different flow states, as well as the transition design between flat high peak channelization schemes, a level crossing variable channelization and signal control coordination system based on video traffic flow detection was completed. And through VISSIM simulation, evaluate the operational efficiency from three aspects: parking frequency, queue length, and delay, and verify its feasibility.

# 2 Variable Channelization and Signal Control Coordination System

## 2.1 Design of Channelization Scheme

Taking the bidirectional four lane intersection as an example, Figs. 1 and 2 respectively show the flat peak channelization and phase schemes and peak peak peak channelization and phase schemes of the designed bidirectional four lane intersection. When the traffic flow is low, adopt the control form of straight left, straight right, two in and two out phases in Fig. 1. As the traffic flow increases, the conflict between direct and left turn traffic continues to increase, resulting in increased delays at intersections and a decrease in traffic efficiency. This is then transformed into a control form of three in, one out, and four phase in Fig. 2. The peak plan mainly controls left turning vehicles to enter the left turning lane through pre signals, so that the lane serves as both the entrance lane and the exit lane.

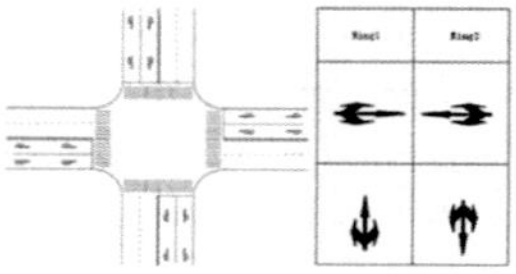

**Fig. 1.** Flat peak channelization scheme

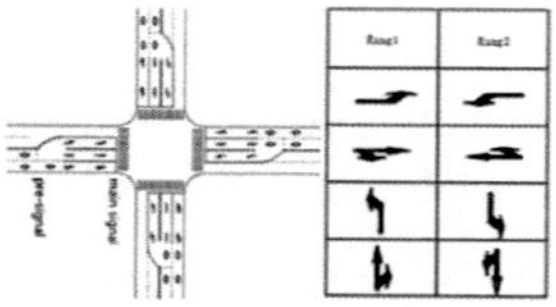

**Fig. 2.** Peak channelization scheme

When shifting from a flat peak plan to a peak peak plan, the setting of the peak channelization plan needs to consider the following three aspects. The length of the 'widened lane', The width of the 'widened lane', Number of 'widened lanes'.

The length of the "widened lane" in the peak channelization scheme cannot be too long or too short. If the lane is too long and exceeds the distance L2 from the stop line to the third group of guide arrows, some left turning vehicles will miss the opportunity to enter the widened lane. At the same time, if the lane is too long, the more time vehicles travel on their roads, the longer the pre signal lights will be closed in advance, resulting in waste of space resources at intersections, increased delays, and reduced Factor of safety. If the length is too short, it is less than the length L1 of the guide line at the intersection entrance, which can easily cause collisions between vehicles entering the left turn lane and the end of the queue of straight vehicles, leading to traffic accidents. Similarly, if the length of the lane is too short to meet the needs of left turning vehicles, it can cause left turning vehicles to queue up at the pre stop line within a certain release time, thereby affecting the overall traffic efficiency of the intersection. Therefore, the length L of the widened lane should be between the length of the intersection guide line and the length of the third group of guide arrows from the stop line to the intersection, as shown in Fig. 3.L1 < L < L2.

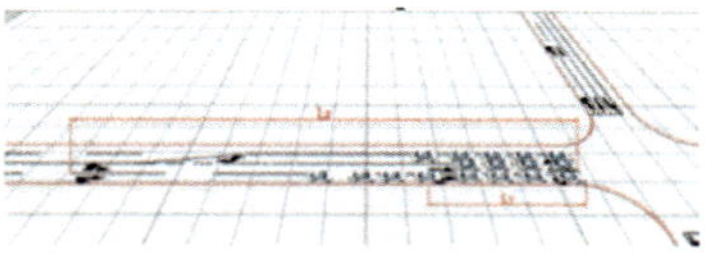

**Fig. 3.** Determination of the length of the left turn lane for peak channelization

Due to the general speed limit of 40 km/h for urban roads, the width of each lane at a two-way four way intersection is 3.5 m, and the speed at the entrance is generally 50–70% of the design speed of each road. Among them, the speed of turning vehicles is taken as 50%, V-turning = 20 km/h, straight vehicles are taken as 70%, and V-straight = 28 km/h. According to the [10], when the speed is 20 km/h, the recommended radius value for curbs is 15 m, so the turning radius for two-way four lane roads is 22 m. Assuming that the vehicle takes 3 s to traverse one lane while driving on the road.

$$L_t = v \times t = \frac{28 \times 3}{3.6} = 23.3\,\text{m} \tag{1}$$

$$L_d = 90 - 25 = 65\,\text{m} \tag{2}$$

In the equation, $L_t$ is the length of the widening gradient section, taken as 25 m, $L_d$ is the length of the widening section, and according to the setting conditions of the left lane in the national standard, it is taken as 65 m.

The width of the lane is mainly influenced by vehicle speed and traffic demand. According to relevant national standards, the width of each lane on the exit of an intersection should not be less than the width of the lane on the road section. In order to ensure the traffic efficiency of the intersection, the width of the widened lane is the same as the width of the exit lane at the intersection, taking 3.5 m.

Considering the matching of the number of lanes at the entrance and exit of the intersection, one widened lane is set up.

At this point, the channelization scheme and supporting phase scheme for Pingfeng are designed, and then the spatiotemporal resource allocation design of the intersection needs to be carried out, including the coordination and coordination between the main and pre signal lights, as well as the timing design under corresponding flow conditions.

## 2.2 Coordination and Cooperation Design of Main Pre signal Lights

Ensure overall signal coordination at the intersection, the main pre signal is set with the same cycle duration.

Prevent conflicts with vehicles entering the lane at the beginning of the next phase due to incomplete platoon of vehicles in the widened lane at the end of the left turn phase, a pre signal light off time should be established. The potential hazards of not turning off the pre signal lights in advance are shown in Fig. 4, where left turn vehicles collide with straight ahead vehicles. The time when the pre signal is closed in advance is mainly related to the length of the widened section and the speed of vehicles driving in the widened section.

$$t = l/v \tag{3}$$

l represents the length of the widened section, taken as 90 m here, v represents the average speed of the vehicle on the widened lane, taken as 40 km/h, t represents the early closing time of the pre signal, calculated as 9 s.

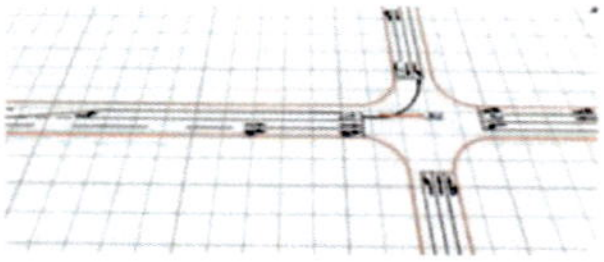

**Fig. 4.** Potential hazards of pre-signal lights not turning off early

Left turn vehicles in the east-west phase were originally able to enter the widened lane and wait when in the north-south straight right phase, but to prevent them from colliding with vehicles in the north-south just ended left turn phase when entering the widened lane, as shown in Fig. 5. Therefore, it is necessary to set a pre signal delay opening time. The delay opening time of the east-west pre signal refers to the time it takes for the last vehicle in the north-south left turn phase to pass through the intersection and leave the widened lane.

$$t_b = \frac{l_z}{v_z} + \frac{v_t - v_z}{a} + \frac{l - \frac{v_t^2 + v_z^2}{2a}}{v_t} \tag{4}$$

$l_z$ refers to the turning length of the vehicle, which can be obtained from data in Vissim simulation, $v_z$ refers to the vehicle's speed when making a turn, obtained by setting a speed detector in vissim, $v_t$ refers to the speed limit of the road, where 40 km/h is taken, a refers to the acceleration, taken as 3 m/s$^2$, $t_b$ refers to the delay opening time of the pre signal, which is calculated to be 15 s.

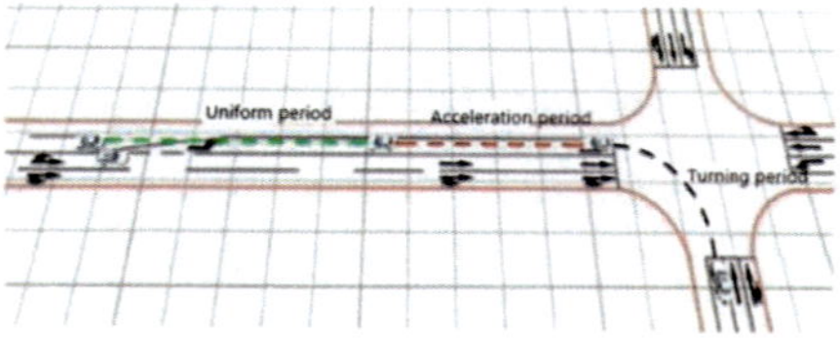


**Fig. 5.** The potential hazards of the pre-signal lamp not delaying the opening

## 3 Design of Peak to Peak Conversion Threshold and VISSIM Evaluation

Under the conditions of 1/4 flow and 1/5 flow in lane groups, 825 pcu/h and 660 pcu/h respectively, the optimal time allocation for two channelization schemes was determined using the Webster time allocation method with a left to right traffic flow ratio of 1:2:1. The time allocation scheme was then kept unchanged, and the traffic flow was increased or decreased by 5% step size. VISSIM was used to evaluate the operational benefits of the two channelization schemes in different traffic flow states from three aspects: parking frequency, delay, and queue length, The results are shown in Figs. 6 and 7.

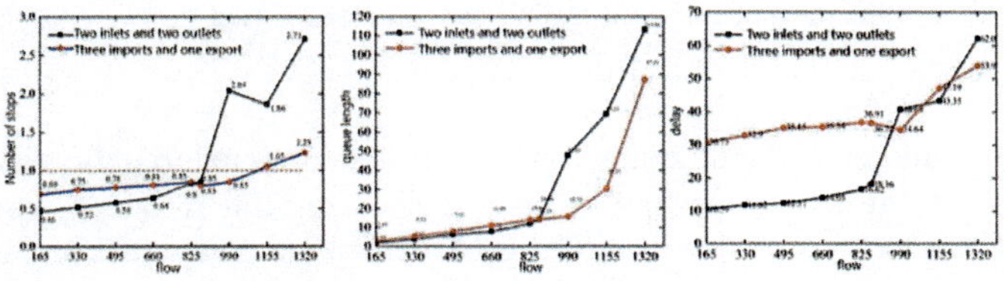


**Fig. 6.** Operation efficiency of different traffic flows in 1/4 flow state

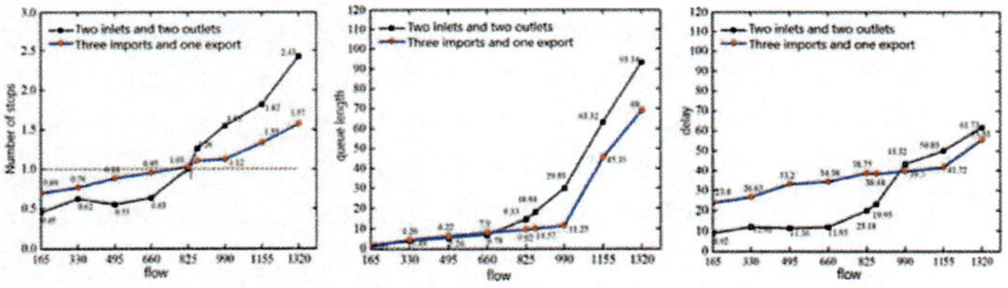


**Fig. 7.** Operation efficiency of different traffic flows in 1/5 flow state

It can be seen that both the 1/4 flow and 1/5 flow have inflection points that happen to occur between 825 and 990 pcu/h in the operating efficiency chart. Then, by expanding this interval and conducting simulation verification again, it was found that the inflection point happens to be around 866 pcu/h. Therefore, the flow data of 866 pcu/h was selected as the flow threshold for variable channelization. When the flow rate is greater than 866 pcu/h, the peak three in and one out channelization scheme is implemented. When the flow rate is less than 866 pcu/h, Implement a channelization scheme of two inputs and two outputs for flat peaks.

## 4 VISSIM Simulation Evaluation

After determining the channelization scheme, the threshold conditions for conversion between channelization schemes, and the coordination and cooperation between the main and pre signal lights, the timing schemes under different flow states are determined. Keep the channelization schemes corresponding to different flows unchanged, calculate their optimal timing separately, and then evaluate and analyze them from three aspects: parking frequency, queue length, and delay [11, 12].

When the flow rate is less than 866 pcu/h, according to the previous text, the channelization scheme of flat peak two in and two out is adopted. Using the Webster timing method, calculate the optimal timing for each traffic state, and use VISSIM simulation to evaluate the operational efficiency, as shown in Fig. 8.

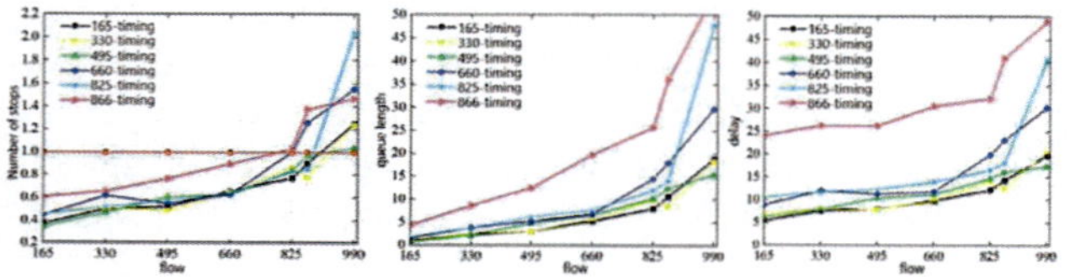

**Fig. 8.** Operation efficiency of different timing schemes under two-in, two-out channelization schemes

It can be found that when the flow rate is less than or equal to 330pcu/h, the corresponding 165 timing scheme performs best in terms of parking frequency, queue length, and delay. The 330-timing scheme performs best when the flow rate is greater than 330 pcu/h and less than or equal to 866 pcu/h. Calculated using the Webster timing method, 165-timing: C = 25 s, g1 = 8 s, g2 = 7 s; 330-timing: C = 30 s, g1 = 10 s, g2 = 10 s.

Similarly, according to the above method for analysis and comparison, when the flow rate is 990 pcu/h, the cycle has already reached 175 s. With the increase of traffic again, the Webster method is no longer used, and at this time, as the cycle increases, the delay increases rapidly, resulting in a decrease in the efficiency of intersection traffic. Therefore, when the flow rate is greater than 866 pcu/h, the optimal timing under the flow states of 825, 866, and 990 was compared, and the results are shown in Fig. 9.

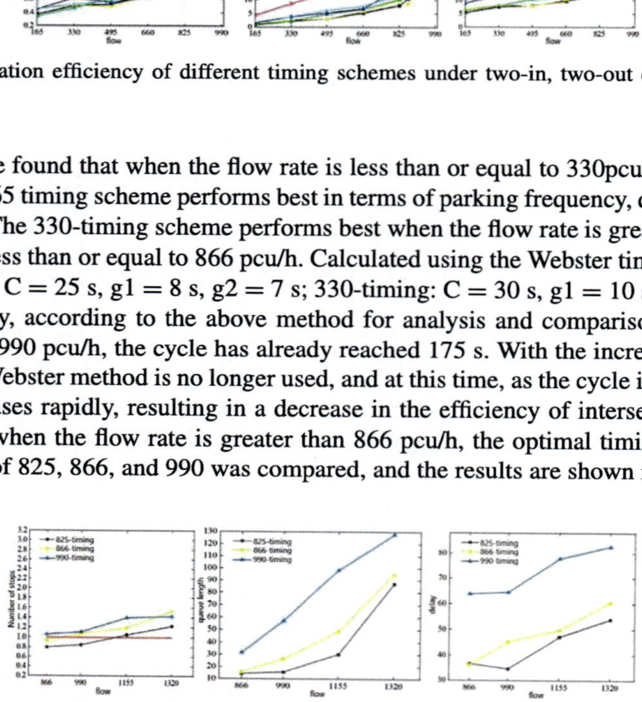

**Fig. 9.** Operation efficiency of different timing schemes under the three-in-one-out channelization scheme

Therefore, it can be seen that when the flow rate is greater than 866 pcu/h, the intersection implements a peak three in and one out channelization scheme in conjunction with the 825-timing scheme C = 105 s, g1 = 17 s, g2 = 25 s, g3 = 18 s, and g4 = 25 s.

In summary, when the flow rate is less than 330 pcu/h, a two in two out channelization scheme is adopted in conjunction with a timing scheme of C = 25 s, g1 = 8 s, g2 = 7 s; When the flow rate is between 330 to 866 pcu/h, a two in two out channelization scheme is used in conjunction with a timing scheme of C = 30 s, g1 = 10 s, and g2 = 10 s; When the flow rate is greater than 866 pcu/h, a three in one out channelization scheme is adopted in conjunction with a timing scheme of C = 105 s, g1 = 17 s, g2 = 25 s, g3 = 18 s, and g4 = 25 s.

Considering the randomness and volatility of traffic flow at intersections, only when there are relatively stable changes in traffic demand at intersections can changing channelization and supporting signal control schemes have better effects. Similarly, due to variable channelization, it changes the road function and redistributes the spatiotemporal resources of intersections, which will have a direct impact on vehicle drivers. The frequency of its changes should be limited to a certain extent. According to [13], the minimum interval between variable channelization and signal control changes at intersections is determined to be 10 min, and a minimum of 3 consecutive cycles are required to reach the threshold condition. When abnormal situations occur at intersections, including traffic accidents, emergency rescue, and other events, detected through video, the cut-off plan can be changed. After the intersection condition is restored, the change can be restarted. As shown in Fig. 10, it is the complete Process flow diagram of variable channelization and signal coordination control at the level crossing.

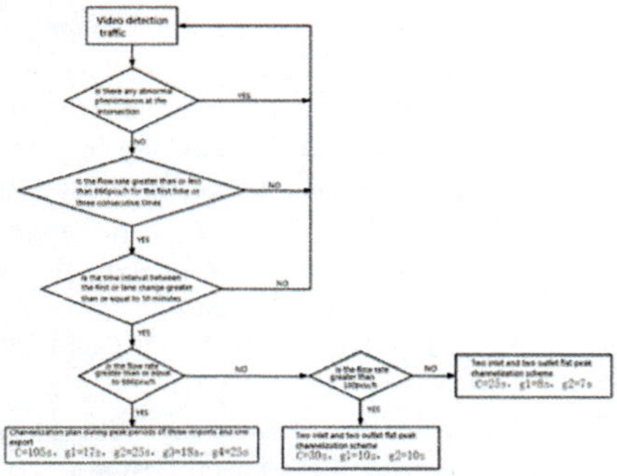


**Fig. 10.** Flow chart of variable channelization and signal coordination control at the level intersection

## 5 Conclusion

This paper takes the more common four-way intersection as an example, obtains the traffic information of the intersection in real time through video, and then analyzes the applicable channelization schemes under different flow states, designs and completes the level intersection variable channelization and signal control coordination control system based on video traffic flow detection, and realizes that the appropriate control scheme can be selected according to the flow Feature selection. And through VISSIM simulation software to build a three-dimensional simulation scene, from three aspects of parking times, queue length, and delay as evaluation indicators, this paper compares and analyzes the operation benefits of different traffic flows in different channelization schemes, explains that different traffic flow states are suitable for different channelization and information control coordination modes, and proves the feasibility of this scheme. We have determined suitable channelization and signal control schemes for different flow states, reducing overall delay at intersections, improving traffic efficiency at intersections, and providing a new approach for improving level crossings.

**Acknowledgments.** This work is supported by the scientific and technological project of 209 provinces, cities and autonomous regions - Research on the Strategic Framework of Transport Development in Ürümqi (2020690002).

## References

1. Alhajyaseen, W.K.M., Najjar, M., Ratrout, N.T., et al.: The effectiveness of applying dynamic lane assignment at all approaches of signalized intersection. Case Stud. Transp. Policy **5**(2), 224–232 (2017). Author, F., Author, S.: Title of a proceedings paper. In: Editor, F., Editor, S. (eds.) Conference 2016, LNCS, vol. 9999, pp. 1–13. Springer, Heidelberg (2016)
2. Sun, W., Wu, X., Wang, Y., et al.: A continuous-flow-intersection-lite design and traffic control for oversaturated bottleneck intersections. Transp. Res. Part C: Emerg. Technol. **56**, 18–33 (2015). Author, F.: Contribution title. In: 9th International Proceedings on Proceedings, pp. 1–2. Publisher, Location (2010)
3. Xuan, Y., Daganzo, C.F., Cassidy, M.J.: Increasing the capacity of signalized intersections with separate left turn phases. Transp. Res. Part B: Methodol. **45**(5), 769–781 (2011)
4. Liang, P.: Design and simulation study on signal timing after setting reverse variable lane at intersection. Beijing Jiaotong University (2017). (in Chinese)
5. Dan, Z., Xuejun, N., Shuhao, Z., Jiaxu, W.: Research and effect evaluation of left turn at plane intersection. J. Syst. Simul. **33**(12), 2975–2982 (2021). (in Chinese)
6. Xiaofang, Y., Ying, W.: Research on optimization of dynamic exit left-turn lane control. Traffic Inf. Safety **39**(05), 85–92 (2021). (in Chinese)
7. Aizeng, L., Wei, W., Wenquan, L.: Highw. Transp. Sci. Technol. **27**(03), 125–129 (2010). (in Chinese)
8. Shidong, L., Lijuan, C., Ying, W., et al.: Optimization design and evaluation analysis of right-straight dynamic lane at signalized intersection. J. Highw. Transp. Sci. Technol. **39**(03), 183–190 (2022). (in Chinese)
9. Xiaoguang, Y., Jing, Z., Ying, Z., Xiaofei, Y.: Study on the influence of short lane on the capacity of signalized intersection. J. Highw. Transp. Sci. Technol. (12), 151–156 (2008).(in Chinese)
10. Industry Standard of the People's Republic of China. Design regulations for urban road intersections: CJJ 152-2010. China Architecture and Building Press, Beijing (2010). (in Chinese)
11. Guo, R., Liu, J., Zhao, Q., et al.: Signal timing and geometric design at contraflow left-turn lane intersections. Int. J. Transp. Sci. Technol. (2021)
12. Calle-Laguna, A.J., Du, J., Rakha, H.A.: Computing optimum traffic signal cycle length considering vehicle delay and fuel consumption. Transp. Res. Interdiscip. Perspect. **3**, 100021 (2019)
13. Jing, Z., Xizhao, Z.: Optimal lane function and signal change method of intersection variable lane. J. Univ. Shanghai Sci. Technol. **38**(04), 380–386 (2016). (in Chinese)

# Research on Safety Assessment Methods for Urban Rail Transit Operation

Zhengjiang Li[1,2](✉), Qianqi Fan[3], and Bin Han[3]

[1] Shanghai Key Laboratory of Rail Infrastructure Durability and System Safety, Tongji University, Shanghai 201804, China
2133349@tongji.edu.cn
[2] College of Transportation Engineering, Tongji University, Shanghai 201804, China
[3] Institute of Rail Transit, Tongji University, Shanghai 201804, China
{2111529,hanbin}@tongji.edu.cn

**Abstract.** With the continuous expansion of urban rail transit operation scale, it presents the characteristics of long operation mileage, complex passenger flow characteristics, and heavy equipment service load, etc., which requires the study of risk generation and evolutionary propagation mechanism in the operation process to guide the active safety control. In this study, Event Evolutionary Graph based Bow-Tie Analysis is proposed. With the help of natural language processing technology to realize the event relationship analysis and automatic extraction, for the text characteristics of urban rail transit field, the sequential relationship to indicate the evolution law of the event, and the bow-tie combination of the cause, effect and control measures of the risk of the path of safety assessment. On the one hand, the assessment is carried out from the evolution path of control measures, and on the other hand, the risk is comprehensively assessed in terms of the possibility of risk occurrence, the possibility of oriented consequences, and the severity of consequences, which provides a reference for the quantification of the safety level. Finally, the application of the method, which is highly operational and can effectively guide the safety management work, is demonstrated by means of a typical scenario of urban rail transit operation in a given location.

**Keywords:** Urban rail transit · Safety assessment · Bow-tie · Event evolutionary graph

## 1 Introduction

With the characteristics of high efficiency, environmental protection and comfort, urban rail transit carries the transportation of a large number of people every day, and shares urban travel services with other transportation modes, which is an important part of building a strong transportation nation and a smart city. However, with the increase of operation kilometers, the number of equipment, and the number of passengers served, the operation safety problem has become more and more prominent. Operational safety accidents not only affect the normal operation of urban transportation, but also threaten the lives and properties of the general public.


Y. Qin et al. (Eds.): EITRT 2023, LNEE 1137, pp. 691–701, 2024.
https://doi.org/10.1007/978-981-99-9311-6_70

It is necessary and important to conduct safety assessment of urban rail transit operations in a timely manner to guide safety management practices and improve social well-being indexes: (1) Preventing disasters, identifying potential risks and hazards in urban rail transit, and taking preventive measures to minimize the occurrence of accidents and protect public safety; (2) Ensure the reliability and stability of the operation service, identify and evaluate the faults and problems in the system, prevent the service interruption caused by potential faults or personnel errors, and improve the operation efficiency; (3) Enhance the emergency response capability, establish and improve the emergency response mechanism according to the ways and means of risk escalation, quickly respond to emergencies, reduce the possibility of risk escalation, and then protect the safety of life and property; (4) Improve the public's trust in urban rail transit, high safety urban rail transit can attract the public to change the travel mode, reduce traffic congestion and environmental pollution.

## 2 Related Work

### 2.1 Safety Assessment

Safety assurance necessitates identifying and managing a broad range of hazards that have the potential to cause harm to individuals, property, and the environment. Safety assessment involves analyzing and evaluating the potential risks and hazards associated with a given system, product, or activity. Safety assessment aims to identify potential hazards, assess the level of risk they pose, and take necessary measures to minimize or eliminate those risks. According to the Oxford Dictionary 3rd edition, risk is defined as: Exposure to chance or circumstance that includes the possibility of loss, injury, or other unfavorable or unwelcome situations. In the safety field, risk refers to the probability and potential harm of a dangerous incident occurring [1]. A risk matrix, which takes both factors into consideration, is widely used for risk assessment [2].

Research on urban rail transit safety approaches accident analysis and risk assessment through two perspectives mainly. The first perspective involves identifying risk sources and proposing safety management strategies based on past accident records. The second perspective involves deducing the law of the evolution of accident risk to guide safety management [3].

In the urban rail transit industry, independent safety assessment (ISA) is a well-established tool that ensures the rail transit system meets the expected safety performance requirements and protects the lives and properties of passengers and staff throughout the design, construction and operation phases [4]. The main task of ISA is to assess the safety management system, design documents, operation procedures, training programs, and other aspects of the rail transit system to determine whether they meet the safety performance requirements. Nevertheless, regional standards for independent safety assessment and audit differ across regions. Unfortunately, inconsistent standards across regions pose a problem for ISA since it is in its nascent stage. Post-operational tracking and monitoring require further research.

Moreover, there is a greater emphasis on researching the construction phase rather than safety management during operation [5]. Various approaches such as the brainstorming method, Delphi method, checklist method, and event tree analysis are currently used for safety assessment during operation. As data accumulates and intelligent monitoring equipment is employed, more detailed and comprehensive assessment methods can be researched.

## 2.2 Event Evolutionary Graph

Knowledge Graph (KG) was proposed by Google in 2012 and is widely used in various fields such as semantic search, recommender systems, and automated Q&A. Generally speaking, KG manages knowledge and information by means of semantic web, which describes entities and attributes by means of relationships in words. Among them, entities are regarded as nodes, relationships between entities are represented as edges, and the basic element unit can be represented as a triple "entity-relationship-entity" or "entity-attribute-attribute value" [6]. A common representation of risk is also a triple (also known as a vector) [7].

$$R = \langle s_i, p_i, x_i \rangle, i = 1, 2, \ldots, N$$

where: $s_i$ is the scenario describing the possible events, $p_i$ is the probability of the scenario, $x_i$ is the outcome of the scenario, and $N$ is the number of scenarios chosen to describe the risk.

This structured representation makes the relationship between knowledge more visible, facilitates knowledge storage, retrieval and reasoning, and both are similar in structure and more portable. KG has been used in railroad systems to describe accidents and hazards through heterogeneous networks and to reveal the patterns of accident occurrence [8].

When the KG represents the semantics of events with dynamic characteristics that are prevalent in the objective world, the static knowledge units and relationships used to represent the semantics are difficult to represent the dynamic logic of the development of things, and the fine-grained entities are difficult to express the rich semantic information of events. In 2017 [9], Liu Ting et al. proposed the Event Evolutionary Graph (EEG), which focuses on expressing the evolutionary laws and patterns of causality, logic, and reasoning relationships between events, to help people better understand and cope with complex events, and is a matter-of-fact logic knowledge base. Among them, predicate phrases are regarded as nodes, and logical relations between events are represented as edges. EEG can effectively capture the diversity and complexity of big data, realize the modeling of event relationships and multi-dimensional correlation, and has been widely used in the study of the evolution process and propagation path of infectious diseases in the medical field, the study of the evolutionary mechanism of market fluctuations and financial crises in the financial field, and so on.

## 2.3 Bow-Tie Analysis

Bow-tie is a method for analyzing and visualizing risk and safety issues in complex systems, presented in the shape of a bow-tie with a well-defined structure [10]. It was first introduced in the field of safety management and has been widely used in risk assessment, safety planning and disaster management. Bow-tie's core idea is to reveal the key elements and vulnerabilities of a system's safety in a comprehensive and intuitive way through the representation of events, causality and controls in the system as a single diagram. The European Organization for the Safety of Air Navigation (EUROCONTROL) defines a hazard as any state, event or circumstance that triggers an accident [11]. Bow-tie is a combination of FTA, which starts with an accident or an undesired top event and identifies the immediate cause of the triggering event, and ETA, which starts with the top event of the FTA and evolves through a series of possible pathways to the possible consequences. By identifying the top event and pathway, critical control points and risk factors of the system can be identified to help predict and prevent potential risks and crises.

The bow-tie abstracts and generalizes complex systems and may not cover all the details and complex relationships; the interpretation and application of the bow-tie model requires a certain level of expertise and domain experience, the involvement of professionals, the construction of which is supported by a large amount of data and information, and the manual identification of which requires a great deal of effort. Consideration can be given to utilizing automated data-driven techniques to reduce the reliance on manual operations and improve the efficiency and accuracy of model construction.

# 3 Methodology

## 3.1 Event Evolutionary Graph Based Bow-Tie Analysis

Event Evolutionary Graph Based Bow-Tie Analysis (EEGBBTA) makes up for the disadvantages of large workload, incomplete analysis and high professional knowledge requirement in bow-tie analysis with the advantage of automatic extraction of event relationships from big data by means of EEG. The integration of the two "node-edge-node" descriptions helps to identify the key events, causes and effects, and gives more scientific and comprehensive safety assessment results.

In accordance with the common paths of accident analysis and risk evaluation, this study mainly mines the evolution pattern from the historical record data. In order to realize the comprehensiveness of the assessment, the data should be as rich as possible when building the bow-tie. The raw data types can include domestic and international research literature, urban rail transit operation accident reports, learning cases of near-miss events in the training of operating companies, operating company operation quality event analysis reports (non-accidental events such as delays), service passenger flow data, equipment failure and maintenance records of various sub-systems of urban rail transit (electromechanical, power supply, rolling stock, communications, signals, etc.), and safety management documents, Emergency management documents, etc., are comprehensively analyzed before, during and after the incident.

EEG's main purpose is to automatically extract event relationships in text data. However, the text corpus in the urban rail transit field lacks obvious logical relationship words making it difficult to identify event relationships, particularly the cause-and-effect relationships. One solution for a corpus without a clear logical relationship is to determine the cause and effect based on the order of events described. That is, the events appearing at the front of the text are considered the causes of the events that appear later, which is consistent with data recording laws.

To achieve a bow-tie analysis, it is necessary to cluster extracted matter-of-fact relationships textually and align entities. Text clustering represents causes, events, and results uniformly. Entity alignment identifies and links instances of the same entity in different data sources, thus discovering events pointing to the same occurrence in multiple sources. A bow-tie analysis graph is drawn using the top event as the central node. This leads to categorization of causes, control measures, remediation measures, consequences, etc. for further safety assessment work.

**Event Evolutionary Graph**

EEG is a tool for analyzing and visualizing the event evolution process. The key operational processes are shown in Fig. 1.

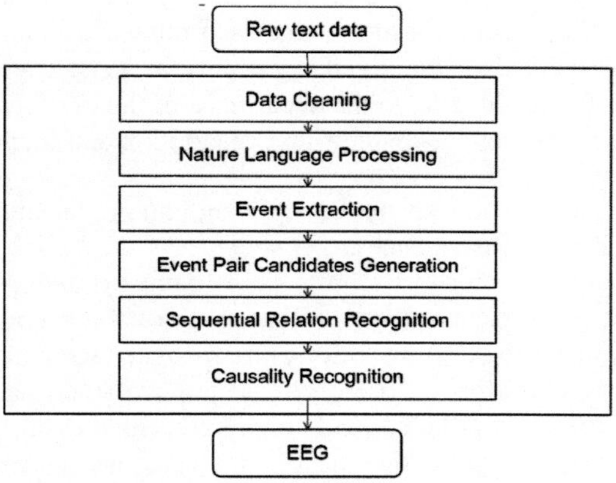


**Fig. 1.** Event evolutionary graph key operational processes.

Data cleaning: removal of blank data, duplicate data, too short descriptions (less than 10 characters for a single record description), mathematical formulas, etc.;

Natural language processing: line by line (one record per line) for word segmentation, entity recognition, semantic dependency analysis;

Event extraction: extract verb phrases as events from semantic dependency analysis;

Event pair candidates generation: take "," as unit, two neighboring events as a candidate event pair;

Sequential relation recognition: extracting event relations in the order in which they appear in the text;

Causality recognition: use logical relationship words to recognize those with obvious logical relationship, and use sequential relationship as causal relationship if there is no clear indication of causal logical words.

**Bow-Tie Analysis**

Bow-tie Analysis is a commonly used risk management tool that can be used to analyze and visualize the causality of complex events, risk propagation pathways, and control measures. The key construction process is shown in Fig. 2.

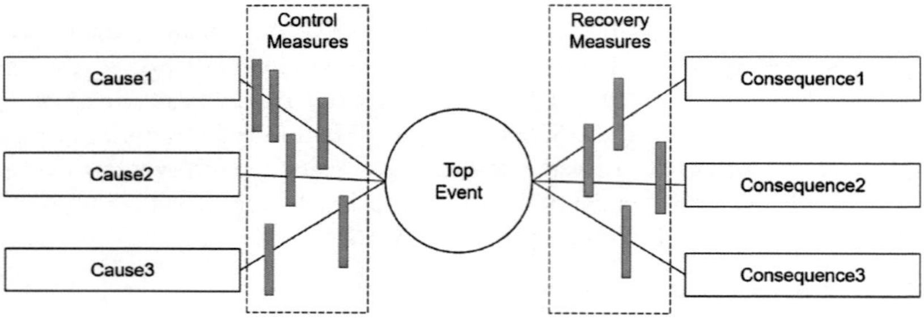


**Fig. 2.** Bow-tie analysis key construction process.

Event identification: identify and determine the key events for analysis, which can be failures of various urban rail transit systems, near misses, etc.

Root cause analysis: For the identified key events, root cause analysis is conducted to find out the main factors leading to the occurrence of the event, which are generally intermediate events in the logical chain of text description and identified as top events in the bow-tie diagram.

Expansion of the top event: Expand the top event into influencing factors, including direct causes and indirect consequences.

Risk Propagation Paths: create risk propagation paths by analyzing the causal relationships between causes and consequences. These pathways can represent the process of event development, including leading factors, intermediate factors, and final outcomes.

Control/Recovery Measures: At each critical point of the risk propagation path, appropriate control measures are identified and implemented to control the risk. These can be preventive measures, protective measures, emergency response and restoration measures.

Monitoring and Evaluation: The implemented control measures are monitored and evaluated to ensure their effectiveness and to make timely adjustments and improvements.

## 3.2 Safety Assessment

Based on the obtained bow-tie diagram, the safety evaluation can be carried out in two ways. On the one hand, the examination of control measures is carried out based on the conduction evolution path of risk events. On the other hand, the assessment of risk causes, consequences and control measures is carried out, drawing on the risk matrix. In particular, a risk event may lead to different consequences, and each chain of direction generates a score, with the maximum value being used as the evaluation criterion for that risk.

$$Risk = Likelihood\ of\ Risk \times Likelihood\ of\ Leading\ to\ Consequence \\ \times Severity\ of\ Consequence$$

**Likelihood of Risk**

Dynamic risk management is the core of life-cycle risk management, which emphasizes that safety risk factors change over time and are subject to uncertainty [12]. Some of the safety risk factors will continue throughout the operation period, and some of the safety risk factors will appear as temporary activities or maintenance and construction work is carried out. Considering the dynamic nature of risk, the entropy weight method is used to assign weights to various causes when evaluating the likelihood of risk occurrence, and then the likelihood of the risk event (Top Event) occurs through the standardized probability calculation of each probability. The probability calculation depends on the actual historical data of the evaluated object. Safety risk factors can be broadly categorized into human factors, mechanical factors, environmental factors and management factors. Among them, human factors use the annual assessment data of the operator; mechanical factors use the failure and maintenance records of various systems and equipment, and more in-depth data analysis can be considered to realize failure prediction after the intelligent monitoring equipment is on line; natural environment factors use the local hydrological and meteorological data, and the working environment adopts the historical records of the assessed party; and the management factors use the 0–1 evaluation of the key aspects of the management and control documents, and based on this, the evaluation is based on personnel interviews, practical records, etc. to assess the implementation. The maximum scale for this item is 5.

**Likelihood of leading to consequences**

The likelihood of leading to consequences is an important part of the risk grading evaluation. According to the bow-tie, a risk event (Top Event) may lead to N consequences. In particular, consequences of the same severity are calculated by taking the path with the highest probability. The expert scoring method is adopted in the actual implementation, and the reference data provided to the experts contain the frequency of industry-related events, the relevant management documents of the remediation measures link of the assessed unit, etc. The maximum scale of this item is 1.

**Severity of consequences**

Generally, the grade of production safety accidents is graded according to the casualties or direct economic losses caused by the accident. According to Heinrich's law, the ratio of the number of accident pieces with serious injury, minor injury and no injury is 1:29:300. Although attempted accidents do not cause personal injury or economic loss, because the cause and development of the accident is consistent with the occurrence of a serious accident or a major accident, if there is no unforeseen event to interrupt the development of the attempted accident, it is highly likely to cause serious injury or a major accident [13]. In order to guide the safety management, this study combines the actual operation of urban rail transit and related policies, and divides the severity of the consequences into 4 levels, with the maximum scale of this item being 4. The grading is shown in Table 1.

**Risk Score Description**

Risks are graded from the risk score, as shown in Table 2.

**Table 1.** Grading of severity of consequences.

| Severity of consequences | Scale | Evaluation criteria (if any) |
|---|---|---|
| Minor | 1 | (1) Sphere of influence: below station level;<br>(2) Other circumstances |
| Marginal | 2 | (1) Scope of impact: station level;<br>(2) Fire in station common areas, major equipment rooms, etc.;<br>(3) Minor equipment failure;<br>(4) Late for 5 min (or more);<br>(5) No casualties, only minor discomfort or minor injuries |
| Critical | 3 | (1) Impact area: line level;<br>(2) Fires in zones, etc.;<br>(3) Failure of key system (vehicle system, signaling system) facilities and equipment requiring repair or replacement;<br>(4) Being late for 30 min (or more);<br>(5) Causing minor injuries to personnel 1–2 |
| Catastrophic | 4 | (1) Scope of impact: line network level;<br>(2) Train derailment, train clashes, train ramming, train crowding switch;<br>(3) Fires in trains, zones, control centers, main substations, vehicle bases, etc.;<br>(4) Passenger stampedes;<br>(5) Flooding and backing up of stations and track areas;<br>(6) Serious deformation and collapse of bridge and tunnel structures, and collapse of road base;<br>(7) Total power outage within a single or more stations, substations, control centers or vehicle bases;<br>(8) Paralysis of communication network;<br>(9) Signal system major failure (unable to monitor train operation or chain failure error lasting for 60 min (inclusive) or more);<br>(10) Contact network breakage or collapse;<br>(11) Major failure of elevator or escalator;<br>(12) Clamping of people and objects on moving trains resulting in passenger casualties;<br>(13) Cyber security incident causing line network level impact;<br>(14) Other operational events resulting in death, serious injury, minor injury to 3 or more persons (inclusive), and one-way traffic disruption at 2 or more stations or intervals on the main line |

**Table 2.** Risk score description.

| Risk score descriptor | Risk score |
|---|---|
| Intolerable | >12 |
| Undesirable | 8–12 |
| Tolerable | 4–8 |
| Negligible | <4 |

## 4 Experiment

### 4.1 Data

The raw data of this experiment contains: (a) a total of 102 publicly reported accident record data (e.g., train derailment, train collision, fire, flood, terrorist attack, and passenger casualties) during urban rail transit operations worldwide; (b) a total of 387 operational quality events originating from scientific research projects and cooperating enterprises; (c) 1,863 records of equipment failure and maintenance of the key systems (rolling stock system, signaling system) of the subject of the evaluation; and (d) 14 documents of the subject of the evaluation, such as maintenance procedures, operation manuals, emergency response plans, and emergency response drills.

### 4.2 EEGBBTA

The raw data belongs to multi-source heterogeneous data, and in order to simplify the natural language processing, the raw data is first cleaned, denoised and reintegrated. The methods used include removing irrelevant data, duplicate data, transcribing text descriptions from images and tables, extracting and transcribing document information, etc., and integrating description information to form a list of event descriptions.

Natural language processing technology is utilized for entity recognition, relationship extraction, etc. for each single item in the event list, and sequential relationships are used as causal relationships to obtain the event-relationship triad. Identify the top event with the key event as the core, take the top event as the intermediate event, the former event as the cause event, and the latter event as the result event, perform text clustering and entity alignment, and unify the representation. Following this, a bow-tie diagram of all the risk events related to operational safety can be constructed. One of the typical scenarios in urban rail transit operation, loss of indication of a switch, is taken as an example for analysis.

### 4.3 Risk Assessment

Figure 3 illustrates the transmission mechanism from the risk source of the switch loss of meter scenario to the incident through path analysis of causes, potential outcomes, and controls. Risk assessment is possible when the paths are independent and the likelihood of the outcome is known.

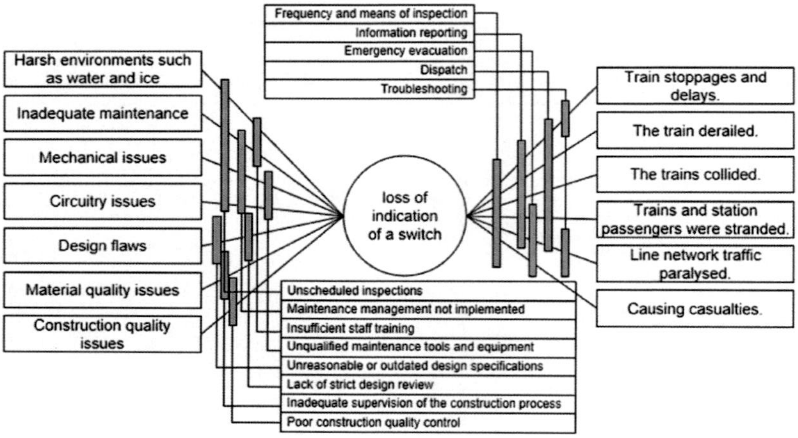


**Fig. 3.** Loss of indication analysis of a switch.

In this case study, the design, materials and construction of the switch were considered to be in accordance with the specifications and the relevant control documents were available. The likelihood of risk is affected by several factors such as the harsh environments, inadequate maintenance, mechanical issues, and circuity issues. Based on records from the past three years, the likelihood of the switch being out of gauge is estimated to be 0.5. The severity of the consequences can be rated between 1 to 4, with a likelihood of 1 being 1, 2 being 0.8, 3 being 0.2, and 4 being 0.1. The calculated risk scores are 0.5, 0.8, 0.3, and 0.2, with the highest score being 0.8 and classified as Negligible. Due to this classification, control measures have been implemented to prevent switch loss on the table, resulting in a high level of safety.

## 5 Conclusion

The EEGBBTA method starts with the logical relationship of event evolution, and employs EEG to automatically analyze textual data. This solves the limitation of manually constructing the bow-tie, resulting in comprehensive safety assessment work. Moreover, the evaluation strategy of the risk scores is more in line with production reality. The example study has achieved better application, and there are extensive application and research prospects.

Urban rail transit is a large, complex system with intricate safety issues. Ensuring operational safety requires joint efforts from scientific research institutions, regulatory authorities, and operating entities. This paper proposes a new method for safety assessment and a strategy for risk score evaluation. However, further research is necessary to determine the efficiency and accuracy of the event relationship extraction and the coupling effect between multiple causes.

## References

1. BS EN 50126-1:1999: Railway applications. The specification and demonstration of Reliability, Availability, Maintainability, and Safety (RAMS). Part 1: basic requirements and generic process. Incorporating corrigendum no. 1. Pg-22
2. Zhang, Y., Fom, D.D.: Hazard identification and risk assessment of Abuja rail mass transit. Transp. Safety Environ. **4**(2) (2022): tdac014
3. Xu, H., Deng, N., Li, Y.: Key risks of rail transit operation based on complex network. In: ICCREM 2021, pp. 155–162 (2021)
4. Wigger, P.: Independent safety assessment-benefits and methodology. Procedia Soc. Behav. Sci. **48**, 3361–3370 (2012)
5. Chai, N., Zhou, W., Xinlei, H.: Safety evaluation of urban rail transit operation considering uncertainty and risk preference: a case study in China. Transp. Policy **125**, 267–288 (2022)
6. Wang, J., et al.: Framework and key technologies of intelligent operation and maintenance of traction transformer based on knowledge graph. In: International Conference on Electrical and Information Technologies for Rail Transportation. Springer Singapore, Singapore (2021)
7. Kaplan, S., John Garrick, B.: On the quantitative definition of risk. Risk Anal. **1**(1), 11–27 (1981)
8. Liu, J., et al.: A knowledge graph-based approach for exploring railway operational accidents. Reliab. Eng. Syst. Safety **207**, 107352 (2021)
9. Ting, L.: From Knowledge Graph to Event Evolutionary Graph. (2017–11–05) [2022–1–25]. http://ir.hit.edu.cn/~zyli/papers/kg2eeg_en.pdf
10. Woods, D.D., Hollnagel, E.: Joint Cognitive Systems: Patterns in Cognitive Systems Engineering. CRC Press (2006)
11. EUROCONTROL. Risk Assessment and Mitigation in ATM. Esarr4 (2001)
12. Ying, X.: Study on Activities Oriented Dynamic Risk Assessment of Metro Shield Construction Safety. China University of Mining and Technology, A thesis (2022)
13. Pengfei, Y., Dou, K.: Discussion about the role of Murphy law and Heinrich theory to control accidents. Safety Health Environ. **08**, 14–15 (2008)

# Study on Optimization of the Adjustment Mechanism for the Allocation of Funds for Transferring Consumption Tax on Refined Oil Products

Ma Yanjun, Cui Min(✉), Luo Shiyi, Huang Liya, Wang Jing, and He Jiayuan

Division of Integrated Transport Research, Transport Planning and Research Institute, Ministry of Transport, Beijing 100028, China
cuimin@tpri.org.cn

**Abstract.** The reform of prices and taxes of refined oil products has great and far-reaching significance in financing the construction and maintenance of transportation infrastructure in accordance with the law and promoting the stable and healthy development of transportation. It has been more than ten years since the reform, and the tax collection method of the existing consumption income of refined oil products used for transportation in China has to be changed, and the distribution mechanism needs to be improved urgently. By sorting out the collection mechanism of the consumption tax on refined oil products used for transportation and the distribution mechanism of the transfer funds, this paper analyzes the problems of this mechanism according to the distribution and use of the transfer funds of the refined oil product tax reform. On this basis, this paper proposes the adjustment plan of the distribution mechanism at the level of the transfer funds supply of the consumption tax on refined oil products, according to the direction of the reform concerning the division of financial affairs.

**Keywords:** Consumption Tax on Refined Oil Products · Transportation · Transfer Payment · Fund Allocation

## 1 The Current System of Allocation Adjustment Mechanism of Transfer Payment of Consumption Tax on Refined Oil Products

In 2008, China implemented the reform of taxes and fees on refined oil products, replacing the original "six fees", such as road maintenance fees, with a consumption tax on refined oil products, and establishing a mechanism for transferring and distributing payments from the central government to localities for the reform of refined oil prices and taxes and fees. The funds transferred from the consumption tax on refined oil products

Fund Project: Strategic Planning Policy Project of the Ministry of Transport (Project No.: 2018-4-4).


Y. Qin et al. (Eds.): EITRT 2023, LNEE 1137, pp. 702–717, 2024.
https://doi.org/10.1007/978-981-99-9311-6_71

refer to the funds allocated by the central financial authorities to localities through standardized financial transfers after the reform of taxes and fees on refined oil products and the corresponding increases in value-added tax, urban maintenance and construction tax and education surcharges, in addition to expenditures such as replacement of maintenance fees for waterways, which are arranged by the central authorities at the local level.

## 1.1 Collection and Management Mechanism of Consumption Tax on Refined Oil Products in China

China's consumption tax on refined oil products refers to the consumption tax paid by consumers when they consume seven types of refined oil products, including gasoline, diesel oil, naphtha, solvent oil, aviation kerosene, lubricating oil and fuel oil. Similar to other consumption taxes, consumption tax on refined oil products also belongs to the central tax, which is uniformly collected by the State Taxation Bureau (the import link is entrusted to the Customs to collect on behalf of the taxpayer). The taxpayers are organizations and individuals who produce, commission processing and import refined oil products in China. The tax is paid at the production stage (including commissioned processing and import stage). The tax is levied from the quantity quota, and the tax is collected within the price. As a kind of consumption tax, the tax revenue from consumption tax on refined oil products is also included in the central general public budget revenue.

China's current consumption tax on refined oil products consists of two parts. One part is the consumption tax on gasoline and diesel fuel between 1994 and 2006, which belonged to the "centralized fixed income" during the tax reform in 1994, and at the same time contains the "two taxes returned" to the local authorities. The other part is the new tax revenue after the reform, in accordance with the Notice of The State Council on the Implementation of the Reform of Refined Oil Prices and Taxes (State Development (2008) No.37) and the Notice of the Ministry of Finance on the Issuance of the Transfer Payment Measures for the Reform of Local refined Oil Prices and Taxes (Financial Advance (2009) No.14) and other documents. The new consumption tax on refined oil products, which replaces the "six fees", has been distributed by the central financial authorities to localities through standardized financial transfers, in order to ensure the maintenance and construction of transportation infrastructure, in addition to the expenditures for the replacement of the maintenance fee for waterways to be arranged at the central level.

## 1.2 Transfer Payment Mechanism of Consumption Tax on Refined Oil Products in China

### 1.2.1 Purpose of the Transfer Payment Funds

The use of funds for the transfer of excise tax on refined oil products can be divided into three categories: first, the maintenance of highways and waterways; second, business expenses for the management of the road and waterway transportation industry; and third, the construction of road and waterway transportation infrastructure (Table 1).

Since the six fees originally belonged to the local special financial revenue, mainly used for the construction, maintenance and management of the local highway and waterway transportation infrastructure, the new tax revenue after the reform maintains the "unchanged attributes of the transportation funds, unchanged uses of the funds, unchanged procedures of the local budget, unchanged local authority". The order of distribution of this part of the tax revenue is:

① The replacement of six charges such as road maintenance fees. The specific amount is based on the six fees such as road maintenance fees in 2007, considering the local actual situation according to a certain growth rate;

**Table 1.** The replaced "six fees" for the original income purposes

| Charging item | Purpose of income | Legal basis |
|---|---|---|
| Highway maintenance fee | In accordance with the principle of "maintaining roads by roads and using special funds", special utility fees for road maintenance, repair, technical transformation, improvement and management are levied on vehicle-owning units and individuals | Former Highway Law of the People's Republic of China, Regulations on Highway Administration |
| Waterway maintenance fee | In accordance with the principle of "using navigation fees to support navigation, and using special funds", special utility fees are levied on the owners or operators of ships, bamboo rafts and floating objects for the maintenance of waterways | Former Regulations of the People's Republic of China on the Administration of Waterways, Measures for the Collection and Use of Maintenance Fees for Inland Waterways |
| Road transportation management fee | Business funds collected from the road transport operators for the administration of the road transport industry | The Original Interim Regulations on the Administration of Highway Transport |
| Highway passenger and freight surcharges | As a highway construction fund, the full amount is used for highway construction | Notice of The State Development Planning Commission, the Ministry of Finance and the Ministry of Communications on Regulating Highway Passenger and Freight surcharges and Increasing Highway Construction Funds (No.1104,1998) |

*(continued)*

**Table 1.** (*continued*)

| Charging item | Purpose of income | Legal basis |
|---|---|---|
| Waterborne transport management fee | To collect business funds from enterprises, units and individuals of domestic commercial waterway transport (service) for the management of the waterway transport industry | The Original Measures for the Collection and Use of Waterway Transportation Management Fee |
| Waterborne passenger and freight transport surcharges | The surcharges are levied on domestic passenger and domestic trade goods carried by ships of shipping enterprises under the Ministry of Transport for the construction of water transport infrastructure | Former Measures for Collection of Waterway Passenger and Freight Surcharges |

② Subsidies for the cancellation of government loan repayment of secondary highway tolls. A certain amount of special subsidy funds is arranged every year, including debt repayment, personnel resettlement, maintenance and management, and highway construction;
③ Increase subsidies for grain farmers, and for some difficult groups and public welfare industries, by taking into account the volume of oil consumption and changes in the price level, and providing subsidies to support them by improving the corresponding supporting subsidies in the price formation mechanism for refined oil products.
④ Incremental funds are allocated according to local fuel consumption, transportation facilities equivalent mileage and other factors, to appropriately reflect the balanced development of national transportation.

### 1.2.2 Distribution Method of Transfer Payment Funds

The total size of the annual transfer of funds is determined on the basis of the total fiscal revenues formed in the year as a result of the implementation of the reform of taxes and fees on refined petroleum products and the proportion of revenues from the six fees in the base period of the reform to the fiscal revenues as measured by the factors related to the reform, which is calculated according to the following formula:

Total amount of transfer payments = fiscal revenues generated by the reform of taxes and fees on refined petroleum products in the current year x the proportion of revenues from the six fees, etc., in the base period of the reform to the fiscal revenues measured according to the factors related to the reform (1.1).

The transfer funds are divided into four categories in terms of distribution, with the scale of funds and channels of payment varying from one to the other (Table 2). Among them, the two funds, the subsidy for the abolition of secondary roads and the subsidy for grain farmers, groups in difficulty and public welfare industries, are mainly in the form of special transfer payments from the central government to local governments, with the size of the funds determined in accordance with the relevant regulations. However, the return of the alternative rebate of "six fees" takes the form of a tax rebate from the central government to the local authorities, while the incremental funds take the form of general transfers, the size of which is determined by the "base plus factor" method, and is divided into two parts, namely, the alternative rebate and the growth subsidy. The alternative rebate of "six fees" refers to the rebate given in lieu of the original revenue base of the "six fees" at the local level. The incremental funds refer to the allocation of incremental funds after deducting the substitution rebate from the total amount of transfer payments for the year, with 60% allocated according to the consumption of refined oil products in each region, and 40% according to the proportion of highway and waterway fees in the base year of the reform, as well as objective factors affecting the maintenance and construction of highways and the maintenance of waterways.

**Table 2.** Distribution of transfer payments for consumption tax on refined oil products

| Projects | Transfer of payment methods | Policy basis | Size of funds | Distribution methods among provinces |
|---|---|---|---|---|
| Elimination of subsidies for the removal of stations on secondary roads | Specialized transfers | Implementation Plan for the Gradual and Orderly Elimination of Tolls on Government-related Secondary Roads | Started at $26 billion per year, then adjusted to about $30 billion per year | Comprehensively consider the 2007 provincial government loan repayment secondary highway toll amount and mileage, regional differences and other factors for quota calculation, according to the east, central and western regions of 40%, 50%, 60% for the cap control |
| Subsidies and grants for grain farmers, needy groups and public welfare industries | Specialized transfers | Notice of the State Council on the Implementation of the Reform of Prices, Taxes and Fees for Refined Products | – | – |

(*continued*)

**Table 2.** (*continued*)

| Projects | Transfer of payment methods | Policy basis | Size of funds | Distribution methods among provinces |
|---|---|---|---|---|
| Replacement of the "six fees" rebate | Tax rebate | Circular on the Issuance of Measures for the Transfer of Payments from the Central Government to Localities for the Reform of Prices of Refined Oil Products and Taxes and Fees | Based on the "six fees" income for 2007, with a growth rate of 10% | Returns are made on a local basis |
| Incremental capital | General transfer payments | Circular on the Issuance of Measures for the Transfer of Payments from the Central Government to Localities for the Reform of Prices of Refined Oil Products and Taxes and Fees | The incremental funds in the total transfers for the year, after deducting the alternative rebates, are subject to the condition that the revenues generated by the reforms increase by more than 10% over the originally measured revenues | From the total transfers for the year, the incremental funds, after deducting the alternative rebates, are subject to the condition that the revenues generated by the reforms increase by more than 10 percent over the originally measured revenues |

## 2 Analysis of the Transfer Payment Mechanism of Consumption Tax on Refined Oil Products

**First, the funds transferred from the consumption tax on refined oil products cannot meet the demand for highway maintenance.** At present, the transfer payment of consumption tax for central oil products accounts for more than 80% of the annual national general road maintenance expenditure (excluding large and medium-sized repairs), and the scale of the base returned part is 153.1 billion yuan per year. In addition to the road maintenance part, it should also be used for waterway maintenance and road transportation management. The incremental funds increased from 63 billion RMB to 69.3 billion RMB from 2013 to 2020, which is limited and not fully used for road maintenance. The base scale and growth of the central transfer payment for refined oil products are much lower than the original road maintenance fees. In addition, even some provinces and regions have the situation that the transfer payment funds cannot be fully allocated, and some of them are balanced by the finance to the general budget or retained by the transportation department as a large and medium-sized repair fund, coupled with the objective factors such as the rising price and wage level year by year, and the increasing maintenance tasks. In recent years, the scale of funding for road maintenance has become more and more insufficient to meet the funding arrangements for maintenance projects and organization operation, and the funding gap has been increasing year by year. It is estimated that the total demand for ordinary highway maintenance funds in China during the "14th Five-Year Plan" period will reach about 2.9 trillion RMB, with an average annual demand of more than 590 billion RMB, while the part of the annual transportation special fund for consumption tax on refined oil products that can be used for highway maintenance is only about 150–180 billion yuan, with a shortfall of as high as 62%. It is expected that 51% of the country's ordinary highways will be trapped in the predicament of "listing maintenance but no money to maintain, and no money to repair for the highways which should be repaired", which will keep about 90,000 km of ordinary national and provincial trunk highways and 1.6 million kilometers of rural highways in the condition of the "medium poor" level for a long time.

**Second, the allocation mode of transfer payment of refined oil consumption tax does not adapt to the structural change of maintenance demand.** In 2008, in order to ensure the smooth progress of the reform of taxes and fees, the base return part of the consumption tax of refined oil products was distributed among provinces according to the "six fees" base in 2007 (base year). This distribution model is far from being able to adapt to the current structural changes in the maintenance demand of each province, and structural problems have appeared in the public service system of the national road network. In particular, at that time, the traffic flow in the central and western provinces of China was low in the early years, and the base of "six fees", which was mainly road maintenance fees, was very low and had a big gap with the eastern provinces. However, with the development so far, the mileage and traffic volume of the road network in the central and western parts of the country have made a big leap, and the demand for funds for the maintenance of its general highways is far greater than that of the eastern part of the country. At this time, it is obviously unreasonable to still adopt the structure of traffic volume of provinces 13 years ago as the allocation of funds for maintenance. The allocation of incremental funds is equally problematic. The allocation of incremental funds nominally takes into account the "road network density", "road condition index" and other industry development indicators of the provinces and regions. However, on the one hand, the total national scale of incremental funds is too small, less than 70 billion RMB nationwide, and the actual amount of funds received by the provinces is limited through the allocation. On the other hand, the algorithmic indicators for incremental funding are not sufficiently comprehensive and detailed, and the algorithm is not transparent, making it difficult to truly reflect the actual development of transportation in each province. Under such an allocation mechanism, the transfer funds for refined oil products per unit mileage of the national ordinary highway vary greatly among provinces, which is a departure from the original intent of the tax and fee reform in the current year. Measured by 2018–2020 data, eastern provinces can receive an average of 56,000 yuan/km of maintenance funds per year, central provinces and regions 38,000 yuan/km, and western provinces and regions only 25,000 yuan/km.

For example, the mileage of ordinary highway in Beijing is about 22,000 km, and the scale of the transfer of consumption tax on refined oil products received is 5.7 billion yuan (of which 4.64 billion yuan is returned from the base and 1.06 billion yuan is incremental funds); corresponding to this, Qinghai Province bears the task of 78,000 km of maintenance of ordinary highway, which is 3.5 times more than that of Beijing, and the scale of the transfer of consumption tax on refined oil products received by Qinghai is only 1.08 billion yuan (of which 4.8 billion yuan is returned from the base and 1.06 billion yuan is incremental funds). Of which 480 million yuan of base return and 600 million yuan of incremental funds), less than one-fifth of Beijing's, and the pressure on maintenance funds far exceeds that of Beijing.

**Third, the consumption tax of refined oil products will be bundled with project funds.** Before the tax reform, according to "Highway Maintenance Fee Use Management Regulations" (Highway Traffic[1987] No. 64), highway maintenance fee in the use of highway maintenance fees need to first ensure that the road maintenance and major repair project fee needs, and then the maintenance of other fees and business development fee needs. Moreover, "highway maintenance fee for maintenance project cost ratio is not less than eighty percent of the total expenditure each year". That is to say, the road maintenance fee is mainly used for "road maintenance" rather than "personnel expenses". After the tax and fee reform, both "people-supporting" funds (including daily cleaning personnel, administration and other fees) and "road-supporting" funds (referring to minor repairs and maintenance fees) are financed by transfers of refined petroleum products. In the case that it is difficult for the supply of funds to cover the demand, highway management agencies at all levels usually put the personnel budget before the daily utility budget and the project expenditure budget in accordance with the principle of "prioritizing basic expenditure" in the preparation of departmental budgets, and do not prepare for deficit budgets. With the increase in personnel expenses, the cost of maintenance project expenditures has been seriously crowded out, and the contradiction between "people eating the road" and "keeping people but not the road" is becoming more and more pronounced in various places. A study found that with some provinces and districts accounting for more than 70% of the personnel expenses for the maintenance of ordinary highways [6].

The fundamental cause of the above problems is that the design of the transfer payment mechanism of refined oil consumption tax does not conform to the principle of matching financial power and administrative power. After the reform of taxes and fees, the original local fees will be turned into the central tax, and the central government will arrange the annual transfer payments and funds in accordance with the principle of "the use of funds, local budget procedures and local authority will remain unchanged", and the local governments will perform their maintenance duties. That is, the financial power of highway maintenance is transferred to the central government, while the administrative responsibility of highway maintenance remains in the local area (see Fig. 1).

In practice, the central finance allocates the funds to the local governments through tax rebates and general transfer payments, not assuming the administrative and expenditure responsibilities of road maintenance. Similarly, the local finance at all levels has no enthusiasm to invest more financial funds in which the highway system funds are included in the financial budget for unified management, the shortage of maintenance

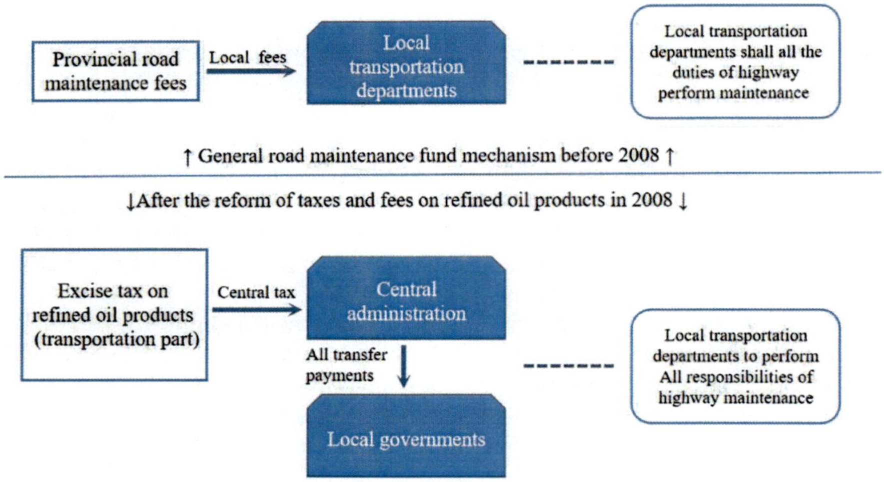


**Fig. 1.** Changes in the fund mechanism of ordinary highway maintenance before and after the tax reform

funds still needs to be borne by the transportation department. Due to the lack of funds, the transportation departments of most provinces and regions have delegated the maintenance authority to various levels, and "construction instead of maintenance" has become more prominent, and the quality of maintenance authority responsibility has been greatly reduced. All levels of finance have no incentive to bear the personnel expenditure of the highway system. From the development trend of more than ten years, this mismatch of financial power does not play the joint force of the central and local governments, transportation and finance to promote the development of highway maintenance.

In addition, from the perspective of the actual use of tax revenue, although the consumption tax on refined oil products is positioned as the central tax, in fact, local governments not only get the tax refund, but also get the transfer payment for the price of refined oil products and tax reform. That is to say, the part of the consumption tax income of refined oil products used for transportation is first collected from the production link as the central tax revenue, and then distributed to the local use by means of tax refund and transfer payment. This system design is different from the current international mainstream practice. According to international experience, the consumption tax on refined oil products (or similar taxes) used for transportation is mostly levied in the sales link, such as the United States, the United Kingdom, France, the Netherlands and other countries. At present, transportation oil accounts for more than half of the total consumption of refined oil products in China. Therefore, the consumption of refined oil products is closely related to the use level of highway and waterway infrastructure. A recent study shows that if the consumption tax for transportation is levied according to the actual consumption of refined oil products in a certain region, it is more reasonable to collect it in the wholesale and sales links [4], and also to avoid the problem of uneven geographical distribution of oil production enterprises. From the perspective of transportation governance and expenditure responsibility division reform, if used for traffic refined oil consumption tax in the wholesale and sales and as a local government tax revenue, to meet the demand of local traffic fiscal governance, make local government better performance in the transportation infrastructure construction and maintenance spending responsibility.

## 3 Suggestions for Optimizing the Allocation Adjustment Mechanism of Consumption Tax on Refined Oil Products

On the one hand, the current state is promoting the reform of fiscal authority, clarifying the responsibilities of the central and local governments, adjusting the central expenditure structure and allocation channels [8], which will bring a series of impacts on the allocation and adjustment mechanism of refined oil consumption tax and transfer payment funds. On the other hand, from the future development trend of China's gasoline and diesel consumption changes, the promotion of new energy vehicles, transportation structure adjustment, the transportation field and the emission peak and deep emission reduction, the incremental income of refined oil consumption tax will be affected to some extent, and there is pressure to reduce revenue. As a result, the future growth rate and scale of refined oil consumption tax transfer payment fund is uncertain. At the supply level of transfer payment funds of refined oil consumption tax, we should focus on the goal of matching the supply side of the capital and the actual capital demand end, adjust the idea of capital supply, and put forward the reform idea of establishing a more scientific, fair and reasonable distribution mechanism.

### 3.1 Maintain the Existing Basic System of Consumption Tax on Refined Oil Products Unchanged, and Adjust the Calculation Method of Transfer Payment Funds

That is, the consumption tax on refined oil products is still the central tax, and the current regulations such as collection and use remain unchanged, and the part of the new tax revenue replacing the original "six fees" is still transferred to local governments for ordinary road maintenance expenses (excluding the operation costs of local road maintenance and management agencies). In this mode, in order to solve the problem of insufficient maintenance funds, should be allocated in the base scale of the capital allocation method, according to the local highway maintenance demand in the allocation of funds, improve the refined oil consumption tax revenue for highway maintenance cutting scale, and develop the corresponding growth mechanism. The core of the adjustment is to make the scale of the transfer payment funds more determined in accordance with the maintenance needs, so as to make the supply side of the transfer payment funds more match with the actual capital demand end. In the "base + factor" calculation adopted by the existing transfer payment mechanism, the funds allocated according to the "base" account for the main body of the transfer payment funds, while the funds allocated according to the "factor" only account for a small part. It means that most of the transfer payment funds are unconditional transfer payment, which can not reflect the development and change of any aspect. Therefore, it is necessary to gradually change the calculation method from the base to the factor, that is, it can more reflect the actual demand.

### 3.1.1 Measurement of the Base Figure

The "base" is retained in the calculation, which reflects the consideration of the original distribution pattern and protects the vested interests of local governments to a certain extent. Even in the extreme case when the consumption tax revenue of refined oil products drops sharply, local governments can still obtain a certain amount of transfer payment funds for the maintenance of transportation infrastructure every year. In the future, if the social and economic conditions reach a certain level, the base can be cancelled and allocated completely according to the factor method.

The calculation of the base can first select a certain year as the base year, and then calculate the total amount of expenditure for maintenance in the actual transfer payment funds of the base year and the target value of the total demand of the maintenance funds in the year as the calculation basis, and the base of transfer payment of refined oil consumption tax income should be calculated. The calculation formula is:

The new national transfer payment base = the original national transfer payment base [1 + lg (the total amount of the total demand target value of conservation funds in the transfer payment of refined oil consumption tax in the base year)].

New local transfer payment base = new national transfer payment base year local maintenance fund demand target value in the proportion of the country.

Make A is the new national consumption tax on refined oil products transfer payment base, Ai's the original national base, $E_0$ the total actual national conservation expenditure in the benchmark year, $D_0$ as the target value of the total demand of the national conservation funds in the benchmark year, a Transfer payment base for j, $d_{0j}$ for the target value of base year j, the above formula can be written as:

$$A = A/ \times \left(1 + \log_{10}\frac{D_0}{E_0}\right) \quad (1)$$

$$as = A \times \frac{d_{oj}}{D_0} \quad (2)$$

### 3.1.2 Calculation of the Total Scale of Transfer Payment Funds

The calculation method of transfer payment funds in a certain year is the growth rate of the total amount of the maintenance funds of the current year and the previous year. Therefore, the calculation formula for the scale of transfer payments in the current year is:

The total amount of national transfer payment funds in the same year = the actual national transfer payment funds in the previous year [(1 + growth rate of national refined oil consumption tax revenue in the previous year) 60% + (1 + growth rate of the target value of national maintenance funds in the same year) 40%].

Let $P_i$ be the total amount of transfer funds in year i, and $I_i$ be the national excise tax revenue from refined oil products in year i, $\Delta I_i$ be the incremental national excise tax revenue from refined oil products in year i, $C_i$ be the national consumption of refined oil products in year i, $D_i$ be the target value of the total national demand for conservation funds in year i, and $\Delta D_i$ be the incremental national demand for conservation funds in year i.

$$
\begin{aligned}
P_i &= P_{i-1} \times \left[ \left(1 + \frac{\Delta I_{i-1}}{I_{i-2}}\right) \times 60\% + \left(1 + \frac{\Delta D_i}{D_{i-1}}\right) \times 40\% \right] \\
&= P_{i-1} \times \left( \frac{I_{i-1}}{I_{i-2}} \times 60\% + \frac{D_i}{D_{i-1}} \times 40\% \right)
\end{aligned} \tag{3}
$$

The total amount of transfer payment funds in a certain place in the same year = the total amount of national transfer payment funds in the same year (the consumption of refined oil in the same year accounted for 60% of the national proportion + the target value of local conservation funds in the same year accounted for 40% of the national proportion).

$p_{ij}$ is the total amount of transfer funds in place j in year i, cij the consumption of refined oil products in place j in year i, and $d_{ij}$ is the target value of conservation fund demand in place j in year i. Then the above equation can be written as:

$$
p_{ij} = P_i \times \left( \frac{c_{i-1,j}}{C_{i-1,j}} \times 60\% + \frac{d_{ij}}{D_{ij}} \times 40\% \right) \tag{4}
$$

among, $P_i \geq 0$,

equal $\Delta I_{i-1} < 0$ When, $P_i = 0$.

When $i = 0$, the $P_0 = A$, $p_{0j} = a_j$.

Among them, in the base year, the scale of transfer payment funds is the transfer payment base, namely:

Total amount of transfer payment funds in the base year = the national transfer payment base.

Transfer payment funds in a place in the base year = transfer payment base in some place.

### 3.1.3 Calculation of Incremental Funds

Based on the calculation method of the total scale and base of transfer payment funds, the calculation formula of incremental funds can be obtained as follows:

The total amount of growth subsidies in the year = the total amount of transfer payment funds in the year-the national transfer payment base.

The local growth subsidy = the total amount of transfer payment funds in the year-the transfer payment base.

Among them, in the base year, the growth subsidy is 0.

Let $B_i$ be the total growth grant for year i and $b_{ij}$ be the growth grant for site j in year i. The above equation can be written as:

$$
B_i = P_i - A \tag{5}
$$

$$
b_{ij} = p_{ij} - a_j \tag{6}
$$

When $i = 0$, the $B_0 = 0$, $b_{0j} = 0$.

#### 3.1.4 Establish a Growth Mechanism

According to the current and future a certain period of the national highway maintenance funds demand growth trend, gradually improve the base calculation formula of the maintenance funds demand target value in the weight, or improve the incremental capital calculation formula in the weight of the weight, establish the mechanism of return funds growth, gradually to the amount of funds allocated to the target value of capital demand.

#### 3.1.5 Establish a Tax Rate Adjustment Mechanism

According to the determined target value of maintenance fund demand in a certain period of time, the growth level of maintenance fund demand will be included in the tax rate adjustment mechanism of refined oil consumption tax as a parameter. If the target value of maintenance fund demand increases by more than 20%, the corresponding tax rate adjustment procedure shall be initiated. The target of the tax rate adjustment is to cover the increase of the maintenance fund demand after the adjustment according to the predicted change trend of refined oil consumption.

### 3.2 Consumption Tax on Refined Oil Products Will Be Changed to Shared Tax

As for the collection of consumption tax on refined oil products in China, the Notice of The State Council on the Implementation of the Reform of refined oil Prices and taxes proposes that "in the improvement of the consumption tax system, actively create conditions, timely move the collection of consumption tax to the wholesale link, and change it to the collection of price". According to the guidance of "taking some central taxes with mature conditions as local income" in the Income division Plan, certain conditions are now met for adjusting the consumption tax on refined oil products from central tax to shared tax.

From the development process of the consumption tax of refined oil products in China, the generation of the transfer payment fund of refined oil consumption tax is from the replacement of the original "six fees" income, and the "six fees" was originally the local fiscal revenue. After the reform of taxes and fees on refined oil products, the newly increased part of the income is returned to local governments through transfer payments. In theory, the new income is still used for the expenditure of local fiscal authority, and mainly for the transportation field, with relatively clear uses.

At present, the consumption of vehicle gasoline and diesel in China accounts for about half of the total consumption of gasoline and diesel. In the current tax system of most countries, the consumption tax on refined oil products used for transportation is higher than that on refined oil products used in agriculture, fishery, forestry and horticulture industries, so as to support specific industries and regulate transportation consumption. At the same time, compared with the refined oil products used in other fields, vehicle gasoline and diesel oil are easier to tax at the consumer end.

Therefore, draw lessons from the experience of the United States, Japan and other countries, can consider the existing oil consumption tax and use mechanism adjustment, cancel the existing oil consumption tax transfer payment mechanism, the refined oil consumption tax from the central tax to Shared tax, by the central and local respectively, and the tax link back, thus better play the role of guiding consumption behavior, but also can avoid the problem of uneven distribution of oil production enterprises.

### 3.2.1 Collection Method

The central government collects a certain amount of consumption tax on all kinds of taxable refined oil in production and import, and the local government collects a certain amount of consumption tax on vehicle gasoline and diesel in wholesale and retail links, including the local fiscal revenue, which is mainly used for the maintenance of roads and channels with local financial authority, which is equivalent to the function of replacing the original "six fees" income.

### 3.2.2 Determine the Base and Tax Rate of Gasoline and Diesel Consumption Tax Levied by Local Governments

Determine the benchmark year and calculate the total amount of transfer payment of consumption tax on refined oil products in the benchmark year, so as the base of gasoline and diesel consumption tax levied by local governments. To calculate the national consumption of gasoline and diesel in the benchmark year, and use the ratio of the tax base to the consumption of gasoline and diesel as the benchmark tax rate of gasoline and diesel consumption tax levied by local governments. The calculation formula is:

Local tax rate of gasoline and diesel consumption tax = tax base year national vehicle gasoline and diesel consumption.

### 3.2.3 Calculate the Tax Rate of Gasoline and Diesel Consumption Tax Collected by the Central Government

The consumption tax on gasoline and diesel levied by the central government is still levied in the production and import links, including vehicle gasoline and diesel and non-vehicle gasoline and diesel. The tax rate is:

Central tax rate of gasoline = (total revenue of consumption tax of gasoline in the base year-local tax rate of consumption tax of gasoline and diesel of automotive gasoline consumption in the base year) total gasoline production and import in the base year.

Central tax rate of diesel = (total revenue of diesel consumption tax in the base year-local tax rate of vehicle diesel consumption of gasoline and diesel consumption tax) Total diesel production and import in the base year.

### 3.2.4 Establish a New Transfer Payment Mechanism

For areas with small tax base and large maintenance demand, the consumption tax revenue of gasoline and diesel levied by local governments is not enough to meet the maintenance demand, so the consumption tax revenue on refined oil products collected by the central government can be transferred to such areas.

## 4 Conclusion

In addition to adjusting the way of supplying funds for the transfer of consumption tax on refined oil products, consideration should also be given to adjusting and optimizing the method of determining the demand for funds for the transfer of consumption tax on refined oil products, scientifically measuring the demand for funds for maintenance, proposing alternative options for funds for maintenance. Moreover, in accordance with the relevant requirements of the reform of the right to fiscal affairs, it is necessary to change the idea of fiscal expenditure for road maintenance from "determining the expenditure on the basis of revenues" to "determining the expenditure on the basis of affairs". In accordance with the relevant requirements of the reform of fiscal authority, it has changed the way of thinking about fiscal expenditure on road maintenance from "setting expenditure on the basis of income" to "setting expenditure on the basis of matters", and has determined the fiscal authority and expenditure responsibility for each specific matter involved in road maintenance. In addition, for the problem of transfer payment mechanism design, it is also necessary to strengthen the research on the basic system level of consumption tax on refined oil products, in accordance with the requirements of deepening the reform of China's fiscal and taxation system, considering the impact of China's "double-carbon" strategy on the consumption tax on refined oil products, and referring to the advanced experience of the developed countries to put forward the direction of adjustment and optimization, so that the system design of consumption tax on refined oil products can be better designed to meet the demands of China's fiscal and taxation system. The system design of the consumption tax can better meet the requirements of its functional position.

## References

1. Notice of The State Council on implementing the Reform of refined Oil Prices and Taxes (State Council (2008) No.37) [EB/OL]. 2008–12–19 [2021–09–11]. http://www.gov.cn/zwgk/2008-12/19/content_1182128.htm
2. Notice of the Ministry of Finance on Printing and Issuing the Transfer Payment Measures for the Reform of local refined oil Prices and taxes (Financial Advance (2009) No.14) [EB/OL]. 2009–2–25 [2021–09–11]. http://yss.mof.gov.cn/ls/czzxzyzf/201109/t20110902_591465.htm
3. Notice of the Ministry of Finance on Issues related to the Management of refined oil price and Tax reform (Ministry of Finance (2009) No.351) [EB/OL]. 2009–8–17 [2021–09–11]. http://yss.mof.gov.cn/zhengceguizhang/200908/t20090825_198300.htm
4. Li, R., Sun, Y.: Reform and improve the consumption tax system of refined oil products. Tax Res. **3**, 15–19 (2019). (in Chinese)
5. Yin, A., Zhang, Y., Meng, Y.: Supply and demand effect of consumption tax reform of refined oil products. Tax Res. **4**, 36–42 (2015). (in Chinese)
6. Wang, S.: Thinking on highway maintenance work after tax and fee reform. Maintenance Manage. **6**, 52–85 (2010). (in Chinese)
7. Bai, Y., Qiu, X., Zhang, J.: Reform scheme selection of current refined oil consumption tax in China. Tax Res. **4**, 52–58 (2015). (in Chinese)
8. Zhang, D.: Consumption tax reform in accelerating the establishment of a new development paradigm. China Financ. Econ. Rev. **11**(2), 117–128 (2023)

9. Zhu, Y., Liu, X.: Satapathy SC. Big data visualization of the quantification of influencing factors and key monitoring indicators in the refined oil products market based on fuzzy mathematics. J. Intell. Fuzzy Syst. **40**(4), 6219–6229 (2021)
10. Han, Y., Zhao, W., Hao, H.: Analysis of income attribution of consumption tax in China's central and local government. Discrete Dyn. Nat. Soc **2021** (2021)
11. Tomoyuki, N., Shuhei, T.: The effectiveness of consumption taxes and transfers as insurance against idiosyncratic risk. J. Money Credit Bank. **52**(2–3), 505–530 (2020)

Check for updates

# A Sparsity-Free Compressed Sensing Method for PHM Data Quality Assurance Using Generative Adversarial Network

Qinglin Xie, Jing Wang, Gongquan Tao, Chenxi Xie, and Zefeng Wen(✉)

State Key Laboratory of Rail Transit Vehicle System, Southwest Jiaotong University, Chengdu, People's Republic of China
zfwen@swjtu.edu.cn

**Abstract.** Recovering missing data to improve data quality is of great importance for prognostics and health management (PHM) of structures and machinery. Existing missing data reconstruction methods are generally limited by strong assumptions such as signal sparsity, coupled with heavy computational burdens, making them poorly adaptable to PHM data. To obtain more reliable PHM results, a novel algorithm is developed to handle missing data based on compressive sensing and enhanced context encoders (CS-ECE). The CS-ECE can reconstruct the detailed characteristics of signals in both the time and frequency domains. The effectiveness of the proposed method is validated using real-world data collected from a high-speed train. The results demonstrate that CS-ECE can obtain realistic and accurate recovered results, thereby providing a high-quality dataset for data-driven PHM studies.

**Keywords:** Compressive Sensing · Context Encoders · Data Quality Assurance · Generative Adversarial Network · Joint Loss Function

## 1 Introduction

The collection and storage of massive data have prompted the study of prognostics and health management (PHM) to enter the era of big data, which brings new opportunities for structure and machinery health monitoring. The complex environment associated with data acquisition, transmission, and storage in modern PHM causes the obtained data have five characteristics: high volume, high variety, high velocity, low-value density, and low veracity, making it more challenging to extract health status information [1]. Particularly, low-quality data significantly reduce the reliability of PHM results. Therefore, data quality assurance has become a new problem for PHM under the background of big data. Generally, data quality refers to four aspects: accuracy, completeness, timeliness, and consistency [2]. Some unavoidable interferences seriously can reduce data quality and increase the challenges associated with PHM studies. Motivated by Big Data, deep learning-based algorithms in PHM exhibit exceptional advantages and substantial advancements. However, these methods are incapable of assessing data quality and suffer from the "garbage in, garbage out" drawback. Conclusions derived from


Y. Qin et al. (Eds.): EITRT 2023, LNEE 1137, pp. 718–726, 2024.
https://doi.org/10.1007/978-981-99-9311-6_72

low-quality PHM data may be incorrect or misleading. Consequently, research on data quality assurance is of great importance to obtain reliable PHM results. Lei et al. [3] pointed out that data cleaning and data recovery are important studies on data quality improvement. In our previous research [1], an adaptive abnormal data detection method was proposed to perform data cleaning tasks. Nevertheless, comprehensive health information cannot be obtained with this cleaning strategy alone, as missing contents may disrupt the temporal structure of the PHM data. Hence, this paper focuses on the issue of missing data recovery to establish a systematic framework for data quality assurance.

As a well-known technique for accurate data representation, compressive sensing (CS) aims to acquire critical information about the original signals from just a few random samples [4]. CS has attracted considerable attention and has been widely applied in PHM data, such as acceleration signals in civil structures [5], acoustic signals [6], road roughness signals [7], ultrasonic signals [8], electromechanical admittance signals [9], and wind speed signals in nature [10]. However, the practical implementation of CS for real-world data is restricted by strong assumptions of sparsity and randomness, as well as the costly reconstruction process [11]. Real-world data may not be naturally sparse due to the complex service environment. Traditional methods for domain transformation are ineffective in conducting signal sparse decomposition. Consequently, CS-based data recovery schemes are unable to be accurate, fast, and robust, especially for high-frequency PHM data.

Fortunately, thanks to the superior end-to-end learning capability of deep learning (DL), maps can be directly acquired from pairs of known low- and high-resolution examples. This can eliminate the need for relying on the unknown sparsity of samples and remove the dependence on random sampling and sparsity. For instance, Kulkarni et al. [12] built a novel DL-based architecture named ReconNet, which takes CS measurements as input and outputs the reconstruction in a non-iterative manner. Experiments confirm that ReconNet can yield better reconstruction quality and time complexity than traditional CS algorithms. Among these DL models, generative adversarial networks (GANs) can derive highly detailed high-frequency textures, thus obtaining the best recovery results. A typical example is the context encoders (CE), which is a feature learning algorithm driven by context-based pixel prediction that shows good performance in missing data recovery [13]. However, most of the DL-based CS studies focus on images and do not address one-dimensional (1-D) signals, such as PHM data. Thus, given the advantages of CE-based CS in data reconstruction in various fields, and their practical limitations in 1-D signals. This paper proposes a missing data recovery model based on CS and enhanced context encoders (CS-ECE) for PHM data quality assurance studies.

## 2 Basic Theory of Compressive Sensing and Context Encoder

### 2.1 Compressive Sensing

The following equation illustrates the basic theory of CS

$$\boldsymbol{y} = \boldsymbol{\Phi x} = \Phi \Psi \boldsymbol{s} = \boldsymbol{A s} \tag{1}$$

where $\boldsymbol{x} \in \mathbb{R}^{N\times1}$ is the original signal to be reconstructed and satisfies the classical Nyquist–Shannon sampling theorem. $\boldsymbol{y} \in \mathbb{R}^{M\times1}$ is the measurement signal. $\boldsymbol{\Phi} \in \mathbb{R}^{M\times N}$ ($M \ll N$), $\boldsymbol{\Psi} \in \mathbb{R}^{N\times N}$, $\boldsymbol{A} \in \mathbb{R}^{M\times N}$ are the measurement matrix, the sparsity transformation matrix, and the sensing matrix, respectively. Let $\boldsymbol{s} \in \mathbb{R}^{N\times1}$ represent the sparse coefficient. According to the CS theory, if a signal $\boldsymbol{x}$ is sparse or if the transformed signal $\boldsymbol{s}$, obtained by applying the transform base $\boldsymbol{\Psi}$ to $\boldsymbol{x}$, exhibits enough sparsity, the original signal $\boldsymbol{x}$ can be first down-sampled by a measurement matrix $\boldsymbol{\Phi}$, and the original signal $\boldsymbol{x}$ can be reconstructed as accurately as possible by solving an optimization problem. However, solving the optimization is an NP-hard problem, and it requires the verification that the sensing matrix $\boldsymbol{A}$ satisfies the conditions of the restricted isometry property, which is usually difficult to meet for missing data recovery tasks in PHM.

### 2.2 Context Encoder

The overall architecture of the classic CE is an intuitive encoder-decoder structure [13]. The encoder takes the incomplete signal as input data and outputs potential information of the incomplete signal. The decoder accepts the feature information and generates the missing region. The training process of the CE involves regression toward the complete original content of the corresponding missing content. It is developed to jointly learn an adversarial discriminative model, denoted as $D$, which provides loss gradients to the generative model, denoted as $G$. Typically, both $G$ and $D$ are DL networks, and $G$ maps correlations form a noise distribution $p_z(z)$ to a data distribution $p_{\text{data}}(\boldsymbol{x})$. The objective of the discriminator is to determine whether the input corresponds to a real or a predicted sample:

$$\min_{G}\max_{D} \mathrm{E}_{\boldsymbol{x}\sim p_{\text{data}}(\boldsymbol{x})}\left[\log D(\boldsymbol{x})\right] + \mathrm{E}_{z\sim p_z(z)}\left[\log(1 - D(G(z)))\right] \tag{2}$$

where E[] averages the values inside the brackets.

## 3 Framework of Data Quality Assurance Method

### 3.1 Data Preprocessing

Figure 1 shows the framework of the proposed CS-ECE. The PHM data fed to the CS-ECE is typically large in volume, and its multifarious information is coupled with each other. It is liable to fall into a dimension disaster in terms of computational efficiency if analyzed directly [14]. Thus, data preprocessing is indispensable. The preprocessing strategy of CS-ECE for PHM data quality assurance mainly consists of abnormal data cleaning, sliding, padding, and masking.

**Sliding** A sliding window with a length of 1024 data points is employed to divide the cleaned data into a series of segments. Note that the arrangement of the window length can be set adaptively according to different requirements.

**Padding** Given the cleaned data, i.e., the compressive signal $\boldsymbol{y}$ is part of the original signal $\boldsymbol{x}$, and the length of $\boldsymbol{y}$ is varied by the compression ratio. It is necessary to pad $\boldsymbol{y}$ with suitable placeholders to ensure that $\boldsymbol{y}$ and $\boldsymbol{x}$ satisfy a uniform scale.

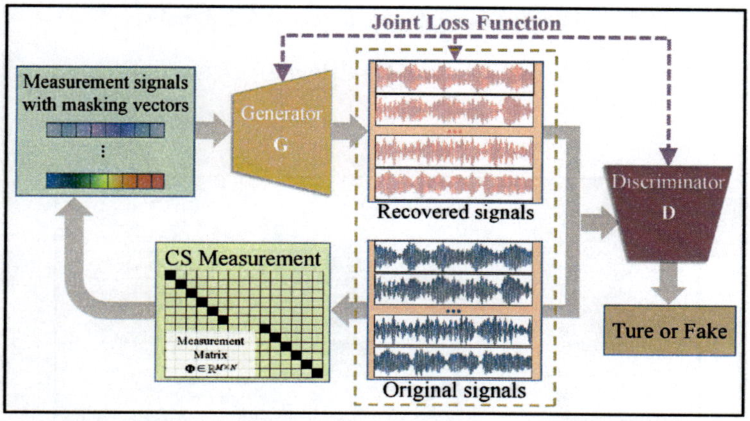


**Fig. 1.** Illustration of the proposed CS-oriented ECE (CS-ECE) model.

**Masking** The CS-ECE may attend to placeholders, so it is important to inform the model of the padded regions in $\boldsymbol{y}$. It is suggested to generate a Boolean mask sequence and append it to the CS-ECE as constraint information.

### 3.2 CS-ECE Architecture

Following data preprocessing, the structured dataset is fed to the CS-ECE for training. It consists of two DL networks: the generator $G$ and the discriminator $D$. $G$ is trained to map the original signal $\boldsymbol{x}$ to the compressive signal $\boldsymbol{y}$ in order to reconstruct a high-quality recovered signal $\hat{\boldsymbol{x}} = G(\boldsymbol{y})$. Conversely, $D$ performs a classification task to discriminate between the $\hat{\boldsymbol{x}}$ and $\boldsymbol{x}$. Figure 2 shows the training scheme of the CS-ECE.

The generator $G$ is an an improved CE. It consists of two modules: an encoder with five stacked convolutional layers and a decoder with five corresponding transposed convolutional layers. An activation function layer and a batch normalization layer are arranged after each convolutional or transposed convolutional layer. The architecture of $G$ is designed to extract high-level abstract features from the input $\boldsymbol{y}$ using the encoder's stacked convolutional layers and then establish the features of the predicted object $\hat{\boldsymbol{x}}$ through the combined decoder. Different from the classical CE, the dimension of the recovered signals $\hat{\boldsymbol{x}}$ is consistent with the original signals $\boldsymbol{x}$, rather than only outputting missing region data. This substantially enhances the continuity of the final reconstructed signal $\hat{\boldsymbol{x}}$. For discriminator $D$, it contains five convolutional layers, followed by a flattened layer and two fully connected layers. $D$ receives the recovery signal $\hat{\boldsymbol{x}}$ or original signal $\boldsymbol{x}$ as input, and a dense layer with the Sigmoid function produces the output for binary classification of whether each input is generated by $G$ or real.

### 3.3 Joint Loss Function

We specify the joint loss function for regulating the learning of network parameters after deploying the architecture of the $G$ and $D$. In CS-ECE, the loss function of $D$ is the cross-entropy between the original signal $\boldsymbol{x}$ and recovered signal $\boldsymbol{y}$.

$$\max_D L_D = \mathrm{E}\big[\log D(\boldsymbol{x})\big] + \mathrm{E}\big[\log(1 - D(G(\boldsymbol{y})))\big] \tag{3}$$

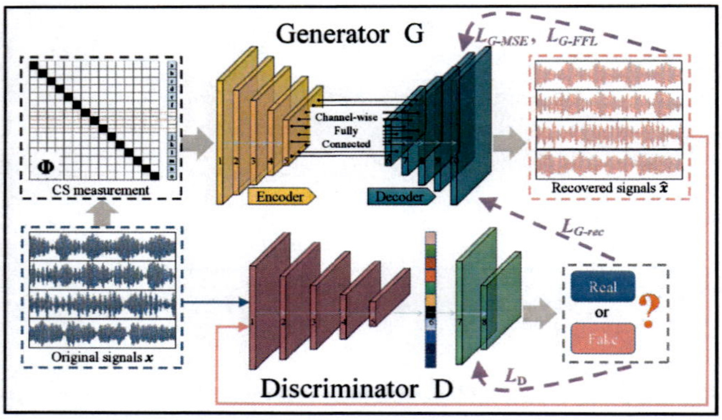


**Fig. 2.** Training scheme and network architecture of the proposed CS-ECE model.

$G$ is trained to minimize the $D$ assigning false labels to recovered signal, i.e., to reconstruct a signal that appears realistic enough to fool $D$ into classifying $\hat{\boldsymbol{x}}$ as real $\boldsymbol{x}$.

$$\min_G L_{G-rec} = \mathrm{E}\big[\log(1 - D(G(\boldsymbol{y})))\big] \tag{4}$$

Although Eq. (4) can give realistic reconstruction results, it may fail to create signals with high accuracy. Previous research [13] suggests that incorporating the error between the recovered signal $\hat{\boldsymbol{x}}$ and original signal $\boldsymbol{x}$ into the loss function is conducive to improving accuracy. Hence, the mean squared error (MSE) is used as an additional loss function $G$ to increase the recovery accuracy, denoted as

$$\min_G L_{G-MSE} = \frac{1}{N}\sum_{i=1}^{N} \left|\boldsymbol{x}_i - \hat{\boldsymbol{x}}_i\right|^2 \tag{5}$$

MSE is commonly used for regression tasks in DL due to its simple optimization. $G$ is compelled to generate $\hat{\boldsymbol{x}}$ close to the true $\boldsymbol{x}$ until $\hat{\boldsymbol{x}} = \boldsymbol{x}$ by minimizing the MSE. For PHM data, errors in the time or spatial domain can be measured using Eq. (5), while disregarding important frequency properties. On the other hand, due to the inherent bias of DL models, $G$ tends to eschew frequency components that are hard to synthesize, i.e., hard frequencies, and converge to an inferior point [15]. To address this issue, we introduce a frequency-level objective function called focal frequency loss (FFL) inspired by hard example mining and focal loss. FFL allows adaptively up-weighting the hard frequencies by a spectrum weight vector $\boldsymbol{\omega}$, as defined in Eq. (6), which is dynamically determined by a non-uniform distribution on the current loss of each frequency during training. The model $G$ then can be forced to rapidly focus on hard frequencies and progressively refine the generated frequencies to ameliorate recovery accuracy. By combining the weight vector $\boldsymbol{\omega}$ and the differentiable Euclidean distance in the frequency domain between $\hat{\boldsymbol{x}}$ and $\boldsymbol{x}$, we define the FFL expression

$$\omega_i = \left|f(\boldsymbol{x}_i) - f(\hat{\boldsymbol{x}}_i)\right|^{\alpha} \tag{6}$$

$$\min_G L_{G-FFL} = \frac{1}{N}\sum_{i=1}^{N} \omega_i \left|f(\boldsymbol{x}_i) - f(\hat{\boldsymbol{x}}_i)\right|^2 \tag{7}$$

where $\alpha$ is the scaling factor for flexibility and is recommended to be set to one [15]. $f$ () refers to the Fourier transform. Finally, the total joint loss function of $G$ is

$$\min_G L_G = \lambda_1 L_{G-rec} + \lambda_2 L_{G-MSE} + \lambda_3 L_{G-FFL} \tag{8}$$

Where $\lambda_1 = 0.01, \lambda_2 = 0.9$, and $\lambda_3 = 0.09$ are determined based on extensive experiments and referring to the settings in [11, 13].

### 3.4 Adversarial CS-ECE Training

The proposed CS-ECE is optimized by alternately maximizing the loss function $L_D$ and minimizing $L_G$ using a gradient descent method. The RMSProp is used to avoid the discontinuity of the discriminator and accelerate convergence during training [16].

## 4 Engineering Application and Experiment

PHM data obtained from a high-speed train were investigated to demonstrate the feasibility and superiority of the proposed method. The tested HSR runs on a specific line in southwestern China. Figure 3 shows the schematic of in-site tests [17].

**Fig. 3.** (a) Test high-speed line, (b) wheel roughness, and (c) axle box vibration measurement.

Figure 4 discloses part of the collected axle box acceleration (ABA) and wheel roughness data. It can be found that several abnormal data exist in ABA, and the $24^{\text{th}}$- and $25^{\text{th}}$-order wheel polygonal wear is significant. The measurement ABA data are divided into 10000 segments of 1024 data points, 80% of the samples are used for training, and the remaining 20% (including abnormal data) are used for testing. We first clean outliers using the proposed abnormal data detection method [1]. Then we obtain the cleaned data (containing missing contents), i.e., compressive signal $\boldsymbol{y}$. Note that the manifestations of abnormal data are random, meaning that they do not occur at a fixed location and scale in PHM data. To simulate this situation, we constructed a slice element containing 205 zero values and rewritten it randomly to each dataset $\boldsymbol{x}$ to acquire the data cleaning results $\boldsymbol{y}$ at a 20% compressive rate. Accordingly, the performance of the proposed CS-ECE can be quantitatively assessed by comparing the recovered with the original $\boldsymbol{x}$. The $R$-square coefficient ($R^2$) is used to evaluate the recovery quality between the original signal $\boldsymbol{x}$ and the recovered signal $\hat{\boldsymbol{x}}$, defined as

$$R^2 = 1 - \frac{\sum_{i=1}^{N}(\boldsymbol{x}_i - \hat{\boldsymbol{x}}_i)^2}{\sum_{i=1}^{N}(\boldsymbol{x}_i - \overline{\boldsymbol{x}}_i)^2} \tag{9}$$

where $\overline{\boldsymbol{x}}$ is the average of $\boldsymbol{x}$. The closer $R^2$ is to one, the higher the fitting degree and the more accurate recovery results. Figure 5 visualizes the output of $G$ at different training stages with a certain sample as input. Initially, $G$ generates completely meaningless noise. As training progresses, the recovered signal $\hat{\boldsymbol{x}}$ gradually approaches the target signal $\boldsymbol{x}$, eventually yielding realistic reconstructions. Figure 5 also illustrates the frequency characteristics of the signals. The frequency properties of the compressive signal $\boldsymbol{y}$ and the original signal $\boldsymbol{x}$ not only exhibit noticeable gaps between the dominant frequencies but also the corresponding energy differences cannot be ignored. Thanks to the benefits of the introduced $L_{G\text{-}FFL}$, the frequency information of the generated signal becomes asymptotically close to the target one, ensuring the reality and accuracy of the recovered signals. Use the trained model to process the abnormal data in Fig. 4. Figure 6 depicts the recovered results of a specific sample containing abnormal data points. It can be observed that not only has the time domain waveform of the measurement ABA signal deviated significantly from the normal level, but its frequency characteristics also fail to identify the excitations of 24$^{\text{th}}$- and 25$^{\text{th}}$-order wheel polygons. Fortunately, the quality of the recovered ABA signal is substantially improved using the trained CS-ECE. The distortion reflected in the time domain is eliminated, and the order information of the wheel polygon can be accurately inferred according to the principal frequency.

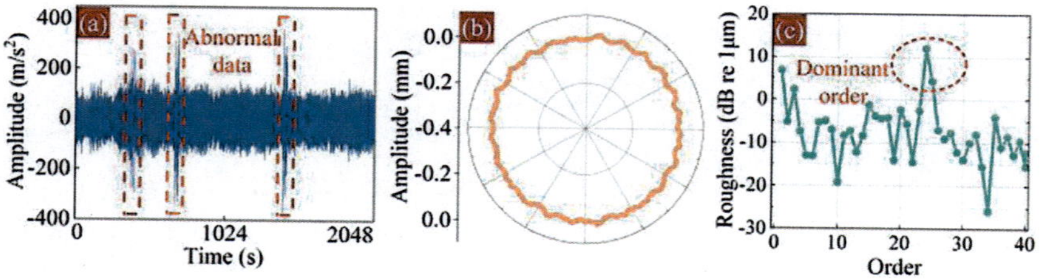


**Fig. 4.** (a) Test high-speed line, (b) wheel roughness and (c) axle box vibration measurement.

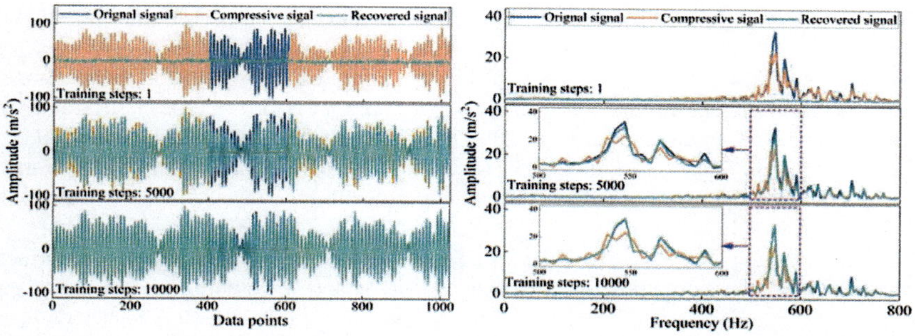


**Fig. 5.** Recovered results under different epoches (left: time domain, right: frequency domain).

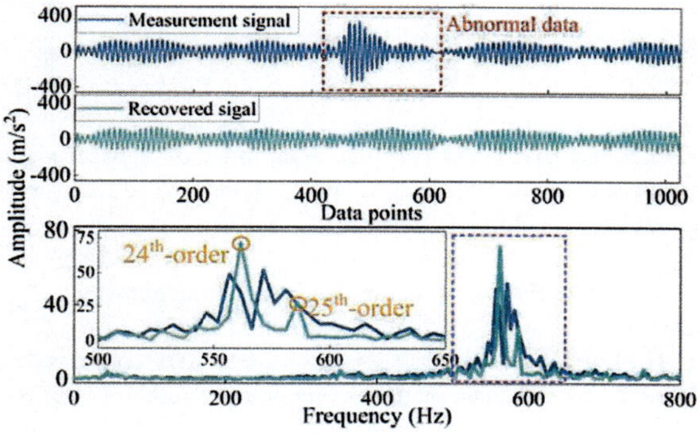


**Fig. 6.** Data quality improvement using the proposed CS-ECE.

## 5 Engineering Application and Experiment

This article develops a novel algorithm named CS-ECE for recovering missing PHM data to improve data quality. CS-ECE can reconstruct the detailed characteristics of target signals in both the time and frequency domains. This approach is not constrained by assumptions of sparsity or random sampling. The availability and superiority of the proposed method were evaluated using real-world data collected from a high-speed train. Results prove that CS-ECE can obtain realistic and accurate recovered results, thereby providing a high-quality dataset for data-driven PHM studies.

**Acknowledgment.** This work was supported by the National Natural Science Foundation (No. U21A20167), the Natural Science Foundation of Sichuan Province (No. 2022NSFSC1914), and the China Postdoctoral Science Foundation (No. 2021T140571).

## References

1. Xie, Q., Tao, G., Xie, C., Wen, Z.: Abnormal data detection based on adaptive sliding window and weighted multiscale local outlier factor for machinery health monitoring. IEEE Trans. Industr. Electron.Industr. Electron. **70**(11), 11725–11734 (2023)

2. Taleb, I., Dssouli, R., Serhani, M.A.: Big data pre-processing: a quality framework. In: IEEE International Congress of Big Data, pp. 191–198, New York, USA (2015)
3. Lei, Y., Xu, X., Cai, X., Li, N., Kong, D., Zhang, Y.: Research on data quality assurance for health condition monitoring of machinery. J. Mech. Eng. **57**(4), 1–9 (2021). (in Chinese)
4. Donoho, D.L.: Compressed sensing. IEEE Trans. Inf. Theory **52**(4), 1289–1306 (2006)
5. Bao, Y., Li, H., Sun, X., Yu, Y., Ou, J.: Compressive sampling-based data loss recovery for wireless sensor networks used in civil structural health monitoring. Struct. Health Monit. **12**(1), 1–18 (2012)
6. Song, S., Zhang, X., Hao, Q., Wang, Y., Feng, N., Shen, Y.: An improved reconstruction method based on auto-adjustable step size sparsity adaptive matching pursuit and adaptive modular dictionary update for acoustic emission signals of rails. Measurement **189**, 110650 (2022)
7. Cheng, Z., Lu, Z.: Two novel reconstruction methods of sparsity adaptive adjustment for road roughness compressive signal based on I-SA and GSM. Mech. Syst. Signal Process. **171**, 108915 (2022)
8. Sawant, S., Banerjee, S., Tallur, S.: Performance evaluation of compressive sensing based lost data recovery using OMP for damage index estimation in ultrasonic SHM. Ultrasonics **115**, 106439 (2021)
9. Li, H., Ai, D., Zhu, H., Luo, H.: An Orthogonal Matching Pursuit based signal compression and reconstruction approach for electromechanical admittance based structural health monitoring. Mech. Syst. Signal Process. **133**, 106276 (2019)
10. Wan, H., Dong, G., Luo, Y.: Compressive sensing of wind speed data of large-scale spatial structures with dedicated dictionary using time-shift strategy. Mech. Syst. Signal Process. **157**, 107685 (2021)
11. Dong, G., Wan, H., Luo, Y., Todd, M.D.: A fast sparsity-free compressive sensing approach for vibration data reconstruction using deep convolutional GAN. Mech. Syst. Signal Process. **188**, 109937 (2023)
12. Kulkarni, K., Lohit, S., Turaga, P., Kerviche, R., Ashok, A.: ReconNet: non-iterative reconstruction of images from compressively sensed measurements. In: 2016 IEEE CVPR, p. 16526936 (2016)
13. Pathak, D., Krähenbühl, P., Donahue, J., Darrell, T., Efros, A.A.: Context encoders: feature learning by inpainting. In: 2016 IEEE CVPR, p. 16526723 (2016)
14. Xie, Q., Tao, G., He, B., Wen, Z.: Rail corrugation detection using one-dimensional convolution neural network and data-driven method. Measurement **200**, 111624 (2022)
15. Jiang, L., Dai, B., Wu, W., Loy, C.C.: Focal frequency loss for image reconstruction and synthesis. In: 2021 IEEE/CVF International Conference on Computer Vision (2022)
16. Sun, Y., Chen, J., Liu, Q., Liu, G.: Learning image compressed sensing with sub-pixel convolutional generative adversarial network. Pattern Recogn.Recogn. **98**, 107051 (2020)
17. Xie, Q., Tao, G., Lo, S. M., Cai, W., Wen, Z.: High-speed railway wheel polygon detection framework using improved frequency domain integration. Veh. Syst. Dyn. (2023). https://doi.org/10.1080/00423114.2023.2235032

# Author Index




Y. Qin et al. (Eds.): EITRT 2023, LNEE 1137, pp. 727–729, 2024.
https://doi.org/10.1007/978-981-99-9311-6